NUTRITION AND HEALTH

Adrianne Bendich, PhD, FACN, SERIES EDITOR

For further volumes:
http://www.springer.com/series/7659

Ronald Ross Watson • Victor R. Preedy • Sherma Zibadi
Editors

Chocolate in Health and Nutrition

Humana Press

Editors
Ronald Ross Watson
Department of Health Promotion Sciences
Health Sciences Center
University of Arizona
Tucson, AZ, USA

Victor R. Preedy
Department of Nutrition and Dietetics
King's College
London, UK

Sherma Zibadi
Mel & Enid Zuckerman College of Public Health
University of Arizona
Tuscon, AZ, USA

ISBN 978-1-4939-5990-7 ISBN 978-1-61779-803-0 (eBook)
DOI 10.1007/978-1-61779-803-0
Springer New York Heidelberg Dordrecht London

Printed on acid-free paper

Humana Press is a brand of Springer
Humana Press is part of Springer Science+Business Media (www.springer.com)

Series Editor Page

The great success of the Nutrition and Health Series is the result of the consistent overriding mission of providing health professionals with texts that are essential, because each includes: (1) a synthesis of the state of the science; (2) timely, in-depth reviews by leading researchers in their respective fields; (3) extensive, up-to-date, fully annotated reference lists; (4) a detailed index; (5) relevant tables and figures; (6) identification of paradigm shifts and the consequences; (7) virtually no overlap of information between chapters, but targeted, inter-chapter referrals; (8) suggestions of areas for future research; and (9) balanced, data-driven answers to patients' as well as health professionals' questions that are based upon the totality of evidence rather than the findings of any single study.

The series volumes are not the outcome of a symposium. Rather, each editor has the potential to examine a chosen area with a broad perspective, both in subject matter as well as in the choice of chapter authors. The editor(s), whose training(s) is (are) both research and practice oriented, have the opportunity to develop a primary objective for their book, define the scope and focus, and then invite the leading authorities to be part of their initiative. The authors are encouraged to provide an overview of the field, discuss their own research, and relate the research findings to potential human health consequences. Because each book is developed de novo, the chapters are coordinated so that the resulting volume imparts greater knowledge than the sum of the information contained in the individual chapters.

Chocolate in Health and Nutrition, edited by Ronald Ross Watson, Ph.D., Victor R. Preedy, Ph.D., D.Sc., FRIPH, FRSH, FIBiol, FRCPath, and Sherma Zibadi, M.D., Ph.D., clearly exemplifies the goals of the Nutrition and Health Series. The major objective of this comprehensive volume is to review the growing evidence that chocolate contains a number of bioactive molecules that can be of value to many aspects of health. However, it would be highly remiss to not first review the complexities of sourcing of cocoa, chocolate manufacture, as well as the physical and biochemical aspects of chocolate components and its by-products so that the clinical studies can be placed in the proper perspective, especially with regard to the potential for comparisons between studies. For example, the volume includes discussions of milk chocolate, dark chocolate, white chocolate, cocoa powder, cocoa butter, and cocoa husks and hulls and comprehensively reviews the effects of temperature and fermentation conditions for the production of each of these "chocolates" on the potential for clinical efficacy. This first comprehensive review of the science behind the active molecules in chocolate and their effects on humans is of great importance to the nutrition community as well as for health professionals who have to answer client questions about this new area of clinical research.

Chocolate in Health and Nutrition represents the first comprehensive compilation of the newest data on the actions of the flavonoids and microorganisms associated with the beneficial effects of chocolate. It is to the credit of Drs. Watson, Preedy, and Zibadi that they have organized this volume so that it provides an in-depth overview of the natural occurrence and biochemistry of relevant molecules in chocolate as well as human exposure to chocolate in its many forms and includes the latest research on the role of chocolate in normal health areas, including mood, pain and weight management, cardiovascular disease and related conditions, as well as their use as adjuncts to therapeutic agents used in the treatment of neurodegenerative diseases. Of importance, this volume includes an in-depth review of the safety of chocolate with emphasis on the mineral content as well as potential for adverse microbial effects. Fortunately, the safety reviews reinforce the confidence in the products sold by major manufacturers.

The volume is organized into five comprehensive sections. The first section comprises five chapters that include reviews of the history of use of cocoa products from the beginning of time up until present times; there are several unique chapters that describe the hundreds of compounds in chocolate and the thousands of microbes that are identified during the fermentation processes. The sensory component of chocolate that most people identify first is its fragrance, and an in-depth chapter examines the volatiles associated with chocolate.

The second section contains six chapters on the composition of chocolate sources and related plant components. There is an important review of the biochemistry of the cacao bean with detailed tables. In the next chapter, there is a comprehensive evaluation of the effects of processing on the bioavailability of the flavonoids in chocolates. The methodologies available for quantifying the bioactive molecules in chocolate products are outlined in the following chapter. Two critical chapters document the exposure to cocoa plants, including the bean and husk, by animals and manufactured products for humans. Cocoa butter's nutritive value is compared to other fat sources in the final, unique chapter of this section.

The third section, composed of four comprehensive reviews, examines the newest data on the metabolism of chocolate, its bioavailability and factors that affect this parameter, and the activities of the bioactives from chocolate in cell culture. The absorption, metabolism, and pharmacokinetics of the polyphenols are emphasized and provide relevant data with regard to the clinical chapters in the next two sections of the volume. Another unique, detailed chapter describes the significant effects of these bioactives on the human gut microbiota. The final chapter in Section C explores the in vitro biological activity of the cacao husk and mass lignin carbohydrate complexes found in by-products from chocolate manufacture.

Half of this volume is devoted to reviews of clinical significance, with the fourth section, containing 11 chapters, examining the evidence for a role of chocolate in prevention and/or treatment of certain chronic diseases. Most of the clinical studies have looked at potential cardiovascular and neurocognitive benefits. Four chapters review the studies of the cardiovascular system and include specific, in-depth examination of the data on endothelial function, arterial disease, coronary heart disease, and hypertension. Extensive tables and figures are included in these chapters. Two other chapters look at the association of chocolate and prevention and treatment of diabetes. There is a single chapter on the possible link between chocolate consumption and cancer risk, as data are emerging and not at the level of research as the cardiovascular area. With regard to cognitive function, three chapters examine the potential of the anti-inflammatory polyphenols in chocolate to reduce the risk of Alzheimer's disease and/or reduce loss of cognition from other causes, including diabetes.

The final section includes 11 chapters – those that examine many diverse aspects of the association of chocolate with human behaviors including the beneficial effects on mood and pain tolerance as well as those that can help the health professional provide data-based, objective responses to clients about the association of chocolate consumption and weight gain (or loss); effects on withdrawal, addiction, and acne; and potential for dental caries and altering food preferences in children. The final chapter provides an enlightened rationale for conducting well-controlled studies in healthy individuals to determine the full scope of actions of chocolate.

The logical sequence of the sections as well as the chapters within each section enhances the understanding of the latest information on the current standards of practice for clinicians and related health professionals including the dietician, nurse, pharmacist, physical therapist, behaviorist, psychologist, and others involved in the team effort required for the successful treatment of cardiovascular and other relevant diseases as well as conditions that adversely affect normal metabolic processes. This comprehensive volume also has great value for academicians involved in the education of graduate students, postdoctoral fellows, medical students, and allied health professionals who plan to interact with patients with disorders that may be beneficially affected by the addition of chocolate products or their constituents to the diet.

Cutting-edge discussions of the roles of signaling molecules, growth factors, hormones, cellular and nuclear receptors, and all of the cells directly involved in chocolate flavonoid metabolism are included in well-organized chapters that put the molecular aspects into clinical perspective. Of great importance, the editors have provided chapters that balance the most technical information with discussions of its importance for clients and patients as well as graduate and medical students, health professionals, and academicians.

The volume contains over 100 detailed tables and figures that assist the reader in comprehending the complexities of the metabolism as well as the biological significance of chocolate for human health. The overriding goal of this volume is to provide the health professional with balanced documentation and awareness of the newest research and therapeutic approaches, including an appreciation of the complexity of this relatively new field of investigation. Hallmarks of the 40 chapters include keywords and bulleted key points at the beginning of each chapter, complete definitions of terms with the abbreviations fully defined for the reader, and consistent use of terms between chapters. There are over 2,500 up-to-date references; all chapters include a conclusion to highlight major findings. The volume also contains a highly annotated index.

This unique text provides practical, data-driven resources based upon the totality of the evidence to help the reader understand the basics, treatments, and preventive strategies that are involved in the understanding of the role chocolate may play in healthy individuals as well as in those with cardiovascular disease, diabetes, or neurocognitive declines. Of equal importance, critical issues that involve patient concerns, such as dental caries and food preferences in children, potential effects on weight gain, and addiction and withdrawal, are included in well-referenced, informative chapters. The overarching goal of the editors is to provide fully referenced information to health professionals so they may have a balanced perspective on the value of various preventive and treatment options that are available today as well as in the foreseeable future.

In conclusion, *Chocolate in Health and Nutrition*, edited by Ronald Ross Watson, Ph.D., Victor R. Preedy, Ph.D., D.Sc., FRIPH, FRSH, FIBiol, FRCPath, and Sherma Zibadi, M.D., Ph.D., provides health professionals in many areas of research and practice with the most up-to-date, well-referenced, and comprehensive volume on the current state of the science and medical uses of chocolate. This volume will serve the reader as the most authoritative resource in the field to date and is a very welcome addition to the Nutrition and Health Series.

Adrianne Bendich, Ph.D., FACN
Series Editor

Preface

Historically, the Mayans used the fruit and seedpods from the cacao tree as a food. This food was one of many brought from the New World by the Spanish. The Aztecs made a drink that was thick, unsweetened, and eaten with a spoon, although now chocolate products are widely used around the world, with many modifications. The first section of the book reviews chocolate's historical use in the Americas and production methods built upon that experience.

The second section then reviews constituents, their analysis, and bioavailability to provide a current understanding of chocolate. These include its role in meeting nutritional requirements, aroma, and the role of chocolate by-products, such as cocoa butter, in health. The third section logically leads to expert reviews of how the human body responds to chocolate and its polyphenol components, particularly their digestion by enzymes and gut microflora, absorption, and metabolic interactions that help define potential and defined health benefits.

The major goal of the third section is to evaluate the variety of clinical benefits of chocolate and especially its polyphenols. Thus, dark chocolate could reduce the risk of heart attack and provide other cardioprotective actions if consumed regularly. Reviews suggest that chocolate may have positive health roles in diabetes, cancer, hypertension, arterial disease, neurocognitive functioning, and modification of brain actions, a developing research and clinical arena.

The fourth section, with the most reviews, builds upon brain stimulation and mood alteration due to chocolate. Several sets of authors evaluate chocolate's biochemistry and chemistry and its effects on the brain, mood, and addiction. Withdrawal, cues that stimulate consumption, pain tolerance, mood, addiction, and flavor are reviewed. Cocoa and chocolate components modulate neurocognitive functioning. Some affect children's food and flavor preferences. There are behavioral, cognitive, and affective consequences of trying to avoid chocolate, including withdrawal, that are included in the book. Two major reported adverse effects of consumption of sweetened chocolate, acne and dental caries, are defined in detail.

Finally, contaminants of the cocoa bean in production and processing are reviewed. Knowledge of some of these is important for protection of the consuming public. Contaminants such as mycotoxin from fungi growing on the beans and pods as well as toxins from other fungal species are described and evaluated for well-known health risks. Chocolate can contain a variety of potential toxic materials with lead absorbed from the environment.

Clearly, chocolate and its many products are a complex and diverse set of foods with wide biological and health effects, both positive and negative. The book has 40 chapters, with world experts reviewing their own and others' research, which dramatically defines chocolate and its human effects. Overall, chocolate is a product with a long history of human use that has developed into a multitude of products, with research now showing that some have disease prevention actions.

Tucson, AZ, USA — Ronald Ross Watson
London, UK — Victor R. Preedy
Tuscon, AZ, USA — Sherma Zibadi

Acknowledgments

This book was made possible by the efforts of Dr. Watson's longtime editorial assistant, Bethany L. Stevens, and Diane Lamsback of Springer. These two editorial experts worked regularly with Dr. Watson and the authors to encourage and assist in the writing and review of 40 chapters for the book. Support for Bethany L. Stevens was generously provided by Southwest Scientific Editing and Consulting. The support of Dr. Adrianne Bendich, the series editor, in the planning as well as the final organization of the book was very useful, leading to an improved book.

Contents

Part I Historical Perspectives and Production

1 **Pre-hispanic Use of Cocoa** 3
Patricia L. Crown

2 **History of the Medical Use of Chocolate** 11
Donatella Lippi

3 **Cocoa and Its By-Products: Identification and Utilization** 23
Emmanuel O.K. Oddoye, Christian K. Agyente-Badu, and Esther Gyedu-Akoto

4 **The Microbiology of Cocoa Fermentation** 39
Dennis S. Nielsen, Michael Crafack, Lene Jespersen, and Mogens Jakobsen

5 **Fungi and Mycotoxin Occurrence in Cocoa** 61
Marina Venturini Copetti, Beatriz Thie Iamanaka, and Marta Hiromi Taniwaki

6 **Nonnutritive Constituents in Chocolate and Cocoa** 73
William I. Manton

7 **Chocolate and Cocoa Aroma** 89
Jürgen Voigt

Part II Composition of Chocolate Sources and Related Plant Components

8 **Composition of Cacao Beans** 105
Antonella Bertazzo, Stefano Comai, Francesca Mangiarini, and Su Chen

9 **Industrial and Home Processing of Cocoa Polyphenols** 119
Rosa M. Lamuela-Raventos, María Izquierdo-Pulido, and Ramón Estruch

10 **Methods in Chocolate Analysis: Use of Atomic Spectrometric Techniques** 125
Denise Bohrer and Carine Viana Silva Ieggli

11 **The Effects of Chocolate and Chocolate by-product Consumption on Wild and Domestic Animals** 135
Brett David Gartrell and Wendi Dianne Roe

12 **Chocolate Bars Based on Human Nutritional Requirements** 143
Anthony A. Robson

13 **Comparing Sunflower Stearins with Cocoa Butter** 149
Joaquín J. Salas, Miguel A. Bootello, Enrique Martínez-Force, and Rafael Garcés

14 **Low-Calorie Chocolates and Acceptability/Sensory Properties** 163
Lauro Melo, Helena Maria André Bolini, and Priscilla Efraim

Part III Chocolate Metabolism and Activities in Culture

15 **Polyphenols in Cocoa: From In Vitro Digestion to In Vivo Bioavailability** 179
Nàdia Ortega Olivé and Maria-José Motilva Casado

16 **The Effect of Chocolate on Human and Gut Microbial Metabolic Interactions: Emphasis on Human Health and Nutritional Status** 189
François-Pierre J. Martin, Sebastiano Collino, Serge Rezzi, and Sunil Kochhar

17 **The Absorption, Metabolism, and Pharmacokinetics of Chocolate Polyphenols** 201
Cesar A. Lau-Cam

18 **Biological Activity of Cacao Husk and Mass Lignin-Carbohydrate Complexes** 247
Hiroshi Sakagami and Tomohiko Matsuta

Part IV Disease-Related Clinical Studies

19 **Clinical Benefits of Cocoa: An Overview** 265
Margarida Castell, Francisco Jose Pérez-Cano, and Jean-François Bisson

20 **Endothelial Control of Vascular Tone by Chocolate and Other Polyphenols** 277
Cyril Auger, Noureddine Idris-Khodja, and Valérie B. Schini-Kerth

21 **Chocolate Flavonoids in the Prevention of Arterial Disease** 289
Nancy J. Correa-Matos and Catherine Christie

22 **Chocolate and Coronary Heart Disease** 303
Imre Janszky and Kenneth J. Mukamal

23 **Dark Chocolate and (Pre-)Hypertension** 313
Karin Ried

24 **Polyphenol-Rich Cocoa and Chocolate: Potential Role in the Prevention of Diabetes** 327
Suzana Almoosawi and Emad Al-Dujaili

25 **Polyphenol-Rich Dark Chocolate in Treatment of Diabetes Mellitus Risk Factors** 341
Emad Al-Dujaili, Catherine Tsang, and Suzana Almoosawi

26 **Chocolate and Cancer Prevention?** 355
Gertraud Maskarinec

27 **The Effects of Cocoa- and Chocolate-Related Products on Neurocognitive Functioning** 369
W. David Crews Jr., David W. Harrison, Kim P. Gregory, Bon Kim, and Allison B. Darling

28 **Flavonoid-Rich Chocolate and Cognitive Improvement** ... 381
Eha Nurk

29 **Cocoa and Chocolate: Chemistry, Biochemistry, and Beneficial Brain Effects** ... 389
Maria Patrizia Carrieri and Joe A. Vinson

Part V Non–Disease-Related Clinical Studies

30 **Chocolate: Mood** ... 409
Marlene M. Millen and Beatrice A. Golomb

31 **Chocolate: Psychopharmacological Aspects, Mood, and Addiction** ... 421
Nuno Rodrigues-Silva

32 **Chocolate and Pain Tolerance** ... 437
Kristina M. Eggleston and Theresa White

33 **Cocoa for Recovery** ... 449
Kelly Pritchett and Phillip A. Bishop

34 **Chocolate and Withdrawal** ... 457
Carla da Silva Benetti and Patrícia Pelufo Silveira

35 **Cues Linked to the Consumption of Chocolate: Orexin and Reward-Based Feeding Behaviors** ... 469
Derrick L. Choi, Jon F. Davis, and Stephen C. Benoit

36 **Behavioral, Cognitive, and Affective Consequences of Trying to Avoid Chocolate** ... 479
James A.K. Erskine and George J. Georgiou

37 **Chocolate and Children's Food and Flavor Preferences** ... 491
Janet E. Standen-Holmes and Djin Gie Liem

38 **Chocolate Use and Knowledge in Oral and Dental Health in Children** ... 505
Emine Efe and Selma Öncel

39 **Acne and Chocolate: Is There Any Evidence of Their Association?** ... 517
Adilson Costa and Maria Carolina Fidelis

40 **Cocoa and Chocolate in Humans: The Rationale for Clinical Studies on Healthy Volunteers** ... 525
Ario Conti, Maria Teresa Pinorini-Godly, and Maria Laura Colombo

About the Editors ... 541

About the Series Editor ... 543

Index ... 545

Contributors

Christian K. Agyente-Badu New Products Development Unit, Department of Cocoa Research, Institute of Ghana, New Tafo-Akim, Eastern Region, Ghana

Emad Al-Dujaili Department of Dietetics, Nutrition and Biological Sciences, Queen Margaret University, Edinburgh, Scotland

Suzana Almoosawi Population Nutrition Research Group, MRC Human Nutrition Research, Elsie Widdowson Laboratory, Cambridge, UK

Cyril Auger UMR CNRS 7213 – Laboratory of Biophotonics and Pharmacology, Strasbourg University, Illkirch, France

Carla da Silva Benetti Hospital de Clínicas de Porto Alegre-HCPA Laboratório de Pediatria Translacional-LPT Largo Eduardo Zaccaro Faraco, Porto Alegre, Rio Grande do Sul, Brazil

Stephen C. Benoit Department of Psychiatry and Behavioral Neuroscience, University of Cincinnati, Cincinnati, OH, USA

Antonella Bertazzo Department of Pharmaceutical Sciences, University of Padova, Padova, Italy

Phillip A. Bishop Department of Kinesiology, The University of Alabama, Tuscaloosa, AL, USA

Jean-François Bisson Department of Cancerology, Human Pathologies & Toxicology, Centre de Recherche ETAP, Vandœuvre-lès-Nancy, France

Denise Bohrer Department of Chemistry, Federal University of Santa Maria, Santa Maria, Brazil

Helena Maria André Bolini Department of Food and Nutrition, School of Food Engineering, State University of Campinas, Campinas, SP, Brazil

Miguel A. Bootello Physiology and Technology of Plant Products, Instituto de la Grasa, Sevilla, Spain

Maria Patrizia Carrieri INSERM, U912 (SESSTIM); Aix Marseille Univ, IRD, UMR-S912; ORS PACA, Observatoire Régional de la Santé Provence Alpes Côte d'Azur, Marseille, France

Maria-José Motilva Casado Department of Food Technology, University of Lleida, Lleida, Spain

Margarida Castell Departament de Fisiologia, Facultat de Farmácia, Universitat de Barcelona, Barcelona, Spain

Su Chen Research and Development, Chainon Neurotrophin Biotechnology Inc, San Antonio, TX, USA

Derrick L. Choi Department of Psychiatry and Behavioral Neuroscience, University of Cincinnati, College of Medicine, Cincinnati, OH, USA

Department of Psychiatry and Behavioral Neuroscience, University of Cincinnati, College of Medicine, UC Reading Campus, Cincinnati, OH, USA

Catherine Christie Brooks College of Health, University of North Florida, Jacksonville, FL, USA

Sebastiano Collino Metabonomics and Biomarkers Group, BioAnalytical Science Department, Nestlé Research Center, Lausanne, Switzerland

Maria Laura Colombo Department of Drug and Science Technology, School of Pharmacy, University of Torino, Torino, Italy

Stefano Comai Department of Pharmaceutical Sciences, University of Padova, Padova, Italy

Ario Conti Alpine Institute of Chemistry and Toxicology, Alpine Foundation for Life Sciences, Olivone, Ticino, Switzerland

Marina Venturini Copetti Department of Food Technology and Science, Universidade Federal de Santa Maria-RS, Santa Maria, Rio Grande do Sul, Brazil

Nancy J. Correa-Matos Department of Nutrition and Dietetics, Brooks College of Health, University of North Florida, Jacksonville, FL, USA

Adilson Costa Dermatology Department, Catholic University of Campinas, Campinas, SP, Brazil

Michael Crafack Department of Food Science, University of Copenhagen, Frederiksberg, Denmark

W. David Crews Jr. Behavioral Neuroscience Laboratory, Department of Psychology, Virginia Polytechnic Institute & State University, Blacksburg, VA, USA

Patricia L. Crown Department of Anthropology, University of New Mexico, Albuquerque, NM, USA

Allison B. Darling Behavioral Neuroscience Laboratory, Department of Psychology, Virginia Polytechnic Institute and State University, Blacksburg, VA, USA

Jon F. Davis Department of Psychiatry and Behavioral Neuroscience, University of Cincinnati, College of Medicine, Cincinnati, OH, USA

Emine Efe Child Health Nursing Department, Akdeniz University, School of Health, Antalya, Turkey

Priscilla Efraim Department of Food Technology, School of Food Engineering, State University of Campinas, Campinas, SP, CEP, Brazil

Kristina M. Eggleston Massachusetts Institute of Technology, Picower Institute of Learning and Memory, Cambridge, MA, USA

James A.K. Erskine School of Population Health Sciences and Education, St. George's, University of London, London, UK

Ramón Estruch Hospital Clinic, Barcelona, Spain

Maria Carolina Fidelis Dermatology Department, Catholic University of Campinas, Campinas, SP, Brazil

Rafael Garcés Physiology and Technology of Plant Products, Instituto de la Grasa, Seville, Spain

Brett David Gartrell Institute of Veterinary, Animal and Biomedical Sciences, Massey University, Palmerston North, New Zealand

George J. Georgiou School of Psychology, University of Hertfordshire, Hatfield, Hertfordshire, UK

Beatrice A. Golomb Department of Medicine and Department of Family and Preventive Medicine, San Diego School of Medicine, University of California, La Jolla, CA, USA

Kim P. Gregory Department of Early Childhood Education, Virginia Western Community College, Roanoke, VA, USA

Esther Gyedu-Akoto New Products Development Unit, Department of Cocoa Research, Institute of Ghana, New Tafo-Akim, Eastern Region, Ghana

David W. Harrison Behavioral Neuroscience Laboratory, Department of Psychology, Virginia Polytechnic Institute & State University, Blacksburg, VA, USA

Beatriz Thie Iamanaka Food Science and Quality Center, Instituto de Tecnologia de Alimentos, Campinas, SP, Brazil

Noureddine Idris-Khodja UMR CNRS 7213 – Laboratory of Biophotonics and Pharmacology, Strasbourg University, Illkirch, France

Carine Viana Silva Ieggli Department of Chemistry, Federal University of Santa Maria, Santa Maria, RS, Brazil

María Izquierdo-Pulido Department of Nutrition and Food Science, School of Pharmacy, University of Barcelona, Barcelona, Spain

Mogens Jakobsen Department of Food Science, University of Copenhagen, Frederiksberg, Denmark

Imre Janszky Public Health Sciences, Karolinska Institute, Stockholm, Sweden

Lene Jespersen Department of Food Science, University of Copenhagen, Frederiksberg, Denmark

Bon Kim Department of Psychology, Liberty University, Lynchburg, VA, USA

Sunil Kochhar BioAnalytical Science Department, Nestlé Research Center, Lausanne, Switzerland

Rosa M. Lamuela-Raventos Department of Nutrition and Food Science, School of Pharmacy, University of Barcelona, Barcelona, Spain

Cesar A. Lau-Cam College of Pharmacy and Allied Health Professions, St. John's University, Jamaica, NY, USA

Djin Gie Liem School of Exercise and Nutrition Science, Deakin Science Group, Deakin University, Burwood, Australia

Donatella Lippi History of Medicine and Medical Humanities, Department of Anatomy, Histology, and Forensic Medicine, University of Florence, Florence, Italy

Francesca Mangiarini Department of Chemistry and Biochemistry, Concordia University, W.Montreal, QC, Canada

William I. Manton Department of Geosciences, University of Texas at Dallas, Dallas, TX, USA

François-Pierre J. Martin Metabonomics and Biomarkers Group, BioAnalytical Science Department, Nestlé Research Center, Lausanne, Switzerland

Enrique Martínez-Force Physiology and Technology of Plant Products, Instituto de la Grasa, Sevilla, Spain

Gertraud Maskarinec University of Hawaii Cancer Center, Honolulu, HI, USA

Tomohiko Matsuta Meikai Pharmaco-Medical Laboratory, Meikai University School of Dentistry, Sakado, Saitama, Japan

Lauro Melo Department of Biochemical Engineering, School of Chemistry, Federal University of Rio de Janeiro Av. Athos da Silveira Ramos, CEP, RJ, Brazil

Marlene M. Millen Department of Internal Medicine, San Diego School of Medicine, University of California, San Diego, CA, USA

Kenneth J. Mukamal Department of Medicine, Beth Israel Deaconess Medical Center, Brookline, MA, USA

Dennis S. Nielsen Department of Food Science, University of Copenhagen, Frederiksberg, Denmark

Eha Nurk Department of Surveillance and Evaluation, National Institute for Health Development, Tallinn, Estonia

Department of Nutrition, Institute of Basic Medical Sciences, University of Oslo, Oslo, Norway

Emmanuel O.K. Oddoye New Products Development Unit, Department of Cocoa Research, Institute of Ghana, New Tafo-Akim, Eastern Region, Ghana

Nàdia Ortega Olivé Research and Development, La Morella nuts, S.A.U., Tarragona, Spain

Selma Öncel Public Health Nursing Department, Akdeniz University, School of Health, Antalya, Turkey

Francisco Jose Pérez-Cano Departament de Fisiologia, Facultat de Farmàcia, Universitat de Barcelona, Barcelona, Spain

Maria Teresa Pinorini-Godly Alpine Institute of Chemistry and Toxicology, Alpine Foundation for Life Sciences, Olivone, Ticino, Switzerland

Kelly Pritchett Department of Nutrition, Exercise, and Health Sciences, Central Washington University, Ellensburg, WA, USA

Serge Rezzi Metabonomics and Biomarkers Group, BioAnalytical Science Department, Nestlé Research Center, Lausanne, Switzerland

Karin Ried Discipline of General Practice, The University of Adelaide, Adelaide, South Australia, Australia

Anthony A. Robson IUEM (UMR CNRS 6539), Université de Bretagne Occidentale, Technopôle Brest-Iroise, Plouzané, France

Nuno Rodrigues-Silva Department of Psychiatry and Mental Health, Vila Nova de Gaia Hospital Center, Vila Nova de Gaia, Portugal

Wendi Dianne Roe Institute of Veterinary, Animal and Biomedical Sciences, Massey University, Palmerston North, New Zealand

Hiroshi Sakagami Division of Pharmacology, Department of Diagnostic and Therapeutic Sciences, and Meikai Pharmaco-Medical Laboratory, Meikai University School of Dentistry, Sakado, Saitama, Japan

Joaquin J. Salas Physiology and Technology of Plant Products, Instituto de la Grasa, Sevilla, Spain

Valérie B. Schini-Kerth UMR CNRS 7213 – Laboratory of Biophotonics and Pharmacology, Strasbourg University, Illkirch, France

Patrícia Pelufo Silveira Hospital de Clínicas de Porto Alegre-HCPA Laboratório de Pediatria Translacional-LPT Largo Eduardo Zaccaro Faraco, Porto Alegre, Rio Grande do Sul, Brazil

Janet E. Standen-Holmes School of Exercise and Nutrition Science, Deakin Science Group, Deakin University, Burwood, Australia

Marta Hiromi Taniwaki Food Science and Quality Center, Instituto de Tecnologia de Alimentos, Campinas, SP, Brazil

Catherine Tsang School of Contemporary Sciences, University of Abertay, Dundee, Scotland

Joe A. Vinson Department of Chemistry, University of Scranton, Scranton, PA, USA

Jürgen Voigt Institute for Biochemistry, Charité – University Medicine Berlin, Berlin, Germany

Theresa White Department of Psychology, Le Moyne College, Syracuse, NY, USA

Part I
Historical Perspectives and Production

Chapter 1
Pre-hispanic Use of Cocoa

Patricia L. Crown

Key Points

- Native to South America, Theobroma cacao was cultivated in Mesoamerica by at least 1800 B.C.E.
- Cacao became a widespread trade item and form of tribute.
- Elites, priests, and warriors consumed a variety of chocolate drinks using distinctive drinking vessels, means of frothing, and additives.
- The healthful properties were recognized, and chocolate was used to treat a variety of ailments before European contact.

Keywords Olmec • Maya • Aztec • Chaco Canyon • Tribute • Currency • Chocolate drinks

Cultures of the New World incorporated cacao into their diet (the foods we eat) and cuisine (how we prepare and eat them) as early as 1800 B.C.E. (Fig. 1.1). The history of cacao use follows a trajectory of use by cultures at an increasing distance from the areas where cacao could be grown, along with increasingly elaborate and distinctive material objects associated with its preparation and consumption. Cacao consumption became an important component in ritual and everyday life, while control over access to cacao fueled economic interactions among New World populations up to (and even after) the arrival of Spanish explorers.

Archeologists rely on multiple lines of evidence in searching for use of plants in the past. Primary evidence may include macrobotanical remains, pollen, phytoliths, impressions of plant parts, and residues. Thus, archeologists traditionally collect soil samples systematically over sites, as well as sampling the interiors of vessels and tools, to search for macrobotanical remains (which often preserve well when they are charred), pollen, and phytoliths. Ancient feces, known as coprolites, can be particularly important in revealing short-term diet and parasite load. Impressions of plant parts in mud, adobe, or ceramics may show the presence of economic plants. Archeologists have employed residue analysis increasingly to determine food remains on vessels or tools, particularly in collaboration with chemists. Gas chromatography–mass spectrometry and high-performance liquid chromatography–mass spectrometry are the techniques most often employed. Some specific plants are more easily

P.L. Crown, M.A., Ph.D. (✉)
Department of Anthropology, University of New Mexico, MSC01 1040, Albuquerque, NM 87131-1086, USA
e-mail: pcrown@unm.edu

R.R. Watson et al. (eds.), *Chocolate in Health and Nutrition*, Nutrition and Health 7,
DOI 10.1007/978-1-61779-803-0_1, © Springer Science+Business Media, LLC 2013

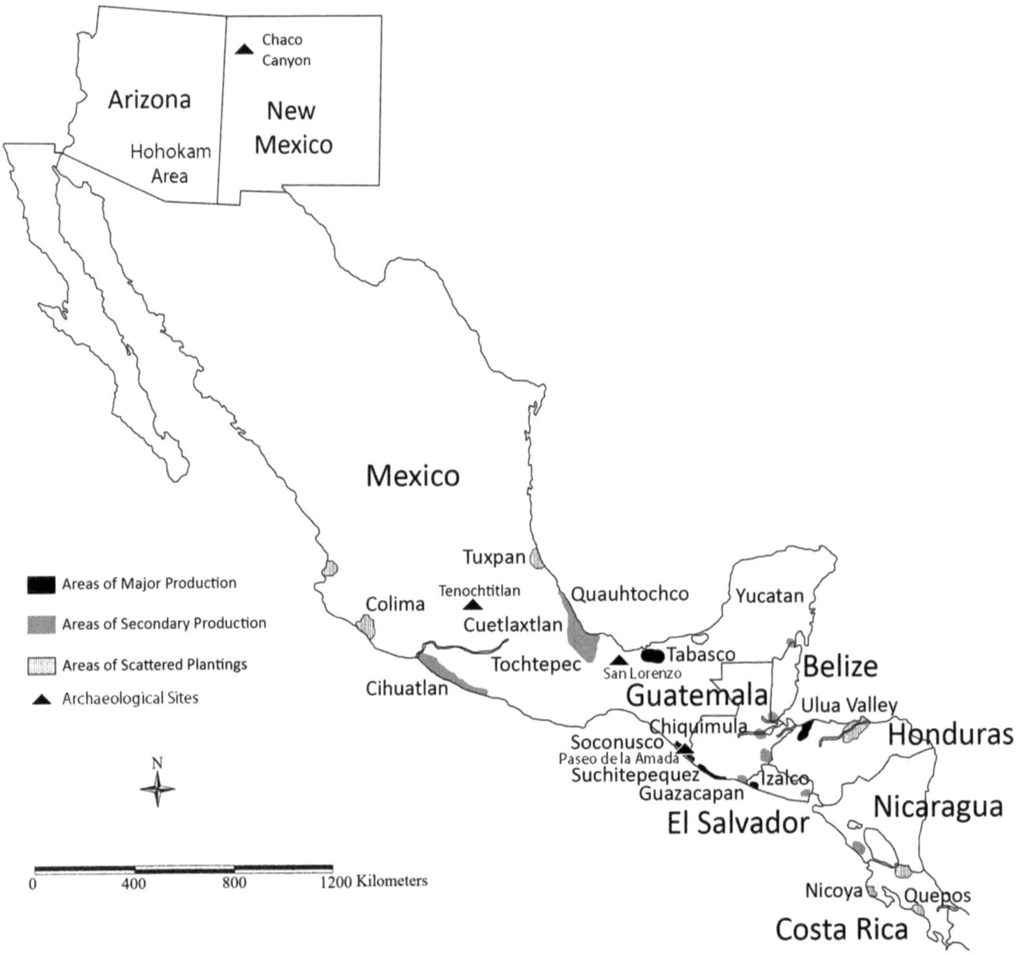

Fig. 1.1 The distribution of cacao cultivation in Central America and Mexico in A.D.1502 and sites/cultures mentioned in the text (Adapted from reference [13])

identified than others because of the presence of compounds, known as biomarkers, that are specific to one or a few plants. Other plants have a dizzying array of compounds, none of which are specific to that plant, so it is not possible to identify them definitively – corn is a good example of a plant known to have been grown and consumed widely in the New World that is difficult to identify as a residue by chemical techniques because there are no lipid biomarkers unique to Zea mays [1].

In addition to the primary means of identifying plants, archeologists employ secondary means of identifying the use of plants. This includes imagery showing plants, documents, and presence of tools known to have been used to prepare or consume specific types of foods. Images painted on pottery or walls or carved into rock may depict plants or foods or even meal preparation and consumption. In more complex societies, written languages document plant and food use. Imagery showing plant use and documents recording plant use in textual form are generally reliable sources, but not always. It is sometimes difficult to determine whether to interpret images literally or whether they may be metaphorical or mythical in meaning. And documents cannot always be trusted as reliable sources of the truth, particularly when written by colonial observers of a culture outside their own. Imagery and texts are very useful, but often archeologists have to rely solely on the presence of tools associated with specific foods. Just as many modern foods and drinks are prepared with tools or served in containers that are specific to that type of food or drink, foods and drinks in the past were often prepared with tools dedicated to that preparation or served in special containers. This is particularly true for foods that anthropologists call *luxury foods,* foods that are desirable but not essential to human nutrition [2]. Such foods are often difficult to obtain, reserved for special occasions, require complex preparation techniques and knowledge, and are often fatty, sweet, or succulent [3]. Use of special containers to serve such foods enhances their distinctive qualities and signals their consumption from a distance. In our own culture, alcoholic drinks are often served in special vessel forms so that it is possible to determine what someone is drinking without asking and from a considerable distance – martini glasses or brandy snifters are good examples. A savvy archeologist in 200 years might interpret the recovery of a martini glass as proof that martinis were consumed without finding martini residues or a charred olive pit! In this same way, archeologists attempt to determine the presence of special tools or containers dedicated to the preparation or consumption of specific foods and drinks. We can argue from such indirect evidence that a specific food or drink was consumed in the past.

Archeologists have employed all of these forms of evidence to search for use of cacao in the past. Working in various parts of the New World, archeologists have recovered macrobotanical remains of cacao, residues, imagery, texts, and tools/containers as evidence of cacao use. Current evidence indicates that from early pre-Hispanic times into the historic period, consumption of chocolate drinks involved special containers, a method for frothing, cacao, and additives. The drinks varied in temperature and additives but shared the four characteristics listed. Research on cacao use is changing rapidly, with yearly publications providing new information. Here I summarize what we know today, with the caveat that new information may alter or supplement this discussion.

The most recent publications indicate that Theobroma cacao is a native Amazonian plant domesticated in Mesoamerica, with possible incipient domestication in Amazonia [4]. At the time of European contact, cultivated cacao trees were noted only in Mesoamerica. Archeological evidence places the earliest cacao use at the Paso de la Amada Site of the Mokaya Culture located on the southern Pacific Coast of Mexico, dating from 1900 to 1500 B.C.E., based on residue analysis of a neckless globular jar form [5]. Cacao residues were also present during the Early Preclassic occupation of the earliest Olmec capital of San Lorenzo at 1800 to 1000 B.C.E. [6], located in southern Veracruz, Mexico. Vessel forms tested confirm that cacao drinks were consumed, but not what these drinks were. They might have been drinks made from cacao nibs or from the sweet pulp surrounding the nibs; a fermented beverage may have been made from the pulp, as has been documented historically for wild Theobroma in South America and cultivated Theobroma among the Highland Maya [6, 7]. The argument for a precedent for a fermented beverage made from the pulp is based on the shape of the vessels containing residues, with earlier bottle forms shifting to later forms more amenable to creating a froth. As the froth is more likely associated with drinks made from cacao nibs, the argument is made that earlier drinks might be of the fermented, nonfrothing variety [7, 8]. The evidence from the Olmec area indicates cacao residues in many distinct vessel forms and ceramic wares, including cups, open bowls, and bottles as well as decorated fine wares and coarse ware [6]. Recovery of residues in such an array of vessel forms and wares suggests that cacao drinks were fairly widely available and consumed, particularly given that 17% of all ceramics tested had cacao residues. Of particular interest, four positive samples from San Lorenzo came from a mass of several hundred broken vessels capping a burial pit containing sacrificial victims; this association is interpreted as indicating the consumption of cacao drinks in "a well-attended, postinternment celebration" [6]. The consumption of cacao drinks in association with ritual thus has a history spanning nearly four millennia. Interestingly, the word "cacao" originated among speakers of an early Mije-Sokean language residing along the Gulf Coast of southern Mexico [9], the Olmec heartland.

Chocolate drinks subsequently became a widespread part of Mesoamerican cultures up to the present day. Consumption of cacao-based drinks moved far outside the area of possible Theobroma tree cultivation, indicating widespread exchange of cacao. These drinks most often were consumed using cacao nibs, a variety of additives, specialized vessel forms, and a method for frothing. Ultimately, cacao became an important part of the economy of Mesoamerica, an economy that was taken over by the Spanish, despite their initial distaste for chocolate drinks. Chocolate drink consumption was too widespread to describe every nuance here, so I provide a brief overview of the best known areas.

Chocolate drinks were probably prepared as they are today through a multistage process. First, growers harvested the cacao pods and removed the nibs from the pods. Then they fermented the cacao beans in the surrounding pulp for one to several days, often on the forest floor. The fermented beans were then removed from the pulp and allowed to dry in the sun for 1–2 weeks. Roasting was the next stage, although sometimes this stage was omitted. Winnowing then involved removing the papery shell to extract the nib. The nibs were then ground into cacao liquor, which might be mixed with a variety of additives and consumed immediately, or formed into tablets for later use. Tablets had a relatively long shelf life, lasting as long as 2 years. Exchange and trade in the past most often involved either the roasted cacao beans or tablets rather than the fruit or liquor.

The Maya occupied areas of the lowlands and highlands of Mexico, Guatemala, and Belize. Maya farmers grew Theobroma cacao trees in the lowlands, and at least some Maya drank chocolate, as attested by residues on vessels, inscriptions, and texts. During the Preclassic, the Maya drank chocolate from spouted vessels, the spout serving not only as a vehicle to pour the drink, but also perhaps as a means to blow air into the chocolate, creating a bubbly froth on the surface [10]. From about A.D. 250 to 900, the Classic Maya built cities with temple-pyramids occupied by kings and elites and supported by a rural population of farmers. The elite consumed chocolate both in drinks and thicker gruels, with a variety of additives; there was no single chocolate drink or recipe. Each concoction had a separate name. There were probably both hot and cold varieties of chocolate drinks, although the historic Maya tended to favor the hot varieties. Chocolate was also used as an additive in drinks and foods with more substantial ingredients, such as corn meal. Depictions of chocolate drink preparation show the Classic Maya using tall cylindrical jars to hold their drinks, a fact supported by inscriptions on the jars that state that they were used to hold chocolate drinks. The drinks were frothed by pouring from one vessel to another from a height, creating a cascading waterfall-like effect with froth bubbling up on the surface of the lower jar. Images confirm preparation of chocolate drinks in formal settings that suggest a theatrical performance surrounding the preparation. Elites and royalty are clearly the primary consumers, but it is unclear if lower classes, including farmers, drank chocolate as well [11].

Cacao was a sign of wealth and power among the Maya, often served at lavish feasts to gain loyalty and obligation. The Theobroma cacao tree was considered sacred and cacao figures prominently in Mayan myths. Cacao was often associated with graves and was considered an important food for the journey to the afterworld. There was a symbolic association among chocolate and blood, rulership, ancestors, and the Underworld [12]. Chocolate drinks were sometimes flavored with achiote/annatto, a red spice that dyed the drinks the color of blood. Cacao was exchanged in wedding ceremonies and was an important component of other rites of passage. Cacao beans were given as gifts and perhaps tribute payments among the Classic Maya [12]. Cacao beans were even counterfeited in clay [11].

By the end of the Classic, some Maya populations had developed widespread trading networks along the Gulf Coast, with large trading canoes. Cacao was both commodity and currency [11]. Cacao was grown wherever it could be, including exploitation of microenvironments in areas where cultivation would otherwise be impossible. Conflict over control of cacao-growing land or access to trade with cacao-growing populations was endemic up to the Conquest.

Cacao had spread to the American Southwest by about A.D. 1000 and perhaps earlier. Current evidence indicates cacao residues on ceramics from Ancestral Pueblo sites in Chaco Canyon, New Mexico, and the Hohokam area of southern Arizona [13, 14]. Ongoing research will likely enhance our understanding of the temporal and spatial extent of cacao use in the American Southwest. Within Chaco Canyon, cacao use is documented in association with cylindrical jars similar in form to those used in the Maya area (Fig. 1.2). However, the Chacoan vessels are locally made and later than the Classic Period Maya jars. Cacao could not be grown in the Chacoan area, so it must have been acquired through one of three mechanisms: by Chacoan people walking south to acquire cacao at the source (a distance of about 1,900 km to the nearest area known to have grown cacao during the historic period), by Mesoamerican traders bringing cacao northward to Chaco, or by a series of shorter exchanges across this distance. Parts of the distance might be traveled by canoe, but Chaco Canyon

Fig. 1.2 Cylinder jar from Chaco Canyon housed in the American Museum of Natural History (Patricia Crown, photographer)

is not located on a navigable watercourse. Other Mesoamerican items occur in some Chacoan sites, including the Scarlet Macaw (another species from the tropics), as well as copper artifacts, pseudo-cloisonné objects, and some species of shells. It is currently unknown whether the cacao came from western Mesoamerica, as did the copper objects and many shell species, or from eastern Mesoamerica.

The specific recipes for consuming chocolate drinks are not known; however, it seems most likely that chocolate was exchanged in prepared bricks rather than as raw beans, particularly because no cacao beans have been recovered in archeological sites in the Southwestern area. The cylinder jars often occur in sets of two to four identical jars, suggesting that preparation included the pouring method to create froth, as was common among both the Maya and later Aztec. Only about 200 of these vessels are known in museum collections and most came from a single large cache within the Chacoan site of Pueblo Bonito.

No records exist to provide information on the use of cacao within Chaco Canyon; however, it seems most likely that chocolate drinks were luxury foods. As defined by anthropologists, luxury foods are foods that are difficult to obtain and nonessential to human nutrition, but desirable. They are often sweet or succulent or fatty, and they often require specialized preparation knowledge [2]. Because the cylinder jars do not occur associated with individual burials, but rather primarily in large caches, it is likely that the jars and chocolate were served in communal feasts or rituals rather than in funeral rites. Serving chocolate drinks at a feast would have signaled wealth and ties to Mesoamerica. It likely created obligations for the guests to reciprocate with subsequent labor or gifts.

Less is known about cacao in the Hohokam area, except that the residues seem to be present on early, shallow oblong bowls that might have been used to grate cacao bricks by about A.D. 900

(ongoing research by Patricia Crown and W. Jeffrey Hurst) and on later beaker-shaped vessels dating to the fourteenth century [14]. The Hohokam area extended from the Arizona/Sonora border north almost to Flagstaff, and they had interaction with Mesoamerica, including macaws, copper bells, and pseudo-cloisonné, just as Chacoans did. The Hohokam particularly crafted items from shells, including many from species found in the Gulf of California, so it is most likely that they exchanged chocolate from West Mexico. Trading canoes from the south might have reached as far as the northern portion of the Gulf of California, with overland trade from there.

Returning to Mesoamerica, the Aztec occupied the Valley of Mexico from the early fourteenth century, conquering a large area by the late fifteenth century. This included the Soconusco area famed for growing high-quality cacao. The large Aztec capital of Tenochtitlan received massive amounts of tribute twice yearly, including cacao and many other luxury items. Aztec rulers occupied huge palaces that included cacao warehouses. In addition to acquiring cacao through tribute, Aztec merchants, called *pochteca*, bought and sold cacao. Cacao was sold in markets within the Aztec Empire. A single *pochteca* would normally carry 24,000 cacao beans in a single backpack [11]. Cacao beans were currency in the Aztec Empire with fixed values, and they were sometimes counterfeited.

The Aztec elite drank chocolate in a variety of elixirs, preferring cool rather than hot drinks. As with the Maya, they mixed chocolate with many different spices and additives. The conquering Spanish recorded various versions of how cacao was prepared, but maize, chili, annatto, honey, and vanilla were common additives, along with several types of powdered flowers. The Aztec consumed chocolate drinks from small, hemispherical bowls of ceramic, gourd, or gold [11]. Such vessels appear on tribute lists along with cacao itself. As with the Maya, the Aztec created a froth by pouring from one vessel to another. Cacao consumption was largely the purview of the elite, warriors, and merchants. Commoners did not have access to chocolate drinks [11].

For the Aztec, cacao was symbolically associated with the south, the Land of the Dead, the color red, and blood [11]. Denied cacao throughout their lives, slaves might taste cacao for the first time if chosen for sacrifice. A drink was even made of chocolate mixed with water containing human blood washed from sacrificial knives to give sacrificial victims courage [11].

Theobroma cacao, the fruit, and the drinks/foods made from it all figured prominently in Mesoamerican myths, songs, poems, and texts. Control over access to cacao created economic partnerships and conflicts throughout much of the millennium before the Spanish entered the New World. The first documented European contact with cacao occurred during Columbus' fourth voyage, when his men encountered a Maya trading canoe filled with trade goods, including cacao [11]. The massive tribute system controlled by the Aztecs, which included cacao, was a critical prize of the Spanish conquest, and although the Spanish did not initially consider chocolate drinks favorably, their success in the New World was at least in part due to their co-opting the cacao trade and tribute.

There is no question then that cacao had symbolic and economic value to many peoples of the New World, but what about medicinal uses for cacao in the past? It is not easy to determine medical practices in the past without texts. Wherever we find residues of cacao, it is possible that the chocolate drinks were viewed as having nutritional and healthful properties, but we cannot prove this. All human societies have healers, and these specialists often use plants to treat disease or to reach altered states of consciousness. Most of what we can glean about the use of cacao in medical practices comes from texts written in the 1500s, and most are specifically about the Aztec practices. The Aztec Emperor maintained a botanical garden to grow and test plants for medicinal purposes [11]. Three manuscripts are particularly useful in understanding Aztec use of plants in medicine: the Badianus manuscript, Florentine Codex, and Princeton Codex [15]. These manuscripts detail several medicinal uses for cacao. Chocolate drinks treated stomach and intestinal problems. When mixed with liquid from silk cotton tree bark, it cured infections [15]. When mixed with rubber, it (not surprisingly) stopped diarrhea [16]. Chocolate ended fever and faintness when combined with ground corn and blended with *Calliandra anomala* [15]. Chocolate drinks mixed with several herbs could also help end coughing. Chocolate was also blended with various medicinal preparations to improve their flavor [15].

Far to the south, a chronicler of the Indies, Gonzalo Fernández de Oviedo y Valdés, visited the Greater Nicoya area between 1527 and 1529. This area encompassed the Pacific coast of Nicaragua and northwestern Costa Rica [17]. Oviedo states that cacao butter was smeared on the skin to protect from sunburn, used as a balm for injury, and that anyone bitten by a snake after drinking cacao would not die [17].

Cacao was a critical commodity in the economy of much of Mesoamerica, an important symbol of status, and used medicinally to cure a variety of ailments. We continue to learn more about the uses of cacao in the past, with new techniques promising to provide important insights into its use and distribution. There are limitations to what we can know from archeological evidence, but those limitations are narrowing with every decade of discovery.

References

1. Reber EA, Evershed RP. How did Mississippians prepare maize? The application of compound-specific carbon isotope analysis to absorbed pottery residues from several Mississippian Valley sites. Archaeometry. 2004;46: 19–33.
2. Van der Veen M. When is food a luxury? World Archaeol. 2003;34:405–27.
3. Hayden B. Feasting in prehistoric and traditional societies. In: Wiessner P, Schiefenhovel W, editors. Food and the status quest: an interdisciplinary perspective. Providence: Berghahn; 1996. p. 127–47.
4. Clement CR, de Cristo-Araujo M, Coppens d'Eeckenbrugge G, Pereira AA, Picanco-Rodrigues D. Origin and domestication of native Amazonian crops. Diversity. 2010;2:72–106.
5. Powis TG, Hurst WJ, del Carmen Rodriguez M, Ponciano Ortíz C, Blake M, Cheetham D, et al. Oldest chocolate in the New World. Antiquity. 2007; 81:314. http://antiquity.ac.uk/projgall/powis/. Accessed 30 June 2011.
6. Powis TG, Cyphers A, Gaikwad NW, Grivetti L, Cheong K. Cacao use and the San Lorenzo Olmec. Proc Natl Acad Sci USA. 2011;108:8595–600.
7. Henderson J, Joyce RA. Brewing distinction. In: MacNeil CL, editor. Chocolate in Mesoamerica: a cultural history of cacao. Gainesville: University Press of Florida; 2006. p. 140–53.
8. Henderson J, Joyce RA, Hall GR, Hurst WJ, McGovern PE. Chemical and archaeological evidence for the earliest cacao beverages. Proc Natl Acad Sci USA. 2007;104:18937–40.
9. Kaufman T, Justeson J. The history of the word for "cacao" and related terms in ancient Meso-America. In: MacNeil CL, editor. Chocolate in Mesoamerica: a cultural history of cacao. Gainesville: University Press of Florida; 2006. p. 118–39.
10. Hurst WJ, Tarka SM, Powis TG, Valdez Jr F, Hester TR. Cacao usage by the earliest Maya civilization. Nature. 2002;418:289–90.
11. Coe SD, Coe MD. The true history of chocolate. New York: Thames & Hudson; 2007.
12. McNeil CL. Introduction. In: MacNeil CL, editor. Chocolate in Mesoamerica: a cultural history of cacao. Gainesville: University Press of Florida; 2006. p. 1–28.
13. Crown PL, Hurst WJ. Evidence of cacao use in the Prehispanic American Southwest. Proc Natl Acad Sci USA. 2009;106:2110–3.
14. Washburn DL, Washburn WN, Shipkova PA. The prehistoric drug trade: widespread consumption of cacao in ancestral pueblo sites in the American Southwest. J Archaeol Sci. 2011;38:1634–40.
15. Dillinger TL, Barriga P, Escárcega S, Jimenez M, Salazar Lowe D, Grivetti LE. Food of the gods: cure for humanity? A cultural history of the medicinal and ritual use of chocolate. J Nutr Suppl. 2000;130:2057S–72.
16. Ortiz de Montellano B. Empirical Aztec medicine. Science. 1975;188:215–20.
17. Steinbrenner L. Cacao in greater Nicoya. In: MacNeil CL, editor. Chocolate in Mesoamerica: a cultural history of cacao. Gainesville: University Press of Florida; 2006. p. 253–70.

Chapter 2
History of the Medical Use of Chocolate

Donatella Lippi

Key Points

- Religious, medical, and cultural controversies have eased off in accepting the therapeutic effectiveness of the various cocoa and chocolate components.
- Recent studies have provided the evidence to a century's established use, supported by empiricism and experience.
- The scientific debate involved the medical class; doctors applied themselves to study the matter in depth.
- Food was considered to be the first medicine, and it was necessary to know chocolate's properties first.
- Some authors supported the adoption of cocoa and chocolate for therapeutic purposes, following the experiences of the people in New Spain.
- The debate involved spices, too, which were added to chocolate, in order to discover the true nature of this beverage.
- In Florence, the custom of drinking chocolate "to the use of Spain" started in1668; then, personalized recipes were elaborated by pharmaceutical factories, which were active in cocoa processing.
- In the eighteenth century, Carl von Linné (Linnaeus) gave a very positive judgment of chocolate's qualities as nourishment and as a therapeutic substance.
- When dietetics separated from medicine, chocolate acquired the role of excipient, being associated with different health problems.

Keywords Exhilarating drinks • Medicine • Dietetics • Sex res non naturals • Hippocratic medicine • Chocolate properties • European recipes • Chocolates de santé • Italian recipes • Florence • Rules of health • Humoral medicine • Religious fast • Hypochondriac melancholy • Madame de Sevigné • Indian chocolate • Santa Maria Novella • Chocolats thérapeutiques du médicinaux

D. Lippi, Ph.D. (✉)
History of Medicine and Medical Humanities, Department of Anatomy, Histology, and Forensic Medicine, University of Florence, Viale Morgagni 85, Florence 50134, Italy
e-mail: donatella.lippi@unifi.it

R.R. Watson et al. (eds.), *Chocolate in Health and Nutrition*, Nutrition and Health 7,
DOI 10.1007/978-1-61779-803-0_2, © Springer Science+Business Media, LLC 2013

The Metaphor

Can chocolate indeed be considered as "manna" or a "cure-all," which was the belief of some seventeenth-century authors? Or does it have to be demonized as a means to "shorten life," as it has been described by detractors who saw in the sweet medicaments and in the foaming ambrosias of "chocomilk" the symbol, perhaps, of a self-conceited and silly century? [1]. More recent studies, which have demonstrated the therapeutic effectiveness of the various cocoa and chocolate components, have come to provide the scientific evidence for centuries of established use, supported by empiricism and experience, but this acknowledgment has not had a straight course, being involved in religious, medical, and cultural controversies.

Christian Europe looked at new, exhilarating drinks, which had revolutionized alimentary habits, with extreme suspicion and often with a condemnatory attitude: from this reaction, the necessity derived to appeal to the reasons of health, with which doctors and scientists hurried to explain that alcohol, tea, coffee, and chocolate were good for the body, and propose thus, in sincere conviction, an intellectual alibi to open the door to desire [2].

The use of chocolate was affirmed, in particular, in Southern European countries, never to become, however, a mass phenomenon, and limiting its own impact to the highest social élite, as it is demonstrated by the alternative recipes which, during the eighteenth century, proposed the substitution of cocoa with other cheaper ingredients, that however offered the same flavor. An example is the work of doctor Saverio Manetti who, in 1765, following the "very big shortage in gender of Wheat of every kind and Fodders, especially of Wheat, happened this year throughout Italy," printed, that same year, a book in which he suggested very economical stratagems: people who could not afford to buy chocolate could use roasted flour, milk, sugar, and egg yolks in order to obtain a beverage which at least looked like chocolate [3]. Is it a sort of chocolate substitute or its…illusion?

Full of metaphors and flavors, seventeenth century's chocolate was about to become a symbol of the weakness and laziness of the aristocrats, as opposed to the activity and rationality of the middle classes, which were associated with coffee consumption. But, in order to allow its official admission in daily use of the middle-high class, it was, however, necessary to receive the viaticum of the acknowledged authority, the medical class, and even in terms of medicine, various mentalities, and various ideologies compared and collided.

Therefore, as famous doctor Pujati claimed,

> If I were to say that Chocolate is only a beverage, that nourishes and refreshes, which is true, the matter would be provided (…) I esteem it to be very useful, and I wish all the Writers and Scholars, that they may drink it with ease, and at their own home, as something, that for long strengthens body and soul, and that in their strenuous tasks, and at times when they cannot, nor should not take sufficient food, it may be of no little help, indeed a Manna. (Fig. 2.1) [4]

Food or Medicament

Cocoa and chocolate represent also today an important occasion of confrontation in terms of medicine and dietetics. In the past, when no effective therapeutic means existed, the only possibility to recover from disease was to use lifestyle and diet as strategies to ensure physical and mental well-being. The concept of a "life regime" in the classical world was expressed by the term díaita/diaeta (which had nothing to do with adjusting rations according to an individual's physical and biometric conditions, as it does today). Its meaning was far broader, encompassing all the areas that were not determined automatically by nature and that humans thus could plan of their own accord such as one's relationship with air and water, food and drink, motion and rest, sleep and wakefulness, dejections and sexuality, love and passion. These rules were collected in the *Regimina Sanitatis*, which were codified during

Fig. 2.1 Cacao arbor. Chiseled, watercolored etching, 350×260 mm. In: Mark Catesby, The natural history of Carolina, Florida, and the Bahama Island. Tomo II. Tav. 6. London, B. White, 1771 (Reprinted with the permission of Biblioteca Nazionale Centrale, Firenze)

the Middle Age: the doctor could, indeed, intervene to cure the sick man, but also the sound, making the required changes to the way of living, using food and drink.

Food was considered to be the first medicine, and it was already contemplated in the Middle Age pharmacopeia the use of animal and vegetable products, to take as food: dietetics became a part of medicine, restricted to rich people, and the alimentary science on the basis of health was designed up to the eighteenth century as a branch of medicine itself, to then reach its own clear identity.

Chocolate as food, then, or as medicament? Surely, it is chocolate as a functional food that recognizes and generates interesting physiological effects, likely to promote or maintain health, and therefore chocolate as medi-food, which exalts its nutritional functions and its therapeutic abilities.

The first authors documenting the adoption of cocoa and chocolate also for therapeutic purposes, referred to the experiences of the people in New Spain, without engaging in the analysis of the alleged reasons for these choices but simply reporting what they had observed.

The Spanish monk Bernardino de Sahagún, for instance [5], referred to the therapeutic purposes that the American indigenous population had for beverages made with cocoa, depending on the type and dosage; he warned against excessive drinking of cocoa prepared from unroasted beans, but recommended it if used in moderation. He stated that drinking large quantities of green cocoa made imbibers confused and deranged, but if taken in moderation, the beverage was invigorating and refreshing.

M. de la Cruz, teacher at Santa Cruz College, founded in Mexico City by the Spanish around 1536, underlined a clear prevalence for the use of cocoa derivatives as nutrients or to be used in case of angina, constipation, dental problems in case of tartar, dysentery, dyspepsia, indigestion, fatigue, gout, and hemorrhoids [6].

If friar Agustín Dávila Padilla reported its use in curing kidney disease [7], it is Francisco Hernandez, in 1577, who identified some fields in which cocoa was used as medicine: without adding any other ingredients, it was used for liver disorders; four grains of cocoa and a dose of gum (holli) toasted and mixed together "restrained dysentery"; the drug called *atextli* was made using a fine paste of cocoa and corn, to which it could be added *mecaxochitl* (*Piper sanctum*) and *tlilxochitl* (*Vanilla planifolia*), as an aphrodisiac [8]. The frequent use of a beverage made with grains of *pochotl* and *cacahoatl*, instead, made one extremely fat and was therefore recommended to slim and asthenic people.

De Usu Et Abusu

When this reflection was transferred to Western Europe, it went beyond the examination of the practical use of chocolate, and it assumed the characteristics of a medico-philosophical debate: therefore, it was necessary to appeal to the tradition of Galenism, which was still an inescapable reference point in the medical practice of that time. The concept of medical science in the seventeenth and eighteenth centuries in fact was still strongly affected by the Hippocratic-Galenic approach. Health was a result of the eukrasìa of the four body fluids, which constituted the human body: blood, phlegm, yellow bile, and black bile. To those corresponded specific dispositions and qualities: hot, cold, humid, and dry.

The therapy was based on an exclusively allopathic system, by which the "hot" illnesses had to be treated with cold substances or foods, as well as dry illnesses needed humid foods and medicaments.

The first debate regarding cocoa, meaning the plant and its berries, and chocolate, product obtained from cocoa in the form of drink, was, therefore, centered on the identification of their respective qualities. The relationship between the balance hot/cold in the ingredients added to the medicinal preparations was raised for the first time by doctor Juan de Cárdenas, in 1591, who underlined the profitable use of cocoa only when it was toasted [9], pointing out the importance of balancing the hot/cold properties of the ingredients added to the medicinal preparations of chocolate.

The reflection on cocoa and chocolate properties and on the various ingredients added to the diverse preparations was connected to the Hippocratic-Galenic system in a systematic way: Santiago de Valverde Turices [10] discriminated the cold quality of cocoa from the hot and dry of chocolate, which had to be then used for the cold and humid illnesses: chocolate had to be considered a medicine because it is able to alter the patient's constitution – useful for breast illnesses if drunk in large quantities, it could also be beneficial for the stomach if taken in small doses. When people were healthy, the use of chocolate was subordinated to the adding of "cold" ingredients to balance its "hot" nature.

In 1618, Bartolomeo Marradon, a Spanish doctor mentioned by Antonio Colmenero de Ledesma in his treatise on nature and quality of chocolate [11], wrote a dialogue that summarized the different uses of cocoa in Spain. The doctor, protagonist of the dialogue made by Marradon, emphasizes how the use of cocoa as currency or as beverage was absolutely restrictive and partial: as "panecitos, tabillas or en coxa como conserva," cocoa could be used appropriately as medicine [12].

This cold drink did not cause drunkenness: in some areas, cocoa was toasted and mixed with water; others mixed it with cooked white corn, adding a kind of water suitable to the pathology that had to be cured. Mixed with sugar, it could be given to the ill "cuando no hay calenture." Cinnamon, sugar, pepper, cloves, vanilla, and anise were the ingredients that were included, as variations, in Marradon's recipe, which provided this prescription:

> 700 (grains of) cocoa, 1 and ½ pounds of white sugar, 2 ounces of cinnamon; 4 grains of Mexico powder, called chile or pimiento; ½ ounce of cloves; 3 little rinds of Campeca (or, instead of this, the weight of 2 *reali* of anis); in the end, 1 hint of "achiote" enough to give it color. Some add almonds, knuckles, and Orange flowers' water.

As Giovanni Battista Anfossi [13] also reminded many years later, this recipe was later changed by Colmenero de Ledesma himself. Here's the recipe:

> For every 100 grains of cocoa are to be mixed 2 grains of Chile, or Mexico pepper, those large grains we said are called chilpatlague (sic) and, in their absence, 2 grains of India's pepper, the larger and less hot, that are found in Spain. 1 handful of anises; 2 of those flowers called mechasuchil, if the belly is hard and tight. In Spain, instead of these latter, it can be added the dust of 6 alexandrine roses, vulgarly called raisins roses; 1 little rind of Campeche; 2 drams of Cinnamon, 1 dozen of almonds and as many "knuckles"; ½ pound of sugar; the amount of achiote enough to give color to the whole. Whenever any of these drugs cannot be found, that it be indeed from the Indies, be it done with our own.

The Anglo-Saxon authors as well intervened in this debate, at the time when Great Britain undertook the conquering of the new continent. Henry Stubbe, starting from the analysis of the previous authors' testimonies, emphasized how in the Indies chocolate was drunk under the doctor's prescription once or twice a day and that it was particularly useful to restore energy "if one is tired through business, and wants speedy refreshment." [14] Stubbe not only told about the use of chocolate in the various sources but proposed his own recipe that closely resembles the one referred by Colmenero de Ledesma:

> For every 100 cocoa seeds, 2 hot red peppers (chile), a handful of anise seeds and *orichelas* (*orejaelas*) and 2 of flowers called *mecasuchill,* 1 vanilla or 6 powdered alexandrine roses, 2 drams of cinnamon, 12 almonds and as many nuts, ½ pound of sugar and as much *achiote* to give color.

In reference to cocoa's nutritional value, Stubbe noticed how the British soldiers stationed in Jamaica, where many plantations had been organized, lived primarily on cocoa paste, mixed with sugar, which was then melted in water: again in subsequent times, chocolate and cocoa will be commonly used by the armies as one of the main genres of livelihood and solace.

Stubbe, Charles the II's doctor and philosopher Thomas Hobbes' friend, dedicated his treaty to famous colleague Thomas Willis, enumerating some cases in which the use of chocolate seemed extremely positive and offering clues of confrontation of great interest: expectorant, diuretic, aphrodisiac. Chocolate was suitable to cure the "hypochondriac melancholy," caused by the veins' obstruction by the black bile, accompanied by stomach weakness and weight loss [15]. Most probably his most important source was Doctor Francisco Ferdinandez, prominent personality in Philip the II's Mexico, to whom William Hughes also referred in his ethnobotanic work on plants that grow in the English plantations in America, in 1672 [16].

During the seventeenth century and in the first half of the next century, this controversy was consumed within the medical context, and it was also subject of some degree thesis at the medical faculty of Paris, between 1684 and 1736 [17].

It seems particularly significant, in this perspective, in Franciscus Foucault's thesis, discussed in 1684 at the medical faculty of Paris, under M. Stephanus Bachot's presidency [18]: after a brief introduction on the different kinds of foods, at Chap. 5, the use of chocolate is described, "massula quaedam, seu confection omnium exquisitissima," prepared with vanilla flavor, "Ambarum Grisium, Ambaro Moschus, Cinnamomum, Zuccarum."

The preparation methods were diverse, just as the forms in which it could be packed ("in rerum omnium simulacra"): chocolate taste was so pleasant that the author concluded, almost in a lapidary way:

> Ergo Chocolatae usus salubris, in quanto calorem nativum & robur viscerum sua caliditate & succitate moderata excitat, firmat, coctionem iuvat, obstructiones solvit, alimenti distributionem, inutilium excretionem promovet, adipem cumulat; non est inimica Cerebro, est amica Veneri, animae & corpori opportunissima.[1]

Still from Paris the voice of Madame de Sevigné had raised, who, after her first enthusiastic adhesion to the consumption of the exotic beverage, suggested categorically to her daughter, who was expecting a baby, to refrain from drinking chocolate during pregnancy because the "marquise de Coetlegon prit tant de chocolat étant grosse l'an passé qu'elle accoucha d'un petit garçon noir comme le diable qui mourut," without thinking to the detail that her unfortunate friend was served chocolate by a black servant [19].

The debate moved, then, on the spices: still according to the *Panckoucke Dictionary* printed in Paris in 1815, it was not correct to call flavorless chocolate *chocolat de santé*, since it was less digestible than the one in which they added vanilla and cinnamon [20].

The "Indian chocolate," which was very spicy, that Neapolitan doctor Giuseppe Donzelli inserted in his book, however, was made by following this recipe:

> You take Cocoa, cleansed of the rind, and well milled, 12 pounds, Cinnamon, beaten and passed through the sieve 1 pound, Anises, scorched, beaten and passed through the sieve ½ pound, Vanillas sticks (these are like pods, long, narrow and thin, of tawny color, of smell like balm, of rather sour taste) scorched, beaten and passed through the sieve at number of 6, of Mecacuce (this they say, that it be a seed like a Nut) 4 handful, or in place of Vanillas, and of Mecacuce, a Nutmeg and 12 cloves of Corn, or Indian Wheat scorched and beaten and gouged out the flower to be sieved 3 pounds, of Aionzoli scorched, and milled 1 pound, or, in place of these Aionzoli of ambrosian Almonds scorched, and well milled ½ pound, of Achiote (this is a blood-red wood, like red sandalwood), ½ ounce, of Sugar, if desired, 4 pounds. You grind Sugar and Achiote, and stir everything diligently, and make bars in Orange leaves. When they are about to use it, those who want to use it, take ½ ounce of the composition thereof and ½ ounce of Sugar, and dissolve them in 6 ounces of hot water, and they drink it as hot as they can endure; many dip biscuits in it [21].

However, not all the doctors agreed on the beneficial effects of chocolate, and its detractors accused it to cause major health problems: in particular this debate, between the favorable and the opposed, started in Florence in the early eighteenth century, where only in 1728 four pamphlets were published that vouch this controversy.

The use of chocolate, in medical field, had also been faced by Antonio Vallisneri, when he had to discuss "of the use and abuse of beverages, and of the hot and cold wetting": he had been requested to judge the practice to consume hot the imported drinks: tea, coffee, and chocolate [22]. Not by chance, the reflection was stimulated by the Spanish doctors' writings, whose metabolic system almost seemed to be different: "If the Spanish bowels are similar to ours, Italian, I know not."

From this purely rhetorical question, Vallisneri got inspiration to confirm the beneficial effects of the hot drinks, compared to the cold ones, especially "on the old people, on the cachectics, on the hypochondriacs, on the weak of stomach," for it "awakens" the bile, which helps digestion; this opinion was supported by Giovan Battista Davini who emphasized the use of hot drinks as "Tea, Coffee, Chocolate," because "Multum habeant spiritosae substantiae, salibusque turgeant volatilibus oleosis, attamen si bibantur frigida & gratiam, & robur videntur amittere nec juvant naturam, ut consuerunt calide hausta.[2]"

Davini invoked the expertise of chemists to confirm that the volatile substances' properties were not enhanced, whenever the beverage was drunk cold. But the discussion was not at all solved.

---

[1] Therefore the use of chocolate is salubrious [for] it excites and strengthens with its warmth mild juiciness the bowels' inborn warmth and strength, it helps digestion, it fosters the spread of food and the secretion of the unnecessary, it accumulates fat, it's not an enemy to the brain, it's Venus' friend and very suitable for body and soul.

[2] Mostly they seem to have a spirituous substance, they seem to be full of volatile and oily salts, however, if one drinks them cold, they seem to lose their grace and strength and they are not a good for nature, like they usually are when drunk hot.

Florence: The Medical Debate

The news about cocoa and chocolate officially got in Florence starting from the end of the sixteenth century. However, the custom of drinking chocolate "to the use of Spain" dated back to 1668, according to what Giovanni Targioni Tozzetti reported, quoting an assertion by Tommaso Rinuccini, reported also by Antonio Cocchi:

> (…) it has been introduced in Florence this year 1668, commonly a beverage to the use of Spain, called Chocolate, and one of the Storekeepers aforesaid sells it in little clay cups, and it seems to be tasty be it hot, or cold [23].

In Florence, the scientific controversy catalyzed between two peculiar personages, Doctor Giovan Battista Felici, great chocolate accuser, considering it "disorder (…) to shorten life" brought by "men's intemperance," and Francesco Zeti, "the Hunchback of Panone," taking his nickname from the coffeehouse where he worked: Zeti was worried that all this "bad talking" about chocolate could provoke a decrease of customers and, fearing to be discharged in his turn, wrote a short book in defense of chocolate (Fig. 2.2).

Cocoa, according to Felici, could not be considered a cold substance, since plants take their qualities from the places in which they grow and, therefore, "it contains a loose, fat and viscous substance, which can easily contain the particles of the heat": the presence of "oil" and its "bitter taste" was a proof of it [24]. It is for this reason that, according to Felici, the custom spread to add some particular flavors in chocolate's composition, such as cinnamon, vanilla, pepper, "cloves, amber and *acciotte* and other similar, very hot spices." Chocolate then had a sort of "slow, long lasting fire," causing in the blood "significant fermentation which can spoil it," since "it causes an extraordinary motion in the

F. Zeti, Altro parere intorno alla natura e all'uso della cioccolata, Firenze 1728; BNCF, Targioni Tozzetti 218.12, frontespizio, particolare.

Fig. 2.2 Detail of the title page from F. Zeti, Altro parere intorno alla natura e all'uso della cioccolata. Firenze, 1728 (Reprinted with the permission of Biblioteca Nazionale Centrale, Firenze)

animal instincts; so, when it gets into the stomach, it makes us more able to perform our doings with vivacity."

In the evaluation of the various substances' nature in this period's medical writings, it is evident the indirect contribution of increasing iatrochemistry, that particular medical movement arising during the 600, on the basis of spagyric tradition and of the figure of Philipp Bombast Theophrast von Hohenheim, called Paracelsus, which started a new approach to the reality of the macrocosm and the human body.

With the suggestions generated from chemistry, the consequences of the use of chocolate even in the decomposition of the different parts of the blood were hypothesized, either for the "viscous nature of cocoa, able to enlarge excessively the body fluids," preventing them from circulating, or for the "obstructions" caused by the impossibility of chocolate to spread "into all parts of the body," still for its viscosity.

Heart palpitations, intermittent pulse, convulsive movements, and apoplexy could be induced by the use of chocolate, to whose negative effects contributed also the ingredients with which it was mixed, as cinnamon and vanilla, which contain "irritant, volatile and stinging substances, which can spoil in thousands of different ways the natural composition of the body fluids," affecting also the nerves and, as a consequence, the animal instinct: equally negative feedback was given to sugar, whose viscosity was demonstrated by the use that painters made of it to blend the colors.

Damages to the fluids and damages to the solid parts of the body: adhering to the inner membrane of the stomach, chocolate inhibited the stomach's gastric juices, corrupting the nerves "texture" as a result of its "constant contractions." The muscles were also affected by this excessive contractile activity and the heart itself suffered: the cases in which the use of chocolate could be of benefit were the hemorrhages, as a proof the fact that "Florentine ladies (…) where healed from the copious blood loss, by the continuous use of chocolate, which they used as desiccant and astringent medicament." These were Doctor Felici's blames: Francesco Zeti's answer was published the following year. He claimed to have commissioned an anonymous doctor (may be Girolamo Giuntini) a sort of defense of chocolate, to protect his own interests: regardless of its alleged quality of hot or cold substance, cocoa was not "replete of oleaginous and sulfurous parts" since it owns also a "milky spirit" which could have beneficial effects on the human body [25]. The fact that cocoa plant grows in warm places could not be a support to its quality, since there exist numerous plants which disprove this relationship: gold caper, Persicaria, calamus aromaticus, etc.

At this point, the pharmaceutical factories, which were particularly active in cocoa processing, elaborating personalized recipes even for cosmetic purposes, were involved in this debate. Among them, the Officina Farmaceutica in Santa Maria Novella was particularly dynamic: right after an inspection at Santa Maria Novella's drugstore on November 23, 1751, by a medical committee in which Giovanni Targioni Tozzetti also participated, chocolate was served in silver bowls in the Green Room.

The different views were summarized, a few years later, by Giovan Battista Anfossi, who examined in a very detailed way the various authors' opinions, referring most of all to the English authors, like Stubbe, who represented an original voice in this debate, coming from a different cultural environment, compared to the Spanish-speaking authors [26].

Through the lines of his dissertation, in addition, new issues came up, compared to the previous authors: enumerating the critics of which chocolate was object, in fact, Anfossi did not limit to mention the "stubborn Galenics," who based their beliefs on the qualities of the various substances, but he referred to the "Professors of more reasonable Systems."

Numerous "systems" came into vogue during the eighteenth century, most of which are of no importance whatever, but three, at least, that came into existence and disappeared during the century are worthy of fuller notice. The Animists had for their chief exponent Georg Ernst Stahl of "phlogiston" fame; the Vitalists were championed by Paul Joseph Barthez (1734–1806); and the third were the Organicists. This last, while agreeing with the other two that vital activity cannot be explained by the laws of physics and chemistry, differed in not believing that life "was due to some spiritual entity" but rather to the structure of the body itself.

The doctors' interest, in fact, was addressed to the elaboration and the study of general theories about the human organism's functioning, with different levels of complexity, related to the influence of the philosophical ideas. This whole of physiological, pathological, and therapeutic principles was deeply influenced by different schools of thought and was very theoretical.

Eventually, the deficiencies and failures of eighteenth-century medical systems became all too apparent: an echo of this debate is felt in Anfossi's pages. It is interesting to note that Anfossi referred to chocolate's implementation even for a topic use as cocoa "butter," solving unequivocally the riddle about the use of chocolate in curing hemorrhoids, which was transmitted by previous authors, apparently colliding with the most elementary precautions to be taken to avoid the annoying inconvenience.

Fin De Siècle

The eighteenth century counts many other statements on the medical use of cocoa and chocolate, not last the one from Carl von Linné (Linnaeus), in which are summarized chocolate qualities, as nourishment and as therapeutic substance (Fig. 2.3) [27].

Fig. 2.3 Chiseled, watercolored etching by Jan Pieter Sluyter, 345×270 mm. In: Maria Sibylla Meria: Dissertatio de generatione et metamorphosibus insectorum surinamensium. Tav 63. Amsterdam, apud Johannem Oosterwyk, 1719 (Reprinted with the permission of Biblioteca Nazionale Centrale, Firenze)

Table 2.1 Medical use of cacao and chocolate

Author	Medical use	Date
Devila Padilla	Kidney disease	1528
de La Cruz	Angina, constipation, dental problems in case of tartar, dysentery, dyspepsia, indigestion, fatigue, gout, hemorrhoids	1552
Hernandez	Liver disease, antipyretic	1577
de Sahagún	Invigorating and refreshing properties	1590
Marradon	Against infections	1618
de Valverde Turices	Against breast and stomach illnesses	1624
de Ledesma	Digestive properties	1631
Stubbe	Expectorant, diuretic, aphrodisiac, suitable to cure the "hypochondriac melancholy"	1662
Foucault	Excites and strengthens with its warmth mild juiciness the bowels' inborn warmth and strength, it helps digestion, it fosters the spread of food and the secretion of the unnecessary, it accumulates fat, it is not an enemy to the brain, it is Venus' friend and very suitable for body and soul	1684
Donzelli	Treats astheny	1696
Davini	If drunk hot, digestive properties	1727
Blancardi	Useful for asthenic people and for the old	1777
Lavedan	Universal medicine. It stimulates natural warmth and the heart, decreasing flatulence, resolving constipations, helping digestion and appetite, increasing virility and slowing down white hair growth, prolonging significantly lifetime	1796

Linnaeus identified three kinds of illnesses in which chocolate could be used appropriately: loss of weight, as a consequence of lung and muscle diseases, hypochondria, and hemorrhoids, adding also that it was an excellent aphrodisiac, confirming a tradition already existing in the pre-Columbian culture.

Later, in Blancardi's work, it is stated that "Chocolata (…) praeparatur cum lacte, vel aqua cocta et servida, atque agitando Chocolata liquescit, et sorbendo calide bibitur. Potus est nutriens, senibus, debilibus utilissimus" [28].

After Lardizabal [29] and Buchan's works [30], the century ended with Antonio Lavedan's work, who claimed, again on the basis of the previous tradition, that chocolate was a sort of universal medicine, since it stimulates natural warmth and the heart, decreasing flatulence, resolving constipations, helping digestion and appetite, increasing virility, and slowing down white hair growth, prolonging significantly lifetime. The history of the medical use of chocolate had reaches its peak (Table 2.1).

In his work, much space is given to the "chocolats de santé" or "chocolats thérapeutiques du médicinaux," [31] with which another chapter in the history of chocolate for a therapeutic use started, destined to relegate it in the role of an excipient.

From being medicine books' subclasses, in fact, cooking books, together with dietetics, progressively emancipate, acquiring their own identity, as a result of the rationalizing effort of French Enlightenment: the road of therapy splits from that of taste, and chocolate will mainly maintain its leading role of excipient, bearing the burden, over time, of a negative valence, generating a *topos* in medical and nonmedical literature, in which it will be associated to obesity, dental problems, unhealthy regimen of life, and so forth.

The rehabilitation of chocolate will occur only in recent times, positively recapturing international scientific magazines' pages and restoring that value that Linnaeus himself credited to chocolate, calling the generous plant *Theobroma Cacao*, food of the gods [32].

References

1. Lippi D. Chocolate and medicine: dangerous liaisons. Nutrition. 2009;25(11–12):1100–3.
2. Dillinger TL, Barriga P, Escarcega S, Jimenez M, Salazar Lowe D, Grivetti LE. Food of the gods: cure for humanity? A cultural history of the medicinal and ritual use of chocolate. J Nutr. 2000;130(Suppl):2057–72.

3. Manetti S. Delle specie diverse di frumento e di pane siccome della panizzazione: memoria del dottor Saverio Manetti; pubblicata sotto gli auspici dell'illustriss. sig. Andrea Ginori … attualmente presidente dell'Imperial Societa fisico-botanica fiorentina. Firenze: Nella Stamperia Moucke: Si vende da Giovacchino Pagani libraio; 1765. (Of the various species of wheat and bread, as well as bread making.)
4. Pujati GA. Della preservazione della salute de' letterati e della gente applicata e sedentaria. Venezia: Presso Antonio Zatta;1762 in-8: 361–367. (How to preserve good health of literates.)
5. Sahagun B. General history of the things of New Spain [Florentine Codex, 1590]. Santa Fe: School of American Research, University of Utah Monographs of the School of American Research, and Museum of New Mexico; 1981.
6. De la Cruz M. The Badianus manuscript, Codex Barberini, Latin 241, Vatican Library; an Aztec herbal of 1552. Baltimore: Johns Hopkins University Press; 1940.
7. Torres EM. La relacio´n fraile-indı´gena en Me´xico. In: A trave´s de las plantas tres ejemplos el tabaco, el cacao y las nopaleras. Rome: Istituto Storico Domenicano; 1997. p. 227–292.
8. Hernàndez F. Historia de las plantas de la nueva España. Mexico: Imprenta Universitaria; 1577.
9. Cardenas J. Problemas y secretos maravillosos de las Indias. Mexico: en casa de Pedro Ocharte; 1591.
10. de ValverdeTurices S. Un discurso del chocolate. Seville: J.Cabrera; 1624.
11. Colmenero de Ledesma A. Curioso tratado de la naturaleza y calidad del chocolate. Madrid: Francisco Martinez; 1631.
12. Marradon B. Curioso tratado de la naturaleza, los dan˜os y provechos que el tempo y experiencia an descubierto de sus effectos, y del chocolate, y otras bevila, que en estos tempo se usan. Interlocutores un medico con un indiano, y un ciudadano. Seville: Gabriel Ramos; 1618.
13. Anfossi GB. Dell'uso ed abuso della cioccolata del dottore Gio. Battista Anfossi romano a sua Eccellenza Pier Vettore Pisani procurator di S. Marco. Venezia: Appresso Francesco Locatelli a S. Bartolommeo; 1779.
14. Stubbe H. The Indian nectar, or, a discourse concerning chocolate [sic]: the nature of the cacao-nut and the other ingredients of that composition is examined and stated according to the judgment and experience of Indian and Spanish writers. London: J.C. for Andrew Crook; 1662.
15. Liebowitz MR, Klein D. Hysteroid dysphoria. Psych Clin North Am. 1979;2:555–75.
16. Hughes W. The American physitian [sic], or a treatise of the roots, plants, trees, shrubs, fruit, herbs etc. growing in the English plantations in America: describing the place, time, names, kindes [sic], temperature, vertues [sic] and uses of them, either for diet, physick [sic], etc. whereunto is added a discourse of the cacao-nut-tree, and the use of its fruit; with all the ways of making of chocolate. The like never extant before 1672. London: J.C. for William Crook the Green Dragon without Temple Bar; 1672.
17. Baron HT. An senibus chocolatae potus? Affirm. Diss. Inaug. Resp. Lud. Guil. Lemonnier, in-4°. Parisiis; 1739.
18. Bachot E. An chocolatae usus salubris? Affirm. Diss. Inaug. Resp. Fr. Foucault, in-4°. Parisiis; 1684.
19. Rabutin Chantal M. Lettres choisies. Paris: Garnier Freres; 1878. (lettres 11.2.1671; 15.4.1671; 13.5.1671; 23.10, 1671).
20. Pancoucke CLF. Dictionnaire des sciences médicales, par une societé de médecins et de chirurgiens. Paris: Pancoucke CLF; 1815.
21. Donzelli G. Teatro farmaceutico, dogmatico, e sparagirico. Roma: Cesaretti; 1677. p. 230–1.
22. Vallisnieri A. Dell'Uso e dell'Abuso delle Bevande e Bagnature calde, o fredde. Terza impressione cui evve annessa Una erudita Dissertazione del Celebre suo Zio, intitolata De potu Vini Calidi, autore Johanne Baptista Davini Serenissimi Raynaldi I. Mutinae, Regii, Mirandulae, & c. Ducis, Medico. Napoli: Felice Mosca; 1727.
23. Targioni Tozzetti G. Notizie degli aggrandimenti delle scienze fisiche in Toscana nel corso di anni IX del secolo XVII. Firenze: Biblioteca Palatina; 1780. p. 221–2.
24. Felici GB. Parere intorno all'uso della cioccolata scritto in una lettera dal conte dottor Gio. Batista Felici all'illustriss. signora Lisabetta Girolami d'Ambra. Firenze: Appresso Giuseppe Manni; 1728.
25. Giuntini G. Altro parere intorno alla natura, ed all'uso della cioccolata disteso in forma di lettera indirizzata all'illustrissimo signor conte Armando di Woltsfeitt. Firenze: si vende allato alla chiesa di Sant'Apollinare; 1728.
26. Anfossi GB. Dell'uso ed abuso della cioccolata del dottore Gio. Battista Anfossi romano a sua Eccellenza Pier Vettore Pisani procurator di S. Marco. Venezia: Appresso Francesco Locatelli a S. Bartolommeo; 1779.
27. von Linne´ (Linneaus) C. Om chokladdryken. Stockholm: Fabel; 1741.
28. Blancardi S. Lexicon medicum tripartitum, renovatum. Lipsiae: Schwickert; 1777.
29. Lardizabal V. Memoria sobre las utilitades de el Chocolate. Pamplona: por Antonio Castilla; 1788.
30. Buchan AP. Medicina Domestica ò Tratado de las Infirmitades Quirurgicas y Cirugìa en General. Madrid: La Emprenta Real; 1792.
31. Lavedan A. Tratado de los usos, abusos, propiendades y virtudes del tabaco, cafe´, te y chocolate. Madrid: Imprenta Real; 1796.
32. Lippi D. Chocolate in health and disease. Maturitas. 2010;67:195–6.

Chapter 3
Cocoa and Its By-Products: Identification and Utilization

Emmanuel O.K. Oddoye, Christian K. Agyente-Badu, and Esther Gyedu-Akoto

Key Points

- Cocoa pulp juice (sweatings) may be made into a fruit drink either alone or in combination with other fruit juices. It may also be used for making jam and marmalade. Moreover, commercial-grade pectin may also be extracted from it.
- Fermentation of the sugars in cocoa pulp juice (sweatings) leads to the production of alcoholic drinks (gin and brandy) and also the production of wine and vinegar.
- Dried cocoa pod husk may be used as a feed ingredient for poultry (10%), pigs (25%), and sheep (40%). Fresh/wet pod husk has been fed to pigs at 300 g/kg of the ration. Fermentation of cocoa pod husk with *Pleurotus ostreatus* improved its feeding value and increased its usage in broiler finisher diets to 20%.
- The ash produced when sun-dried cocoa pod husk is burnt contains about 40% potash, which can be used as the alkali for the making of soft soap and liquid soap. The ash may also be converted into a potassium-rich fertilizer by adding starch and then pelletizing the mixture.
- Cocoa butter can be extracted from discarded cocoa beans and may be used in the production of toilet soap, soft soap, and body pomade.
- A feasibility study, conducted as part of the ICCO/CFC/COCOBOD-funded cocoa by-products project, indicated that there is the potential for cocoa farmers to enhance their incomes through the processing of cocoa waste into the developed by-products.

Keywords Cocoa pulp juice (sweating) • Cocoa pod husk • Fruit drink • Jam • Marmalade • Pectin • Gin • Brandy • Wine • Vinegar • Animal feed • Liquid soap • Soft soap • Toilet soap

E.O.K. Oddoye, Ph.D. (✉) • C.K. Agyente-Badu, M.Phil. • E. Gyedu-Akoto, Ph.D.
New Products Development Unit, Department of Cocoa Research,
Institute of Ghana, 8, New Tafo-Akim, Eastern Region, Ghana
e-mail: emmanuel_oddoye@yahoo.com; kagyente@yahoo.com; akua_akoto2004@yahoo.co.uk

R.R. Watson et al. (eds.), *Chocolate in Health and Nutrition*, Nutrition and Health 7,
DOI 10.1007/978-1-61779-803-0_3, © Springer Science+Business Media, LLC 2013

Introduction

Over the years cocoa has been cultivated mainly for the beans, with the rest of the fresh fruit being discarded. The beans make up about 33% of the fruit by weight (Fig. 3.1) [1]. There are many other by-products of cocoa that could be generated from the rest of the fruit and that could form the basis of small- and medium-scale industries in cocoa-producing countries.

A. W. Knapp [2] suggested that from its composition, cocoa pulp juice (sweatings) could be used to produce soft drinks, industrial alcohol, citric acid, vinegar, and cocoa jelly. In the chocolate manufacturing countries, where dry beans were on hand in large quantities, utilization of the shell wastes was the subject of much research [3]. In Brazil the use of cocoa sweatings for jelly manufacture started as a small-scale industry several decades ago [4], while the fresh pod husk waste was tested for its suitability as cattle feed in the early 1950s. In Ghana, a very old rural soft soap industry based on cocoa pod husk (CPH) potash has gained greater importance as an export trade in and beyond West Africa over the past decade [4].

The Cocoa Research Institute of Ghana (CRIG) initiated research into cocoa by-products, in mid-1965, by setting up a committee of experts, with representatives from the University of Ghana, to identify by-products that could be produced from cocoa. From 1982 to 1986, and in collaboration with the Departments of Animal Science of the University of Science and Technology and the University of Ghana as well as the Animal Research Institute (ARI), poultry and livestock feeding trials were carried out. Further collaboration between CRIG and the Mechanical Engineering Department of the University of Science and Technology led to the production and utilization of CPH as a source of potash fertilizer for different crops.

In 1992 cocoa by-products research received a further boost with the establishment of the New Products Development Unit of CRIG. The unit was to be responsible for the development of new products, assessment of the economic viability of new products, and promotion of the use of cocoa products generally. In 1993 the unit received financial support through the International Cocoa Organisation (ICCO)/Common Fund for Commodities (CFC)/Ghana Cocoa Board (COCOBOD)-funded project to develop by-products from cocoa.

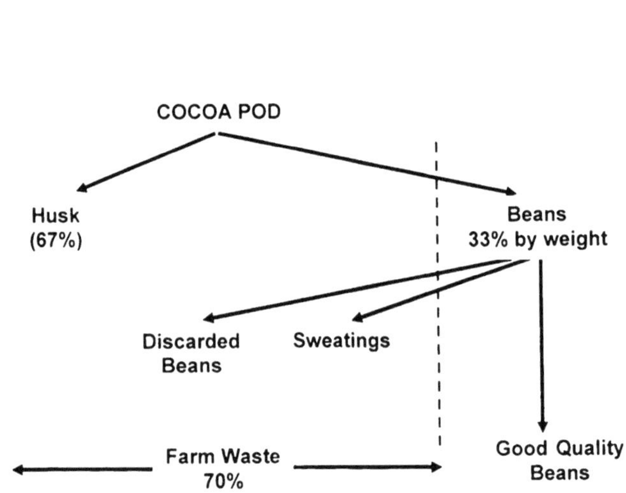

Fig. 3.1 Relative distribution of cocoa by-products that go to waste on cocoa farms (Courtesy of Dr. M.R. Appiah)

By-Products from Cocoa Pulp Juice (Sweatings)

Description

Cocoa beans are surrounded by an aromatic pulp (Fig. 3.2) that arises from the bean teguments. The mucilaginous pulp is composed of spongy parenchymatous cells containing cell sap rich in sugars, citric acid, and salts [5]. After harvesting, the ripe pods are broken open and the beans removed and placed in baskets for transport to fermentaries (Fig. 3.3). The bruising of the pulp during handling and the pressure of the weight of the mass release some of the pulp juices, which trickle from the collected bean mass as a slightly turbid, whitish liquid. This liquid is known in the industry as cocoa pulp juice or sweatings. It is a common practice for laborers and children to collect these drippings in makeshift containers and use as a refreshing drink.

Collection

In the traditional method of cocoa fermentation, pulp juice is allowed to drain to waste from fermentation baskets, sweat boxes, and heap fermentation under gravity. Dwapanyin [6] had shown that the collection of some sweatings from the beans before fermentation did not affect the bean quality. Further work by Adomako and Takrama [7] led to the development of an improved method for the collection of large volumes of sweatings within a short time.

Polytanks, normally used for water storage are modified by cutting open the top and perforating the bottom. The modified polytank is mounted on a wooden platform with a hole in the middle, into which the perforated end of the polytank fits. An assembly of perforated polyvinyl chloride (PVC) pipes are placed vertically in the tank to aid the flow of pulp juice. The tank is filled with fresh cocoa beans and covered with a circular wooden board. Some pressure is exerted by means of weights to facilitate the flow of pulp juice. The juice is collected through a funnel into a plastic receptacle placed

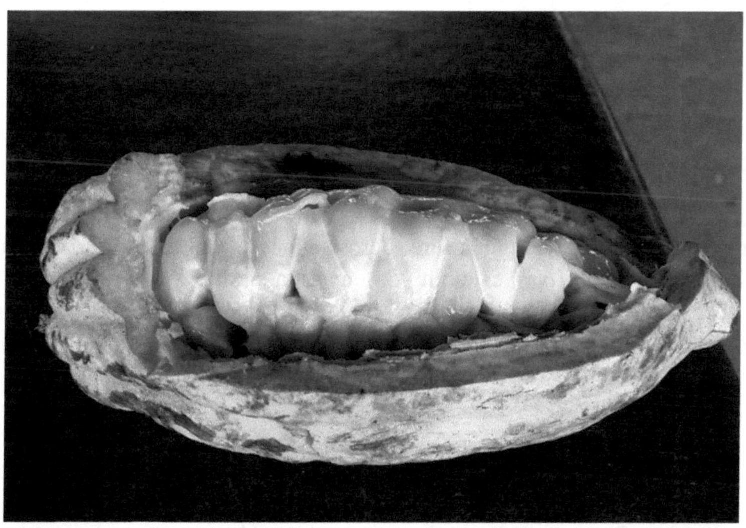

Fig. 3.2 A cocoa pod broken open to show the beans surrounded by mucilaginous pulp

Fig. 3.3 Cocoa beans ready to be transported to the fermentary

below the tank (Figs. 3.4, 3.5, 3.6, and 3.7). The pulp juice is immediately dosed with 0.15 g/L of potassium metabisulphite to prevent early deterioration due to microbial action. Using this method, 100–150 L of sweating may be collected per ton of wet cocoa beans.

Gyedu and Oppong [8] used a screw press extractor to extract cocoa pulp juice. With this method, the yield of pulp juice was 175.5 L/t.

Physicochemical Analyses of Fresh Cocoa Pulp Juice

The main sugars present in cocoa pulp juice are sucrose, glucose, and fructose. Predominant minerals include potassium (K), sodium (Na), calcium (Ca), and magnesium (Mg) (Table 3.1) [8, 9]. Pectin obtained from cocoa pulp juice compares favorably to that from apples, lemon, and pawpaw, which are the main sources of commercial pectin [4].

Usage

In late 1969, following Greenwood-Barton's review paper [10] on the occurrence of substantial amounts of pectin and neutral polysaccharides in cocoa pod husk, CRIG received enquiries from

Fig. 3.4 A polytank

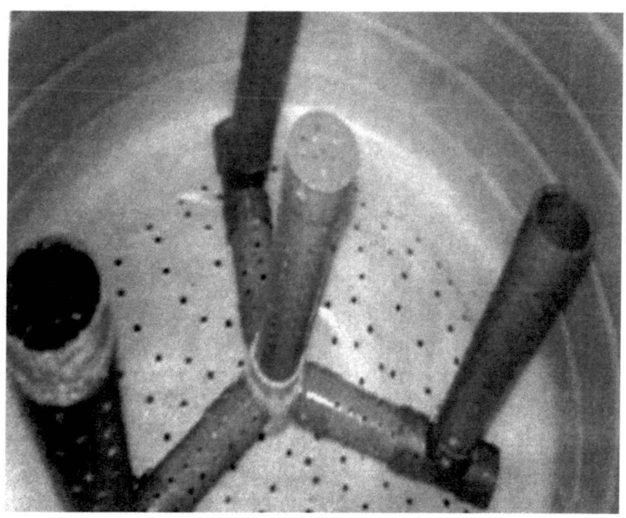

Fig. 3.5 A polytank with a perforated *bottom* fitted with an assembly of perforated polyvinyl chloride (*PVC*) pipes

Fig. 3.6 A polytank setup being filled with fresh cocoa beans

Europe concerning cocoa pectin. Results of the pectin studies showed that the pectin in cocoa pod husk was of low grade, very much like sugar beet pectin, which has little or no commercial value. In contrast, the pectin from cocoa pulp juice (sweatings) was of high grade and similar to lemon and apple pectin used in commerce [11, 12].

Gyedu and Oppong [8] reported on cocoa sweatings as a potential raw material for the development of soft drinks. A natural shelf-stable, ready-to-serve fruit juice was prepared using cocoa sweatings, sugar, and water. The authors also reported on the production of shelf-stable jam and marmalade using cocoa sweatings as one of the ingredients. Selamat et al. [13] also reported on the development of juice from cocoa pulp as well as mixed juices from cocoa sweatings with either mango, guava, or coconut. The combination of cocoa sweatings and mango proved to be the best.

Cocoa pulp juice contains between 10% and 18% of fermentable sugars [14, 15]. The pulp juice can therefore serve as a suitable medium for fermentation. Fermentation of the sugars in cocoa sweatings by naturally occurring yeasts led to the production of alcohol, which was distilled (triple distillation) and blended into gin and brandy of the finest quality [16]. Wine was developed from cocoa sweatings by

Fig. 3.7 Collection of cocoa pulp juice by the polytank method

pasteurizing the sweatings to destroy these natural yeasts, and introducing a specific wine yeast [17]. The second stage of this fermentation process (fermentation with *acetobacter*) led to the production of vinegar that was light brown/orange in color, had a pH of 3.07, specific gravity of 1.029, and an acetic acid content of 6.54% (measured as percentage of total acids w/v) [18, 19].

By-Products from Cocoa Pod Husk

Description

Cocoa pod husk may be described as the leftover pod material of the matured cocoa fruit, after the wet cocoa beans, sweatings, and placenta have been removed. It is often discarded as waste/residue during the processing of raw cocoa beans on the farm.

Table 3.1 Physicochemical composition of cocoa pulp juice (sweating)

	Composition	
Parameter	Gyedu and Oppong [8]	Anvoh et al. [9]
PH	3.66	3.75
Ref. index	1.358	
Spec. gravity	1.067	
Moisture (%)		85.30
Ash (%)		3.76
Fat (%)		3.54
Soluble solids (%)	16.0	
Total soluble solids (EBrix)		16.17
Total solids (%)	18.2	
Protein (%)	0.41	
Total proteins (g L^{-1})		7.2
Total sugars (%)	7.5	
Citric acid (%)	0.65	
Citric acid (mg L^{-1})		9.14
Titratable acidity (meq L^{-1})		170
Malic acid(mg L^{-1})		3.6
Acetic acid (mg L^{-1})		2.28
Oxalic acid (mg L^{-1})		1.27
Lactic acid (mg L^{-1})		1.23
Fumaric acid (mg L^{-1})		0.02
Vitamin C (mg L^{-1})		18.3
Citric acid (mg L^{-1})		9.14
Malic acid (mg L^{-1})		3.6
Sucrose (mg ml^{-1})	107.6	
Glucose	23.3 (mg/ml)	214.24 (g L^{-1})
Fructose (mg ml^{-1})	10.6	
Saccharose (g L^{-1})		21.31
Pectin (%)	1.33	
K	0.090 (mg 100 ml^{-1})	950.0 (mg L^{-1})
Na	0.085 (mg 100 ml^{-1})	30.5 (mg L^{-1})
P	0.023 (mg 100 ml^{-1})	62.47 (mg L^{-1})
Ca	0.089 (mg 100 ml^{-1})	171.5 (mg L^{-1})
Mg	0.086 (mg 100 ml^{-1})	82.5 (mg L^{-1})

Preparation and Storage of CPH for Animal Feed

Preparation of CPH is an important determinant of its final quality. The husk needs to be dried as quickly as possible to prevent deterioration due to attack by fungi. At CRIG, this is done by first passing fresh pods through a slicing machine to increase surface area before sun drying. After 24 h, when moisture is about 65%, the partially dry material is passed through a mill, which grinds and pellets the husk. The pelleted, wet husk is then further sun-dried for an additional day or two, depending on the weather, to a moisture content of about 10% and stored for future use. Where such machinery is not available, the key to producing quality CPH is to dry it as quickly as possible by increasing the surface area through slicing or chopping the material with a knife.

Table 3.2 Chemical composition of cocoa pod husk

Component	Falaye and Juancey [20] (%)	Sobamiwa [21] (%)	Donkoh et al. [22] (%)	Okai et al. [23] (%)	Alemanor et al. [24] ($g\ kg^{-1}$)
Dry matter	89.1	–	94.60	–	890
Ash	9.80	9.14	10.10	10.0	91
Calcium	–	0.32	0.81	0.88	8.14
Phosphorus	–	–	0.44	0.48	4.39
Magnesium	–	–	0.69	0.71	
Iron	–	–	0.03	0.04	
Potassium	–	–	7.18	7.36	
Manganese	–	–	0.02	0.01	
Crude protein	7.80	5.94	7.66	7.40	91.4
Crude fiber	33.40	22.59	32.50	–	357
Ether extract	1.80	1.24	4.37	8.20	99.6
Nitrogen-free extract	35.30	62.17	–	–	93.6
Metabolizable energy	–	–	4.72 MJ/kg	–	
Gross energy	–	–	–	–	
NDF	–	–	52.20	64.4	598
ADF	–	–	41.4	56.4	470
Lignin	–	14	–	–	
Acid-insoluble lignin					209
Cellulose	–	35	–	–	262
Hemicelluloses	–	11	10.80	8.0	128
Pectin	–	6	–	–	
Nitrogen	–	–	–	–	
Theobromine	0.30	–	–	–	
Tannin					11.88

NDF neutral detergent fiber, *ADF* acid detergent fiber

Nutrient Analysis and Antinutritional Factors

Several studies have been made on the composition of CPH. Table 3.2 summarizes some of these studies [20–24]. Cocoa pod husk also contains a mixture of condensed or polymerized flavonoids (anthocyanidins, catechins, and leukoanthocyanidin), sometimes linked with glucose [25]. Tannins (1–4%) have been detected in CPH [24, 26]. Theobromine, the main alkaloid in CPH, has been detected at 0.3–0.4%, but this is negligible compared to the proportions in cocoa bean cake (1.5–3.0%) and cocoa bean shell (0.7–1.9%) [20, 27].

Use of CPH as an Animal Feed Ingredient

Several poultry and livestock feeding trials were conducted by CRIG, in collaboration with the Department of Animal Science of the University of Ghana, Legon and the Kwame Nkrumah University of Science and Technology, Kumasi, as well as the Animal Research Institute of the Council for Scientific and Industrial Research. The trials also studied digestibility, egg laying, and carcass quality of animals fed with rations containing CPH. Results of these studies indicated that CPH can be incorporated into animal feeds at the following optimum levels: poultry, 10%; pigs, 25%; sheep, 40% [22, 28–36].

Similar results were obtained in other countries [37–44]. The use of CPH in diets for tilapia has also been reported [20]. In Cameroon the usual blend of tilapia feed (blend of cornmeal, wheat bran, and rice), which costs 35 US cents/kg, was reduced to 2 US cents/kg when CPH replaced 200 g/kg of feed [45].

Improving the Feeding Value of CPH

The addition of alkalis to fibrous feedstuffs to improve their feeding value is a well-known technique. There are, however, contrasting reports as to its effectiveness for CPH. Tuah and Orskov [46] observed a slight decline in dry-matter disappearance of CPH (40.6–39.4%) following ammonia (3.5%) treatment and subsequent 48-h incubation in the rumen of cannulated sheep using the nylon bag technique. Tuah [47] also observed an insignificant increase in the in vitro dry-matter digestibility of CPH after NaOH (5%) pretreatment. Smith et al. [39] used different concentrations of CPH ash solutions as their alkaline agent to treat CPH and observed a linear increase in the rumen degradability of CPH. The authors also reported that the improvement in rumen degradability was similar to that obtained by using NaOH solutions of equivalent alkalinity.

A series of experiments by Alemawor et al. [48, 49] have indicated the possibility of using extracellular enzymes and bioconversion with *Pleurotus ostreatus* (oyster mushroom) to enhance the feeding value of CPH. The optimum level of inclusion of *Pleurotus ostreatus*-fermented CPH and enzyme-supplemented CPH in broiler finisher diets was 200 and 100 $g\,kg^{-1}$, respectively [24].

Wet Feeding of CPH

Sun drying is the usual method of drying CPH as other drying methods have proven expensive, given the low nutrient content of CPH. This led to experiments by Oddoye et al. [50] to investigate the possibility of feeding CPH fresh or wet. The authors included fresh/wet CPH in growing pig feeds at up to 300 $g\,kg^{-1}$ (on air-dry weight basis) of the diet without any deleterious effects.

CPH as a Source of Alkali for Soap Making

The ash produced when sun-dried CPH is burnt in a kiln contains about 40% potash [4, 51]. This potash is made up of about 43% potassium carbonate and 27% potassium hydroxide and can be used as the alkali for the making of soft soap, traditionally known as *alata samina*. Water is percolated through the ash to dissolve the potassium salts. The resulting solution is evaporated to dryness and the potash salt produced stored for later use. Soap making uses a traditional method. A mixture of vigorously boiling cocoa butter and palm kernel oil in the ratio 1:10 is saponified with potash salt (potash salt dissolved in water) in a galvanized pot placed over a heat source. The contents of the pot are continuously stirred and beaten until a thick soap mass is produced. The pot is temporarily covered, and carbon dioxide, generated from the carbonate component during the saponification reaction, causes the soap mass to rise in the form of a foam. The soap foam is scooped and air-dried after which it is packaged [52].

In the production of liquid soap, the initial potash solution obtained after percolating water through the CPH ash is not evaporated to dryness. Instead calcium hydroxide is added to the solution, converting the carbonate portion of the potassium salt to hydroxide. Calcium carbonate, formed as a result, is allowed to sediment. The supernatant is drained off and used in liquid soap production. A blend of coconut and palm kernel oils is heated to about 80°C. The caustic potash solution is added a little at a time while stirring until an emulsion is formed. The emulsion is allowed to boil for about 2 h with the regular addition of water to reduce the rate of reaction. The heating process is stopped when a clear emulsion is obtained. The liquid soap is allowed to cool overnight, which also allows unsaponifiable matter to sediment. This product is still under development.

CPH as a Potash Fertilizer

Most farmers leave the empty pods to rot on their farms. While this should have been a good practice, as it would return nutrients to the soil, the pods also serve as a reservoir for the causative agent of the cocoa black pod disease. As a result, farmers are being encouraged to remove the pods from their farms after harvesting. CPH ash has, however, been converted into a potassium-rich fertilizer by the addition of starch and then pelletizing the mixture. The results, when tested on farm, were encouraging [53]. Reports from Nigeria also indicated the use of cocoa pod husk as a fertilizer for maize production [54].

By-Products from Discarded Cocoa Beans

Cocoa butter extracted from discarded cocoa beans, with little or no commercial value, has been used in the production of toilet soap and body pomade [55, 56]. It has also been used in the production of soft soap [52]. Useable beans are selected from the discards. These are then roasted and milled. Water is added to the milled material and boiled. The cocoa butter floats to the top and is skimmed off, washed, dried, and stored for further use.

By-Products from Cocoa Bean Shell

During processing of cocoa beans, the seed coat or testa is removed. This is known as the cocoa bean shell. Ayinde et al. [57] reported that untreated cocoa bean shell could be included in rabbit diets at a level of 100 g/kg. After treatment with hot water, however, it could be included at 200 g/kg. The major use of cocoa bean shell is as a mulch. This mulch contains approximately 2.5% nitrogen, 1% phosphate, and 3% potash as well as a natural gum that is activated when watered. This enables the cocoa shell mulch to slow soil moisture loss through evaporation as well as retarding weed growth [58]. The texture of the cocoa shell also deters slugs and snails and helps prevent plant damage [59]. Cocoa bean shell mulch has been implicated in methylxanthine toxicosis in dogs. Methylxanthine is a breakdown product of theobromine, one of the main alkaloids in cocoa [60].

Other By-Products of Cocoa

At the International Workshop on Cocoa By-Products held in Ghana in 2003, the delegation from Malaysia indicated other by-products that they had developed from cocoa. These by-products are listed in Table 3.3 [61].

Table 3.3 Other by-products that may be produced from cocoa waste

By-product	Source
Cocoa gum	Pod husk, leaves, and chupons
Activated carbon	Pod husk
Dietary fiber	Shells
Pectic enzyme	Pod husk
Methylxanthine	Shells, leaves, beans, and chupons
Theobromine	Shells and beans
Polyphenols	Shells, leaves, pod, and branches
Particle boards	Shells
Essential oils	Leaves, chupons, flowers, and pod husk
Compost	Pod husk
Biogas	Pod husk

By-Products as a Way of Increasing Farmer Income and Alleviating Farmer Poverty

There has been a growing concern among governments of cocoa-producing countries in the world, particularly in Africa, on the increasing poverty among cocoa farmers. The low yield obtained by the cocoa farmer has resulted in some social and economic consequences such as low incomes, poor farm maintenance arising from high incidence of pests and diseases, and increased deforestation. As a result, the youth from the cocoa-producing areas are not finding cocoa farming attractive enough and are thus drifting into the urban areas in search of nonexisting jobs.

A feasibility study, conducted as part of the ICCO/CFC/COCOBOD-funded cocoa by-products project, indicated that there is the potential for cocoa farmers to enhance their incomes through the processing of cocoa waste into the developed by-products in all cases except the production of industrial alcohol, which only results in an unattractive 3% enhanced income. The other products result in an enhancement of revenues ranging between 13% and 18% [62].

Problems with Utilization of Cocoa By-Products

The logistics involved in the evacuation of cocoa by-products (fresh pods, pulp juice) from the farm gate to a processing factory in an urban center are quite enormous. Cocoa farming, as practiced in Ghana, is made up of several small farmers scattered all over the six cocoa growing regions, and this makes the collection of by-products a headache for any potential investor. Such small farmers would probably not generate enough by-products. A more realistic approach would be to process at the district level (farmers' cooperative) or on large plantations. Cooperatives could also organize central processing of raw materials collected from, or delivered by, their members and successfully run a profitable enterprise, generating additional revenue for member farmers in the process. Products derived solely from cocoa would, perhaps, be most suitable for such strategies rather than those which combine cocoa by-products with other inputs. Products that would fall into this category, either as products suitable for retail or as intermediate products for sale to other processing industries, would include pure CPH animal feed, potash, alcohol, pectin, and cocoa butter extracted from cocoa beans discards.

It has not been easy getting farmers to adopt new methods that will allow for the collection of raw/waste materials, for example, cocoa sweating. Furthermore, apart from the production of alcohol, all other products require that the sweatings be quite fresh and not have undergone deterioration. It is a major challenge collecting and keeping the sweating fresh until they are processed.

Another major problem has to do with the drying of cocoa pod husk for animal feed. Researchers have relied on sun drying as a means of reducing cost as the cost of fossil fuels or electricity for drying is rather prohibitive. For large volumes of material, however, this may not be wholly appropriate.

Cocoa pod husk is very fibrous, which limits its use in monogastric diets. With the development of the use of enzyme preparations in feed formulations and the use of bioconversion, future work will have to examine whether these techniques would allow a higher proportion of CPH to be used in monogastric feeds.

References

1. Appiah MR. Introduction to technical presentations by Cocoa Research Institute of Ghana. Paper presented at: International Workshop on Cocoa By-Products in Ghana; 2003 July 14–16; La-Palm Royal Hotel, Accra. p. 22–5.
2. Knapp AW. Cocoa and chocolate: their history from plantation to consumer. London: Chapman and Hall; 1920.
3. Knapp AW, Churchman A. Cocoa shell and its use as accessory fodder. J Soc Chem Ind Lond. 1937;56:29–33.
4. Adomako D. A review of researches into the commercial utilization of cocoa by-products with particular reference to the prospects in Ghana. CMB Newsl. 1975;61:12–20.
5. Lopez AS, Pereira HIS, Llamosas A, et al. Present status of cacao by-products utilization in Brazil. In: Proceedings of the 9th International Cocoa Research Conference; 1984; Lome. p. 425–32.
6. Dwapanyin AO, Adomako D, Tetteh JP. The sugar content of cocoa sweatings and the effect of pressing the sweatings prior to fermentation on bean quality. Ghana J Biochem Mol Biol. 1991;1:109–20.
7. Adomako D, Takrama JFK. Large–scale collection of cocoa bean pulp juice sweatings. In: Proceedings of the 12th International Cocoa Research Conference; 1996; Salvador, Bahia; 1999. p. 1033–40.
8. Gyedu E, Oppong H. Cocoa pulp juice (sweatings) and its potential for soft drink, jam and marmalade production. Paper presented at: International Workshop on Cocoa By-products in Ghana; 2003 July 14–16; La-Palm Royal Hotel, Accra. p. 37–40.
9. Anvoh KYB, Zoro BA, Gnakri D. Production and characterization of juice from mucilage of cocoa beans and its transformation into marmalade. Pak J Nutr. 2009;8(2):129–33.
10. Greenwood-Barton L. Cocoa beans and cocoa pods: have they any unconventional uses? In: FAO technical working party on cocoa: production and projection. 1st Session. Rome, 1964 Sept 7–11. Working Paper 67.
11. Adomako D. Cocoa pod husk pectin. Phytochemistry. 1972;11:1145–8.
12. Adomako D. Chemical characterization of cocoa pectin. Chem Ind. 1974;1974:873–4.
13. Selamat J, Yosof S, Jimbuh M, et al. Development of juice from cocoa pulp. In: Proceedings of the 2nd Malaysian International Cocoa Conference; 1994. p. 351–57.
14. Adomako D. Cocoa products and by-products research: utilisation of cocoa pulp juice. ANN.REP, Cocoa Research Institute of Ghana. 1972/73, 1975. p. 41–142.
15. Agyeman KOG. Pilot plant studies on collection of cocoa sweatings and production of fermentation products from cocoa sweatings [M.Sc. thesis]. Kumasi: University of Science and Technology; 1976.
16. Takrama JFK, Adomako D. Production of cocoa gin and brandy. Paper presented at: International Workshop on Cocoa By-products in Ghana; 2003 July 14–16; La-Palm Royal Hotel, Accra. p. 64–9.
17. Agyente-Badu CK, Adomako D, Asante EG. Development of wine from cocoa pulp juice. Paper presented at: International Workshop on Cocoa By-products in Ghana; 2003 July 14–16; La-Palm Royal Hotel, Accra. p. 48–52.
18. Agyente-Badu CK, Takrama JFK, Appiah MR. Development of pilot-scale process for the production of vinegar from cocoa pulp juice (sweating). Technical report submitted to FAO, Rome. 2001
19. Agyente-Badu CK, Adomako D, Asante EG. Production of vinegar from cocoa pulp juice. Paper presented at: International Workshop on Cocoa By-products in Ghana; 2003 July 14–16; La-Palm Royal Hotel, Accra. p. 53–7.
20. Falaye AE, Jauncey K. Acceptability and digestibility by tilapia (*Oreochromis niloticus*) of feeds containing cocoa husk. Acquacult Nutr. 1999;5:157–61.
21. Sobamiwa O. Use of cocoa pod husk in poultry feeds: a particular reference to the Nigerian situation. In: Proceedings of the 11th International Cocoa Research Conference; 1993; Yamoussoukro. p. 797–801.
22. Donkoh A, Atuahene CC, Wilson BN, et al. Chemical composition of cocoa pod husk and its effects on growth and food efficiency in broiler chicks. Anim Feed Sci Technol. 1991;35:161–9.
23. Okai DB, Easter RA, Frank GR. The nutritive value of some non-conventional Ghanaian feed ingredients; nutrient composition and effects on performance of weanling rats. World Rev Anim Prod. 1984;20(23):11–6.

24. Alemawor A, Oddoye EOK, Dzogbefia VP, Oldham JH, Donkoh A. Broiler performance on finisher diets containing different levels of either *Pleurotus ostreatus*-fermented dried cocoa pod husk or dried cocoa pod husk supplemented with enzymes. Trop Anim Health Prod. 2010;42:933–9.
25. Kimura K. Manufacturing procedure of natural pigment from cacao bean. Japanese Patent no. Showa 54–10567 1979.
26. Adamafio NA, Dosoo DK, Aggrey EC, et al. Effect of corn stalk ash extract on oxalate and tannin levels in crop residues. In: Proceedings of the 13th Faculty of Science Colloquium, University of Ghana; 2003.
27. Figueira A, Janick J, BeMiller JN. New products from *Theobroma cacao*: seed pulp and pod gum. In: Janick J, Simon JE, editors. New crops. New York: Wiley; 1993. p. 475–8.
28. Atuahene CC, Adams C, Adomako D. Cocoa pod husk in diets of broiler chickens. In: Proceedings of the 9th International Cocoa Research Conference; 1984; Lome; 1985. p. 495–500.
29. Barnes AR, Amega WK, Manu M, et al. Utilization of cocoa husk meal by growing-finishing pigs. In: Proceedings of the 9th International Cocoa Research Conference; 1984; Lome; 1985. p. 449–54.
30. Boa-Amponsem K, Agudu EW, Manu M. Effect of cocoa pod husk on broiler performance. In: Proceedings of the 9th International Cocoa Research Conference; 1984; Lome; 1985. p. 501–03.
31. Okai DB, Osafo ELK, Adomako D. Effects of different levels of cocoa pod husk on performance and carcass characteristics of finisher pigs. In: Proceedings of the 9th International Cocoa Research Conference; 1984; Lome; 1985. p. 455–59.
32. Tuah AK, Dzoagbe S, Adomako D. Evaluation of cocoa pod husk as feed ingredient for sheep in Ghana. In: Proceedings of the 9th International Cocoa Research Conference; 1984; Lome; 1985. p. 505–10.
33. Adomako D, Tuah AK. Digestibility of cocoa pod husk pectin polysaccharides by sheep fed rations containing cocoa pod husk. In: Proceedings of the 10th International Cocoa Research Conference; 1987; Santa Domingo, Dominican Republic; 1988. p. 853–56.
34. Osei SA, Atuahene CC, Heathcote D, et al. Cocoa pod and husk meal as a feed ingredient in layer diets. Anim Feed Sci Technol. 1991;35:283–8.
35. Osei SA, Sarpong PK, Adomako D, et al. Response of laying chickens fed graded levels of cocoa pod and husk meal. J Univ Sci Technol. 1991;15:5–8.
36. Otchere EO, Musah IA, Bafi-Yeboah M. The digestibility of cocoa husk based diets fed to sheep. Trop Anim Prod. 1983;8:33–8.
37. Olubamiwa O, Otun AR, Longe OG. Dietary inclusion rate of cocoa husk for starter cockerels. Int J Poult Sci. 2002;1(5):133–5.
38. Smith OB, Adegbola AA. Studies on the feeding value of agro-industrial by-products for livestock. III. Digestibility of cocoa pod and cocoa pod-based diets by ruminants. Anim Feed Sci Technol. 1985;13:249–54.
39. Smith OB, Osafo ELK, Adegbola AA. Studies on the feeding value of agro-industrial by-products: strategies for improving the utilization of cocoa pod-based diets by ruminants. Anim Feed Sci Technol. 1988;20(3):189–201.
40. Sobamiwa O. Use of cocoa pod husk in poultry feeds: a particular reference to the Nigerian situation. In: Proceedings of the 11th International Cocoa Research Conference; 1993; Yamoussoukro; 1994. p. 797–801.
41. Sobamiwa O. Performance and egg quality of laying hens fed cocoa husk-based diets. Niger J Anim Prod. 1998;25:22–4.
42. Sobamiwa O, Longe OG. Utilization of coca pod pericarp fractions in broiler chick diets. Anim Feed Sci Technol. 1998;37:237–44.
43. Afolami CA, Eruvbetine D. Economics of substituting cocoa pod husk for maize in livestock feeds: the Nigerian case. In: Proceedings of the 11th International Cocoa Research Conference; 1993; Yamoussoukro; 1994. p. 803–07.
44. Alawa JP, Umunna NN. Alternative feed formulation in the developing countries: prospects for utilization of agro-industrial by-products. J Anim Prod Res. 1993;13:63–98.
45. More to cocoa than the bean. http://www.new-agri.co.uk/99-2/focuson/focuson6.html. Accessed 13 June 2011.
46. Tuah AK, Orskov ER. A study on the degradation of untreated, ammonia-treated, sodium hydroxide-treated and water soaked corncob and cocoa pod husk in the rumen using the nylon bag technique. In: Proceedings of the 4th African Research Network for Agricultural By-products Workshop; 1987; Bamenda, Cameroon; 1989. p. 363–70.
47. Tuah AK. Utilization of agricultural by-products for village and commercial production of sheep rations in Ghana. Proceedings of the 1st Joint Pastures Network for Eastern and Southern Africa/African Research Network for Agricultural By-products (PANESA/ARNAB) Workshop; 1988; Lilongwe, Malawi. http://www.ilri.org/Infoserv/Webpub/Fulldocs/X5536e/x5536e04.htm. Accessed 15 June 2011.
48. Alemawor F, Dzogbefia VP, Oddoye EOK, et al. Effect of *Pleurotus ostreatus* fermentation on cocoa pod husk composition: influence of fermentation period and Mn^{2+} supplementation on the fermentation process. Afr J Biotechnol. 2009;8(9):1950–8.
49. Alemawor F, Dzogbefia VP, Oddoye EOK, et al. Enzyme cocktail for enhancing poultry utilization of cocoa pod husk. Sci Res Essay. 2009;4(6):555–9.

50. Oddoye EOK, Rhule SWA, Agyente-Badu K, et al. Fresh cocoa pod husk as an ingredient in the diets of growing pigs. Sci Res Essay. 2010;5(10):1141–4.
51. Adomako D. Recent developments in cocoa by-products research in Ghana. Proceedings of the 6th International Cocoa Research Conference; 1997; Venezuela; 1998. p. 706–18.
52. Agyente-Badu CK, Gyedu E, Oppong H. Production of local soft soap (*Alata samina*) from cocoa pod husk. Paper presented at: International Workshop on Cocoa By-products in Ghana; 2003 July 14–16; La-Palm Royal Hotel, Accra. p. 41–3.
53. Ahenkorah Y, Appiah MR, Halm BJ, et al. Production and utilization of cocoa pod husk as a source of potash for fertilizer. In: Proceedings of the 10th International Cocoa Research Conference; 1987; Santa Domingo; 1988. p. 843–48.
54. Oladokun MAO. Use of cocoa pod husk as fertilizer for maize production. Niger J Agron. 1986;1:103–9.
55. Agyente-Badu CK, Adomako D, Asante EG. Development of cocoa butter-based toilet soap. Paper presented at: International Workshop on Cocoa By-products in Ghana; 2003 July 14–16; La-Palm Royal Hotel, Accra. p. 44–7.
56. Agyente-Badu CK, Esther Gyedu, Adomako D, et al. Development of cocoa butter-based pomade. Paper presented at: International Workshop on Cocoa By-products in Ghana; 2003 July 14–16; La-Palm Royal Hotel, Accra. p. 35–6.
57. Ayinde OE, Ojo V, Adeina AA, et al. Economics of using cocoa bean shell as feed supplement for rabbits. Pak J Nutr. 2010;9(2):195–7.
58. Sukha DA. Potential value added products from Trinidad and Tobago cocoa. In: Proceedings of the Seminar/Exhibition on the Revitalization of the Trinidad and Tobago Cocoa Industry – targets, problems and options, The University of the West Indies, St. Augustine; 2003 Sept 20. p. 69–73.
59. Vitasoil. (2003). Cocoa shell mulch. http://www.vitasoil.com.cocoa%20main.htm. Accessed 15 June 2011.
60. Hansen S, Trammel H, Dunayer E, et al. Cocoa bean pulp as mulch: methylxanthine toxicosis in dogs. ASPCA Animal Poison Control Center, Urbana, IL. http://www.aspcapro.org/mydocuments/cocoa-mulch.pdf. Accessed 15 June 2011.
61. Sabariah S, Fisal A, Yap SKC. R & D on cocoa by-product development in Malaysian Cocoa Board. Paper presented at: International Workshop on Cocoa By-products in Ghana; 2003 July 14–16; La-Palm Royal Hotel, Accra. p. 80–7.
62. Brewer A. Summary report on the feasibility study on pilot plants to process cocoa by-products. Paper presented at: International Workshop on Cocoa By-products in Ghana; 2003 July 14–16; La-Palm Royal Hotel, Accra. p. 70–9.

Chapter 4
The Microbiology of Cocoa Fermentation

Dennis S. Nielsen, Michael Crafack, Lene Jespersen, and Mogens Jakobsen

Key Points

- Fermentation is an essential step in the development of "cocoa" or "chocolate" flavor.
- The fermentation is carried out by a complex consortium of microorganisms.
- Yeast, lactic acid bacteria, and acetic acid bacteria are involved.
- Pulp sugars are metabolized into ethanol and acetic acid through exothermal processes.
- Acetic acid and ethanol penetrate the beans and in combination with the heat generated kill the germ and break down cell walls inside the beans, initiating the processes that lead to "cocoa" flavor.
- Proper fermentation practices are important to avoid formation of mycotoxins.

Keywords Fermentation • Yeast • Lactic acid bacteria • Acetic acid bacteria • Microbial interactions • Cocoa quality

Primary Processing of Cocoa

Raw cocoa has an astringent unpleasant taste and has to be fermented, dried, and roasted to obtain the characteristic "cocoa" taste and flavor. The fermentation and drying steps are often referred to simply as "curing." The cocoa beans originate as seeds in fruit pods of the tree *Theobroma cacao*. Each fruit pod contains 30–40 beans embedded in a mucilaginous pulp. The pulp is rich in glucose, fructose, and sucrose (total content 10–15%), and the initial pH is relatively low (pH = 3.3–4.0), primarily due to a high concentration of citric acid (1–3%). A relatively high content of pectin and other polysaccharides (1–2%) makes the pulp viscous, limiting diffusion of air [1–4]. The actual fermentation takes place in the pulp, where microbiological activity in the pulp leads to the initiation of various biochemical processes important for taste and flavor development inside the beans, processes that continue during the drying step. A further purpose of the fermentation is to facilitate

D.S. Nielsen, M.Sc., Ph.D. (✉) • M. Crafack, M.Sc. • L. Jespersen, M.Sc., Ph.D. • M. Jakobsen, DVM
Department of Food Science, University of Copenhagen,
Rolighedsvej 30, Frederiksberg C 1958, Denmark
e-mail: dn@life.ku.dk; mcrafack@life.ku.dk; lj@life.ku.dk; moj@life.ku.dk

R.R. Watson et al. (eds.), *Chocolate in Health and Nutrition*, Nutrition and Health 7,
DOI 10.1007/978-1-61779-803-0_4, © Springer Science+Business Media, LLC 2013

Fig. 4.1 (**a**) Heap under construction; the beans are placed on plantain leaves. (**b**) Heap covered with plantain leaves (Picture courtesy of Susanne Hønholt)

removal of the mucilaginous pulp surrounding the beans as the pulp inhibit drying of the beans to a microbiological stable water content [5–7].

It is common practice in many cocoa-producing regions to harvest the pods over several days before the collected pods are transported to a place at the farm suitable for subsequent handling. Storing the pods for some days before opening is considered beneficial for the fermentation, as it results in a more rapid increase in temperature during fermentation – and thus a faster fermentation – presumably because sucrose is converted to glucose and fructose [8–11]. The pods are broken open with, for example, a cutlass or any other convenient tool, and the beans scooped out of the broken pod [6, 12]. It is important to avoid damaging the beans during opening, as damaged beans are more prone to mold attacks, leading to formation of mycotoxins such as ochratoxin A [13, 14].

Fermentation

Following opening of the pods, the cocoa beans are spontaneously inoculated with a variety of microorganisms [5, 15–18]. During the fermentation, various yeasts, lactic acid bacteria (LAB), acetic acid bacteria (AAB), and possibly *Bacillus* spp. develop in a form of succession, carrying out the fermentation [1, 2, 4, 5, 16–24]. The microorganisms metabolize the fermentable pulp sugars to ethanol. Subsequently, some of the ethanol is further oxidized to acetic acid through an exothermal process, causing an increase of the temperature of the fermenting mass to 45–50°C. The ethanol and acetic acid penetrate the beans. This, in combination with the heat produced, kills the germ and breaks down the cell walls in the bean, initiating the processes leading to well-fermented beans [2, 5, 16, 18, 21]. More details on the involvement and roles of different groups of microorganisms in the fermentation of cocoa are presented in other sections of this chapter.

Various cocoa fermentation systems have been developed. The heap fermentation system dominates in Ghana and other West African countries [6, 12, 25]. In a heap fermentation, the beans are piled on and covered with plantain leaves (Fig. 4.1) or plastic tarpaulin. The cover protects the fermenting mass against insects and conserves heat [6, 25].

Fig. 4.2 (**a**) Turning of a heap fermentation. (**b**) Heap fermented for 48 h; the beans of half the heap have been pushed aside. Three zones with varying degree of fermentation marked *1* (*outer zone*), *2* (*middle zone*), *3* (*center zone*) (Pictures courtesy of D.S. Nielsen)

It is recommended to turn/mix the heaps every 24–72 h during the fermentation to ensure a uniform fermentation, enhance growth of beneficial microorganisms, and limit the growth of unwanted microorganisms [12, 26]. However, turning of the heaps by hand is tedious (Fig. 4.2a); according to Baker et al. [12] 57% of the Ghanaian farmers do not turn their heaps. A more recent investigation by Aneani and Takrama [25] confirmed this finding, as they reported that 62% of Ghanaian farmers do not turn their heaps.

The fermentation progresses faster in the outer, well-aerated parts of the fermenting mass, as illustrated in Fig. 4.2b, where a heap fermentation has been opened after 48 h of fermentation and the beans of half the fermentation pushed aside. Three zones with varying degrees of fermentation can be recognized: (1) a thin, outer layer, where the beans are almost fully fermented and the pulp surrounding the beans has been broken down and drained away; (2) a middle zone, where the fermentation has progressed some; and (3) a central zone, where the beans are surrounded by a white mucilaginous pulp and have the same appearance as when they were removed from the pod. Turning is, in other words, necessary to achieve a final product with a uniform degree of fermentation.

To circumvent the laborious process of turning the heaps, an experimental tray-based system claimed to give high-quality beans in shorter time than the traditional heap system has been developed at the Cocoa Research Institute of Ghana (CRIG). In the tray system, the raw cocoa beans are placed in 10-cm deep trays, and 8–10 trays are stacked on top of each other, as illustrated in Fig. 4.3a. Air is allowed to circulate between the trays, ensuring aeration of the fermenting mass without turning the beans [27, 28]. Until recently, the tray system remained an experimental system, but during recent years, the system has caught renewed interest from chocolate producers and has now been distributed to several hundred farmers in Ghana [29]. The tray system has been found to change the aroma of the produced beans in a more fruity direction compared to traditional heap fermentations [30].

A third system widely used in Brazil, Indonesia, Malaysia, and to some extent in West Africa, for example, is the box fermentation system [6]. As illustrated in Fig. 4.3b, a number of boxes (here, three) are arranged in a stair-wise manner on top of each other. Following pod breaking, the beans are placed in the top box. After 1–2 days of fermentation, the beans are moved to the middle box and, finally, after 2–3 days of further fermentation into the lowest. The box fermentation systems facilitate turning, as the movement of the beans is aided by gravity and cocoa of good quality can be produced [6, 26]. However, a problem occasionally encountered during box fermentation is uneven temperature and oxygen distribution through the fermenting mass with the corners and areas around aeration holes

Fig. 4.3 (**a**) Tray fermentation. (**b**) Box fermentation of cocoa beans (Pictures courtesy of D.S. Nielsen)

being better aerated and occasionally colder and less acidic than the rest of the fermenting mass, enabling molds to grow abundantly. Care must be taken with respect to adequate design and fermentation practice to avoid this [4, 26, 31, 32].

Several other fermentation systems, including fermentation on drying platforms, in bags, barrels, baskets, and holes dug in the ground, have been used over the last century [6, 8, 18, 26]. However, these methods are not widely used.

Unfortunately, a firm and easy-to-follow definition of when to stop the fermentation has never been developed. As a consequence, fermentation time varies widely from country to country and even from farmer to farmer. As an example, it can be mentioned that in Ghana, some farmers ferment their heap fermentations for 3 days, whereas others ferment for up to 7 days [12, 25].

However, even though no strict definition of when to terminate the fermentation has been developed, the experienced farmer still has a good idea of when to stop fermenting and start drying. This is based on the smell of the fermenting mass (development of acetic odor from the activity of acetic acid bacteria), the internal and external appearance of the beans, and falling temperature of the fermenting mass [33, 34].

Drying

The fermentation is stopped by drying the beans. The moisture content of the beans must be brought from the initial 40–60% to 6–7% to avoid growth of molds. Furthermore, biochemical processes important for flavor and color development of the cocoa beans take place during drying, and the drying process is thus essential for the production of high-quality cocoa. From the point of avoiding mold growth, it is desirable to dry as fast as possible, whereas from the point of view of proper flavor development, drying should not be too fast. It has been established that drying should take at least 48 h to allow proper flavor development [6, 18, 35, 36].

Two systems are used for drying: sun drying and mechanical drying. During sun drying, the beans are spread in a thin layer on, for example, bamboo mats raised from the ground (Fig. 4.4a), plastic sheds, or concrete floors. It is important to ensure uniform drying by mixing the beans regularly, breaking up clumps of beans, and so forth. Furthermore, the beans should be protected from rewetting by dew during the night and rain by covering the beans adequately (Fig. 4.4b) or by collecting and moving the beans indoor during rain and at night [6, 18].

Under sunny conditions, the beans dry within a week, but under cloudy or rainy conditions, drying may take up to 3–4 weeks. Obviously, prolonged drying increases the risk of mold growth and spoilage [6, 18].

Mechanical drying is another option. Generally, hot-air dryers driven by wood or oil are employed. Numerous designs have been developed, but normally, indirect heating using heat exchangers is preferred. The initial drying rate must be slow and with frequent mixing to achieve uniform removal of water and allow time for the flavor and aroma-precursor producing biochemical reactions to complete. This is achieved by keeping the temperature at 60°C or lower and drying for at least 48 h [6, 18].

Sun drying is the most widely used method for several reasons: First and foremost, sun drying is cheap, with no need to invest in expensive equipment and consumables (fuel). Furthermore, cocoa is very prone to contamination with smoke. If smoke from the heating source of the mechanical dryer reaches the cocoa owing to badly constructed or poorly maintained equipment, it will result in cocoa with smoky off-flavors, which will severely limit the value of the product [6, 18].

Further Handling

Following proper drying, the beans are ready for sale. Normally, the beans are stored and transported in jute bags or as bulk goods. Some cocoa-producing countries have established industry for further processing of the beans into cocoa butter, cocoa powder, cocoa mass, and chocolate, but typically, the fermented and dried beans are exported directly to Europe, the United States, Japan, and other parts of the industrialized world [6, 37]. Here the beans are roasted, de-shelled, and further processed into cocoa butter (the fat-containing part), cocoa powder (the defatted part), or used directly for production of chocolate.

Fig. 4.4 (**a**) Sun drying on bamboo mats. (**b**) Beans covered to protect against rain (Pictures courtesy of D.S. Nielsen)

The Microbiology of Cocoa Fermentation

Early investigations by Preyer-Buitenzorg [84], Koeppen [48], and Bainbridge and Davies [44] all showed that microorganisms were involved in the fermentation of cocoa. Over the last century, a number of studies have investigated the microbiology of cocoa fermentations, and especially during the last decade, our understanding of the process has greatly expanded with the publication of a number of studies taking advantage of molecular biology–based methods.

Origin of Inocula

The fermentation of cocoa is a spontaneous microbiological process. The interior of undamaged, healthy pods are sterile or almost sterile, containing no more than a few hundred microorganisms per gram [5, 17, 38, 39].

Faparusi investigated the occurrence of yeasts associated with cocoa at different stages from flower to ripe pods. A number of yeasts often present in high numbers during the actual fermentation, such as *Candida krusei* (imperfect form of *Pichia kudriavzevii*), *Kloeckra apiculata* (imperfect form of *Hanseniaspora uvarum*), and *Pichia membranifaciens,* were detected at the different stages of maturation [38]. The incidence of yeasts isolated from mature, aseptically opened pods differed from farm to farm. The incidence of *P. membranifaciens*–infected pods differed from 25% to 50% among the five farms investigated, whereas the incidence of *Saccharomyces* spp.–infected pods ranged from 2% to 7.5% of the pods [38]. Maravalhas [31] mentions the occasional isolation of *C. krusei* from the interior of aseptically opened pods, and Jespersen et al. [17] reported that Ghanaian cocoa pods contained around 10^2 yeast/g with *P. kudriavzevii* (then named *Issatchenkia orientalis* [40]) constituting 80% of the isolates and *P. membranifaciens* 20% of the isolates.

A wide range of yeasts have been isolated from the surface of Ghanaian cocoa pods, with *Hanseniaspora guilliermondii, Hanseniaspora opuntiae,* and *Pichia guilliermondii* being among the most frequently isolated [15, 17].

Rombouts [47], Ostovar and Keeney [43], Jespersen et al. [17], and Daniel et al. [15] all isolated a wide range of yeasts involved in the fermentation of cocoa from equipment such as knives used for pod breaking, fermentation boxes, pod surfaces, dried pulp, and from workers' hands. Furthermore, plantain leaves used for covering the fermenting mass have been suggested as a source of inoculation as well [41].

The fruit fly *Drosophila melanogaster* and other insects such as ants are another possible (and possibly underestimated) source of inoculation [42, 43]. During cocoa fermentations, *D. melanogaster* is present in numbers so high that it is even referred to as "the cocoa fly" in a few early publications [44, 45]. Early experiments by Nicholls [45] suggested that *D. melanogaster* plays an important role in inoculating the cocoa pulp with yeasts, and Bainbridge and Davies [44] state that *D. melanogaster* is the main agent responsible for inoculating the fermenting mass with acetic acid bacteria. Ostovar and Keeney [43] isolated three different AAB species, four different *Bacillus* spp., four different LAB species, and various yeasts from four fruit flies collected at a cocoa farm on Trinidad. All isolated species were isolated from fermenting cocoa at the farm as well. In Brazil, it has been found that *Drosophila* spp. normally carries *P. kudriavzevii* (*C. krusei*) and is an important vector for transferring microorganisms between ecological niches [31, 42, 46].

Overall Microbial Development During Fermentation

During the initial phases of the fermentation, growth of yeasts is favored due to the high sugar content, low pH (due to the relatively high content of citric acid), and limited oxygen availability in the pulp [18]. During the first 24–36 h of fermentation, the yeast population increase to 10^7–10^8 CFU/g, normally followed by a steady decline throughout the rest of the fermentation [1, 2, 4, 21, 22, 47]. Climatic conditions have been reported to influence the progress of fermentation [6, 21, 47, 48]. As an example, Jespersen et al. [17] reported that during some Ghanaian cocoa fermentations carried out in December, one of the dry months in Tafo, Ghana [6], yeast growth was slower, not peaking until after 48–72 h of fermentation.

Table 4.1 Growth [log(CFU/g), standard deviations in brackets] of yeast, lactic acid bacteria (LAB), acetic acid bacteria (AAB), and *Bacillus* spp. during fermentation of cocoa in the outer (15 cm from the surface) and central parts of a large (500 kg) heap fermentation, a small (50 kg) heap fermentation (15 cm from surface), and a tray fermentation. All fermentations carried out in October. The large heap fermentation was turned after 48 and 96 h

	Fermentation time (hours)												
	0	12	24	36	48	60	72	84	96	108	120	132	144
Yeast, log(CFU_{yeast}/g)													
Large heap,	6.97	7.24	6.60	4.90	5.38	5.08	6.87	6.67	7.32	5.40	7.86	5.65	5.55
outer part	(0.04)	(0.03)	(0.21)	(0.43)	(0.25)	(0)	(0.02)	(0.16)	(0.34)	(0.02)	(0.12)	(0.14)	(0.14)
Large heap,	7.23	7.47	5.73	6.34	5.19	5.25	4.76	5.82	4.30	5.48	4.40	5.85	4.30
central part	(0.21)	(0.12)	(0.02)	(0.06)	(0.16)	(0.07)	(0.06)	(0.06)	(0)	(0.04)	(0.71)	(0.10)	(0)
Small heap	7.38	8.03	6.68	5.53	5.88	3.75	5.96	4.99	4.30				
	(0)	(0.05)	(0.10)	(0.04)	(0.14)	(0.21)	(0.03)	(0.55)	(0)				
Tray	7.45	7.57	7.18	6.33	6.49	4.25	5.55	6.58	4.29				
	(0)	(0.10)	(0.38)	(0.17)	(0.16)	(0.07)	(0.07)	(0.03)	(0.12)				
LAB, log(CFU_{LAB}/g)													
Large heap,	5.99	8.09	8.86	9.08	9.03	8.47	9.09	9.25	9.23	6.72	9.16	8.73	8.88
outer part	(0.12)	(0.04)	(0.15)	(0.11)	(0.10)	(0.02)	(0.27)	(0.03)	(0.12)	(0.03)	(0.06)	(0.07)	(0.03)
Large heap,	5.53	7.02	9.16	9.18	9.41	9.40	9.29	8.17	6.76	4.90	6.43	8.29	5.58
central part	(0.04)	(0.03)	(0.13)	(0.04)	(0.05)	(0.01)	(0.08)	(0)	(0.05)	(0)	(0.05)	(0.12)	(0.04)
Small heap	6.25	8.35	8.36	9.06	9.15	8.93	8.60	6.37	N.D.				
	(0.07)	(0.21)	(0.18)	(0.16)	(0.18)	(0.18)	(0.60)	(0.10)					
Tray	7.23	8.62	9.71	9.54	9.33	8.27	9.07	8.45	8.58				
	(0.21)	(0.03)	(0.10)	(0.34)	(0.04)	(0.04)	(0.18)	(0.04)	(0.19)				
AAB, log(CFU_{AAB}/g)													
Large heap,	5.60	N.D.	7.14	7.33	7.76	6.94	No data[a]	7.54	7.38	5.41	8.14	5.84	5.99
outer part	(0)		(0.02)	(0.17)	(0.02)	(0.01)		(0.06)	(0)	(0.05)	(0.09)	(0.34)	(0.12)
Large heap,	N.D.	N.D.	5.60	6.99	4.60	7.13	6.29	4.90	4.60	6.47	3.70	N.D.	N.D.
central part			(0)	(0.12)	(0)	(0.02)	(0.12)	(0.43)	(0)	(0.04)	(0)		
Small heap	N.D.	6.19	7.76	7.88	7.60	6.26	6.25	6.60	N.D.				
		(0.16)	(0.02)	(0.06)	(0)	(0.04)	(0.07)	(0)					
Tray	N.D.	N.D.	7.05	7.70	7.34	7.49	5.96	3.99	N.D.				
			(0.21)	(0.14)	(0.06)	(0.16)	(0.03)	(0.55)					
Bacillus spp. log($CFU_{Bacillus}$/g)[b]													
Large heap,	N.D.	N.D.	N.D.	N.D.	5.49	7.74	N.D.	3.60	9.10	5.75	7.99	7.75	7.81
outer part					(0.15)	(0.09)		(0)	(0.71)	(0.21)	(0.12)	(0.04)	(0.19)
Large heap,	N.D.	N.D.	N.D.	N.D.	N.D.	7.84	N.D.	4.14	4.66	6.96	5.57	7.43	6.83
central part						(0.34)		(0.09)	(0.08)	(0.26)	(0.10)	(0.10)	(0.32)
Small heap	N.D.	N.D.	N.D.	N.D.	N.D.	4.05	N.D.	5.75	5.75				
						(0.21)		(0.21)	(0.21)				

Adapted from reference [2]
N.D. none detected
[a] No data, plates partly overgrown by slimy non-AAB. CFU_{AAB} approximately as for 60 h of fermentation
[b] No *Bacillus* spp. detected during tray fermentation

Aeration by turning of the heap influences yeast growth [2], as seen from Table 4.1, where a 500-kg heap is turned after 48 and 96 h of fermentation, resulting in a decrease in the yeast cell count in the outer part of the fermenting mass after turning and followed by renewed growth.

The primary activity of the yeasts is production of ethanol from carbohydrates. Consequently, a sharp increase in the ethanol concentration and a decrease in the concentration of fermentable sugars are observed during the first 24–36 h of fermentation [1, 2, 4, 7, 16, 21]. Maximum pulp ethanol concentrations as high as 6.5% [1] and as low as 1% or less [4, 22] have been reported. Recent studies by

Nielsen et al. [2], Camu et al. [16], Camu et al. [49], and Papalexandratou all report maximum ethanol values in the pulp around 2%. In general, the fermentation progresses slower in the center compared to the outer parts of heaps [2, 22].

The conversion of glucose and fructose to ethanol is an exothermic process producing 93.3 kJ/mol reaction [34, 50, 51]. Consequently, the production of ethanol in the initial phases (~first 24 h) of fermentation is accompanied by a moderate increase in temperature to around 35°C [2].

During the early phases of fermentation abundant growth of LAB also fermenting the sugars producing lactic acid is observed [1, 4, 18, 21, 24]. Lactic acid bacteria have been reported to reach 10^8–10^9 CFU/g during the first 24–48 h of fermentation [1, 2, 4, 16, 22, 49]. In Brazilian, Indonesian, and Trinidadian box fermentations, LAB have been found to decrease to low (10^4 CFU/g) or nondetectable numbers during the later stages of fermentation [1, 4, 47], whereas other studies report that LAB constitute a significant part of the micropopulation throughout Ghanaian heap and tray fermentations [2, 16, 22, 49], Dominican box fermentations [52], and Brazilian box fermentations [53].

The role of yeast and LAB during the fermentation of cocoa is not restricted to producing ethanol and lactic acid. Breakdown of the pulp through pectinolytic activity by the yeast, assimilating citric acid by the yeast and LAB, and reduction of fructose to mannitol by some heterofermentative LAB such as *Lb. fermentum* are other important functions [18, 24, 54–58]. The assimilation of citric acid causes the pulp pH to increase with 0.3–0.8 units from the onset usually around pH 3.5–4 [2, 16, 52].

Ardhana and Fleet [1] found that highly pectinolytic filamentous fungi grew well during the first 36 h of fermentation and suggested that filamentous fungi play a key role in the degradation of pulp pectin during the early phases of fermentation. On the contrary, Roelofsen and Giesberger [21] stated that molds play no role during normal cocoa fermentations on Java (Indonesia).

The breakdown of the pulp causes some of the pulp to drain away and increase aeration of the fermenting mass, favoring growth of aerobic AAB [4]. In general, AAB have been reported to grow from low or even nondetectable levels at the onset of fermentation to levels anywhere between 10^5 and 10^9 CFU/g, often followed by a decline toward the end of fermentation [1, 2, 4, 16, 18, 21, 22, 43, 47, 52, 53]. As an example, Nielsen et al. [2] found that AAB grew from low or nondetectable levels at the onset of fermentation to 10^7–10^8 CFU/g after 36 h of fermentation. Aeration of the fermenting mass by turning has been found to positively influence AAB growth [2, 22, 49].

In general, it is believed that AAB growth requires oxygen, but some strains are seemingly capable of growing under conditions with very low oxygen tension, as Carr and Davies [23] found that AAB from cocoa fermentations could be isolated from plates incubated anaerobically. The ability of AAB to spoil bottled wine further underlines their potential for growing under very limited oxygen conditions [59].

The AAB metabolizes the ethanol initially formed by the yeasts to acetic acid through an exothermal process (496 KJ/mol) [34]. Furthermore, some AAB species are capable of completely oxidizing acetic acid to CO_2 and H_2O through an exothermal process [34, 60]. The AAB activity leads to a further increase in temperature, reaching 45–50°C or even higher [2, 4, 16, 21, 22, 34, 51, 61–63]. In general, the temperature development in the center of large heap fermentations is relatively slow the first 36–48 h of fermentation compared to the outer part of large heaps and smaller fermentations with better aeration [2, 21, 22, 61, 64, 65].

The ethanol and acetic acid diffuse into the beans. This in combination with the heat produced by the activities of the AAB leads to bean death and induces the biochemical changes leading to well-fermented cocoa beans [18, 63].

The high temperatures in combination with the ethanol and acetic present are a severe stress factors, limiting the growth or killing vegetative cells of many microorganisms [18, 21, 26]. In the later phases of fermentation, environmental conditions with increasing pH and aeration become favorable for growth of spore-forming *Bacillus* spp., often reaching 10^8 CFU/g or more in the later stages of the fermentation [1, 2, 4, 22, 47].

The role of *Bacillus* spp. in the fermentation of cocoa has not been fully elucidated. However, due to their high enzymatic activity and production of, for example, short-chain fatty acids, pyrazines, and 2,3-butanediol, the growth and activity of *Bacillus* spp. is likely to influence the final product, possibly causing off-flavors [66–69], but *Bacillus* spp. have also been suggested to play a positive role as producers of pectinolytic enzymes during the fermentation [70, 71]. However, as *Bacillus* spp. in general are present in relatively low numbers during the first days of fermentation where pectin breakdown is most important, it can be argued that even though *Bacillus* spp. isolated during cocoa fermentation produce pectinolytic enzymes, their actual importance for the process is negligible.

Toward the end of fermentation, filamentous fungi grow well in the cooler and more aerated parts of the fermenting mass [26, 31]. Mold growth has been associated with internal molding of the beans (a serious commercial defect), various off-flavors such as "moldy odor," and increased free fatty acids levels [72–77]. Furthermore, mold growth may lead to production of mycotoxins such as ochratoxin A, aflatoxins, and citrinin [78, 79]. Ochratoxin A can be produced not only during the drying step but also during the actual cocoa fermentation [14, 80, 81]. It is furthermore likely that excessive mold growth during the later parts of fermentation will enhance mold growth and thus potential mycotoxin production during the subsequent drying phase [5, 75, 76, 82].

Yeast Involved in Fermentation

The distinct smell of alcohol apparent during the early stages of cocoa fermentation soon led early investigators of the cocoa fermentation to conclude that yeasts were involved in the process [45, 83–85]. These early studies reported the occurrence of a *Saccharomyces*-like yeast for which Preyer-Buitenzorg [84] and later Nicholls [45] suggested the name *Saccharomyces theobromae*. However, as stated by Rombouts [85], it is most likely that *Sc. theobromae* was a mixture of different yeasts species. Furthermore, a fission yeast, possibly *Schizosaccharomyces pombe*, and an apiculate yeast, possibly *K. apiculata* and *Saccharomyces* spp., were reported frequently [44, 45, 83–85].

The involvement of *Saccharomyces* spp., *Schizosaccharomyces* spp., and apiculate yeasts in the fermentation of cocoa in all the important cocoa-producing regions of the world, including South America, Indonesia, and West Africa, was confirmed by later studies [73, 86–88]. According to Knapp [50], H.A. Dade isolated a number of yeasts from Gold Coast (Ghana) cocoa fermentations and deposited these isolates at the Centralbureau voor Schimmelcultures (CBS), Holland, where they were later identified as *K. apiculata*, *C. krusei* (*P. kudriavzevii*), *Pichia fermentans*, *Candida mycoderma* (now *Candida vini* [89]), and *Saccharomyces cerevisiae* [50, 85].

Roelofsen and Giesberger [21] carried out a comprehensive study on box fermentation of cocoa in Java (Indonesia). Unfortunately, except for the identification of a few isolates to genus level (*Pichia* spp. and *Saccharomyces* spp.), no serious attempts were made to identify the yeasts isolated during the study. Rombouts [47] studied box fermentations on Trinidad and found that yeasts dominated the fermentations during the first 24 h. The isolates were not identified in the original publication, but according to Roelofsen [5], the isolates were later identified and *C. krusei* was found to be the predominating yeast. Among others, *Sc. cerevisiae* (then named *Sc. cerevisiae* var. *ellipsoideus* [89]), *K. apiculata,* and *P. membranifaciens* were isolated frequently as well [5].

Studies of box and heap fermentations in Belize, Brazil, Côte d'Ivoire, The Dominican Republic, Ghana, Malaysia, and Indonesia have confirmed that the yeast community associated with cocoa fermentations is complex, involving a number of species with *Sc. cerevisiae*, *K. apis*, *K. apiculata*, *P. membranifaciens*, *P. fermentans*, *P. kudriavzevii* (and the imperfect form *C. krusei*), and various *Candida*, *Torulopsis,* and *Schizosaccharomyces* spp. being isolated frequently [1, 4, 18, 22, 23, 43, 90–94].

More recent studies have used molecular biology–based methods for accurate identification and culture-independent characterization of the yeast communities associated with cocoa fermentation.

Table 4.2 Composition of the yeast community sampled 15 cm from the surface (outer part) and in the center of a large (500 kg) heap cocoa fermentation turned after 48 and 96 h

	Fermentation time (hours)												
	0	12	24	36	48	60	72	84	96	108	120	132	144
Outer part % of yeast population													
H. opuntiae [a]	79	67	71										
S. crataegensis	7												
P. pijperi	9	2											
C. awuaii	3												
P. sporocuriosa	2												
C. zemplinina		19	14										
C. michaelii		3											
C. diversa		9	14										
P. membranifaciens			Det.[b]	100	70	50	100	100	100	90	100	100	100
C. ethanolica					15						Det.	Det.	
Sc. cerevisiae					15	17							
Schiz. pombe						33							
P. kudriavzevii										10			
Central part, % of yeast population													
H. opuntiae [a]	78	53	21		25								
P. kluyveri	11												
C. diversa	11				13								
C. awuaii	Det.	14											
S. crataegensis	Det.												
C. zemplinina		33	57										
P. membranifaciens				9	13	11	31	97	50	100	100	67	100
Sc. cerevisiae			22	91	25	89	45						
P. kudriavzevii				Det.									
P. occidentalis				Det.									
C. ethanolica					13			3	50			33	
T. delbreuckii					13								
Schiz. pombe						Det.	24						

Adapted from references [90] and [2]

H. Hanseniaspora, S. Saccharomycopsis, P. Pichia, C. Candida, Sc. Saccharomyce, Schiz. Schizosaccharomyces

[a] Reported as *H. guilliermondii* in [2], but later actin gene sequencing showed that the isolates are the closely related *H. opuntiae*

[b] Detected sporadically

These studies confirmed that the yeast communities associated with cocoa fermentation are complex, with often 3–6 different yeast species detected during the first 24 h of fermentation and members of the *Hanseniaspora guilliermondii/opuntiae* complex in general being dominant (Table 4.2) [2, 15, 17, 95]. After the first 24 h of fermentation, the *Hanseniaspora guilliermondii* complex members in general have been found to die out [2], which has been explained by the relatively low ethanol tolerance of *Hanseniaspora* spp. and the imperfect form, *Kloeckera* [1, 4]. The increasing temperatures during the fermentation probably influence the growth and death of the yeast species present during the fermentation as well [1, 89]. Later in the fermentation, species such as *P. kudriavzevii, S. cerevisiae,* and *P. membranifaciens* in general dominate the yeast flora [2, 15, 17, 89, 95] but in lower numbers than what is seen in first phases of fermentation (see Table 4.1) [2]. In a recent study, Papalexandratou and De Vuyst showed that members of the *Hanseniaspora guilliermondii/opuntiae* complex seem to be indigenous to cocoa fermentations worldwide (South America, West Africa, and Asia) [96].

Investigation of chromosome length polymorphism (CLP) using pulsed field gel electrophoresis (PFGE) of a number of *H. guilliermondii*, *P. membranifaciens*, *P. kluyveri*, *Sc. cerevisiae,* and

P. kudriavzevii strains isolated during cocoa fermentations revealed that CLP was evident within all species investigated [17]. This finding further adds to the complexity of cocoa fermentations, as not only is a number of different species involved in the fermentation, but within the different species, different strains are involved as well.

Lactic Acid Bacteria Involved in Fermentation

The involvement of LAB in the fermentation of cocoa was established relatively late compared to the other major microbiological groups involved. Using microscopy, Busse et al. [88] and Ficker and von Lilienfeld-Toal [86] both observed LAB-like microorganisms in the pulp from fermenting cocoa, but LAB was not cultivated from fermenting cocoa until the work by Roelofsen and Giesberger [21] and Rombouts [47], where Roelofsen and Giesberger found *Lb. fermentum* to dominate the LAB community during Indonesian cocoa fermentations [88].

Carr et al. [22] and Carr and Davies [23] found *Lactobacillus collinoides*, *Lactobacillus fermentum*, *Lactobacillus mali,* and *Lactobacillus plantarum* to be the dominant LAB during Ghanaian heap fermentations. However, only a limited number of isolates were fully identified [23]. Later, Ardhana and Fleet [1] found *Lactobacillus cellobiosus* (later synonym of *Lb. fermentum* [97]) and *Lb. plantarum* to be the dominant LAB involved in Indonesian box fermentations.

The LAB population of Brazilian and Caribbean box fermentations have been reported to be considerably more complex involving a number of *Lactobacillus*, *Leuconostoc*, *Lactococcus,* and *Pediococcus* species, represented by species such as *Lb. fermentum*, *Lactobacillus brevis*, *Lb. plantarum*, *Leuconostoc mesenteroides*, *Lactococcus lactis,* and *Pediococcus acidilactici* [18, 24, 43, 52, 98].

Using a combination of culture-dependent and culture-independent methods and taking advantage of modern molecular biology–based methods the LAB community associated with the fermentation of cocoa has been established in a handful of more recent publications (Table 4.3) [2, 16, 20, 49, 53, 99–103]. Taken together, these studies show that a number of different LAB species (mainly *Lactobacillus*, *Weisella, Leuconostoc,* and *Fructobacillus* spp.) are associated with the fermentation of cocoa, but a clear pattern is also emerging, showing that *Lb. plantarum* and especially *Lb. fermentum* are indigenous to the fermentation of cocoa worldwide and are likely to play important roles during the fermentation.

Recently, successful attempts have been made to include *Lb. fermentum* in a starter culture consortium for consistent fermentation of cocoa [55, 104].

Acetic Acid Bacteria Involved in Fermentation

The ethanol formed by the yeasts in the early phase of fermentation is metabolized to acetic acid by AAB as mentioned previously. The smell of acetic acid associated with fermenting cocoa is easily recognizable and soon led early investigators to conclude that acetic acid–producing microorganisms were involved in and significant contributors to the fermentation of cocoa [44, 45, 50, 83, 105].

The first comprehensive study on AAB associated with cocoa was carried out by Eckmann [106], who grew AAB cultures in Germany from pulp sweatings, dry beans, and fresh fruits transported from the tropics. Eckmann [106] mainly isolated *Gluconacetobacter xylinus* (then named *Bacterium xylinus* [107]), *Acetobacter orleanensis* (then named *Bacterium orleanese* [107]), and *Acetobacter pasteurianus* (then named *Bacterium ascendens* [107]). However, as the samples were transported for

Table 4.3 Composition of the lactic acid bacteria (LAB) community during a large (500 kg) heap fermentation (samples were taken from two positions: 15 cm from the surface [outer part] and in the center of the fermenting mass), a small (50 kg) heap fermentation (samples taken 15 cm from the surface), and a tray fermentation. The large heap fermentation was turned after 48 and 96 h

	Fermentation time (hours)												
	0	12	24	36	48	60	72	84	96	108	120	132	144
Large heap, outer part, % of LAB population													
Lb. plantarum	60	11	5		9		9				21	85	63
Lb. fermentum	40	81	76	100	86	100	81	89	92	69	74	15	37
Lb. ghanensis		8						11	8		5		
Lc. pseudoficulneum			19										
Lb. Hilgardii[a]					5								
Pd. acidilactici										31			
Large heap, center, % of LAB population													
Lc. cpseudomesenteroides	82												
Lb. rossii[b]	18												
Lb. fermentum		100	98	47	87	100	100	100	100	100	52	75	4
Lc. pseudoficulneum			2										
Lb. plantarum				53	9						42	25	76
Lb. ghanensis					4								
Pd. acidilactici											6		
Lb. hilgardii[b]													20
Small heap, % of LAB population													
Lc. pseudomesenteroides	100	83		5									
Lb. fermentum		17	63	90	77	100	100	100					
Lb. plantarum			2										
Lc. pseudoficulneum			28										
Lac. lactis			7										
Pd. acidilactici				5	23								
Tray, % of LAB population													
Lb. plantarum	89	30	35	5					10				
Lc. pseudoficulneum	11		3										
Lb. fermentum		64	57	95	100	100	100	89	90				
Lc. pseudomesenteroids		6	2										
Lb. brevis			3										
Pd. acidilactici								11					

Adapted from references [2] and [102]

Lb. Lactobacillus, Lc. Leuconostoc, Pd. Pediococcus

[a] Isolates did not grow upon purification. Identified directly by rep-PCR grouping and 16S rRNA gene sequencing (see reference [2] for details)

several weeks before being examined, it is impossible to conclude anything about the significance of the isolated species on actual fermentations.

Roelofsen and Giesberger [21] found that *A. pasteurianus* (then named *Acetobacter rancens* [107]) and *Gluconobacter oxydans* (then named *Acetobacter melanogenum* [107]) were the dominant AAB during Indonesian (Java) cocoa fermentations, and in a later study on Trinidadian box fermentations, Ostovar and Keeney [43] found that *Gluconobacter oxydans* (then named *Acetobacter suboxydans* [107]) was the predominant AAB during the first 24–32 h of fermentation, whereas *A. pasteurianus* predominated the AAB community from 40 h of fermentation and onward.

Gluconobacter oxydans, *A. aceti*, and *A. pasteurianus* have been found to dominate the AAB community in a number of other studies on cocoa box fermentations in Malaysia, Belize, Brazil, and

Table 4.4 Composition of the acetic acid bacteria (AAB) community during a large (500 kg) heap fermentation (samples were taken from two positions: 15 cm from the surface [outer part] and in the center of the fermenting mass), a small (50 kg) heap fermentation (samples taken 15 cm from the surface), and a tray fermentation. The large heap fermentation was turned after 48 and 96 h

	Fermentation time (hours)												
	0	12	24	36	48	60	72	84	96	108	120	132	144
Large heap, outer part, % of AAB population							No data						
A. pasteurianus	100			55		22	–			62			
A. syzygii			57			33	–	72					
A. tropicalis			29	45	100	45	–	14	100	38	100	100	100
A. malorum			14				–						
G. oxydans							–	14					
Large heap, center, % of AAB population													
A. pasteurianus			100			22	80			13			
A. syzygii				100		45	20			34			
A. tropicalis					100	33		100	100	53	100		
Small heap, % of AAB population													
A. syzygii		100	83	89	100	100	22	75					
A. pasteurianus			17				33	25					
A. malorum				11									
A. tropicalis							45						
Tray, % of AAB population													
A. syzygii			50	100									
A. malorum			50										
A. pasteurianus					73	100							
A. tropicalis					27		100	100					

Adapted from reference [2]
A. Acetobacter, G. Gluconobacter

Ghana [18, 22, 23, 108, 109]. More recently, Ardhana and Fleet [1] isolated *A. pasteurianus* and *Acetobacter aceti* during Indonesian cocoa fermentations, but interestingly, no *Gluconobacter* spp. were isolated in this study.

During recent years, the composition of the AAB microbiota associated with the fermentation of cocoa worldwide has been investigated in detail in a number of studies taking advantage of modern molecular biology–based methods, including culture-independent methods [2, 16, 20, 49, 53, 100, 101, 103]. In the study by Nielsen et al. [2], *Acetobacter pasteurianus*, *Acetobacter syzygii*, and *Acetobacter tropicalis* were the dominating AAB during the fermentations, with occasionally isolation of *A. malorum* and *G. oxydans* detected as well (Table 4.4) [2].

All in all recent years' advancement of our understanding of the AAB microbiota associated with fermentation of cocoa strongly indicates that *A. pasteurianus* is indigenous to cocoa fermentations around the world. Recently, *A. pasteurianus* has been included in a starter culture consortium successfully used for controlled fermentation of cocoa [55, 104].

Bacillus spp. Involved in Fermentation

Bainbridge and Davies [44] were the first to observe growth of aerobic spore-forming bacteria during the later stages of cocoa fermentation and identified the bacteria as being of the *Bacillus subtilis* type.

Table 4.5 Composition of the *Bacillus* spp. community during a large (500 kg) heap fermentation and a small (50 kg) heap fermentation. The large heap fermentation was turned after 48 and 96 h and samples were taken from two positions: 15 cm from the surface (outer part) and in the center of the fermenting mass. From the small heap fermentation, samples were taken 15 cm from the surface

	Fermentation time (hours)												
	0	12	24	36	48	60	72	84	96	108	120	132	144
Large heap, outer part, % of *Bacillus spp.* population													
B. licheniformis					100				50	100	72	96	80
B. cereus						Det.[a]							
B. pumilus											18	4	11
B. megaterium						100		100	50		10		
B. subtilis													9
Large heap, center, % of *Bacillus spp.* population													
B. licheniformis								29	35	17	68	60	34
B. pumilus						Det.			22	66	32	40	66
B. megaterium						100		71					
B. sphaericus										17			
B. subtilis									43				Det.
Small heap, % of *Bacillus spp.* population													
B. megaterium						100							
B. licheniformis								100	100				

Adapted from reference [2]
B. Bacillus
[a] Detected sporadically

Later, H.A. Dade (mentioned in a private communication to Knapp, 1937) isolated *Bacillus undulatus* (*B. undulatus* is not a validly published name; furthermore, it is not clear what the present name of *Bacillus undulates* would be [107]) and *Bacillus megaterium* during Gold Coast (Ghana) cocoa fermentations. In the studies by Carr et al. [22] and Carr and Davies [23], the *Bacillus* community associated with cocoa fermentation was found to be relatively simple, as all identified isolates from Ghanaian heap fermentations were identified as *B. subtilis* and only *Bacillus licheniformis* and *Bacillus subtilis* were isolated during Malaysian box fermentations. In a later study carried out in Indonesia, Ardhana and Fleet [1] isolated *Bacillus pumilus* at each of three estates investigated and in general found that the *Bacillus* community diversity was relatively restricted, with 2–4 different species being detected at each of the sites investigated. On the contrary, Ostovar and Keeney [43] and Schwan et al. [68] reported that the *Bacillus* community associated with box fermentations in Trinidad and Brazil was complex, involving 8–14 different *Bacillus* spp.

Nielsen et al. [2] only isolated *Bacillus* spp. occasionally and in relatively low numbers during the first 2–3 days of fermentation. During later stages of fermentation, *Bacillus* spp. were detected in higher number (see Table 4.1), probably because *Bacillus* spp. are able survive the high temperatures resulting from AAB activity and in general thrive better with the increasing oxygen availability in the later stages of fermentation, caused by the breakdown of pulp. *Bacillus licheniformis* and *B. megaterium* were dominating the *Bacillus* microflora in the fermentations investigated by Nielsen et al. [2] (Table 4.5).

Whether *Bacillus* spp. play an important role during cocoa fermentation or not remains controversial. In several studies, they are not detected at all [16, 100], and in other studies, they are only detected more occasionally and often late in the fermentation [2, 20]. On the other hand, Ouattara et al. [70] reported that *Bacillus* communities were present and generally in significant numbers throughout Ivorian box and heap fermentations. Ouattara et al. [70, 71] argue that *Bacillus* spp. might play an important role due cocoa fermentation due to their production of pectinolytic enzymes.

Fig. 4.5 Visible mold growth on the surface of an unturned Ghanaian heap fermentation (Picture courtesy of Susanne Hønholt)

Other Bacteria Involved in Fermentation

A wide range of bacteria other than LAB, AAB, and *Bacillus* spp. have been isolated during cocoa fermentations, but the importance of any given species is difficult to judge, as also stated by Lehrian and Patterson [26]. However, some species are encountered frequently and probably play some role in the fermentation. In Indonesia, Ardhana and Fleet [1] isolated *Micrococcus kristinae* and various *Staphyloccus* spp. during cocoa fermentations carried out at three different plantations, and *Pseudomonas cepacea* was isolated from two out of three fermentations. Also, Roelofesen and Giesberger [21] and Ostovar and Keeney [43] mention the occasional isolation of *Micrococcus* spp. during Indonesian and Trinidadian box fermentations.

During an investigating of cocoa box fermentations carried out at two different estates in Trinidad, Ostovar and Keeney [43] observed that *Zymomonas mobilis* constituted 40–80% of the micropopulation after 16–48 h of fermentation in the central part of a box fermentation on an estate where the beans were tightly packed in the boxes and the oxygen availability thus limited. On the contrary, *Z. mobilis* accounted for a much smaller percentage of the microorganisms at the other estate, where the beans were packed more loosely and the oxygen availability higher. *Zymomonas mobilis* grows well under anaerobic conditions and is an efficient producer of ethanol from glucose, fructose, and sucrose. Consequently, it has been suggested that *Z. mobilis* along with the yeasts play a role in converting sugars to ethanol during the first days of fermentation [43, 110].

More recently, various Enterobacteriaceae, such as *Tatumella saanichensis* and *Tatumella punctata*, have been detected during the first days of cocoa fermentation, and it has been suggested that they might play a role in pectin degradation and citric acid metabolism during the early phase of fermentation [100, 103].

Molds Involved in Fermentation

Filamentous fungi can be readily observed on the surface of fermenting cocoa if left undisturbed (i.e., no turning) for a few days, as seen in Fig. 4.5. Fungal growth is rarely observed below the outer layers of the fermenting mass [5, 21, 26, 31, 72, 82], with the exception of the study by Ardhana and Fleet [1], who observed fungal growth in samples taken from the center of Indonesian box fermentations during the first 36 h of fermentation.

According to Bunting [82] and Dade [72], the molds most frequently isolated during Gold Coast (Ghana) cocoa heap fermentations were members of the *Aspergillus fumigatus* group and *Mucor* spp. Maravalhas [31] observed that molds appear in succession with *A. fumigatus* appearing first followed by *Mortierella gamsii* (then named *Mortierella spinosa* [111]) and later *Paecilomyces variotii*. In the later stages of fermentation, the temperature of the fermenting mass decreases, and this is according to Maravalhas [31] accompanied by growth of *Penicillium citrinum* and *Aspergillus glaucus*. The occurrence of *P. citrinum* was later confirmed by Ardhana and Fleet [1], who furthermore reported growth of *Penicillium purpurogenum*, *Penicillium ochrochloron*, *Aspergillus versicolor*, and *Aspergillus wentii* during the fermentations.

It is of concern that mycotoxins (ochratoxin, citrinin) have been detected and mycotoxin-producing molds such as *Aspergillus carbonarius* and *Aspergillus niger* have been isolated from fermented cocoa in frequent numbers [14, 79, 81]. However, according to Mounjouenpou et al., ochratoxin (and probably mycotoxins as such) contamination of cocoa beans can be avoided or at least minimized by adhering to good cocoa-processing practices [13, 14].

Microbial Growth During Drying

The ultimate goal of the drying process is to obtain microbiologically stable cocoa beans by drying to an extent where microbial growth becomes impossible. A water content of 7.2% has been established as a safe limit [5, 21]. As the water availability decreases during the drying process, fewer and fewer microorganisms are capable of growing, and a pronounced reduction in the number of viable microorganisms is observed [22, 24]. At the onset of drying, yeast, AAB, LAB, and *Bacillus* spp. were isolated by Carr et al. [22] in Ghana. During the solar drying process, all but spore-forming *Bacillus* spp. died out. A similar picture has been observed in Brazil [24].

Many molds are capable of growing at low water activities, and if the reduction in moisture content during drying is too slow, molds will grow [5, 6]. Various *Aspergillus* spp., *Penicillium* spp., and *Mucor* spp. have been frequently isolated from moldy cocoa beans. Mold growth during drying should be avoided, because it can result in the formation of off flavors, internal molding of the beans, and formation of mycotoxins [5, 6, 13, 14, 72, 75, 76, 82].

Use of Starter Cultures for Fermentation of Cocoa

Soon after it was realized that microorganisms were responsible for the fermentation of cocoa beans, the first attempts of controlling the fermentation using starter cultures were made, as Preyer-Buitenzorg [84] inoculated cocoa fermentations with his so-called *Sc. theobromae* (see the section Yeast Involved in Fermentation). According to Preyer-Buitenzorg [84], the resulting cocoa was of a better quality than cocoa fermented without the addition of *Sc. theobromae*. Subsequently, a number of attempts have been made to improve or control the quality of fermented cocoa by using yeast as a starter culture

[21, 50, 86, 94, 112, 113], and more recently, an interesting study using a hybrid *Kluyveromyces marxianus* strain with increased pectinolytic activity as starter culture resulting in chocolate more acceptable to a sensorical panel was published [114].

According to Busse et al. [88], the quality of Cameroonian cocoa suffered from being too acidic, and it was unsuccessfully attempted to retard the growth of AAB by adding *Lactobacillus delbrueckii* (then named *Bacillus delbrückii* [107]).

As evident from the previous sections of this chapter, cocoa fermentations are carried out by a consortium of microorganisms, including yeast, LAB, and AAB. The first attempt at using a consortium of all these groups of microorganisms for fermentation of cocoa was first published by Schwan [109]. Schwan [109] used a defined microbial inoculum consisting of a mixture of *Sc. cerevisiae*, *Lb. plantarum*, *Lactobacillus delbrueckii* subsp. *lactis* (then named *Lactobacillus lactis* [107, 115]) *A. aceti,* and *G. oxydans*. The first 2–3 days of fermentation the inoculum dominated the fermentation, but during the later phases of fermentation, the inoculum was outgrown by naturally occurring microorganisms. The chocolate produced from the inoculated bean was of a quality comparable to chocolate produced from a spontaneously inoculated control fermentation. In a couple of recent studies, Lefeber et al. [55, 101, 104] used a systematic approach for screening isolates for interesting cocoa fermentation starter culture properties and identified *L. fermentum* and *A. pasteurianus* as the essential bacteria for the process.

So far, the aim of the different attempts to use starter cultures for the fermentation of cocoa has been to improve the quality of the resulting beans in terms of flavor potential, absence of excess acidity, and/or to increase the speed of fermentation. The work of Masoud et al. and Masoud and Kaltoft points at another potential use of artificial inoculation of cocoa fermentations, as strains of *Pichia anomala*, *P. kluyveri,* and *H. uvarum* isolated from coffee fermentations have been found to inhibit growth and ochratoxin A production by *Aspergillus ochraceus* [116, 117]. The primary processing of cocoa and coffee has a number of similarities [117], and it is indeed possible that the potential of some yeasts to inhibit mold growth and mycotoxin production could be exploited in the primary processing of cocoa as well.

References

1. Ardhana M, Fleet GH. The microbial ecology of cocoa fermentations in Indonesia. Int J Food Microbiol. 2003;86:87–99.
2. Nielsen DS, Teniola OD, Ban-Koffi L, Owusu M, Andersson T, Holzapfel WH. The microbiology of Ghanaian cocoa fermentations analysed using culture dependent and culture independent methods. Int J Food Microbiol. 2007;114:168–86.
3. Pettipher GL. Analysis of cocoa pulp and the formulation of a standardised artificial cocoa pulp medium. J Sci Food Agric. 1986;37:297–309.
4. Schwan RF, Rose AH, Board RG. Microbial fermentation of cocoa beans, with emphasis on enzymatic degradation of the pulp. J Appl Bacteriol Suppl. 1995;79:96S–107.
5. Roelofsen PA. Fermentation, drying, and storage of cacao beans. Adv Food Res. 1958;8:225–96.
6. Wood GAR, Lass RA. Cocoa. London: Longman; 1985.
7. Lopez AS, Dimick PS. Cocoa fermentation. In: Reed G, Nagodawithana TW, editors. Enzymes, biomass, food and feed, vol. 9. 2nd ed. Weinheim: VCH; 1995. p. 561–77.
8. Rohan TA. Processing of raw cocoa for the market. Rome: FAO; 1963.
9. Berbert PRF. Contribuicão para o conhecimento dos acúcares components da amêndoa e do mel de cacau. Rev Theobroma. 1979;9:55–61.
10. Dougan J. Cocoa fermentation in Ghana and Malaysia (part 2): methods for the analysis of cocoa pulp and cotyledons. Bristol: Long Ashton Research Station; 1980.
11. Tomlins KI, Baker DM, Daplyn P. Effect of fermentation and drying practices on the chemical and physical profiles of Ghana cocoa. Food Chem. 1993;46:257–63.

12. Baker DM, Tomlins KI, Gay C. Survey of Ghanian cocoa farmer fermentation practices and their influence on cocoa flavor. Food Chem. 1994;51:425–31.
13. Mounjouenpou P, Gueule D, Ntoupka M, Durand N, Fontana-Tachon A, Guyot B, et al. Influence of post-harvest processing on ochratoxin A content in cocoa and on consumer exposure in Cameroon. World Mycotoxin J. 2011;4:141–6.
14. Mounjouenpou P, Gueule D, Fontana-Tachon A, Guyot B, Tondje PR, Guiraud JP. Filamentous fungi producing ochratoxin a during cocoa processing in Cameroon. Int J Food Microbiol. 2008;121:234–41.
15. Daniel HM, Vrancken G, Takrama JF, Camu N, de Vos P, De Vuyst L. Yeast diversity of Ghanaian cocoa bean heap fermentations. FEMS Yeast Res. 2009;9:774–83.
16. Camu N, De Winter T, Verbrugghe K, Cleenwerck I, Vandamme P, Takrama JS, et al. Dynamics and biodiversity of populations of lactic acid bacteria and acetic acid bacteria involved in spontaneous heap fermentation of cocoa beans in Ghana. Appl Environ Microbiol. 2007;73:1809–24.
17. Jespersen L, Nielsen DS, Hønholt S, Jakobsen M. Occurrence and diversity of yeasts involved in fermentation of West African cocoa beans. FEMS Yeast Res. 2005;5:441–53.
18. Thompson SS, Miller KB, Lopez AS. Cocoa and coffee. In: Doyle MJ, Beuchat LR, Montville TJ, editors. Food microbiology – fundamentals and frontiers. 2nd ed. Washington, DC: ASM; 2001. p. 721–33.
19. Nielsen DS, Nielsen PS, van den Berg F. Investigating the fermentation of cocoa by correlating denaturing gradient gel electrophoresis profiles and near infrared spectra. Int J Food Microbiol. 2008;125:133–40.
20. Papalexandratou Z, Camu N, Falony G, De Vuyst L. Comparison of the bacterial species diversity of spontaneous cocoa bean fermentations carried out at selected farms in Ivory Coast and Brazil. Food Microbiol. 2011;28:964–73.
21. Roelofsen PA, Giesberger G. Onderzoekingen over cacaobereidung. Archiv voor de koffiecultur in Nederlandisch-Indie. 1947;16:1–159.
22. Carr JG, Davies PA, Dougan J. Cocoa fermentation in Ghana and Malaysia. London: Tropical Products Institute; 1979.
23. Carr JG, Davies PA. Cocoa fermentation in Ghana and Malaysia (part 2): further microbiological methods and results. Bristol: Long Ashton Research Station; 1980.
24. Schwan RF, Wheals AE. The microbiology of cocoa fermentation and its role in chocolate quality. Crit Rev Food Sci Nutr. 2004;44:205–22.
25. Aneani F, Takrama JF. Practices that influence the quality of cocoa in Ghana: perceptions of cocoa farmers. In: Proceedings of the 15th international cocoa research conference, San Jose, Costa Rica; 2006.
26. Lehrian DW, Patterson GR. Cocoa fermentation. In: Reed G, editor. Food and feed production with microorganisms, vol. 5. 1st ed. Weinheim: Verlag; 1983. p. 529–75.
27. Allison HWS, Rohan TA. A new approach to the fermentation of West African Amelonado Cocoa. Trop Agric StAugustine. 1958;35:279–88.
28. Allison HWS, Kenten RH. Tray fermentation of cocoa. Trop Agric StAugustine. 1963;40:15–24.
29. Anonymous: http://www.toms.dk/Default.aspx?ID=1336 (2011).
30. Owusu M, Petersen MA, Heimdal H. Aroma of chocolate produced from tray-fermented cocoa at different stages of fermentation. In: Ho C-T, Mussinan CJ, Shahidi F, Contis ET, editors. Recent advances in food and flavor chemistry. London: RSC; 2010. p. 221–3.
31. Maravalhas N. Microbiological deterioration of cocoa beans during fermentation and storage in Bahia. Revue Internationale de la Chocolaterie. 1966;21:375–8.
32. Senenayake M, Jansz ER, Buckle KA. Effect of different mixing intervals on the fermentation of cocoa beans. J Sci Food Agric. 1997;74:42–8.
33. Forsyth WGC, Quesnel VC. Variations in cacao preparation. In: Sixth interamerican technical cacao committee; 1956, p. 157–168.
34. Forsyth WGC, Quesnel VC. The mechanisms of cacao curing. Adv Enzymol. 1963;25:457–92.
35. Faborode MO, Favier JF, Ajayi OA. On the effects of forced air drying on cocoa quality. J Food Eng. 1995;25:455–72.
36. Nganhou J, Njomo D, Benet JC, Augier F, Berthomieu G. Perfecting a method of micro-analysis of water and acetic acid in a cocoa bean in the course of drying: applying to determine transportation coefficients. Heat Mass Trans. 2003;39:797–803.
37. Anonymous. ICCO annual report 2009/2010. International Cocoa Organization, 2011. http://www.icco.org/about/anualreport.aspx
38. Faparusi SI. The yeasts associated with cacao (*Theobroma cacao*) pods. Egypt J Microbiol. 1974;9:105–15.
39. Hoynak S, Polansky TS, Stone RW. Microbiological studies of cacao fermentation. Food Res. 1941;6:471–9.
40. Kurtzman CP, Robnett CJ, Basehoar-Powers E. Phylogenetic relationships among species of *Pichia*, *Issatchenkia* and *Williopsis* determined from multigene sequence analysis, and the proposal of *Barnettozyma* gen. nov., Lindnera gen. nov and *Wickerhamomyces* gen. nov. FEMS Yeast Res. 2008;8:939–54.

41. Grimaldi J. Les possibilités d'amélioration des techniques d'écabossage et de fermentation dans le processus artisinal de la préparation du cacao. Café Cacao Thé. 1978;22:303–16.
42. Gilbert DG. Dispersal of yeasts and bacteria by *Drosophila* in a temperate forest. Oecologia. 1980;46:135–7.
43. Ostovar K, Keeney PG. Isolation and characterization of microorganisms involved in the fermentation of Trinidad's cacao beans. J Food Sci. 1973;38:611–7.
44. Bainbridge JA, Davies SH. The essential oil of cocoa. J Chem Soc. 1912;101:2209–20.
45. Nicholls L. The fermentation of cacao. In: Smith HH, editor. The fermentation of cacao. 1st ed. London: Bale; 1913. p. 221–351.
46. Dobzhansky T, Da Cunha AB. Differentiation of nutritional preferences in Brazilian species of *Drosophila*. Ecology. 1955;36:34–9.
47. Rombouts JE. Observations on the microflora of fermenting cacao beans in Trinidad. Proc Soc Appl Bacteriol. 1952;15:103–11.
48. Koeppen. Bericht über die Dienstreise des Herrn Koeppen, Chemiker an des Landes-Versuchsanstalt zu Victoria, Kamerun, nach Fernando Poo und Sao Tome zum Studium der Kakaoaufbereitungsmethoden. Tropenpflanzer. 1907;11:566–72.
49. Camu N, Gonzáles Á, De Winter T, Van Schoor A, De Bruyne K, Vandamme P, et al. Influence of turning and environmental contamination on the dynamics of populations of lactic acid and acetic acid bacteria involved in spontaneous cocoa heap fermentation in Ghana. Appl Environ Microbiol. 2008;74:86–98.
50. Knapp AW. Cacao fermentation – a critical survey of its scientific aspects. London: Bale & Curnow; 1937.
51. Carr JG. Cocoa. In: Rose AH, editor. Fermented foods, Economic microbiology, vol. 7. 1st ed. London: Academic; 1982. p. 275–92.
52. Galvez SL, Loiseau G, Paredes JL, Barel M, Guiraud JP. Study on the microflora and biochemistry of cocoa fermentation in the Dominican Republic. Int J Food Microbiol. 2007;114:124–30.
53. Papalexandratou Z, Vrancken G, De BK, Vandamme P, De VL. Spontaneous organic cocoa bean box fermentations in Brazil are characterized by a restricted species diversity of lactic acid bacteria and acetic acid bacteria. Food Microbiol. 2011;28:1326–38.
54. Sanchez J, Guiraud JP, Galzy P. A study of the polygalacturonase activity of several yeast strains isolated from cocoa. Appl Microbiol Biotechnol. 1984;20:262–7.
55. Lefeber T, Janssens M, Camu N, De Vuyst L. Kinetic analysis of strains of lactic acid bacteria and acetic acid bacteria in cocoa pulp simulation media toward development of a starter culture for cocoa bean fermentation. Appl Environ Microbiol. 2010;76:7708–16.
56. Hugenholtz J. Citrate metabolism in lactic acid bacteria. FEMS Microbiol Rev. 1993;12:165–78.
57. Schwan RF, Rose AH. Endopolygalacturonase production by *Kluyveromyces marxianus*: effect of medium composition. J Appl Bacteriol. 1994;76:62–7.
58. Schwan RF, Cooper RM, Wheals AE. Endopolygalacturonase secretion by *Kluyveromyces marxianus* and other cocoa pulp-degrading yeasts. Enzyme Microb Technol. 1997;21:234–44.
59. Drysdale GS, Fleet GH. Acetic acid bacteria in winemaking: a review. Am J Enol Viticult. 1988;39:143–56.
60. Kersters K, Lisdiyanti P, Komagata K, Swings J. The family Acetobacteraceae: the genera *Acetobacter, Acidomonas, Asaia, Gluconacetobacter, Gluconobacter*, and *Kozakia*. In: Dworkin M, Falkow S, Rosenberg E, Schleifer K-H, Stackebrandt E, editors. The Prokaryotes, an evolving electronic ressource for the microbiological community; 2003. http://www.prokaryotes.com
61. Biehl B. Über die Bildung flüchtiger Säuren in der Pulpa, und in den Kakaokernen während der Kakaofermentation in Bahia. Zeitschrift für Lebensmitteluntersuchung und Forschung. 1969;139:76–85.
62. Kenten RH, Powell BD. Production of heat during fermentation of cacao beans. J Sci Food Agric. 1960;7:396–400.
63. Quesnel VC. Agents inducing the death of cacao seeds during fermentation. J Sci Food Agric. 1965;16:441–7.
64. Rohan TA. Observations on the fermentation of West African Amelonado Cocoa. London, UK: Cocoa conference report; 1957. p. 203–207.
65. Passos FML, Lopez AS, Silva DO. Aeration and its influence on the microbial sequence in cacao fermentations in Bahia, with emphasis on lactic acid bacteria. J Food Sci. 1984;49:1470–4.
66. Lopez A, Quesnel VC. Volatile fatty acid production in cacao fermentation and the effect on chocolate flavor. J Sci Food Agric. 1973;24:319–26.
67. Romanczyk LJ, McClelland CA, Post LS, Aitken WM. Formation of 2-acetyl-1-pyrroline by several *Bacillus cereus* strains isolated from cocoa fermentation boxes. J Agric Food Chem. 1995;43:469–75.
68. Schwan RF, Vanetti MCD, Silva DO, Lopez A, De Moraes CA. Characterization and distribution of aerobic, spore-forming bacteria from cacao fermentations in Bahia. J Food Sci. 1986;51:1583–4.
69. Zak DL, Ostovar K, Keeney PG. Implication of *Bacillus subtilis* in the synthesis of tetramethylpyrazine during fermentation of cocoa beans. J Food Sci. 1972;37:967–8.
70. Ouattara HG, Koffi BL, Karou GT, Sangare A, Niamke SL, Diopoh JK. Implication of *Bacillus* sp. in the production of pectinolytic enzymes during cocoa fermentation. World J Microbiol Biotechnol. 2008;24:1753–60.

71. Ouattara HG, Reverchon S, Niamke SL, Nasser W. Molecular identification and pectate lyase production by *Bacillus* strains involved in cocoa fermentation. Food Microbiol. 2011;28:1–8.
72. Dade HA. Internal molding of prepared cacao. Dept Agr Gold Coast Year Book Bull. 1982;16:74–100.
73. Ciferri R. Studies on cacao. J Dept Agric (Puerto Rico). 1931;15:223–86.
74. Hansen AP, Welty RE, Shen R. Free fatty acid content of cacao beans infested with storage fungi. J Agr Food Chem. 1973;21:665–70.
75. Anonymous. Joint Caobisco/ECA/FCC updated position on Ochratoxin A in cocoa and chocolate products. Paper presented by Caobisco, ECA and FCC to the EU Commission, 2003.
76. Anonymous. Ochratoxin A in the cocoa supply chain. Presentation to the EU Commission by Cabisco, ECA and FCC, 2006 Jan 12.
77. Ciferri R. Untersuchungen über den muffigen Gerauch der Kakaobohnen. Centralblatt für Bakteriologie, Parasitenkunde und Infektionskrankheiten, Zweite Abteilung. 1927;71:80–93.
78. Höhler D. Ochratoxin A in food and feed: occurence, legislation and mode of action. Z Ernahrungswiss. 1998;37:2–12.
79. Meister U. New method of citrinin determination by HPLC after polyamide column clean-up. Eur Food Res Technol. 2004;218:394–9.
80. Petithuguenin P. Causes and development of ochratoxin A on cocoa beans. In: International ZDS symposium, 2002 Dec 10–12; Solingen.
81. Copetti MV, Pereira JL, Iamanaka BT, Pitt JI, Taniwaki MH. Ochratoxigenic fungi and ochratoxin A in cocoa during farm processing. Int J Food Microbiol. 2010;143:67–70.
82. Bunting RH. Fungi occurring in cacao beans. Dept Agr Gold Coast Year Book Bull. 1928;16:44–57.
83. Loew O. Fermentation of cocoa with some comparative notes on the fermentation of tea and coffee. Annual Report of the Porto Rico agricultual experiment station, 1907.
84. Preyer-Buitenzorg A. Über Kakaofermentation. Tropenpflanzer. 1901;5:157–73.
85. Rombouts JEA. Critical review of the yeast species previously described from cacao. Trop Agr. 1953;30:34–41.
86. Ficker M, von Lilienfeld-Toal OA. Beobachtungen über die Fermentation des Kakaos in Brasilien. Tropenpflanzer. 1930;33:408–52.
87. Steinmann A. De Biologie der Cacao-gisten. Archief voor de Cacao en Andere kleine Cultures in Nederlandisch Indie. 1928;3:71–80.
88. Busse W, Henneberg W, Zeller T. Neue Untersuchungen und Versuche über Fermentation des Kakaos. Tropenpflanzer. 1929; 26, Beihefte 1:1–87
89. Kurtzmann CP, Fell JW, Boekhout T. The yeasts: a taxonomic study. New York: Elsevier; 2011.
90. Nielsen DS, Jakobsen M, Jespersen L. *Candida halmiae* sp. nov., *Geotrichum ghanense* sp. nov. and *Candida awuaii* sp. nov., isolated from Ghanaian cocoa fermentations. Int J Syst Evol Microbiol. 2010;60:1460–5.
91. Gálvez SL, Loiseau G, Paredes JL, Barel M, Guiraud J-P. Study on the microflora and biochemistry of cocoa fermentation in the Dominican Republic. Int J Food Microbiol. 2007;114:124–30.
92. Ravelomanana R, Guiraud JP, Vincent JC, Galzy P. The yeast flora of cocoa bean fermentation in the Ivory Coast. World J Microbiol Biotechnol. 1985;1:319–26.
93. Gauthier B, Guiraud J, Vincent JC, Parvais JP, Galzy P. Remaeques sur la Flore de Levures de la Fermentation traditionelle du Cacao en Cote d'Ivorie. Revue des Fermentations et des Industries Alimentaires (Bruxelles). 1977;32:160–3.
94. Sanchez J, Daguenet G, Guiraud JP, Vincent JC, Galzy P. A study of the yeast flora and the effect of pure culture seeding during the fermentation process of cocoa beans. Lebensm Wiss Technol. 1985;18:69–75.
95. Nielsen DS, Hønholt S, Tano-Debrah K, Jespersen L. Yeast populations associated with Ghanaian cocoa fermentations investigated using denaturing gradient gel electrophoresis (DGGE). Yeast. 2005;22:271–84.
96. Papalexandratou Z, De Vuyst L. Assessment of the yeast species composition of cocoa bean fermentations in different cocoa-producing regions using denaturing gradient gel electrophoresis. FEMS Yeast Res. 2011;11:564–74.
97. Dellaglio F, Torriani S, Felis GE, et al. Reclassification of *Lactobacillus cellobiosus* Rogosa et al. 1953 as a later synonym of *Lactobacillus fermentum* Beijerinck 1901. Int J Syst Evol Microbiol. 2004;54:809–12.
98. Passos FML, Silva DO, Lopez A, Ferreira CLL, Guimarães WV. Characterization and distribution of lactic acid bacteria from traditional cocoa bean fermentations in Bahia. J Food Sci. 1984;49:205–8.
99. Kostinek M, Ban-Koffi L, Ottah-Atikpo M, Teniola D, Schillinger U, Holzapfel WH, et al. Diversity of predominant lactic acid bacteria associated with cocoa fermentation in Nigeria. Curr Microbiol. 2008;56:306–14.
100. Garcia-Armisen T, Papalexandratou Z, Hendryckx H, Camu N, Vrancken G, De Vuyst L, et al. Diversity of the total bacterial community associated with Ghanaian and Brazilian cocoa bean fermentation samples as revealed by a 16 S rRNA gene clone library. Appl Microbiol Biotechnol. 2010;87:2281–92.
101. Lefeber T, Gobert W, Vrancken G, Camu N, De Vuyst L. Dynamics and species diversity of communities of lactic acid bacteria and acetic acid bacteria during spontaneous cocoa bean fermentation in vessels. Food Microbiol. 2011;28:457–64.

102. Nielsen DS, Schillinger U, Franz CMAP, Bresciani J, Amoa-Awua W, Holzapfel WH, et al. *Lactobacillus ghanensis*; a novel motile lactic acid bacteria isolated from Ghanaian cocoa fermentations. Int J Syst Evol Microbiol. 2007;57:1468–72.
103. Papalexandratou Z, Falony G, Romanens E, Jimenez JC, Amores F, Daniel HM, et al. Species diversity, community dynamics, and metabolite kinetics of the microbiota associated with traditional Ecuadorian spontaneous cocoa bean fermentations. Appl Environ Microbiol. 2011;77:7698–714.
104. Lefeber T, Janssens M, Moens F, Gobert W, De Vuyst L. Interesting starter culture strains for controlled cocoa bean fermentation revealed by simulated cocoa pulp fermentations of cocoa-specific lactic acid bacteria. Appl Environ Microbiol. 2011;77:6694–8.
105. Ashby SF. The sweating of cacao. Trop Agric (Trinidad). 1925;2:99–101.
106. Eckmann C. Über Essigbakterien aus der Kakaogärung. Dissertation, Universität Kiel. Universitätsverlag von Robert Noske, Borna-Leipzig, 1928.
107. Euzéby JP. List of names with standing in nomenclature. 2011. http://www.bacterio.cict.fr
108. Passos FML, Passos FJV. Descricao e classificacão de bactérias acéticas isoladas da fermentacão do cacau, com base em uma analise numérica. Revista de Microbiologia. 1985;16:290–8.
109. Schwan RF. Cocoa fermentations conducted with a defined microbial cocktail inoculum. Appl Environ Microbiol. 1998;64:1477–83.
110. Lin Y, Tanaka S. Ethanol fermentation from biomass resources: current state and prospects. Appl Microbiol Biotechnol. 2006;69:627–42.
111. Gams W. A key to the species of *Mortierella*. Persoonia. 1977;9:381–91.
112. Dzogbefia VP, Buamah R, Oldham JH. The controlled fermentation of cocoa (*Theobroma cacao* L) using yeasts: enzymatic process and associated physico-chemical changes in cocoa sweatings. Food Biotechnol. 1999; 13:1–12.
113. Samah OA, Putih MF, Selamat J. Biochemical changes during fermentation of cocoa beans inoculated with *Saccharomyces cerevisiae* (Wild strain). J Food Sci Technol (Mysore). 1992;29:341–3.
114. Leal GA, Gomes LH, Efraim P, Tavares FCD, Figueira A. Fermentation of cacao (*Theobroma cacao* L.) seeds with a hybrid *Kluyveromyces marxianus* strain improved product quality attributes. FEMS Yeast Res. 2008; 8:788–98.
115. Weiss N, Schillinger U, Kandler O. *Lactobacillus lactis*, *Lactobacillus leichmannii* and *Lactobacillus bulgaricus* subjective synonyms of *Lactobacillus delbrueckii* subsp. *lactis* comb. nov. and *Lactobacillus delbrueckii* subsp. *bulgaricus* comb. nov. System. Appl Microbiol. 1983;4:552–7.
116. Masoud W, Poll L, Jakobsen M. Influence of volatile compounds produced by yeasts predominant during processing of Coffea arabica in East Africa on growth and ochratoxin A (OTA) production by *Aspergillus ochraceus*. Yeast. 2005;22:1133–42.
117. Masoud W, Kaltoft CH. The effects of yeasts involved in the fermentation of Coffea arabica in East Africa on growth and ochratoxin A (OTA) production by *Aspergillus ochraceus*. Int J Food Microbiol. 2006;106:229–34.

Chapter 5
Fungi and Mycotoxin Occurrence in Cocoa

Marina Venturini Copetti, Beatriz Thie Iamanaka, and Marta Hiromi Taniwaki

Key Points

- Fungi can grow and produce mycotoxin during cocoa processing.
- Mycotoxins are stable to most food processing and remain in cocoa by-products and chocolate.
- There are worldwide reports of mycotoxin occurrence in cocoa products and chocolate, especially ochratoxin A and aflatoxins.
- Ochratoxin A is nephrotoxic, while aflatoxins are hepatotoxic and hepatocarcinogenic.
- Chocolate is a minor source of ochratoxin A and aflatoxin in diet.

Keywords Fungi • Ochratoxin A • Aflatoxins • Chocolate • Cocoa by-products • Food safety

Introduction

Chocolate quality may be influenced by a variety of environmental, agronomic, and technological factors to which the cocoa beans are subjected from the opening of the fruit until the end of industrial processing. To achieve the chocolate flavor, the cocoa beans have to go through a microbial curing process that starts at fermentation. After fermentation, beans are transferred to sun drying platforms where a gradual reduction in moisture content occurs. When moisture content reaches values sufficiently low to avoid microbial spoilage, beans are transferred to storage rooms, then later bagged and marketed. At the manufacturing stage, the dried fermented cocoa beans are submitted to a similar

M.V. Copetti, Ph.D. (✉)
Department of Food Technology and Science, Universidade Federal de Santa Maria-RS,
Av Roraima 1000, Predio 42, Sala 3210, Campus UFSM, Santa Maria, Rio Grande do Sul, 97105-900, Brazil
e-mail: mvc@smail.ufsm.br

B.T. Iamanaka, Ph.D.
Food Science and Quality Center, Instituto de Tecnologia de Alimentos,
Av. Brasil, 2880, Campinas, SP, 13070-178, Brazil
e-mail: beatriz@ital.sp.gov.br

M.H. Taniwaki, Ph.D.
Food Science and Quality Center, Instituto de Tecnologia de Alimentos,
Av. Brasil, 2880, Campinas, SP, 13070-178, Brazil
e-mail: marta@ital.sp.gov.br

R.R. Watson et al. (eds.), *Chocolate in Health and Nutrition*, Nutrition and Health 7,
DOI 10.1007/978-1-61779-803-0_5,

treatment of roasting and grinding at the beginning and then follow two distinctive processing lines giving as final product powdered cocoa or chocolate.

Research has shown a steady and marked participation of filamentous fungi during the fermentation process and during the drying and storage steps [1–4]. When present, they occur with higher intensity in the last days of fermentation and are usually related to the formation of off flavors and deterioration of cocoa beans.

Besides the deteriorative potential and consequent influence on sensorial quality of cocoa and chocolate, the presence of fungi in food is also a public health issue owing to the possibility of mycotoxin formation. *Aspergillus*, *Penicillium*, and *Fusarium* are the main genera capable of producing toxic secondary metabolites, many of which have mutagenic, teratogenic, and carcinogenic effects on humans and animals. Consequently, this chapter will focus mainly on the factors affecting the development of filamentous fungi and production of mycotoxins, as well as the reports of mycotoxins in cocoa and chocolate.

Fungi in Cocoa and Cocoa Products

Fungi are able to grow in all kinds of food, and there is general agreement that an increase in fungal activity in food can affect both its chemical composition and nutritional value. The fungal development can result in various forms of deterioration: loss of flavor, aroma and nutrients, toxin formation, discoloration, decay and formation of allergenic and pathogenic propagules.

The occurrence of alterations in sensory properties of food is often due to the synthesis of exoenzymes during fungal growth. A large number of enzymes can be produced (e.g., lipases, proteases, and carbohydrases), which can cause changes in protein and fat content while increasing the level of fatty acids and nonreducing sugars. Volatile substances can also be produced (such as dimethyl sulfides and 2-methyl-isoborneol) during fungal development, which affect the quality of food and beverages even when present in very small amounts. However, fungal contamination in food can also be of concern because several fungi are capable of producing a wide variety of toxic substances.

It should be noted that the presence of toxigenic fungi does not necessarily indicate the production of mycotoxins, since there are several factors that influence the production of mycotoxins by a fungal species. Nor does the absence of fungi ensure the absence of mycotoxins, since these toxins can persist for long periods even after fungal death.

Considering the very different parameters present in the cocoa beans according to the processing steps where the samples are collected for evaluation, a quite significant difference in the fungal species presence is expected. Figure 5.1a–h illustrates the succession on fungal contamination during the processing of cocoa beans, from cocoa pod to storage. Detailed studies on cocoa mycobiota were published elsewhere [5].

Both the cocoa beans and pulp inside a not injured cocoa pod are microbiologically sterile but are subsequently contaminated with microorganisms when the fruit is opened. These microorganisms will contribute to the subsequent spontaneous fermentation process.

The role of filamentous fungi during cocoa fermentation is not totally understood. It is known that these microorganisms are able to cause hydrolysis of the pulp, produce acids, off flavors, and alter the taste of the cocoa beans [6], and it is believed that an extensive fungal development at the end of fermentation causes increased deterioration in the consecutive phase of drying [7].

During the cocoa fermentation, high amounts of alcohol and lactic and acetic acid are produced by the fermentative yeasts and bacteria. These products of metabolism acting together with other factors such as the low pH, elevation of temperature produced by exothermic metabolic reactions, the competitive environment dominated by bacteria and yeasts in high water activity substrates and the microaerophilic conditions, will restrict the species able to develop.

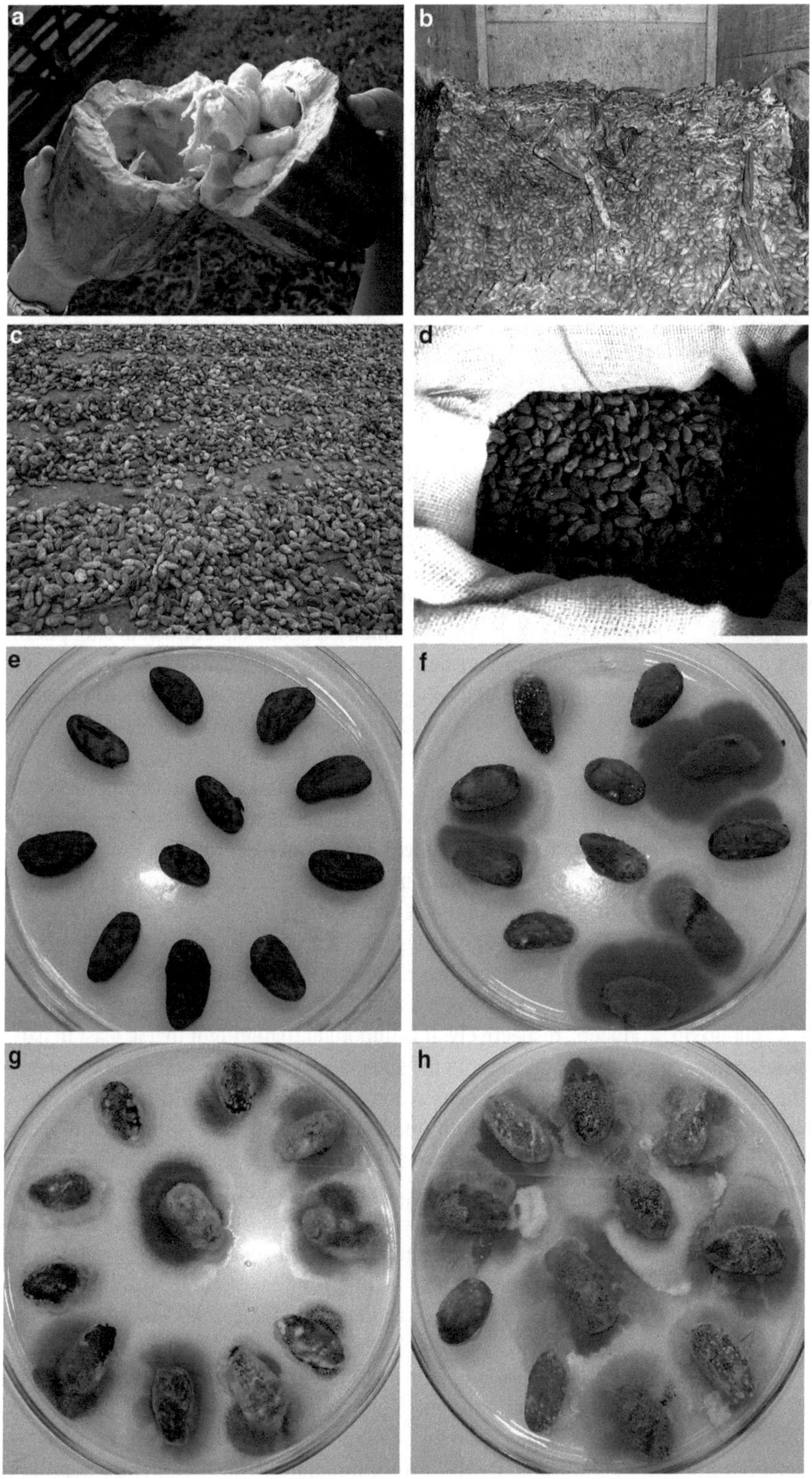

Fig. 5.1 Cocoa pod (**a**), fungal presence during cocoa fermentation (**b**), drying (**c**) and storage (**d**), mycological evaluation of cocoa beans sampled before fermentation (**e**), at fermentation (**f**), at drying (**g**), and at storage (**h**)

In an investigation carried out in Brazil [5], of 18 species of filamentous fungi isolated in samples from fermentation, only three species (*Monascus ruber*, *Penicillium paneum*, and *Geotrichum candidum*) were present in more than 20% of the 51 samples evaluated during this processing step. Considering some physiologic aspects of these more prevalent fungi isolated at fermentation, Pitt and Hocking [8] remark that *G. candidum* is able to grow under microaerophilic conditions and high a_w and *M. ruber* and *P. paneum* are also adapted to environments of low oxygen pressure. *P. paneum* is a species closely related to *P. roqueforti,* which has an optimum pH for growth between 4.0 and 5.0, and tolerates high levels of CO_2, as well as the organic acids commonly found in the cocoa fermentation. Ribeiro et al. [1] also reported the presence of some acetic acid-tolerant species, *P. variotii* and *T. ethaceticus.*

A low occurrence of potentially toxigenic isolates of *A. flavus*, *A. parasiticus* (aflatoxin producers), *A. niger*, and *A. carbonarius* (ochratoxin producers) has been reported in samples from fermentation [3, 9, 10].

At the end of fermentation, the fermented cocoa beans are submitted to a drying step to reduce their water activity (a_w) and consequently avoid microbial spoilage at storage. This process can be carried out in artificial driers or on sun drying platforms. Sun drying is the most common method and usually takes 7–14 days or more, depending on weather conditions.

The a_w is one of the most important parameters regarding the control of microorganism growth. The decrease of a_w during the drying step reduces the number of competitors due to the higher sensitivity of bacteria and yeasts to low water availability [11], so filamentous fungi become the predominant population. In addition, the dispersal of fermented beans on wooden drying platforms facilitates contact with fungal spores which can act as inoculums for future developing species. This procedure on the platform also increases the oxygen tension essential for growth of these fungi and facilitates the volatilization of some inhibitory acids produced during the fermentation, especially the acetic acid.

Sun drying can be considered a transition stage where the water content, competitors' microorganisms, and inhibitory substances (e.g., alcohol and acetic acid) will decrease gradually, allowing a better development of filamentous fungi. The fungal species predominant in the first days tend to be the ones that were more prevalent at fermentation, and in the course of time, it changes to species adapted to intermediate moisture foods, like the genera *Aspergillus* and *Penicillium.*

An investigation carried out in Brazil [12] demonstrates a visible elevation of the occurrence of fungal species in this phase of cocoa processing. The accentuated increase in the presence of *A. flavus*, *A. parasiticus*, *A. niger*, and *A. carbonarius* is a cause for concern since these species are potentially aflatoxin and ochratoxin producers, respectively. It is important to highlight that at least in part of the drying stage these species can find enough water in the substrate to allow for later mycotoxin production.

When the fermented and dried cocoa beans reach about 6–7% of humidity, they are considered "cured" and are stored in bags at the farm until marketed. The species isolated from stored samples tend to be similar to those present at drying, since the fungal spores can remain viable even if there is not enough water at this stage to sustain the fungal growth. Nevertheless, an increase in the occurrence of xerophilic species is expected and was reported [5], especially concerning the genus *Eurotium* (*E. amstelodami*, *E. chevalieri*, and *E. rubrum*) and *A. penicillioides*. These species are adapted to grow under conditions of reduced a_w and are responsible for considerable economic losses in stored grains, nuts, spices, and cereal products [8], although they have little importance regarding mycotoxin production.

Even if the cocoa beans have a water content low enough to avoid any microbial growth when bagged and stored; storage under poor conditions of high humidity can provide suitable conditions for spore germination, fungal growth, and spoilage. Cocoa beans are hygroscopic, so when cocoa is stored under high humidity conditions, it will absorb moisture from the environment until reaching equilibrium. For this reason, Wood [13] recommends that the storage of cocoa in tropical countries should not exceed 2–3 months if not carefully carried out.

When marketed, the cured beans are subjected to an initial cleaning at the manufacturing plant to remove residues of fibers, insects, and stones previous to the industrial processing that starts with a thermal treatment.

Most cocoa contaminants are present in the testa (shell), which corresponds to about 12% of a cured cocoa bean and is almost completely removed by winnowing after the preroasting of beans. These steps are extremely important to reduce the level of contaminants present in the beans before the subsequent processing stages.

Roasting is essential to complete the chemical reactions responsible for development of the chocolate flavor and at the same time reducing the microbial contaminants. Roasting (treatments of 15 min to 2 h at 105–150°C) is considered the only step in the chocolate production allowing for complete destruction of vegetative microorganisms [14].

After the steps of roasting and alkalization, the cocoa by-products used to be considered free of fungi; nevertheless, some fungal species can be isolated from cocoa products [5]. These contaminants probably originate from a postprocess contamination by microorganisms released in the previous manufacturing procedures and which remain present in the processing equipment and environment.

Chocolate is considered a product microbiologically stable because of its low a_w that is not conducive to microbial growth and spoilage. Nevertheless, there are reports of development of xerophilic fungi such as *Bettsia alvei*, *Chrysosporium xerophilum*, and *Neosartorya glabra* in deteriorated chocolate and chocolate confetti [14, 15]. These episodes occur due to inadequate storage problems leading to the formation of an environment with increased availability of water at the interface of the packaging and chocolate when it is stored in environments with high relative humidity.

Mycotoxins in Cocoa and Cocoa Products

Mycotoxins are substances of low molecular weight synthesized during the growth of certain fungi. Some toxins remain restricted to the fungal mycelium while the majorities are secreted into the substrate. It is known that most mycotoxins are extremely resistant to chemical and physical treatments and, once present in a raw material, remain stable through the steps of food processing and storage.

The fungal toxins are variable in their chemical composition, biological action, and toxicological effects. The most expressive effects are the induction of various types of tumors and immune suppression [16]. When ingested as a contaminant of food, part can be fixed to the tissues, despite the majority being metabolized and excreted. An important feature is their ability to affect specific organs without causing apparent changes in the others. The severity of symptoms depends on the toxicity of the mycotoxin; the level of exposure; age, sex, and nutritional status; and the possible synergistic effects with other chemicals [17, 18].

Human exposure to mycotoxins through consumption of contaminated food is a public health issue worldwide. Programs for monitoring the levels of food contamination by mycotoxins are essential for setting priorities in health surveillance. The aflatoxins and ochratoxin A are among the most known mycotoxins. Special attention will be given to these two toxins due to their toxicity, high presence in the food chain, and recent reports of occurrence in cocoa and chocolate.

Ochratoxin A

Ochratoxin A (OTA) is a toxin produced by several species of *Aspergillus* and *Penicillium*. These filamentous fungi are capable of growing in different climates and on different substrates, allowing a worldwide contamination of food crops with OTA. There is a general consensus that ochratoxigenic

Aspergillus species are responsible for accumulation of OTA in tropical regions, while in temperate climates such a problem is attributed to the occurrence of *Penicillium* species, such as *P. verrucosum* and *P. nordicum* [8].

The toxicology and human health risks of OTA have been assessed by the Joint FAO/WHO Expert Committee on Food Additives (JECFA), which has established tolerable intakes of OTA from food [19]. These tolerable intakes have been used in many countries as a basis for regulation of OTA in some foods, and nowadays, the provisional tolerable weekly intake (PTWI) for OTA is stated at 100 ng/kg of body weight [20].

OTA is known for its carcinogenic, nephrotoxic, teratogenic, and immunotoxic effects in animal cells [16]. The International Agency for Research on Cancer [16] has classified OTA as a possible carcinogen to humans (Group 2B), since there are a limited number of studies. Humans are more sensitive to the nephrotoxic effects [21], where this mycotoxin has a plasma half-life of 35 days [16]. Assuming that the period necessary to achieve the theoretical value of zero is eight times, a detectable serum level could still be found in humans after 280 days from the first exposure [22].

The main dietary sources of OTA are cereals, followed by wine, spices, coffee, grape juice, cocoa, beer, and dried fruits [16, 23]. OTA has an extremely high affinity for serum albumin and other blood macromolecules, with a long half-life in mammals. Because of this, it has also been found in animal tissues, especially in pork meat and derivatives [24].

Ochratoxin A in Cocoa and Chocolate

The occurrence of OTA in cocoa powder and chocolate has been reported around the world (Table 5.1). It is estimated that cocoa responds to 5% of OTA intake in a diet [29]. The number of investigations regarding the occurrence of ochratoxin A in cocoa and by-products has increased in the last decade powered by international discussions supported by the Joint FAO/WHO Expert Committee on Food Additives (JECFA); European Food Safety Authority (EFSA); Association of Chocolate, Biscuit and Confectionery Industries of the European Union (CAOBISCO); European Cocoa Association (ECA); and Federation for Cocoa Commerce (FCC). In 2011 the Codex Committee on Contaminants in Food (CCCF) elaborated a discussion paper on ochratoxin A in cocoa [35]. The main objective of these studies was to try to understand the main factors involved in OTA production during the different steps of cocoa processing at farm and the possibilities of reducing the contamination in the final products through technological processes. Once possessing this information, a code of practice to reduce and manage the incidence of OTA in cocoa could be formulated.

Gilmour and Lindblom [7] carried out extensive research in an attempt to elucidate the main factors involved in the occurrence of OTA in cocoa, which was focused mainly in the Ivory Coast. Preliminary results of this study revealed that the contamination already starts between harvest and fermentation; damaged pods appear to be critical for OTA accumulation in cocoa.

In an investigative study carried out on Brazilian farms during the period of 2006–2008 [9], no OTA was detected in the cocoa samples collected from the pods (previous to the beginning of fermentation), and only low levels of this toxin were detected during the fermentation time. Higher numbers of samples were detected with OTA during sun drying and storage of cocoa beans, with about 50% of samples contaminated, although the OTA levels found in Brazilian cocoa were low (mean of 0.13 and 0.10 μg/kg at sun drying and storage, respectively). This study also correlated the occurrence of OTA in the samples with the presence of OTA-producing species, concluding that *A. carbonarius* was the main OTA-producing species involved in OTA contamination in cocoa.

Experiments carried out in Brazil [5] demonstrated the influence of fermentative practices in the accumulation of ochratoxin A in cocoa artificially contaminated with *A. carbonarius* as well as the inhibitory effect of the organic acids present in the cocoa fermentation and pH on the growth and ochratoxin A production by *A. carbonarius* and *A. niger*, highlighting the importance of conducting an adequate fermentation to minimize the development of ochratoxigenic fungi during the drying step.

Table 5.1 Ochratoxin A (OTA) occurrence in chocolate worldwide

	Level of OTA contamination in the samples							
	Number of samples			μg/kg				
Product	Total	>LOQ	>2μg/kg	Max.	Median	Mean	Origin	Reference
Cocoa powder	31	29 (94%)	17 (55%)	4.4	–	2.41	Various	[25]
Cocoa powder	21	21 (100%)	–	–	0.24	–	Spain	[26]
Cocoa powder	5	5 (100%)	–	–	0.17	–	Various	[26]
Cocoa powder	1189	1094 (92%)	143 (12%)	–	–	1	Various	[7]
Cocoa powder	18	9 (50%)	0	0.77	–	0.43	Italy	[27]
Cocoa powder	20	19 (95%)	0	1.1	–	0.68	United Kingdom	[28]
Cocoa powder	20	20 (100%)	–	2.4	–	1.67	United Kingdom	[28]
Cocoa powder	96	91 (95%)	0	1.8	–	0.38	Germany	[29]
Cocoa powder	40	39 (98%)	–	2.4	–	1.2	United Kingdom	[29]
Cocoa powder	40	40 (100%)	0	1.8	0.59	–	Germany	[30]
Cocoa powder	25	25 (100%)	0	0.92	–	0.39	Brazil	[31]
Cocoa powder	40	40 (100%)	0	1.82	–	0.55	Italy	[32]
White chocolate	5	5 (100%)	–	–	0.03	–	Spain	[26]
White chocolate	9	8 (88%)	–	–	0.03	–	Various	[26]
White chocolate	25	23(92%)	–	0.05	–	0.03	Brazil	[31]
Milk chocolate	47	47 (100%)	–	–	0.12	–	Spain	[26]
Milk chocolate	122	122(100%)	–	–	0.1	–	Spain	[26]
Milk chocolate	228	52 (23%)	–	–		0.16	Various	[7]
Milk chocolate	39	36 (92%)	0	0.41	0.08		Germany	[30]
Milk chocolate	25	25 (100%)	0	0.45	–	0.15	Brazil	[31]
Milk chocolate	7	5 (71%)	0	0.19	–	0.11	Canada	[33]
Milk chocolate	78	21 (27%)	0	0.26	–	0.15	Italy	[32]
Dark chocolate	35	52 (100%)	–	–	0.25	–	Spain	[26]
Dark chocolate	52	52 (100%)	–	–	0.27	–	Various	[26]
Dark chocolate	536	300 (60%)	–	–	–	0.26	Various	[7]
Dark chocolate	25	25 (100%)	0	0.87	–	0.34	Brazil	[31]
Dark chocolate	14	14 (100%)	0	0.88	–	0.38	Canada	[33]
Dark chocolate	120	92 (77%)	0	0.74	–	0.20	Italy	[32]
Dark/Bitter chocolate	78	78 (100%)	0	0.66	0.14	–	Germany	[30]
Bitter chocolate	41	27 (66%)	0	0.94		0.35	Japan	[34]
Bitter chocolate	25	25 (100%)	0	0.92		0.39	Various	[31]

With regard to the secondary processing steps, studies demonstrate that most OTA is concentrated in the shell fraction and just a small part of the toxin contaminates the nibs [7, 36–38]. About 50% of the toxin is physically removed by industrial shelling [7], while handmade shelling reduces between 50% and 100% of OTA contamination [36, 38]. In the next processing steps, the shelled cocoa (cocoa nibs) can follow two different ways, resulting in chocolate or powdered cocoa. Investigations demonstrate that OTA remains bound to the nonfatty fraction of cocoa (see Table 5.1), so levels found in the butter are very low and the more cocoa solids a processing product has, the higher will be the amounts of OTA on it [7, 25, 37].

Usually a high number of chocolate samples are positive to OTA (40–100%), but generally the OTA levels present in these products are low (see Table 5.1). As pointed out for cocoa products, these values are directly related to the amount of cocoa solids used in the chocolate formulation. The low amounts of cocoa solids used in chocolate and chocolate derivatives manufacture can explain the low levels of OTA in these products, outlining a relatively low contribution of cocoa and chocolate for human exposure to OTA in food intake.

Aflatoxins B_1, B_2, G_1, and G_2

Aflatoxins are found as contaminants of food after the development of fungi in both pre- and postharvest, the level of contamination depending on many environmental conditions and substrate factors. Although a wide range of foods are susceptible to aflatoxin contamination, these toxins have been most commonly associated with groundnuts and groundnut products, dried fruit, tree nuts, spices, figs, crude vegetable oils, cocoa beans, maize, rice, cottonseed, and copra [17]. An inhibition of their production in some substrates has been related to the presence of some food components such as caffeine and polyphenols [39–41].

The International Agency for Research on Cancer (IARC) has recognized the carcinogenicity of aflatoxins since 1976 as a Group 1 carcinogen [16]. Given the proven knowledge of aflatoxin B_1 as a primary carcinogen for humans, there are no recommended values of tolerable weekly intake (PTWI), since the presence of these contaminants in the diet should be as low as possible [17].

The effects of aflatoxins are dependent on the dose and exposure time. An acute exposure results in acute hepatotoxicity, with a mortality rate of approximately 25%. On the other hand, the occurrence of hepatocellular carcinoma due to chronic exposure of aflatoxin has been widely documented, usually with a higher incidence in areas where infection with hepatitis B is endemic or in association with risk factors [42, 43].

Aflatoxins in Cocoa and Chocolate

There are few data on the natural occurrence of aflatoxins in cocoa, although in the 1970s methods were described for extraction and determination of aflatoxins in cocoa beans [44, 45].

Campbell [46] reported the presence of aflatoxins at levels up to 17 mg/kg in two of nine samples of cocoa beans analyzed, but in general this product has been set apart from the concerns of accumulation of aflatoxins because of the existing theory that caffeine inhibits the production of this mycotoxin [47].

In an extensive study carried out in Brazil [10] despite the high number of aflatoxigenic fungi isolated during processing of cocoa on the farm, fortunately the levels of aflatoxins found in the samples evaluated were low, suggesting the existence of antitoxigenic properties in cocoa, limiting the accumulation of aflatoxins in this product. This inhibition could be related to the high presence of polyphenols in the cocoa. The action of polyphenols on the synthesis and accumulation of aflatoxins in food was observed by Molyneux et al. [41] in oilseeds (walnuts, almonds, and pistachios). The researchers found up to 99.8% inhibition of the synthesis of aflatoxins induced by the presence of these antioxidants [41].

Concerning the presence of aflatoxins in cocoa by-products, a project developed in Germany showed that 73.5% of the 334 products examined contained traces of aflatoxins [48].

Only two reports on the presence of aflatoxins in chocolate are available on literature (Table 5.2). The survey carried out on 42 samples of bitter chocolate in Japan [34] revealed 22 (52%) samples contaminated with mean aflatoxin B1 concentrations of 0.18 μg/kg and a maximum of 0.6 μg/kg. Aflatoxin contamination was also reported by Copetti et al. [31] when evaluating the co-occurrence of ochratoxin A and aflatoxins in 125 samples of chocolate marketed in Brazil (powdered and bitter, dark, milk, and white bars, 25 samples each). The analyses revealed the co-occurrence of these two mycotoxins in 80% of the samples, with presence of aflatoxins at mean levels of 0.53, 0.66, 0.43, 0.08, and 0.01 μg/kg in 96% of powdered, 100% of bitter, 100% of dark, 72% of milk, and 20% of white chocolate samples, respectively. In the case of chocolate, besides concerns about aflatoxins coming from cocoa fractions, an additional contamination from cocoa-butter substitutes is possible [49].

Table 5.2 Reports on occurrence of total aflatoxins in chocolate

	Level of aflatoxin contamination in the samples						
	Number of samples			μg/kg			
Product	Total	>LOQ	>1 μg/kg	Max.	Mean	Origin	Reference
Cocoa powder	25	24 (96%)	1	1.70	0.53	Brazil	[31]
Bitter chocolate	42	22 (52%)	0	0.6	0.18	Japan	[34]
Bitter chocolate	25	25 (100%)	4	1.65	0.66	Various	[31]
Dark chocolate	25	25 (100%)	0	0.91	0.43	Brazil	[31]
Milk chocolate	25	18 (72%)	0	0.32	0.08	Brazil	[31]
White chocolate	25	5 (20%)	0	0.10	0.01	Brazil	[31]

Mycotoxin Regulation in Cocoa and Cocoa By-products

Most of countries have not set regulatory limits for ochratoxin A and/or aflatoxins in cocoa beans and cocoa products. Considering that cocoa products and chocolate are also consumed by children, and taking into account the studies carried out by Copetti et al. [9, 10], the Brazilian Sanitary Surveillance Agency (ANVISA) decided to set limits for cocoa beans and cocoa products in 2011. The set limits were 10 μg/kg for cocoa beans and 5 μg/kg for cocoa products and chocolate, both for ochratoxin A, and aflatoxins B_1, B_2, G_1, and G_2 [50].

Summary

The presence of the mycotoxins ochratoxin A and aflatoxin in chocolate is quite common, but fortunately the levels reported in this product are low. Chocolate appears not be a major source of ochratoxin A and aflatoxins in the diet, although one concern is the fact that products containing chocolate are widely consumed by children; thus, constant monitoring of their occurrence in these products is necessary.

References

1. Ribeiro NCA, Bezerra JL, Lopez A. Micobiota na fermentação do cacau no estado da Bahia, Brasil. Revista Theobroma. 1986;16:47–55.
2. Ardhana MM, Fleet GH. The microbial ecology of cocoa bean fermentations in Indonesia. Int J Food Microbiol. 2003;86:87–99.
3. Mounjouenpou P, Gueule D, Fontana-Tachon A, Guyot B, Tondje PR, Guiraud JP. Filamentous fungi producing ochratoxin A during cocoa processing in Cameroon. Int J Food Microbiol. 2008;128:234–41.
4. Sanchez-Hervas M, Gil JV, Bisbal F, Ramon D, Martinez-Culebras PV. Mycobiota and mycotoxin producing fungi from cocoa beans. Int J Food Microbiol. 2008;125(3):336–40.
5. Copetti MV, Iamanaka BT, Pereira JL, Frisvad JC, Taniwaki MH. Mycobiota of cocoa: from farm to chocolate. Food Microbiol. 2011;28(8):1499–504.
6. Schwan RF, Wheals AE. The microbiology of cocoa fermentation and its role in chocolate quality. Crit Rev Food Sci Nutr. 2004;44:1–17.
7. Gilmour M, Lindblom M. Management of ochratoxin A in the cocoa supply chain: a summary of work by the CAOBISCO/ECA/FCC Working Group on Ochratoxin A. In: Leslie JF, Bandyopadhyay R, Visconti A, editors. Mycotoxins: detection methods, management, public health and agricultural trade. Wallingford: CABI; 2008. p. 231–43.
8. Pitt JI, Hocking AD. Fungi and food spoilage. 3rd ed. New York: Springer; 2009.

9. Copetti MV, Pereira JL, Iamanaka BT, Pitt JI, Taniwaki MH. Ochratoxigenic fungi and ochratoxin A in cocoa during farm processing. Int J Food Microbiol. 2010;143:67–70.
10. Copetti MV, Iamanaka BT, Pereira JL, Fungaro MH, Taniwaki MH. Aflatoxigenic fungi and aflatoxin in cocoa. Int J Food Microbiol. 2011;148(2):141–4.
11. Beuchat LR. Influence of water activity on sporulation, germination, outgrowth, and toxin production. In: Rockland LD, Beuchat LR, editors. Water activity: theory and applications to food. New York: Marcel Dekker; 1987. p. 153–72.
12. Copetti MV, Iamanaka BT, Pereira JL, Frisvad JC, Taniwaki MH. The effect of cocoa fermentation and weak organic acids on growth and ochratoxin A production by *Aspergillus* species. Int J Food Microbiol. 2012a;155: 158–164.
13. Wood GAR. From harvest to store. In: Wood GAR, Lass RA, editors. Cocoa. New York: Longman Scientific and Technical; 1985. p. 444–504.
14. ICMSF (International Commission on Microbiological Specifications for Foods). Cocoa, chocolate, and confectionery. In: ICMSF, Microorganisms in Food. Vol 6. Microbial Ecology of Food Commodities. New York: Blackie Academic and Professional; 2005. p. 467–79.
15. Kinderlerer JL. *Chrysosporium* species, potential spoilage organisms of chocolate. J Appl Microbiol. 1997; 83:771–8.
16. International Agency for Research on Cancer (IARC). IARC monographs on the evaluation of the carcinogenic risks to humans. Some naturally occurring substances: food items and constituents, heterocyclic aromatic amines and mycotoxins, vol. 56. Lyon: IARC Press; 1993.
17. Joint FAO/WHO Expert Committee on Food Additives (JECFA). Safety evaluation of certain food additives and contaminants – aflatoxins. Geneva: World Health Organization; 1998.
18. Peraica M, Radic B, Lucic A, Pavlovic M. Toxic effects of mycotoxins in humans. Bull WHO. 1999;77(9): 754–66.
19. Joint FAO/WHO Expert Committee on Food Additives (JECFA). Safety evaluation of certain mycotoxins in food, WHO Technical Report Series, vol. 47. Geneva: World Health Organization; 2001.
20. Joint FAO/WHO Expert Committee on Food Additives (JECFA). JECFA/68/SC. Summary and conclusions. Geneva: World Health Organization; 2007.
21. Biancardi A, Riberzani A. Determination of ochratoxin A in cereals and feed by SAX-SPE clean up and LC fluorimetric detection. J Liq Chromatogr R T. 1996;19:2395–401.
22. Castegnaro M, Plestina R, Dirheimer R, Chernozemsky IN, Bartsch H. Mycotoxins, endemic nephropathy and urinary tract tumors. IARC Scientific Publications no. 115. Lyon: World Health Organisation/International Agency for Research on Cancer; 1991.
23. Joint FAO/WHO Expert Committee on Food Additives (JECFA). Evaluation of certain mycotoxins in food, WHO Technical Report Series, vol. 906. Geneva: World Health Organization; 2002.
24. Galtier P, Alvinerie M, Charpenteau JL. The pharmacokinetic profiles of ochratoxin A in pigs, rabbits and chickens. Food Cosmet Toxicol. 1981;19:735–8.
25. Bonvehi JS. Occurrence of ochratoxin A in cocoa products and chocolate. J Agric Food Chem. 2004;52: 6347–52.
26. Burdaspal PA, Legarda TM. Ocratoxina A en distintos tipos de chocolate y preparados de cacao en polvo comercializados en espana y en otros países extranjeros. Alimentaria. 2003;347:143–53.
27. Tafuri A, Ferracane R, Ritieni A. Ochratoxin A in Italian marketed cocoa products. Food Chem. 2004;88:487–94.
28. Ministry of Agriculture and Fisheries and Food (MAFF UK). Survey of Aflatoxins and ochratoxin A in cereals and retail product. 1998 Food Surveillance Information Sheet number 185, 1999. http://archive.food.gov.uk/maff/archive/food/infsheet/1999/no185/185ochra.htm. Accessed 1 July 2011.
29. Miraglia M, Brera C. Reports on tasks for scientific cooperation, task 3.2.7. Available at: http://europa.eu.int/comm/food/fs/scoop/3.2.7_en.pdf (2002). Accessed 23 Apr 2006.
30. Engel G. Ochratoxin A in sweets, oil seeds and dairy products. Arch lebensmittelhyg. 2000;51:81–128.
31. Copetti MV, Iamanaka BT, Pereira JL, Lemes DP, Nakano F, Taniwaki MH. Co-occurrence of ochratoxin A and aflatoxins in chocolate marketed in Brazil. Food Control. 2012b;26: 36–41.
32. Brera C, Debegnach F, De Santis B, Iafrate E, Pannunzi E, Berdini C, et al. Ochratoxin A in cocoa and chocolate products from the Italian Market: occurrence and exposure assessment. Food Control. 2011;22:1663–7.
33. Turcotte AM, Scott PM. Ochratoxin A in cocoa and chocolate sampled in Canada. Food Addit Contam. 2011;28(6):762–6.
34. Kumagai S, Nakajima M, Tabata SE, Ishikuro E, Tanaka T, Norizuki H, et al. Aflatoxin and ochratoxin A contamination of retail food and intake of these mycotoxins in Japan. Food Addit Contam. 2008;25(9):1101–6.
35. Codex Alimentarius. Discussion paper on ochratoxin A in cocoa. (CAC/REP 11-2011). Joint FAO/WHO Food Standards Program, FAO, Rome. 2011.
36. Amezqueta S, Gonzalez-Penas E, Murillo M, Lopez de Cerain A. Occurrence of ochratoxin A in cocoa beans: effect of shelling. Food Addit Contam. 2005;22(6):590–6.

37. Copetti MV. Micobiota do cacau: fungos e micotoxinas do cacau ao chocolate. Ph.D. thesis in Food Science. State University of Campinas, Brazil. 2009.
38. Manda P, Dano DS, Kouadio JH, Diakite A, Sangare-Tigori B, Ezoulin MJM, et al. Impact of industrial treatments on ochratoxin A content in artificially contaminated cocoa beans. Food Addit Contam. 2009;26(7):1081–8.
39. Lenovich LM. Effect of caffeine on aflatoxin production on cocoa beans. J Food Sci. 1981;46:655–7.
40. Hasan HAH. Role of caffeine and tannin in anti-toxigenic properties of coffee and tea. Cryptogamie Mycol. 1999;20(1):17–21.
41. Molyneux RJ, Mahoney N, Kim JH, Campbell BC. Mycotoxins in edible tree nuts. Int J Food Microbiol. 2007; 119:72–8.
42. Henry SH, Bosch FX, Bowers JC. Aflatoxin, hepatitis and worldwide liver cancer risks. In: DeVries JW, Trucksess MW, Jackson LS, editors. Mycotoxins and food safety. New York: Kluwer Academic/Plenum Publications; 2002. p. 229–320.
43. Omer RE, Kuijsten A, Kadaru AM, Kampman E, el Khidir IM, Fedail SS, et al. Population-attributable risk of dietary aflatoxins and hepatitis B virus infection with respect to hepatocellular carcinoma. Nutr Cancer. 2004;48(1):15–21.
44. Scott PM, Przybylski W. Collaborative study of a method for the analysis of cocoa beans for aflatoxins. J Assoc Off Ana Chem Int. 1971;54(3):540–4.
45. Hurst WJ, Lenovich LM, Martin Jr RA. Liquid-chromatographic determination of aflatoxins in artificially contaminated cocoa beans. J Assoc Off Ana Chem Int. 1982;65(4):888–91.
46. Campbell TC. 83rd Annual meeting of the AOAC. Report presented to the Joint AOAC-AOCS Committee on Aflatoxins, October, 13–16, Washington, DC, 1969.
47. Buchanan RL, Lewis DF. Caffeine inhibition of aflatoxin synthesis: probable site of action. Appl Environ Microb. 1984;47(6):1216–20.
48. Raters M, Matissek R. Vorkommen der Mykotoxine Aflatoxin B1, B2, G1, G2 und Ochratoxin A in Kakao und Kakaoprodukten. Projekt Nr. 9 der Stiftung der Deutschen Kakao und Schokoladenwirtschaft, Hamburg. 2000. http://www.kakao-stiftung.de/menu4_frameset_e.html. Accessed 1 July 2011.
49. Kershaw SJ. Occurrence of aflatoxins in oilseeds providing cocoa-butter substitutes. Appl Environ Microb. 1982;43(5):1210–2.
50. ANVISA (Agência Nacional de Vigilância Sanitária). Resolução RDC nº 7, de 18 de fevereiro de 2011. Dispõe sobre limites máximos tolerados (LMT) para micotoxinas em alimentos, 2011.

Chapter 6
Nonnutritive Constituents in Chocolate and Cocoa

William I. Manton

Key Points

- Contaminants include the toxic metals lead, cadmium, and nickel; the pesticides; and ochratoxin A, a mold-produced toxin.
- Lead, ochratoxin A, and probably pesticides are associated with the shells and enter cocoa and chocolate through imperfect shell removal.
- Lead, though present in soil, is not taken up by the plant. What lead is found in cocoa and chocolate results from beans being dried on bare soil, which adheres to the shell or from which it is absorbed by the shell.
- Cadmium and nickel are taken up by the plant from the soil, and the concentration in the bean depends on soil type. Fertilizer use may increase the cadmium content of the soil.
- Ochratoxin A concentration in cocoa powder is greater than expected from simple inclusion of shell fragments, as if, like lead, it is concentrated during processing.
- Concentrations of contaminants are generally within mandated guidelines, and the fraction of chocolate products in the diet is so small that what contaminants are present will have little effect on overall intake.

Keywords Cocoa beans • Cocoa powder • Chocolate • Lead • Cadmium • Nickel • Pesticides • Ochratoxin A

Introduction

Having been consumed for centuries without apparent ill effect, cocoa and chocolate were until recently subjected to little scrutiny or regulation with regard to their content of toxic metals. In 1981, for example, the World Health Organization [1] set the standards for arsenic (As), lead (Pb), and copper (Cu) in cocoa powder at what seem today the extraordinarily high levels of 1, 2, and 50 mg/kg, respectively, and publications for the next 20 years were entirely restricted to reporting regional levels

W.I. Manton, Ph.D. (✉)
Department of Geosciences, University of Texas at Dallas,
800 W. Campbell Road, 75080-3021 Richardson, TX, USA
e-mail: manton@utdallas.edu

R.R. Watson et al. (eds.), *Chocolate in Health and Nutrition*, Nutrition and Health 7,
DOI 10.1007/978-1-61779-803-0_6,

of these elements [2–5]. Even market basket surveys that showed chocolate and chocolate-containing products to have cadmium (Cd) and Pb levels conspicuously greater than those of most foods [6, 7] excited little interest from the viewpoint of public health. The situation changed in 1986 with the passage of California's Safe Drinking Water and Toxic Enforcement Act, commonly referred to as Proposition 65 [8], which allows private parties acting in the public interest to bring civil suits against businesses that market products containing toxic substances introduced by human activity. Under this statute, the American Environmental Institute in 2002 brought suit against the principal chocolate manufacturers doing business in the United States on the grounds that Pb contained in chocolate was adversely affecting children's health [9]. Though it was later dropped, the suit drew attention to the potential health effects of metals in cocoa, and, after a number of instances of Pb-contaminated candy were widely reported in the press, the US Food and Drug Administration set a guideline of 0.1 mg/kg for Pb in chocolate [10, 11]. It is important to note that they addressed only the finished product and advised the manufacturers to "source their raw materials appropriately." International standards for Pb in cocoa products are still set by the United Nations Codex Alimentarius Commission. In 2000, they proposed to reduce the standard for cocoa to 1 mg/kg and proposed applying the same value to cocoa mass and chocolate. For cocoa butter, they proposed 0.1 mg Pb/kg and, only for chocolate, standards of 0.5 mg As/kg and 20 mg Cu/kg [12]. None of these values has yet been formally adopted, however. The inclusion of As and Cu is questionable as it has been many years since Pb arsenate was last used as a pesticide and Cu is not generally considered toxic. Although concern has been expressed over cadmium (Cd) and nickel (Ni) in cocoa, the commission has not yet addressed the question of setting standards for them, but some countries have acted independently and set their own. Current interest in toxic metals in cocoa and chocolate is reflected in the number of publications that have appeared in recent years [13–21].

Quite apart from the problem of metals is that of insecticides and fungicides which make for the never-ending experiment of adding toxic substances to the environment and finding out the rate at which they have to be replaced either because target species become immune to them or because the ecosystem cannot withstand them. The first insecticides to be introduced were the organochlorines with familiar names like DDT, dieldrin, lindane, and chlordane. With generally low volatility (lindane, however, is quite volatile) and chemical stability, they were both efficient and cost-effective, but their persistence in the food chain and accumulation in higher animals brought about their ban in developed countries. In the 1950s, they were replaced by the organophosphates, of which parathion and malathion are the best known and by the carbamates, typified by Sevin and propoxur. As both types are toxic to humans, they have been in part replaced by the pyrethrins which have low toxicity, are readily metabolized, and break down in the environment. In the 1990s, neonicotinoids were introduced and are favored by having lowest mammalian toxicity, but the fact that they act systemically creates a new set of problems in that they may kill pollinators [22]. In its native environment in the Amazon basin, the cocoa tree is not immune to pests and disease and, after having been introduced into Africa and Asia, has been beset by species indigenous to those regions. As a consequence, it has been and is vigorously treated with insecticides and fungicides. However, in contrast to all other major crops, 95% of cocoa production is in the hands of smallholders living in rural areas who may apply pesticides inconsistently or inappropriately or may use up old stocks of banned materials [22–24]. Another source of pesticides are the fumigants that have to be applied to piles of sacked beans awaiting shipment or sprayed into containers during transport to avoid spoilage by warehouse moths and beetles. Cocoa beans can also be contaminated by the herbicides used to control weeds in the plantations [22].

Finally, molds growing on beans while they are drying after fermentation may produce toxic metabolites (mycotoxins) that on a weight for weight basis are more toxic than pesticides. These include the aflatoxins and ochratoxins, of which ochratoxin A is of most concern.

Contaminants, Toxic Effects, and Regulatory Values

Metals

Lead

Lead occurs in house dust and soil and in the diet. It is absorbed through the gut, and upon entering plasma, about 50% is sequestered in bone and more than 99% of the remainder becomes bound to the red cells. Clinical Pb poisoning in adults typically manifests itself as peripheral neuropathy resulting in wrist drop and, in children, as encephalopathy. Such cases seldom occur today thanks to vigorously enforced regulations and overall awareness of the danger through aggressive advertising. Of interest here is subclinical poisoning, the sequelae to sustained exposure to low levels of Pb. In children, deficits in IQ are well documented [25, 26], and a blood Pb concentration of 10 μg/dL is taken as the threshold of harmful neurodevelopmental effects [27], though some would have that figure reduced [28]. Blood Pb concentration has been shown to correlate with blood pressure in adult men [29], in black as opposed to white men and women [30], and in perimenopausal and postmenopausal women [31]. Lead is excreted through the kidney and follows first-order kinetics with a half-life of about 20 days so that blood Pb concentration can be calculated for a given intake and body mass. Difficulties arise because the absorption of Pb varies in the presence of Ca and because Pb may be almost completely absorbed if it is ingested on an empty stomach, as happens when children mouth soiled fingers and hands. To account for variations in day-to-day exposure, the concept of a provisional tolerable weekly intake (PTWI) was introduced by the United Nations' Joint Expert Committee on Food Additives (JECFA) and set at 0.025 mg/kg body weight. This figure, however, has been revoked on the grounds that it is associated with a deficit in IQ of three points in children and an increase of 3 mmHg in systolic blood pressure in adults and so cannot be considered as health protective [32].

Cadmium

Plants readily take up Cd that is naturally present in soil [33, 34], rendering exposure universal whether through the diet or through smoking. Absorbed through the gut, it is detoxified by binding to the cysteine-rich protein metallothionein, which is synthesized de novo in the liver, and is ultimately stored in the cortex of the kidney where it has a half-life of about 15 years. Occupational exposure may produce renal tubular damage usually monitored by the excretion of the low-molecular-weight protein β2 microglobulin. There is evidence, however, that low-level environmental exposures damage the kidney [35–37] and cause bone loss in women [38, 39] and possibly also breast cancer [40]. In view of the long half-life of Cd in the body, JECFA proposed that exposure should be averaged over a period of at least a month and implemented a provisional tolerable monthly intake (PTMI) of 0.025 mg/kg body weight [32], superceding the PTWI of 0.007 mg/kg body weight. In an independent review, the European Food Safety Authority concurred with this figure [41].

Nickel

The diet contains large amounts of Ni which plants take up to different degrees from the soil. Gastrointestinal absorption is low at about 1%. The principal health effect of nickel is allergy which affects 10–20% of the population. Sensitized individuals will develop a skin rash after contact with nickel or nickel-plated objects such as buttons or jewelry. Occasionally ingestion of food items such as chocolate with a high nickel content results in eczema [42, 43]. There is no PTWI for nickel, only a tolerable daily intake (TDI) of 0.012 mg/kg body weight [44].

Pesticides

Urban dwellers encounter agricultural pesticides as residues in food, and most have detectable quantities of the environmentally more persistent varieties in their adipose tissue. The organochlorines are of particular concern in that they are potential endocrine disruptors and carcinogens [45]. That they might cause breast cancer has been examined by several groups over the period 1995–2005 without any clear-cut positive correlation being found [46–52]. On the other hand, an association has been found with non-Hodgkin's lymphoma and the organochlorine heptachlor epoxide and, at the highest concentrations in adipose tissue, with dieldrin and chlordane [53]. Blood serum levels of β hexachlorocyclohexane (beta-HCH), which constituted 5–14% of technical grade lindane, have been found associated with Parkinson's disease [54, 55]. Up to this point, the residues of a few pesticides in cocoa beans have been regulated. The Codex Alimentarius set maximum residue limits (MRLs) of 0.2 mg/kg for endosulphan (an organochlorine insecticide) and metalaxyl (a systemic fungicide) and 0.01 mg/kg for the fumigant phosphine (hydrogen phosphide) [56]. The US Food and Drug Administration placed an action level of 1 mg/kg for DDT, TDE, and DDE and 0.5 mg/kg for benzene hexachloride and lindane [57]. Most recently, the European Union has published MRLs for a large number of pesticides in fermented cocoa beans [58], a decision that has caused some concern in the cocoa industry [22]. The US EPA lists tolerances for some substances in cocoa beans and cocoa products [59]. These include pyriproxyfen (0.02 mg/kg), a pesticide that acts on larvae, pyrethrins (1 mg/kg), and the newly introduced insecticide chlorantraniliprole (0.08 mg/kg) [60]. It lists tolerances for several herbicides including glyphosate (0.2 mg/kg) and paraquat (0.05 mg/kg), for the fungicide chlorothalonil (0.05 mg/kg), and for residues of the fumigants phosphine (0.1 mg/kg) and sulfuryl fluoride (0.2 mg/kg). It should be mentioned that Bateman [22] does not record pyriproxyfen, chlorantraniliprole, or chlorothalonil as being used on African plantations but notes that the herbicide 2,4-D, which for cocoa is omitted from the EPA list, is widely used.

Mycotoxins

Aflatoxins

Having caused widespread poultry poisoning in the 1960s through consumption of contaminated feed [61], the aflatoxins have been scrutinized for their potential of causing human disease, chiefly through the consumption of maize and peanuts. Although in large doses they cause liver cancer in rats, they have not been conclusively shown to be carcinogenic in humans [62]. During the fermentation, drying, and storage of cocoa, the mycotoxin-producing *Aspergillus* species flourish, but for reasons not clearly understood, little aflatoxin is produced [63].

Ochratoxin A

Ochratoxin A, which is also produced by *Aspergillus* species, is of widespread occurrence in grains and nuts and, if contaminating animal feed, in meat and milk and thus also in breast milk.

It is primarily nephrotoxic to all species and in animals has been shown be teratogenic and carcinogenic, inducing renal tumors [64, 65]. It has been suggested that it may be a factor in testicular cancer [66], the incidence of which has been increasing in Europe [67, 68], and exposure to chocolate in utero and in early life has been put forward as a factor [69]. The PTWI is set at 100 ng/kg bw/week [70], but it has been proposed that this is too high by a factor of 10 and that a virtually safe dose of 1.8 ng/kg bw/day should be adopted [65].

Agricultural Practices and Cocoa Bean Processing

Growing and Harvesting

Distribution

Cocoa is a subtropical crop that is grown across the world in regions within 10° of the equator. Production is dominated by the African countries of Ivory Coast and Ghana which together produced about 59% of that grown in 2005/2006. Indonesia, principally the island of Sulawesi, was second with 13%. The remaining production comes from the African countries of Nigeria, Cameroon, and Madagascar; Malaysia and various Pacific Islands like New Britain, Papua New Guinea, Vanuatu, and Hawaii; Central and South American countries including Mexico, Ecuador, Venezuela, Panama, Peru, and Brazil; and the Caribbean islands of Grenada and Hispaniola. Thus, cocoa is grown on soils ultimately derived from a variety of igneous rock types: granites in the case of the African countries, Brazil, Peru, and Venezuela and basaltic to andesitic lavas in the Pacific Islands and the Caribbean. Each soil will have a different content of trace elements.

Fertilizing

To obtain maximum production, both shade and nutrition must be managed [71]. For mature plantings, nutrition requirements are 438 kg/ha for nitrogen, 48 for phosphorus, and 633 for potassium. The need for phosphorus often exceeds its availability in soil despite the fact that about 50% is returned through leaf litter and, if yields are to be increased, it is found that nitrogen applications must be accompanied by phosphorus.

Diseases

Cocoa is affected by a number of diseases including those caused by fungi – black pod rot which is found worldwide [72], witches' broom and frosty pod rot which are restricted to the Americas [73], and vascular streak dieback restricted to SE Asia [74]. Quarantine of young plants, sanitation, shade reduction, and canopy management help reduce the incidence of infection, and applications of Cu as oxide or hydroxide or of metalaxyl are the most commonly used treatments. Fungal diseases cause the greatest crop losses. Next are losses caused by insects, in particular the sap-sucking capsids or mirids found in all growing countries and the cocoa pod borer restricted to SE Asia. These have been treated in the past by the whole spectrum of insecticides. Other diseases include swollen shoot virus which is restricted to W Africa. Being transmitted by mealy bugs [75], it is also controllable by insecticides.

Harvesting

Pods grow on the trunk and branches of the tree. When ripe, they are harvested with machetes and are cut open. The beans, covered by a white pulp or mucilage, are then scooped out, pooled, and fermented. Smaller growers do this in the field by piling the beans into heaps which they then cover with banana or plantain leaves. Others may use baskets. Larger growers will use wooden boxes. During fermentation, bacteria consume the glucose- and fructose-rich pulp and raise the temperature of the mass to 50°C. The process produces watery fluids or sweatings that soak the shells. The beans must then be dried and again the procedure varies. In the simplest case, they are spread out on the ground or on mats to dry in the sun. More elaborate is to place them on wooden racks. In climates with an insufficient number of sunny days, they may be force-dried by heated air. To prevent damage by mold or moths, piles of sacks or containers must be fumigated. Some processing may be done in the country of origin because partially processed material is cheaper to ship, being lighter than whole beans of which about 10% may be shell.

Processing

Upon arrival at the processing plant, unwanted debris are removed by means of vibrating screens, and the beans are washed to remove any soil particles. High-velocity impaction against steel plates is used to break the shells free from the kernels or nibs as they are known in the industry. Shell fragments are separated from the nibs by winnowing. Total separation is unachievable, and up to 1.7% of shell material may legally be processed with the nibs but, in practice, 1% is usually achieved. Nibs are then ground to less than 30 μm at which point the fat-containing vesicles rupture and a viscous fluid variously called cocoa mass or cocoa liquor is produced. At some point, the material is sterilized by roasting. This may be done at the beginning before removal of the shells or after production of the liquor. The liquor may be retained for chocolate manufacture or be hydraulically pressed to remove the fat or cocoa butter, leaving behind a solid mass termed cocoa cake. At some point, either the nibs or the cake may be treated with alkali to develop color or reduce acidity. Grinding of the cake produces cocoa powder which is used in baked products or chocolate-flavored drinks. In the manufacture of chocolate, the liquor is further ground to the point where the surface area of the solid material has reached the point where it can no longer be coated by the available fat and the material becomes powdery. Cocoa butter and sugar are added to make dark chocolate and these ingredients plus milk powder to make milk chocolate. All must be ground to less than 30 μm and thoroughly mixed in a process called conching to ensure even coating of the ingredients with fat. The liquid mixture is then poured into molds where it solidifies at room temperature. Detailed descriptions of the chocolate making process can be found in Beckett et al. [76].

Contamination Pathways

Metals

Potential sources of metal contamination are physiological uptake from the soil, airborne fallout on beans while drying or being transported, adsorption of ions on to shell from soil while drying, adhesion of soil to beans, incorporation of shell fragments into cocoa liquor, contamination from grinding equipment, and airborne contamination of chocolate during processing.

Physiological Uptake

While plants naturally take up Cd and Ni from the soil and translocate them to their tissues [33, 34], whatever Pb may be taken up tends to stay in the roots [77]. This is true for cocoa, because the Pb content of the shells and nibs of freshly harvested and conservatory grown beans are about 1 μg/kg [15, 19]. The Cd content of cocoa is reported to vary according where it is grown, Mounicou et al. [13] stating that South American varieties except Brazil have higher contents than those from West Africa and Malaysia. In geological terms, this implies that cocoa grown on the volcanic rocks of the Andes has a higher Cd content than that grown on the Precambrian granites of Brazil and West Africa. The Malaysian samples referred to above may have been derived from granitic soils of eastern Peninsular Malaysia [78] because the soils in the principal growing region of Tawau in northern Borneo are volcanic [79] and cocoa grown on them has a high Cd content [5]. Tropical soils are typically acid, resulting in aluminum toxicity, and are leached of nutrients. They are usually remediated by liming and fertilization [80]. Fertilizers derived from rock phosphate typically have high Cd contents and their use can increase the Cd content of crops including cocoa [81, 82].

Airborne Fallout

Phase out of leaded gasoline began in the United States in 1975 and was completed in 1996 as was the case for most countries of Europe. The rest of the world followed more slowly. and a compilation made in 2005 indicated that 81 countries had not banned its use, including the cocoa producers of Cameroon, Indonesia, Ivory Coast, Madagascar, Sao Tome and Principe, Vanuatu, and Venezuela. Peru had banned it the year earlier. In 2011 the list had fallen to six, none of which are cocoa producers [83]. Contamination of beans could occur during drying by direct atmospheric fallout or from contact with previously contaminated soil.

Ion Adsorption

In common with other biomass, cocoa shells have the ability to adsorb metal ions from solution. Meunier et al. [84] state that from strongly acid solutions, the adsorption for the metals of interest is Pb>Cd=Cu>Ni. Thus, if adsorption occurs at neutral pH, it is possible that the shells will take up these metals while drying on the soil [19]. In the case of Pb, both naturally occurring and contaminant Pb may be simultaneously adsorbed. Soil Cd, whether natural or from fertilizer, and Ni may likewise be adsorbed and increase the amount naturally present in the shell.

Contamination by Soil

If wet beans are dried on bare earth, soil particles will stick to the shells. Beans are washed at the processing plant, but it is possible that some soil particles will not be removed and act as a source of trace metals. The content of ash insoluble in HCl in cocoa is permitted to be 0.3% or 3 g/kg and may have been put in place to limit the quantity of soil.

Shell Fragments

The winnowing process is about 99% efficient in separating shells from nibs, and some shell material is inevitably ground with the nibs. Industry standards allow 1.7% of shell material in cocoa liquor. The shells are the principal source of Pb in cocoa products, but it is unclear how much of the metal is bound to the shell and how much is merely in soil adhering to the shell [19]. The extent to which the presence of shell affects the concentration of Cd and Ni in cocoa depends on the relative concentration in the shell and the nib.

Contamination by Grinding Equipment

Nibs are ground to size in a series of mills, usually first an impact mill followed by either ball or disc mills. The materials used, hardened steel, or silicon carbide should add no more contaminants than are found in other mechanically ground foods.

Contamination During Chocolate Manufacture

In chocolate manufacture sugar – and milk powder if milk chocolate is to be made – is ground and added to cocoa mass and cocoa butter. Then follows the mixing process of conching. Today, this is done in enclosed rotary conches but in the past was carried out in so-called long conches, which could be open or partially open vessels into which ingredients were shoveled manually. Under such conditions, it would seem there was much opportunity for contamination by airborne particulate, especially as conching could take 72 h.

Pesticides

Pesticides historically used on cocoa were applied topically and killed the target pest by direct or secondary contact or through ingestion [22]. Experiments with radiolabeled propoxur showed that penetration from the surface of the pod to the beans does not occur [85], raising the question of how beans become contaminated if they are not directly sprayed. To answer this, it seems inevitable that when handling pods, which are not washed, workers' hands will become contaminated by pesticide residues that they will transfer to the beans when scooping them out of the pods. Organochlorines have very low solubilities in water and tend to bind to soil particles where they persist for long periods even in subtropical climates [86]. Thus, if beans are set out to dry on bare earth, it is likely that they will also acquire pesticides from the soil that inevitably sticks to them. Another possible route is from pallets that may have been vigorously treated against termites [22].

Mycotoxins

The development of molds and concomitant production of ochratoxin A in cocoa is greatest if broken pods have been stored before being fermented [87], but only about half of the beans that are fermented are affected [88]. Ochratoxin A is located on the shells of the beans [89]. It is stable with a half-life of 200 min at 150°C [90] and will largely survive the cocoa roasting process which is carried out between 110°C and 140°C for up to 45 min and 1 h [76].

Current Status of Contaminants

Metals

Lead

Recent studies [15, 19] have demonstrated that the shells of cocoa beans may contain between 0.9 and 3 mg/kg of Pb, while the nib contains about 1 μg/kg. With such a disparity, it is near impossible to prepare a nib sample that has not been to some extent contaminated by the shell, and it is not surprising that earlier studies reported nib concentrations too high by factors of 10–1,000 [2, 4]. The incorporation of up to 1.7% of shell material in cocoa mass, though small, will greatly increase the Pb content over that in the nibs alone, and concentrations in cocoa powder ranging from 0.03 to 0.65 mg/kg have been reported [15, 19]. Even so, the actual content of Pb in cocoa powder is possibly as much as 10 times higher than expected, leading Manton [19] to propose that during winnowing, a metal-rich dust is formed that contaminates the surfaces of the nibs. The origin of the Pb on the shell has been disputed. Rankine et al. [15] postulated it was entirely the result of contamination by leaded gasoline during harvesting and transportation. Manton [19], on the other hand, showed that the shells also contained the nonessential lanthanide elements neodymium (Nd) and samarium (Sm), which, taken with the presence of insoluble silicate mineral residues, implied that the Pb at least in part was derived from that naturally present in soil. From the apparent age of the soil protolith (determined from Sm-Nd isotope systematics), it was possible to identify where the cocoa was grown, and it turned out that cocoa originating from Sulawesi had on average twice the Pb content of that from Africa. From Pb isotope systematics, it could then be shown that the Pb in the Sulawesi samples was an equal mixture of natural Pb and Australian Pb, which was probably derived from gasoline, as Indonesia was one of the last countries to phase out leaded fuel. In contrast, the Pb contained in African cocoa appeared entirely derived from that naturally present in soil.

In reporting the Pb content of chocolate, it is conservative to report that of dark chocolate which may contain up to 70% cacao rather than milk chocolate which is diluted by powdered milk and sugar. For US-produced chocolate, concentrations between 0.012 and 0.275 mg/kg have been reported [11, 15, 19] and for Europe a range of 0.017 to 0.510 mg/kg [19]. Concentrations that are similar to US values are reported from Saudi Arabia [20], but from other countries, much higher concentrations have recently appeared in the literature. For example, Dahiya et al. [14] found dark chocolate sold in Mumbai to contain between 0.24 and 8.0 mgPb/kg, and from Pakistan, Jalbani et al. [18] report chocolate containing 1.11–48 mg Pb/kg. In Turkey Duran et al. [17] report concentrations between 0.84 and 2.06 mgPb/kg in "cocoa-based candies," while Iwegbue [21] reports Nigerian chocolate having concentrations from less than 0.4 to 2.3 mg/kg with a mean value of 0.8 mg Pb/kg.

Cadmium

The Cd content of cocoa varies greatly, Mounicou et al. [13] reporting a 20-fold range from 0.094 to 1.83 mg/kg. A similar variability occurs in chocolate. Srogi [16] reported a range from 0.01 to 0.05 mg/kg for Polish chocolate, and Iwegbue [21], a mean value of 0.07 mg/kg for Nigeria. These values are similar to the 0.028 mg/kg obtained for milk chocolate analyzed in recent US total diet surveys [91]. Higher values are reported from other countries. For example, from Mumbai, Dahiya et al. [14] report a mean of 0.24 mg/kg; from Turkey Duran et al., a mean of 0.24 mg/kg [17]; from Malaysia Lee and Low [3], a mean of 0.32 mg/kg; and from Pakistan Jalbani et al. [18], a mean of 0.35 mg/kg.

The reason for the variation is usually ascribed to andesites having a higher Cd content than other igneous rocks, but there are too few published analyses to say whether this is generally true. In Malaysia, however, where Fauziah et al. [5] carried out a detailed study of the distribution of Cd, Cu, Ni, and Zn in tissues of the cocoa tree, it is well established that soil developed on andesite contains far more Cd than the median concentration of 0.06 mg/kg reported for world soils [92]. Working with alluvial and volcanic soils that ranged in Cd content from 0.3 to 1.9 mg/kg, they found that the uptake of Zn and Cu did not vary with soil content but that of Cd did. Furthermore, extractable Zn and Cd correlated with available P, indicating that those elements had been added by fertilizer derived from rock phosphate.

Nickel

Nickel concentrations have been reported only from chocolate, and, while there may be a large range from sample to sample, average values for different countries differ by a factor of 5. The lowest of 1.05 mg/kg is reported by Duran et al. [17] from Turkey followed by Scrogi et al.'s [16] values of 1.–2.2 mg/kg from Poland. These are similar to the US milk chocolate value of 1.0 mg/kg [91]. Means of samples in Mumbai, Pakistan, and Nigeria were, respectively, 2.8, 4.3, and 5.4 mg/kg [14, 18, 21]. The last is unexpected because the Ni content of granite is one-tenth that of basalt [93] and the Nigerian samples, grown on granite, should be among the lowest. Fauziah et al. [5] reported concentrations of both Cd and Ni in cocoa, and if their measurements are plotted, a straight line passing through the origin results. The slope of the line is ten which is similar to the Ni/Cd ratios of chocolates from Mumbai and Pakistan and may indicate that they were made from Malaysian beans.

Pesticides

The content of pesticides has become tightly regulated with the implementation of Regulation (EC) No 396/2005 of the European Parliament [94] and the US Food Safety Modernization Act [95], which was enacted into law early in 2011. The European law sets out maximum residue limits (MRLs) for pesticides in foods imported into the European Union. The US law is far reaching and calls for inspections of foreign facilities where fumigation is carried out, places responsibility upon importers for certifying that foods they import are safe, and requires traceability and extensive record keeping. Though pesticides residues are of great concern, it is difficult to find published studies beyond the listing for milk chocolate in the USFDA total diet survey where lindane was found present in all 44 samples analyzed in concentrations ranging from 0.1 to 6.0 μg/kg with a mean value of 1.7 μg/kg [96]. Other pesticides appeared in less than half the samples and maximum concentrations were generally less than 10 μ(mu) g/kg. The figure for lindane should be viewed in the light of the MRL of 1 mg/kg in the European Union database [58] and 0.5 mg/kg for the USFDA action level [57]. In FY 2006, the FDA reports that for 20 samples of imported "candy, chocolate, and cocoa products," none was out of compliance [97]. If pesticides are associated with the shells, the level in cocoa can be expected to be low.

Fumigation of stored cocoa is necessary to avoid infestations of pests like warehouse moths and beetles, and in the past, methyl bromide and phosphine have been used. Both are highly toxic gases, and the first was banned in developed countries in 2004 under the Montreal Protocol for its potential damage to the ozone layer. However, the US EPA is withdrawing its replacement sulfuryl fluoride, which it approved in 2004, on the grounds of children's heath, because the residues could result in excessive exposures to fluoride [98].

Ochratoxin A

Serra Bonvehí [99] carried out an industrial scale experiment examining the distribution of ochratoxin A in the various products of processing. The highest mean concentration of 11 μg/kg was found in roasted shell, while those in nibs and cocoa butter were below the detection limit of 0.10 μg/kg. The mean concentration in cocoa powder was 2.41 μg/kg which is at least ten times greater than would be expected if the powder contained 1% by weight of shell. The situation is thus similar to that reported for Pb [19].

The regulation proposed by the European Union to set maximum levels of ochratoxin A in imported cocoa and chocolate, respectively, at 2 and 1 μg/kg would have caused about 40% of imported cocoa cake to be rejected [99], and these levels have since been withdrawn. The proposed regulation did stimulate interest in the problem, and ochratoxin A concentrations in cocoa and chocolate have been published for Canada [100] and Italy [101]. For Canada, most samples analyzed fell below the limits originally proposed, and in the case of Italy, all did. On average, alkalized Canadian cocoa had a higher ochratoxin A concentration than untreated powder, an observation for which there is no ready explanation.

Strategies to Reduce Contaminants

Lead contamination and soilborne pesticide contamination can be essentially reduced to zero if beans are not placed on bare earth to dry. Drying on tarpaulins or plastic sheets may have the advantage of reducing ochratoxin A since it is known that wooden racks tend to harbor *Aspergillus* species, and if coffee is any guide, drying on tarpaulins with frequent raking will reduce production of the toxin [102]. The use of galvanized wire racks is probably inadvisable as the shells may take up Pb and Cd from them. As there seems to be an unintended amplification of Pb and ochratoxin A during processing (and probably also pesticides), it is desirable to keep contamination of the shell as low as possible. Contamination with Cd and Ni is unavoidable as they are naturally taken up from the soil. If soil derived from andesitic lavas is avoided, Cd contents can be minimized but the use of phosphate fertilizer will add Cd to the soil. The work of Lee and Low [3] indicates that the Cd content of nibs is half that of the shells. In view of this, drying beans on the ground will probably have little effect on the Cd content of cocoa. As granite contains little Ni, cocoa from Africa and Brazil is expected to be lower than that from other regions, and cocoa grown on basalt or serpentine will be the highest. Modern processing equipment will add little in the way of contamination, but there may be countries where the old-fashioned long conches are still used. If so, there would be a possibility of contamination from airborne Pb. Enough seems to be known of the level of Pb that can be acquired during harvesting that the levels of Pb recently reported in chocolate from India, Turkey, and Pakistan [14, 17, 18] must be ascribed to post-harvesting sources. The problem of residues from chemical fumigation could be solved by turning to physical techniques. Finkelman et al. [103], for example, showed that after 3 days storage at 100 mmHg and 30°C, eggs of moths and beetles were no longer viable.

In the final analysis, it must be remembered that chocolate products constitute a minor part of the diet. In the European diet, they total 4% of the total weekly dietary intake [101], and in the US diet, the characteristic Pb isotopic signature of cocoa and chocolate is overwhelmed by those from other sources [19]. On the other hand, chocolate products are the major source of some flavonoids which may reduce the chance of heart disease [104].

References

1. FAO/WHO. Codex Alimentarius. Current Food Standards. Codex Stan 105-1981. Codex standard for cocoa powders (cocoas) and dry cocoa-sugar mixtures. http://www.codexalimentarius.net/web/standard_list.jsp. Accessed 11 Aug 2011.
2. Knesevic G von. Schwermetalle in lebensmitteln. 2. Mitteilung: Über den Gehalt an Blei in Rohkakao und in Kakao-Halb- und Fertigprodukten. (Heavy metals in food. Part 2. Lead content in unrefined cocoa and in semi-finished and finished cocoa products.) Deut Lebensm-Rundsch.1982;78:178–80.
3. Lee CK, Low KS. Determination of cadmium, lead, copper and arsenic in raw cocoa, semifinished and finished chocolate products. Pertanika. 1985;8:243–8.
4. Prugarová A, Kovác M. Lead and cadmium content in cocoa beans. Nahrung. 1987;31:635–6.
5. Fauziah CI, Rozita O, Zauyah S, et al. Heavy metals content in soils of peninsular Malaysia grown with cocoa and cocoa tissues. Malaysian J Soil Sci. 2001;5:47–58.
6. U.S. Department Health and Human Services. Food and drug administration. Total diet study-analytical results. Summary of toxic and nutritional elements found in TDS foods: 1991–2005. http://www.fda.gov/Food/FoodSafety/FoodContaminantsAdulteration/TotalDietStudy/ucm184293.htm. Accessed 4 Aug 2011.
7. Pedersen GA, Mortensen GK, Larsen EH. Beverages as a source of toxic trace element intake. Food Addit Contam. 1994;11:351–63.
8. Ca.Gov. Office of Environmental Health Hazard Assessment. Safe drinking water and toxic enforcement Act of 1986. http://www.oehha.org/prop65/law/p65.html. Accessed 4 Aug 2011.
9. American Environmental Safety Institute. Lead in chocolate: the impact on children's health. Fact Sheet-May 2002. www.wcaslab.com/tech/FactSheet. Accessed 4 Aug 2011.
10. U.S. Department Health and Human Services. Food and drug administration. Lead in candy likely to be consumed frequently by small children: recommended maximum level and enforcement policy. http://www.fda.gov/Food/GuidanceComplianceRegulatoryInformation/GuidanceDocuments/ChemicalContaminantsandPesticides/ucm077904.htm. Accessed 11 Aug 2011.
11. U.S. Department Health and Human Services. Food and drug administration. Supporting document for recommended maximum level for lead in candy likely to be consumed frequently by small children. Docket No. 2005D-0481. http://www.fda.gov/Food/FoodSafety/FoodContaminantsAdulteration/Metals/Lead/ucm172050.htm. Accessed 11 Aug 2011.
12. FAO/WHO Food Standards Programme. Codex Alimentarius Commission. Report of the eighteenth session of the Codex Committee on cocoa products and chocolate. Fribourg, Switzerland, 2–4 Nov 2000. Appendices II-V. http://www.fao.org/docrep/meeting/005/x8943e/x8943e00.htm#Contents. Accessed 5 Aug 2011.
13. Mounicou S, Szpunar J, Andrey D, et al. Concentrations and bioavailability of cadmium and lead in cocoa powder and related products. Food Addit Contam. 2003;20:343–52.
14. Dahiya S, Karpe R, Hegde AG, et al. Lead, cadmium and nickel in chocolates and candies from suburban areas of Mumbai, India. J Food Comp Anal. 2005;18:517–22.
15. Rankin CW, Nriagu JO, Aggarwal JK, et al. Lead contamination in cocoa and cocoa products: isotopic evidence of global contamination. Environ Health Perspect. 2005;113:1344–8.
16. Srogi K. Assessment of selected heavy metal contents in medicinal plants, tea leaves and chocolate using atomic absorption spectrometry. Acta Toxicol. 2006;14:117–28.
17. Duran A, Tuzen M, Soylak M. Trace metal contents in chewing gums and candies marketed in Turkey. Environ Monit Assess. 2009;149:283–9.
18. Jalbani N, Kazi TG, Hassan I, et al. Determination of toxic metals in different brand of chocolates and candies, marketed in Pakistan. Pak J Anal Environ Chem. 2009;10:48–52.
19. Manton WI. Determination of the provenance of cocoa by soil protolith ages and assessment of anthropogenic lead contamination by Pb/Nd and lead isotope ratios. J Agric Food Chem. 2010;58:713–21.
20. Al Othman ZA. Lead contamination in selected foods from Riyadh City market and estimation of the daily intake. Molecules. 2010;5:7482–97.
21. Iwegbue CMA. Concentrations of selected metals in candies and chocolates consumed in southern Nigeria. Food Addit Contam: Part B: Surveillance. 2011;4:22–7.
22. Bateman RP. Pesticide use in cocoa. A guide for training administrative and research staff. 2nd Ed 2009 7. http://www.icco.org/sps/manual.asp. Accessed 11 Aug 2011.
23. Clarke EEK, Levy LS, Spurgeon A, et al. The problems associated with pesticide use by irrigation workers in Ghana. Occup Med. 1997;47:301–8.
24. Tutu AS, Yeboah PO, Golow AA, et al. Organochlorine pesticides residues in the breast milk of some primiparae mothers in La community, Accra, Ghana. Res J Environ Earth Sci. 2011;3:153–9.
25. Pocock SJ, Smith M, Baghurst P. Environmental lead and children's intelligence: a systematic review of the epidemiological evidence. Br Med J. 1994;309:1189–97.

26. Schwartz J. Low-level lead exposure and children's IQ: a meta-analysis and search for a threshold. Environ Res. 1994;65:42–55.
27. U.S. Department of Health and Human Services, Public Health Service, Centers for Disease Control. Preventing lead poisoning in young children. Publication date: 10/01/1991. http://wonder.cdc.gov/wonder/prevguid/p0000029/p0000029.asp. Accessed 11 Aug 2011.
28. Canfield RL, Henderson CR, Cory-Slechta DA, et al. Intellectual impairment in children with blood lead concentrations below 10 24 μg per deciliter. N Engl J Med. 2003;348:1517–26.
29. Hu H, Aro A, Payton M, et al. The relationship of bone and blood lead to hypertension. The Normative Aging Study JAMA. 1996;275:1171–6.
30. Vupputuri S, He J, Muntner P, et al. Blood lead level is associated with elevated blood pressure in blacks. Hypertension. 2003;41:463–8.
31. Nash D, Magder L, Lustberg M, et al. Blood lead, blood pressure, and hypertension in perimenopausal and postmenopausal women. JAMA. 2003;289:1523–32.
32. FAO/WHO, Joint FAO/WHO Expert Committee on food additives, seventy-third meeting. summary and conclusions. Geneva, Switzerland, 2010, pp. 12–13. www.fao.org/ag/agn/agns/.../JECFA73%20Summary%20Report%20Final.p... Accessed Aug11 2011.
33. Smolders E. Cadmium uptake by plants. Int J Occup Med Environ Health. 2001;14:177–83.
34. Luo J, Zhang H, Zhao F-J, et al. Distinguishing diffusional and plant control of Cd and Ni uptake by hyperaccumulator and nonhyperaccumulator plants. Environ Sci Technol. 2010;44:6636–41.
35. Järup L, Hellström L, Alfvén T, et al. Low level exposure to cadmium and early kidney damage. Occup Environ Med. 2000;57:668–72.
36. Satarug S, Haswell-Elkins MR, Moore MR. Safe levels of cadmium intake to prevent renal toxicity in human subjects. Br J Nutr. 2000;84:791–802.
37. Järup L, Åkesson A. Current status of cadmium as an environmental health problem. Toxicol Appl Pharmacol. 2009;238:201–8.
38. Alfvén T, Elinder C-G, Carlson MD, et al. Low-level cadmium exposure and osteoporosis. J Bone Miner Res. 2000;15:1579–86.
39. Schutte R, Nawrot TS, Richart T, et al. Bone resorption and environmental exposure to cadmium in women: a population study. Environ Health Perspect. 2008;116:777–83.
40. Gallagher CM, Chen JJ, Kovach JS. Environmental cadmium and breast cancer risk. Aging (Albany NY). 2010;2:804–14.
41. EFSA Panel on Contaminants in the Food Chain (CONTAM); Scientific Opinion on tolerable weekly intake for cadmium. EFSA Journal 2011;9(2):1975. [p 19] doi:10.2903/j.efsa.2011.1975. Available online: http://www.efsa.europa.eu/cs/Satellite. Accessed Aug 11 2011.
42. Sharma AD. Relationship between nickel allergy and diet. Indian J Dermatol Venereol Leprol. 2007;73:307–12.
43. Krecisz B, Chomiczewska D, Kiec-Swierczynska M, et al. Systemic contact dermatitis to nickel present in cocoa in 14-year-old boy. Pediatr Dermatol. 2011;28:335–6.
44. World Health Organization. Nickel in drinking-water. Background document for preparation of WHO Guidelines for drinking-water quality. Geneva: world health organization (WHO/SDE/WSH/04.08/55); 2004.
45. Longnecker MP, Rogan WJ, Lucier G. The human health effects of DDT (dichlorodiphenyltrichloroethane) and PCBs (polychlorinated biphenyls) and an overview of organochlorines in public health. Annu Rev Public Health. 1997;18:211–44.
46. Adami HO, Lipworth L, Titus-Ernstoff L, et al. Organochlorine compounds and estrogen-related cancers in women. Cancer Causes Control. 1995;6:551–66.
47. Ahlborg UG, Lipworth L, Titus-Ernstoff L, et al. Organochlorine compounds in relation to breast cancer, endometrial cancer, and endometriosis: an assessment of the biological and epidemiological evidence. Crit Rev Toxicol. 1995;25:463–531.
48. Safe SH, Zacharewski T. Organochlorine exposure and risk for breast cancer. Prog Clin Biol Res. 1997;396:133–45.
49. Høyer AP, Jørgensen T, Rank F, et al. Organochlorine exposures influence on breast cancer risk and survival according to estrogen receptor status: a Danish cohort-nested case-control study. BMC Cancer. 2001;1:8.
50. Zheng T, Holford TR, Mayne ST, et al. Risk of female breast cancer associated with serum polychlorinated biphenyls and 1,1-dichloro-2,2'-bis(p-chlorophenyl)ethylene. Cancer Epidemiol Biomarkers Prev. 2000;9:167–74.
51. Laden F, Collman G, Iwamoto K, et al. 1,1-Dichloro-2,2-bis(p-chlorophenyl)ethylene and polychlorinated biphenyls and breast cancer: combined analysis of five U.S. studies. J Natl Cancer Inst. 2001;93:768–76.
52. Raaschou-Nielsen O, Pavuk M, Leblanc A, et al. Adipose organochlorine concentrations and risk of breast cancer among postmenopausal Danish women. Cancer Epidemiol Biomarkers Prev. 2005;14:67–74.
53. Quintana PJE, Delfino RJ, Korrick S, et al. Adipose tissue levels of organochlorine pesticides and polychlorinated biphenyls and risk of non-Hodgkin's lymphoma. Environ Health Perspect. 2004;112:854–61.

54. Richardson JR, Shalat SL, Buckley B, et al. Elevated serum pesticide levels and risk of Parkinson disease. Arch Neurol. 2009;66:870–5.
55. Richardson JR, Roy A, Shalat SL, et al. β-Hexachlorocyclohexane levels in serum and risk of Parkinson's disease. Neurotoxicology. 2011;32(5):640–5.
56. FAO/WHO Food Standards, Pesticide Residues in Food. Maximum residue limits; extraneous maximum residue limits. http://www.codexalimentarius.net/mrls/pestdes/jsp/pest_q-e.jsp. Accessed Aug 11 2011.
57. US Food and Drug Administration. Inspections, compliance, enforcement, and criminal investigations. CPG Sec. 575.100 Pesticide residues in food and feed – enforcement criteria. 1995: http://www.fda.gov/iceci/compliancemanuals/compliancepolicyguidancemanual/ucm123236.htm. Accessed Aug 11 2011.
58. European Union Pesticides Database. Available online: http://ec.europa.eu/sanco_pesticides/public/index.cfm?event=commodity.resultat. Accessed Aug 11 2011.
59. Code of Federal Regulations. Title 40, Part 180.Tolerances and exemptions for pesticide chemical residues in food. Electronic code of federal regulations. http://ecfr.gpoaccess.gov/cgi/t/text/text-idx?c=ecfr&tpl=/ecfrbrowse/Title40/40cfr180_main_02.tpl. Accessed Aug 11 2011.
60. Lahm GP, Cordova D, Barry JD. New and selective ryanodine receptor activators for insect control. Bioorg Med Chem. 2009;17:4127–33.
61. Kotsonis FN, Burdock GA, Flamm WG. Food toxicology. In: Claasen CD, editor. Casarett and Doull's toxicology: the basic science of poisons. 6th ed. New York: McGraw-Hill; 2001. p. 1049–88.
62. World Health Organization. International agency for research on cancer. Some traditional herbal medicines, some mycotoxins, naphthalene and styrene. IARC Monogr Eval Carcinog Risks Hum. 2002;82:1–556.
63. Copetti MV, Iamanaka BT, Pereira JL, et al. Aflatoxigenic fungi and aflatoxin in cocoa. Int J Food Microbiol. 2011;148:141–4.
64. O'Brien E, Dietrich DR. Ochratoxin A: the continuing enigma. Crit Rev Toxicol. 2005;35:33–60.
65. Pfohl-Leszkowicz A, Manderville RA. Ochratoxin A: an overview on toxicity and carcinogenicity in animals and humans. Mol Nutr Food Res. 2007;51:61–99.
66. Schwartz GG. Hypothesis: does ochratoxin A cause testicular cancer? Cancer Causes Control. 2001;13:91–100.
67. Huyghe E, Matsuda T, Thonneau P. Increasing incidence of testicular cancer worldwide: a review. J Urol. 2003;170:5–11.
68. Huyghe E, Plante P, Thonneau PF. Testicular cancer variations in time and space in Europe. Eur Urol. 2007; 51:621–8.
69. Giannandrea F. Correlation analysis of cocoa consumption data with worldwide incidence rates of testicular cancer and hypospadias. Int J Environ Res Public Health. 2009;6:568–78.
70. WHO/FAO. Evaluation of certain food additives and contaminants. Sixty-eighth report of the joint FAO/WHO expert committee on food additives. WHO technical report series No. 947. Geneva, Switzerland, 2007.
71. Wessel M. Shade and nutrition. In: Wood GAR, Lass RA, editors. Cocoa. 4th ed. Oxford: Blackwell Science; 1985. p. 166–94.
72. Evans HC. Cacao diseases-the trilogy revisited. Phytopathology. 2007;97:1460–3.
73. Aime MC, Phillips-Mora W. The causal agents of witches' broom and frosty pod rot of cacao (chocolate, Theobroma cacao) form a new lineage of Marasmiaceae. Mycologia. 2005;97:1012–22.
74. Guest D, Keane P. Vascular-streak dieback: a new encounter disease of cacao in Papua New Guinea and Southeast Asia caused by the obligate basidiomycete Oncobasidium theobromae. Phytopathology. 2007;97:1654–7.
75. Dongo LN, Orisajo SB. Status of cocoa swollen shoot virus disease in Nigeria. Afr J Biotech. 2007;6:2054–61.
76. Beckett ST, Harding J, Freeman B. The science of chocolate. 2nd ed. Cambridge: The Royal Society of Chemistry; 2008.
77. Pourrut B, Shahid M, Dumat C, et al. Lead uptake, toxicity, and detoxification in plants. Rev Environ Contam Toxicol. 2011;213:113–36.
78. Ng SK, Thamboo S, Cheah TE. Soil and leaf nutritional status of cacao on the east and west coasts of Malaya. Malaysian Agric J. 1970;47:390–408.
79. Musta B, Hennie FW, Soehady E, et al. Geochemical characterization of volcanic soils from Tawau, Sabah. Geol Soc Malaysia. 2008;54:33–36.
80. Baligar VC, Fageria NK. Soil aluminum effects on growth and nutrition of cacao. Soil Sci Plant Nutr. 2005;51:709–13.
81. Syers JK, Mackay AD, Brown MW, et al. Chemical and physical characteristics of phosphate rock materials of varying reactivity. J Sci Food Agric. 1986;37:1057–64.
82. Lugon-Moulina N, Ryan L, Doninia P, et al. Cadmium content of phosphate fertilizers used for tobacco production. Agron Sustain Dev. 2006;26:151–5.
83. Taylor R, Gethin-Damon Z. Countries where leaded petrol is possibly still sold for road use as at 17th June 2011. The Lead Group Inc. Available at http://www.lead.org.au/fs/fst27.html. Accessed Aug 11 2011.
84. Meunier N, Laroulandie J, Blais JF, et al. Cocoa shells for heavy metal removal from acidic solutions. Bioresource Technol. 2003;90:255–63.

85. Adomako D, Owusu-Manu E. Propoxur residues in cocoa beans. In: Radiotracer studies of pesticide residues in edible oil seeds and related products. Vienna: IAEA; 1979: 51–57. www.iaea.org/inis/collection/NCLCollectionStore/_Public/.../10455193.pdf.
86. Samuel T, Agarwal HC, Pillai MKK. Persistence and binding of DDT and gamma-HCH in a sandy loam soil under field conditions in Delhi, India. Pest Sci. 1988;22:1–15.
87. Mounjouenpou P, Gueule D, Fontana-Tachon A, et al. Filamentous fungi producing ochratoxin A during cocoa processing in Cameroon. Int J Food Microbiol. 2008;121:234–41.
88. Copetti MV, Pereira JL, Iamanaka BT, et al. Ochratoxigenic fungi and ochratoxin A in cocoa during farm processing. Int J Food Microbiol. 2010;143:67–70.
89. Amézqueta S, González-Peñas E, Murillo M, et al. Occurrence of ochratoxin A in cocoa beans: effect of shelling. Food Addit Contam. 2005;22:590–6.
90. Boudra H, Le Bars P, Le Bars J. Thermostability of ochratoxin A in wheat under two moisture conditions. Appl Environ Microbiol. 1995;61:1156–8.
91. U.S. Department Health and Human Services. Food and drug administration. Total diet study-analytical results. Summary of toxic and nutritional elements found in TDS foods: 2006–2008. http://www.fda.gov/Food/FoodSafety/FoodContaminantsAdulteration/TotalDietStudy/ucm184293.htm. Accessed Aug 6 2011.
92. Helmke PA. The chemical composition of soils. In: Sumner ME, editor. Handbook of soil science. Boca Raton: CRC Press; 2000. p. B3–24.
93. Callender E. Heavy metals in the environment – historical trends. In: Sherwood Lollar B, editor. Environmental geochemistry, Treatise on geochemistry, vol. 9. Oxford: Elsevier; 2005. p. 67–105.
94. Europa. EUR-lex. Regulation (EC) No 396/2005 of the European Parliament and of the Council of 23 Feb 2005 on maximum residue levels of pesticides in or on food and feed of plant and animal origin and amending Council Directive 91/414/EECText with EEA relevance. http://eurlex.europa.eu/LexUriServ/LexUriServ.do?uri=CELEX:32005R0396:EN:NOT. Accessed 6 Aug 2011.
95. U.S. Department Health and Human Services. Food and drug administration. The new FDA food safety modernization Act (FSMA). http://www.fda.gov/food/foodsafety/fsma/default.htm. Accessed 6 Aug 2011.
96. U.S. Department Health and Human Services. Food and drug administration. Total diet study-analytical results. Summary of pesticide residues and industrial chemicals found in TDS foods: sorted by food. http://www.fda.gov/Food/FoodSafety/FoodContaminantsAdulteration/TotalDietStudy/ucm184293.htm. Accessed 6 Aug 2011.
97. U.S. Department Health and Human Services. Food and drug administration. Pesticide residue monitoring program results and discussion FY 2006 http://www.fda.gov/Food/FoodSafety/FoodContaminantsAdulteration/Pesticides/ResidueMonitoringReports/ucm125187.htm. Accessed 11 Aug 2011.
98. U.S. Environmental Protection Agency. Pesticides: registration review. EPA proposes to withdraw sulfuryl fluoride tolerances. http://www.epa.gov/pesticides/sulfuryl-fluoride/evaluations.html. Accessed 11 Aug 2011.
99. Serra Bonvehí J. Occurrence of ochratoxin A in cocoa products and chocolate. J Agric Food Chem. 2004; 52:6347–52.
100. Turcotte A-M, Scott PM. Ochratoxin A in cocoa and chocolate sampled in Canada. Food Add Contam. 2011; 28:762–6.
101. Brera C, Debegnach F, De Santis B, et al. Ochratoxin A in cocoa and chocolate products from the Italian market: occurrence and exposure assessment. Food Control. 2011;22:1663–7.
102. Velmourougane K, Bhat R, Gopinandhan TN. Impact of drying surface and raking frequencies on mold incidence, ochratoxin A contamination, and cup quality during preparation of arabica and robusta cherries at the farm level. Foodborne Pathog Dis. 2010;7:1435–40.
103. Finkelman S, Navarro S, Rindner M, et al. Effect of low pressures on the survival of three cocoa pests at 30°C. J Stored Prod Res. 2004;40:499–506.
104. Ding DL, Hutfless SM, Ding X, et al. Chocolate and prevention of cardiovascular disease: a systematic review. Nutr Metab (Lond). 2006;3:2. Published online 3 Jan 2006. doi: 10.1186/1743-7075-3-2

Chapter 7
Chocolate and Cocoa Aroma

Jürgen Voigt

Key Points

- Chocolate- and cocoa-specific aroma is formed during roasting of fermented, but not of unfermented, cocoa beans.
- Essential components of the chocolate- and cocoa-specific aroma are generated by heat-induced Maillard reactions of free amino acids and peptides with reducing sugars.
- Essential aroma precursors are derived from the vicilin (7S)-class globular storage protein of the cocoa beans by acid-induced successive degradation by endogenous aspartic endoprotease and carboxypeptidase activities.
- Effects of the time course of acidification and the final pH values of cocoa fermentations on the aroma potential of the resulting raw cocoa are due to the differential pH optima of the aspartic endoprotease and the carboxypeptidase of the cocoa beans.
- Genotype-dependent differences in the aroma potential of cocoa clones are at least partly due to varying contents of the vicilin (7S)-class globular storage protein and/or the amounts and activities of the aspartic endoprotease and carboxypeptidase of the cocoa beans.

Keywords Aroma components • Aspartic endoprotease • Carboxypeptidase • Cocoa beans • Free amino acids • Maillard reactions • Peptides • Storage protein • Vicilin (7S)-class globulin

Introduction

Chocolate has a distinctive aroma character that is strongly dependent on its proportion of cocoa powder and the quality of raw cocoa used for its production. The aroma quality of raw cocoa is determined by bean genotype, growth conditions, and processing factors [1–3]. Fermentation is a key processing stage that causes the death of the beans and facilitates the removal of the pulp and subsequent drying. During this stage, there is initiation of aroma formation, color development, and a significant reduction in bitterness. Essential precursors of the chocolate- and cocoa-specific aroma notes are generated during this stage by acid-induced proteolysis of the vicilin-type (7S) globular

J. Voigt, Dipl.-Biol., Dipl.-Chem. (✉)
Institute for Biochemistry, Charité – University Medicine Berlin, Charité- Platz 1, Virchowweg 24, Berlin D-10117, Germany
e-mail: juergen.voigt@charite.de

R.R. Watson et al. (eds.), *Chocolate in Health and Nutrition*, Nutrition and Health 7,
DOI 10.1007/978-1-61779-803-0_7,

storage protein of the cocoa beans [4, 5]. During roasting of the fermented cocoa beans, the typical aroma notes of chocolate and cocoa are formed by heat-induced chemical reactions of these proteolysis products with reducing sugars, including Maillard reactions [4]. These cocoa-specific aroma notes, which become volatile at temperatures around 30°C, are predominant in dark, bitter-tasting chocolate, less pronounced in fine chocolates, almost absent in milk chocolates, and completely absent in white chocolate [6]. The specific aroma notes of fine chocolates are formed in the pulp and are taken up by the beans during fermentation [5].

Cocoa-Specific Aroma Components of Dark Chocolates and Cocoa Powder

The typical chocolate aroma is characteristic for dark, bitter-tasting chocolate and depends on its high proportion of cocoa powder. Dark chocolate and cocoa powder contain several hundred volatile constituents, including pyrazines, thiazoles, oxazoles, pyrrole derivatives, pyridines, and furans [6–8], and it is still difficult to assess which components really contribute to the specific chocolate aroma. Candidates are those compounds which are more or less prominent in both cocoa powder and dark chocolate but are considerably less prominent in the volatile fraction of milk chocolate. Key aroma compounds have been identified in milk and dark chocolates as well as in cocoa powder using GC–mass spectrometry and GC–olfactometry [9–16]. Aroma extract dilution analysis of the volatile fractions of dark chocolate revealed that three constituents had a strong chocolate aroma: 2-methylpropanal, 2-methylbutanal, and 3-methylbutanal (Table 7.1). Many other components of the volatile fractions of

Table 7.1 Aroma components with cocoa- and chocolate-related sensorial attributes detected in raw cocoa or dark chocolate

Compound[a]	Sensorial attribute	Reference
Aldehydes		
2-Methylpropanal	Chocolate	[12, 16]
2-Methylbutanal	Chocolate	[12, 16]
3-Methylbutanal	Chocolate	[12, 16]
2-Phenyl-2-butenal	Cocoa, roasted	[12, 16]
4-Methyl-2-phenyl-2-pentenal	Cocoa	[8]
5-Methyl-2-phenyl-2-hexenal	Cocoa	[8]
Pyrazines		
2-Methylpyrazine	Nutty, cocoa	[8]
2-Ethyl-5-methylpyrazine	Cocoa, roasted	[16]
2,3-Dimethylpyrazine	Caramel, cocoa	[8]
2,3-Dimethyl-5-ethylpyrazine	Cocoa, chocolate	[16]
2,5-Dimethylpyrazine	Cocoa, roasted nuts	[8]
3,5-Dimethyl-2-ethylpyrazine	Cocoa, chocolate	[16]
2,3,5-Trimethylpyrazine	Cocoa, nutty	[8, 12, 16]
2,3,5,6-Tetramethylpyrazine	Cocoa, coffee	[8]
3,5-Diethyl-2-methylpyrazine	Cocoa, chocolate	[16]
Furans		
3-Phenylfuran	Cocoa, green, mint	[12, 16]
Pyrroles		
1-(2-Furanylmethyl)-1*H*-pyrrole	Roasted, chocolate, green	[12, 16]
2,3-Dimethyl-1*H*-pyrrole	Cocoa, chocolate	[16]
3-Ethyl-2,5-dimethyl-1*H*-pyrrole	Cocoa, hazelnut, coffee	[12, 16]

[a]After vacuum distillation and liquid–liquid extraction, the volatile fractions of cocoa powder [8] or dark chocolates [12, 16] were analyzed by gas chromatography–olfactometry and gas chromatography–mass spectrometry. Only those compounds are listed which revealed chocolate- or cocoa-related sensorial attributes

dark chocolate and cocoa mass were characterized by cocoa/nutty/roasted/coffee notes like several pyrazines and some pyrrole derivatives (see Table 7.1). However, other authors have reported different aroma notes for these components [17, 18]. Some additional compounds with cocoa-like aroma characters have been found, whose chemical nature is still unknown [12]. All these cocoa- and chocolate-specific aroma compounds are largely missing in the volatile fractions of milk chocolates [6, 11]. In roasted fine cocoa (Criollo), which usually has a rather low cocoa-typical aroma but instead contains some special aroma notes [5, 6], at least some of the cocoa- and chocolate-related aroma components of dark chocolate have been detected: 2-methylbutanal, 3-methylbutanal, 3,5 dimethyl-2-ethylpyrazine, and 2,3,5-trimethylpyrazine [17].

The characteristic cocoa and chocolate aroma notes are generated during the roasting process by Maillard or nonenzymatic browning reactions, which involve the reactions of free amino groups of amino acids and/or peptides with reducing sugars [6, 10, 11, 19]. Such reactions also occur during roasting of meat and other foods [20]. Some of the pyrazine derivatives occurring in dark chocolates with cocoa-related sensory attributes (see Table 7.1) are also formed in roasted beef [9], posing the question as to which these particular compounds are really characteristic for chocolate and cocoa aroma.

Fermentation of Cocoa Beans Is Required for the Formation of Cocoa-Specific Aroma Precursors

The typical cocoa aroma is formed during roasting of fermented but not of unfermented cocoa beans [4, 21]. Therefore, essential aroma precursors are generated during the fermentation process (Fig. 7.1). During fermentation, sugars in the pulp are transformed to acetic and lactic acid by environmental microorganisms, causing acidification of the cocoa beans [22]. This acidification, but not the presence of microorganisms, is required for the formation of the cocoa-specific flavor precursors as revealed by bean incubations under aseptic conditions [23]. Therefore, raw cocoa quality is strongly dependent

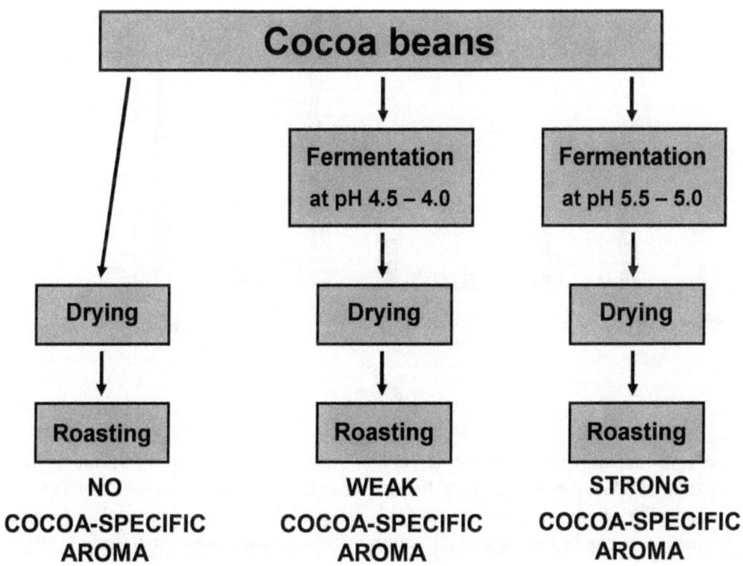

Fig. 7.1 Cocoa- and chocolate-specific aroma precursors are generated during fermentation (Adapted from Botanica Acta, 108, J. Voigt, B. Biehl, Precursors of the cocoa-specific aroma components are derived from the vicilin-class (7 S) globulin of the cocoa seeds by proteolytic processing, 283, Copyright 1995, with permission from Wiley-Blackwell)

on the degree and time course of acidification of the cotyledons during the fermentation process as well as on the duration of the fermentation process and the final pH [4, 6, 24].

Fine cocoas (Arriba, Criollo, National, Trinitario) contain additional aroma notes (floral, fruity, nutty, spicy, winy), which originate from the pulp and are strongly influenced by the genotype and growth conditions [6]. In fine cocoas, the level of basic cocoa- and chocolate-specific aroma components is usually low due to a rather short fermentation duration (2–4 days) as compared to bulk cocoa (5–7 days).

Essential Precursors of Cocoa-Specific Aroma Notes Are Generated by Acid-Induced Proteolytic Processes in Cocoa Beans

The level of free amino acids is considerably higher in fermented than in unfermented cocoa beans [23, 25–27]. Together with reducing sugars, amino acids are important precursors of Maillard reactions, leading to various aroma components [6, 9–11, 19]. Liberation of free amino acids was not only observed after fermentation of cocoa beans but also after fermentation-like incubations of cocoa beans under axenic, moderate acidic conditions [25, 26] and after autolysis of acetone-dry powder prepared from unfermented cocoa beans [28]. Hydrophobic amino acids were predominantly accumulated under all these conditions, and essentially, no differences in the patterns of free amino acids liberated under these different conditions were observed (Fig. 7.2). These findings clearly show that the liberation of predominantly hydrophobic amino acids during cocoa fermentation is due to an acid-induced

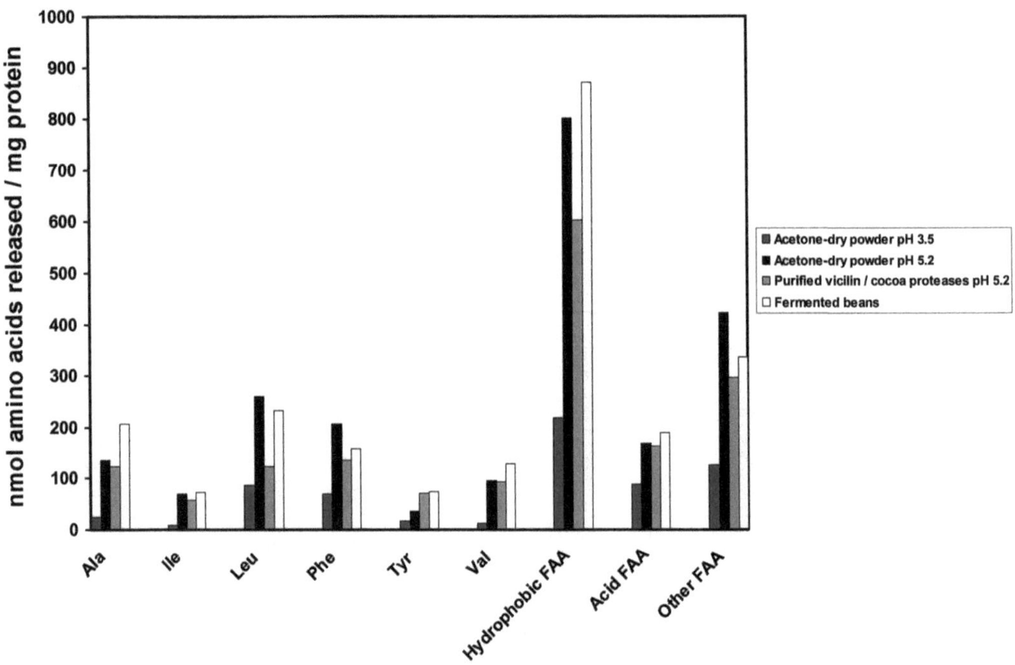

Fig. 7.2 Free amino acids accumulated in fermented cocoa beans and during in vitro proteolysis of beans proteins. FAA, free amino acids; acid FAA, Asp+Asn+Glu; hydrophobic FAA, Ala+Ile+Leu+Phe+Tyr+Val; other FAA, Ar g+Gly+Lys+Met+Ser+Thr; *dark gray bars*, amino acids released during autolysis at pH 3.5 of acetone-dry powder prepared from unfermented cocoa beans; *black bars*, amino acids released during autolysis at pH 5.2 of acetone-dry powder prepared from unfermented cocoa beans; *light gray bars*, amino acids released during incubation of purified cocoa vicilin (7)-class globulin with cocoa proteases at pH 5.2; *white bars*, fermented cocoa beans (This figure is based on results published in references [28] and [50])

Table 7.2 Aroma notes detected upon in vitro roasting of cocoa proteolysis products in the presence of reducing sugars[a, b]

Sample	Aroma notes[c]
Precursor extract from fermented beans[d]	Cocoa, chocolate
Acdp[e] autolysis products (pH 5.2)	Cocoa, chocolate
Acdp[e] autolysis products (pH 3.5)	Acid, herbaceous, vegetable
Acdp[e] autolysis products (pH 3.5) plus mixture of free amino acid	Herbaceous, floral, acid
Mixture of free amino acids[f]	Cereal, floral
Acdp[e] autolysis products (pH 3.5) posttreated with carboxypeptidase A	Cocoa, chocolate

[a]Data obtained from reference [28]

[b]Mixture of 0.13 g of glucose and 0.37 g of fructose dissolved/suspended in 0.15 ml distilled water was added to 0.25 g of precursor sample and ground in a mortar and pestle with 2.5 g of deodorized cocoa butter and 1.0 g silicate. The mixture was transferred to a glass Petri dish and heated for 10 min at 130°C

[c]Aroma notes were determined by sniffing of the roasted samples preheated to 37–40°C

[d]Aroma precursors were extracted with 70% (v/v) aqueous methanol. After evaporation of the solvent, the residue was suspended in water and freeze-dried

[e]AcDP, acetone-dry powder of cocoa seeds prepared by thorough extraction with aqueous acetone of the defatted seed powder

[f]Mixture of free amino acids adjusted to the composition of the free amino acids of fermented cocoa beans (see Fig. 7.2)

proteolysis of bean proteins by endogenous proteases. During fermentation, these proteolytic processes start after 1–3 days, when the beans are killed by testa opening at the micropyle caused by acidification of the pulp and increased temperature. This aperture of the testa enables the penetration of the nib by acetic and lactic acid [5]. In the case of bulk cocoa, when a strong cocoa-specific roasting aroma is desired, fermentation is usually stopped after 5–7 days [6, 24]. In the case of fine cocoas, however, fermentation is stopped earlier to avoid masking of the particular aroma notes by the typical cocoa-specific roasting aroma [5].

Questions arose about whether the generation of the chocolate- and cocoa-specific aroma components during roasting of the fermented cocoa beans is merely dependent on the particular pattern of free amino acids released during the fermentation process (see Fig. 7.2) or whether specific peptides are also required, as suggested by Mohr and coworkers [29]. A suitable experimental approach to verify this question was the in vitro formation and subsequent organoleptic analysis of the aroma notes [28]. It has been shown that cocoa-specific aroma precursors are generated in vitro when polyphenol-free acetone-dry powder prepared from unfermented cocoa beans was subjected to autolysis at pH 5.2, i.e. under conditions of optimal fermentation (Table 7.2). The proteolysis products obtained under these conditions consisted of hydrophilic peptides and hydrophobic free amino acids [28]. When the same material was incubated at pH 3.5, no cocoa-specific aroma precursors were formed (see Table 7.2). Under the latter conditions, rather low amounts of liberated free amino acids were observed, and hydrophobic instead of hydrophilic peptides were detected in the autolysis products generated by the action of endoprotease activities [28]. These hydrophobic peptides could be transformed to hydrophilic peptides by treatment with commercial carboxypeptidase A at pH 5.2 accompanied by liberation of hydrophobic free amino acids [28]. This mixture of hydrophilic peptides and hydrophobic free amino acids revealed typical cocoa aroma when roasted in the presence of reducing sugars and deodorized cocoa butter (see Table 7.2). No typical cocoa or chocolate aroma was obtained, however, when synthetic amino acid mixtures adapted to the spectrum of free amino acids present in fermented cocoa seeds (see Fig. 7.2) were roasted in the presence of reducing sugars (see Table 7.2). The latter findings are in agreement with earlier observations of Mohr and coworkers [29] that, in addition to free amino acids, an oligopeptide fraction isolated from fermented and air-dried cocoa beans is absolutely required to obtain cocoa and chocolate aroma upon roasting.

Proteases of Cocoa Beans

Aspartic endoprotease, carboxypeptidase, and aminopeptidase activities have been found in ungerminated cocoa beans [28, 30]. For the aspartic endoprotease (EC 3.4.23) partially purified from cocoa beans, activity optima were measured at 55°C and pH 3.5 [30]. Two cocoa cDNA species, TcAP1 and TcAP2, respectively, encoding different polypeptides of the plant aspartic endoprotease family, have been cloned and sequenced [31]. Both genes are induced early in seed development and show decreased expression as the beans reach maturity [31]. These data are in agreement with the results of protein blot analyses, indicating an accumulation of aspartic endoprotease protein during seed development up to an optimum level reached after ripening [32]. Cocoa beans have unusually high levels of such aspartic endoprotease activity as compared to seeds of other plants [33]. Guilloteau et al. [34] assume that physical and biochemical properties of the active cocoa TcAP2 aspartic endoprotease complex are novel, suggesting the highly expressed gene product may represent a previously unknown activity. A purified TcAP2 gene product efficiently degrades cocoa seed vicilin into low molecular products including di- and tripeptides, implying that this gene product may play an important role during fermentation.

Aspartic endoproteases like pepsin and renin cleave protein substrates preferentially at hydrophobic amino acid residues. This holds true also for the aspartic endoprotease of ungerminated cocoa beans, although its exact cleavage specificity has not yet been characterized. However, it has been reported that this particular protease generates peptides with hydrophobic amino acid residues at their C-terminal ends, which can be cleaved off by commercial carboxypeptidase A [28].

As described above, cocoa-specific aroma precursors can be generated in vitro from acetone-dry powder prepared from unfermented cocoa beans by either autolysis at pH 5.2 or by posttreatment with commercial carboxypeptidase A of the mixture of hydrophobic peptides formed at pH 3.5 (see Table 7.2). Under both conditions, essentially, the same mixture of free amino acids was obtained [28]. These findings show that the cocoa-specific aroma precursors are generated by cooperation of the aspartic endoprotease and the carboxypeptidase of the cocoa beans [28, 35, 36].

The carboxypeptidase present in ungerminated cocoa beans preferentially liberates hydrophobic amino acids, whereas acid amino acids are released very slowly [37]. Peptides with carboxyterminal arginine, lysine, or proline residues are resistant to degradation by this carboxypeptidase [37]. The rate of hydrolysis is not only dependent on the carboxyterminal but also affected by the neighboring amino acid residue as also reported for the carboxypeptidases from other organisms [38–40]. The carboxypeptidase of the cocoa beans has a pH optimum between pH 4.8 and 5.8, depending on the substrate [28, 30, 37].

Proteolytic Precursors of Chocolate- and Cocoa-Specific Aroma Components Originate from the Globular Storage Protein of Cocoa Beans

Cocoa beans contain two predominant storage proteins: a 21-kDa albumin with homology to the Kunitz trypsin inhibitors [41–45] and a vicilin (7S)-class globulin [43, 46–48]. It has been shown that under optimal fermentation conditions, some predominant polypeptides of the cocoa beans are selectively degraded [49]. These selectively degraded polypeptides have been later identified as the polypeptide subunits of the vicilin (7S)-class globulin of the cocoa beans [43, 50]. Typical cocoa and chocolate aroma was obtained when the vicilin (7S)-class globulin was successively degraded by the cocoa aspartic endoprotease and carboxypeptidase activities and the resulting proteolysis products subsequently roasted in the presence of reducing sugars and deodorized cocoa butter (Table 7.3; see Voigt, Heinrichs, Voigt, et al. [50]). No cocoa-specific aroma precursors were obtained when the 21-kDa

Table 7.3 Proteolytic formation of the essential precursors of the cocoa- and chocolate-specific aroma notes depends on the structure of the globular storage protein of the cocoa beans[a, b]

Sample	Aroma notes[c]
Precursor extract from fermented beans[d]	Cocoa, chocolate
In vitro proteolysis products of cocoa albumin with proteases from cocoa beans (pH 5.2)	Cereal, herbaceous
In vitro proteolysis products of cocoa globulin with proteases from cocoa beans (pH 5.2)	Cocoa, chocolate
In vitro proteolysis products of coconut globulin with proteases from cocoa beans (pH 5.2)	Herbaceous, cereal
In vitro proteolysis products of hazelnut globulin with proteases from cocoa beans (pH 5.2)	Cereal, nutty
In vitro proteolysis products of sunflower seed globulin with proteases from cocoa beans (pH 5.2)	Cereal, herbaceous

[a]Data obtained from reference [51]

[b]Proteolysis products were formulated with reducing sugars and deodorized cocoa butter and roasted as described in the legend of Table 7.2

[c]Aroma notes were determined by sniffing of the roasted samples preheated to 37–40°C

[d]Aroma precursors were extracted with 70% (v/v) aqueous methanol. After evaporation of the solvent, the residue was suspended in water and freeze-dried

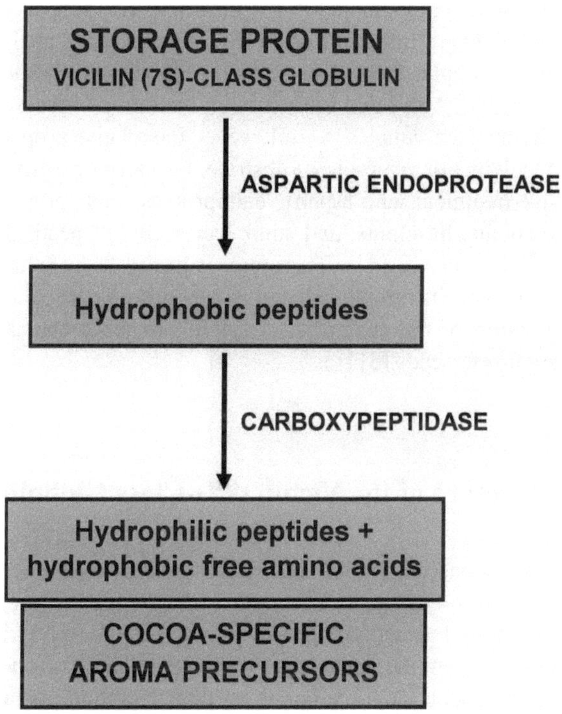

Fig. 7.3 Proteolytic formation of the cocoa- and chocolate-specific aroma precursors (Adapted from Botanica Acta, 108, J. Voigt, B. Biehl, Precursors of the cocoa specific aroma components are derived from the vicilin-class (7 S) globulin of the cocoa seeds by proteolytic processing, 283, Copyright 1995, with permission from Wiley-Blackwell)

albumin was degraded by the same proteolytic enzymes. When the cocoa vicilin(7S)-class globulin was merely incubated with the aspartic endoprotease purified from ungerminated cocoa beans, a mixture of more or less hydrophobic peptides was obtained (Fig. 7.3; see Heinzler and Eichner [58]). These particular hydrophobic peptides did not reveal cocoa- or chocolate-specific aroma upon roasting in the presence of reducing sugars even in the presence of a mixture of free amino acids adapted to the pattern of free amino acids detected in well-fermented cocoa beans [50]. These hydrophobic peptides were transformed to hydrophilic peptides and hydrophobic free amino acids rather similar to the pattern of proteolysis products found in fermented cocoa beans [28, 50]. These data indicate that the cocoa-specific aroma precursors are derived from the vicilin (7S)-class globulin by the cooperative action of the aspartic endoprotease and the carboxypeptidase present in ungerminated, ripe cocoa beans (see Fig. 7.3). In the first step, the vicilin (7S)-class globulin is degraded by the aspartic endoprotease, which generates a complex mixture of more or less hydrophobic oligopeptides, as judged by reversed-phase HPLC. By the action of the carboxypeptidase, these hydrophobic oligopeptides are transformed to a mixture of hydrophilic peptides and hydrophobic free amino acids containing the essential precursors of the typical components of the cocoa and chocolate flavor (see Fig. 7.3). Therefore, the hydrophobic properties of the oligopeptides generated by the aspartic endoprotease are largely due to the hydrophobic amino acid residues at their carboxyterminal ends.

When the globulins isolated from cocoa beans, coconuts, hazelnuts, and sunflower seeds were successively degraded with the aspartic endoprotease and the carboxypeptidase partially purified from ripe, ungerminated cocoa beans and the obtained proteolysis products subsequently roasted in the presence of reducing sugars and deodorized cocoa butter, the typical cocoa- and chocolate-specific aroma was obtained exclusively in the case of the cocoa globulin (see Table 7.3; see Voigt, Wrann, Heinrichs, et al. [51]). The globular fraction of coconuts only contain a vicilin (7S)-class globulin as the cocoa beans [51], and those from hazelnuts and sunflower seeds largely consist of legumin (11S)-class globulin [51, 52]. The amino acid sequences of legumin (11S)-class and vicilin (7S)-class globulins are rather different [53]. Comparative analyses of the free amino acids released from all these globular storage proteins by the action of aspartic endoprotease and carboxypeptidase from cocoa beans revealed a preferential liberation of the hydrophobic amino acids Ala, Leu, Phe, Tyr, and Val in all the samples [51]. However, the relative proportions of these amino acids were rather different depending on the protein substrate. Furthermore, different peptide patterns were obtained by successive treatment with aspartic endoprotease and carboxypeptidase of the globulins from cocoa beans, coconuts, hazelnuts, and sunflower seeds, respectively, as revealed by comparative reversed-phase HPLC analyses [51]. Therefore, it has to be concluded that the proteolytic formation of essential cocoa-specific aroma precursors depends on particular amino acid sequences of the vicilin (7S)-class globulin of the cocoa beans lacking in the globular storage proteins of coconuts, hazelnuts, and sunflower seeds [51].

Structure of the Vicilin (7S)-Class Globulin of Cocoa Beans

The cocoa vicilin (7S)-class globulin consists of overlapping polypeptides with apparent molecular masses of 47 kDa, 31 kDa, and 14.5 kDa [28, 46, 47]. These polypeptide subunits are generated by differential proteolytic processing of a common 66-kDa precursor, whose amino acid sequence was derived from the nucleotide sequences of the corresponding cDNA [47] and gene [48]. Localization of the cocoa vicilin subunits on their common precursor by N-terminal sequencing was, however, impossible because all these subunits were found to be resistant against Edman degradation, presumably due to N-terminal protection. MALDI-ReTOF-MS analyses of the tryptic fragments of the purified 47-kDa subunit revealed that this particular subunit was derived from the C-terminal domain of the 66-kDa precursor [54]. By the same approach, the other constituents of the mature cocoa vicilin-class globulin were localized in the same precursor region. The results were further corroborated by epitope mapping of polyclonal antibodies raised against the particular polypeptide subunits using 185 overlapping pentadecapeptides covering the whole sequence of the precursor [54].

Two-dimensional gel electrophoresis of cocoa seed proteins revealed heterogeneities in the patterns of the vicilin (7S)-class globulin subunits [54, 55]. The same patterns of globular storage proteins were found in cocoa cotyledons from various genetic origins both after SDS-PAGE and 2D-electrophoresis, respectively [55]. By IEF, the 47-kDa subunit was split into six spots with isoelectric points (IEPs) between 5.2 and 6.4 [54]. The 31-kDa and 28-kDa components were split into several spots with IEPs between 6.1 and 7.3. These heterogeneities could be either due to specific posttranslational modifications of particular amino acid residues or to the existence of a gene family generated by gene duplications followed by mutations of the individual members of this hypothetic gene family. In both cases, the formation of additional tryptic peptide fragments is expected, whose amino acid sequence differs from the sequence of the 66-kDa precursor. The lack of such additional fragments in the MALDI-ReTOF spectra clearly showed that despite the heterogeneities with respect to their isoelectric points, all the components of the vicilin (7S)-class globulin of the cocoa beans are encoded by a single gene and that there is no specific posttranslational modification of specific amino acid residues [54]. Calculation of the theoretical isoelectric points of the 47-kDa and 31-kDa subunits revealed values that roughly corresponded to those experimentally determined by two-dimensional electrophoresis for the corresponding subspecies with highest pI values [54]. The occurrence of subspecies with lower pI values could be explained by statistical transformation of Asn or Gln to Asp or Glu residues, respectively, due to the action of a protein deaminase during maturation. Such an enzyme was found in germinating wheat grains [56].

Particular Cleavage Specificity of the Cocoa Aspartic Endoprotease Is Essential for Generation of Cocoa-Specific Aroma Precursors

Apart from the particular amino acid sequence of the vicilin (7S)-class globulin of the cocoa beans, the formation of the cocoa- and chocolate-specific aroma precursors is also strongly dependent on the substrate and cleavage specificities of the aspartic endoprotease of the ripe, ungerminated cocoa beans, which is not sufficiently studied to date. Presently, it is only known that this particular enzyme cleaves protein substrates at hydrophobic amino acid residues to produce oligopeptides with hydrophobic amino acid residues at their carboxyterminal ends [28]. Similar substrate and cleavage specificities have been reported for chymotrypsin and pepsin. The cocoa bean globulin was efficiently degraded not only by the aspartic endoprotease from ripe, ungerminated cocoa beans but also by chymotrypsin and pepsin. As revealed by reversed-phase HPLC analysis, similarities were observed for the oligopeptide patterns generated by pepsin and cocoa aspartic endoproteinase, whereas the pattern of chymotryptic peptides was completely different [57]. After posttreatment with carboxypeptidase, pepsin revealed similar patterns of free amino acids and oligopeptides as the cocoa aspartic endoprotease. Indeed, after roasting in the presence of reducing sugars, a weak cocoa and chocolate flavor was observed in the case of the proteolysis products of cocoa vicilin (7S)-class globulin generated by pepsin and carboxypeptidase, but not for the proteolysis products obtained by successive treatments with chymotrypsin and carboxypeptidase [57]. These findings indicate that the particular specificity of the cocoa aspartic endoprotease is important for the formation of the cocoa-specific aroma precursors [57], in addition to the amino acid sequence of the protein substrate.

Summary

Chocolate- and cocoa-specific aroma is obtained by roasting of fermented, but not of unfermented, cocoa beans (see Fig. 7.1). Therefore, essential precursors are generated during fermentation, which are transformed to the chocolate- and cocoa-specific aroma components during the roasting process by complex Maillard or nonenzymatic browning reactions. These are initiated by reactions of free amino groups of amino acids or peptides with the carbonyl groups of reducing sugars [9, 10, 58–60]. Indeed, accumulation of free amino acids has been observed both during fermentation and fermentation-like incubations of cocoa beans under moderate acidic conditions [23, 25–27]. During the fermentation process, the cocoa pulp is degraded by various microorganisms accompanied by the accumulation of acetic and lactic acid [22, 24]. This acidification causes the death of the beans as well as acid-induced proteolysis of bean proteins [23, 25–27]. It is well known that fermentations accompanied by a strong acidification of the cotyledons (pH < 5.0) result in poor aroma potential of the resulting raw cocoa (see Fig. 7.1). This fact is obviously due to the pH dependency of the carboxypeptidase, whose activity is optimal around pH 5.5 and strongly decreases at pH values below 5.0 [28, 30, 37]. At pH values greater than 5.5, the aspartic endoprotease is inactive [28, 30]. These findings provide an explanation for the dependency of the aroma formation on the degree of acidification of the cocoa beans during the fermentation process (see Fig. 7.1).

The specific mixture of free amino acids and peptides present in well-fermented cocoa beans and required for the formation of the cocoa-typical aroma notes during roasting depends on both the particular cleavage specificity of the aspartic endoprotease and the amino acid sequence of the vicilin (7S)-class globular storage protein of the cocoa beans (see Fig. 7.3) [50, 51, 57]. The aspartic endoprotease of the cocoa beans cleaves its protein substrate preferentially at hydrophobic amino acid residues [50, 51]. However, its exact cleavage specificity is still unknown and has to be analyzed in the future. The pattern of free amino acids liberated from cocoa vicilin (7S)-class globulin by the cooperation of the aspartic endoprotease and the carboxypeptidase is characteristic for well-fermented cocoa beans (see Fig. 7.2). However, this particular mixture of free amino acids is not sufficient for the formation of the typical cocoa-specific aroma during roasting in the presence of reducing sugars (see Table 7.2). Both free amino acids and hydrophilic peptides proteolytically derived from the cocoa vicilin (7S)-class globulin are essential for the formation of the cocoa-specific aroma components during the roasting process (see Table 7.2 and Fig. 7.3). These peptides exclusively occur in the proteolysis products of the cocoa vicilin (7S)-class globulin (see Table 7.3). Their amino acid sequences are predetermined by both the amino acid sequence of the cocoa vicilin (7S)-class globulin [54, 61] and the particular cleavage specificity of the cocoa aspartic endoprotease. However, their number and particular amino acid sequences are still unknown and have to be analyzed in the future.

Another important aspect is the ensuring of raw cocoa quality in the future. In recent years, raw cocoa production has been threatened by diverse cocoa diseases caused by different fungal pathogens and by certain mirids (Heteroptera; Miridae), particularly by the cocoa pod borer, *Conopomorpha cramerella*, in Asia [62]. Most of these biotic constraints are presently restricted to certain geographic areas. All together, they reduce yields by about 20% [62] but could cause far greater losses if certain highly damaging pathogens were to become more widely distributed. This risk is increased by monocultures of pathogen-susceptible clones. Therefore, national and international breeding programs are presently underway to improve resistance of cocoa trees to various biotic constraints. In the past, traditional cocoa breeding was not very efficient owing to the long generation time of the trees. However, application of modern molecular genetic techniques like marker-assisted selection to cocoa breeding are expected to greatly improve the rate of the genetic gain and result in the selection of numerous new, productive, disease-resistant cultivars [63, 64]. Such breeding programs could result in cocoa clones with successively decreasing aroma potential when selection is restricted to pathogen resistance and yield. This undesirable effect is well known from breeding of other crops. Genotype influences both aroma quality and intensity [13, 14, 65, 66] likely by determining the quantities of precursors and the activities of enzymes involved in the formation of aroma components and aroma precursors, respectively. Reineccius [15] concluded that the known varietal differences are primarily due to quantitative rather than to qualitative differences in the aroma precursor and polyphenol contents. A comparative study of cocoa beans of various clones revealed relevant differences both in the contents of the vicilin (7S)-class globulin and the aroma potential [67]. Furthermore, a strict correlation of cocoa aroma with the amount of vicilin (7S)-class globulin in the beans was observed [67]. These findings show that comparative analysis of the new, pathogen-resistant clones for the levels of both this particular globular storage protein and the proteolytic enzymes involved in aroma precursor formation is requested for ensuring the aroma potential of raw cocoas.

Apart from the basic cocoa- and chocolate-specific aroma notes, cocoa and chocolate quality is also influenced by other aroma compounds, part of which (like linalool) are generated in the pulp and enter the nib during cocoa fermentation after opening of the testa caused by acidification and swelling of the beans [5, 68]. Both the patterns and amounts of the latter aroma components are also dependent on the genetic background and can be strongly changed in new clones selected for pathogen resistance and yield. Linalool content in both the pulp and in the fermented beans is a suitable marker for some flavor grade cocoas [69].

References

1. Clapperton JF. A review of research to identify the origins of cocoa flavour characteristics. Cocoa Growers Bull. 1994;48:7–16.
2. Beckett ST. Is the taste of British milk chocolate different? Int J Dairy Technol. 2003;56:138–42.
3. Whitefield R. Making chocolates in the factory. London: Kennedy's Publications Ltd; 2005.
4. Voigt J, Biehl B. Precursors of the cocoa specific aroma components are derived from the vicilin-class (7S) globulin of the cocoa seeds by proteolytic processing. Bot Acta. 1995;108:283–9.
5. Voigt J, Lieberei R. Biochemistry of cocoa fermentation. In: Schwan RF, Fleet GH, editors. Cocoa and coffee fermentations. Boca Raton: CRC Press; 2013.
6. Afoakwa EO, Paterson A, Fowler M, et al. Flavor formation and character in cocoa and chocolate: a critical review. Crit Rev Food Sci Nutr. 2008;48:840–57.
7. Ziegleder G, Biehl B. Analysis of cocoa flavour components and flavour precursors. In: Linskens HF, Jackson JF, editors. Modern methods of plant analysis, New Series, vol. 8. New York: Springer; 1988. p. 321–93.
8. Bonvehí JS. Investigation of aromatic compounds in roasted cocoa powder. Eur Food Res Technol. 2005;221:19–29.
9. Cerny C, Grosch W. Precursors of ethylpyrazine isomers and 2,3-diethyl-5-methylpyrazine formed in roasted beef. Z Lebensm Unters Forsch. 1994;198:210–4.
10. Amrani-Hemaimi M, Cerny C, Fay LB. Mechanism of formation of alkylpyrazine in the Maillard reaction. J Agric Food Chem. 1995;43:2818–22.
11. Schnermann P, Schieberle P. Evaluation of key odorants in milk chocolate and cocoa mass by aroma extract dilution analysis. J Agric Food Chem. 1997;45:867–72.
12. Counet C, Callemien D, Ouwerx C, et al. Use of gas chromatography-olfactometry to identify key odorant compounds in dark chocolate. Comparison of samples before and after conching. J Agric Food Chem. 2002; 50:2385–91.
13. Taylor AJ. Food flavour technology. Sheffield: Sheffield Academic Press; 2002.
14. Taylor AJ, Roberts DD. Flavour perception. Oxford: Blackwell Publishing; 2004.
15. Reineccius G. Flavour chemistry and technology. 2nd ed. Boca Raton: CRC Press; 2006.
16. Afoakwa EO, Paterson A, Fowler M, et al. Matrix effects on flavour volatiles release in dark chocolates varying in particle size distribution and fat content using GC-mass spectrometry and GC-olfactometry. Food Chem. 2009;113:208–15.
17. Frauendorfer F, Schieberle P. Changes in key aroma compounds of Criollo cocoa beans during roasting. J Agric Food Chem. 2008;56:10244–51.
18. Huang Y, Barringer SA. Alkylpyrazines and other volatiles in cocoa liquor at pH 5 to 8, by selected ion flow tube-mass spectrometry (SIFT-MS). J Food Sci. 2010;75:C121–7.
19. Granvogel M, Bugan S, Schieberle P. Formation of amines and aldehydes from parent amino acids during thermal processing of cocoa and model systems: new insights into pathways of the Strecker reaction. J Agric Food Chem. 2006;54:1730–9.
20. Ledl F, Schleicher E. New aspects of the Maillard reaction in foods and in the human body. Angew Chem Int Ed Engl. 1990;29:565–94.
21. Rohan TA. The precursors of chocolate aroma: a comparative study of fermented and unfermented cocoa beans. J Sci Food Agric. 1964;29:456–9.

22. Ostovar K, Keeney PG. Isolation and characterization of microorganisms involved in the fermentation of Trinidad cocoa beans. J Food Sci. 1973;39:611–7.
23. Biehl B, Brunner E, Passern D, et al. Acidification, proteolysis and flavour potential in fermenting cocoa beans. J Sci Food Agric. 1985;36:583–98.
24. Schwan RF, Wheals AE. The microbiology of cocoa fermentation and its role in chocolate quality. Crit Rev Food Sci Nutr. 2004;44:205–21.
25. Kirchhoff PM, Biehl B, Crone G. Peculiarity of the accumulation of free amino acids during cocoa fermentation. Food Chem. 1989;31:295–311.
26. Kirchhoff PM, Biehl B, Ziegeler-Berghausen H, et al. Kinetics of the formation of free amino acids in cocoa seeds during fermentation. Food Chem. 1989;34:161–79.
27. Rohsius C, Matissek R, Lieberei R. Free amino acid amounts in raw cocoas from different origins. Eur Food Res Technol. 2006;222:432–8.
28. Voigt J, Biehl B, Heinrichs H, et al. In-vitro formation of cocoa-specific aroma precursors: aroma-related peptides generated from cocoa seed protein by co-operation of an aspartic endoprotease and a carboxypeptidase. Food Chem. 1994;49:173–80.
29. Mohr W, Landschreiber E, Severin T. Zur Spezifität des Kakaoaromas. Fette Seifen Anstrichmittel. 1976; 78:88–95.
30. Biehl B, Heinrichs H, Ziegeler-Berghausen H, et al. The proteases of ungerminated cocoa seeds and their role in the fermentation process. Angew Bot. 1993;67:59–65.
31. Laloi M, McCarthy J, Morandi O, et al. Molecular and biochemical characterization of two proteinases TcAP1 and TcAP2 from *Theobroma cacao* seeds. Planta. 2002;215:754–62.
32. Voigt J, Kamaruddin S, Heinrichs H, et al. Developmental stage-dependent variation of the levels of globular storage protein and aspartic endoprotease during ripening and germination of *Theobroma cacao* seeds. J Plant Physiol. 1995;145:299–307.
33. Voigt G, Biehl B, Heinrichs H, et al. Aspartic proteinase levels in seeds of different angiosperms. Phytochemistry. 1997;44:389–92.
34. Guilloteau M, Lalaoi M, Michaux S, et al. Identification and characterization of the major aspartic endoprotease activity in *Theobroma cacao* seeds. J Sci Food Agric. 2005;85:549–62.
35. Hansen CE, del Olmo M, Burri C. Enzyme activities in cocoa beans during fermentation. J Sci Food Agric. 1998;77:273–81.
36. Hansen CE, Manez A, Burri C, et al. Comparison of enzyme activities involved in flavour precursor formation in unfermented beans of different cocoa genotypes. J Sci Food Agric. 2000;80:1193–8.
37. Bytof G, Biehl B, Heinrichs H, et al. Specificity and stability of the carboxypeptidase activity in ripe, ungerminated seeds of *Theobroma cacao* L. Food Chem. 1995;54:15–21.
38. Neurath H. Carboxypeptidase A. In: Boyer PD, Lardy HA, Myrbäck K, editors. The enzymes, vol. 4. New York: Academic; 1960. p. 11–24.
39. Ambler RP. Enzymatic hydrolysis with carboxypeptidases. Methods Enzymol. 1972;25:143–54.
40. Breddam K, Sœrensen SB, Svendson I. Primary structure and enzymatic properties of carboxypeptidase II from wheat bran. Carlsberg Res Commun. 1987;52:297–311.
41. Spencer ME, Hodge R. Cloning and sequencing of the cDNA encoding the major albumin of *Theobroma cacao*. Planta. 1991;183:528–35.
42. Tai H, McHenry L, Fritz PJ, et al. Nucleic acid sequence of a 21 kDa cocoa seed protein with homology to the soybean trypsin inhibitor (Kunitz) family of protease inhibitors. Plant Mol Biol. 1991;16:913–5.
43. Voigt J, Biehl B, Kamaruddin S. The major seed proteins of *Theobroma cacao* L. Food Chem. 1993;47:145–51.
44. Kochhar S, Gartenmann K, Juillerat MA. Primary structure of the abundant seed albumin of *Theobroma cacao* by mass spectrometry. J Agric Food Chem. 2000;48:5593–9.
45. Kochhar S, Gartenmann K, Gilloteau M, et al. Isolation and characterization of 2 S cocoa seed albumin storage polypeptide and the corresponding cDNA. J Agric Food Chem. 2001;49:4470–7.
46. Pettipher GL. The extraction and partial purification of cocoa storage proteins. Café Cacao Thé. 1999;34:23–6.
47. Spencer ME, Hodge R. Cloning and sequencing of a cDNA encoding the major storage proteins of *Theobroma cacao*. Identification of the proteins as members of the vicilin class of storage proteins. Planta. 1992;186:567–76.
48. McHenry L, Fritz PJ. Comparison of the structure and nucleotide sequence of vicilin genes of cocoa and cotton raise questions about vicilin evolution. Plant Mol Biol. 1992;18:1173–6.
49. Biehl B, Wewetzer C, Passern D. Vacuolar (storage) proteins and their degradation during germination and fermentation. J Sci Food Agric. 1982;33:1291–304.
50. Voigt J, Heinrichs H, Voigt G, et al. Cocoa-specific aroma precursors are generated by proteolytic digestion of the vicilin-like globulin of the cocoa seeds. Food Chem. 1994;50:177–84.
51. Voigt J, Wrann D, Heinrichs H, et al. The proteolytic formation of essential cocoa-specific aroma precursors depends on particular chemical structures of the vicilin-class globulin of the cocoa seeds lacking in the globular storage proteins of coconuts, hazelnuts, and sunflower seeds. Food Chem. 1994;51:197–205.

52. Derbyshire E, Wright DJ, Boulter D. Legumin and vicilin, storage proteins of legume seeds. Phytochemistry. 1976; 15:3–24.
53. Borroto K, Dure L. The globulin seeds storage proteins of flowering plants are derived from two ancestral genes. Plant Mol Biol. 1987;8:113–31.
54. Kratzer U, Frank R, Kalbacher H, et al. Subunit structure of the vicilin-like globular storage protein of cocoa seeds and the origin of cocoa- and chocolate-specific aroma precursors. Food Chem. 2009;113:903–13.
55. Amin I, Jinap S, Jamilah B, et al. Analysis of vicilin (7S)-class globulin in cocoa cotyledons from various genetic origins. J Sci Food Agric. 2002;82:728–32.
56. Vaintraub IA, Kotova LV, Shaha R. Protein deamidase from germinating wheat grains. FEBS Lett. 1992;302: 169–71.
57. Voigt J, Voigt G, Heinrichs H, et al. In vitro studies on the proteolytic formation of the characteristic aroma precursors of fermented cocoa seeds: the significance of endoprotease specificity. Food Chem. 1994;51:7–14.
58. Heinzler M, Eichner KV. Verhalten von Amadori-Verbindungen während der Kakaoverarbeitung. Bildung von Aromastoffen unter Röstbedingungen. Z Lebensm Unters Forsch. 1991;192:445–50.
59. Garbe LA, Wuertz A, Piechotta CT, et al. The peptide-catalyzed Maillard reaction. Characterization of $^{13°}$C reductones. Ann NY Acad Sci. 2008;1126:244–7.
60. Van Lancker F, Adams A, De Kimpe N. Formation of pyrazines in Maillard model systems of lysine-containing dipeptides. J Agric Food Chem. 2010;58:2470–8.
61. Voigt J, Kratzer U, Biehl B, et al. The origin of cocoa- and chocolate-specific aroma precursors. Malays Cocoa J. 2009;5:68–72.
62. Ploetz RC. Cacao diseases: important threats to chocolate production worldwide. Phytopathology. 2007;97:1634–9.
63. Schnell RJ, Brown JS, Kuhn DN, et al. Development of a marker assisted selection program for cacao. Phytopathology. 2007;97:1664–9.
64. Schnell RJ, Kuhn DN, Shapiro HY, et al. New tools to increase the efficiency of traditional cacao breeding. Malays Cocoa J. 2009;5:24–36.
65. Luna F, Crouzillat D, Cirou L, et al. Chemical composition and flavor of Ecuadorian cocoa liquor. J Agric Food Chem. 2002;50:3527–32.
66. Counet C, Ouwerx C, Rosoux D, et al. Relationship between procyanidin and flavour contents of cocoa liquors from different origins. J Agric Food Chem. 2004;52:6243–9.
67. Khairul Bariah S, Amin I, Furtek D. Correlation of cocoa flavor with the concentration of vicilin (7S)-class globulin in embryos. In: Proceedings of the 2007 Conference on Plantation Commodities; Kuala Lumpur; 2007. p. 55–58.
68. Pino J, Roncal E. Linalool content in roasted cocoa butter as characteristic of several flavour grade cocoas. Die Nahrung. 1992;36:299–301.
69. Ziegleder G. Linalool contents as characteristic of some flavour grade cocoas. Z Lebensm Unters Forsch. 1990; 191:306–9.

Part II
Composition of Chocolate Sources and Related Plant Components

Chapter 8
Composition of Cacao Beans

Antonella Bertazzo, Stefano Comai, Francesca Mangiarini, and Su Chen

Key Points

- The physics and chemistry of cocoa beans are very complex and change throughout the life of the bean, mainly depending on the processing it receives and on geographical origin.
- The main component of cocoa beans is lipid fraction, approximately 50%, mainly constituted by neutral lipids, with a predominant fraction of triglyceride molecules.
- Protein fraction constitutes 10–15% of the dry weight of cocoa seeds, and it is composed of 52% and 43% of albumin and globulin fractions, respectively. Other proteins, such as glutelins and prolamins, are present in lower concentrations.
- Cocoa beans contain stimulant substances, such as theobromine, caffeine, and theophylline, named purinic alkaloids, which affect the central nervous system.
- Various bioactive compounds with vasoactive effects have also been reported in cocoa beans.

Keywords *Theobroma cacao* • Cocoa beans • Nutrient composition • Geographic influence • Purinic alkaloids

---

A. Bertazzo, Ph.D. (✉)
Department of Pharmaceutical Sciences, University of Padova, Via Marzolo 5, Padova 35131, Italy
e-mail: antonella.bertazzo@unipd.it

S. Comai, Ph.D.
Department of Pharmaceutical Sciences, University of Padova, Via Marzolo 5, Padova 35121, Italy
e-mail: stefano.comai@unipd.it

F. Mangiarini, Ph.D.
Department of Chemistry and Biochemistry, Concordia University,
7141 Sherbrooke St, W.Montreal, QC H4B 1R6, Canada
e-mail: fmangiar@live.concordia.ca

S. Chen, Ph.D.
Research and Development, Chainon Neurotrophin Biotechnology Inc,
12436 Vance Jackson, Suite 721, San Antonio, TX 78230, USA
e-mail: su@chainonbio.com

R.R. Watson et al. (eds.), *Chocolate in Health and Nutrition*, Nutrition and Health 7,
DOI 10.1007/978-1-61779-803-0_8, © Springer Science+Business Media, LLC 2013

Introduction

Cacao beans are derived from the tropical tree *Theobroma cacao L* (Sterculiaceae family).

It originated in Central and South America, and it is cultivated extensively for its seeds, which are the source of cocoa, chocolate, and cocoa butter.

The cacao tree is a wide-branched evergreen and currently grows within 20°C latitude of the equator. Being a tropical species, it lives in regions characterized by warm temperature (18/21–30/32°C), high humidity (70–100%), and at elevations up to 600 m. Trees are planted under the shade of taller trees because of their sensitivity to sun and wind. The standard height of the cocoa tree is kept at 2–5 m on plantations, and it bears seedpods, named cabosse, up to 30 cm long and 10 cm wide. The fruit is harvested twice a year, before and after the rainy seasons. Cocoa pods have different shapes and sizes. Pods contain 20–50 almond-sized seeds that are about 2 cm long and 1 cm wide and are surrounded by mucilaginous pulp. The cocoa bean weighs around 1 g after drying and is composed of an inner nib portion covered by an outer shell (Fig. 8.1) [1].

Three important varieties of cocoa exist: Criollo, Forastero, and Trinitario. Criollo trees provide highly aromatic cocoa beans but are more sensitive to climatic changes. For this reason, the Criollo can be attacked by diseases and pests and are lower in yield, compared with Forastero trees. Approximately 95% of the world's cocoa product is made of the Forastero, owing to its disease resistance and high productivity. However, its organoleptic qualities are not as highly regarded as the Criollo variety. Trinitario is a crossbreed among Criollo and Forastero, and it displays characteristics of both varieties.

At harvest, the fully ripe pods are opened and the seeds with surrounding pulp are removed. The seeds are usually subjected to fermentation for 2–8 days, leading to a moisture content of about 6–8% after drying. After this step, they are called cocoa beans, which can be exported as raw materials for the production of cocoa derivatives. Further processing is usually performed by consuming countries (i.e., roasting at 120–130°C but not exceeding 150°C, for 10–35 min). Roasting parameters depend on variety, moisture content, ripening, and measure of cacao beans. In this phase, the humidity is lowered down to 2% or 3%. It results in further oxidation of phenolic compounds and the removal of acetic acid, volatile esters, and other undesirable aroma components. During drying and roasting, peptides and free amino acids, together with reducing sugars also present in fermented cocoa beans, undergo a Maillard reaction, which is responsible for the typical cocoa aroma. This complex and multistep process, referred to as nonenzymatic browning reaction [2], is subdivided into three main phases. The first one, called the initial phase, is partly reversible and produces compounds that later react, giving rise to additional products; in the second phase, a variety of highly reactive carbonyl compounds, which are responsible for color formation and flavor development, are formed; and in the final phase, the production of a heterogenous group of substances (e.g., melanoidins) occurs that determines the deep-brown color.

Chemical Compositions of Cocoa Beans

The physics and chemistry of cocoa beans are very complex and change throughout the life of the bean, mainly depending on the processing it receives. The chemical composition of cocoa beans also differs according to varieties, as indicated in Table 8.1.

Fig. 8.1 Cocoa pods and beans

Table 8.1 Composition (%) of cacao beans of various geographical origin[a]

Constituent	Trinidad	Giava	Arriba	Caracas
Moisture	6.34	5.12	5.90	6.63
Lipids	43.66	45.50	43.31	36.81
Cellulose	13.01	13.85	14.07	16.35
Pigments	8.31	8.90	9.00	12.72
Albumin	11.90	9.25	10.14	11.09
Starch	4.98	5.17	6.37	3.81
Glucose	1.38	1.23	0.42	2.76
Sucrose	0.32	0.51	1.58	1.56
Theobromine	0.85	1.16	0.86	1.13
Ash	3.60	3.31	8.73	4.36

[a]Adapted with permission from reference Arlorio [31]

Protein and Amino Acids

Proteins represent about 60% of the total nitrogen content of fermented beans and constitute 10–15% of the dry weight of cocoa seeds, which are the second most abundant constituent after the cocoa fat.

The nonprotein nitrogen of fermented beans is found in the form of amino acids; about 0.3% presents in amide form and 0.02% as ammonia, which is formed during fermentation of the beans and methylxanthines such as theobromine and caffeine.

In the fresh cacao beans, a lot of enzymes can be detected, but most of them are inactivated as a result of the production process. They include (beta)β-glucosidase, (beta)β-fructosidase, (alpha)α-amylase, (beta)β-gactosidase, proteinase, alkaline and acid phosphatases, lipase, polyphenol oxidase, pectinesterase, catalase, and peroxidase.

Cocoa cotyledon proteins can be fractioned into albumin (water soluble), globulins (salt soluble), glutelins (soluble in dilute acids and alkali), and prolamins (alcohol soluble) [3].

Voigt et al. [4] suggested that the total seed protein content is composed of 52% and 43% of albumin and globulin fractions, respectively. The former is described as a 21-kDa storage protein with trypsin inhibitory properties [5], whereas the latter is referred as a vicilin-like globular storage protein, consisting of three subunits with molecular weights around 47, 31, and 15 kDa [4, 6], which are derived from a common precursor with 66 kDa.

Glutelins and prolamins are present in a lower concentration; the former representing 5% and the latter being about 1%. Fermentation process affects the alteration of protein concentrations (as reported in Table 8.2), leading to the increase of albumin and glutelin from 52% to 79% and the decrease of globulins level from 43% to 8.3%.

Cocoa albumin, amino acid content, and their nutritional values are all highly affected by the extent of roasting [7]. Globulins also undergo extensive degradation already during fermentation [8], leading to the production of hydrophobic amino acids and peptides, considered cocoa-specific flavor precursors [9]. The increase of hydrophobic free amino acids concentrations, such as leucine, alanine, phenylalanine, and tyrosine, is explained by the activity of two cocoa proteases: the aspartic endoprotease, which attacks the proteins preferentially at sites of hydrophobic amino acids, and the carboxypeptidase, which releases single hydrophobic amino acids [9, 10].

The content and distribution of free amino acids in fermented cocoa beans from different origins vary greatly (5–25 mg/g fat-free dry matter), and in some cases, geographical region-specific differences were apparent (Table 8.3) [11, 12].

Thus, fermentation, roasting, and drying, as well as type of soil, climate, and harvest conditions, greatly affect cocoa characteristics. In particular, fermentation involves microbiological and enzymatic

Table 8.2 Proportions of the different solubility classes of seed proteins in unfermented and fermented cocoa seeds[a]

Seed protein fraction (%)	Unfermented seeds	Fermented seeds
Albumin	52±3.3	79±10.1
Globulin	43±3.1	8.3±3.7
Prolamin	<1	<1
Glutelin	5±1.0	12.8±3.7

[a]Reprinted from reference Voigt et al. [4], Copyright 1993, with permission from Elsevier

Table 8.3 Amino acid profiles (mg/g crude protein) of unfermented and fermented cocoa seeds[a]

Amino acid	Unfermented nibs	Fermented nibs	Mean	SD	CV%
Lys[b]	42.0±0.02	52.6±0.02	47.3	7.50	15.9
His[b]	20.0±0.00	23.3±0.02	21.7	2.33	10.7
Arg[b]	43.6±0.01	51.4±0.20	47.5	5.52	11.6
Asp	100±0.10	82.5±0.11	91.3	12.4	13.6
Thr[b]	29.9±0.03	23.3±0.10	26.6	4.67	17.6
Ser	23.7±0.01	32.6±0.03	28.2	6.29	22.3
Glu	128±0.20	153±0.40	141	17.7	12.6
Pro	12.5±0.02	12.5±0.03	12.5	0.00	–
Gly	20.5±0.01	32.0±0.02	26.3	8.13	30.9
Ala	29.8±0.20	40.1±0.03	35.0	7.28	20.8
Cys	7.8±0.01	6.9±0.02	7.35	0.64	8.71
Val[b]	32.1±0.10	35.1±0.02	33.6	2.12	6.31
Met[b]	9.9±0.01	8.0±0.00	8.95	1.34	15.0
Ile[b]	21.4±0.02	29.3±0.20	25.4	5.59	22.0
Leu[b]	72.2±0.30	62.4±0.20	67.3	6.93	10.3
Tyr	18.6±0.02	27.0±0.01	22.8	5.94	26.1
Phe[b]	28.6±0.01	36.3±0.02	32.5	5.44	16.7
Try[b]	–[c]	–	–	–	–
Crude protein (g/100 g)	13.6±0.30	15.2±0.21			

[a]Adapted from reference Adeyeye et al. [13], Copyright 2010, with permission from Elsevier
[b]Essential amino acids
[c]Not determined

reactions that lead to extensive breakdown of the cocoa proteins with an improvement of free amino acids concentration that, together with oligopeptides and reducing sugars, are considered cocoa aroma precursors [13].

Lipid Content

The total fat content of the whole cocoa bean on a dry basis is approximate 50%. The composition of total lipid extract from the cocoa beans was 98% of neutral lipids and 2% of polar lipids [14]. Of the neutral lipids [15], it is composed of predominantly over 75% of triglyceride molecular species with oleic acid esterified at the *sn*-2 position [16]. Among the polar lipids [14], phospholipids and glycolipids contain approximately 30% and 70%, respectively. Both triglycerides and phospholipids are the major lipids in the cocoa bean and that will be discussed in this chapter.

Table 8.4 Percentage of triglyceride molecular species from raw cocoa beans[a]

Geographical origin	LOO	PLO	PLP	OOO	SLO	POO	PLS	POP	POP	SLS	SOP	SOS	Other
Ivory Coast	0.73	0.94	2.33	0.61	0.58	2.47	3.23	15.5	3.71	1.54	37.7	27.2	4.20
Nigeria	0.34	0.58	1.92	0.45	0.39	2.56	3.03	16.1	3.2	1.41	39.7	27.5	2.82
Brazil (Bahia)	0.94	1.66	2.35	1.23	1.16	5.93	3.31	14.9	7.85	1.47	34.4	22.7	2.10
Brazil (Rd)	0.34	0.58	1.84	0.61	0.51	2.72	2.99	16.7	3.81	1.33	38.9	25.9	3.77
Brazil (Pará)	0.57	0.69	1.73	0.85	0.63	4.81	3.2	16.1	6.58	1.57	36.4	23.7	3.17
Ecuador (Guayaquil)	0.42	0.96	1.89	0.58	0.68	4.34	3.35	15.6	6.1	1.27	36.3	25.0	3.51
Indonesia	0.22	0.42	1.81	0.37	0.34	1.9	3.07	15.7	3.82	1.59	39.3	28.4	3.06
Malaysia	0.14	0.36	1.44	0.22	0.28	1.8	2.63	14.3	3.25	1.27	40.0	31.0	3.31
Guinea	0.30	0.91	2.08	0.52	0.81	3.89	3.48	15.3	5.82	1.73	37.3	25.4	2.46

P palmitic acid, *S* stearic acid, *O* oleic acid, *L* linoleic acid
[a]Adapted from reference Hernandez et al. [15]

Molecular Species of Triglycerides

Triglycerides have a glycerol backbone to which three fatty acids are esterified. Mono- or diglycerides are closely related to this structure but contain only one or two fatty acids. Triglycerides, diglycerides, and monoglycerides consist of a variety of molecular species. A molecular species of triglycerides contains three saturated and/or unsaturated fatty acids that are esterified at the *sn-1*, the *sn-2*, and the *sn-3* positions of the glycerol backbone. Therefore, a molecular species of diglycerides is based on the two saturated/unsaturated fatty acids, which are esterified either at the *sn-1*/the *sn-3* positions (or the *sn-3*/the *sn-1*) or the *sn-1*/the *sn-2* (or the *sn-2*/the *sn-3*). Monoglycerides contain only one saturated or unsaturated fatty acid that is linked to either at the *sn-1* (or the *sn-3*) or the *sn-2* position of the glycerol backbone.

The percentage of molecular species of the triglycerides extracted from cocoa beans from different geographical origins was analyzed by a reversed-phase HPLC method. The environmental temperature, climate, rainfall, and sunshine during the growth and ripening of cocoa fruit, genetics of the cocoa tree, and post-harvesting and processing conditions can affect the lipid composition of cocoa [17–19]. Table 8.4 shows the molecular species percentage of triglycerides in cocoa beans from Ivory Coast, Nigeria, Brazil, Ecuador, Indonesia, Malaysia, and Guinea [15]. The major fatty acids in molecular species of the triglycerides are palmitic acid (16:0; P), stearic acid (18:0; S), oleic acid (18:1; O), and linoleic acid (18:2; L). There is no study, so far, regarding the presence of ether linkage triglyceride molecular species in cocoa beans (Table 8.4).

Structures and Percentages of Phospholipids

A phospholipid class, such as lecithin or phosphatidylcholine, also consists of a mixture of molecular species. Their structural diversity is due to (1) a variety of fatty acid chains esterified at the *sn-1* and the *sn-2* positions of the glycerol backbone; (2) locations of the double bond(s) (between 1 and 6) with unsaturated fatty acids, which are usually located at the *sn-2* position, with a number of carbon atoms (between 16 and 22); and (3) a polar head carried at the *sn-3* position, such as phosphocholine, phosphoethanolamine, phosphoinositol, phosphoserine, phosphatidic acid, phosphoglycerol, as well as related lysophospholipids. Figure 8.2 shows phospholipid classes extracted in cocoa beans from different places [14].

So far, the molecular species of cocoa bean phospholipids have not been reported. Furthermore, there is no study regarding the presence of ether linkage molecular species of phospholipids in cocoa beans.

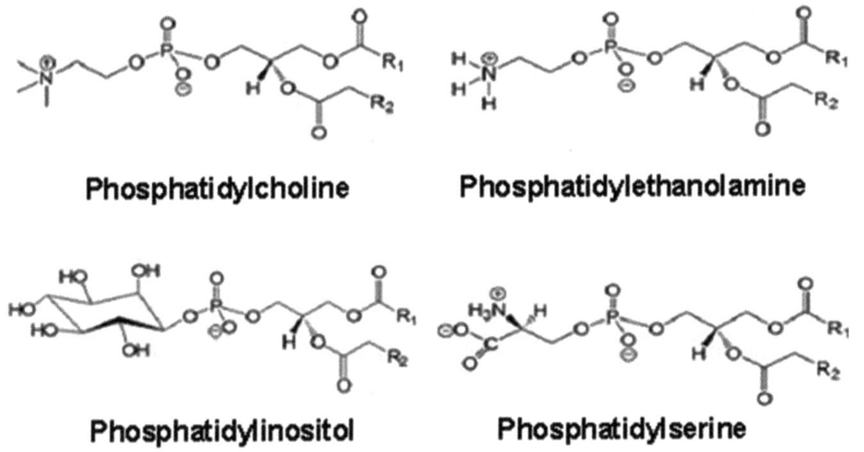

Phospholipid	Percentage (%)[a]		
	Accra	Arriba	Bahia
Phosphatidylcholine	36±0.4	40±1	36±1.3
Phosphatidylinositol	26±1	28±1.2	29±1.4
Phosphatidylethanolamine	14±0.3	11±0.7	19±1.3
Lysophosphatidylcholine	9±1.1	7±0.3	5±0.5
Phosphatidylserine	3±0.6	2.3±0.6	3.3±1
Phosphatidic acid	10±1	7.5±0.5	7±0.5
Unidentified phospholipids	2	4.2	0.7

[a]Quantitatively analyzed by TLC.

Fig. 8.2 Structures and percentage of phospholipids from raw cocoa beans. (Adapted from Parsons et al. [14], with permission of John Wiley and Sons)

Carbohydrates

In cocoa beans, mono-, oligo-, and polysaccharides are present. Starch is the major digestible polysaccharide, ranging from 3% to 7%. Cellulose is about 12% in fermented/dried cocoa beans, and it is one of the predominant components of the cell wall polysaccharides, together with pectic polysaccharides that are made up of a heterogeneous mixture of rhamnogalacturonans with variable degrees of branching. Lesser amounts of hemicellulose, which consists of a mixture of a fucosylated xyloglucan, galactomannans, and glucuronoarabinoxylan, are also isolated from cocoa shells [20]. The soluble carbohydrates found in fermented cacao beans are glucose, sucrose, raffinose, fructose, stachyose, and verbascose, and they range from 0.39% to 3.48% [21]. Major sugars are fructose and sucrose [22]. The variable concentration of the two sugars is probably caused by different conditions

Table 8.5 Changes to monosaccharides and oligosaccharides during roasting of cocoa beans[a]

	Glucose (mg/20 g)	Fructose (mg/20 g)	Sucrose (mg/20 g)	Raffinose (mg/20 g)	Stachyose (mg/20 g)	Verbascose (mg/20 g)
Ghana						
Non-roasted	12.4	83.6	31.6	4.0	15.1	1.9
Roasted	0.9	11.9	28.2	5.4	10.6	1.3
Ivory Coast						
Non-roasted	15.9	56.0	31.0	3.0	12.2	0.9
Roasted	1.0	8.7	40.5	5.5	14.4	0.9
Ecuador						
Non-roasted	16.8	34.4	96.6	12.2	20.1	0.8
Roasted	2.1	12.1	124.8	14.9	22.2	0.9

[a]Adapted from reference Redgwell et al. [21], Copyright 2003, with permission from Elsevier

of the fermentation that markedly affect the amount of sucrose in cocoa beans. Sucrose hydrolysis, which occurs during fermentation of the beans, provides the reducing sugar pool important for aroma formation during roasting process. While fermentation is essential to the formation of flavor precursors, the typical aroma of chocolate does not develop until the cocoa beans have been roasted. Indced, a nearly complete destruction of reducing sugars occurs during the roasting process (Table 8.5) since they are involved in the formation of volatile compounds from nonenzymatic browning reactions. On the contrary, there are no decreases in the concentration of nonreducing sugars, sucrose, raffinose, stachyose, and verbascose, as they are unable to undergo Maillard reaction.

Polyphenols

The stringent phenolic compounds contained in the pigment cells of cocoa cotyledons are defensive protections that the plant uses to repel animals and microbes. However, these compounds are responsible for the bitter and astringent taste of cocoa developed during the processing steps, such as fermentation. In the unfermented cocoa beans the amount of polyphenols is around 2 wt.% [23], whereas fermented cocoa beans contain 6% of phenolic compounds.

Among the polyphenols, the main compounds are catechins (e.g., (−)-epicatechin, (+)-catechin, (+)-gallocatechin, and (−)-epigallocatechin), anthocyanins (e.g., cyanidin-3-(alpha)α-L-arabinoside and cyanidin-3- (beta)β-D-galactoside), and proanthocyanidins (e.g., flavan-3,4-diols). In particular, the latter can form oligomers via condensation with the carbons C-4 and C-8 or C-4 and C-6 (Fig. 8.3). Moreover, some hydroxybenzoic and hydroxycinnamic acids are present in cocoa beans, and their concentration depends on the temperature reached during the roasting step. In the final product, chocolate, the amount of polyphenols depends on the percentage of nonfat cocoa solids [24]. Indeed, polyphenols are present in high concentrations in dark chocolates, in which the content of cocoa solid is higher than in milk chocolate.

The amount of polyphenolic compounds also depends on the provenience of the cocoa beans and can vary dramatically from region to region. However, during fermentation, the content of polyphenols is greatly modified. In fact, polyphenols diffuse in the cells, and they are degraded by polyphenol oxidase, with a consequent decrease of the amount of flavonoids. Moreover, the degradation of anthocyanins during the fermentation modifies the color of the beans, which is initially purple and then becomes brown. The amount of phenols also decreases with roasting steps at high temperature and long times.

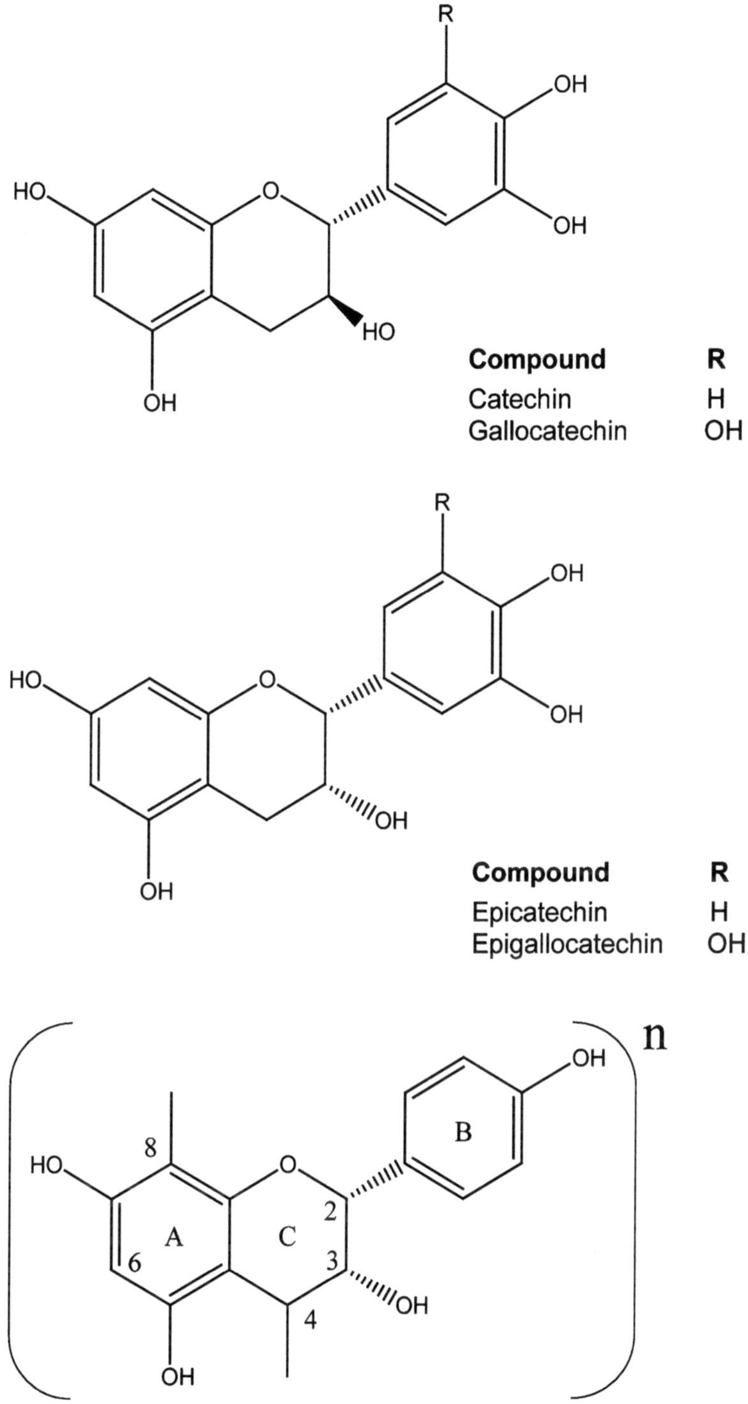

Fig. 8.3 Structures of main polyphenols present in cocoa

Organic Acids

The kind and amount of organic acids contained in cocoa beans depend on the maturation and fermentation stages. Also, geographical origin is another important factor to determine a total and/or single acid composition of the beans (Table 8.6).

The most common organic acids are citric, oxalic, malic, acetic, and formic. The most important is acetic acid because of its influence on the taste of cocoa (Fig. 8.4). Indeed, fermentation and drying duration affect the amount of acetic acid, which is produced by the fermentation of lactic acid and ethanol. During this step, acetic acid destroys the cells of the cotyledons and diffuses into the beans, making possible the reactions between phenolic compounds, proteins, and oxygen, which lead to less bitter complexes and develop the flavor of cocoa [25].

However, organic acids represent the antinutritional compounds contained in cocoa, and, according to their quantity, they may have adverse health effects.

The amount of phytic acid is reduced during the processing of cocoa, and its concentration depends on the type of cocoa and the strength of the roasting step. Phytic acid may have effects at the intestinal level, where it forms insoluble complexes with Ca^{2+}, preventing its absorption. On the other hand, oxalic acid (0.3–0.5% in cocoa powder) produces insoluble oxalates that bind to calcium and, as a consequence, inhibits its absorption.

Table 8.6 Organic acids levels in cocoa beans from different geographic regions[a,b]

Origin	Acetic acid (g/kg)	Citric acid (g/kg)	Lactic acid (g/kg)	Oxalic acid (g/kg)	Total acid (g/kg)
West Africa	3.0	6.1	1.2	4.5	14.8
Indonesia	5.9	4.6	3.6	3.5	17.6
Philippines	6.2	7.2	4.6	3.8	21.8
Malaysia	6.6	5.8	4.6	4.3	21.4
South Pacific	7.1	4.7	23.0	2.3	17.7
Papua New Guinea (plantation)	8.3	3.0	5.7	4.8	33.3
Papua New Guinea (smallholder)	7.1	3.6	4.0	4.9	18.1

[a]Adapted from reference Holm et al. [25], with permission from John Wiley and Sons
[b]For each region, values are the means from samples collected in different locations within the area

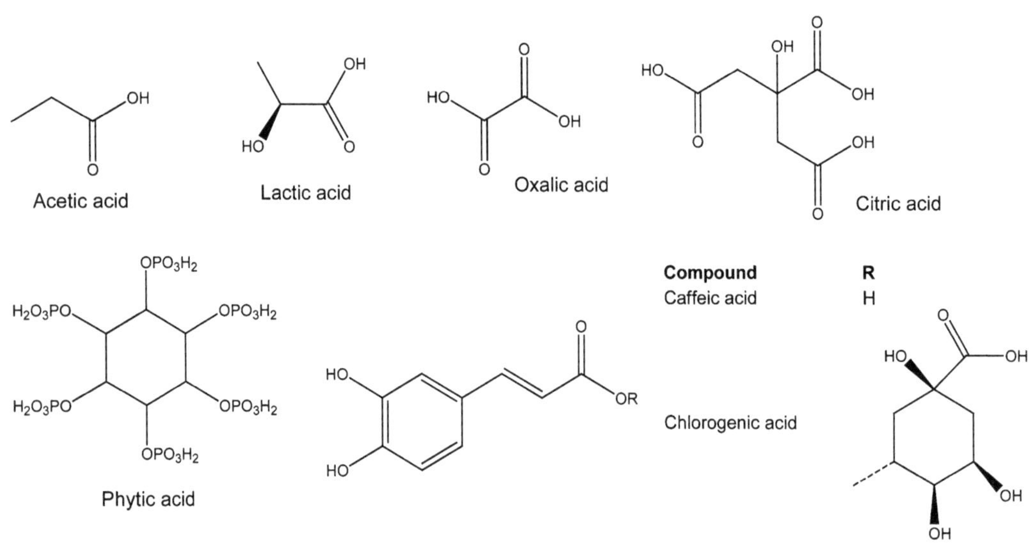

Fig. 8.4 Structures of organic acids present in cocoa

Table 8.7 Amount of caffeic, chlorogenic, and vanillic acids in different varieties of cocoa[a]

Variety	Caffeic acid (mg/kg)	Chlorogenic acid (mg/kg)	Vanillic acid (mg/kg)
Arriba	109.5	8.80	4.14
Avorio	13.92	17.47	0.98
Ghana	27.03	14.38	10.57

[a]Adapted with permission from reference Arlorio [31]

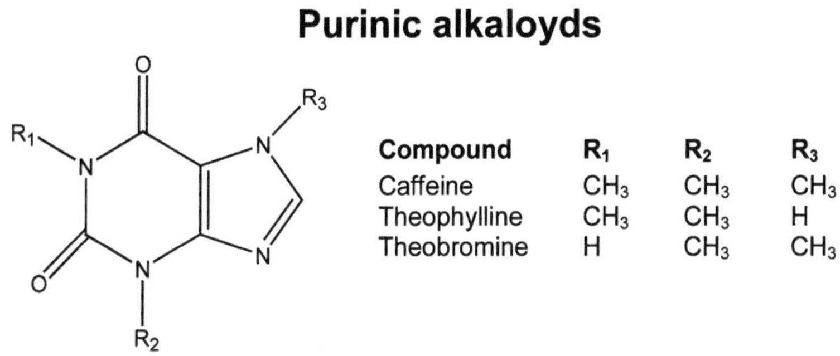

Fig. 8.5 Structures of purinic alkaloids

Chlorogenic acid (5-caffeoil-quinic acid) contained in cocoa beans varies from 8.8 to 17.5 mg/kg, according to the origin of cocoa. Its quantity decreases with the roasting step due to thermal degradation (Table 8.7). In water solution, chlorogenic acid liberates caffeic acid and quinic acid since it reacts with quinines and proteins, limiting, as a consequence, the bioavailability of amino acids and enzymes. In particular, caffeic acid inhibits the activity of thiamine (vitamin B1) and thus its absorption.

Purinic Alkaloids

Cocoa beans contain stimulant substances, purinic alkaloids, such as caffeine, theobromine, and theophylline (Fig. 8.5). While theobromine (whose name derives from *Theobroma cacao*) is present in a concentration between 2% and 3%, cocoa contains quantities of caffeine lower than coffee (<1%). However, theobromine has a weaker effect on the nervous system compared to caffeine. In general, the effects of these substances on the nervous system include increased concentration and attention and vasodilatation with consequent increase in diuresis [26].

Biogenic Amines

2-Phenylethylamine, tryptamine, and tyramine are the most common amines in cocoa (Fig. 8.6). Their concentration increases after roasting step, probably due to thermal decarboxylation of free amino acids. In general, they derive from microbial decarboxylation of amino acids (phenylalanine, tryptophan, and tyrosine, respectively) via decarboxylase-positive bacteria. Biogenic amines are vasoactive compounds that, if taken at appropriate doses, especially in susceptible persons, can provoke blushing, headaches, and blood pressure variation up to fatal effects, such as cardiovascular shock. These effects are due to the inhibition of enzymes in the human body, which catabolize biogenic amines such as the monoamine oxidase (MAO) and diaminoxidase (DAO). For this reason, foods containing these biogenic amines should be avoided for patients with the treatment of MAO and DAO inhibitors.

2-Phenylethylamine, a molecule with a structure similar to the one of amphetamines, is found in cocoa in discrete quantities. Due to this structural similarity, it can activate brain receptors of dopamine

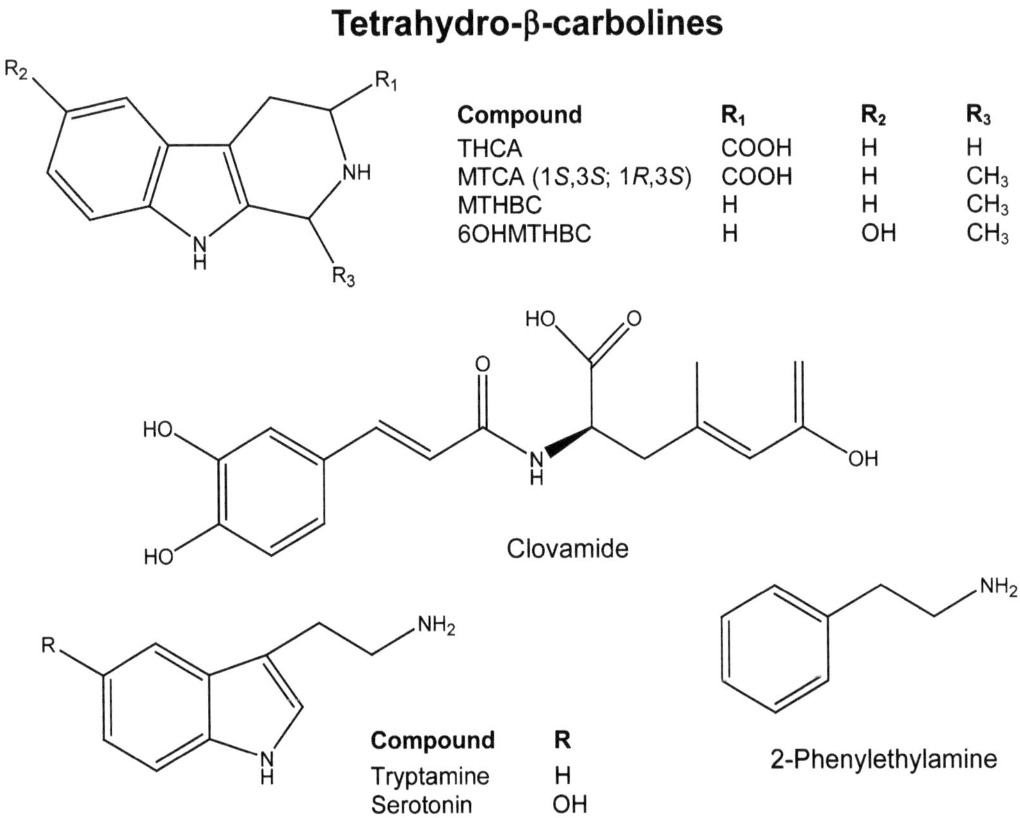

Fig. 8.6 Structures of biogenic amines present in cocoa

and noradrenalin (e.g., in delaying the appearance of fatigue). Moreover, 2-phenylethylamine is considered to be the bioactive compound responsible of chocolate craving [27, 28]. The quantity of the amine in cocoa is quite contradictory since concentrations between 1.8 and 22.0 mg/kg have been reported. But some authors were not able to detect this amine in cocoa. The sense of euphoria and satisfaction consequent to chocolate consumption derives from anandamide, an endogenous fat that can bind to the receptor of cannabinoids [29].

Another class of important compounds present in cocoa beans are tetrahydro-beta-carboline derivatives (0.18 μg/g) (see Fig. 8.6), indolic natural alkaloids produced by condensation of indolamine and aldehydes. Serotonin (1.25 μg/g) and tryptamine (0.69 μg/g) are also present in cocoa and are the precursors of tetrahydro-beta-carboline alkaloids [30]. These compounds appear to have neurologically active roles since they can bind to the receptor of benzodiazepines and modulate the uptake and release of serotonin, thus inhibiting MAOs.

Antioxidants, such as clovamide (see Fig. 8.6), have also been found in cocoa liquor. They are molecules analogous of rosmarinic acid, and they have an antioxidant effect similar to that of ascorbic acid.

References

1. Belitz HD, Grosch W, Schieberle P. Food chemistry. 4th ed. Berlin/Heidelberg: Springer; 2009.
2. Maillard LC. Action of amino acids on sugars. Formation of melanoidins in a methodical way. Compt Rendus Acad Sci. 1912;154:66–8.

3. Zak DL, Keeney PG. Extraction and fractionation of cocoa proteins as applied to several varieties of cocoa beans. J Agric Food Chem. 1976;24:479–83.
4. Voigt J, Biehl B, Kamaruddin S. The major seed proteins of *Theobroma cacao* L. Food Chem. 1993;47:145–51.
5. Kochhar S, Gartenmann K, Juillerat MA. Primary structure of the abundant seed albumin of *Theobroma cacao* by mass spectrometry. J Agric Food Chem. 2000;48:5593–9.
6. Spencer ME, Hodge R. Cloning and sequencing of a cDNA encoding the major storage proteins of *Theobroma cacao*. Planta. 1992;186:567–76.
7. Abecia-Soria L, Pezoa-Garcia NH, Amaya-Farfan J. Soluble albumin and biological value of protein in cocoa (*Theobroma cacao* L.) beans as a function of roasting time. J Food Sci. 2005;70:S294–8.
8. Amin I, Jinap S, Jamilah B. Vicilin-class globulins and their degradation during cocoa fermentation. Food Chem. 1997;59:1–5.
9. Voigt J, Biehl B, Heinrichs H, Kamaruddin S, Marsoner GG, Hugi A. In-vitro formation of cocoa-specific aroma precursors: aroma-related peptides generated from cocoa-seed protein by co-operation of an aspartic endoprotease and a carboxypeptidase. Food Chem. 1994;49:173–80.
10. Hashim P, Selamat J, Syed Muhammad SK, Ali A. Changes in free amino-acid, peptide-N, sugar and pyrazine concentration during cocoa fermentation. J Sci Food Agric. 1998;78:535–42.
11. Rohsius C, Matissek R, Lieberei R. Free amino acid amounts in raw cocoas from different origins. Eur Food Res Technol. 2006;222:432–8.
12. Bertazzo A, Comai S, Brunato I, Zancato M, Costa CVL. The content of protein and non-protein (free and protein-bound) tryptophan in *Theobroma cacao* beans. Food Chem. 2011;124:93–6.
13. Adeyeye EI, Akinyeye RO, Ogunlade I, Olaofe O, Boluwade JO. Effect of farm and industrial processing on the amino acid profile of cocoa beans. Food Chem. 2010;118:357–63.
14. Parsons JG, Keeney PG, Patton S. Identification and quantitative analysis of phospholipids in cocoa beans. J Food Sci. 1969;34:497–9.
15. Hernandez B, Castellote AI, Permanyer JJ. Triglyceride analysis of cocoa beans from different geographical origins. Food Chem. 1991;41:269–76.
16. Pittenauer E, Allmaier G. The renaissance of high-energy CID for structural elucidation of complex lipids: MALDI-TOF/RTOF-MS of alkali cationized triacylglycerols. J Am Soc Mass Spectrom. 2009;20:1037–47.
17. Lehrian DW, Keeney PG, Butler DR. Triglyceride characterizatics of cocoa butter from cocoa fruit matured in a microclimate of elevated temperature. J Am Oil Chem Soc. 1980;57:66–9.
18. Chin AHG, Zainuddin N. Characteristics of Malaysian cocoa butter. In: Proceedings of the 1984 international conference on cocoa and coconuts. Kuala Lumpur; 1994. p. 1.
19. Chalseri S, Dimick PS. Cocoa butter: its composition and properties. Manuf Confect. 1987;68:115–22.
20. Redgwell RJ, Trovato V, Merinat S, Curti D, Hediger S, Manez A. Dietary fibre in cocoa shell: characterisation of component polysaccharides. Food Chem. 2003;81:103–12.
21. Redgwell RJ, Trovato V, Curti D. Cocoa bean carbohydrates: roasting-induced changes and polymer interactions. Food Chem. 2003;80:511–6.
22. Reineccius GA, Andersen DA, Kavanagh ET, Keeney PG. Identification and quantification of the free sugars in cocoa beans. J Agric Food Chem. 1972;20:199–202.
23. Porter LJ, Ma Z, Chan BG. Cacao procyanidins: major flavonoids and identification of some minor metabolites. Phytochemistry. 1991;30:1657–63.
24. Miller KB, Stuart DA, Smith NL, Lee CY, McHale NL, Flanagan JA, et al. Antioxidant activity and polyphenol and procyanidin contents of selected commercially available cocoa-containing and chocolate products in the United States. J Agric Food Chem. 2006;54:4062–8.
25. Holm CS, Aston JW, Douglas K. The effects of the organic acids in cocoa on the flavour of chocolate. J Sci Food Agric. 1993;61:65–71.
26. Smit HJ. Theobromine and the pharmacology of cocoa. Handb Exp Pharmacol. 2011;200:201–34.
27. Bruinsma K, Taren DL. Chocolate: food or drug? J Am Diet Assoc. 1999;99:1249–56.
28. di Tomaso E, Beltramo M, Piomelli D. Brain cannabinoids in chocolate. Nature. 1996;382:677–8.
29. Granvogl M, Bugan S, Schieberle P. Formation of amines and aldehydes from parent amino acids during thermal processing of cocoa and model systems: new insights into pathways of the strecker reaction. J Agric Food Chem. 2006;54:1730–9.
30. Herraiz T. Tetrahydro-beta-carbolines, potential neuroactive alkaloids, in chocolate and cocoa. J Agric Food Chem. 2000;48:4900–4.
31. Arlorio M. Alimenti nervini: cacao, caffè, tè. In: Cabras P, Martelli A, editors. Chimica degli Alimenti. Padova: Piccin; 2004. p. 291–354.

Chapter 9
Industrial and Home Processing of Cocoa Polyphenols

Rosa M. Lamuela-Raventos, María Izquierdo-Pulido, and Ramón Estruch

Key Points

- Cocoa and chocolate increase flow-mediated dilatation.
- Fermentation and roasting decrease phenol bioavailability.
- Milk may drive the phase II generation of sulfated metabolites.
- Milk does not appear to exert the same effect in beverage matrices as in confections.
- Flavonoids that have not been absorbed in the small intestine and reach the colon intact could also suffer metabolization.

Keywords Bioavailability • Cocoa • Processing • Polyphenols • Flavonoids milk effect • Fermentation • Roasting

Introduction

Since the discovery, 15 years ago, by Waterhouse [1] that cocoa can inhibit LDL oxidation, many studies have reported the health benefits of cocoa polyphenols for humans [2–5], and nowadays, the consumption of dark cocoa is recommended by dietitians and nutritionists as an polyphenol-rich food. In chronic studies, chocolate and cocoa have been found to increase flow-mediated dilatation (FMD) by 1.45% (95% CI: 0.62%, 2.28%; 2 studies) and to reduce both systolic (by 5.88 mmHg; 95% CI: _9.55,_2.21; 5 studies; P for heterogeneity _ 0.0003, I2 _ 81%) and diastolic (by 3.30 mmHg; 95% CI: _5.77, _0.83; 4 studies; P for heterogeneity _ 0.009, I2 _ 70%) blood pressure. In acute studies, only chocolate or cocoa significantly improved FMD (3.99%; 95% CI: 2.86, 5.12; 6 studies, P for heterogeneity _ 0.1, I2 _ 46%; 70–177 mg epicatechin/d, at 90–149 min), compared with wine or tea [6].

Cocoa is very rich in polyphenols, mainly flavan-3-ols. Phenolic compounds comprise 12–18% of the total weight of dried cocoa nibs [7]. Approximately 35% of the total content of polyphenols in

R.M. Lamuela-Raventos, M.S., Ph.D. (✉) • M. Izquierdo-Pulido, M.S., Ph.D.
Department of Nutrition and Food Science, School of Pharmacy, University of Barcelona, Av. Joan XXIII s/n, Barcelona 08028, Spain
e-mail: lamuela@ub.edu; maria_izquierdo@ub.edu

R. Estruch, M.D., Ph.D.
Hospital Clinic, Calle Villarroel 170, Barcelona 08036, Spain
e-mail: restruch@clinic.ub.es

R.R. Watson et al. (eds.), *Chocolate in Health and Nutrition*, Nutrition and Health 7,
DOI 10.1007/978-1-61779-803-0_9, © Springer Science+Business Media, LLC 2013

Table 9.1 Levels of flavonoids found in cocoa nibs

Compound	Cocoa nibs (mg/100 g)
Catechin	72.4
Epicatechin	328.1
Procyanidin B1	16.6
Procyanidin B2	92.1
Quercetin	2.5
Isoquercitrin	11.0
Hyperoside	9.0
Quercetin 3-O-arabinoside	16.5
Apigenin	0.5
Vitexin	0.4
Isovitexin	0.4
Luteolin	0.5
Luteolin-7-O-glucoside	1.2

non-fermented cocoa nibs belonging to the Forastero variety is epicatechin (Table 9.1). The epicatechin content in non-fermented cocoa nibs of different varieties ranges between 34.65 and 43.27 mg/g (defatted samples). Compared with other flavonoid-rich food sources, such as green and black tea, red wine, and so forth, flavonoids from chocolate or cocoa seem to be more active in reducing cardiovascular risk factors [6].

Effect of Fermentation and Roasting

Fermentation results in a substantial loss of flavonoids (80–94%), depending on its duration [8, 9]. To evaluate the effect of fermentation and roasting on the bioavailability of phenolic compounds, Tomas-Barberán et al. [10] performed a randomized double-blind crossover study with six volunteers, in which they compared unfermented and unroasted cocoa, treated thermally by blanching with water at an internal temperature of 95°C for 5 min (to inactivate oxidative enzymes such as polyphenol oxidase), with regular cocoa. Unfermented cocoa contained eight times more epicatechin and procyanidin B2 than conventional powder. Upon consumption of unfermented cocoa, the content of epicatechin glucuronide in plasma was five times higher, and levels of methyl epicatechin sulfate in urine were also two- to twelve-fold higher, depending on the metabolite, than after ingestion of regular cocoa. However, microbiota metabolites were not evaluated in this study. Strategies that increase polyphenolic or counteract processing effects on polyphenol bioavailability are needed in developing-country settings.

Biological Properties of Cocoa Flavonoids Are Conditioned by Their Bioavailability

Different studies have proven the absorption of catechin, epicatechin, and dimeric procyanidins after the intake of different cocoa by-products by animals and humans by reporting an increase in their plasmatic concentrations [11–15]. The absorption of these flavonoids across the human intestinal epithelial membrane varies between aglycones and polymeric forms. However, in contrast to what was initially thought, neither dimeric nor polymeric forms (procyanidins) are hydrolyzed at a gastric level, thus arriving intact to the small intestine [16]. The aglycones, using the Caco-2 cellular model

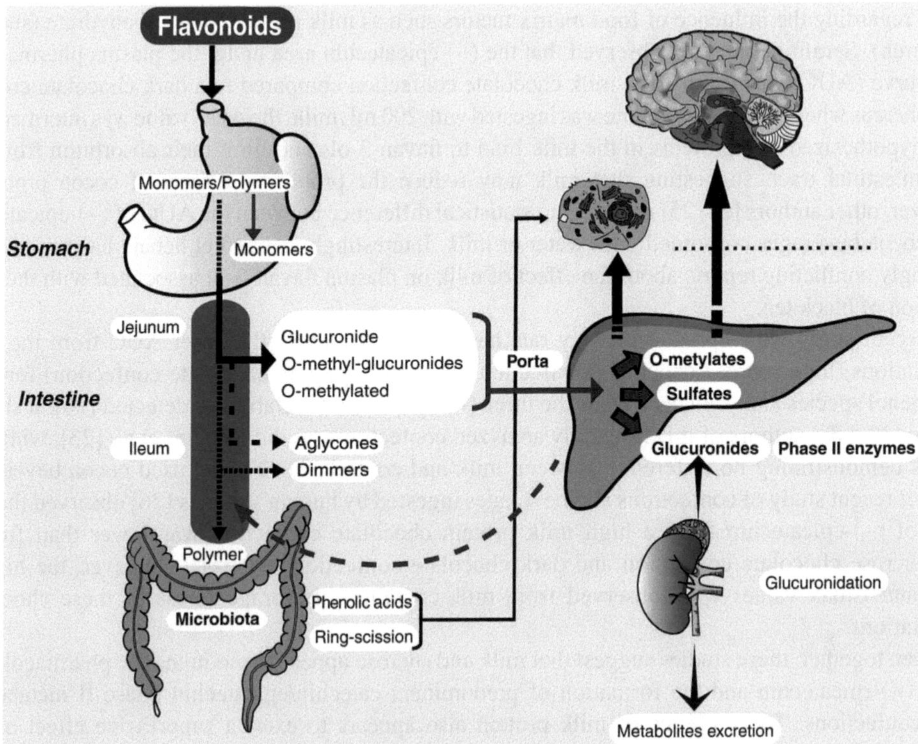

Fig. 9.1 Bioavailability of cocoa flavonoids in the human body

of human intestinal absorption, pass through the intestinal cells by means of a multispecific organic anion transporter [17]. Procyanidin absorption depends on its molecular weight. Dimers and probably trimers are absorbed in the small intestine, although less efficiently (under 0.5%) than epicatechin and catechin monomers, whose absorption level is in the 22–55% range [11–13, 18]. Epicatechin is much more efficiently absorbed than catechin [4], possibily due to sterochemical differences that result in varying hydrophobicity [15]. Data on oral absorption percentages of these compounds are very limited. Added to this great interindividual variability, polyphenol-containing diet constituents and their interaction in the food matrix can also have an important effect on their oral bioavailability. In contrast, polymeric procyanidins, given their high molecular weight, are scarcely absorbed. However, results for cocoa procyanidin activity in in vitro assays, especially the data obtained with long-chain oligomers, should be treated carefully, because their potential activity in vivo is conditioned by poor bioavailability; however, these procyanidins are metabolized by microbiota and absorbed in the form of small molecules that may be biologically active.

Aglycones are extensively metabolized during their movement through the epithelial cells of the small intestine, resulting in the appearance of glucuronide conjugates (mainly), sulfate conjugates, sulfoglucuronide conjugates, and methylated conjugates of corresponding aglycones [13, 19–21]. These metabolites could be responsible, at least in part, for the effects observed on the central nervous system after cocoa and chocolate consumption. Specifically, this has been seen in laboratory animals where epicatechin metabolites (glucuronide and 3-O-methylglucuronide) can cross the blood–brain barrier and act at a cerebral level (Fig. 9.1) [22].

Cocoa and its by-products are usually consumed with milk. Studies on the overall bioavailability of (+) and (−) -catechin and (−)-epicatechin from chocolate in human subjects report conflicting results regarding the influence of food matrix factors such as milk protein and carbohydrate (sucrose and starch). Serafini et al. [23] observed that the (−)-epicatechin area under the plasma pharmacokinetic curve (AUC) was lower for a milk chocolate confection compared to a dark chocolate confection, whereas when the dark chocolate was ingested with 200 mL milk, the AUC value was intermediate. They hypothesized that proteins in the milk bind to flavan-3-ols and limit their absorption from the gastrointestinal tract, suggesting that milk may reduce the protective effects of cocoa products. However, other authors [24, 25] reported no statistical difference between the AUC of (−)-epicatechin from cocoa beverages consumed with water or milk. Interestingly, a parallel debate has arisen amid seemingly conflicting reports about the effect of milk on plasma flavan-3-ol associated with the consumption of black tea.

A recent study with Sprague-Dawley rats has found a significantly lower AUC from the milk formulations (high sucrose chocolate confection > high milk protein chocolate confection) for total polyphenol species analyzed, as well as the three (−)-epicatechin metabolites detected [26]. It should be noted that the authors of this last study analyzed confections, as did Serafini et al. [23], while the studies demonstrating no difference between milk and control [21, 27, 28] used cocoa beverages. Another recent study of confections and beverages ingested by human subjects [26] observed that the AUC of (−)-epicatechin from a high milk protein chocolate confection was lower than from a high sucrose chocolate confection and dark chocolate confection (control). However, the highest AUC and Cmax values were observed from milk-containing beverage forms of these chocolate formulations.

Taken together, these studies suggest that milk and sucrose appear to modulate the pharmacokinetics of (−)-epicatechin and the formation of predominant catechin/epicatechin phase II metabolites from confections. The presence of milk protein also appears to exert a suppressive effect on the bioavailability of these compounds from confections. In fact, milk may drive the phase II generation of sulfated metabolites [20], which are rapidly excreted to urine, thus reducing the formation of glucuronides and O-Me glucuronides. In another study, milk markedly reduced the concentration of cocoa flavan-3-ol metabolites appearing in urine but did not affect either gastric emptying or the time for the passage of the meal through the gastrointestinal tract. Thus, the effect of the milk is more likely to be a consequence of the binding of its components directly to flavan-3-ols or to an interference with the mechanism involved in their transport through the wall of the small intestine into the portal vein. However, milk does not appear to exert these effects to the same extent in beverage matrices as in confections.

The flavonoids that are not absorbed through the gut barrier (mainly long-chain procyanidins) or that are excreted in the bile reach the colon where they can interact with the existing microbiota. Gut microflora catalyze the breakdown into smaller molecules, producing phenolic acids, which have been described as microbial degradation products of flavanols [29–33].

This metabolization has already been seen in vitro [34]. Regarding the organisms responsible for this conversion, Schneider et al. [35] identified two types of bacteria in human feces capable of metabolizing these compounds: *Enterococcus casseliflavus,* which would only react with the sugar molecule, and *Eubacterium ramulus*, which is capable of degrading the aromatic ring, thus transforming flavonoids into lower molecular weight phenols such as hydroxyphenylacetic acids. This implies that monomeric flavonoids that have not been absorbed in the small intestine and reach the colon intact could also suffer metabolization. All these phenolic acids of low molecular weight, and with antioxidant activity could also have biological activity, have a local effect in the colon or be absorbed, thus contributing to the biological effects observed after the consumption of cocoa and its derivatives. The main microbial-derived metabolites of flavanols, such as 5-(dihydroxyphenyl)-(gamma)γ-valerolactone and 5-(hydroxymethoxyphenyl)-(gamma)γ-valerolactone, were also detected in their glucuronide and sulfate forms after cocoa intake [30].

References

1. Waterhouse AL, Shirley JR, Donovan JL. Antioxidants in chocolate. Lancet. 1996;348(9030):834.
2. Khan N, Monagas M, Andres-Lacueva C, Casas R, Urpí-Sardà M, Lamuela-Raventós RM, et al. Regular consumption of cocoa powder with milk increases HDL cholesterol and reduces oxidized LDL levels in subjects at high-risk of cardiovascular disease. Nutr Metab Cardiovasc Dis. 2011;1–8. [Epub ahead of print].
3. Visioli F, Bernaert H, Corti R, Ferri C, Heptinstall S, Molinari E, et al. Chocolate, lifestyle, and health. Crit Rev Food Sci Nutr. 2009;49(4):299–312.
4. Vázquez-Agell M, Urpi-Sarda M, Sacanella E, Camino-López S, Chiva-Blanch G, Llorente-Cortés V, et al. Cocoa consumption reduces NF-κB activation in peripheral blood mononuclear cells in humans. Nutr Metab Cardiovasc Dis. 2011;1–7. [Epub ahead of print].
5. McShea A, Ramiro-Puig E, Munro SB, Casadesus G, Castell M, Smith MA. Clinical benefit and preservation of flavonols in dark chocolate manufacturing. Nutr Rev. 2008;66(11):630–41.
6. Hooper L, Kroon PA, Rimm EB, Cohn JS, Harvey I, Le Cornu KA, et al. Flavonoids, flavonoid-rich foods, and cardiovascular risk: a meta-analysis of randomized controlled trials. Am J Clin Nutr. 2008;88(1):38–50.
7. LamuelaRaventos RM, RomeroPerez AI, Andres-Lacueva C, Tornero A. Review: health effects of cocoa flavonoids. Food Sci Technol Int. 2005;11(3):159–76.
8. Andres-Lacueva C, Monagas M, Khan N, Izquierdo-Pulido M, Urpi-Sarda M, Permanyer J, et al. Flavanol and flavonol contents of cocoa powder products: influence of the manufacturing process. J Agric Food Chem. 2008;56(9):3111–7.
9. Payne MJ, Hurst WJ, Miller KB, Rank C, Stuart DA. Impact of fermentation, drying, roasting, and Dutch processing on epicatechin and catechin content of cacao beans and cocoa ingredients. J Agric Food Chem. 2010;58(19):10518–27.
10. Tomas-Barberan FA, Cienfuegos-Jovellanos E, Marin A, Muguerza B, Gil-Izquierdo A, Cerda B, et al. A new process to develop a cocoa powder with higher flavonoid monomer content and enhanced bioavailability in healthy humans. J Agric Food Chem. 2007;55(10):3926–35.
11. Baba S, Osakabe N, Natsume M, Muto Y, Takizawa T, Terao J. Absorption and urinary excretion of (−)-epicatechin after administration of different levels of cocoa powder or (−)-epicatechin in rats. J Agric Food Chem. 2001;49(12):6050–6.
12. Baba S, Osakabe N, Natsume M, Yasuda A, Takizawa T, Nakamura T, et al. Cocoa powder enhances the level of antioxidative activity in rat plasma. Br J Nutr. 2000;84(5):673–80.
13. Baba S, Osakabe N, Yasuda A, Natsume M, Takizawa T, Nakamura T, et al. Bioavailability of (−)-epicatechin upon intake of chocolate and cocoa in human volunteers. Free Radic Res. 2000;33(5):635–41.
14. Holt RR, Lazarus SA, Sullards MC, Zhu QY, Schramm DD, Hammerstone JF, et al. Procyanidin dimer B2 [epicatechin-(4beta-8)-epicatechin] in human plasma after the consumption of a flavanol-rich cocoa. Am J Clin Nutr. 2002;76(4):798–804.
15. Keen CL, Holt RR, Oteiza PI, Fraga CG, Schmitz HH. Cocoa antioxidants and cardiovascular health. Am J Clin Nutr. 2005;81(1 Suppl):298S–303.
16. Rios LY, Bennett RN, Lazarus SA, Remesy C, Scalbert A, Williamson G. Cocoa procyanidins are stable during gastric transit in humans. Am J Clin Nutr. 2002;76(5):1106–10.
17. Vaidyanathan JB, Walle T. Transport and metabolism of the tea flavonoid (−)-epicatechin by the human intestinal cell line caco-2. Pharm Res. 2001;18(10):1420–5.
18. Donovan JL, Crespy V, Oliveira M, Cooper KA, Gibson BB, Williamson G. (+)-Catechin is more bioavailable than (−)-catechin: Relevance to the bioavailability of catechin from cocoa. Free Radic Res. 2006;40(10):1029–34.
19. Roura E, Almajano MP, Bilbao ML, Andres-Lacueva C, Estruch R, Lamuela-Raventos RM. Human urine: epicatechin metabolites and antioxidant activity after cocoa beverage intake. Free Radic Res. 2007;41(8):943–9.
20. Roura E, Andres-Lacueva C, Estruch R, Lourdes Mata Bilbao M, Izquierdo-Pulido M, Lamuela-Raventos RM. The effects of milk as a food matrix for polyphenols on the excretion profile of cocoa (−)-epicatechin metabolites in healthy human subjects. Br J Nutr. 2008;100(4):846–51.
21. Roura E, Andres-Lacueva C, Estruch R, Mata-Bilbao ML, Izquierdo-Pulido M, Waterhouse AL, et al. Milk does not affect the bioavailability of cocoa powder flavonoid in healthy human. Ann Nutr Metab. 2007;51(6):493–8.
22. Abd El Mohsen MM, Kuhnle G, Rechner AR, Schroeter H, Rose S, Jenner P, et al. Uptake and metabolism of epicatechin and its access to the brain after oral ingestion. Free Radic Biol Med. 2002;33(12):1693–702.
23. Serafini M, Bugianesi R, Maiani G, Valtuena S, De Santis S, Crozier A. Plasma antioxidants from chocolate. Nature. 2003;424(6952):1013.
24. van het Hof KH, Kivits GA, Weststrate JA, Tijburg LB. Bioavailability of catechins from tea: the effect of milk. Eur J Clin Nutr. 1998;52(5):356–9.
25. Serafini M, Ghiselli A, Ferro-Luzzi A. In vivo antioxidant effect of green and black tea in man. Eur J Clin Nutr. 1996;50(1):28–32.

26. Neilson AP, Sapper TN, Janle EM, Rudolph R, Matusheski NV, Ferruzzi MG. Chocolate matrix factors modulate the pharmacokinetic behavior of cocoa flavan-3-ol phase II metabolites following oral consumption by Sprague-Dawley rats. J Agric Food Chem. 2010;58(11):6685–91.
27. Schramm DD, Karim M, Schrader HR, Holt RR, Kirkpatrick NJ, Polagruto JA, et al. Food effects on the absorption and pharmacokinetics of cocoa flavanols. Life Sci. 2003;73(7):857–69.
28. Schroeter H, Holt RR, Orozco TJ, Schmitz HH, Keen CL. Nutrition: milk and absorption of dietary flavanols. Nature. 2003;426(6968):787–8. discussion 788.
29. Urpi-Sarda M, Llorach R, Khan N, Monagas M, Rotches-Ribalta M, Lamuela-Raventos R, et al. Effect of milk on the urinary excretion of microbial phenolic acids after cocoa powder consumption in humans. J Agric Food Chem. 2010;58(8):4706–11.
30. Urpi-Sarda M, Monagas M, Khan N, Lamuela-Raventos RM, Santos-Buelga C, Sacanella E, et al. Epicatechin, procyanidins, and phenolic microbial metabolites after cocoa intake in humans and rats. Anal Bioanal Chem. 2009;394(6):1545–56.
31. Gonthier MP, Verny MA, Besson C, Remesy C, Scalbert A. Chlorogenic acid bioavailability largely depends on its metabolism by the gut microflora in rats. J Nutr. 2003;133(6):1853–9.
32. Rios LY, Gonthier MP, Remesy C, Mila I, Lapierre C, Lazarus SA, et al. Chocolate intake increases urinary excretion of polyphenol-derived phenolic acids in healthy human subjects. Am J Clin Nutr. 2003;77(4):912–8.
33. Selma MV, Espin JC, Tomas-Barberan FA. Interaction between phenolics and gut microbiota: role in human health. J Agric Food Chem. 2009;57(15):6485–501.
34. Deprez S, Brezillon C, Rabot S, Philippe C, Mila I, Lapierre C, et al. Polymeric proanthocyanidins are catabolized by human colonic microflora into low-molecular-weight phenolic acids. J Nutr. 2000;130(11):2733–8.
35. Schneider H, Schwiertz A, Collins MD, Blaut M. Anaerobic transformation of quercetin-3-glucoside by bacteria from the human intestinal tract. Arch Microbiol. 1999;171(2):81–91.

Chapter 10
Methods in Chocolate Analysis: Use of Atomic Spectrometric Techniques

Denise Bohrer and Carine Viana Silva Ieggli

Key Points

- Chocolate analysis for metal determination is an analytical challenge owing to matrix characteristics.
- Sample pretreatment is an important step for successful chocolate analysis.
- Atomic spectrometric techniques enable one to determine major and minor metallic elements in chocolate samples.

Keywords Metal determination • Atomic spectrometry • Sample pretreatment • Chocolate • Cocoa beverages

Chocolate is consumed all over the world in all segments of society and by people of all ages. Nowadays, the consumer is more and more concerned with the nutritional status of foodstuff, and considering that chocolate may be an extremely rich source of many essential minerals, it can contribute to a healthy diet. Nevertheless, the evaluation of nutrient ingestion is a very complex task [1].

The determination of metals in food has become an important field in food analysis. However, the accurate determination of metals in chocolate is still an analytical challenge owing to difficulties arising from matrix characteristics. Because chocolate is a complex matrix and metallic constituents are present in ppm (mg/kg) or ppb (µg/kg) range, a successful determination of elements is a commitment between both the technique for sample decomposition and for metal measurement.

Official methods of analysis are designed for groups of samples classified according to their nature or for a specific analyte [2–6]. Among them, those for solid samples and metal determination using atomic spectrometric techniques that could be used for chocolate analysis are listed in Table 10.1. All

D. Bohrer, Ph.D. (✉)
Department of Chemistry, Federal University of Santa Maria,
Av. Roraima, 1000, Campus Camobi, Santa Maria 97105 900, Brazil
e-mail: ndenise@quimica.ufsm.br

C.V.S. Ieggli, Ph.D.
Department of Chemistry, Federal University of Santa Maria,
Avenida Roraima, 1000, Santa Maria, RS 971110-970, Brazil
e-mail: carineviana@yahoo.com.br

R.R. Watson et al. (eds.), *Chocolate in Health and Nutrition*, Nutrition and Health 7,
DOI 10.1007/978-1-61779-803-0_10, © Springer Science+Business Media, LLC 2013

Table 10.1 Overview of published methods for extraction and digestion of solid samples for inorganic analysis that could be applied for chocolate samples

Method	Scope	Uses	Reagents and conditions	Elements	Suitable for
EPA 3031	Acid digestion of oils for metal analysis by atomic absorption or ICP spectrometry	Oils, oil sludges, waxes, viscous petroleum products	0.5 g sample mixed with 0.5 g potassium permanganate and 10 mL conc. H_2SO_4 after reaction finished 2 mL conc. HNO_3 and 10 mL conc. HCl	Sb, As, Ba, Be, Cd, Cr, Co, Cu, Pb, Mo, Ni, Se, Ag, Tl, V, Zn	FAAS ICP-AES
EPA 3040	Dissolution procedure	Dissolution of oils, greases, and waxes	2 g sample + solvent: xylene, kerosene, or methyl isobutyl ketone	Sb, As, Ba, Be, Cd, Cr, Cu, Fe, Pb, Mn, Ni, V	FAAS ICP-AES
EPA 3050	Acid digestion	Sediments, soils, sludges	1-2 g (wet) or 2 g (dry) sample + conc. HNO_3 (repeated additions) + H_2O_2	As, Be, Cd, Cr, Co, Fe, Pb, Mo, Se, Tl	GFAAS
			1-2 g (wet) or 2 g (dry) sample + conc. HNO_3 (repeated additions) + H_2O_2 + conc. HCl	Al, Sb, Ba, Be, Cd, Ca, Cr, Co, Cu, Fe, Pb, Mg, Mn, Mo, Ni, K, Ag, Na, Tl, V, Zn	FAAS ICP-AES
EPA 3051	Microwave-assisted acid digestion	Sediments, soils, sludges, oils	0.25–0.50 g sample + 10 mL conc. HNO_3	Al, Sb, As, Ba, Be, B, Cd, Ca, Cr, Co, Cu, Fe, Pb, Mg, Mn, Hg, Mo, Ni, K, Se, Ag, Na, Sr, Tl, V, Zn	FAAS GFAAS ICP-AES
			0.25–0.50 g sample + 9 mL conc. HNO_3 + 3 mL conc HCl	Al, Sb, Ba, Be, Cd, Ca, Cr, Co, Cu, Fe, Pb, Mg, Mn, Mo, Ni, K, Ag, Na, Tl, V, Zn	FAAS ICP-AES

Observation: AOAC International and CEN present methods under "chocolate" but none of them contemplates metal determination

Table 10.2 Performance parameters of flame atomic absorption spectrometry, graphite furnace atomic absorption spectrometry, and inductively coupled plasma atomic emission spectrometry[a]

Parameter	FAAS	GFAAS	ICP-AES
Linear range	One order of magnitude	One order of magnitude	More than one order of magnitude
Concentration range	mg/L	μg/L	mg/L
Typical detection limit	0.1 mg/L	0.1–10 μg/L	0.1 mg/L
Precision (relative standard deviation)	1–2%	1–5%	<1%
Interferences	Not a problem	Need to be corrected	Need to be corrected
Sample volume	1–10 mL	10–100 μL	1–10 mL
Multielement analysis	Sequential	Sequential	Simultaneous
Time required for one run (one measurement)	<1 min	2–5 min	<1 min
Operating skills	Low	High	High
Maintenance	Low cost	Expensive	Expensive

[a]Data collected from reference [7]

methods require sample decomposition, and the reagents are selected depending on the metal and spectrometric technique used.

Some features of the atomic techniques are displayed in Table 10.2. Flame atomic absorption spectrometry (FAAS) is a simple and powerful detection technique for determining metallic elements. The sample is aspirated into a flame (temperature about 2,300–2,700°C) where the element to be determined is atomized. As the radiation of the wavelength that the atoms are able to absorb is supplied through the flame, the absorbed radiation is proportional to the number of atoms of the element present in the flame. Although FAAS is a powerful technique, only about 5% of the aspirated sample solution reaches the flame, which reduces the sensitivity allowing measurements in the mg/L range. The advantages of FAAS include well-characterized interferences, low operator skill required for operation, and comparatively low cost of instrumentation and maintenance. In the graphite furnace atomic absorption spectrometry (GFAAS) technique, the sample is introduced in discrete amounts into a small graphite tube, which is heated to a determined temperature to ensure the atomization of the element to be determined. As in FAAS, radiation of adequate wavelength crosses the tube and is absorbed by the atoms of the element of interest atomized inside the tube. Because the atoms have a longer residence time in the optical path (radiation path) than in the flame, a higher peak concentration of atoms is obtained and, therefore, enhanced sensitivity. Detection capability of GFAAS is in the range of μg/L [8].

Techniques using atomic emission spectrometry (AES), also called optical emission spectrometry (OES), use an inductively coupled plasma (ICP) to produce excited atoms and ions that emit electromagnetic radiation at wavelengths characteristic of a particular element. The wavelength of the atomic spectral line gives the identity of the element, while the intensity of the emitted light is proportional to the number of atoms of the element. Advantages of ICP-AES are excellent limit of detection, wide linear dynamic range, and multielement capability. On the other hand, owing to the multielement detection, it is possible that spectral interferences (due to the existence of several emission lines) can occur [9].

The literature reports methods for chocolate sample treatment involving mostly wet digestion, as described in EPA 3050 method (see Table 10.1). The procedures listed in Table 10.3 include conventional and microwave heating and high-pressure systems for wet digestion. In all cases, nitric acid is used, combined either with acids such as perchloric acid or hydrogen peroxide. Wet digestion allows for the subsequent metal measurement by flame, graphite furnace, and atomic emission spectrometry.

Table 10.3 Selected methods for metal determination in chocolate- and cocoa-derived products

Treatment/sample	Sample size	Reagents	Procedure	Technique	Metals	Reference
Wet digestion						
Chocolate	5 g	$HNO_3/HClO_4$ 3:1	16 h heating	FAAS	Pb, Cd, Ni	[10]
Chocolate beverage	1.5 g	15 mL HNO_3 2 mL 30% H_2O_2	15 min microwave	ICP-OES	Na, K, Ca, Mg, S, P, Fe, Cu, Zn, Se, Mo, Cr, Mn	[11]
Chocolate	0.25 g	5 mL HNO_3 2 mL 35% (v/v) H_2O_2	59 min microwave at 35 bar and 100-195°C	FAAS GFAAS	Ni, Cd, Pb, Fe, Cu, Zn, As, Hg	[12]
Powdered cocoa beverage	0.5 g	25 mL HNO_3 4 mL $HClO_4$ 2 mL H_2SO_4	Gently heated until no trace of carbon	FAAS	Ca, Fe, Pb, Cu, Cr	[13]
Dry ashed						
Cocoa beverage	n.i.	1 M HNO_3 for ash dissolution	Muffle furnace at 550°C	FAAS	Cd, Co, Cr, Cu, Fe, Ni, Pb, Zn	[14]
Chocolate	2 g	1 mL water + 1.5 mL HNO_3 Ash dissolution in 0.5 mL HNO_3 + 5 mL water	Heated to dryness Heating overnight at 450°C muffle furnace	GFAAS	Al	[15]
Slurry sampling						
Chocolate	0.05–0.20 g	2 mL 1 M HNO_3/1.5 M H_2O_2 + 1 mL 0.1% Triton X-100	20 min ultrasonic bath	GFAAS	Cd, Co, Cr, Cu, Fe, Ni, Pb	[16]
Powdered chocolate	150 mg dried sample (particle size ≅ 23 μm)	50 mL 2.0 M HNO_3	15 min ultrasonic bath	FAAS	Mn, Zn	[17]
Powdered chocolate	200 mg dried sample (particle size ≅ 15 μm)	50 mL 2.0 M HNO_3		FAAS	Cu	[18]

Cloud point extraction						
Powdered chocolate	100 mg	For sample digestion 2.0 mL HNO_3 1.0 mL 30% H_2O_2 for extraction pH adjusted to 8.6 Volume made up to 50 mL + 720 µL PAN in Triton X-114 + 750 µL 1.5 mM Tris buffer (pH 8.6) 300 µL 1% HNO_3 in ethanol	12 h at 150°C in a closed oven 15 min heated at 40°C 10 min centrifugation at 2500 rpm 10 min cooling in ice bath Separation of surfactant-rich phase Addition to the phase	FAAS	Cu, Zn	[19]
Emulsion						
Chocolate	0.02–0.2 g	1 g Tween 80 in 50 mL final volume	Surfactant, chocolate samples and hot water (75°C) were mixed with continuous agitation. After, magnetic stirring was maintained during 15 min at 22°C	GFAAS	Al, Mn, Cu	[20]
Chocolate	0.1–2.0 g	2 g Triton X-100 or Tween 80 + 2 g octyl stearate in 50 mL final volume	Surfactant, chocolate samples and hot water (75°C) were mixed with continuous agitation. After, magnetic stirring was maintained during 15 min at 22°C	FAAS	Na, K, Ca, Mg, Zn, Fe	[21]

n.i. not informed. Where acid concentration is not informed = concentrated acid

Another conventional sample treatment is dry ashing, which was employed by Onianwa et al. [14] for the determination of several metallic elements in cocoa beverages and by Sepe et al. [15] for the determination of Al in different types of chocolate. In these procedures, after ashing, the residue is dissolved in diluted nitric acid and the elements measured by either FAAS or GFAAS.

Nonconventional treatments applied to chocolate samples include slurry sampling, cloud point extraction, and emulsification.

Slurry sampling coupled to FAAS or GFAAS has been employed for powdered chocolate analysis. In a common sense, slurry is a thick, stable suspension of a solid in a liquid phase. From the analytical point of view of sample preparation, slurry is a dispersion thin enough to ensure its transportation as a solution, that is, to be pipetted or aspirated from one reservoir into another. Slurry sampling offers potential advantages in analytical speed compared with wet digestion because the solid sample does not need to be decomposed by the action of acids or other oxidizing agents. The solid sample, in the form of a fine powder, is only suspended in a solvent. Important attributes that contribute to slurry stability are particle size, solvent, stabilizing agent, and solid mass to total slurry volume [22]. Nevertheless, drawbacks associated with slurry sampling arise from weighing small amounts of sample, instrument calibration, and slurry inhomogeneity. Therefore, the main difficulties for slurry sampling are to find (1) optimal mixture for slurry preparation, (2) instrumental parameters for AAS determination, and (3) conditions for calibration. Features of slurry sampling methods for the analysis of chocolate and chocolate-related products are listed in Table 10.3.

A general problem associated with the measurement of slurries by GFAAS is the production of a carbonaceous residue into the tube. Karadjova et al. [16] observed that hydrogen peroxide used for slurry preparation assists in the decomposition of organic matrix by converting the charring step into an oxidative decomposition process. They observed that a concentration of 0.1–0.2 mol/L hydrogen peroxide in the slurry is able to prevent the buildup of carbonaceous residues.

Another approach for sample pretreatment was proposed by Ferreira et al. [19] using cloud point extraction (CPE) for pre-concentration of copper and zinc from powdered chocolate with subsequent determination by FAAS. Cloud point extraction is a separation technique based on the phenomenon in which an aqueous solution of surfactant becomes turbid (cloud point) at a specific temperature, with the following separation of two distinct phases: one rich in surfactant containing the metallic ions and the other with the aqueous matrix. In order to be extracted into the micellar phase (surfactant), metallic species should be in the form of hydrophobic chelates [23]. Nevertheless, because CPE is a separation technique that requires the analyte to be in solution, chocolate samples must be digested prior to separation. This approach allowed for enhancement factors of 36 and 32 for Cu and Zn, respectively; therefore, lower limits of detection than obtained with direct FAAS measurement could be attained [19].

Direct emulsification with surfactants is another procedure for sample preparation that does not require destruction of the organic matrix [24]. It simply reduces the viscosity and the organic content of the sample, making the properties of the chocolate sample close to those capable of being analyzed by any spectrometric technique, while maintaining the system homogeneity and stability.

Emulsions and microemulsions are thermodynamically stable systems composed of water, oil, and surfactant. They differ in appearance, with the former usually being milky white and the latter being transparent. Both emulsion types, however, are obtained by the dispersion of two immiscible liquids with the concomitant fragmentation of one phase into the other with the aid of emulsifiers to enhance the stability of the emulsion [25].

To utilize micellar media (as an emulsion or microemulsion) in the sample preparation for AAS measurements, some criteria should be taken into account: (1) the media should contain only the minimum components necessary to stabilize the sample (i.e., sample, surfactant, and water), (2) the emulsion stability should be maintained for a time interval long enough to complete the analysis, (3) the media should have low viscosity to allow correct sample aspiration, and (4) all components should have low metal contamination and low background during the AAS measurement.

Table 10.4 Summarized results of elements in chocolate- and cocoa-derived products

Element	Sample	Amount (μg/g) range		Measurement	Reference
Aluminum	White chocolate	10.6	21.0	GFAAS	[20]
	White chocolate	1.2	1.9	GFAAS	[15]
	Milk chocolate	17.1	22.7	GFAAS	[20]
	Milk chocolate	4.1	8.5	GFAAS	[15]
	Dark chocolate	27.2	92.1	GFAAS	[20]
	Dark chocolate	9.9	30.1	GFAAS	[15]
	Cocoa powder	24.9	89.0	GFAAS	[15]
Arsenium	Chocolate with pistachio	0.004	0.02	GFAAS	[12]
Cadmium	Chocolate with pistachio	0.01	0.03	GFAAS	[12]
	Cocoa-based chocolate	0.01	2.73	GFAAS	[10]
	Milk-based chocolate	0.01	0.85	GFAAS	[10]
Calcium	White chocolate	3,203.4	4,533.7	FAAS	[21]
	Milk chocolate	1,546.2	2,523.3	FAAS	[21]
	Dark chocolate	324.4	2,069.5	FAAS	[21]
	Chocolate beverages	116	2,540	ICP-OES	[11]
	Powdered cocoa beverages	149	5723	FAAS	[13]
Copper	White chocolate	<LOD	0.2	GFAAS	[25]
	Milk chocolate	1.5	2.3	GFAAS	[25]
	Dark chocolate	4.3	12.4	GFAAS	[25]
	Powdered chocolate	26.6	31.5	FAAS	[24]
	Chocolate with pistachio	9.15	10.61	FAAS	[20]
	Chocolate beverages	2.6	6.6	ICP-OES	[19]
	Powdered cocoa beverages	3.9	25.8	FAAS	[21]
	Powdered chocolate	3.3 ±0.1[a]		FAAS	[13]
Chromium	Chocolate beverages	< LOD	0.8	ICP-OES	[11]
	Powdered cocoa beverages	< LOD	3.2	FAAS	[13]
Iron	White chocolate	1.2	3.0	FAAS	[21]
	Milk chocolate	14.7	27.4	FAAS	[21]
	Dark chocolate	35.8	140.8	FAAS	[21]
	Chocolate with pistachio	2.31	3.67	FAAS	[12]
	Chocolate beverages	44.3	93	ICP-OES	[11]
	Powdered cocoa beverages	95	652	FAAS	[13]
Lead	Chocolate with pistachio	0.001	0.04	GFAAS	[12]
	Cocoa-based chocolate	0.236	8.04	GFAAS	[10]
	Milk-based chocolate	0.234	2.62	GFAAS	[10]
	Powdered cocoa beverages	0.8	3.8	FAAS	[13]
Magnesium	White chocolate	325.7	496.8	FAAS	[21]
	Milk chocolate	1,546.2	2,523.3	FAAS	[21]
	Dark chocolate	324.4	2,069.5	FAAS	[21]
	Chocolate beverages	363	1,090	ICP-OES	[11]
Manganese	White chocolate	0.8	2.7	GFAAS	[20]
	Milk chocolate	3.6	5.2	GFAAS	[20]
	Dark chocolate	8.4	17.3	GFAAS	[20]
	Powdered chocolate	42.8	52.7	FAAS	[17]
	Chocolate beverages	2.4	35.9	ICP-OES	[11]
Mercury	Chocolate with pistachio	0.008	0.02	FAAS	[12]
Molybdenum	Chocolate beverages	0.3	2.6	ICP-OES	[11]

(continued)

Table 10.4 (continued)

Element	Sample	Amount (μg/g) range		Measurement	Reference
Nickel	Chocolate with pistachio	0.33	1.52	GFAAS	[12]
	Cocoa-based chocolate	0.049	8.29	GFAAS	[10]
	Milk-based chocolate	0.137	8.288	GFAAS	[10]
Phosphorus	Chocolate beverages	517	2950	ICP-OES	[11]
Potassium	White chocolate	2,745.3	3,952.1	FAAS	[21]
	Milk chocolate	2,595.6	3,671.5	FAAS	[21]
	Dark chocolate	3,653.9	6,361.6	FAAS	[21]
	Chocolate beverages	1,340	6,620	ICP-OES	[11]
Selenium	Chocolate beverages	3.1	7.6	ICP-OES	[11]
Sodium	White chocolate	921.0	1,411.2	FAAS	[21]
	Milk chocolate	450.1	932.2	FAAS	[21]
	Dark chocolate	59.8	509.8	FAAS	[21]
	Chocolate beverages	49	1,770	ICP-OES	[11]
Sulfur	Chocolate beverages	198	2,480	ICP-OES	[11]
Zinc	White chocolate	10.3	13.5	FAAS	[21]
	Milk chocolate	7.5	10.7	FAAS	[21]
	Dark chocolate	12.1	23.3	FAAS	[21]
	Powdered chocolate	88.6	102.4	FS-FAAS	[17]
	Chocolate with pistachio	14.05	16.68	FAAS	[12]
	Chocolate beverages	3.9	119	ICP-OES	[11]
	Powdered chocolate	9.0 ±0.1[a]		FAAS	[19]

[a]Mean ± standard deviation

Spontaneous formation/stabilization of emulsions or microemulsions rarely occurs because this process requires a sequence of steps. The formation of droplets of the internal phase must occur first and be followed by the stabilization of these droplets in the external phase. These two steps should occur before the internal phase begins to coalesce and generally requires heating and mechanical stirring. Stabilization and the coalescence speed depend mainly on the stirring time and on the temperature, thereby requiring optimal parameters for the formation of a stable microemulsion [26].

An increased temperature reduces interfacial tension but can also favor the separation of the phases due to the increased kinetic energy of the droplets. The emulsification temperature as well as cloud point is a characteristic of each surfactant. By the addition of other substances, this temperature can be increased or reduced; thus, for each sample/emulsifier combination, an optimization of the temperature is required. In general, emulsions are prepared at temperatures between 70°C and 80°C. In the initial period of stirring, the droplets necessary for emulsification are formed. If the stirring exceeds the necessary period for ideal stability, adhesion can take place due to droplet collisions. This period of time is usually determined empirically.

Table 10.4 summarizes the results found in the studies listed in Table 10.3. Differences between the levels found for some elements are probably correlated to sample composition, which varies from one sample to another. Among the elements recorded in Table 10.4 are essential and those considered toxic for humans. Although some toxic elements are present in chocolate- or cocoa-derived samples, their levels are in ng/g, whereas the essential elements are in the range of mg/kg.

References

1. Borchers AT, Keen CL, Hannum SM, Gershwin ME. Cocoa and chocolate: composition, bioavailability, and health implications. J Med Food. 2000;3:77–103.
2. United States Environmental Protection Agency: http://www.epa.gov/fem/methcollectns.htm. Accessed 2012.
3. AOAC International Official Methods: http://www.eoma.aoac.org/. Accessed 2012.
4. European Committee for Standardization (CEN): http://www.cen.eu/. Accessed 2012.
5. The National Institute for Occupational Safety and Health (NIOSH): http://www.cdc.gov/niosh/. Accessed 2012.
6. National Environmental Methods Index: https://www.nemi.gov/. Accessed 2012.
7. Ingle JD, Crouch SR. Spectrochemical analysis. Upper Saddle River: Prentice Hall; 1988.
8. Welz B, Sperling M. Atomic absorption spectrometry. 3rd ed. Weinheim: Wiley-VCH; 1999.
9. Nölte J. Inductively coupled plasma (ICP) emission spectrometry: a practical guide. Weinheim: Wiley-WCH; 2003.
10. Dahiya S, Karpe R, Hegde AG, Sharma RM. Lead, cadmium and nickel in chocolates and candies from suburban areas of Mumbai, India. J Food Compos Anal. 2005;18:517–22.
11. Pedro NAR, Oliveira E, Cadore S. Study of the mineral content of chocolate flavoured beverages. Food Chem. 2006;95:94–100.
12. Güldaş M. Comparison of digestion methods and trace elements determination in chocolates with pistachio using atomic absorption spectrometry. J Food Nutr Res. 2008;47:92–9.
13. Shittu TA, Badmus BA. Statistical correlations between mineral element composition, product information and retail price of powdered cocoa beverages in Nigeria. J Food Compos Anal. 2009;22:212–7.
14. Onianwa PC, Adetola IG, Iwegbue CMA, Ojo MF, Tella OO. Trace heavy metals composition of some Nigerian beverages and food drinks. Food Chem. 1999;66:275–9.
15. Sepe A, Costantini S, Ciaralli L, Ciprotti M, Giordano R. Evaluation of aluminium concentrations in samples of chocolate and beverages by electrothermal atomic absorption spectrometry. Food Addit Contam. 2001; 18:788–96.
16. Karadjova I, Girousi S, Iliadou E, Stratis I. Determination of Cd, Co, Cr, Cu, Fe, Ni and Pb in milk, cheese and chocolate. Mikrochim Acta. 2000;134:185–91.
17. Silva EGP, Santos ACN, Costa ACS, Fortunato DMN, José NM, Korn MGA, et al. Determination of manganese and zinc in powdered chocolate samples by slurry sampling using sequential multi-element flame atomic absorption spectrometry. Microchem J. 2006;82:159–62.
18. Santos WNL, Silva EGP, Fernandes MS, Araujo RGO, Costa ACS, Vale MGR, et al. Determination of copper in powdered chocolate samples by slurry-sampling flame atomic-absorption spectrometry. Anal Bioanal Chem. 2005;382:1099–102.
19. Ferreira HS, Santos ACN, Portugal LA, Costa ACS, Miró M, Ferreira SLC. Pre-concentration procedure for determination of copper and zinc in food samples by sequential multi-element flame atomic absorption spectrometry. Talanta. 2008;77:73–6.
20. Ieggli CVS Bohrer D, Do Nascimento PC, De Carvalho LM, Gobo LA. Determination of aluminum, copper and manganese content in chocolate samples by graphite furnace atomic absorption spectrometry using a microemulsion technique. J Food Compos Anal. 2011;24:465–8.
21. Ieggli CVS, Bohrer D, Do Nascimento PC, De Carvalho LM. Determination of sodium, potassium, calcium, magnesium, zinc and iron in emulsified chocolate samples by flame atomic absorption spectrometry. Food Chem. 2011;124:1189–93.
22. Ferreira SLC, Miró M, da Silva EGP, Matos GD, Reis PS, Brandão GC, et al. Slurry sampling: an analytical strategy for the determination of metals and metalloids by spectroanalytical techniques. Appl Spectrosc Rev. 2010;45:44–62.
23. Bezerra MA, Ferreira SLC. Extração em ponto de nuvem: princípios e aplicações em química analítica. Edições da UESB (2006)
24. Sanz-Medel A, De La Campa MRF, Gonzalez EB, Fernandez-Sanchez ML. Organised surfactant assemblies in analytical atomic spectrometry. Spectrochimi Acta B. 1999;54:251–87.
25. Prista N, Alves AC, Morgado R, Lobo JS. Tecnologia farmacêutica, vol. 1. 6th ed. Lisbon: Editora Fundação Calouste Guebenkian; 2003.
26. Lachman L, Lieberman HA, Kanig JL. Teoria e prática na indústria farmacêutica, vol. 2. Lisbon: Editora Fundação Calouste Gulbenkian; 2001.

Chapter 11
The Effects of Chocolate and Chocolate by-product Consumption on Wild and Domestic Animals

Brett David Gartrell and Wendi Dianne Roe

Key Points

- Modern, high-yield cocoa plantations have a negative impact on biodiversity, and there is a pressing need for financial incentives to encourage use of less destructive farming systems.
- The use of low-value cocoa by-products, particularly cocoa husk, in animal feed has potential economic benefits for many cocoa-producing countries.
- Inclusion of chocolate and its by-products in animal feed has led to fatal toxicity in pigs, poultry, cattle, and horses.
- Theobromine is the main toxin responsible for accidental poisoning in domestic and wild animals. Horses, pigs, chickens, and dogs are particularly sensitive.
- Chocolate is one of the most common causes of accidental poisoning in dogs.
- Chocolate toxicity is rarely encountered in wildlife, but it does occur. Chocolate has been used to poison coyotes.

Keywords Cocoa by-products • Toxicity • Theobromine • Animal feed • Chocolate poisoning

Introduction

In this chapter we present an overview of the effects of chocolate on animals, starting with the effects of cacao plantations on biodiversity, the use of cocoa by-products in domestic animal husbandry, and the reported and varied toxicity of chocolate in both domestic and wild animals.

B.D. Gartrell, BVSc., Ph.D., MACVSc. (Avian Health) (✉)
W.D. Roe, B.Sc., BVSc., MACVSc. (Internal Medicine), DipACVP.
Institute of Veterinary, Animal and Biomedical Sciences, Massey University,
Palmerston North 4410, New Zealand
e-mail: B.Gartrell@massey.ac.nz; W.D.Roe@massey.ac.nz

R.R. Watson et al. (eds.), *Chocolate in Health and Nutrition*, Nutrition and Health 7,
DOI 10.1007/978-1-61779-803-0_11, © Springer Science+Business Media, LLC 2013

Cacao Plantations and Biodiversity

Cocoa is a crop from the cacao tree *Theobroma cacao,* which is produced in lowland tropical areas such as western Africa (Côte d'Ivoire and Ghana), South and Latin America, and Indonesia [1]. These areas are also regions of high biodiversity, and tropical forest loss to agriculture is occurring at alarming rates [2]. The current protection for natural forest in these areas is inadequate to protect their biodiversity; increasingly, conservationists are looking for ways to include the adjacent agricultural areas in the conservation of the natural resources of the tropical forests [3]. Shaded cacao plantations have been held up as an example of how agroforestry practices can be integrated with conservation [2]. It is widely acknowledged that shaded plantation systems contain more biodiversity than monocultures or alternative land uses [1, 2, 4–6]. However, recent trends in cacao farming practices and wider studies of the impact of these farms on biodiversity suggest that integrating conservation and farming may not be as simple as first thought [3, 7].

A range of farming systems are used for farming cocoa. Traditionally, in a rustic cacao system, small-scale producers clear the understory and plant cacao beneath thinned primary or old secondary forest. These systems require low input, with limited use of chemicals, but are very low yield [2]. At the other end of the management spectrum, some farms clear-fell and burn the forest before planting either a managed variety of shade trees or, more recently, modern varieties of cacao that produce higher yields without shade trees and are managed as monocultures. These systems require higher input in both labor and chemicals but produce higher yields [3]. There are dramatic reductions in biodiversity of plants and animals in these systems.

Rustic cacao systems occupy large areas of the lowland agroforests – for example, 40% of remaining Atlantic Forest in Brazil [8]. While these farming systems contain more biodiversity than other land uses, they are still heavily modified landscapes that have much lower biodiversity than native forests, including large-bodied mammals, primates, bats, reptiles, and birds [2, 3, 6–11]. Biodiversity in shade farms also dramatically declines when they are distant from natural forest remnants, suggesting that they cannot be considered the sole answer to conservation in these regions [10]. Conservationists currently suggest that there need to be market incentives for the use of shade systems of farming over more managed, higher-production systems.

Finally, cacao farming is globally affected by boom and bust cycles. In times of poor prices, many farms are left fallow or converted to pasture. In times of high demand, natural forest remnants are cleared and converted to cacao [1]. Improved sustainability of farming systems and conservation solutions that better address the needs of local farmers and global markets are urgently needed.

The Use of Cocoa By-Products in Animal Feed

The manufacture of chocolate results in several by-products that are of low economic value, and some of these have been used, both historically and more recently, in animal feed. During processing, cocoa beans are roasted, fermented, and ground to form a mixture that can then be pressed to separate out the cocoa oils. The remaining residue is known as press cake, and it can be ground to form cocoa powder. Cocoa shell, the thin outer layer of the bean, is usually removed before pressing and is a waste product of cocoa bean processing. While cocoa shell and cocoa meal (surplus press cake) are potential feed components that are produced in countries where cocoa beans are processed, cocoa-growing countries have access to a further by-product that has been found to be a useful animal feed. This product is cocoa husk, the fibrous outer layer of the cocoa pod that encloses the cocoa beans embedded in a mucilaginous pulp. In many cocoa-growing countries, cocoa husks are removed on site during harvesting, and were previously left to rot. In countries such as sub-Saharan Africa and the Pacific Islands, however, where animal feed can be costly or seasonally scarce, cocoa husk can provide a low-cost, locally produced source of feed for livestock [12–14]. Although they are comparatively low in energy and protein content, husks can be dried and ground to provide a high-fiber compound that can be safely used in ruminant and rabbit feed at up to 45% of total diet without adversely affecting digestibility or palatability [13, 15, 16].

Transport of cocoa husk to other countries is not economically viable, but in Europe, the outbreak of World War II led to an increased demand for cheap livestock feed components. During this time, there was a high demand for production of cocoa butter, generating large amounts of cocoa shell by-product as well as a surplus of cocoa cake [17]. These products began to be incorporated into animal feed in the early 1940s [18], and while this initially seemed to be a cost-effective means of supplementing animal feed rations, it was not long until problems with palatability and toxicity began to be reported [17–19]. In fact, deaths in horses fed on cocoa shell had first been described in 1919, and decreased weight gain and food intake had been demonstrated in pigs, cattle, and chickens fed on various cocoa by-products [17]. Theobromine was identified as the toxic principle present, and horses, poultry, and swine were identified as being particularly susceptible to poisoning by this alkaloid [17].

Subsequent studies have shown that theobromine levels vary between cocoa genotypes as well as between parts of the fruit itself, resulting in wide variations in concentration for various by-products. Theobromine levels are higher in African cocoa plants [20] and are much higher in the kernel than in other parts of the unfermented bean [21]. During fermenting and roasting, however, theobromine leaches from the kernel into the shell [18], and potentially toxic levels can be reached in shells that are removed after these processes are completed.

In general, theobromine levels in cocoa husk are comparatively low, and the low economic value of this waste product has made it an appealing option for use in animal feeds in cocoa-growing developing countries. Husks are low in protein, but high in fiber, and have been suggested to be similar in nutritive value to grass hay [22, 23]. Several studies have shown no loss in production parameters when husks are incorporated in feeds at low rates, although higher rates tend to result in decreased digestibility of nutrients or decreased food intake due to low palatability [12–16, 22]. Interestingly, in 1978 an application was lodged at the United States patent office to use cocoa products as an appetite stimulant in ruminants, with the applicants citing a stimulatory effect on food intake for lambs and rats [24, 25].

Toxicity is more likely to be encountered when products containing higher levels of theobromine, such as cocoa meal and cocoa shell, are included in the diet. Alexander et al. [26] reviewed the literature on adverse effects of theobromine in feed and found that decreased food intake and decreased production (reduced milk yield or reduced weight gain) were the earliest observed effects and occurred at low concentrations in pigs, poultry, and horses, and at higher levels in ruminants and rabbits. More dramatic clinical signs of toxicity, including sudden deaths, have been reported in pigs, poultry, and horses that were given feed containing cocoa shell or meal [17, 18].

Chocolate in Feed

While the use of by-products from growing and processing cocoa beans in animal feed has obvious economic advantages, a less obvious source of theobromine is the incorporation of waste chocolate into feed rations. Cramshaw [27] estimated that approximately 20,000 T of chocolate is fed to animals in the United Kingdom each year. The majority of this chocolate is sourced from rejected product that fails to meet quality-control standards or following seasonal overproduction of chocolates at Easter and Christmas time. This chocolate can be incorporated into feed in solid form for use in ruminants and pigs or supplied in a comparatively concentrated liquid form for feeding to pigs.

The theobromine content of feed containing chocolate depends on the ratio of chocolate to other components as well as on the nature of the chocolate used (dark, milk, solid vs soft-centered chocolate). Several cases of poisoning resulting from inclusion of chocolate in the feed ration have been reported and are typified by the following report. In 1971 a veterinarian was called to examine a group of six calves that were showing signs of excitability and seizures, were sweating, and had rapid respiratory and heart rates. One calf died during a seizure. Investigation of the feeding history found that the calves were being fed a ration containing waste chocolate. Once this was removed from the diet, the remaining five calves recovered uneventfully [28].

The incorporation of chocolate in feed has also had unanticipated effects in the horse-racing industry. Theobromine, caffeine, and theophylline, all of which are found in chocolate, are included in the list of substances banned from use in racehorses owing to their ability to artificially enhance performance through increased heart rate and respiratory rate. In order to police this ban, urine or serum from a randomly selected sample of horses is tested after racing, and assays are performed to detect the presence of banned substances. High levels of theobromine were commonly found in the urine of horses up until the 1980s and were due to the inclusion of cocoa shell and cocoa meal in pelleted horse feed. Because of this, urinary theobromine levels of up to 300 ppm were considered legal in Europe for over 50 years. In 1979, however, a diuretic containing theobromine was released for medical use in horses, and numerous positive urine tests were found to be due to use of this drug. The acceptable urinary concentration of theobromine was subsequently lowered to 2 μg/mL [29, 30]. A number of high-profile cases of apparently accidental positive tests have occurred since this time, including one case of a horse being fed a Mars bar, several that had been given chocolate peanuts as treats, and a horse that had been fed pelleted feed containing bakery waste that included chocolate [31, 32].

Chocolate Toxicity in Dogs and Cats

Dogs are well known as being fairly indiscriminate and opportunistic eaters, and chocolate is a ubiquitous compound in households, so one might expect that the two would often "meet." In 1983, however, Glauber and Blumenthal [33] reported a case of fatal chocolate toxicity in a dog and noted that there were very few similar reports in the literature. While chocolate poisoning of dogs does not seem common based on the lack of case reports in the veterinary literature, the picture is slightly different when information from animal poison control centers and veterinary emergency centers is taken into account [34, 35]. The ASPCA center in Illinois, for example, found that chocolate toxicity was the second most common animal toxicosis, based on the number of calls they received from veterinarians and dog owners between 2001 and 2005 [35]. The Dove-Lewis Emergency Center in Portland, Oregon, found that chocolate poisoning accounted for 25% of their poisoning admissions and was by far the most commonly diagnosed toxin [36]. In contract, no cases of toxicity have been reported in cats, which tend to be more discerning in their eating habits.

Chocolate toxicity can occur when owners feed chocolate to their dogs as treats or when dogs help themselves to chocolate they find around the house. Poisoning has also been reported in dogs that have eaten organic garden mulch made from cocoa beans and shells [37–39]. Gans et al. [40] undertook toxicity studies in crossbred dogs and found that the earliest signs of poisoning were restlessness, panting, and muscle tremors. Further reported clinical signs include vomiting, diarrhea, seizures, coma, and sudden death [33, 39, 41, 42]. Although theobromine is believed to be the most important toxin in chocolate poisoning, caffeine may also contribute to the clinical signs, and vomiting and diarrhea have been attributed to the fat present in ingested chocolate products [43].

The severity of the clinical signs depends on the type and amount of chocolate ingested and the size of the dog. Gans et al. [40] also found that there was some individual variation in response to doses of theobromine. Most chocolate products contain far higher levels of theobromine than caffeine, with unsweetened dark chocolate having the highest levels and white chocolate the lowest. Cocoa powder, which has been reported as causing the death of two dogs who shared a 225-g tin between them [42], also has high levels of theobromine [40]. As a general rule, as little as 125 g of unsweetened baking chocolate is potentially fatal for a 20-kg dog.

Although there is no specific antidote for methylxanthine poisoning, if chocolate toxicity is diagnosed early, it can be successfully treated. In cases where dogs are observed to eat chocolate,

Gwaltney-Brant [43] recommends calculating the likely total methylxanthine intake by adding the amount of theobromine and caffeine ingested using known values for various types of chocolate. When the exact nature of the chocolate is not known, it is safest to use a worst-case-scenario approach by assuming the highest theobromine concentration (unsweetened chocolate). Mild signs of toxicosis are expected at levels of 20 mg/kg, severe signs at 40–50 mg/kg, and seizures at 60 mg/kg. Induction of vomiting can decrease absorption of toxins if done within 1–4 h of ingestion, and veterinary attention should be sought promptly.

Chocolate Toxicity in Birds

The reports of chocolate toxicity in captive birds are sketchy and anecdotal and much less frequent than in dogs. It has been suggested that birds are less likely to eat toxic amounts of chocolate but that they may be more sensitive to its effects, as all reported clinical cases have resulted in the death of the bird. Clinical signs reported before death have included depression or hyperactivity, cardiac arrhythmias, regurgitation, convulsions, and diarrhea [44]. There is a single non-peer-reviewed case report of chocolate toxicity in an African gray parrot (*Psittacus erithacus*) that ate a chocolate doughnut and died showing liver, lung, and renal congestion as its main pathological findings [45].

Chocolate Toxicity in Wild Animals

Chocolate and chocolate analogs have been used to deliberately poison coyotes (*Canis latrans*). The authors of this study did not list doses of the ingredients used, just a 5:1 ratio of theobromine: caffeine, but claimed that all coyotes that ingested even small quantities of bait died [46].

There is only one report of accidental poisoning of wild mammals in the scientific literature. In this study from Sweden, a red fox (*Vulpes vulpes*) and a European badger (*Meles meles*) were both found dead, with postmortem findings of chocolate waste from a nearby farm in the gastrointestinal tract. The paper mentions that a number of other foxes had been seen dead in the area, which prompted the investigation. High-performance liquid chromatography detected theobromine and caffeine in both the gastric contents and livers of both animals. The authors speculate that as canids, foxes may share domestic dogs' susceptibility to chocolate intoxication [47].

The only report of chocolate toxicity in wild birds is of the death of an adult male kea (*Nestor notabilis*) in good body condition at Mt. Cook village, in the Southern Alps of New Zealand [48]. The bird had been previously recorded as being able to open rubbish bins. The crop contained 20 g of what appeared to be dark chocolate; a conservative estimate of the dose of methylxanthines ingested by the bird was 250 mg/kg theobromine, 20 mg/kg caffeine, and 3 mg/kg theophylline. Histopathological examination revealed acute degenerative changes to hepatocytes, renal tubules, and cortical neurons. The postmortem diagnosis was acute combination methylxanthine toxicity after opportunistic ingestion of chocolate [48].

Summary

Humanity's love affair with chocolate has had profound implications for biodiversity in the lowland tropics, but there is hope that with global support and appropriate market incentives, local farmers can make a living from cocoa farming and still support the conservation of wild animals in their regions. The use of cocoa products in livestock feeds is limited by the toxic effects of theobromine, and while it may be useful as a feed additive in cocoa-producing areas to enhance the value of otherwise worthless by-products, it is unlikely to find great favor in overseas markets. The toxic effects of chocolate on domestic dogs are underreported in the scientific literature but widely recognized by practicing veterinarians and poison information centers. Other domestic animals are more rarely affected. Wildlife fatalities due to accidental poisoning with chocolate and chocolate waste products are rare but do occur, especially in inquisitive and scavenging species, and care should be taken in the disposal of chocolate waste.

References

1. Franzen M, Mulder MB. Ecological, economic and social perspectives on cocoa production worldwide. Biodivers Conserv. 2007;16(13):3835–49.
2. Rice RA, Greenberg R. Cacao cultivation and the conservation of biological diversity. Ambio. 2000; 29(3):167–73.
3. Anglaaere LCN, Cobbina J, Sinclair FL, McDonald MA. The effect of land use systems on tree diversity: farmer preference and species composition of cocoa-based agroecosystems in Ghana. Agrofor Syst. 2011;81(3):249–65.
4. Feintrenie L, Schwarze S, Levang P. Are local people conservationists? Analysis of transition dynamics from agroforests to monoculture plantations in Indonesia. Ecol Soc.. 2010;15(4):37.
5. Laird SA, Awung GL, Lysinge RJ. Cocoa farms in the Mount Cameroon region: biological and cultural diversity in local livelihoods. Biodivers Conserv. 2007;16(8):2401–27.
6. Oliveira LC, Hankerson SJ, Dietz JM, Raboy BE. Key tree species for the golden-headed lion tamarin and implications for shade-cocoa management in southern Bahia, Brazil. Anim Conserv. 2010;13(1):60–70.
7. Cassano CR, Kierulff MCM, Chiarello AG. The cacao agroforests of the Brazilian Atlantic forest as habitat for the endangered maned sloth Bradypus torquatus. Mamm Biol. 2011;76(3):243–50.
8. Faria D, Laps RR, Baumgarten J, Cetra M. Bat and bird assemblages from forests and shade cacao plantations in two contrasting landscapes in the Atlantic Forest of southern Bahia, Brazil. Biodivers Conserv. 2006;15(2): 587–612.
9. Clough Y, Faust H, Tscharntke T. Cacao boom and bust: sustainability of agroforests and opportunities for biodiversity conservation. Conserv Lett. 2009;2(5):197–205.
10. Faria D, Baumgarten J. Shade cacao plantations (Theobroma cacao) and bat conservation in southern Bahia, Brazil. Biodivers Conserv. 2007;16(2):291–312.
11. Wilsey CB, Temple SA. The effects of cropping systems on avian communities in cacao and banana agro-forestry systems of Talamanca, Costa Rica. Biotropica. 2011;43(1):68–76.
12. Adamfio N, Afeke I, Wepeba J, Ali E, Quaye F. Biochemical composition and in vitro digestibility of cocoa (*Theobroma cacao*) pod husk, cassava (*Manihot esculenta*) peel and plantain (*Musa paradisiaca*) peel. Ghana J Sci. 2004;44:29–38.
13. Aregheore E. Chemical evaluation and digestibility of cocoa (*Theobroma cacao*) by-products fed to goats. Trop Anim Health Prod. 2002;34:339–48.
14. Owusu-Domfeh K. The future of cocoa and its by-products in the feeding of livestock. Ghana J Agric Sci. 1972;5:57–64.
15. Ridzwan B, Fadzli M, Rozali M, Chin D, Ibrahim B, Faridnordin B. Evaluation of cocoa-pod husks on performance of rabbits. Anim Feed Sci Technol. 1993;40:267–72.
16. Smith O, Adegbola A. Studies on the feeding value of agro-industrial by-products. III. Digestibility of cocoa-pod and cocoa-pod based diets by ruminants. Anim Feed Sci Technol. 1985;13:249–54.
17. Blakemore F, Shearer G. Suspected chocolate poisoning of calves. Vet Rec. 1943;55(14):165.
18. Black D, Barron N. Observations on the feeding of a cacao waste product in poultry. Vet Rec. 1943;55(15): 166–7.
19. Woodman HE. The cocoa bean in animal feeding. J Ministry Agric. 1941;48:45–7.
20. Matissek R. Evaluation of xanthine derivatives in chocolate – nutritional and chemical aspects. Z Lebens Unters Forsch Food Res Technol. 1997;205(3):175–84.
21. Sotelo A, Alvarez RG. Chemical composition of wild theobroma species and their comparison to the cacao bean. J Agric Food Chem. 1991;39(11):1940–3.
22. Greenwood-Barton L. Utilisation of cocoa by-products. Food Manuf. 1965;50:52–6.
23. Dittmar HFK. The analysis of cacao pod shells. Gordian. 1958;58(1387):48–9.
24. Tarka S, Zoumas B, Trout G. Examination of the effect of cocoa shells and theobromine in lambs. Nutr Rep Int. 1978;18(3):301–12.

25. Trout GA, Zouman BL, Tarka SM. Method of stimulating appetite in ruminants and ruminant feed containing appetite stimulant. US patent application no. 711912. 1978.
26. Alexander J, Benford D, Cockburn A, et al. Scientific opinion of the panel on contaminants in the food chain on a request from the European Commission on theobromine as undesirable substances in animal feed. Eur Food Safety Auth. 2008;725:1–66.
27. Crawshaw R. Co-product feeds: animal feeds from the food and drinks industries. Nottingham: Nottingham University Press; 2001.
28. Curtis P, Griffiths J. Suspected chocolate poisoning of calves. Vet Rec. 1972;90:313–4.
29. Moss MS. A common sense change in doping rules. Equine Vet J. 1988;20(2):84.
30. Delbeke FT, Debackere M. Urinary excretion of theobromine in horses given contaminated pelleted food. Vet Res Commun. 1991;15(2):107–16.
31. Hintz HF. Theobromine. Equine Pract. 1987;9(4):6–7.
32. Dyke T, Sams R. Detection and determination of theobromine and caffeine in urine after administration of chocolate-coated peanuts to horses. J Anal Toxicol. 1998;22(2):112–6.
33. Glauberg A, Blumenthal HP. Chocolate poisoning in the dog. J Am Anim Hosp Assoc. 1983;19(2):246–8.
34. Hornfeldt C. Chocolate toxicity in dogs. Mod Vet Pract. 1987;68:552–4.
35. Meadows I, Gwaltney-Brant S. The 10 most common toxicoses in dogs. Vet Med. 2006;101(3):142–8.
36. Cope RB, White KS, More E, Holmes K, Nair A, Chauvin P, et al. Exposure-to-treatment interval and clinical severity in canine poisoning: a retrospective analysis at a Portland Veterinary Emergency Center. J Vet Pharmacol Ther. 2006;29(3):233–6.
37. Drolet R, Arendt TD, Stowe CM. Cacao bean shell poisoning in a dog. J Am Vet Med Assoc. 1984;185(8):902.
38. Hansen S, Trammell H, Dunayer E, Gwaltney S, Farbman D, Khan S. Cocoa bean mulch as a cause of methylxanthine toxicosis in dogs. J Toxicol Clin Toxicol. 2003;41(5):720.
39. Hovda L, Kingston R. Cacao bean mulch poisoning in dogs. Vet Hum Toxicol. 1994;36(4):357.
40. Gans JH, Korson R, Cater MR, Ackerly CC. Effects of short-term and long-term theobromine administration to male dogs. Toxicol Appl Pharmacol. 1980;53(3):481–96.
41. Stidworthy MF, Bleakley JS, Cheeseman MT, Kelly DF. Chocolate poisoning in dogs. Vet Rec. 1997;141(1):28.
42. Strachan ER, Bennett A. Theobromine poisoning in dogs. Vet Rec. 1994;134(11):284.
43. Gwaltney-Brant S. Chocolate intoxication. Vet Med. 2001;96(2):108–11.
44. Lightfoot TM, Yaeger JM. Pet bird toxicity and related environmental concerns. Vet Clin North Am Exot Anim Pract. 2008;11:229–59.
45. Cole G, Murray M. Suspected chocolate toxicosis in an African Grey parrot (Psittacus erithacus). In: Proceedings of the Annual Conference of the Association Of Avian Veterinarians; Monterey: Association of Avian Veterinarians; 2005. p. 339–40.
46. Johnston JJ. Evaluation of cocoa- and coffee-derived methylxanthines as toxicants for the control of pest coyotes. J Agric Food Chem. 2005;53(10):4069–75.
47. Jansson DS, Galgan V, Schubert B, Segerstad CHA. Theobromine intoxication in a red fox and a European badger in Sweden. J Wildlife Dis. 2001;37(2):362–5.
48. Gartrell BD, Reid C. Death by chocolate: a fatal problem for an inquisitive wild parrot. N Z Vet J. 2007; 55(3):149–51.

Chapter 12
Chocolate Bars Based on Human Nutritional Requirements

Anthony A. Robson

Key Points

- The nutritional value of chocolate bars should be based on the nutritional value of the low energy dense late Paleolithic human diet to help reduce mental ill health, obesity, and other postprandial insults.
- Current chocolate bars have a high energy density (>2 kcal/g).
- Cocoa can be sweetened by the addition of calorie-free Purefruit™ (Tate & Lyle) monk fruit (*Siraitia grosvenorii*) extract. PUREFRUIT™ is approximately 200 times sweeter than sugar and has exceptional stability.
- The energetic cost of the assimilation of chocolate can be increased by increasing its protein and fibre content.
- Self-assembled, water-filled, edible nanotubes that self-organise into a more complex structure, possibly a 3D network of nanocellulose, could be incorporated into chocolate bars to lower their energy density to <1.6 kcal/g.
- Durethan® KU 2-2601 packaging film enables the water content of chocolate bars to be increased without reducing product shelf life.
- Aquatic biotechnology can provide all the nutrients needed to make chocolate really nutritious.

Keywords Diet • Food • Disease • Energy density • Nutrition • Food technology • Biotechnology Nanocellulose

Cocoa is a source of health-promoting polyphenols including procyanidins, but they do not make chocolate a healthy food item [1]. Chocolate bars have a high energy density (>2 kcal/g) [2] and are not as nutrient dense as wild foods (Tables 12.1 and 12.2) consumed by humans during the late Paleolithic [3]. Degenerative non-communicable diseases are rare or nonexistent in hunter-gatherers eating a late Paleolithic diet, that is, a low energy dense diet with a wild plant-to-animal energy intake ratio ~1:1, with fish and shellfish providing a significant proportion of the animal component [3]. Nearly all the genes and epigenetic regulatory mechanisms humans carry today were originally selected for behaviorally modern humans who appeared in Africa between 100,000 and 50,000 years

A.A. Robson, B.Sc., Ph.D. (✉)
IUEM (UMR CNRS 6539), Université de Bretagne Occidentale, Technopôle Brest-Iroise, Rue Dumont d'Urville, 29280 Plouzané, France
e-mail: a.a.robson@hotmail.com

R.R. Watson et al. (eds.), *Chocolate in Health and Nutrition*, Nutrition and Health 7,
DOI 10.1007/978-1-61779-803-0_12, © Springer Science+Business Media, LLC 2013

Table 12.1 Energy density, water, protein, carbohydrate, and fiber content of a selection of food and drink (value per 100 g)[a]

	Energy (kcal)	Water (g)	Protein (g)	Carbohydrate (g)	Fiber (g)
Chocolate, dark (19904)	598	1.37	7.79	45.90	10.9
Hot chocolate (14194)	55	86.34	0.92	11.54	0.5
Oyster meat, cooked (15231)	163	64.12	18.90	9.90	0.0
Broccoli, cooked (11091)	35	89.25	2.38	7.18	3.3
Raspberries (35202)	62	84.48	1.12	13.85	7.5

[a]Entries retrieved from the USDA National Nutrient Database for Standard Reference, Release 23 (2010) and are identified by a five-digit nutrient database number in parentheses

ago [3]. Thus, it has been argued that the typical diet, physical activity patterns, and body composition of late Paleolithic humans remain normative for contemporary humans – and models for disease-prevention recommendations [3]. Indeed, an article in *Nature* highlighted, "It is difficult to refute the assertion that if modern populations returned to a hunter-gatherer state then obesity and diabetes would not be the major public health threats that they now are" [4]. A double-blind, placebo-controlled, randomized trial failed to find any beneficial effects of dark chocolate and cocoa consumption on neuropsychological or cardiovascular health-related variables [5]. Furthermore, an increase in body weight has been found owing to the consumption of just 25 g of chocolate per day [6], and dark chocolate and cocoa consumption has been associated with significantly higher pulse rates [5].

Human food production should be linked to human nutritional requirements as its first priority [7, 8]. Thus, the high energy density and low nutrient density that characterize the modern diet must be overcome simultaneously [7–9]. Overweight and obese people can develop paradoxical nutritional deficiency from eating high-energy-dense foods with a poor nutrient content. The finding that people with a low-energy-dense diet (<1.6 kcal/g) have the lowest total intakes of energy, even though they consume the greatest amount of food, has important implications for promoting compliance with a healthy diet [2]. A processed food that is not both low energy dense and high nutrient dense is of poor dietary quality compared to the low-energy-dense foods of high nutrient density that humans should eat i.e. the most nutritious cooked wild plant and animal foods for humans [3, 10–12].

In 2009, Barry Callebaut AG, Switzerland, reported that they had reduced the energy density of their chocolate by aerating it; however, the resulting product still had a deleteriously high energy density (~4.3 kcal/g) [13]. Chocolate bars have a high energy density principally because they have a low water content [9, 14]. Self-assembled, water-filled, edible nanotubes that self-organise into a more complex structure, possibly a 3D network of nanocellulose, could be incorporated into chocolate bars to lower their energy density to <1.6 kcal/g [14, 15]. Nanocellulose is composed of nanosized cellulose fibrils (fibre diameter: 20–100 nm), has a water content of up to 99 per cent and the same molecular formula as plant cellulose [16]. The water inside the nanosized cellulose fibrils could contain flavour with few calories e.g. a cup of tea without milk = 0.01 kcal g^{-1}. The shape and supramolecular structure of the nanocellulose can be regulated directly during biosynthesis to produce fleeces, films/patches, spheres and tubes [17]. Other edible materials can strongly adhere to the surface and the inside of nanocellulose structures such as fleeces to form edible composites [18]. Taste sensation per mouthful could be improved by adding flavouring substances processed on the nanoscale (increased surface area in contact with taste and smell receptors) to edible composites (Ultrafine food technology: Eminate Limited, Nottingham, United Kingdom). Durethan® KU2-2601 packaging film produced by Bayer Polymers, Germany, is a nanocomposite film enriched with silicate nanoparticles which is designed to prevent the contents from drying out and prevent the contents coming into contact with oxygen and other gases. Durethan® KU 2-2601 can prevent food spoilage [19] and thus the water content of chocolate bars can be increased without reducing product shelf life. Therefore, nanocellulose is expected to be widely used a as a nature-based food additive [17, 18].

Table 12.2 Nutrient density of a selection of food and drink (value per 100 g)[a]

	DHA	AA	I[b]	Fe[c]	Cu	Zn	Mn	Se	Vitamin					
	(g)	(g)	(µg)	(mg)	(mg)	(mg)	(mg)	(µg)	A (µg_RAE)	B12 (µg)	B6 (mg)	C (mg)	D[l] (µg)	Folate (µg)
Chocolate, dark (19904)	0.000	0.001	2.3	11.90	1.766	3.31	1.948	6.8	2	0.28	0.038	0.0	0	0.0
Hot chocolate (14194)	0.000	0.000	1.3	0.17	0.048	0.21	0.037	0.7	0	0.05	0.016	0.0	0	1
Oyster meat, cooked (15231)[d]	0.500	0.076	200	9.20	2.679	33.24	1.222	154.0	146	28.80	0.090	12.8	0.14	15
Broccoli, cooked (11091)	0.000	0.000	0.4	0.67	0.061	0.45	0.194	1.6	77	0.00	0.200	64.9	0	108
Raspberries (35202)	0.000	0.000	0.5	0.64	0.097	0.47	0.368	0.2	2	0.00	0.104	26.4	0	5

[a]Entries retrieved from the USDA National Nutrient Database for Standard Reference, Release 23 (2010) and are identified by a five-digit nutrient database number in parentheses

[b]Data from the Australian Food, Supplement and Nutrient Database (AUSNUT) 2007 (Available at: http://www.foodstandards.gov.au/)

[c]Two billion people, over 30% of the world's population, are anemic, mainly because of iron deficiency (World Health Organization. Micronutrient deficiencies: Iron deficiency anemia. Available at: http://www.who.int/nutrition/topics/ida/en/print.html; 2009)

[d]One hundred grams provide 100% or more of the adult RDA for iodine (Institute of Medicine. Dietary reference intakes for vitamin A, vitamin K, boron, chromium, copper, iodine, iron, manganese, molybdenum, nickel, silicon, vanadium, and zinc. Food and Nutrition Board. Washington, DC: National Academy Press; 2001)

Chocolate may appear to be high in fiber, but it is not when its energy density is reduced to a level similar to that of a typical fruit or vegetable (see Table 12.1). The energetic cost of the assimilation of chocolate can be increased by adding more protein and fiber to it [9]. Protein has more than three times the thermic effect of either fat or carbohydrate [20], and because it has a greater satiety value than fat or carbohydrate [20, 21], a high-protein diet (protein and carbohydrate intake both being approximately one-third of total energy intake) is of vital importance as a weight-loss strategy for the overweight or obese and for weight maintenance [22, 23]. Clinical trials have shown that calorie-restricted, high-protein diets are more effective than are calorie-restricted, high-carbohydrate diets in promoting [24–26] and maintaining [27] weight loss in overweight subjects, while producing less hunger and more satisfaction [28]. Furthermore, high-protein diets have been shown to improve metabolic control in patients with type 2 diabetes [29–31]. Some protein-based nanotubes are food-grade materials [32] and can increase protein consumption at the expense of lowered carbohydrate. Chocolate bars must simultaneously satisfy the human "sweet tooth" and almost completely remove added sugars such as glucose, fructose and sucrose from the diet [3]. Cocoa can be sweetened by the addition of calorie-free Purefruit™ (Tate & Lyle) monk fruit (Siraitia grosvenorii) extract [14]. PUREFRUIT™ is approximately 200 times sweeter than sugar and has exceptional stability.

The modern diet is too acidic [33], and chocolate is an acid-producing food. The known health benefits of a net base-yielding diet include preventing and treating osteoporosis [34, 35], age-related muscle wasting [36], calcium kidney stones [37, 38], hypertension [39, 40], exercise-induced asthma [41], and the progression of age- and disease-related chronic renal insufficiency [42]. Adding sufficient amounts of potassium bicarbonate to cocoa and chocolate will make them net base-yielding foods. The nutrition facts labels for Good Cacao™ "Thc World's Finest Superfood Chocolate®" (1) lemon ginger and (2) coconut chocolate bars show that they contain no vitamin A or vitamin C [43]. Vitamin A is particularly important during periods of rapid growth, both during pregnancy and in early childhood. Vitamin A derivatives, retinoids, control the differentiation of neurones, and a role has been suggested in memory, sleep, depression, Parkinson's disease, and Alzheimer's disease [44]. Furthermore, vitamin A plays a critical role in visual perception, and a deficiency causes blindness [45]. Vitamin C deficiency causes scurvy. Chocolate contains practically no folate (see Table 12.1). Folate deficiency causes neural tube defects [46], which produce malformations of the spine, skull, and brain and neurological disorders, such as depression and cognitive impairment [47]. It may be misleading to describe any current chocolate product as a superfood when its nutritional value is compared to that of an oyster (see Tables 12.1 and 12.2).

The bioavailable nutrient content including cofactors of chocolate bars should be based on the nutritional value of the most nutritious cooked wild foods for humans and can be increased using existing bioactive encapsulation [9]. Aquatic biotechnology can provide chocolate manufactures with sufficient amounts of all the nutrients needed to make chocolate really nutritious, including protein, docosahexaenoic acid (DHA), eicosapentaenoic acid (EPA), arachidonic acid (AA), vitamins, minerals, and fiber [48–52]. Reducing particle size using nanotechnology can further improve the properties of bioactive compounds (e.g., DHA and EPA), such as delivery, solubility, prolonged residence time in the gastrointestinal tract, and efficient absorption through cells [53].

Summary

The nutritional value of chocolate bars should be based on human nutritional requirements to help reduce mental ill health, obesity, and other postprandial insults [7–9]. In the near future, food technology could allow people to get their chocolate fix, while simultaneously and significantly increasing nutrient intake and reducing energy intake per day. Nanocellulose and calorie-free monk fruit extract could be used to lower the energy density of chocolate, and aquatic biotechnology can provide the nutrients needed to make it really nutritious.

References

1. Rimbach G, Egert S, de Pascual-Teresa S. Chocolate:(un) healthy source of polyphenols? Genes Nutr. 2011;6:1–3.
2. Ledikwe JH, Blanck HM, Kettel Khan L, et al. Dietary energy density is associated with energy intake and weight status in US adults. Am J Clin Nutr. 2006;83(6):1362–8.
3. Eaton SB, Konner MJ, Cordain L. Diet-dependent acid load, paleolithic nutrition, and evolutionary health promotion. Am J Clin Nutr. 2010;91(2):295–7.
4. O'Rahilly S. Human genetics illuminates the paths to metabolic disease. Nature. 2009;462(7271):307–14.
5. Crews WD, Harrison DW, Wright JW. A double-blind, placebo-controlled, randomized trial of the effects of dark chocolate and cocoa on variables associated with neuropsychological functioning and cardiovascular health: clinical findings from a sample of healthy, cognitively intact older adults. Am J Clin Nutr. 2008;87(4):872–80.
6. Desch S, Kobler D, Schmidt J, et al. Low vs. higher-dose dark chocolate and blood pressure in cardiovascular high-risk patients. Am J Hypertens. 2010;23(6):694–700.
7. Robson AA. Preventing the epidemic of diet induced disease: an overview. In: Bioactive foods in chronic disease states (Vol. Bioactive food as dietary interventions for liver and gastrointestinal disease). San Diego: Academic Press; 2013 (in press); edited by Victor Preedy, King's College London, School of Biomedical and Health Sciences, London.
8. Robson AA. Preventing the epidemic of mental ill health: an overview. In: Bioactive foods in chronic disease states (Vol. Bioactive food as dietary interventions for the aging population). San Diego: Academic Press; 2013 (in press); edited by Victor Preedy, King's College London, School of Biomedical and Health Sciences, London.
9. Robson AA. Food nanotechnology: water is the key to lowering the energy density of processed foods. Nutr Health. 2011;20(3 and 4):231–6.
10. Robson A. Shellfish view of omega-3 and sustainable fisheries. Nature. 2006;444(7122):1002.
11. Marean CW, Bar-Matthews M, Bernatchez J, et al. Early human use of marine resources and pigment in South Africa during the Middle Pleistocene. Nature. 2007;449:905–8.
12. Wang Y, Lehane C, Ghebremeskel K, et al. Modern organic and broiler chickens sold for human consumption provide more energy from fat than protein. Public Health Nutr. 2010;13:400–8.
13. Tschofen G. 'Volcano': aerated, crunchy chocolate that melts in your mouth, not in your hands. Factsheet: Volcano chocolate, p. 1–3. Available at: http://www.barry-callebaut.com/84?release=5256. Accessed 16 Jan 2011.
14. Robson, AA. Tackling obesity: can food processing be a solution rather than a problem? Agro Food Industry Hi-Tech. 2012;23(2 supplement):10–11.
15. Norton J, Fryer P, Parkinson J, Cox P. Development and characterisation of tempered cocoa butter emulsions containing up to 60% water. Journal of Food Engineering. 2009;95(1):172–8.
16. Klemm D, Schumann D, Kramer F, Heßler N, Hornung M, Schmauder H-P, et al. (Eds.). *Nanocelluloses as innovative polymers in research and application. Polysaccharides II* (Vol. 205): Springer Berlin/Heidelberg. 2006.
17. Klemm D, Kramer F, Moritz S, Lindström T, Ankerfors M, Gray D, et al. Nanocelluloses: a new family of nature-based materials. *Angewandte Chemie International Edition*. 2011;50(24):5438–66.
18. Chang S-T, Chen L-C, Lin S-B, Chen H-H. Nano-biomaterials application: Morphology and physical properties of bacterial cellulose/gelatin composites via crosslinking. Food Hydrocolloids. 2012;27(1):137–44.
19. Neethirajan S, Jayas DS. Nanotechnology for the food and bioprocessing industries. *Food and Bioprocess Technology*. 2011;4(1):39–47.
20. Crovetti R, Porrini M, Santangelo A, et al. The influence of thermic effect of food on satiety. Eur J Clin Nutr. 1998; 52(7):482–8.
21. Stubbs RJ. Appetite, feeding behaviour and energy balance in human subjects. Proc Nutr Soc. 1998;57(3): 341–56.
22. Veldhorst M, Smeets A, Soenen S, et al. Protein-induced satiety: effects and mechanisms of different proteins. Physiol Behav. 2008;94(2):300–7.
23. Robson AA. Preventing diet induced disease: bioavailable nutrient-rich, low-energy-dense diets. Nutr Health. 2009;20(2):135–66.
24. Baba NH, Sawaya S, Torbay N, et al. High protein vs high carbohydrate hypoenergetic diet for the treatment of obese hyperinsulinemic subjects. Int J Obes. 1999;23(11):1202–6.
25. Skov AR, Toubro S, Ronn B, et al. Randomized trial on protein vs carbohydrate in ad libitum fat reduced diet for the treatment of obesity. Int J Obes. 1999;23(5):528–36.
26. Layman DK. The role of leucine in weight loss diets and glucose homeostasis. J Nutr. 2003;133(1):261S–7.
27. Westerterp-Plantenga MS, Lejeune M, Nijs I, et al. High protein intake sustains weight maintenance after body weight loss in humans. Int J Obes. 2004;28:57–64.
28. Johnston CS, Tjonn SL, Swan PD. High-protein, low-fat diets are effective for weight loss and favorably alter biomarkers in healthy adults. J Nutr. 2004;134(3):586–91.

29. Seino Y, Seino S, Ikeda M, et al. Beneficial-effects of high protein-diet in treatment of mild diabetes. Hum Nutr Appl Nutr. 1983;37(3):226–30.
30. Odea K. Marked improvement in carbohydrate and lipid-metabolism in diabetic Australian aborigines after temporary reversion to traditional lifestyle. Diabetes. 1984;33(6):596–603.
31. Odea K, Traianedes K, Ireland P, et al. The effects of diet differing in fat, carbohydrate, and fiber on carbohydrate and lipid-metabolism in type-II diabetes. J Am Diet Assoc. 1989;89(8):1076–86.
32. Graveland-Bikker JF, De Kruif CG. Unique milk protein based nanotubes: food and nanotechnology meet. Trends Food Sci Technol. 2006;17(5):196–203.
33. Cordain L, Eaton SB, Sebastian A, et al. Origins and evolution of the Western diet: health implications for the 21st century. Am J Clin Nutr. 2005;81(2):341–54.
34. Bushinsky DA. Metabolic alkalosis decreases bone calcium efflux by suppressing osteoclasts and stimulating osteoblasts. Am J Physiol Renal Physiol. 1996;271(1):F216–22.
35. Sebastian A, Harris ST, Ottaway JH, et al. Improved mineral balance and skeletal metabolism in postmenopausal women treated with potassium bicarbonate. N Engl J Med. 1994;330(25):1776–81.
36. Frassetto L, Morris RC, Sebastian A. Potassium bicarbonate reduces urinary nitrogen excretion in postmenopausal women. J Clin Endocrinol Metab. 1997;82(1):254–9.
37. Pak CYC, Fuller C, Sakhaee K, et al. Long-term treatment of calcium nephrolithiasis with potassium citrate. J Urol. 1985;134(1):11–9.
38. Preminger GM, Sakhaee K, Skurla C, et al. Prevention of recurrent calcium stone formation with potassium citrate therapy in patients with distal renal tubular-acidosis. J Urol. 1985;134(1):20–3.
39. Sharma AM, Kribben A, Schattenfroh S, et al. Salt sensitivity in humans is associated with abnormal acid–base regulation. Hypertension. 1990;16(4):407–13.
40. Morris RC, Sebastian A, Forman A, et al. Normotensive salt sensitivity – effects of race and dietary potassium. Hypertension. 1999;33(1):18–23.
41. Mickleborough TD, Gotshall RW, Kluka EM, et al. Dietary chloride as a possible determinant of the severity of exercise-induced asthma. Eur J Appl Physiol. 2001;85(5):450–6.
42. Alpern RJ, Sakhaee K. The clinical spectrum of chronic metabolic acidosis: Homeostatic mechanisms produce significant morbidity. Am J Kidney Dis. 1997;29(2):291–302.
43. Good Cacao. Retail price schedule. Nutrition facts. Available at: http://www.goodcacao.com/pdf/GOOD_CACAO.pdf. Accessed 14 July 2011.
44. Tafti M, Ghyselinck NB. Functional implication of the vitamin A signaling pathway in the brain. Arch Neurol. 2007;64(12):1706–11.
45. Benton D. Micronutrient status, cognition and behavioral problems in childhood. Eur J Nutr. 2008;47 (Suppl. 3):38–50.
46. Shaw GM, Schaffer D, Velie EM, et al. Periconceptional vitamin use, dietary folate, and the occurrence of neural tube defects. Epidemiology. 1995;6(3):219–26.
47. Gomez-Pinilla F. Brain foods: the effects of nutrients on brain function. Nat Rev Neurosci. 2008;9(7):568–78.
48. Harun R, Singh M, Forde GM, et al. Bioprocess engineering of microalgae to produce a variety of consumer products. Renew Sustain Energy Rev. 2010;14(3):1037–47.
49. Ortiz J, Romero N, Robert P, et al. Dietary fiber, amino acid, fatty acid and tocopherol contents of the edible seaweeds Ulva lactuca and Durvillaea antarctica. Food Chem. 2006;99(1):98–104.
50. Clarke GL. Plankton as a food source for man. Science. 1939;89(2322):602–3.
51. Clarke GL, Bishop DW. The nutritional value of marine zooplankton with a consideration of its use as an emergency food. Ecology. 1948;29(1):54–71.
52. Liu L, Liu CC, Li JL. Comparison of biochemical composition and nutritional value of Antarctic krill (*Euphausia Superb*) with several species of shrimps. Adv Mater Res. 2012;361–363:799–803.
53. Chen L, Remondetto GE, Subirade M. Food protein-based materials as nutraceutical delivery systems. Trends Food Sci Technol. 2006;17(5):272–83.

Chapter 13
Comparing Sunflower Stearins with Cocoa Butter

Joaquín J. Salas, Miguel A. Bootello, Enrique Martínez-Force, and Rafael Garcés

Key Points

- High-oleic high-stearic sunflower mutant is one of the new commercial varieties of this oil crop that have been developed by breeding and mutagenesis from common sunflower lines. The oil extracted from this line contains high levels of stearic acid in a high-oleic background and could be a source of stearate-rich fats in the future.
- Dry fractionation is a process by which the triacylglycerols with a high degree of saturation of an oil or fat are selectively crystallized and separated from the liquid matrix without the addition of any solvent to yield a solid fraction enriched in saturated fatty acids and a liquid fraction richer in unsaturated fatty acids. This method is used for palm oil fractionation.
- Solvent fractionation involves the addition of solvent to oil as a step previous to its fractionation. The resulting micelle is fractionated by crystallization of the species of triacylglycerols with higher levels of saturated fatty acids. The resulting solid phase is separated from the liquid one and washed with fresh solvent. This method is more efficient than dry fractionation, although it requires higher operation costs. It is often used to obtain highly valuable stearate-rich butters from tropical fats.
- Stearin is the name given to the solid fraction obtained in the fractionation of an oil or fat. It contains higher levels of saturated fatty acids than the initial oil. Its characteristics and composition will depend on the composition of the initial oil and the method and conditions of fractionation.
- A cocoa butter equivalent is a vegetable fat or a mixture of vegetable fats with similar physical and chemical properties to those of cocoa butter. It has to be compatible in mixtures with cocoa butter and display similar polymorphism. Cocoa butter equivalents are produced by mixing stearate-rich tropical butters and fractions resulting from palm oil fractionation.

Keywords High-stearic–high-oleic sunflower • Dry fractionation • Solvent fractionation • Stearin • Cocoa butter equivalent

---

J.J. Salas, Ph.D., Chemistry (✉) • M.A. Bootello • E. Martínez-Force, Ph.D., in Biological Science
R. Garcés, Ph.D., Biology
Physiology and Technology of Plant Products, Instituto de la Grasa,
Av. Padre García Tejero, 4, Sevilla, Sevilla 41012, Spain
e-mail: jjsalas@cica.es; bootello@cica.es; emforce@cica.es; rgarces@cica.es

R.R. Watson et al. (eds.), *Chocolate in Health and Nutrition*, Nutrition and Health 7,
DOI 10.1007/978-1-61779-803-0_13,

Introduction

Cocoa butter (CB) is the fat accumulated in the seeds from the *Theobroma cacao* tree. This fat possesses a unique triacylglycerol (TAG) composition, with the disaturated species 1,3-dipalmitoyl-2-oleoyl glycerol (POP), 1-palmitoyl-3-stearoyl-2-oleoyl glycerol (POSt), and 1,3-distearoyl-2-oleoyl glycerol (StOSt) predominant (Table 13.1), accounting for about 80–85% of the total TAGs of this fat [1]. There are slight differences in the composition of the CB depending on the country of origin. Cocoa butters from South America (or soft CBs) have a higher content of unsaturated fatty acids (oleic and linoleic) and palmitic acid, at the expense of the concentration of stearic and arachidic acid (Table 13.2). Cocoa butters from Ivory Coast and Malaysia (hard CBs) are richer in stearic acid. This composition of fatty acids is reflected in their TAG composition (see Table 13.1). This TAG composition is responsible for the characteristic melting profile of this fat, a process that takes place between 35°C and 38°C in the form of a steep step [2]. This melting temperature interval matches the human body temperature, which makes CB-based products, such as chocolate, melt quickly in the mouth, releasing aroma and flavor and causing a refreshing sensation [3]. The TAG composition of CB involves that this butter contain a high level of saturated fatty acids (up to 55%). However, unlike other saturated rich fats, CB did not cause negative effects on cardiovascular health [4]. The reason for this is the high level of stearic acid, the only saturated fatty acid that does not increase the levels of blood cholesterol, and the distribution of the saturated fatty acids that are esterified to the *sn*-1 and *sn*-3 positions of the glycerol backbone almost in their totality, minimizing their absorption rates in the small intestine. Moreover, this fat contains high levels of oleic acid in the *sn*-2 position of the TAG molecules. These facts, in addition to the positive effects of the antioxidants present in cocoa powder [5, 6], have led to black chocolate being considered a healthy food from the point of view of cardiovascular disease and cancer prevention.

Among the minor components of the CB, polar lipids play an important role in the crystallization of this fat. Phospholipids account for 0.8–0.9 wt.% of the CB [7]. Arruda and Dimick [8] associated the nucleation rate of the CB with the polarity of the phospholipids. Thus, polar phospholipids are related to slow-nucleating CBs, whereas rapid-nucleating CBs are rich in phosphatidylcholine and phosphatidylethanolamine, which provide nuclei in the early stages of crystallization. Furthermore, polar lipids delay the (alpha) α-to-β′ (beta) polymorphic transition and affect the kinetic of TAG crystallization of CB [9].

The high popularity of chocolate in developed countries, its perception as a healthy food, and its increasing consumption in developing countries have boosted world demand for CB in the first decade of the twenty-first century. However, the production of CB is constrained by the difficulties inherent to cocoa tree cultivation. Thus, this species can only be grown in tropical areas with a specific rainfall and altitude. Moreover, the cocoa tree is less productive than other oil crops (500 kg beans/Ha approx.), and production is often affected by pest attacks and climatology [3]. These limitations in the production usually result in market tensions and seriously compromise its supply in the future. In this regard, alternative fats with characteristics similar to CB for the production of confectionary products have been sought by food scientists [10]. Since there is no other species with a TAG composition similar to CB, the alternatives have been often selected as a function of their melting profiles. A group of fats used in confectionary as an alternative to CB have been the cocoa butter substitutes (CBSs), which are based in lauric fats [11]. These CBSs are prepared from coconut and palm kernel fats rich in laurate and myristate. These fats are fractionated to yield stearins richer in saturated fatty acids or treated by hydrogenation and transesterification to produce fats displaying melting profiles similar to those from CB. CBSs have been broadly used in confectionary formulations. However, they display low compatibility with CB owing to the blends of both fats, which give rise to eutectics that resulted in important decreases of the melting point [10]. The use of CBSs in confectionary also has the inconvenience of the negative effects of these fats on human health, caused by its high levels of lauric and myristic fatty acids, which have been reported to increase the levels of blood cholesterol and the incidence of cardiovascular disease [12]. Similar physical properties and better compatibility with CB present the so-called cocoa

Table 13.1 Triacylglycerol composition of cocoa butter from different countries

	Triacylglycerols (%, w/w)[a]																	
Cocoa butter	PPP	PStP	POP	PLP	PStSt	POSt	POO	PLSt	POL	StStSt	StOSt	StOO	StLSt	OOO	StOL	AOSt	OOA	SUS[b]
Ivory Coast	0.1	0.5	14.0	1.2	0.9	39.2	1.7	3.1	0.3	0.6	30.8	2.6	2.0	0.3	0.5	1.8	0.3	92.0
Malaysia	0.2	0.6	13.8	1.1	0.9	39.0	1.8	2.7	0.4	0.5	31.1	3.1	1.8	0.3	0.4	1.8	0.4	91.2
Ecuador	0.2	0.5	16.7	1.8	0.7	38.6	3.2	3.7	0.4	0.3	25.0	4.1	2.0	0.7	0.7	1.3	0.3	88.9
Cuba	0.3	0.4	15.1	1.5	0.5	38.7	3.1	3.4	0.3	0.5	26.7	4.3	2.0	0.5	0.8	1.5	0.5	88.8

P palmitic; *O* oleic; *St* stearic; *L* linoleic; *A* arachidic

[a]Triacylglycerols were named with three letters. The first and the third corresponded to fatty acids esterified in positions 1 and 3; the second, to the fatty acid esterified in position 2. Peaks accounting for less than 0.1% of total triacylglycerols were not integrated

[b]Represents total content of disaturated triacylglycerols

Table 13.2 Fatty acid composition of cocoa butter from different countries

	Fatty acids (%, w/w)[a]				
Cocoa butter	16:0	18:0	18:1	18:2	20:0
Ivory Coast	25.5	37.8	32.7	2.7	1.3
Malaysia	24.5	38.4	33.2	2.5	1.4
Ecuador	27.9	33.7	34.1	3.4	1.0
Cuba	26.2	35.0	34.4	3.3	1.1

[a]Trace fatty acids (less than 0.2%) were not integrated. 16:0 palmitic acid; 18:0 stearic acid; 18:1 oleic acid; 18:2 linoleic acid; 20:0 arachidic acid

butter replacers (CBRs), which are butters obtained by hydrogenation of common vegetable oils. CBRs represent an accessible source of high melting point confectionary fats. However, as happens with other hydrogenated fats, they contain trans-fatty acids, which have been demonstrated to raise the levels of LDH cholesterol at the expense of the HDH form, promoting arteriosclerosis [13]. The negative impact of trans-fatty acids on human health is now being considered even by food legislation. Currently, the only known healthy alternatives to CB are the fats defined as cocoa butter equivalents (CBEs). These are defined as non-lauric fats with a similar melting profile, composition, and polymorphisms as CB, which should be compatible with CB without presenting any eutectic point [10]. These fats are usually prepared by blending tropical fats or fat fractions rich in stearic acid and palm mid fractions. Since there is not a natural source of POS alternative to CB, these blends cannot exactly match CB composition, but they displayed a similar melting profile. The European legislation allows the addition of CBE to regular chocolate to a limit of the 5% of the total fat of the formulation. The tropical fats used in these formulations come from exotic species like shea, kokum, or illipe, which are wild trees growing in the rainforest and involve hand harvest and artisanal processing, so its production is often neither reliable nor predictable and is subjected to incidences and variations. In this regard, it would be of interest having a trusted source of natural stearate-rich butters to allow the production of healthy confectionary products. Sunflower is a well-established annual oil crop that can be cultivated in large areas within a temperate climate. This plant has been demonstrated to have great plasticity in its fatty acid metabolism. Thus, it was possible to produce new lines with altered fatty acid composition by techniques of breeding and mutagenesis, which did not involve any transgenesis and so avoid any discussion on the proliferation of genetically modified organisms in agriculture [14]. Among the different phenotypes of the sunflower lines obtained, the lines displaying high content of saturated fatty acids in high-oleic background are especially remarkable. In this regard, the high-oleic–high-stearic (HOHS) sunflower mutant is of special interest [15]. This was obtained by transferring the high-stearic phenotype obtained by chemical mutagenesis of common sunflower to high-oleic sunflower backgrounds. These lines displayed contents of stearic acid ranging between 15% and 25%, which contrasted with the common sunflower, which contains no more than 5% of this fatty acid. The oleic acid content of this mutant is also high, ranging from 60% to 70% of total fatty acids, and it also displays very low amounts of polyunsaturated fatty acids (from 2% to 6%). Therefore, HOHS sunflower could be a reliable source of stearic acid–rich fats for confectionary. However, with reference to the TAG composition of this oil, its content of disaturated TAGs is far from that necessary for the production of CBEs. So the melting curve of these oils displayed high contents of solids than common and high-oleic sunflower oils but far less than CB or shea stearins. So, the use of this oil as a source of stearate-rich fat would require the previous concentration of disaturated TAGs by means of fat fractionation techniques.

Oil Fractionation Technology

Oil or fat fractionation is a process in which the TAGs of the initial fat will be separated as a function of their melting point. It involves the crystallization of the TAGs with a higher melting point followed by a step of separation. The solid crystallized fat is called stearin, whereas the resulting liquid fraction is called olein. This method allows low-cost production of fats with specific properties and applications and avoids any chemical transformation. Therefore, it is widely used, mainly by the palm oil industry, to produce palm stearins, mid fractions, and oleins, whose use is quite extensive within the food industry. Two main techniques are used to fractionate oils: dry or solvent fractionation [16]. Dry fractionation involves the crystallization of TAGs with the higher degree of saturated fatty acids by cooling down the melted fat without the addition of any solvent. The filtration step is usually fulfilled at medium or high pressures to remove most of the olein remaining in the solid phase [17]. The solvent fractionation process requires the addition of an organic solvent, such as hexane or acetone, and the crystallization of the TAGs with higher levels of saturation within the resulting micelle [18]. The resulting solids are vacuum-filtered. Dry fractionation is less efficient, and often several steps are required to reach the desired results, whereas solvent fractionation is more efficient, producing high enrichment of saturated TAG species in a single step. However, the solvent management increases the cost of the solvent fractionation process, so it is usually applied for the production of valuable fats and butters, as is the case of shea stearin used for preparation of CBE blends and confectionary products. The HOHS sunflower oil has been fractionated by both dry and solvent procedures, yielding fractions of different degree of saturation in function of the conditions of the process.

Dry Fractionation of High-Oleic High-Stearic Sunflower Oil

These oils can be fractionated without the addition of any solvent, as was depicted by Bootello et al. [19]. The influence of temperature and seeding on the kinetics of the process and the final composition of the fractions were carefully studied in that work. Moreover, the crystallization kinetics of these fats was studied applying the mathematical models of Gompertz [20] and Avrami [21]. Crystallization is the determinant step in fat fractionation. This involves a first step of nucleation, which is the process, whereby the first stable crystals are formed in the overcooled liquid matrix of the melted fat [22]. It can be either homogeneous, if the nuclei are spontaneously formed by direct interaction of the TAGs with a higher melting point, or heterogeneous, when the process takes place on the surface of particles or on the walls of the container. The second step within crystallization is crystal growth that takes place when the most saturated TAGs are incorporated in the formed nuclei, increasing the size of the crystals until their concentration reaches the value of their solubility in the liquid matrix at the crystallization temperature [18]. For an optimum fractionation, the formation of stable crystals of a big size is required. This would allow an easier step of filtration and a better compression of the filtered cake to remove the trapped liquid phase or olein. Oils and fats can crystallize in several forms. The most unstable crystals are those of (alpha)α form, which occurs when the fat is cooled down quickly. The more stable forms (beta)β' and (beta)β are much more adequate for fractionation [23], so they should be promoted by applying slower cooling rates and appropriate seeding.

HOHS sunflower oil can only be fractionated within a short range of temperatures, ranging from 17°C to 19°C [19]. The effect of the addition of seeding crystals on the crystallization kinetics of this oil was therefore studied at a temperature of 18°C. The model of Gompertz parameterized the kinetics of crystallization by fitting the curve of crystalline fraction formed to a sigmoidal curve. Therefore, the curve is determined by the parameters (lambda)λ, which is related with the lag time or time necessary for the crystallization to start; (mu)μ, which is the maximum crystallization rate; and a_c, which is the maximum amount of crystals formed. The equation of Avrami corresponded to a logarithmic fitting that give notice of the mechanism of crystallization.

When crystallization is fulfilled without the addition of crystal seeding, there were longer lag times and higher maximum crystallization rates. The specificity of the crystallization was generally poorer, yielding softer stearins with lower contents of saturated fatty acids and disaturated TAGs. The addition of high melting point stearin powder as a seeding agent induced a reduction in the crystallization lag time but a decrease of the maximum crystallization rate, so it did not change the total crystallization time. Moreover, the parameterization of the crystallization curve through the Avrami equation allowed discovery of a change in the crystallization mechanism from spontaneous nucleation with spherical growth to instantaneous nucleation with spherical growth at seeding additions of 0.25% of stearin or higher. With regard to the final composition of the resulting stearins, the addition of seeding crystals induced a higher enrichment of saturated fatty acids in the final fractions, which ranged around 30% of disaturated TAGs, with yields that varied from 10% to 12%. Therefore, the addition of seeding crystals improved the specificity of the crystallization of HOHS oil. Furthermore, studies on the influence of temperature on the crystallization of this oil indicated that temperatures in the lower range of the narrow interval in which this oil can be fractionated (17.5°C) accelerated the process of crystallization, whereas higher ones retarded the process. Temperature affected mainly to maximum crystallization rates and not the lag time of crystallization. Lower fractionation temperatures produced higher yields of stearin with lower levels of saturated fatty acids. On the other hand, stearins obtained at 19°C contained higher amounts of disaturated TAGs.

Characteristics of HOHS Sunflower Stearins Obtained by Dry Fractionation

Dry fractionation produced stearin fractions enriched in saturated fatty acids from HOHS sunflower oils (Tables 13.3 and 13.4). This enrichment was about threefold for stearic acid and fivefold for longer chained fatty acids like arachidic or behenic. However, these fractions were far from the content of saturated fatty acids and disaturated TAGs found in CB or CBE formulations. The result was that they usually did not display the content of solid fat required by confectionary fats, which is estimated to be 40% at a temperature of 25°C [24]. This can be observed in Fig. 13.1, where the melting profiles of the original HOHS oil, some stearins from dry fractionation of this oil or sunflower soft stearins (SSS1 and SSS2 from Tables 13.3 and 13.4), a typical CB, and two other tropical fats (mango and shea) are shown. The SSS fractions displayed melting profiles intermediate between the original oil and CB, with lower amounts of solids than shea and mango fat. Since SSSs have less than 50% of solids at 20°C, they can be used as filling fats [25]. Nevertheless, the dry fractionation of this oil was an easy, cheap, and straightforward method to increase the content of disaturated TAGs of this oil. Thus, it could be used as an intermediate step within industrial processing aimed at the production of sunflower stearins with a higher melting point. Moreover, the SSSs obtained by dry fractionation displayed a melting profile appropriate for other confectionary uses, such as plastic fat products, structured lipids, fillings, and shortenings, with the advantage of being a healthy, natural fat rich in stearic and oleic fatty acids.

Solvent Fractionation of HOHS Sunflower Oil

Solvent fractionation involves the crystallization of TAGs with a higher degree of saturation by cooling down a micelle made up with the initial fat dissolved into an organic solvent (usually hexane or acetone). The fat crystals produced in this process are vacuum-filtered. Solvent fractionation is more efficient than dry fractionation and produces a higher enrichment of disaturated TAGs species in a single step [16]. This fractionation method has the additional advantage of inducing β-crystallization of TAGs, which increases selectivity and crystal stability. The main disadvantage of solvent fractionation was the operating costs, which could be tenfold higher than those involved with the dry fractionation processes. Therefore, this technique is usually applied in the production of valuable fractions, such as stearate-rich butters used in CBE or confectionary formulation.

Table 13.3 Triacylglycerol composition of high-oleic–high-stearic sunflower oils, sunflower soft stearins (SSS) obtained by dry fractionation, and sunflower hard stearins (SHS) obtained by solvent fractionation

	Triacylglycerols (%, w/w)[a]																	
Fraction	POP	POSt	POO	POL	PLL	StOSt	StOO	StLSt	OOO	StOL	OOL	OLL	AOSt	OOA	OLA	BOSt	OOB	OLB	SUS[b]
N17	0.3	2.2	7.2	1.0	0.3	4.0	32.5	0.0	31.9	5.2	5.6	0.7	0.7	2.8	0.3	0.7	4.2	0.4	7.9
N20	0.6	4.4	8.7	0.8	0.1	7.5	35.8	0.0	23.5	3.9	3.1	0.4	1.5	3.1	0.3	1.4	4.9	0.0	15.4
SSS1	0.5	4.6	6.6	0.5	0.0	14.2	27.2	0.0	30.7	2.6	3.3	0.7	2.0	2.0	0.0	2.0	3.1	0.0	23.3
SSS2	0.5	5.8	5.7	0.4	0.0	20.9	24.0	0.0	26.4	2.2	3.0	0.4	3.0	1.8	0.0	3.1	2.7	0.0	33.4
SHS1	1.1	12.1	4.8	0.4	0.0	28.0	21.7	0.5	12.1	2.3	1.5	0.1	5.1	1.8	0.0	5.6	2.9	0.0	52.4
SHS2	0.8	11.7	3.5	0.3	0.0	37.7	16.4	0.5	9.2	1.6	1.1	0.0	6.3	1.3	0.0	7.5	2.1	0.0	64.5
SHS3	0.6	10.7	1.2	0.1	0.0	57.5	6.5	0.4	2.9	0.5	0.3	0.0	8.0	0.5	0.0	10.3	0.7	0.0	87.5
SHS4	0.4	9.1	0.8	0.1	0.0	63.2	4.2	0.5	1.8	0.5	0.2	0.0	7.8	0.3	0.0	10.5	0.4	0.0	91.5

P palmitic; *O* oleic; *St* stearic; *L* linoleic; *B* behenic; *A* arachidic

[a] Triacylglycerols were named with three letters. The first and the third corresponded to fatty acids esterified in positions 1 and 3; the second, to the fatty acid esterified in position 2. Peaks accounting for less than 0.1% of total triacylglycerols were not integrated

[b] Represents total content of disaturated triacylglycerols

Table 13.4 Fatty acid composition of high-oleic–high-stearic sunflower oils, sunflower soft stearins (SSS) obtained by dry fractionation, and sunflower hard stearins (SHS) obtained by solvent fractionation

	Fatty acids (%, w/w)[a]					
Fraction	16:0	18:0	18:1	18:2	20:0	22:0
N17	4.1	16.8	72.0	4.4	1.3	1.6
N20	5.4	20.4	66.3	3.3	1.7	2.8
SSS1	4.7	22.9	65.0	3.4	1.6	2.4
SSS2	4.8	27.1	61.1	2.8	1.8	2.5
SHS1	7.1	33.8	52.3	1.8	2.2	2.9
SHS2	6.4	39.0	47.7	1.4	2.3	3.1
SHS3	5.1	50.1	38.2	0.5	2.6	3.6
SHS4	4.3	52.9	36.4	0.4	2.5	3.5

[a] Trace fatty acids (less than 0.2%) were not integrated. 16:0 palmitic acid; 18:0 stearic acid; 18:1 oleic acid; 18:2 linoleic acid; 20:0 arachidic acid

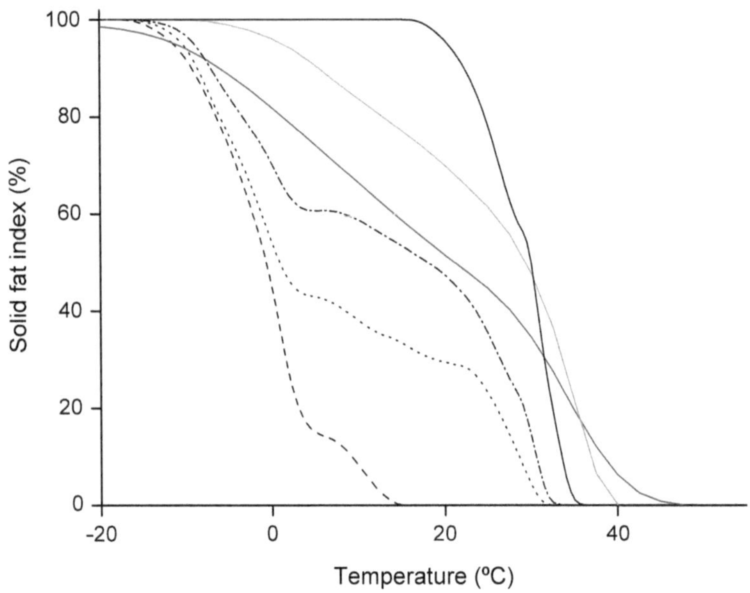

Fig. 13.1 Solid fat indexes calculated by differential scanning calorimetry of initial high-oleic–high-stearic oil (*dashed line*), soft stearins SSS1 (*dotted line*), and SSS2 (*dashed-dotted line*; see Tables 13.3 and 13.4), mango fat (*dark gray line*), shea (*gray line*), and cocoa butter (*solid line*). Data corresponding to stearins and initial oil were taken from reference [19]

Two HOHS oils were tested to obtain sunflower hard stearins (SHS) by fractionation with hexane [26]. These oils were within the possible variation on the content of saturated fatty acids found in these sunflower mutants. One of the oils, named N17, displayed a stearic acid content of 16.8%, and the other, named N20, displayed a stearic acid content of 20% (see Table 13.4). These differences in the content of stearic acid considerably altered the TAG composition of these oils (see Table 13.3), so the N20 oil contained amounts of disaturated TAGs that were almost double those in the N17 oil. The pattern of the solvent fractionation of these oils was investigated at the laboratory scale and was demonstrated to strongly depend on the fractionation temperature and the ratio oil/solvent (O/S ratio) applied. In general, higher proportions of solvent and fractionation temperatures decreased the rate of crystallization of this oil, which also depended on its initial content of stearic acid. Thus, the N20 oil was fractionated in times ranging 20–40 h, whereas the N17 oil generally required longer fractionation times (up to 60 h). Fractionation temperatures and O/S ratios exerted a strong influence on the composition of the precipitates or stearins resulting from fractionation. Thus, higher O/S rations give way to a higher yield of stearins displaying lower contents of saturated fatty acids. In a similar way, lower fractionation temperatures also diminished the specificity of fractionation, producing higher recoveries of stearin. Furthermore, the composition of the fractions obtained from different oils was essentially the same, the only difference being that the yield of stearin was higher when the initial amount of disaturated TAGs was also higher. All these studies indicated that increasing the content of disaturated TAGs induced lower recoveries of the stearin product and that the composition of the final stearin can be increased or decreased by manipulating the O/S ratio and the fractionation temperature. Solvent fractionation induced a much higher enrichment of disaturated TAG species in the stearin than dry fractionation, reaching levels up to 85–90% of these TAGs in a single step of fractionation (see Table 13.3).

Chemical Composition of High Melting Point Sunflower Stearins

The levels of disaturated TAGs species found in hard stearins from HOHS sunflower obtained by solvent fractionation were at the same level as those found in CB. However, the composition of those butters was different than CB. Thus, the levels of palmitic acid in these stearins were lower than those typically found in CB. This was caused by the low palmitic/stearic ratio that is found in the HOHS oils and by the fact that the TAG POP did not precipitate in the conditions in which the oil fractionation took place. The levels of POSt were higher than those of the POP, although they were still far from those found in CB. Sunflower hard stearins displayed high contents in stearic acid (see Table 13.4), and their most abundant TAG was StOSt, which ranged from 30% to 63% in the stearins produced (see Table 13.3), so these fats were similar in composition to tropical butters used for CBE formulation, which are fats or fractions of fats rich in stearic acid, as in the case of shea stearin, which is prepared by solvent fractionation from shea butter. Another factor that distinguishes these new hard stearins from sunflower is the relatively high content of very-long-chain arachidic (20:0) and behenic (22:0) fatty acids. These fatty acids are produced by extraplastidial elongation of stearic acid [27] and are present in common sunflower in small amounts (around 1% the sum of both). The content of very-long-chained fatty acids increased in the HOHS line, reaching levels of between 3% and 5% (see Table 13.4). These fatty acids have a high melting point, so the TAGs carrying these moieties, mainly 1-behenoyl-3-stearoyl-2-oleoyl glycerol (BOSt) and 1-arachidoyl-3-stearoyl-2-oleoyl glycerol (AOSt), tended to crystallize quickly in the conditions of the fractionation, so they are concentrated in the resulting stearins to reach levels of 10.5% in the case of BOSt and 8% for AOSt in the stearins displaying the higher level of saturated fatty acids (see Table 13.3). These high levels of very-long-chained fatty acids are not present in CB or in the stearins from tropical species like shea or mango and could confer to them different properties. Thus, TAGs carrying these fatty acids are used to improve the properties of confectionary formulations because they prevent blooming, an effect caused by the loss of the template or crystalline form of the fat in chocolates caused by high temperatures [28]. Blooming typically produces whitish spots that are rejected by consumers. Owing to their high melting point, TAGs containing behenic acid act as nuclei for recrystallization and so help to keep the appropriate crystalline form of the product ((beta) β in the case of chocolate). Moreover, owing to their high melting point, these very-long-chained fatty acids presented a low bioavailability and are poorly absorbed, especially when they are esterified top the external *sn*-1 and *sn*-2 positions of the TAGS, as in the case of sunflower stearins.

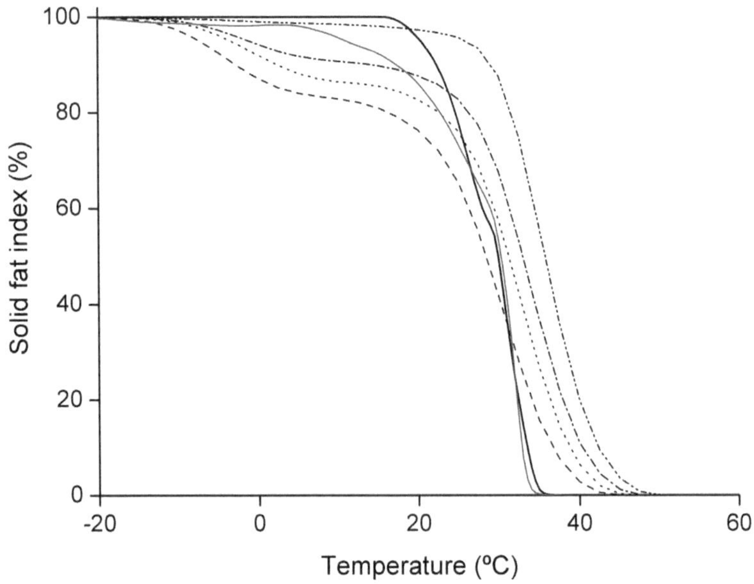

Fig. 13.2 Solid fat indexes calculated by differential scanning calorimetry of hard stearins SHS1 (*dashed line*), SHS2 (*dotted line*), SHS3 (*dashed-dotted line*), SHS4 (*dashed double dotted line*; see Tables 13.3 and 13.4), sunflower cocoa butter equivalent (50% of SHS4 and 50% palm mid fraction; *gray line*), and cocoa butter (*solid line*). Data corresponding to hard stearins were taken from reference [26]

Physical Properties of Sunflower Hard Stearins: Melting Curves and Compatibility with CB

Although the chemical compositions of the stearins yielded by HOHS sunflower were different than those found in CB, their potential uses as a CB alternative and as confectionary fats were studied by analyzing their melting and crystallization profiles [26]. These were carried out by differential scanning calorimetry techniques on samples crystallized to favor the formation of stable (beta)β and β′ crystals. This method involved the continuous integration of the differential scanning calorimetry (DSC) melting peaks, which yielded data of the index of solids in function of the time. The melting curves of different stearins were compared with CB, which typically displayed a sharp melting interval from 29°C to 40°C. The melting profiles of some stearins obtained by solvent fractionation are shown in Fig. 13.2. Stearins with a content of disaturated TAGs lower than 65% of SUS displayed melting profiles similar to those from soft stearins obtained by dry fractionation, which are not appropriate for confectionary fats. Furthermore, stearins between 65% and 75% of disaturated TAGs melting curves are similar to CB, although none of them exactly fit the line corresponding to that curve. Thus, stearins SHS1 and SHS2 displayed lower amounts of solids in the range of lower temperatures (0–30°C) than CB but fit the curve of this fat almost exactly in the higher temperature range (30–45°C). On the other hand, stearins of the type of SHS3 or SHS4 fitted better in the region of lower temperatures but displayed more solid content at higher temperatures. This effect is caused by the different TAG composition of both fats (see Fig. 13.2). It was also remarkable that in all cases the content of disaturated TAGs of these stearins was lower than that usually found in CB, which was caused by the occurrence of TAGs with a higher melting point in these fractions. In this regard, stearins containing up to 80% of disaturated TAGs displayed higher amounts of solids than CB, and similar to those corresponding to high melting point stearins used for CBE formulations, like shea butter stearins. Thus, CBEs are prepared by blending high-stearic butters and palm mid fractions (PMFs) as

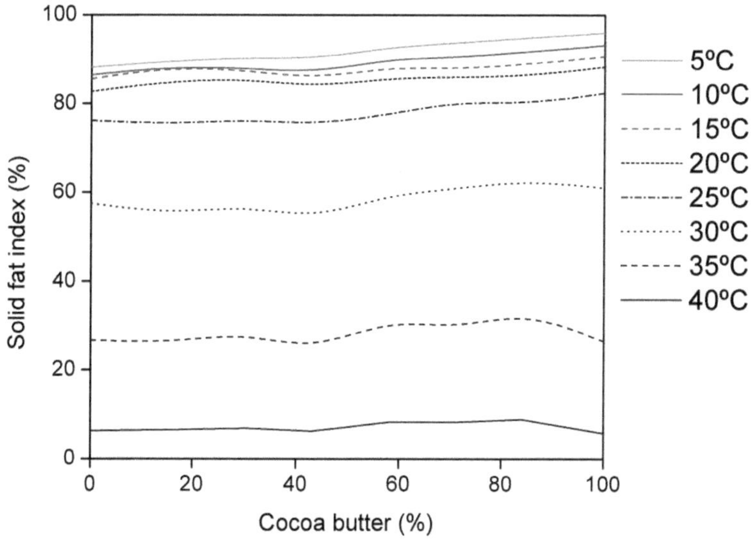

Fig. 13.3 Isothermal diagram obtained from solid fat indexes calculated by differential scanning calorimetry from mixtures of cocoa butter and a hard sunflower stearin at different temperatures. Data were taken from reference [26]

a source of POP, which can be made up with some of the stearins prepared from HOHS sunflower oil stearins. In this regard, Fig. 13.2 presented the melting profile corresponding to a blend of 50% PMF and 50% of a sunflower stearin containing 90% of disaturated TAGs (SHS4) and displaying a melting curve similar to CB, making it adequate for confectionary use. In this regard, Timms et al. [24] set in 1980 an empirical rule to determine whether a fat was adequate for confectionary uses, which was that the fat should display a value of at least 40% solids at 25°C. This criterion was accomplished by all hard stearins having 65% disaturated TAGs or more presented in this work.

Confectionary fats, on the other hand, should be compatible in mixtures with CB to be adequate for CBE formulation. This means that they should not display any eutectic effect when blended with CB, as happens in the case of CBSs and CBRs. The stearin SHS3, which displays a melting profile similar to CB, was used to prepare several mixtures with CB, and the amount of solid at different temperatures was estimated by DSC techniques. Results in Fig. 13.3 showed the parallel lines typical of a compatible blend of fats, with no eutectic effect observed in the whole diagram. So only a slight effect of softening (at high temperature) and hardening (at low temperature) took place with a high proportion of CB in the blend. This result demonstrated that the stearins produced from HOHS sunflower are compatible with CB and appropriate for CBE formulations.

Summary

The main conclusion of the studies on the properties of HOHS sunflower oils and fractions is that new fats rich in stearic acid can be produced from a reliable source through simple physical methods not involving chemical transformations. Depending on the method and conditions of fractionation, it was possible to produce soft stearins with melting profiles that could be appropriate for fillings or bakery use, being rich in oleic and stearic acid. This fat would be a stearic-based natural and healthier alternative to palm oil and fractions.

Moreover, solvent fractionation of this oil produced a single harder stearin with a high content of disaturated TAGs, mainly StOSt, POSt, BOSt, and AOSt, which displayed melting profiles similar to CB. The quality of the final stearin can be controlled in function of the fractionation parameters and included total amounts of disaturated TAGs ranging from 65% to 91%. All were appropriate for use as confectionary fats, and in the case of those containing higher levels of saturated fatty acids, they could be used for CBE formulation, just like other hard stearins obtained from tropical species. This was confirmed by fulfilling studies of compatibility of these stearins with CB, which resulted in the absence of any eutectic in the resulting isothermal diagram. Moreover, the presence of high melting point TAGs containing very-long-chained fatty acids could help to prevent blooming in the confectionary products made up with these fractions. Thus, HSHO sunflower hard stearins are excellent candidates to be a reliable and healthy source of stearate-rich butter for the formulation of CBE and confectionary fats in the immediate future.

References

1. Gunstone FD, Harwood JL, Dijkstra DJ. The lipid handbook. 3rd ed. Boca Raton: CRC Press; 2007.
2. Shukla VKS. Cocoa butter properties and quality. Lipid Technol. 1995;5:54–57.
3. Knapp AW. Cocoa and chocolate: their history from plantation to consumer. Whitefish: Kessinger Publishing LLC; 2007.
4. Cooper KA, Donovan JL, Waterhouse AL, Williamson G. Cocoa and health: a decade of research. Br J Nutr. 2008;99:1–11.
5. Steinberg FM, Bearden MM, Keen CL. Cocoa and chocolate flavonoids: implications for cardiovascular health. J Am Diet Assoc. 2003;103:215–23.
6. Weisburger JH. Chemopreventive effects of cocoa polyphenols on chronic diseases. Exp Biol Med. 2001;226:891–7.
7. Savage CM, Dimick PS. Influence of phospholipids during crystallization of hard and soft cocoa butters. Proc Penn Manuf Confect Assoc. 1994;49:54–9.
8. Arruda DH, Dimick PS. Phospholipid composition of lipid seed crystal isolates from Ivory Coast cocoa butter. J Am Oil Chem Soc. 1991;68:385–90.
9. Toro-Vazquez JF, Rangel-Vargas E, Dibildox-Alvarado E, Charó-Alonso MA. Crystallization of cocoa butter with and without polar lipids evaluated by rheometry, calorimetry and polarized light microscopy. Eur J Lipid Sci Technol. 2005;107:641–55.
10. Smith KW. Cocoa butter and cocoa butter equivalents. In: Gunstone FD, editor. Structured and modified lipids. New York: Marcel Dekker; 2001. p. 401–22.
11. Pease JJ. Confectionery fats from palm oil and lauric oil. JAOCS. 1985;62:426–30.
12. Katan MB, Zock PL, Mensink RP. Dietary oils, serum lipoproteins, and coronary heart disease. Am J Clin Nutr. 1995;61:1368S–73.
13. Ascherio A, Willett WC. Health effects of *trans* fatty acids. Am J Clin Nutr. 1997;66:1006S–10.
14. Hazebroek JP. Analysis of genetically modified oils. Prog Lipid Res. 2000;39:477–506.
15. Fernández-Moya V, Martínez-Force E, Garcés R. Oils from improved high stearic acid sunflower seeds. JAOCS. 2005;53:5326–30.
16. Gibon V. Fractionation of lipids for use in food. In: Gunstone FD, editor. Modifying lipids for use in food. Boca Raton: CRC Press; 2006. p. 201–33.
17. Deffense E. Dry multiple fractionation: trends in products and applications. Lipid Tech. 1995;7:34–8.
18. Timms RE. Fractionation. In: Gunstone FD, editor. Lipid technologies and applications. New York: Marcel Dekker; 1997. p. 199–222.
19. Bootello MA, Garcés R, Martínez-Force E, Salas JJ. Dry fractionation and crystallization kinetics of high-oleic high-stearic sunflower oil. JAOCS.2011;88:1511–9.
20. Zwietering MH, Jongenburger I, Rombouts FM, van't Riet K. Modeling of the bacterial growth curve. Appl Environ Microbiol. 1990;1990(56):1875–81.
21. Chong CL. Crystallization of palm oil products. In: Widlak N, Hartel RW, Narine S, editors. Crystallization and solidification properties of lipids. Champaign: AOCS; 2001. p. 53–78.
22. Metin S, Hartel RW. Crystallization of fats and oils. In: Shahidi F, editor. Bailey's industrial oil and fat products, vol. 6. 6th ed. London: Wiley; 2005.
23. Sato K. Crystallization behavior of fats and lipids—a review. Chem Eng Sci. 2001;56:2255–65.

24. Timms RE. The phase behaviour and polymorphism of milk fat, milk fat fractions and fully hardened milk fat. Aust J Dairy Technol. 1980;35:47–53.
25. Talbot G. Fats for confectionery coatings and fillings. In: Talbot G, editor. Science and technology of enrobed and filled chocolate, confectionery and bakery products. Boca Raton: CRC Press; 2009.
26. Salas JJ, Bootello MA, Martínez-Force E, Garcés R. Production of stearate-rich butters by solvent fractionation of high stearic–high oleic sunflower oil. Food Chem. 2011;124:450–8.
27. Salas JJ, Martínez-Force E, Garcés R. Very long chain fatty acid synthesis in sunflower kernels. J Agric Food Chem. 2005;53:2710–6.
28. Tietz RA, Hartel RW. Effect of minor lipids on crystallization of milk fat—cocoa butter blends and bloom formation in chocolate. JAOCS. 2000;77:763–71.

Chapter 14
Low-Calorie Chocolates and Acceptability/ Sensory Properties

Lauro Melo, Helena Maria André Bolini, and Priscilla Efraim

Key Points

- Introduction to sensory evaluation applied to the development of diabetic and reduced-calorie milk chocolates
- Assessing high-intensity sweetness equivalence in milk chocolates using time-intensity methodology
- Sensory profile, acceptability, and drivers of liking for diabetic and reduced-calorie milk chocolates
- Storage time of diabetic and reduced-calorie milk chocolates studied by sensory properties
- Expectations and acceptability of diabetic and reduced-calorie milk chocolates among nondiabetics and diabetics

Keywords Sensory evaluation • High-intensity sweeteners • Fat replacer • Time intensity • QDA • Consumer testing

Introduction

There is an increased interest in low-calorie foods and beverages [1]. Food cravings are extremely common, particularly among women. Cravings are frequently reported for specific types of foods, including chocolate and foods high in both sugar and fat. Cravings for specific macronutrients, such

L. Melo, Ph.D. (✉)
Department of Biochemical Engineering, School of Chemistry, Federal University of Rio de Janeiro Av. Athos da Silveira Ramos, 149 - Bloco E - Sala E203 - Centro de Tecnologia, Cidade Universitária, Ilha do Fundão, CEP 21941-909, Rio de Janeiro, RJ, Brazil
e-mail: lauro@eq.ufrj.br

H.M.A. Bolini, Ph.D.
Department of Food and Nutrition, School of Food Engineering, State University of Campinas, CP 6121, Campinas, SP 13083-862, Brazil
e-mail: hellini@fea.unicamp.br

P. Efraim, Ph.D. in Food Technology
Department of Food Technology, School of Food Engineering, State University of Campinas, Rua Monteiro Lobato, nº 80, Cidade Universitária, Campinas, SP, CEP 13035-388, Brazil
e-mail: efraim@fea.unicamp.br

R.R. Watson et al. (eds.), *Chocolate in Health and Nutrition*, Nutrition and Health 7,
DOI 10.1007/978-1-61779-803-0_14,

as carbohydrate, have been postulated to result from a physiological need to alter neurotransmitters in such states as eating disorders, affective disorders, or obesity [2]. Among women, the consumption of high-fat sweet products may be a factor in understanding obesity. Furthermore, the observation of high consumption of these foods among obese women is consistent with measured preferences for these high-fat sweet foods [3]. However, larger portion sizes of foods that are low in fat and commercial energy-dense foods and beverages could be important factors in maintaining a high energy intake, causing overconsumption and enhancing the prevalence of obesity in the population. In light of this development, portion size ought to take central place in dietary guidelines and public campaigns [4].

Chocolate is known as a food that promotes well-being and sensory pleasure. Chocolate cravings have not yet been scientifically confirmed, but it is commonly reported for women [5]. Chocolate, despite being a food enjoyed by people of varied ages and countries, in general, contains high levels of fats, especially cocoa butter, and carbohydrates, mainly sucrose. High levels of these nutrients contribute with the high caloric content of the chocolates.

Alternatives to sucrose serve a number of purposes. They are used to expand food and beverage choices for those who must or who want to control caloric, carbohydrate, or sugar intake, assist control of dental caries, enhance the usability of pharmaceuticals and cosmetics, provide sweetness when sugar is not available, and assist the cost-effective use of limited resources. It is important that the use of intense sweeteners instead of sucrose does not cause significant changes in the sensory characteristics of the product [6].

For instance, the salivary habituation may be affected more by the sensory characteristics of the food than by the differences in calorie intake [7]. The women's accurate perception of the sugar content of the chocolate puddings played a primary role in determining their hedonic and other perceptual responses [8]. Sugar and fat have an effect on the texture, mouthfeel, and flavor of milk chocolate [9]. Chocolate sensory attributes, a characteristic flavor, and a unique mouthfeel (attributable to a narrow melting range near body temperature) are extremely appealing [10].

Among several options of high-intensity sweeteners, stevioside is an example of a natural sweetener and sucralose, an artificial one. Stevioside is a natural high-intensity sweetener extracted from the leaves of the plant *Stevia rebaudiana* (Bert.) Bertoni, which is indigenous to Paraguay. The leaves contain a complex mixture of sweet diterpene glycosides, including stevioside, steviolbioside, rebaudiosides A, B, C, D, E, and dulcoside A. The sweetness of "pure" stevioside (about 90%) is quoted as about 300 times that of sugar [11]. Sucralose is produced from ordinary table sugar, sucrose, by a process involving selective chlorination at the 4, 1′, and 6′ positions of the sugar molecule. Sucralose offers a unique combination of features, including a clean, sweet taste and no undesirable off-flavors. It is a versatile ingredient, very stable under the processing and storage conditions used for food products, and it does not present any problem of interaction with other food components. It is odorless, with an intense sweet taste approximately 600 times that of sugar (compared with a 5% sugar solution) [12].

Another important group of sucrose substitutes is the bulking agents, such as polydextrose, lactitol, and maltitol, among others. Polydextrose is made up from glucose and sorbitol and also contains minor residues of citric acid. It has a clean, mildly sweet flavor and can be used in chocolate when combined with other bulk sweeteners – for example, lactitol and an intense sweetener [13]. The FDA has recognized that polydextrose contains less than 1 Kcal/g, compared with sucrose and other carbohydrates that contain 4 Kcal/g [14]. One example of a polyalcohol used for diet chocolates is lactitol, which is obtained by the hydrogenation of lactose. It has a sweetening power of about 40% of that of sucrose and a caloric value of 2.0–2.4 Kcal/g, depending on the specific legislation of each country or union of countries [13].

In addition, it is important to study sweeteners in each food that they could be used in because their sweetness potencies depend on the dispersion matrix where they are inserted. If sucrose in cocoa-containing or any other foodstuff is to be partially or completely replaced by other types of sugar or by sugar substitutes, theoretical models are at present unable to predict the sweetness of the product. These will only provide rough guidelines, and it is always necessary to carry out sensory evaluations on actual samples of the product. In order to obtain the best results, the formulations will normally need adapting, and a series of samples should be manufactured [15]. Despite high lipid and sugar contents, chocolate consumption makes a positive contribution to human nutrition through provision of antioxidants, principally polyphenols including flavonoids such as epicatechin, catechin, and notably the procyanidin [16].

Assessing High-Intensity Sweeteners Equivalence in Milk Chocolates

In the case of chocolate, it is not just the quality of the ingredients (whether determined by location of origin or some other means), but the proportions in which they are mixed and the ways in which they are processed, that determine quality [17].

Perception of aroma, taste, flavor, and texture in foods is a dynamic, not static, phenomenon. In other words, the perceived intensity of the sensory attributes changes during the consumption of a food since start until extinction. In particular to chocolate, the time evolution of the perceived flavor and texture occurs during fat melting and sugar dissolution [18]. It is widely believed that the consumer acceptability of different high-intensity sweeteners depends on the similarity of their time profile to that of sucrose [19]. Single-point sensory response measurement techniques such as magnitude estimation and category scaling only contain a limited amount of information. Time-intensity (T-I) data collection techniques provide much more information, addressing rate-related and duration aspects as well as intensity quantification [20]. Thus, time-intensity method scaling has increased in popularity as an applied sensory evaluation method [21]. Time-intensity data may also better approximate conditions of consumption and be more suitable for examining the functionality of these sweeteners [22]. Thus, the potency of the sweeteners could be predicted from the temporal profiles of several concentrations of both test and reference sweeteners [23].

With all this in mind, Melo et al., [24] using a trained panel, determined equisweetness of the high-intensity sweeteners sucralose and stevioside in diabetic milk chocolates, as sensorially near to the conventional product as possible, mainly with respect to temporal profile. Initially, following the concept that sweetness acceptance is the limiting factor for acceptability of chocolate, [25] consumer testing was applied to determine which sample of conventional chocolates – varying sucrose concentration – was the most accepted. Figure 14.1 shows the results of the hedonic rating for five different

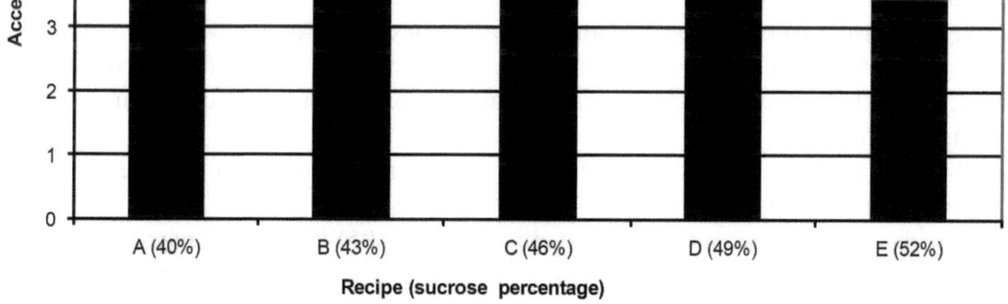

Fig. 14.1 Acceptance of each milk chocolate formulation ($F_{4,116}$ value: 3.64; p value: 0.0079). Same letters indicate that there is no significant difference ($p<0.05$) (Reprinted from De Melo LLMM, Bolini HMA, Efraim P. Equisweet milk chocolates with intense sweeteners using time-intensity method. J Food Qual. 2007;30(6):1056–67. With permission from John Wiley and Sons)

Table 14.1 Means for parameters of T-I curves (Reprinted from De Melo LLMM, Bolini HMA, Efraim P. Equisweet milk chocolates with intense sweeteners using time-intensity method. J Food Qual. 2007;30(6):1056–67. With permission from John Wiley and Sons)

Chocolate	Imax	T-Imax	Area	Ttot
Sucrose	5.86[a]	17.23[a]	104.91[a]	32.26[a]
Sucralose	5.68[a]	17.82[a]	110.07[a]	35.16[a]
Stevioside	4.36[b]	17.49[a]	80.65[b]	31.54[a]
$F_{2,66}$ value	6.04	0.17	4.72	1.58
p value	0.009	0.8417	0.021	0.230

Imax maximum intensity, *T-Imax* time to maximum intensity, *Area* area under the curve, *Ttot* total time of stimuli duration

Note: Same letters in a column indicate that there is no significant difference ($p<0.05$)

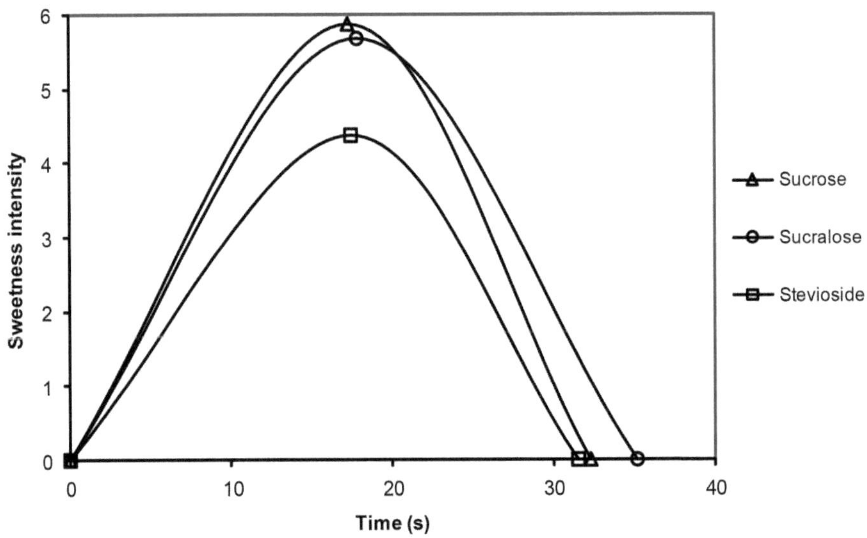

Fig. 14.2 Time-intensity sweetness curves of milk chocolates formulated with different sweeteners (sucrose, sucralose, and stevioside) in equisweet equivalence (Reprinted from De Melo LLMM, Bolini HMA, Efraim P. Equisweet milk chocolates with intense sweeteners using time-intensity method. J Food Qual. 2007;30(6):1056–67. With permission from John Wiley and Sons)

formulations with sucrose. Formulation B (43% of sucrose) had the best acceptance mean for Brazilian consumers; however, it was not statistically different ($p>0.05$) from those of formulations A and C. Also, formulation B was significantly different ($p<0.05$) from formulations D and E, whereas formulations A and C were not. Besides, formulation B was the sample that received the biggest percentage of positive purchase intent (63.3% of "definitely would buy" and "probably would buy" answers). Therefore, the formulation B (43% sucrose milk chocolate) was considered the chocolate with ideal sweetness, and diabetic chocolates with sucralose and stevioside were produced aiming for the same T-I sweetness profile.

Table 14.1 shows parameters of T-I curves (Fig. 14.2) for conventional and diabetic milk chocolates (sucralose and stevioside concentrations that presented best results). Diabetic milk chocolate sweetened with sucralose was not statistically different ($p>0.05$) from the standard (formulation B) for all parameters of curves: maximum intensity (Imax), time to maximum intensity (T-Imax), area under the curve (Area), and total time of stimuli duration (Ttot). The formulation with sucralose that matched formulation B's sweetness was produced considering sucralose 700 times sweeter than sucrose. Chocolate with stevioside presented two parameters of T-I curves not statistically different

($p>0.05$) from the formulation with sucrose (ideal sweetness) – T-Imax and Ttot – although Ttot tended to be longer for the chocolate with stevioside ($p=0.0528$). To achieve these results, stevioside was considered to be 200 times sweeter than sucrose. Results showed that it was possible to produce milk chocolates with sucrose and sucralose very similar between each other with regards to temporal sweetness profile. Chocolates produced with stevioside presented profile not so similar (statistically different in Imax and Area). However, the formulation with stevioside was considered equisweet to the one with sucrose because an additional stevioside concentration increase would probably increase bitter aftertaste, as bitterness is a stevioside characteristic frequently reported in literature. A new alternative has been tested to solve the undesirable aftertaste, with the use of another molecule of the plant *Stevia Rebaudiana* (Bert.) Bertoni, the rebaudioside.

However, it was important to confirm these results by evaluating acceptance (with consumers) and sensory profile (with trained panel) of the diabetic milk chocolates and by checking if they are similar to the acceptability and sensory profile of the product with sucrose.

Sensory Profile and Acceptability for Diabetic and Reduced-Calorie Milk Chocolates

A very effective way of comparing conventional food is to perform sensory evaluations, such as descriptive analysis and consumer affective testing. Descriptive techniques are frequently used in product development to measure how close a new introduction is to the target or to assess suitability of prototype products. In consumer sensory analysis, the investigator is interested in whether the consumer likes the product, prefers it to another product, or finds the product acceptable based on its sensory characteristics [19].

Sensory properties are some of the most important factors on consumer liking and preference; thus, it is very important to determine factors affecting the product attributes, acceptance, and preference, especially for foods and drinks [26]. Understanding what sensory properties drive consumer liking is critical for maximum market share [27]. Consumers would not be interested in consuming a functional beverage if the ingredients caused noticeable off-flavors that consumers found unpleasant, despite the added health advantages [28]. However, conflicting research does exist. A study performed with elderly consumers demonstrated that sensory appeal was less important than health perception and fat content, with regard to the purchase intent of fat-modified foods. Although high-intensity sweeteners are essentially calorie-free, some of these sweeteners impart undesirable flavors and aftertastes, such as bitterness, that can limit their applications in foods and beverages [29].

During the rush of publicity of the new nutritional recommendations in the early 1980s, the first strategy to evolve was simply to remove fat from standard products, such as milk or meat, without any attempt to address the organoleptic changes resulting from the reduction in fat. Such a strategy is not feasible for most other food products (as milk chocolate) because physical stability, functional properties, and microbiological stability may be adversely affected.

Considering this challenge, Melo et al. [30] investigated the influence of sucrose and fat replacement and the correlation between consumer testing results and descriptive sensory data. The major challenge in the development of reduced-fat foods is to achieve fat reduction while matching as closely as possible the flavor and mouthfeel of traditional full-fat products [31]. Whey protein-based fat replacers can mimic fat in terms of texture and flavor retention [32]. Conventional milk chocolate was compared with lab-developed prototypes of diabetic and diabetic/reduced calorie chocolates as well as with a diabetic but not reduced-calorie commercial product (see Table 14.2, for details on samples' formulations).

The lab-developed prototype of conventional milk chocolate was prepared with sucrose (Sucro). Diabetic prototypes were prepared substituting sucrose with high-intensity sweeteners, sucralose

Table 14.2 Formulations of diabetic and reduced-calorie milk chocolates (Reprinted from Melo LLMM, Bolini HMA, Efraim P. Sensory profile, acceptability, and their relationship for diabetic/reduced calorie chocolates. Food Qual Prefer. 2009;20/2:138–43. Copyright 2009, with permission from Elsevier)

	Sample				
	Conventional	Diabetic		Diabetic/reduced calorie	
Ingredient (%)	Sucro	Sucra	Ste	Sucra/WPC	Ste/WPC
Sucrose	43	–	–	–	–
Sucralose	–	0.061	–	0.061	–
Stevioside	–	–	0.22	–	0.22
Polydextrose[a]	–	25.8	25.8	25.8	25.8
Lactitol[a]	–	17.2	17.2	17.2	17.2
Cocoa mass	14	14	14	14	14
Cocoa butter	21.4	21.4	21.4	15.8	15.8
WPC[b]	–	–	–	5.6	5.6
Powdered milk	12	12	12	12	12
Skim powdered milk	9	9	9	9	9
Soy lecithin	0.5	0.5	0.5	0.5	0.5
Vanilla flavor	0.1	0.1	0.1	0.1	0.1

[a]Bulking agents
[b]Whey protein concentrate (partial fat replacer)

(Sucra) or stevioside (Ste), and a polydextrose/lactitol (60/40) blend as a bulking agent. Polydextrose (a fermentable nonstarch polysaccharide fiber) and lactitol (a digestible, fermentable hydrogenated disaccharide sugar alcohol) are synthetic low-digestible carbohydrates [33] and are usually well tolerated but may also have some dose-related undesirable effects owing to their natural osmotic potential and/or excessive fermentation [34]. In practice, the bulking agent must have no negative physical, chemical, sensory, or economical impact on the product – in other words, it should be tasteless, inert, and inexpensive [35]. Polydextrose has been reported as a good option as bulking agent to improve the overall acceptability of low-sugar milk chocolates [36]. Low doses of lactitol can beneficially affect the fecal flora without eliciting gross symptoms of intolerance, and lactitol can be classified as a prebiotic [37].

Both conventional (Sucro) and diabetic (Sucra and Ste) chocolates were formulated to be equisweet at the most acceptable sweetness intensity as determined by the time-intensity methodology. Diabetic/reduced-calorie milk chocolates were prepared through substitution of sucrose in the same way as diabetic chocolates but with partial replacement of cocoa butter with whey protein concentrate (WPC) in order to reduce the calorie content by 25% (Sucra/WPC and Ste/WPC). In theory, reduced-fat products can be highly acceptable, and nutritionally modified food products with good initial acceptance may largely retain their high acceptance under conditions of repeated use over time [38]. A sucrose-free milk chocolate bar formulated with Stevia as a sweetening agent and inulin and polydextrose as bulking agents was successfully developed, but the use of whey proteins for partial fat replacement in sucrose-free chocolate resulted in unacceptable rheological and sensory properties [39].

About 500 volatile compounds have been detected in cocoa, and each may react differently with milk fat and fat replacers [40]. Therefore, a concern with replacing fat with WPC was that an undesirable increase in powdered milk flavors characteristic of milk chocolate could result. In addition, replacing the sugar with various combinations of sugar substitutes in order to produce low-sugar chocolate may have any influence on the rheological properties [41]. Milk ingredients influence consumer liking of milk chocolate through the quality-driving parameters of particle size/sandiness, viscosity/melting, mouthfeel, and milk flavor [42]. In addition, texture was more important than flavor in determining overall acceptability of the low-fat foods [43].

Preference mapping is a sensory tool to accomplish integration between consumer reactions and descriptive data [44]. By relating consumer data with descriptive data, the researcher can discover the relationships between product attributes and the ultimate bottom line, consumer acceptance [45]. One of the more recent topics in product research is the notion of "drivers of liking" or the nature of sensory features that drive acceptance [46].

As main results, Melo et al. [30] found a perceptible difference between conventional milk chocolate and their diabetic/reduced-calorie counterparts (Fig. 14.3). On an overall basis, consumers preferred the sensory characteristics of conventional chocolate to their alternative counterparts when tasted without product information such as health benefits. This result could only be attributed to the sensory attributes associated with the alternative chocolate options since no information was provided to influence preference. The crucial attributes that determined consumer acceptability in the chocolate samples were sweet aroma, melting rate, and sweetness, whereas bitterness, bitter aftertaste, adherence, and sandiness were drivers of disliking. Therefore, the sucrose substitution by high-intensity sweeteners, sucralose and stevioside, in conjunction with bulking agents and partial fat replacement by WPC, have potential as a palatable food in the formulation of diabetic/reduced-calorie milk chocolates. The major problems remain the increase in bitterness and bitter aftertaste (due to high-intensity sweeteners), increase in hardness, sandiness, and adherence, and decrease in melting rate (due to bulking agents and partial fat replacement).

An investigation on similarities and differences in sensory properties of control chocolate chip cookies and cookies made with two levels of selected protein, lipid, and carbohydrate-based fat replacers showed that carbohydrate-based fat replacers produced cookies with greater textural differences from the control than the protein or lipid-based replacers [47]. The physical properties of powdered cocoa drink mixtures prepared from two cocoa powders with various fat contents and different sweeteners, as well as the bioactive content and sensory properties of cocoa drinks prepared from them, were investigated, and results of the sensory evaluation showed that there was a preference for the cocoa drinks made with sweeteners (aspartame/acesulfame K and stevia extract), and there was a significant difference in the sensory attributes between the experimental mixtures and the control. The displayed results indicate the significant potential of using alternative sweeteners for the preparation of cocoa drink mixtures, which may provide good physical and sensory properties and also enhance the already existing beneficial effects of cocoa [48].

In order to avoid possible problems in relating descriptive data from trained panels with affective responses from consumers, a study suggested the application of "check-all-that-apply" (CATA) question to understand how consumers perceive chocolate milk desserts with different sugar and cocoa concentration and suggested that this methodology was able to detect differences in consumer perception of the desserts [49].

Storage Time of Diabetic and Reduced-Calorie Milk Chocolates

Some microbiologically stable food products, such as chocolate, must have their shelf lives determined by sensory analyses because, in many cases, sensory changes occur before microbiological changes. In such cases, the quality of food stored might not be acceptable after a period of time [50].

Little information is available on the sensory evaluation of chocolates, and few studies have been published on the effect of storage time on the sensory quality of alternative chocolates, such as diabetic and reduced-calorie chocolates made with sucralose or stevioside as high-intensity sweeteners, polydextrose/lactitol as bulking agent, and WPC as partial fat replacer. This is very important, especially for growing markets that need these products, such as diabetic people and people under reduced-calorie treatment diets. The effect of storage time (at 20°C) on the sensory characteristics of conventional, diabetic, and reduced-calorie chocolates was studied by Melo et al. [50] using sensory

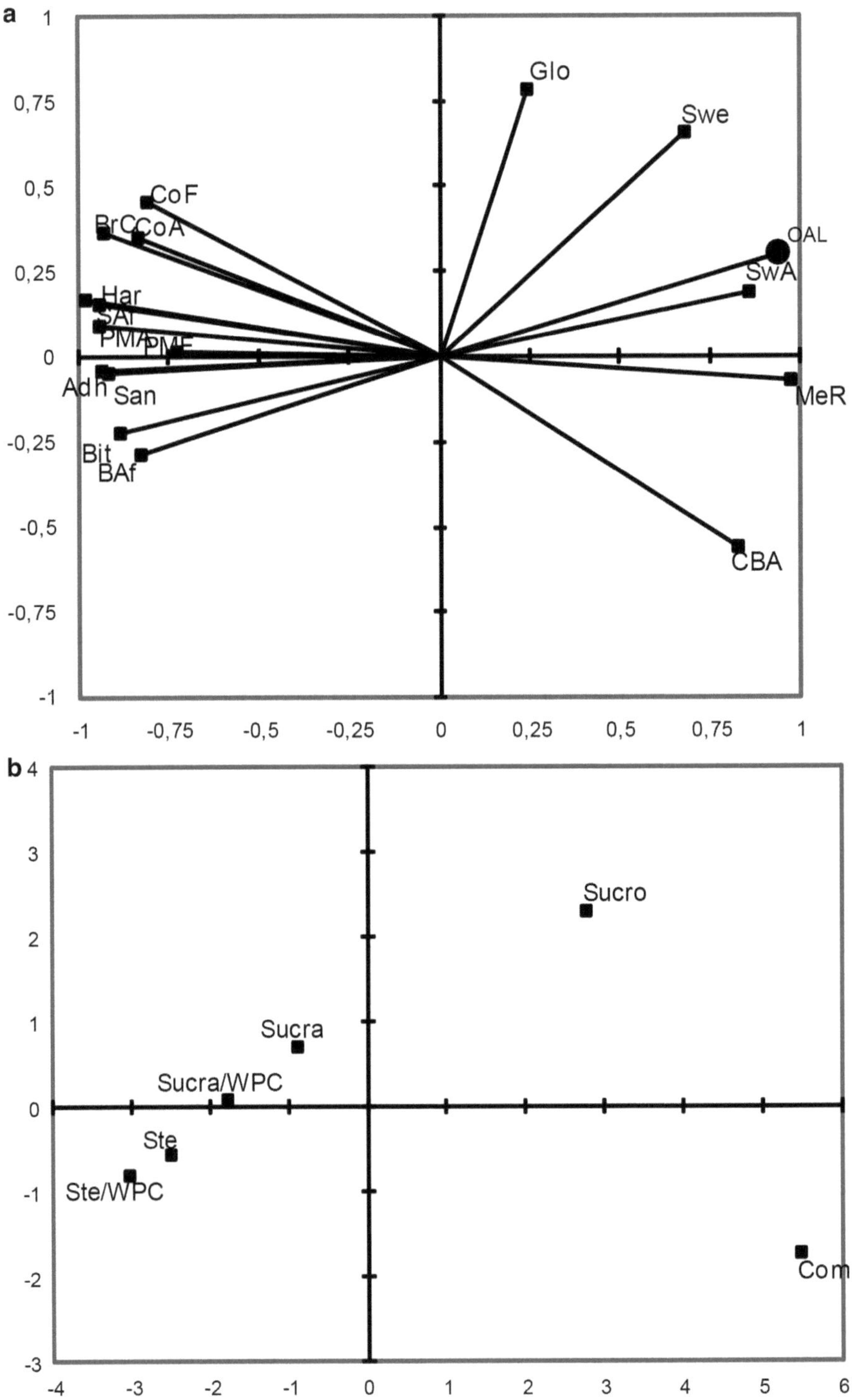

Fig. 14.3 (**a**) Relation of acceptance (overall liking, OAL, with consumers) to descriptive data (with trained panel), and (**b**) the positions of chocolate samples, using PLS regression. Abbreviations: *Adh* adherence, *Baf* bitter aftertaste, *Bit* bitterness, *BrC* brown color, *CBA* cocoa butter aroma, *CoA* cocoa aroma, *CoF* cocoa flavor, *Glo* gloss, *Har* hardness, *MeR* melting rate, *PMA* powdered milk aroma, *PMF* powdered milk flavor, *Saf* sweet aftertaste, *San* sandiness, *SwA* sweet aroma, *Swe* sweetness (Reprinted from Melo LLMM, Bolini HMA, Efraim P. Sensory profile, acceptability, and their relationship for diabetic/reduced calorie chocolates. Food Qual Prefer. 2009;20(2):138–43. Copyright 2009, with permission from Elsevier)

evaluation through descriptive analysis and acceptability testing (Fig. 14.4). All milk chocolates retained acceptable quality characteristics after 9 months of storage (20°C). Although the diabetic and reduced-calorie chocolates with sucralose and stevioside (high-intensity sweeteners), polydextrose and lactitol (bulking agents), and WPC (fat replacer) presented some sensory property changes, these changes were not enough to interfere in the acceptability among consumers for a long storage period, driving decisions on production and storage of diabetic and reduced-calorie chocolates.

The effects of 18°C and 30°C storage temperatures on texture, polymorphic structure, bloom formation, and sensory attributes of dark chocolate stored for 8 weeks were studied, and the sensory evaluation indicated that storage at 18°C was better than 30°C [51], and temperature fluctuations during storage had more influence on texture perception than storage at high temperatures or high relative humidity [52]. Storage time of dietary chocolates up to 1 year had statistically highly significant ($p < 0.01$) effects on the sensory attributes of chocolate, as well as on instrumentally measured color parameters [53].

Expectations and Acceptability of Diabetic and Reduced-Calorie Milk Chocolates Among Nondiabetics and Diabetics

Despite the importance of all previous results, optimal sensory quality, on its own, will not guarantee success. Consumer perception about the safety, cost, and risk/benefits associated with novel technologies can negatively influence consumer choice and purchase decisions. Industry should utilize the predictive power of conjoint analysis for the design of new food products, and this consumer-based, predictive research should be followed up with validity testing in test markets that mimic the consumer segments used in generating the conjoint predictions [54]. An optimized product formulation is necessary for a successful innovation; however, consumers are also influenced by extrinsic product information such as brand, price, or labeling. Understanding the relative importance of product attributes influencing food choice at the point of sale is important to the success of new product development [55].

Conjoint analysis is extensively used in marketing research to evaluate industrial products and services and is being increasingly used in the study of food choices by consumers [56]. It examines trade-offs consumers make at point of purchase to determine what features are most valued, allowing the researcher to evaluate the effectiveness of different communications, and the ways in which they interact and affect the consumers [57]. Marketing researchers have made considerable use of conjoint analysis to estimate the impact of selected product or service characteristics on consumer preference [58]. Conjoint measurement comprises a set of approaches, whereby concepts are created by experimental design, reacted to by consumers, with the reactions then being traced to the presence/absence of specific concept elements. The stimuli or the raw material for a conjoint measurement study comprises single, stand-alone elements. These elements of phrases are combined with other phrases to form a concept [59], enabling a number of products to be compared and ranked based on their attributes [60–62]. For example, the developer might decide to incorporate certain ingredients with known health benefits or create the product under a specified brand name [63].

Foods with low-calorie or reduced-calorie sweeteners can have fewer calories than foods made with sugar and other caloric sweeteners. Such foods can also be appealing and healthful for weight loss or weight maintenance for nondiabetic consumers. These products often also have less carbohydrate, which can be helpful in managing blood glucose levels for diabetic consumers. If sucrose in cocoa-containing foods is to be partially or completely replaced by other types of sugar or by sugar substitutes, theoretical models are at present unable to predict acceptance [15]. Melo et al. [64] compared attitudes and acceptance between two groups of consumers in the United States, diabetics and nondiabetics, of sugar and calorie reductions in a widely accepted product: milk chocolate. Conjoint

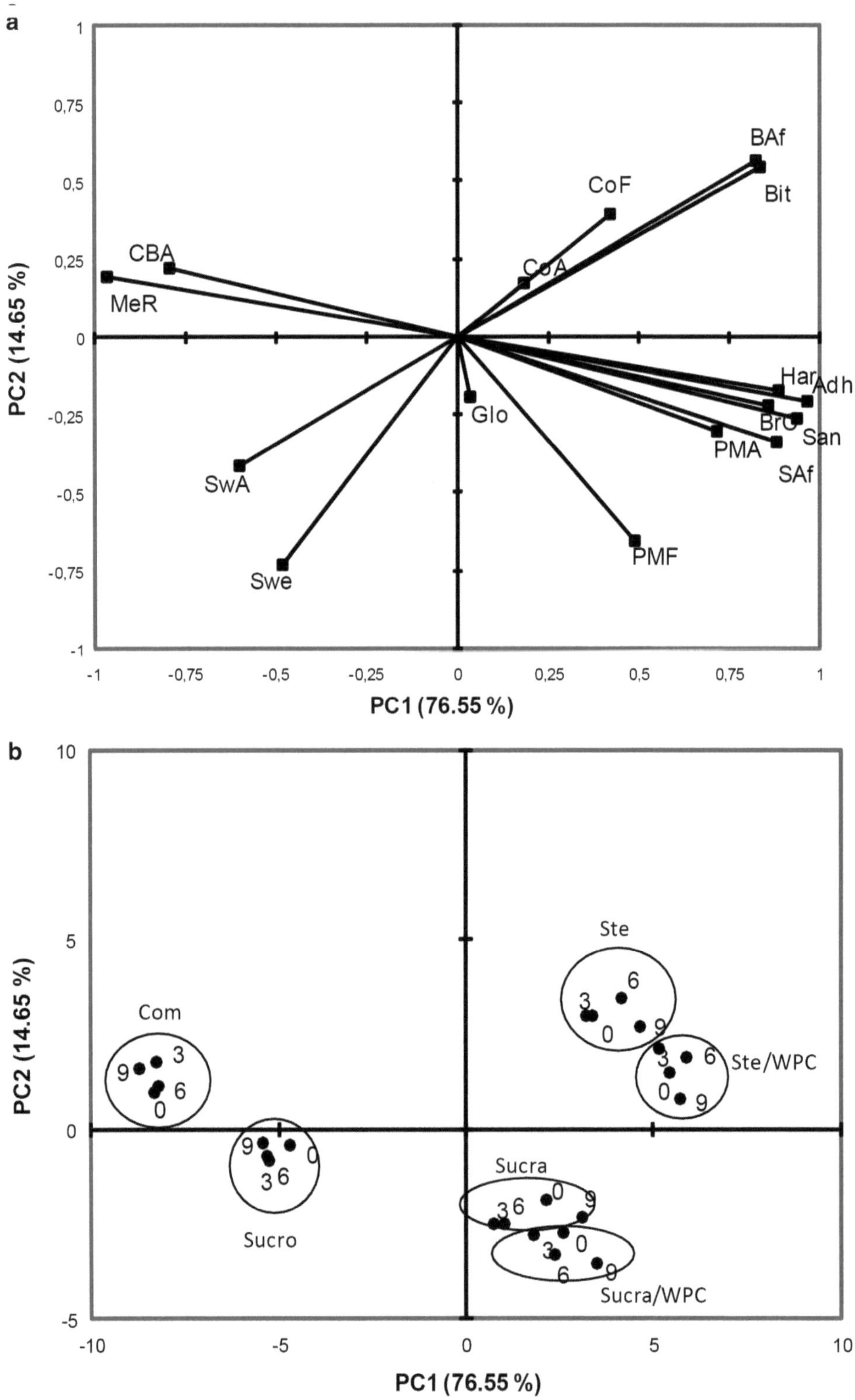

Fig. 14.4 Principal component analysis (PCA) loading (**a**) descriptors and (**b**) chocolate samples during storage time (number into the circlus: 0, 3, 6, and 9 months). Abbreviation: *Com* diabetic commercial sample (Reprinted from Melo LLMM, Bolini HMA, Efraim P. Storage time study of sugar-free and reduced calorie milk chocolates. J Food Qual. 2009;32(5):577–89. With permission from John Wiley and Sons)

Table 14.3 Relative importance and average zero-centered utility scores for conjoint analysis levels among diabetic and nondiabetics for milk chocolates (Reprinted from Melo L, Childs JL, Drake M, Bolini HMA, Efraim P. Expectations and acceptability of diabetic and reduce-calorie milk chocolates among nondiabetics and diabetics in the USA. J Sensory Stud. 2010;25:133–152. With permission from John Wiley and Sons)

	Relative importance (%)			Average zero-centered utility scores		
Attribute	Diabetics (N=68)	Nondiabetics (N=103)	p values	Level	Diabetics (N=68)	Diabetics (N=68)
Sugar claim	34.6	28.7	0.048	Absence of sugar claim	−1.57^{c}	−0.24^{c}
				Reduced sugar	0.58^{a}	0.88^{a}
				Sugar-free	0.88^{a}	0.19^{b}
				Diabetic	0.13^{b}	−0.83^{d}
Sweetener type	43.7	46.9	0.40	Artificial	−1.92^{c}	−2.50^{c}
				Natural	2.06^{a}	2.75^{a}
				Combination	−0.15^{b}	−0.25^{b}
Calorie reduction	21.7	24.4	0.35	0%	−1.19^{c}	−1.50^{c}
				20%	0.40^{b}	0.61^{b}
				25%	0.79^{a}	0.89^{a}

Note: Same letters in a column and within an attribute indicate that there is no significant difference ($p<0.05$)

Table 14.4 Acceptance means for chocolates with different sweeteners among nondiabetic and diabetic consumers (Reprinted from Melo L, Childs JL, Drake M, Bolini HMA, Efraim P. Expectations and acceptability of diabetic and reduce-calorie milk chocolates among nondiabetics and diabetics in the USA. J Sensory Stud. 2010;25:133–52. With permission from John Wiley and Sons)

		Attribute				
Samples	Consumer	Overall liking	Appearance	Aroma	Flavor	Texture
Sucro	Nondiabetic	5.3^{A}	6.5^{AB}	6.0^{AB}	5.3^{A}	5.6^{A}
	Diabetic	–	–	–	–	–
Sucra	Nondiabetic	5.0A^{Bb}	6.5^{Aa}	6.0^{Ab}	5.3^{Ab}	5.1^{Bb}
	Diabetic	6.3^{Aa}	6.9^{Aa}	6.6^{Aa}	6.3^{Aa}	6.4^{Aa}
Ste	Nondiabetic	3.5^{Cb}	6.3^{ABb}	5.4^{Ca}	3.4^{Cb}	3.9^{Cb}
	Diabetic	4.9^{Ba}	6.2^{Ba}	5.9^{BCa}	4.8^{Ba}	5.3^{Ba}
Sucra/WPC	Nondiabetic	4.9^{ABb}	6.2^{Ba}	6.0^{ABa}	5.0^{ABb}	4.2^{Cb}
	Diabetic	5.8^{Aa}	6.8^{Aa}	6.3^{ABa}	5.8^{Aa}	5.6^{Ba}
Ste/WPC	Nondiabetic	4.0^{Cb}	6.3^{ABa}	5.6^{BCa}	3.7^{Cb}	4.1^{Cb}
	Diabetic	4.7^{Ba}	6.3^{Ba}	5.8^{Ca}	4.9^{Ba}	5.1^{Ba}
Com	Nondiabetic	4.7^{Bb}	6.4^{ABa}	5.8^{ABCb}	4.6^{Bb}	5.6^{Aa}
	Diabetic	6.1^{Aa}	6.7^{Aa}	6.4^{Aa}	5.9^{Aa}	6.3^{Aa}

Note: Means in column and within a consumer group (nondiabetic or diabetic) followed by different capital letters are different ($p<0.05$). Means in a column and within a sample followed by different lowercase letters are different ($p<0.05$) across consumer groups (nondiabetics and diabetics)

analysis was used to determine which levels of sugar claim, sweetener type, and calorie reduction in chocolate were more important for nondiabetic and diabetic consumers. Additionally, acceptance of actual laboratory-developed chocolates was evaluated among these two groups of consumers to determine if actual products matched consumer expectations.

Nondiabetics and diabetics had different expectations regarding sugar claim, sweetener type, and calorie reduction (Table 14.3), which were likely due to their different nutritional necessities. In addition, consumer testing showed different acceptance means for conventional, diabetic, and reduced-calorie milk chocolates among nondiabetics, diabetics, and between them (Table 14.4). A natural sweetener was appealing to both consumer groups but did not meet consumer expectations of either group when tasted in chocolates. Conventional, diabetic chocolate made with sucralose (Sucra) and diabetic/reduced-calorie chocolate made with sucralose and WPC (Sucra/WPC) were the most accepted among nondiabetics, as were Sucra and Sucra/WPC chocolates among nondiabetics, who did not evaluate the conventional sample due to its sucrose content. Diabetic consumers were more accepting than nondiabetic consumers of diabetic and diabetic/reduced-calorie milk chocolates.

Alternative products must be developed and labeled according to the specific consumer groups they are intended to address. Future studies should also evaluate consumer acceptance of products with and without nutritional information (sugar claim, sweetener type, and calorie reduction) since this information may influence acceptance and purchase intent. For instance, pleasantness and buying probability of chocolate bars were rated higher with regular than with reduced-fat information [65].

References

1. Parpinello GP, Versari A, Castellari M, et al. Stevioside as a replacement of sucrose in peach race: sensory evaluation. J Sens Stud. 2001;16(5):471–84.
2. Yanovski S. Sugar and fat: cravings and aversions. J Nutr. 2003;133(3):835S–7.
3. Macdiarmid JI, Vail A, Cade JE, et al. The sugar-fat relationship revisited: differences in consumption between men and women of varying BMI. Int J Obes. 1998;22(11):1053–61.
4. Matthiessen J, Fagt S, Biltoft-Jensen A, et al. Size makes a difference. Publ Health Nutr. 2003;6(1):65–72.
5. Lambert JP. Nutrition and health aspects of chocolate. In: Beckett ST, editor. Industrial chocolate manufacture and use. London: Wiley-Blackwell; 2009. p. 623–35.
6. Cardello H, Da Silva M, Damasio MH. Measurement of the relative sweetness of stevia extract, aspartame and cyclamate/saccharin blend as compared to sucrose at different concentrations. Plant Foods Hum Nutr. 1999;54(2):119–30.
7. Epstein LH, Caggiula AR, Rodefer JS, et al. The effects of calories and taste on habituation of the human salivary response. Addict Behav. 1993;18(2):179–85. Mar-Apr.
8. Geiselman PJ, Smith CF, Williamson DA, et al. Perception of sweetness intensity determines women's hedonic and other perceptual responsiveness to chocolate food. Appetite. 1998;31(1):37–48.
9. Guinard JX, Mazzucchelli R. Effects of sugar and fat on the sensory properties of milk chocolate: descriptive analysis and instrumental measurements. J Sci Food Agric. 1999;79(11):1331–9.
10. Hoskin JC. Sensory properties of chocolate and their development. Am J Clin Nutr. 1994;60(6):S1068–70.
11. Higginbotham JD. Recent developments in non-nutritive sweeteners. In: Grenby TH, Parker KJ, Lindley MG, editors. Progress in sweeteners. London: Applied Science Publishers; 1983. p. 119–55.
12. Jenner MR. Sucralose: unveiling its properties and applications. In: Grenby TH, editor. Progress in sweeteners. London: Elsevier Applied Science; 1989. p. 121–41.
13. Krüger C. Sugar and bulk sweeteners. In: Beckett ST, editor. Industrial chocolate manufacture and use. London: Wiley-Blackwell; 2009. p. 48–75.
14. Liebrand J, Smiles R. Polydextrose for reduced calorie confections. Manuf Confect. 1981;61(11):35–40.
15. Krüger CH. Sugar. In: Beckett TH, editor. Industrial chocolate manufacture and use. London: Black Academic & Professional; 1994. p. 25–42.
16. Afoakwa EO, Paterson A, Fowler M. Factors influencing rheological and textural qualities in chocolate - a review. Trends Food Sci Technol. 2007;18(6):290–8.
17. Cidell JL, Alberts HC. Constructing quality: the multinational histories of chocolate. Geoforum. 2006;37(6):999–1007.
18. Ziegler GR, Mongia G, Hollender R. The role of particle size distribution of suspended solids in defining the sensory properties of milk chocolate. Int J Food Prop. 2001;4(2):353–70.
19. Lawless HT, Heymann H. Sensory evaluation of food – principles and practices. Gaithersburg: Aspen Publishers Inc; 1999.
20. Lee III WE. Single-point versus time-intensity sensory measurements: an informational entropy analysis1. J Sens Stud. 1989;4(1):19–30.
21. Clark CC, Lawless HT. Limiting response alternatives in time-intensity scaling - an examination of the halo-dumping effect. Chem Senses. 1994;19(6):583–94.
22. Ayya N, Lawless HT. Quantitative and qualitative evaluation of high-intensity sweeteners and sweetener mixtures. Chem Senses. 1992;17(3):245–59.

23. Calviño A, Garrido D, García M. Potency of sweetness of aspartame, d-tryptophan and thaumatin evaluated by single value and time-intensity measurements. J Sens Stud. 2000;15(1):47–64.
24. De Melo LLMM, Bolini HMA, Efraim P. Equisweet milk chocolates with intense sweeteners using time-intensity method. J Food Qual. 2007;30(6):1056–67.
25. San Juan EM, Edra EV, Fadrigalan EN. Consumer-based optimization of peanut-chocolate bar using response surface methodology. J Food Process Pres. 2005;29(3-4):208–27.
26. Dos A, Ayhan Z, Sumnu G. Effects of different factors on sensory attributes, overall acceptance and preference of Rooibos (Aspalathus linearis) tea. J Sens Stud. 2005;20(3):228–42.
27. Thompson JL, Drake MA, Lopetcharat K, et al. Preference mapping of commercial chocolate milks. J Food Sci. 2004;69(9):S406–13. Nov-Dec.
28. Luckow T, Delahunty C. Consumer acceptance of orange juice containing functional ingredients. Food Res Int. 2004;37(8):805–14.
29. Zhao LQ, Tepper BJ. Perception and acceptance of selected high-intensity sweeteners and blends in model soft drinks by propylthiouracil (PROP) non-tasters and super-tasters. Food Qual Prefer. 2007;18(3):531–40.
30. Melo LLMM, Bolini HMA, Efraim P. Sensory profile, acceptability, and their relationship for diabetic/reduced calorie chocolates. Food Qual Prefer. 2009;20(2):138–43.
31. Jones SA. Issues in fat replacement. In: Roller S, Jones SA, editors. Handbook of fat replacers. Boca Raton: CRC Press; 1996. p. 3–26.
32. Prindiville EA, Marshall RT, Heymann H. Effect of milk fat, cocoa butter, and whey protein fat replacers on the sensory properties of lowfat and nonfat chocolate ice cream. J Dairy Sci. 2000;83(10):2216–23.
33. Grabitske HA, Slavin JL. Gastrointestinal effects of low-digestible carbohydrates. Crit Rev Food Sci Nutr. 2009;49(4):327–60.
34. Marteau P, Flourie B. Tolerance to low-digestible carbohydrates: symptomatology and methods. Br J Nutr. 2001;85(1):S17–21.
35. Bolenz S, Amtsberg K, Schape R. The broader usage of sugars and fillers in milk chocolate made possible by the new EC cocoa directive. Int J Food Sci Technol. 2006;41(1):45–55.
36. Farzanmehr H, Abbasi S. Effects of inulin and bulking agents on some physicochemical, textural and sensory properties of milk chocolate. J Texture Stud. 2009;40(5):536–53.
37. Finney M, Smullen J, Foster HA, et al. Effects of low doses of lactitol on faecal microflora, pH, short chain fatty acids and gastrointestinal symptomology. Eur J Nutr. 2007;46(6):307–14.
38. Stubenitsky K, Aaron JI, Catt SL, et al. Effect of information and extended use on the acceptance of reduced-fat products. Food Qual Prefer. 1999;10(4–5):367–76.
39. Shah AB, Jones GP, Vasiljevic T. Sucrose-free chocolate sweetened with *Stevia rebaudiana* extract and containing different bulking agents - effects on physicochemical and sensory properties. Int J Food Sci Technol. 2010;45(7):1426–35.
40. Prindiville EA, Marshall RT, Heymann H. Effect of milk fat on the sensory properties of chocolate ice cream. J Dairy Sci. 1999;82(7):1425–32.
41. Abbasi S, Farzanmehr H. Optimization of the formulation of prebiotic milk chocolate based on rheological properties. Food Technol Biotechnol. 2009;47(4):396–403.
42. Bolenz S, Thiessenhusen T, Schape R. Influence of milk components on properties and consumer acceptance of milk chocolate. Eur Food Res Technol. 2003;216(1):28–33.
43. Devereux HM, Jones GP, McCormack L, et al. Consumer acceptability of low fat foods containing inulin and oligofructose. J Food Sci. 2003;68(5):1850–4.
44. Geel L, Kinnear M, de Kock HL. Relating consumer preferences to sensory attributes of instant coffee. Food Qual Prefer. 2005;16(3):237–44.
45. Meilgaard M, Civille GV, Carr BT. Sensory evaluation techniques. Boca Raton: CRC Press; 1999.
46. Moskowitz H, Gofman A, Beckley J. Using high-level consumer-research methods to create a tool-driven guidebook and database for product development and marketing. J Sens Stud. 2006;21(1):54–100.
47. Armbrister WL, Setser CS. Sensory and physical-properties of chocolate chip cookies made with vegetable shortening or fat replacers at 50 and 75-percent-levels. Cereal Chem. 1994;71(4):344–51.
48. Belscak-Cvitanovic A, Benkovic M, Komes D, et al. Physical properties and bioactive constituents of powdered mixtures and drinks prepared with cocoa and various sweeteners. J Agric Food Chem. 2010;58(12):7187–95.
49. Ares G, Barreiro C, Deliza R, et al. Application of a check-all-that-apply question to the development of chocolate milk desserts. J Sens Stud. 2010;25:67–86.
50. Melo LLMM, Bolini HMA, Efraim P. Storage time study of sugar-free and reduced calorie milk chocolates. J Food Qual. 2009;32(5):577–89.
51. Ali A, Selamat J, Man YBC, et al. Effect of storage temperature on texture, polymorphic structure, bloom formation and sensory attributes of filled dark chocolate. Food Chem. 2001;72(4):491–7.
52. Andrae-Nightingale LM, Lee SY, Engeseth NJ. Textural changes in chocolate characterized by instrumental and sensory techniques. J Texture Stud. 2009;40(4):427–44.

53. Popov-Raljic JV, Lalicic-Petronijevic JG. Sensory properties and color measurements of dietary chocolates with different compositions during storage for up to 360 days. Sensors. 2009;9(3):1996–2016.
54. Cardello AV, Schutz HG, Lesher LL. Consumer perceptions of foods processed by innovative and emerging technologies: a conjoint analytic study. Innov Food Sci Emerg Tech. 2007;8(1):73–83.
55. Enneking U, Neumann C, Henneberg S. How important intrinsic and extrinsic product attributes affect purchase decision. Food Qual Prefer. 2007;18(1):133–8.
56. Haddad Y, Haddad J, Olabi A, et al. Mapping determinants of purchase intent of concentrated yogurt (Labneh) by conjoint analysis. Food Qual Prefer. 2007;18(5):795–802.
57. Paulus KO, Aarts P, Moskowitz HR, et al. Purchasing habits and sensory preferences for cola consumers. Fruit Process. 2003;13(6):409–15.
58. Cheng HW, Clarke AD, Heymann H. Influence of selected marketing factors on consumer response to restructured beef steaks: a conjoint analysis. J Sens Stud. 1990;4(3):165–78.
59. Moskowitz H, Silcher M, Beckley J, et al. Sensory benefits, emotions and usage patterns for olives: using Internet-based conjoint analysis and segmentation to understand patterns of response. Food Qual Prefer. 2005;16(4):369–82.
60. Beckley J, Ashman H, Maier A, et al. What features drive rated burger craveability at the concept level? J Sens Stud. 2004;19(1):27–47.
61. Kupiec B, Revell B. Measuring consumer quality judgements. Br Food J. 2001;103(1):7–22.
62. Lee CM, Lee SY. Consumer insights on healthy breakfast cereal - a focus group research. J Sens Stud. 2007;22(4):417–32.
63. Gofman A. Emergent scenarios, synergies and suppressions uncovered within conjoint analysis. J Sens Stud. 2006;21(4):373–414.
64. Melo L, Childs JL, Drake M, et al. Expectations and acceptability of diabetic and reduced-calorie milk chocolates among nondiabetics and diabetics in the USA. J Sens Stud. 2010;25:133–52.
65. Kahkonen P, Tuorila H. Consumer responses to reduced and regular fat content in different products: effects of gender, involvement and health concern. Food Qual Prefer. 1999;10(2):83–91.

Part III
Chocolate Metabolism and Activities in Culture

Chapter 15
Polyphenols in Cocoa: From In Vitro Digestion to In Vivo Bioavailability

Nàdia Ortega Olivé and Maria-José Motilva Casado

Key Points

- In vitro digestion models permit the characterization of cocoa procyanidins during digestion under physiological conditions, caused by alimentary enzymes.
- Cultured cell models facilitate studies of small intestinal absorption and metabolism, as an important step towards elucidating the potential impact of these cocoa procyanidins on human health.
- Characterization of the digestive species of cocoa procyanidins forming as consequence of colonic fermentation provides useful information for future studies in vivo.
- The in vitro digestion models could act as a first directional influence on the reformulation of new cocoa products in order to achieve the required "nutrient profiles."

Keywords In vitro digestion • Colonic fermentation • Cellular models • Procyanidin metabolism • Bioavailability

Introduction

Studies on the health benefits of cocoa and cocoa products have been conducted over the past decade, with a major focus on degenerative diseases. These benefits have been attributed mainly to phenolic compounds located in the pigment cells, also called polyphenol-storage cells, of the cotyledons of cocoa beans [1]. The main groups of cocoa polyphenols are catechins or flavan-3-ols (catechin and epicatechin), proanthocyanidins, and anthocyanins in low proportion. Besides flavonoids, cocoa and cocoa products are also rich in methylxanthines, namely, caffeine, theobromine, and theophylline [2]. Theobromine is the major methylxanthine present in cocoa, constituting about 4% on a fat-free basis.

N.O. Olivé, Ph.D. (✉)
Research and Development, La Morella nuts, S.A.U.,
Camí Ample, s/n, Castellvell del Camp, Tarragona 43392, Spain
e-mail: n.ortega@morellanuts.com

M.-J.M. Casado, D.Pharm., Ph.D.
Department of Food Technology, University of Lleida,
Alcade Rovira Roure 191, Lleida 25198, Spain
e-mail: motilva@tecal.udl.es

R.R. Watson et al. (eds.), *Chocolate in Health and Nutrition*, Nutrition and Health 7,
DOI 10.1007/978-1-61779-803-0_15, © Springer Science+Business Media, LLC 2013

The mechanisms of action of polyphenols in vivo are not fully elucidated. They are believed to act mainly as free-radical scavengers and/or chelators of transition metals (minerals or trace elements). But it is unlikely that their free-radical scavenging capacity is sufficient to explain their antioxidant action in vivo, due to their relatively low bioavailability within the digestive tract [3]. Also after digestion, naturally occurring polyphenols can undergo significant modifications that can result in a diverse family of bioactive molecules. Flavanols and procyanidins are relatively stable in stomach acid and during gastric transit. During digestion and transfer across the small intestine, and in the liver, flavanols are rapidly metabolized in phase I and II biotransformations to various O-sulfated, O-glucuronidated, and O-methylated forms [4].

Considering the important transformations of polyphenols during the digestion, in addition to their low bioavailability, their mechanisms of action are probably linked to the activation/repression of particular genes. Recent investigations have shown that they can also act on gene expression via transcription factors (e.g., Nrf2), activating the antioxidant response element [5], which can lead to the transcription of antioxidant compounds such as glutathione (GSH) or enzymes of glutathione metabolism [6]. Therefore, efforts to understand the transformations of polyphenols during digestion and their bioavailability will be useful to elucidate in the future the mechanisms of action of polyphenols in the prevention of diseases related to oxidative stress and degenerative diseases.

Bioavailability refers to the fraction of any compound ingested and made available for utilization, metabolism, and/or storage by the organism. Different factors have been suggested to influence the bioavailability of cocoa polyphenols, but the results are not always in agreement, probably due to the use of different study designs, cocoa sources, or food matrices between and within studies. With evidence that cocoa polyphenols are absorbable in humans, it is necessary to elucidate the factors that impact intestinal uptake and potential bioavailability. It is likely that the general mechanism of polyphenol absorption would follow routes similar to those taken by other xenobiotic compounds that require consideration of (1) efficient release of the polyphenol from the food matrix highly conditioned by food processing and food matrix composition (percentage of fat, protein, or carbohydrates), (2) ingested dose, (3) stability to gastric and small intestinal digestive conditions, (4) uptake by small intestinal absorptive epithelial cells, and (5) secretion into circulation.

Taking into account the multiple factors that determine the bioavailability of cocoa polyphenols, the selection of a cocoa derivate formula or food processing through in vivo bioavailability studies is unfeasible. In this sense, the in vitro models are a useful tool for the early stages of development of new processes and new food formulations.

In Vitro Digestion Models

Studies of food digestibility in animal or human models (in vivo methods) should provide more accurate results, but they are time consuming and costly. The in vitro digestion models are developed to mimic the conditions of reactions that occur in the mouth, stomach, intestines, and colon. The aim of performing these models is to characterize the changes in food components due to the physiological conditions. Mainly the first part of digestion (mouth, stomach, and small intestine) is focused on digested fraction, while the nondigestion fraction is transformed in the colon.

The food industry is focused on developing innovative microstructure and texture of food products as well as formulating enhanced and balanced structured matrices that are rich in nutrients, especially in bioactive compounds, using technologies such as extrusion, encapsulation, and so on. In that line, it is important to acquire enough knowledge of the real contribution of each specific nutrient in a food product in order to maximize the health benefits to consumers. For that reason, recently, the in vitro digestion models have also focused on the study of how the food matrix affects changes in food components. The food matrix effect can be defined as the influence on the release of nutrients contained into a larger continuous medium from different cellular origins such as fruits, vegetables, processed foods, and so on [7]. In this sense, the in vitro digestion models are perfectly suited to predict the outcome of in vivo experiments and also to enable study of underlying mechanisms of the release and bioavailability of nutrients in different food matrices. These models should replicate closely the conditions that occur in vivo.

The two main types of digestion models are static and dynamic. Static models are models that do not mimic the physical processes that occur in vivo. In these models, the food remains immobile, and they do not reflect the physical conditions of the digestion process. Indeed, concentrations and activities of the digestive components, as well as the passage of the food ingredients through the gastrointestinal tract, are dynamic processes. Some static models were improved with the incorporation of a dialysis step that enables the estimation of the bioavailable fraction [8–10]. Even though they have certain limitations, static models could be considered as good indicators for simple matrices and useful for effective release studies of the nutrients from the food matrix. However, complex matrices should not be enough simulated by simply static model. For that reason, some digestion models have been improved and overcome as dynamic models. Dynamic models include physical processing (e.g., mixing, diffusion, particle size reduction) that is useful to mimic the in vivo conditions. Recently, several commercial gastrointestinal models such as the TIM model and the dynamic digestion model have been developed by TNO Centre and the Institute of Food Research in Norwich [11, 12], respectively. The advantage of both models is the wide application in pharmacological and food testing and the established correlation of in vitro-in vivo observed [13].

In order to evaluate the digestion models, several concepts pertaining to the different steps of a digested nutrient have been defined. A commonly used term is "bioaccessibility," which is defined as the fraction of an ingested nutrient from food matrix that is available for absorption [14]. Another term is "bioavailability," which is considered the fraction of ingested nutrient that is absorbed and becomes available for the action of physiologic functions. In vitro digestion models provide information about both bioaccessibility (digestion process) and bioavailability (absorption process).

In Vitro Digestion Models Applied to Cocoa Matrices

Due to limited knowledge of typology and quantities of polyphenols that are really absorbed, several studies have performed digestion models to better understand the bioaccessibility and bioavailability of polyphenols from cocoa matrices. Of particular interest is the knowledge of the absorption and uptake of the cocoa oligomeric procyanidins and whether such polyphenol complexes are stable in the acidic environment of the stomach after consumption, prior to absorption. Spencer et al. [15] studied the effects of the acidic environment, as found in the gastric milieu, on procyanidin oligomers ranging from a dimer to hexamer isolated from *Theobroma cacao*, as well as on the monomer epicatechin. Procyanidin oligomers (100 mM) were incubated in acid (pH 2.0) and in simulated gastric juice (pH 2.0) at 37°C for up to 3.5 h and the products analyzed and identified by HPLC. Results of this study showed that under these conditions there is a time-dependent decomposition of each oligomer (trimer to hexamer) with the progressive appearance of dimer and monomer over the period of incubation. Decomposition of the oligomers during the first 90 min was more pronounced ($60 \pm 80\%$) than during the subsequent equivalent period and virtually complete by 3.5 h. The results show that under gastric conditions the procyanidin oligomers (trimer to hexamer) are hydrolyzed to mixtures of epicatechin monomer and dimer, thus enhancing the potential for their absorption in the small intestine. In the same trend, Zhu et al. [16] investigated the stability of the cocoa monomers, epicatechin and catechin, and the dimers, epicatechin-(4(beta)β-8)-epicatechin (dimer B2) and epicatechin-(4(beta)β-6)-epicatechin (dimer B5), in simulated gastric and intestinal juice and at different pH values. Epicatechin and catechin were stable in simulated gastric juice. In contrast, dimer B2 and dimer B5 were unstable producing epicatechin; at the same time, incubation of dimer B2 and dimer B5 in simulated gastric juice resulted in isomerization to dimer B5 and dimer B2, respectively. When incubated in simulated intestinal juice (alkaline pH), all four compounds degraded almost completely within several hours. These results suggest that the amount and type of flavanols and procyanidins in the gastrointestinal tract following the consumption of cocoa can be influenced by the stability of these compounds in both acidic and alkaline environments. Additionally, the potential degradation of dimers in both the stomach and the intestine helps to explain their low concentrations in plasma. In that sense, Neilson et al. [17] monitored the native catechin degradation and simultaneously characterized the structural variation of the resulting catechin degradation products using an in vitro gastric and small intestinal digestion model to simulate preabsorptive digestion events. To perform the study, catechin, epicatechin, epigallocatechin, epigallocatechin gallate, and epicatechin gallate were tested. Results showed that simple catechins were much more stable under digestive conditions than the corresponding gallocatechins and catechin gallates that were significantly degraded, with losses between 60% and 100%. The main contribution of this study is the discovery that both homo- and heterocatechin autoxidation dimers are formed under simulated digestive conditions, which suggests that luminal alteration of ingested catechin profiles in vivo warrants further consideration. This work facilitates future in vivo studies of catechin digestive behavior and bioavailability by providing analytical methods specifically designed to identify and characterize catechin autoxidation dimers.

Matrix effects are becoming increasingly important for food as EU legislation Directive 2006/1924/EC [18] on nutrition and health claims made on foods asks for evidence between the food labeling and its beneficial compound which must be able to show that it is bioavailable from that food product and is also effective in its implied benefit. Cocoa, unlike other vegetal foods, is not consumed as such but as an ingredient in food formulations, such as chocolate, dark cocoa power, cocoa drinks, or as an ingredient in coating or in cookies. For example, the basic ingredients required for the manufacture of chocolate are cocoa nibs or cocoa liquor, sugar and other sweeteners, cocoa butter or butter fat, milk powder, and emulsifiers, among others. Therefore, foods that usually include cocoa are complex mixtures. In this sense, besides the studies of stability of cocoa polyphenols under digestion conditions, it is of special interest to evaluate the effect of the food matrix composition on the stability and digestibility of these compounds.

As mentioned previously, the food matrix is composed mainly for different ratios and typology of macronutrients such as carbohydrates, fat, and protein, which could modify the bioaccessibility of cocoa polyphenols and therefore their bioavailability. Indeed, the know-how of this influence could help to improve and enhance the formulation of new cocoa food products in order to maximize the utilization of the health-promoting nutrients such as the polyphenols. In order to assess the matrix effect on the bioaccessibility of cocoa polyphenols, Ortega et al. [19] developed an in vitro digestion model for the evaluation of fat content on the digestibility and bioaccessibility of cocoa polyphenols. In this study, two main cocoa ingredients used in the confectionary industry were studied – liquor and powder cocoa (containing 50% and 15% of fat, respectively). This work tried to show that a higher fat content seems to have a protective role for the polyphenols, mainly due to a better micellarization that enhances their stability during digestion conditions. This conclusion was later demonstrated by the evaluation of the presence in the food matrix of soluble water fraction (soluble fiber and soluble sugars) and/or hazelnut oil (rich in polyunsaturated fatty acids) on the digestibility and stability of carob flour phenols using an in vitro digestion model [20]. The presence of the soluble dietary fiber enhanced the stability of the phenolic compounds during the duodenal digestion phase. Similarly, the hazelnut oil showed a protective effect on the stability of the phenolic compounds during duodenal digestion. The presence of lipids in the digestion mixture could suggest a micellarization that could establish interactions with certain polyphenols, enhancing their recovery and stability during digestion. These results may be useful for the food industry by suggesting that an emulsion food matrix structure could be able to give the desired protection during digestion process and achieve health benefits.

A recent comprehensive study [21] assessed changes in the polyphenol fraction of water-insoluble cocoa fraction by in vitro digestion simulating the human gastrointestinal process. Results of the study showed that the digestive process solubilizes a significant part of the bound polyphenols (high-degree polymerized procyanidins), and it increases their potential bioaccessibility. All samples obtained after protease digestion (pepsin, pancreatin, and pronase) showed the presence of catechin, epicatechin, and procyanidin B2, while B1 dimer was released only by pepsin hydrolysis. After the action of Viscozyme L (multienzyme complex containing a wide range of carbohydrases, including arabanase, cellulase, (beta)β-glucanase, hemicellulase, and xylanase), epicatechin (but no catechin or procyanidins) was found in the digested fraction. These data suggest that the catechin and dimeric procyanidins, particularly B1, were mainly linked to the protein moiety of the cocoa-insoluble material, while a minor part of epicatechin was associated with the polysaccharide moiety. Following digestion in the small intestine, about 75% of the cocoa-insoluble fraction remained undigested, so it might reach the colon, where bacteria play a major role in fermenting it.

In Vitro Colonic Fermentation

Mainly monomers (catechin and epicatechin) and dimers have been detected to be able to cross the gut barrier in the small intestine. Procyanidic fraction that is not absorbed in the small intestine reaches the colon where it exerts its action or is degraded by intestinal or colonic bacteria before absorption or excretion of the resulting metabolites. In that line, several in vitro gut models have been developed in order to assess the changes in the higher polyphenol unit's fraction in the last step of the digestion process (colon model). The amount of substrate, the fecal microbiota, and substrate-to-inoculum ratio are some crucial factors in fermentation studies that are able to show discordance results.

In a previously described study by Flogiano et al. [21], the cocoa-insoluble fraction that remained undigested after in vitro gastrointestinal digestion was submitted to a three-stage continuous system-gut model simulating the proximal, transverse, and distal colons. The application of the colonic fermentation model revealed that microflora action led to the solubilization of a further 17.8% of this undigested fraction that is potentially absorbable through the colon. Results showed that the microflora extensively metabolized the procyanidins into various aromatic acids. 3-hydroxyphenylpropionic, 3-hydroxyphenylacetic, and 3,4-dihydroxybenzoic were the phenolic acids detected in the three vessels, showing that phenolic acids are the major metabolites of cocoa polyphenols.

Previous studies on cocoa procyanidins showed similar results. Rios et al. [22] reported that the consumption of procyanidin-rich chocolate by humans is associated with the production of several aromatic acids, including derivatives of phenylpropionic, phenylacetic, and benzoic acids. A study by Gonthier et al. [23] compared the metabolism of procyanidin dimer B3, trimer C2, and polymer isolated from willow tree catkins to that of the catechin monomer administered in the rat diet. Sixteen metabolites of microbial origin were detected and identified as phenylvaleric, phenylpropionic, phenylacetic, and benzoic acid derivatives. Their total yields significantly decreased from the catechin monomer to the procyanidin dimer, trimer, and polymer, confirming a higher ability of rat fecal microflora to degrade catechin monomer in relation with more complex procyanidins. Therefore, the degree of procyanidin polymerization has a major impact on their fate in the body characterized by a poor absorption through the gut barrier and a limited metabolism by the intestinal microflora as compared to catechin.

A recent work by Serra et al. [24] studied the colonic metabolic pathways of procyanidins (catechin, epicatechin, epicatechin gallate, epigallocatechin gallate, and dimer B2) using rat colonic microflora. Results showed an intense metabolism of all procyanidins tested after 24–48 h of fermentation. Differences in the fermentation of the stereoisomers catechin and epicatechin were observed. When epicatechin was fermented, a lower number of fermentation products were obtained,

5-(hydroxyphenyl)- γ(gamma)-valerolactone being the main fermentation product. However, when catechin was fermented, a lower concentration of (hydroxyphenyl)- γ(gamma)-valerolactone was produced; phenylacetic acid was the main fermentation product. The metabolism of galloylated epicatechins, epicatechin gallate and epigallocatechin gallate, resulted in a high number of metabolites, and p-hydroxybenzoic, protocatechuic, phenylacetic, 2-hydroxyphenylacetic, and 4-hydroxyphenylacetic acids were the common metabolites. In contrast with what was expected, the fermentation of the dimer B2 did not lead to the formation of the same metabolic products of epicatechin and phenylacetic, and 4-hydroxyphenylacetic acids were the only common metabolites. The dimer B2 was hydrolyzed to epicatechin during the first period of incubation. After the first hydrolysis of the dimer B2, several derivatives of benzoic and phenylacetic acids with different patterns of hydroxylation and protocatechuic acids were identified.

The intense colonic metabolism of monomers and the limited metabolism of procyanidins with high polymerization degree will have to be considered to explain the health effects of cocoa procyanidins. In fact, some microbial metabolites may have a physiologic effect; for example, hydroxyphenylacetic acids have been suggested to inhibit platelet aggregation [25]. Besides, among the wide array of aromatic acids with low molecular weight, some may be used as biomarkers for polyphenol intake [26, 27]. Thus, the identification and quantification of microbial metabolites of monomeric units of flavan-3-ol (e.g., catechin, and epicatechin) constitute an important field for future research.

Caco-2 Cellular Models: Phase II Metabolism

The intestinal barrier is also an active member that modifies the native polyphenols that are extensively altered during first-pass metabolism, so the molecular forms that reach the peripheral circulation and tissues are often different from those in foods [28]. During the course of absorption, polyphenols are conjugated in the small intestine and later in the liver. Cell culture models have been also taken into account as one step of the in vitro digestion model. In particular, the Caco-2 cell (the name of polarized human colon carcinoma cell line) model has been widely used as a predictive tool for the absorption of food nutrients. In this model, the amount of nutrient after the Caco-2 step is designed as the bioavailable fraction. However, there are no specific studies related to the evaluation of the cocoa procyanidins metabolism using Caco-2 cell models.

Human intestinal epithelial Caco-2 cells grown on permeable inserts attain many of the morphological and functional characteristics of intestinal enterocytes. The low bioavailability of green tea catechins, including epigallocatechin gallate, epigallocatechin, epicatechin gallate, and epicatechin, is low in both animals and humans. The contribution of intestinal efflux to this low bioavailability has been suggested. The Caco-2 cell monolayer has been used to investigate the kinetics of efflux transport (basal-to-apical transport) of the tea catechins [29]. Kinetic studies indicated that active and saturable efflux transport of epicatechin took place in Caco-2 cells, suggesting the importance of efflux transporters and their role in the limited bioavailability of epicatechin. By contrast, no saturation could be observed for the efflux transport of epicatechin gallates even at concentrations up to about 200 μM, which may be due to their low affinity towards the transporters.

The gut absorption of proanthocyanidins and of the related catechin monomer was investigated with colonic carcinoma (Caco-2) cells of a human origin, grown in monolayers on permeable filters. Permeability of various radiolabeled proanthocyanidins differing in their molecular weight was compared with that of the radiolabeled catechin [30]. Catechin, dimer, and trimer had similar permeability coefficients close to that of mannitol, a marker of paracellular transport. In contrast, permeability of a proanthocyanidin with an average polymerization degree of six was ten times lower. These results suggest that dimers and trimers could be absorbed in vivo and that polymer bioavailability is limited to the gut lumen. Radiolabeled monomers, dimers, and trimers were transported across a layer of Caco-2 cells, in contrast to oligomers of six units, which were transported approximately tenfold less [30].

The coculture of Caco-2 with hepatocytes has been shown to be a feasible system for assaying the conjugation mechanisms in the small intestine and the potential bioactivity of a procyanidin-rich extract derived from grape seeds [31]. The main metabolites resulting from the Caco-2 metabolism corresponded to sulfate derivatives of catechin and epicatechin. Also glucuronidated forms were quantified in lower concentration. By contrast, dimer and trimer were quantified in cellular medium in their native forms. This demonstrates the inability of enterocytes to the conjugation of molecular structures more complexes that monomeric units of flavan-3-ol as catechin and epicatechin.

Bioavailability In Vivo

The bioavailability of polyphenols greatly depends on the chemical structure, mainly due to the glycosylation, conjugation, and polymerization form. Thus, the forms circulating in blood are mainly glucuronidated, sulfated, and methylated derivatives of the parent compounds. Additionally, the inter-subject variability has to be taken into account when bioavailability trials are performed with humans which sometimes may obscure the true meaning of the obtained results. For that reason, when it is feasible, it is desired to measure the bioavailability in vivo by two approaches – the determination of polyphenol concentration in plasma and in urine after an ingestion of a known amount of polyphenols in a determined food matrix. In general, plasma metabolites showed maximum concentration peak (C_{max}) after 1–3 h of food intake; meanwhile, the urine metabolites are detected after 24 h, suggesting that enterohepatic circulation occurs in the liver, producing mainly sulfate conjugates, whereas the first metabolism produces the glucuronide conjugates in the gastrointestinal cell's tract.

The bioaccessibility of monomers (catechin, epicatechin, and gallates) and dimers from several cocoa sources has been shown in the previous section through in vitro digestion models. There is extensive literature on in vivo studies of bioavailability of cocoa polyphenols that has confirmed the presence of cocoa polyphenols in plasma at levels often ranges between nanomolar and low micromolar concentration. [32, 33] Richelle et al. [34] were one of the first studies aimed at the bioavailability of cocoa polyphenols in humans. This study analyzed the plasma of healthy men after a single dose of 40–80 g of dark chocolate. Indeed, in this study, results indicated that epicatechin was quickly absorbed. Schroeter et al. [35] observed similar rate of absorption (T_{max}) to those reported by Richelle et al. [34] even though for this study a high-flavonol cocoa drink was assayed and the peak serum levels (C_{max}) quantified were around 2.75 nM. Several further studies confirmed the rapid absorption of epicatechin into plasma, with plasma concentrations peaking at two or three hours after ingestion [36]. The chiral nature of polyphenols is also seen as a variable that could influence their bioavailability. In that sense, Donovan et al. [37] observed that the intestinal absorption of (−)-catechin was lower than the intestinal absorption of (+)-catechin. After perfusion of 10, 30, and 50 μmol/L of each catechin enantiomer ((−) and (+)) in the jejunum and ileum from a rat, plasma concentrations of (−)-catechin were significantly reduced compared to (+)-catechin, suggesting that (+)-catechin enantiomer is more bioavailable than (−)-catechin enantiomer. Indeed, these in vivo studies confirm the functional properties of cocoa polyphenols based on targets such as plasma antioxidant status and oxidative stress, endothelial function and flow-mediated dilation, the levels and oxidation of plasma lipids, and nitric oxide, among others.

Studies related to the bioavailability of oligomeric and polymeric procyanidins, performed either with animals and humans, have indicated that these structures are not absorbed. Therefore, it can be supposed that most polymeric units pass unaltered to the large intestine where they are catabolized by the colonic microflora, overcoming a diversity of phenolic acids and several metabolites [38]. Rios et al. [22] verified in an in vivo experiment with humans the stability of cocoa procyanidins during the gastric transit suggested by in vitro studies [15]. However, minor quantities of procyanidin dimers B1 and B2 were detected in human plasma after the consumption of a flavan-3-ol-rich cocoa [36]. A trend is observed in the available studies: the smaller the polyphenol structure, the higher the concentration in plasma, which gives it a greater chance of reaching its target organs in the body.

Bioavailability can also be affected by the matrix in which the cocoa polyphenols are delivered. To overcome this item, several trials have been focused on the assessment of the matrix effect in the bioavailability of polyphenols. For example, Schram et al. [39] examined the absorption of flavonoid aglycones after a carbohydrate-rich meal (bread, sucrose). This study showed an enhanced bioavailability of flavonols that could be affected by the carbohydrate source. Likewise, Neilson et al. [40] also observed that bioavailability of cocoa polyphenols was influenced by the sucrose content as well as by the physical matrix form that evolved the cocoa polyphenols, both affecting mainly the rate of absorption (T_{max}) and the peak serum levels (C_{max}).

High molecular weight polyphenols such as procyanidins may interact with protein, indicating that the addition of milk could reduce the polyphenol bioavailability. Previous studies showed contradictory results about the milk effect on bioavailability of polyphenols in other foodstuffs. Catechin was reported in plasma at low levels after the intake of tea with milk in relation to those levels detected after the intake of black tea [41]. In contrast, the addition of milk to English and Indian black tea showed comparable bioaccessibility of total catechins in relation to both tea controls [42]. Indeed, in this study, no influence on catechin bioaccessibility was observed depending on the type of milk used (skimmed or full-fat milk). Due to the controversial results, several works have been aimed at the milk's effect on flavonol cocoa absorption. In this sense, Keogh et al. [43] did not observe any correlation between flavonoid absorption and the presence of milk. In the same light, Tomás-Barberán et al. [44] and Roure et al. [45] did not perceive any influence on the polyphenol absorption concentration in relation to the presence of milk in the cocoa beverage intake. These studies indicated that milk did not seem to influence the bioavailability of cocoa flavonols (mainly monomers and dimers).

Summary

The studies of bioavailability of phytochemicals carried out in animals or human subjects are complex, expensive, and lengthy. In vitro digestion models permit the characterization of cocoa procyanidins during digestion under physiological conditions, caused by alimentary enzymes.

In combination with a cultured cell model that facilitates studies of small intestinal absorption and metabolism, it is possible to determine the relationship between the molecular structures, gastrointestinal stability, conjugation, and extent of absorption for cocoa procyanidins as an important step towards elucidating the potential impact of these compounds on human health.

Characterization of the digestive species of cocoa procyanidins forming as a consequence of hydrolysis, conjugation, and colonic fermentation will provide valuable insight regarding the potential alteration of these compounds during digestive transit and will provide a framework for future animal and human studies in vivo.

The in vitro digestion models allow the screening of multiple samples and may provide data on the relative bioavailability of different type of procyanidins. Moreover, these models could act as a first directional influence on the reformulation of new cocoa products in order to achieve the required "nutrient profiles," in terms of sugar content and fatty acids, allowing the effect on potential bioaccessibility of procyanidins related with the food matrix composition.

References

1. Wollgast J, Anklam E. Review on polyphenols in *Theobroma cacao*: changes in composition during the manufacture of chocolate and methodology for identification and quantification. Food Res Int. 2000;33:423–47.
2. Jalil AMM, Ismail A. Polyphenols in cocoa and cocoa products: is there a link between antioxidant properties and health? Molecules. 2008;13:2190–219.

3. Scalbert A, Williamson G. Dietary intake and bioavailability of polyphenols. J Nutr. 2000;130:2073S–85.
4. Hackman RM, Polagruto JA, Zhu QY, Sun B, Fujii H, Keen CL. Flavanols: digestion, absorption and bioactivity. Phytochem Rev. 2008;7:195–208.
5. Na HK, Surh YJ. Intracellular signaling network as a prime chemopreventive target of (–)-epigallocatechin gallate. Mol Nutr Food Res. 2006;50:152–9.
6. Myhrstad MCW, Carlsen H, Nordstrom O, Blomhoff R, Moskaug JO. Flavonoids increase the intracellular glutathione level by transactivation of the gamma-glutamylcysteine synthetase catalytical subunit promoter. Free Radic Biol Med. 2002;32:386–93.
7. Parada J, Aguilera JM. Food microstructure affects the bioavailability of several nutrients. J Food Sci. 2007;72:21–32.
8. Gil-Izquierdo A, Zafrilla P, Tomás-Barberán F. An in vitro method to simulate phenolic compounds release from the food matrix in gastrointestinal tract. Eur Food Res Technol. 2002;214:155–9.
9. Bermúdez M, Tomás-Barberan F, García M. Stability of polyphenols in chokeberry (*Aronia melanocarpa*) subjected to an in vitro gastric and pancreatic digestion. Food Chem. 2007;68:1285–94.
10. Ortega N, Reguant J, Romero MP, Macia A, Motilva MJ. Effect of fat content on the digestibility and bioaccessibility of cocoa polyphenol by an in vitro digestion model. J Agric Food Chem. 2009;57:5743–9.
11. Wickham M. Plant Bioscience Ltd., Norwich Research Institute, World Intellectual Property Organization, WO 2007/010238.
12. Wickham M, Faulks R, Mills C. In vitro digestion methods for assessing the effect of food structure on allergen breakdown. Mol Nutr Food Res. 2009;53:952–8.
13. Kong F, Singh RP. A model stomach system to investigate disintegration kinetics of solid foods during gastric digestion. J Food Sci. 2008;73:202–10.
14. Hendrén E, Mulokozi G, Svanberg U. In vitro bioaccessibility of carotenoids from green leafy vegetables cooked with sunflower oil or red palm oil. Int J Food Sci Nutr. 2002;53:445–53.
15. Spencer JPE, Chaudrya F, Pannalaa AS, Sraib SK, Debnamb E, Rice-Evans C. Decomposition of cocoa procyanidins in the gastric milieu. Biochem Biophys Res Commun. 2000;272:236–41.
16. Zhu Q, Holt R, Lazarus S, Ensunsa J, Hammerstone J, Schmitz H. Stability of the flavan-3-ols epicatechin and catechin and related dimeric procyanidins derived from cocoa. J Agric Food Chem. 2002;50:1700–5.
17. Neilson A, Hopf A, Cooper B, Pereira M, Bomser J, Ferruzzi M. Catechin degradation with concurrent formation of homo- and heterocatechin dimers during in vitro digestion. J Agric Food Chem. 2007;55:8941–9.
18. Regulation (EC) No. 1924/2006 of the European Parliament of the Council of December 2006 on nutrition and health claims made on foods. Official Journal L. 30.12.2006; 404:9.
19. Ortega N, Reguant J, Romero MP, Macià A, Motilva MJ. Effect of fat content on the digestibility and bioaccessibility of cocoa polyphenol by an in vitro digestion model. J Agric Food Chem. 2009;57:5743–9.
20. Ortega N, Macia A, Romero MP, Reguant J, Motilva MJ. Matrix composition effect on the digestibility of carob flour phenols by an in-vitro digestion model. Food Chem. 2011;124:65–71.
21. Fogliano V, Corollaro M, Vitaglione P, Napolitano A, Ferracane R, Travaglia F, et al. In vitro bioaccessibility and gut biotransformation of polyphenols present in the water-insoluble cocoa fraction. Mol Nutr Food Res. 2011;55:44–55.
22. Rios L, Gonthier M, Rémésy C, Mila I, Lapierre C, Lazarus S, et al. Chocolate intake increases urinary excretion of polyphenol-derived phenolic acids in healthy human subjects. Am J Clin Nutr. 2003;73:912–8.
23. Gonthier M, Donovan J, Teixer O, Felgines C, Remesy C, Scalbert A. Metabolism of dietary procyanidins in rats. Free Radic Biol Med. 2003;35:837–44.
24. Serra A, Macià A, Romero MP, Anglés N, Morelló JR, Motilva MJ. Metabolic pathways of the colonic metabolism of procyanidins (monomers and dimers) and alkaloids. Food Chem. 2011;126:1127–37.
25. Kim DH, Jung EA, Sohng IS, Han JA, Kim TH, Han MJ. Intestinal bacterial metabolism of flavonoids and its relation to some biological activities. Arch Pharm Res. 1998;21:17–23.
26. Clifford MN, Copeland EL, Bloxsidge JP, Mitchell LA. Hippuric acid as a major excretion product associated with black tea consumption. Xenobiotica. 2000;30:317–26.
27. Graefe EU, Veit M. Urinary metabolites of flavonoids and hydroxycinnamic acids in humans after application of a crude extract from Equisetum arvense. Phytomedicine. 1999;6:239–46.
28. Manach C, Williamson G, Morand C, Scalbert A, Rémésy C. Bioavailability and bioefficacy of polyphenols in humans. I. Review of 97 bioavailability studies. Am J Clin Nutr. 2005;81(1 Suppl):230S–42.
29. Chan KY, Zhang L, Zuo Z. Intestinal efflux transport kinetics of green tea catechins in Caco-2 monolayer model. J Pharm Pharmacol. 2007;59:395–400.
30. Deprez S, Mila I, Huneau J-F, Tome D, Scalbert A. Transport of proanthocyanidin dimer, trimer, and polymer across monolayers of human intestinal epithelial Caco-2 cells. Antioxid Redox Signal. 2001;3:957–67.
31. Castell-Auví A, Motilva MJ, Macià A, Torrell H, Bladé C, Pinent M, et al. Organotypic co-culture system to study plant extract bioactivity on hepatocytes. Food Chem. 2010;122:775–81.
32. Cooper K, Donovan J, Waterhouse A, Williamson G. Cocoa and health: a decade of research. Br J Nutr. 2008;99:1–11.

33. Rimbach G, Melchin M, Moehring J, Wagner A. Polyphenols from cocoa and vascular health – a critical review. Int J Mol Sci. 2009;10:4290–309.
34. Richelle M, Tavazzi I, Enslen M, Offord E. Plasma kinetics in man of epicatechin from black chocolate. Eur J Clin Nutr. 1999;53:22–6.
35. Schroeter H, Heiss C, Balzer J, Kleinbogard P, Keen CL, Hollenberg NK, et al. (−)-Epicatechin mediates beneficial effects of flavonol rich cocoa on vascular functions in humans. Proc Natl Acad Sci USA. 2006;103:1024–9.
36. Holt RR, Lazarus SA, Sullards MC, Zhu QY, Schramm DD, Hammerstone JF, et al. Procyanidin dimer B2 [epicatechin-(4β-8)-epicatechin] in human plasma after the consumption of a flavanol-rich cocoa. Am J Clin Nutr. 2002;76:798–804.
37. Donovan J, Crespy V, Oliveira M, Cooper K, Gibson B, Williamson G. (+)-Catechin is more bioavailable than (−)-catechin: relevance to the bioavailability of catechin from cocoa. Free Radic Res. 2006;40:1029–34.
38. Selma MV, Espín SC, Tomás-Barberán FA. Interaction between phenolics and gut microbiota: role in human health. J Agric Food Chem. 2009;12:6485–64501.
39. Schramm D, Karim M, Schrader H, Roberta R, Kirkpatrick N, Polagruto J, et al. Food effects on the absorption and pharmacokinetics of cocoa polyphenols. Life Sci. 2003;73:857–69.
40. Neilson AP, Sapper TN, Janle EM, Rudolph R, Matusheski NV, Ferruzzi MG. Chocolate matrix factors modulate the pharmacokinetic behavior of cocoa flavan-3-ol phase II metabolites following oral consumption by Sprague–Dawley rats. J Agric Food Chem. 2010;58:6685–91.
41. Reddy V, Vidya S, Sreeramulu D, Venu L, Raghunath M. Addition of milk does not alter the antioxidant activity of black tea. Ann Nutr Metab. 2005;49:189–95.
42. Van der Burg-Koorevaar M, Miret S, Duchateau G. Effect of milk and brewing method on black tea catechin bioaccessibility. J Agric Food Chem. 2011;59:7752–8.
43. Keogh JB, McInerney J, Clifton PM. The effect of milk protein on the bioavailability of cocoa polyphenols. J Food Sci. 2007;72:S230–3.
44. Tomas-Barberán FA, Cienfuegos-Jovellanos E, Marín A, Muguerza B, Gil-Izquierdo A, Cerdá B, et al. A new process to develop a cocoa powder with higher flavonoid monomer content and enhanced bioavailability in healthy humans. J Agric Food Chem. 2007;55:3926–35.
45. Roure E, Lacueva A, Estruch R, Mata M, Izquierdo M, Waterhouse A, et al. Milk does not effect the bioavailability of cocoa powder. Ann Nutr Metab. 2008;51:493–8.

Chapter 16
The Effect of Chocolate on Human and Gut Microbial Metabolic Interactions: Emphasis on Human Health and Nutritional Status

François-Pierre J. Martin, Sebastiano Collino, Serge Rezzi, and Sunil Kochhar

Key Points

- One of the greatest scientific challenges in modern nutrition is to decipher the key metabolic interactions between food and human metabolism and to understand its role in health and wellness.
- Despite the growing evidence on the health benefits associated with chocolate, the mechanisms of action of chocolate bioactive components at the molecular level are poorly understood.
- Recent analytical advances now provide comprehensive approaches for monitoring the nutritional effects at the levels in multiple biochemical pathways using integrative systemic metabolic and microbiome profiling.
- Through the rigorous characterization of interactions between the diet, human metabolism, and the microbiota, metabonomics is providing new ventures for modulating the microbiota toward the improvement of human health.

Keywords Chocolate • Cocoa • Gut microbiota • Host metabolism • Metabonomics

Introduction

Over the past years, changes in diet and recognition of health benefits of specific food components have evolved rapidly, having a direct influence on the well-being and nutritional status of individuals. Dietary preferences and habits, which are predominantly cultural in origin, affect the health of both individuals and populations. Food selection initiates from a decision process that integrates multiple biological, socioeconomic, psychological, and behavioral determinants. For instance, food intake is a complex process under the influence of both physiological (e.g., hormonal regulation of hunger and

F.-P.J. Martin, Ph.D. • S. Collino, Ph.D. • S. Rezzi, Ph.D.
Metabonomics and Biomarkers Group, BioAnalytical Science Department,
Nestlé Research Center, Vers-chez-les-Blanc, 44, Lausanne VAUD 1000, Switzerland
e-mail: francois-pierre.martin@rdls.nestle.com; sebastiano.collino@rdls.nestle.com; serge.rezzi@rdls.nestle.com

S. Kochhar, Ph.D. (✉)
BioAnalytical Science Department, Nestlé Research Center, Vers-chez-les-Blanc, 44,
Lausanne VAUD 1000, Switzerland
e-mail: sunil.kochhar@rdls.nestle.com

R.R. Watson et al. (eds.), *Chocolate in Health and Nutrition*, Nutrition and Health 7,
DOI 10.1007/978-1-61779-803-0_16, © Springer Science+Business Media, LLC 2013

satiety) and subjective aspects [1, 2]. The latter encapsulates senses (e.g., taste, smell, palatability, or texture), cognitive perception (e.g., preferences, aversions), and the postprandial feeling of wellness perceived by individuals after the consumption of the product [3]. In particular, genetically mediated taste response to specific foods or ingredients is ultimately reflected in various dietary preferences [4, 5]. Moreover, consumers have specific ideas about the product features that influence the perceived satiety level, which may subsequently affect their food preference [6]. For instance, products perceived as fat, high in proteins, with a savory taste are expected to have a higher level of satiety compared to sweet products. Dietary composition impacts the metabolism of an individual, and the dynamic responses of the metabolic phenotype to dietary modulation are well documented [2, 7, 8]. Long-term dietary preferences associated with, for example, high levels of saturated fat or carbohydrate consumption are strongly associated with obesity and cardiovascular disease [9, 10]. There are also clear metabolic differences associated with the alternation of vegetarian and omnivorous diets [8]. In addition, there are also strong interactions between dietary composition and gut microbial metabolic activities that might contribute to nutritional and health status of an individual. Recent studies have also highlighted the importance of altered gut microbiota as a possible major contributor to calorific harvest [11] and to obese phenotypes in animal [12] and human models [13].

Recent applications of top-down system biology approaches provided the ability to measure subtle changes in biological processes as a result of different nutritional effects [14–16], and described the importance of gut microbiota modulation in complex organisms, including host lipid, carbohydrate, and amino acid metabolism at a panorganismal scale [17–20]. Modern nutrition research has indeed promoted the use of metabonomics to gain a deeper understanding of the interactions between nutrition, health, and physiological processes (Fig. 16.1) [2]. Nutrimetabonomics, or nutritional metabonomics, is today a well-established approach for the analysis of physiological regulatory processes via the quantitative measurement of nutrition-induced metabolic variations [2]. Metabolites are the products of the many intricate biosynthetic and catabolic pathways existing in all living systems. Biological systems that exist at steady state maintain relatively constant concentrations of metabolic intermediates. Specific physiological states, gene expression, and environmental stimuli can cause changes in existing homeostatic conditions and reaction fluxes. All these fluxes across cells, tissues, and organs of the organism are thus reflected ultimately in the metabolic composition of the different biological compartments. Therefore, monitoring the resulting metabolic variations provides unique insights into intra- and extracellular regulatory processes, and it allows the characterization of individual metabolic phenotypes [21–24]. In particular, metabolic profiles encapsulate information on the metabolic activity of symbiotic partners, that is, gut microbiota, of complex organisms, which represent a major determinant in nutrition and health.

Perhaps one of the greatest challenges in modern nutrition is to interrogate and classify the critical metabolic interactions between the complex food matrices, which contain a wide range of biologically active compounds, and human system metabolism and to understand their role in human diseases. The main goal of nutrimetabonomic studies is therefore to study the effects of selected ingredients and foods in individuals. The sheer complexity of a food matrix, such as dark chocolate, may determine a large variety of biological outcomes. Polyphenols are of great interest in nutrition owing to their potential antioxidant capacity and possible protective effects in reducing the risk of diseases, such as cardiovascular diseases. Cocoa or chocolate contains flavanols that form a unique class of polyphenols, including monomers (epicatechin and catechin), oligomers, and polymers (proanthocyanidins) [25]. Different studies showed that cocoa flavanols prevent LDL oxidation, enhance endothelial functions, or modulate cytokine transcription in peripheral blood mononuclear cells [26, 27]. Moreover, cocoa polyphenols may increase the concentration of HDL cholesterol, whereas chocolate fatty acids may modify the fatty acid composition of LDL and make it more resistant to oxidative damage. In a recent study, the serum concentration of HDL cholesterol was increased by 4% in healthy subjects following dark chocolate consumption [28]. Another study described that the daily consumption of 75 g of dark chocolate or dark chocolate enriched with cocoa polyphenols

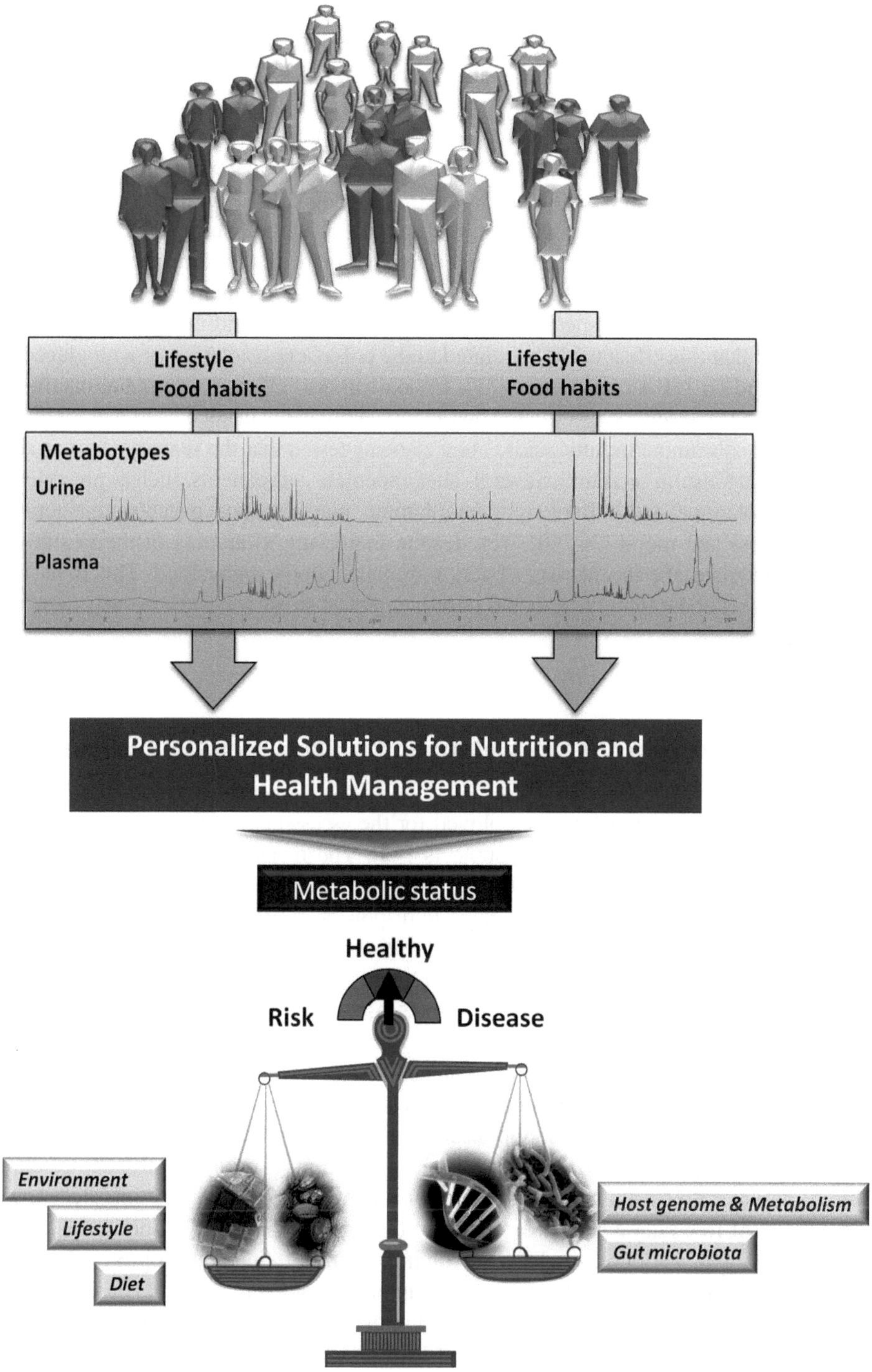

Fig. 16.1 Conceptualization of nutritional metabonomics for health and risk management. The metabolic status of individuals results from the continuous interaction of genes, environment, lifestyle, food, and gut microbiota. Different metabotypes, as reflected in urine or plasma biochemical composition, for instance, are under homeostatic regulations. Metabolic profiling can be used to measure homeostasis status, nutritional effects, and likelihood to develop specific metabolic deregulations. Nutritional metabonomics aims at optimizing nutrition for health maintenance and disease prevention (Images © Nestlé S.A.)

increased serum HDL cholesterol (11.4% and 13.7%, respectively) over a period of 3 weeks [29]. Cocoa and chocolate products have been reported to exhibit a total antioxidant capacity often exceeding other antioxidant-rich foods such as red wine and green or black tea [30]. Great attention in nutrition research has been mostly directed to the antioxidant ability of cocoa and chocolate and their eventual protective effects against cardiovascular diseases [31]. Systolic blood pressure can be, for instance, lowered by 5.6% with dark chocolate [32]. In addition, several benefits of cocoa on improvement of insulin sensitivity and glucose tolerance were reported [33, 34]. For instance, insulin sensitivity was higher in healthy volunteers consuming 100 g dark chocolate during 15 days compared to white chocolate [35]. It has also been demonstrated that tea increases in vitro insulin activity in the insulin-potentiating epididymal fat-cell assay with rat adipocytes and that this potentiating activity was mainly due to polyphenols (especially epigallocatechin gallate and epicatechin gallate) [36]. It was hypothesized that this effect could be related to the polyphenols interfering with glucose transport via GLUT 1 and GLUT 4 transporters [37]. Cocoa can also affect postprandial insulinemia in lean young adults [38]. The authors postulated that many components in cocoa could elicit this effect, from caffeine and theobromine to fatty acids. They also suggested that the sensory characteristics of cocoa might potentiate insulin secretion. Several other chocolate components, such as phenylethylamine, N-oleoyl-ethanolamine, and N-linoleoyl-ethanolamine, may also have psychoactive activity that could modulate stress and mood [39, 40]. Yet, despite increasing awareness of the health benefits associated with chocolate, the mechanism of action remains poorly understood. This is particularly the case for benefits related to brain health and improvement of stress states where only symptomatic data are available [41].

Specific Food Habits Imprint the Metabolic Profile of Healthy Individuals

Recently, a metabonomics approach was employed for the assessment of the metabolic response to chocolate consumption in 22 healthy men subjects [42]. The metabolic profiles generated by ^{1}H nuclear magnetic resonance (NMR) spectroscopy [43] and mass spectrometry (MS) [44] on both plasma and urine samples showed distinct metabolic responses in subjects according to their chocolate liking which was determined on the basis of a questionnaire [42]. This clinical trial consisted of a study period of 5 days with two test days on which blood and 24-h urine were collected before and after intervention. The authors demonstrated how spectroscopically generated metabolic phenotypes can be correlated with behavioral/psychological dietary preference for chocolate products. The metabolic profiles collected on both test days thus indicated that "chocolate liking" is a measurable individual preference class, although dietary chocolate itself is likely to be only one indicator of a more complex dietary pattern. The main metabolic determinants in these metabotypes were a differentiating signature in the plasma concentrations of LDL and albumin, even in the absence of chocolate stimulation. Albumin, the most abundant plasma protein, is involved in many physiological processes, including transport of free fatty acids generated from the lipoprotein lipase action in the lipoprotein metabolism. In particular, a relationship between albumin concentration and apoprotein B-100 secretion and LDL catabolism has also been reported on a human hepatocyte model [45]. Furthermore, the class-specific metabolic phenotype was also encoded in the urinary profiles, suggesting both distinct energy and microbiota metabolism. The increased urinary excretion of phenylacetylglutamine and citrate in the chocolate "likers" suggested a different modulation of the tricarboxylic acid cycle [46], associated with changes in the levels of carnitine, N-acetyl-carnitine, and ketone bodies that could be related to a different basal energy metabolism [47, 48]. Urinary metabolic profiles also encapsulate information on the gut microbial metabolic activities, as observed through the excretion of polar microbial-mammalian cometabolites [7, 18, 20], including hippurate, 2-hydroxyhippurate, phenylacetylglutamine, p-cresol sulfate, indole-3-acetate, and 3-hydroxyisovalerate [18, 49–52].

These metabolic variations reflect different metabolism of dietary aromatic amino acids and polyphenols by colon microorganisms [53]. The modulated urinary excretion of phenolic and indolic compounds reveals variations in gut microbial activities in relation to nutritional competition [54, 55], indoleacetate being reported, for instance, to inhibit the growth and survival of lactobacilli [55]. In particular, the conversion of 4-hydroxyphenylacetate to 4-cresol exhibits a negative correlation in the urine profiles of chocolate "likers," which corresponds to a molecular process that has been associated with the presence of *Clostridium difficile* but also other bacterial strains [49]. This observation suggests a differential management of the metabolic pool of the precursor 4-hydroxyphenylacetate by gut bacteria. Interestingly, the correlation of 2-hydroxyhippurate with hippurate, two metabolites deriving from the metabolism of dietary polyphenols by colon microorganisms [53], and the particular correlation of trigonelline with 2-hydroxyhippurate might indicate a specific microbial modulation of dietary flavonoids and niacin metabolism in "chocolate likers." The production of phenylacetate and indoleacetate has been restricted to a certain taxonomic group of gut bacteria, including Bacteroides, Clostridia, and *E. coli*, which count among the dominant species in humans [54, 56]. The differential urinary excretion of 2-hydroxyhippurate and hippurate indicates further a specific microbial modulation of dietary flavonoids in each group of subjects. Altogether, these findings demonstrate imprinted differences in the gut microbiotal metabolic activities of the individuals that appear to be dependent on the dietary consumption habits. This imprinting was reported independently of the ingested food, as food challenge had no direct effects.

Chocolate, Energy, Gut Microbiota, and Stress-Related Metabolism in Free-Living Subjects

In a follow-up study, ^{1}H NMR spectroscopy [43] and mass spectrometry [44] were employed as complementary analytical techniques to capture a global view of the metabolic events associated with chocolate consumption in healthy and free-living men and women [57], ^{1}H NMR spectroscopy [43] and mass spectrometry [44] were employed as complementary analytical techniques [57]. A clinical trial was conducted on a population of thirty human subjects, who were classified in low- and high-anxiety trait groups using validated psychological questionnaires. Subjects maintained their usual lifestyle and nutrition while consuming dark chocolate twice a day for 2 weeks. Biological fluids (urine and blood plasma) were collected at baseline, 1 week, and 2 weeks after daily consumption of chocolate. The metabonomics analysis of blood and urine samples revealed variations in relation to trait anxiety levels even prior to nutritional intervention [57]. For instance, stress leads to elevation of hormones, such as the glucocorticoids and catecholamines [58, 59]. Although the presence of these hormones is essential for modulating acute stress, there are valid associations between long-term exposures to high levels of these stress hormones and the development of multiple forms of chronic illnesses [60, 61]. In particular, the individual response to chronic stress is tightly connected to the hypothalamic–pituitary–adrenal metabolic axis and the sympathoadrenal system, which in turn determine the overall homeostatic balance [58]. The reported results described systemic changes in hormonal metabolism of high trait anxiety individuals when compared to low trait anxiety subjects, as observed with a higher urinary excretion of adrenaline, 3,4-dihydroxyphénylalanine (DOPA) and 3-methoxy-tyrosine, two intermediates in dopamine synthesis [62]. Additional metabolic changes were consistent with the stress-mediated modulation of gluconeogenesis by catecholamines [58], as noted with the metabolic changes observed in several pathways involving the tricarboxylic acid cycle (citrate, succinate, aconitate) [46], gluconeogenetic pathways (lactate), and urea cycle (urea, proline). Nowadays, there is compelling evidence that life stress impacts directly upon gastrointestinal health in animals and humans via modulation of key functional parameters, such as intestinal permeability and secretion and release of biological mediators [60, 63, 64]. Changes of gastrointestinal function are intimately linked to gut microbial populations and activities [65], and metabolic monitoring of urinary excretion of many mammalian microbial cometabolites – namely, phenolics, indoles, and benzoyl derivatives, methylamines, short-chain fatty acids – provides information on the gut microbial metabolic activities [49, 52, 66]. Differences in trait anxiety levels were associated to differential urinary excretion of p-cresol sulfate and hippurate, two metabolites associated with gut microbial metabolism of aromatic amino acids [20, 49]. Certain aromatic compounds, such as benzoate and phenylacetate, that can be coprocessed by the gut microbiota, are well-characterized agents reducing the levels of circulating glycine and glutamine [46, 67]. Both NMR- and MS-based metabolic profiling of urine revealed relatively higher excretion of glycine in high trait anxiety individuals, with inferred relationships with amino acid interconversion and benzoate metabolism.

The metabolic response to chocolate intervention revealed that a daily intake of dark chocolate resulted in subtle and cumulative metabolic effects of the urinary excretion of gut microbial cometabolites over a 2-week period. Increased levels of m-hydroxyphenylacetate and decreased content of phenylacetylglutamine and p-cresol sulfate can result from the adaptation of gut microbiota to process dark chocolate constituents, such as phenylethylamine, N-oleoyl-ethanolamine and N-linoleoyl-ethanolamine, theobromine, and flavonoids (epicatechins, catechins, and its oligomers) [49, 52, 53]. In particular, urinary excretion of m-hydroxyphenylacetate was previously ascribed to the intake of polyphenol-rich products, such as chocolate [25]. These observations are therefore complementary to our preliminary investigations of metabolic imprinting associated to chocolate dietary habits [42]. Moreover, these observations indicated that the metabolic impact of a daily intake of dark chocolate was strongly dependent on the dispositional stress state of the individuals, as noted with statistically significant metabolic effects only in subjects with inherent high trait anxiety. Consumption of dark chocolate resulted in the decrease of the urinary levels of catecholamines (adrenaline, noradrenaline, normetanephrine), corticosterone, and stress hormone cortisol in subjects with high trait anxiety. Chronic stress is correlated with increases in stress hormones cortisol and catecholamines [68–71], suggesting potential beneficial implications of dark chocolate for reduction of mental and/or physical stress and improvement of the metabolic response to stress. At the same time, chocolate consumption partially normalized stress-related differences in energy metabolism (glycine, citrate, trans-aconitate, proline, β-alanine) and gut microbial activities (hippurate and p-cresol sulfate).

This study demonstrated the ability of metabonomics to measure the metabolic signatures of stress in healthy individuals under free-living conditions while assessing their metabolic changes to daily consumption of dark chocolate. The outcomes of this study provided additional evidence that consumption of dark chocolate may beneficially impact on stress-associated metabolism as observed through a partial normalization of stress-related differences in energy metabolism and gut microbial activities.

The Beneficial Effects of Chocolate and Cocoa Bioactives Are Partially Mediated by Gut Microbiota

The complex association of gut bacterial cells extends the processing of undigested food to the benefit of the host through metabolic capacities not encoded in mammalian genomes [65, 72]. In mammals, microbial communities differ in composition from the stomach to the colon, where the competition for space and nutrients in the large bowel determines the microbial composition of this internal ecosystem. In particular, the colon is one of the most populated sections of the gastrointestinal tract, with high density of bacterial cells of up to 10^{12} per mL [73], and gut bacteria may represent up to 60% of the total fecal mass [65, 74]. The main human intestinal bacteria, which are predominantly species of the *Bacteroides, Clostridium, Lactobacillus, Eubacterium, Faecalibacterium,* and *Bifidobacterium* groups, therefore coexist in a dynamic ecological equilibrium together with various yeasts and other microorganisms [75], the whole ecosystem varying in addition with the host's age, diet, and health status.

The members of the gut microbiota consortium are diverse and provide the host with specific capacities ranging from dietary energy recovery from nutrient load, generating digestible carbohydrates, amino acids, and vitamins, and ultimately determine the concentrations of bioactive and bioavailable polyphenols to the host metabolism [25, 76]. Indeed, despite polyphenols are of great interest in nutrition and medicine, it is often assumed from in vitro studies that cocoa polyphenols are bioavailable and reach the target inner tissues. However, proanthocyanidins are poorly absorbed through the gut barrier because of their high molecular weight [77–79]. The formation of several phenolic acids from proanthocyanidins in in vitro studies with human fecal microflora [78] suggests that they could be absorbed through the colon barrier and contribute to the biological effects of chocolate polyphenols observed in vivo. Rios et al. [25] analyzed the phenolic acids recovered in the urine of human subjects collected for 2 days after the consumption of polyphenol-rich chocolate. They reported that m-hydroxyphenylpropionic acid, m-hydroxyphenylacetic acid, and m-hydroxybenzoic acid urinary excretions, which are likely to arise from the microbial metabolism of catechin and proanthocyanidins in chocolate, increase after chocolate consumption, which are likely to arise from the microbial metabolism of catechin and proanthocyanidins in chocolate. The delayed excretion of these phenolic acids (9–48 h after the test meal) indicates their microbial origin. It is possible that the biological effects ascribed to polyphenols may not be due to a direct action of proanthocyanidins themselves but to an effect of some more readily absorbed low-molecular-weight metabolites. Circulating levels of some phenolic acids formed in the colon exceed those of the parent flavonoids after regular consumption of a flavonoid-rich diet [80], highlighting potential implication of microbial metabolites in antioxidant protection and possible beneficial health effects of cocoa polyphenols.

Health effects of cocoa flavanols depend on their bioavailability, which is strongly influenced by the food matrix and the degree of flavanol polymerization. There is increasing awareness for instance that milk significantly impact the bioavailability of cocoa flavonoids [81]. For instance, when participants consumed 40 g of cocoa powder dissolved either in 250 mL of whole milk or in 250 mL of water, the urinary concentration of 3,4-dihydroxyphenylacetic, protocatechuic, p-hydroxybenzoic, p-hydroxyhippuric, hippuric, caffeic, and ferulic acids diminished after the intake of cocoa with milk, whereas urinary concentrations of vanillic and phenylacetic acids increased. In conclusion, milk may partially affect the formation of microbial phenolic acids derived from the colonic degradation of procyanidins and other compounds present in cocoa powder [81].

The antioxidant and biological effects of chocolate are therefore not exclusively related to the absorption of catechin monomers, but also by the absorption of microbial metabolites [25]. Generally, the daily intake of dark chocolate resulted in cumulative changes on the urinary excretion of chocolate compounds and gut microbial cometabolites, mainly phenolics, which result from the digestion of chocolate constituents, as well as methylxanthines and theobromine, major active molecules in chocolate. These features reflect the adaptation of gut microbiota to process dark chocolate polyphenols, mainly flavan-3-ol epicatechins and procyanidins, that are known to reach the intestine intact before being degraded into bioavailable low-molecular-weight phenolic acids produced by the microbiota [25].

Modulation of Gut Microbial Activities to Provide Health Benefits to the Host Metabolism

Nowadays, foods are commonly used to modulate the composition of the gut microbiota contributing to the maintenance of the host health and prevention of diseases. The effects of consuming live microbial supplements (probiotics) on the microbial ecology and on human health and nutritional status have been extensively investigated [82]. Probiotic supplementation aims at replacing or reducing the number of potentially pathogenic bacteria in the intestine by enriching the populations of gut microbiota that ferment carbohydrates and that have little proteolytic activity [82]. Probiotics need to resist the manufacturing process and storage in order to be viable in the commercial product until the end of the shelf life but they also need to be resistant to the gastrointestinal environment and to compete with the resident intestinal microbiota [83]. Lately, microencapsulation in sealed capsules of different materials, which release their content under specific environmental conditions, is being investigated as a mean to improve delivery and assure better protection of probiotics. The lipid fraction of cocoa butter was shown to be protective for bifidobacteria [84], and more recently, chocolate was also evaluated as a potential protective carrier for oral delivery of a microencapsulated mixture of *Lactobacillus helveticus* and *Bifidobacterium longum* [85]. A sequential in vitro setup was used to evaluate the protection of the probiotics during passage through the stomach and small intestine, when embedded in dark and milk chocolate or liquid milk. Both chocolates offered superior protection (91% and 80% survival in milk chocolate for *L. helveticus* and *B. longum*, respectively, compared to 20% and 31% found in milk, or 69% and 58% in dark chocolate). This latter observation may be explained by the almost fivefold higher polyphenol content in the dark chocolate, as these cocoa compounds can exert antimicrobial effects [86, 87]. These data indicate that the coating of the probiotics in chocolate is an excellent solution to protect them from environmental stress conditions and for optimal delivery. The simulation with a gastrointestinal model showed that the formulation of a probiotic strain in a specific food matrix could offer superior protection for the delivery of the bacterium into the colon.

As an alternative, prebiotics, nondigestible food ingredients, generally oligosaccharides, modify the balance of the intestinal microbiota by stimulating the activity of health beneficial bacteria, such as lactobacilli and bifidobacteria [82, 88]. Today, there is considerable evidence that manipulation of the gut microbiota by prebiotics could beneficially influence the host's health [88–94]. Eventually, the combined use of prebiotics and probiotics may offer superior effects in health maintenance through modulating the microbial functional ecology [82, 95].

It has been reported that the flavanol monomer (+)-catechin significantly increases the growth of the *Clostridium coccoides–Eubacterium rectale* group, *Bifidobacterium* spp., and *Escherichia coli* and significantly inhibits the growth of the *Clostridium histolyticum* group [96]. These data suggest that the consumption of flavanol-rich foods may have the potential to support gut health through their ability to exert a nutritional pressure that favors certain groups of bacteria. In a randomized, double-blind, crossover, controlled intervention study, 22 healthy human volunteers were randomly assigned to either a high-cocoa flavanol group (494 mg cocoa flavanols per day) or a low-cocoa flavanol group (23 mg cocoa flavanols per day) for 4 weeks of intervention with a 4-week washout period. Fecal samples were collected before and after each intervention, and bacterial numbers were measured by fluorescence in situ hybridization. The daily consumption of the high-cocoa flavanol drink increased the bifidobacteria and lactobacilli populations. In addition, the high dose of polyphenols also induced a significant decrease in the *C. histolyticum* group, which suggested that cocoa-derived flavanols also have an inhibitory effect on the growth of this bacterial group, which includes *Clostridium perfringens*. These microbial changes were associated with significant reductions in plasma triacylglycerol, and C-reactive proteins suggesting the potential benefits associated with the dietary inclusion of flavanol-rich foods may be mediated by the gut microbial activities. Sequential in vitro digestion of the water-insoluble cocoa fraction with gastrointestinal enzymes and bacterial fermentation in a human colonic model system were carried out to investigate bioaccessibility and biotransformation of this source of polyphenols, respectively [97]. On one hand, in vitro digestion solubilized 38.6% of the water-insoluble cocoa and released 51% of the total phenols from the insoluble material. However, this release of phenols was not linked to any reduction in the total antioxidant capacity of the digestion-resistant material. On the other hand, bacterial fermentation of the insoluble material was associated with an increase of bifidobacteria and lactobacilli as well as butyrate production. Flavanols were converted into phenolic acids by the microbiota following a concentration gradient resulting in high concentrations of 3-hydroxyphenylpropionic acid in urine. The association between fermentable polysaccharides and some flavonoids, such as the catechins, may be very effective in the modification of microflora metabolic functions.

Summary

The perspective of inducing changes in the host metabolism triggered by unique combinations of polysaccharides and some flavonoids establishes an important step forward in the efforts to develop tailored nutritional solutions at an individual level. Overall, these examples demonstrated the possibility to characterize the biochemical effects of nutritional modulation of gut bacterial populations as well as depicting their metabolic activities. In turn, the displayed microbial changes were associated with the modulation of a range of host metabolic pathways. Recent analytical advances now provide comprehensive approaches for monitoring the nutritional effects in multiple metabolic pathways using integrative systemic metabolic and microbiome profiling. Finally, through the rigorous characterization of interactions between the diet and the microbiota, metabonomics is providing new ventures for modulating the microbiota toward the improvement of human health. The prospective of preventing the progression of human metabolic disorders using specific nutritional intervention programs could benefit from the applications of metabonomics for therapeutic surveillance and assessment of nutrition efficacy. Undeniably, the elucidation of these host gut microbial metabolic interactions will bring at the design of novel and enhanced functional foods for health promotion.

References

1. Boelsma E, Brink EJ, Stafleu A, Hendriks HF. Measures of postprandial wellness after single intake of two protein-carbohydrate meals. Appetite. 2010;54(3):456–64.
2. Rezzi S, Ramadan Z, Fay LB, Kochhar S. Nutritional metabonomics: applications and perspectives. J Proteome Res. 2007;6(2):513–25.
3. Sorensen LB, Moller P, Flint A, Martens M, Raben A. Effect of sensory perception of foods on appetite and food intake: a review of studies on humans. Int J Obes Relat Metab Disord. 2003;27(10):1152–66.
4. Drewnowski A. Taste preferences and food intake. Annu Rev Nutr. 1997;17:237–53.
5. Drewnowski A, Henderson SA, Levine A, Hann C. Taste and food preferences as predictors of dietary practices in young women. Public Health Nutr. 1999;2(4):513–9.
6. Bilman EM, van Trijp JC, Renes RJ. Consumer perceptions of satiety-related snack food decision making. Appetite. 2010;55(3):639–47.
7. Dumas ME, Barton RH, Toye A, Cloarec O, Blancher C, Rothwell A, et al. Metabolic profiling reveals a contribution of gut microbiota to fatty liver phenotype in insulin-resistant mice. Proc Natl Acad Sci USA. 2006;103(33):12511–6.
8. Stella C, Beckwith-Hall B, Cloarec O, Holmes E, Lindon JC, Powell J, et al. Susceptibility of human metabolic phenotypes to dietary modulation. J Proteome Res. 2006;5(10):2780–8.
9. Lichtenstein AH, Appel LJ, Brands M, Carnethon M, Daniels S, Franch HA, et al. Diet and lifestyle recommendations revision 2006: a scientific statement from the American Heart Association Nutrition Committee. Circulation. 2006;114(1):82–96.
10. Kuller LH. Nutrition, lipids, and cardiovascular disease. Nutr Rev. 2006;64(2 Pt 2):S15–26.
11. Martin FP, Dumas ME, Wang Y, Legido-Quigley C, Yap IK, Tang H, et al. A top-down systems biology view of microbiome-mammalian metabolic interactions in a mouse model. Mol Syst Biol. 2007;3:112.
12. Turnbaugh P, Ley R, Mahowald M, Magrini V, Mardis E, Gordon J. An obesity-associated gut microbiome with increased capacity for energy harvest. Nature. 2006;444:1027–31.
13. Ley R, Turnbaugh P, Klein S, Gordon J. Human gut microbes associated with obesity. Nature. 2006;444:1023–4.
14. Nicholson JK, Lindon JC. Systems biology: metabonomics. Nature. 2008;455(7216):1054–6.
15. Holmes E, Wilson ID, Nicholson JK. Metabolic phenotyping in health and disease. Cell. 2008;134(5):714–7.
16. Nicholson JK, Wilson ID. Opinion: understanding "global" systems biology: metabonomics and the continuum of metabolism. Nat Rev Drug Discov. 2003;2(8):668–76.
17. Martin FP, Sprenger N, Yap IK, Wang Y, Bibiloni R, Rochat F, et al. Panorganismal gut microbiome-host metabolic crosstalk. J Proteome Res. 2009;8(4):2090–105.
18. Nicholson JK, Holmes E, Wilson ID. Gut microorganisms, mammalian metabolism and personalized health care. Nat Rev Microbiol. 2005;3(5):431–8.

19. Martin F-PJ, Wang Y, Sprenger N, Yap IK, Rezzi S, Ramadan Z, et al. Top-down systems biology integration of conditional prebiotic transgenomic interactions in a humanized microbiome mouse model. Mol Syst Biol. 2008;4:205.
20. Martin FP, Wang Y, Sprenger N, Yap IK, Lundstedt T, Lek P, et al. Probiotic modulation of symbiotic gut microbial-host metabolic interactions in a humanized microbiome mouse model. Mol Syst Biol. 2008;4:157.
21. Rezzi S, Vera FA, Martin FP, Wang S, Lawler D, Kochhar S. Automated SPE-RP-HPLC fractionation of biofluids combined to off-line NMR spectroscopy for biomarker identification in metabonomics. J Chromatogr B Analyt Technol Biomed Life Sci. 2008;871(2):271–8.
22. Nicholson JK, Sadler PJ, Bales JR, Juul SM, MacLeod AF, Sonksen PH. Monitoring metabolic disease by proton NMR of urine. Lancet. 1984;2(8405):751–2.
23. Bales JR, Higham DP, Howe I, Nicholson JK, Sadler PJ. Use of high-resolution proton nuclear magnetic resonance spectroscopy for rapid multi-component analysis of urine. Clin Chem. 1984;30(3):426–32.
24. Montoliu I, Martin FP, Collino S, Rezzi S, Kochhar S. Multivariate modeling strategy for intercompartmental analysis of tissue and plasma (1)H NMR spectrotypes. J Proteome Res. 2009;8(5):2397–406.
25. Rios LY, Gonthier MP, Remesy C, Mila I, Lapierre C, Lazarus SA, et al. Chocolate intake increases urinary excretion of polyphenol-derived phenolic acids in healthy human subjects. Am J Clin Nutr. 2003;77(4):912–8.
26. Serafini M, Bugianesi R, Maiani G, Valtuena S, De SS, Crozier A. Plasma antioxidants from chocolate. Nature. 2003;424(6952):1013.
27. Mathur S, Devaraj S, Grundy SM, Jialal I. Cocoa products decrease low density lipoprotein oxidative susceptibility but do not affect biomarkers of inflammation in humans. J Nutr. 2002;132(12):3663–7.
28. Wan Y, Vinson JA, Etherton TD, Proch J, Lazarus SA, Kris-Etherton PM. Effects of cocoa powder and dark chocolate on LDL oxidative susceptibility and prostaglandin concentrations in humans. Am J Clin Nutr. 2001;74(5):596–602.
29. Mursu J, Voutilainen S, Nurmi T, Rissanen TH, Virtanen JK, Kaikkonen J, et al. Dark chocolate consumption increases HDL cholesterol concentration and chocolate fatty acids may inhibit lipid peroxidation in healthy humans. Free Radic Biol Med. 2004;37(9):1351–9.
30. Scalbert A, Williamson G. Dietary intake and bioavailability of polyphenols. J Nutr. 2000;130(8):2073S 85.
31. Steinberg FM, Bearden MM, Keen CL. Cocoa and chocolate flavonoids: implications for cardiovascular health. J Am Diet Assoc. 2003;103(2):215–23.
32. Grassi D, Lippi C, Necozione S, Desideri G, Ferri C. Short-term administration of dark chocolate is followed by a significant increase in insulin sensitivity and a decrease in blood pressure in healthy persons. Am J Clin Nutr. 2005;81(3):611–4.
33. Taubert D, Roesen R, Lehmann C, Jung N, Schomig E. Effects of low habitual cocoa intake on blood pressure and bioactive nitric oxide: a randomized controlled trial. JAMA. 2007;298(1):49–60.
34. Schroeter H, Heiss C, Balzer J, Kleinbongard P, Keen CL, Hollenberg NK, et al. (−)-Epicatechin mediates beneficial effects of flavanol-rich cocoa on vascular function in humans. Proc Natl Acad Sci USA. 2006; 103(4):1024–9.
35. Grassi D, Necozione S, Lippi C, Croce G, Valeri L, Pasqualetti P, et al. Cocoa reduces blood pressure and insulin resistance and improves endothelium-dependent vasodilation in hypertensives. Hypertension. 2005;46(2): 398–405.
36. Anderson RA, Polansky MM. Tea enhances insulin activity. J Agric Food Chem. 2002;50(24):7182–6.
37. Strobel P, Allard C, Perez-Acle T, Calderon R, Aldunate R, Leighton F. Myricetin, quercetin and catechin-gallate inhibit glucose uptake in isolated rat adipocytes. Biochem J. 2005;386(Pt 3):471–8.
38. Brand-Miller J, Holt SH, de Jong V, Petocz P. Cocoa powder increases postprandial insulinemia in lean young adults. J Nutr. 2003;133(10):3149–52.
39. Parker G, Parker I, Brotchie H. Mood state effects of chocolate. J Affect Disord. 2006;92(2–3):149–59.
40. di Tomaso E, Beltramo M, Piomelli D. Brain cannabinoids in chocolate. Nature. 1996;382(6593):677–8.
41. McDonald I. Gordon Holmes lecture: Gordon Holmes and the neurological heritage. Brain. 2007;130(Pt 1): 288–98.
42. Rezzi S, Ramadan Z, Martin FP, Fay LB, van Bladeren P, Lindon JC, et al. Human metabolic phenotypes link directly to specific dietary preferences in healthy individuals. J Proteome Res. 2007;6(11):4469–77.
43. Nicholson JK, Holmes E, Lindon JC. Metabonomics; understanding the consequences of xenobiotic induced metabolic dysfunction using NMR spectroscopy of the biofluids, multivariate statistics and expert system. Xenobiotica. 1999;29(11):1181–9.
44. Wilson ID, Plumb R, Granger J, Major H, Williams R, Lenz EM. HPLC-MS-based methods for the study of metabonomics. J Chromatogr B Analyt Technol Biomed Life Sci. 2005;817(1):67–76.
45. Cianflone K, Vu H, Zhang Z, Sniderman AD. Effects of albumin on lipid synthesis, apo B-100 secretion, and LDL catabolism in HepG2 cells. Atherosclerosis. 1994;107(2):125–35.
46. Yang D, Brunengraber H. Glutamate, a window on liver intermediary metabolism. J Nutr. 2000;130(4S Suppl): 991S–4.

47. Bales JR, Bell JD, Nicholson JK, Sadler PJ. 1H NMR studies of urine during fasting: excretion of ketone bodies and acetylcarnitine. Magn Reson Med. 1986;3(6):849–56.
48. Brass EP. Supplemental carnitine and exercise. Am J Clin Nutr. 2000;72(2 Suppl):618S–23.
49. Selmer T, Andrei PI. p-Hydroxyphenylacetate decarboxylase from *Clostridium difficile*. A novel glycyl radical enzyme catalysing the formation of p-cresol. Eur J Biochem. 2001;268(5):1363–72.
50. Gostner A, Blaut M, Schaffer V, Kozianowski G, Theis S, Klingeberg M, et al. Effect of isomalt consumption on faecal microflora and colonic metabolism in healthy volunteers. Br J Nutr. 2006;95(1):40–50.
51. Wang Y, Holmes E, Nicholson JK, Cloarec O, Chollet J, Tanner M, et al. Metabonomic investigations in mice infected with *Schistosoma mansoni*: an approach for biomarker identification. Proc Natl Acad Sci USA. 2004;101(34):12676–81.
52. Zeisel SH, Blusztajn JK. Choline and human nutrition. Annu Rev Nutr. 1994;14:269–96.
53. Rechner AR, Kuhnle G, Bremner P, Hubbard GP, Moore KP, Rice-Evans CA. The metabolic fate of dietary polyphenols in humans. Free Radic Biol Med. 2002;33(2):220–35.
54. Smith EA, Macfarlane GT. Enumeration of human colonic bacteria producing phenolic and indolic compounds: effects of pH, carbohydrate availability and retention time on dissimilatory aromatic amino acid metabolism. J Appl Bacteriol. 1996;81(3):288–302.
55. Nowak A, Libudzisz Z. Influence of phenol, p-cresol and indole on growth and survival of intestinal lactic acid bacteria. Anaerobe. 2006;12(2):80–4.
56. Xu ZR, Hu CH, Wang MQ. Effects of fructooligosaccharide on conversion of L-tryptophan to skatole and indole by mixed populations of pig fecal bacteria. J Gen Appl Microbiol. 2002;48(2):83–90.
57. Martin FP, Rezzi S, Pere-Trepat E, Kamlage B, Collino S, Leibold E, et al. Metabolic effects of dark chocolate consumption on energy, gut microbiota, and stress-related metabolism in free-living subjects. J Proteome Res. 2009;8:5568–79.
58. Noble RE. Diagnosis of stress. Metabolism. 2002;51(6 Suppl 1):37–9.
59. Eisenhofer G, Kopin IJ, Goldstein DS. Catecholamine metabolism: a contemporary view with implications for physiology and medicine. Pharmacol Rev. 2004;56(3):331–49.
60. Santos J, Perdue MH. Stress and neuroimmune regulation of gut mucosal function. Gut. 2000;47(Suppl 4): iv49–51.
61. Santos J, Benjamin M, Yang PC, Prior T, Perdue MH. Chronic stress impairs rat growth and jejunal epithelial barrier function: role of mast cells. Am J Physiol Gastrointest Liver Physiol. 2000;278(6):G847–54.
62. Hyland K, Surtees RA, Rodeck C, Clayton PT. Aromatic L-amino acid decarboxylase deficiency: clinical features, diagnosis, and treatment of a new inborn error of neurotransmitter amine synthesis. Neurology. 1992;42(10): 1980–8.
63. Caso JR, Leza JC, Menchen L. The effects of physical and psychological stress on the gastro-intestinal tract: lessons from animal models. Curr Mol Med. 2008;8(4):299–312.
64. Barclay GR, Turnberg LA. Effect of psychological stress on salt and water transport in the human jejunum. Gastroenterology. 1987;93(1):91–7.
65. Tannock GW. New perceptions of the gut microbiota: implications for future research. Gastroenterol Clin North Am. 2005;34(3):361–82, vii.
66. Yap IK, Li JV, Saric J, Martin FP, Davies H, Wang Y, et al. Metabonomic and microbiological analysis of the dynamic effect of vancomycin-induced gut microbiota modification in the mouse. J Proteome Res. 2008;7(9):3718–28.
67. Beliveau GP, Brusilow SW. Glycine availability limits maximum hippurate synthesis in growing rats. J Nutr. 1987;117(1):36–41.
68. Seematter G, Binnert C, Tappy L. Stress and metabolism. Metab Syndr Relat Disord. 2005;3(1):8–13.
69. Goldstein DS. Catecholamines and stress. Endocr Regul. 2003;37(2):69–80.
70. Goldstein DS, Dionne R, Sweet J, Gracely R, Brewer Jr HB, Gregg R, et al. Circulatory, plasma catecholamine, cortisol, lipid, and psychological responses to a real-life stress (third molar extractions): effects of diazepam sedation and of inclusion of epinephrine with the local anesthetic. Psychosom Med. 1982;44(3):259–72.
71. Berk LS, Tan SA, Berk D. Cortisol and catecholamine stress hormone decrease is associated with the behavior of perceptual anticipation of mirthful laughter. FASEB J. 2008;22(1_MeetingAbstracts):946.
72. Gilmore MS, Ferretti JJ. The thin line between gut commensal and pathogen. Science. 2003;299:1999–2002.
73. Tuohy KM, Gougoulias C, Shen Q, Walton G, Fava F, Ramnani P. Studying the human gut microbiota in the trans-omics era–focus on metagenomics and metabonomics. Curr Pharm Des. 2009;15(13):1415–27.
74. Dethlefsen L, Fall-Ngai M, Relman DA. An ecological and evolutionary perspective on human-microbe mutualism and disease. Nature. 2007;449(7164):811–8.
75. Eckburg PB, Bik EM, Bernstein CN, Purdom E, Dethlefsen L, Sargent M, et al. Diversity of the human intestinal microbial flora. Science. 2005;308(5728):1635–8.
76. Xu J, Bjursell MK, Himrod J, Deng S, Carmichael LK, Chiang HC, et al. A genomic view of the human-Bacteroides thetaiotaomicron symbiosis. Science. 2003;299:2074–6.

77. Donovan JL, Manach C, Rios L, Morand C, Scalbert A, Remesy C. Procyanidins are not bioavailable in rats fed a single meal containing a grapeseed extract or the procyanidin dimer B3. Br J Nutr. 2002;87(4):299–306.
78. Deprez S, Brezillon C, Rabot S, Philippe C, Mila I, Lapierre C, et al. Polymeric proanthocyanidins are catabolized by human colonic microflora into low-molecular-weight phenolic acids. J Nutr. 2000;130(11):2733–8.
79. Scalbert A, Deprez S, Mila I, Albrecht AM, Huneau JF, Rabot S. Proanthocyanidins and human health: systemic effects and local effects in the gut. Biofactors. 2000;13(1–4):115–20.
80. Rechner AR, Kuhnle G, Hu H, Roedig-Penman A, van den Braak MH, Moore KP, et al. The metabolism of dietary polyphenols and the relevance to circulating levels of conjugated metabolites. Free Radic Res. 2002; 36(11):1229–41.
81. Urpi-Sarda M, Llorach R, Khan N, Monagas M, Rotches-Ribalta M, Lamuela-Raventos R, et al. Effect of milk on the urinary excretion of microbial phenolic acids after cocoa powder consumption in humans. J Agric Food Chem. 2010;58(8):4706–11.
82. Collins MD, Gibson GR. Probiotics, prebiotics, and synbiotics: approaches for modulating the microbial ecology of the gut. Am J Clin Nutr. 1999;69(5):1052S–7.
83. Fooks LJ, Gibson GR. Probiotics as modulators of the gut flora. Br J Nutr. 2002;88(Suppl 1):S39–49.
84. Lahtinen SJ, Ouwehand AC, Salminen SJ, Forssell P, Myllarinen P. Effect of starch- and lipid-based encapsulation on the culturability of two Bifidobacterium longum strains. Lett Appl Microbiol. 2007;44(5):500–5.
85. Possemiers S, Marzorati M, Verstraete W, van de Wiele T. Bacteria and chocolate: a successful combination for probiotic delivery. Int J Food Microbiol. 2010;141(1–2):97–103.
86. Percival RS, Devine DA, Duggal MS, Chartron S, Marsh PD. The effect of cocoa polyphenols on the growth, metabolism, and biofilm formation by *Streptococcus mutans* and *Streptococcus sanguinis*. Eur J Oral Sci. 2006;114(4):343–8.
87. Datla KP, Zbarsky V, Rai D, Parkar S, Osakabe N, Aruoma OI, et al. Short-term supplementation with plant extracts rich in flavonoids protect nigrostriatal dopaminergic neurons in a rat model of Parkinson's disease. J Am Coll Nutr. 2007;26(4):341–9.
88. Gibson GR, Roberfroid MB. Dietary modulation of the human colonic microbiota: introducing the concept of prebiotics. J Nutr. 1995;125(6):1401–12.
89. Parracho H, McCartney AL, Gibson GR. Probiotics and prebiotics in infant nutrition. Proc Nutr Soc. 2007;66(3):405–11.
90. Rastall RA. Modulation of the microbial ecology of the human colon by probiotics, prebiotics and synbiotics to enhance human health: an overview of enabling science and potential applications. FEMS Microbiol Ecol. 2005;52(2):145–52.
91. Sartor RB. Therapeutic manipulation of the enteric microflora in inflammatory bowel diseases: antibiotics, probiotics, and prebiotics. Gastroenterology. 2004;126(6):1620–33.
92. Delzenne NM, Kok N. Effects of fructans-type prebiotics on lipid metabolism. Am J Clin Nutr. 2001;73 (2 Suppl):456S–8.
93. Roberfroid MB. Prebiotics and synbiotics: concepts and nutritional properties. Br J Nutr. 1998;80(4):S197–202.
94. Lim CC, Ferguson LR, Tannock GW. Dietary fibres as "prebiotics": implications for colorectal cancer. Mol Nutr Food Res. 2005;49(6):609–19.
95. Schrezenmeir J, de Vrese M. Probiotics, prebiotics, and synbiotics – approaching a definition. Am J Clin Nutr. 2001;73(2 Suppl):361S–4.
96. Tzounis X, Rodriguez-Mateos A, Vulevic J, Gibson GR, Kwik-Uribe C, Spencer JP. Prebiotic evaluation of cocoa-derived flavanols in healthy humans by using a randomized, controlled, double-blind, crossover intervention study. Am J Clin Nutr. 2011;93(1):62–72.
97. Fogliano V, Corollaro ML, Vitaglione P, Napolitano A, Ferracane R, Travaglia F, et al. In vitro bioaccessibility and gut biotransformation of polyphenols present in the water-insoluble cocoa fraction. Mol Nutr Food Res. 2011;55(Suppl 1):S44–55.

Chapter 17
The Absorption, Metabolism, and Pharmacokinetics of Chocolate Polyphenols

Cesar A. Lau-Cam

Key Points

- Cacao products such as chocolate and cocoa are major food sources of polyphenolic compounds in the form of monomeric catechins and oligomeric procyanidins.
- These compounds display a myriad of health-related properties that are only realized after they have transited through the digestive tract, interacted with the intestinal microflora, undergone presystemic biotransformations, and become absorbed from the intestinal milieu.

Keywords Chocolate • Cocoa • Polyphenols • Oral absorption • Metabolism • Excretion • Pharmacokinetics • Dietary factors • Chocolate matrix effects

Cocoa Beans and Their Products

A review of the scientific literature on human and animal studies linking the consumption of vegetables, fruits, and certain plant-derived beverages with a reduction in the risk of coronary artery disease [1–3], ischemic stroke [4, 5], hypertension [6], maculopathy [7], chronic obstructive pulmonary disease [8], gallstones [9], certain types of cancer [10, 11], and possibly type 2 diabetes [12, 13] suggests that although the extent of the association is sometimes weak [11], uncertain [14, 15], or unclear [16], in most cases the health benefits that these dietary components can offer may be due, to a great extent, to their high content in antioxidant polyphenols such as flavonoids, procyanidins, lignans, and stilbenes [17, 18].

Cacao beans, the unprocessed seeds (beans) from the fruits (pods) of *Theobroma cacao*, along with black and green teas, red wines, ground cloves, nuts, cereals, legumes, and several types of berries, fruit juices, and chocolates, have been identified as foods with a high content in antioxidant polyphenols [19, 20], especially flavonoids [21]. However, owing to a natural unpleasant astringency, acidity, and bitterness, the cacao beans need to undergo several manufacturing steps aimed at making them palatable and exhibiting the characteristic aroma and taste of cocoa and chocolate through a

C.A. Lau-Cam, B.S., M.S., Ph.D. (✉)
College of Pharmacy and Allied Health Professions, St. John's University,
8000 Utopia Parkway, Jamaica, NY 11439, USA
e-mail: claucam@usa.net

R.R. Watson et al. (eds.), *Chocolate in Health and Nutrition*, Nutrition and Health 7,
DOI 10.1007/978-1-61779-803-0_17, © Springer Science+Business Media, LLC 2013

series of complex chemical reactions. Following the opening of the pods, the beans are exposed to the enzymatic action of microbes present in the sugar-rich, acidic (pH 3.3–4.0), and relatively anaerobic testa (pulp) surrounding the seeds. This fermentation process, which normally takes place between 5 and 7 days, will chemically break down the pulp, induce chemical changes within the seed cotyledons, and allow the shells to release the seeds, which are then dried and roasted to promote the development of the flavor qualities associated with cacao products [22, 23]. Further flavors may develop when the roasted cacao beans are subjected to winnowing, a violent vibratory process that removes the woody husks (or shells) and breaks the cacao beans into small irregular fragments known as cacao nibs, and a second roasting of the cacao nibs at 120–150°C. Milling of the roasted nibs at a constant temperature of just above 32°C (90°F), followed by expression, separates the cacao solids from the liquefied fatty components of cacao, designated as cacao (or chocolate) liquor, the main component of commercial bitter chocolates [24, 25] containing cocoa solids and fatty components (or cocoa butter) in roughly equal proportions [26]. To make cocoa, the cacao liquor is expressed with a hydraulic press, an operation that removes all but about 10–12% of cocoa butter present in chocolate [27], to leave a cake that is then pulverized to cacao powder [24]. In turn, cocoa can be of two kinds: natural cocoa if it has not been mixed with alkali during the manufacturing process as a further modifier of the organoleptic properties of the product and Dutch-processed (alkalinized) cocoa if it has been treated with an alkaline solution, usually 2–3% in strength [24].

Polyphenols in Cacao, Cocoa, and Chocolates

Total polyphenols in unfermented cacao beans range from 12 to 18% of the dry weight of the bean (with the predominant forms being the epimeric flavan-3-ol monomers (+)-catechin and (−)-epicatechin (Fig. 17.1) and oligomeric to polymeric proanthocyanidins, which are exclusively represented by procyanidins (Fig. 17.2) [23, 28]. In general, monomers comprise 5–10% of the total polyphenols in cacao and oligomers and polymers ≥90% [27, 29].

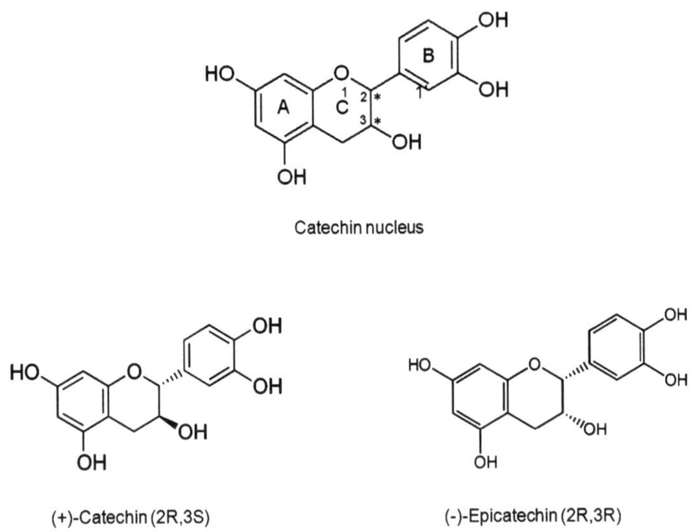

Fig. 17.1 Monomeric catechins found in cacao and its products

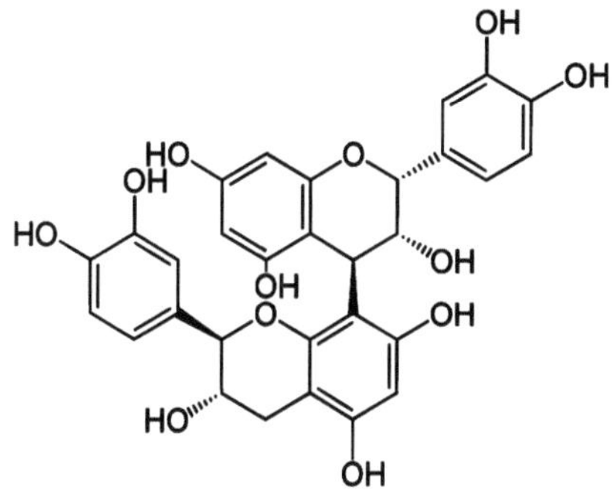

Procyanidin B1

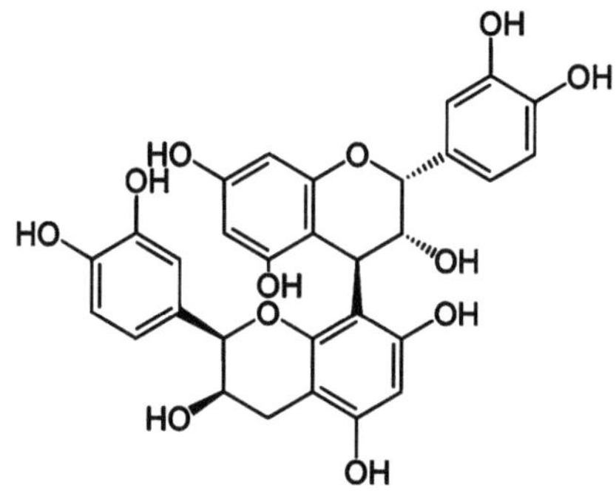

Procyanidin B2

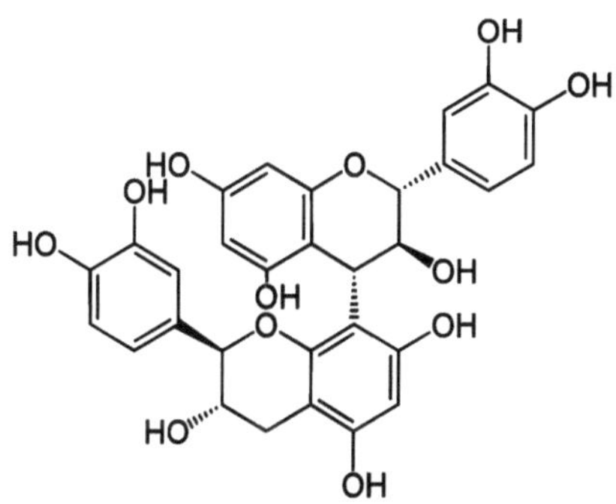

Procyanidin B3

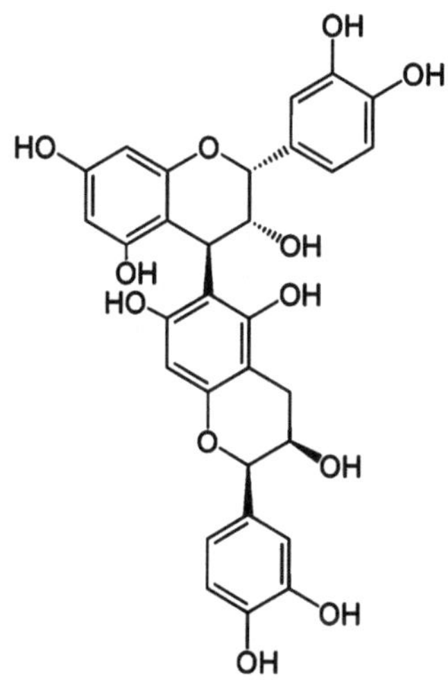

Procyanidin B5

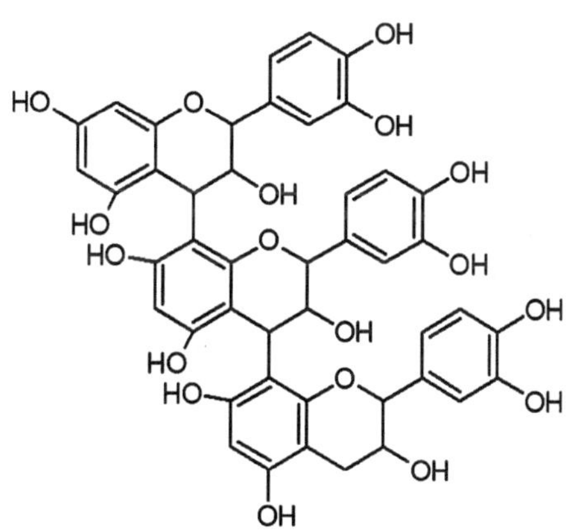

Procyanidin C1

Fig. 17.2 Procyanidins found in cacao and its products

The content of polyphenols in fresh cacao beans varies quite widely depending on their geographic origin [30, 31], and changes rather rapidly under the influence of fermentation [32]. For a sample of fresh unfermented West Indian cacao beans, the proportion of flavan-3-ol monomers to procyanidins was found to be ~75/31, but after 8 days of fermentation, it had increased to ~83/10 [32]. It was also reported that the levels of the various polyphenols in unfermented beans amounted to ~37% flavan-3-ols, ~4% anthocyanins, and ~58% procyanidins [28, 34]. Among these polyphenols, (−)-epicatechin appears to be the major flavan-3-ol, with one study reporting a content of 34.65–43.27 mg per g (or up to 35% of the total phenolics) of defatted freshly harvested cacao beans, with (+)-catechin occurring in smaller amounts and (+)-gallocatechin and (−)-epigallocatechin being only present in traces amounts [33, 34].

The procyanidins found in cacao-derived products are represented by oligomers and polymers of (−)-epicatechin such as procyanidins B1, B2, B5, and C1 (see Fig. 17.1) [29]. Procyanidins are particularly abundant in cacao beans, where types B and C are present in much higher concentrations than type A. While in types B and C the monomers are linked by a *4β→8* carbon to carbon bond to form a dimer such as procyanidin B1 (epicatechin-(*4β→8*)-catechin), procyanidin B2 (epicatechin-(*4β→8*)-epicatechin), or a trimer such as procyanidin C1 (epicatechin-(*4β→8*)-epicatechin-(*4β→8*)-epicatechin), in type A the monomers are linked by both a *4β→8* carbon bond and a *2β→O-7* ether bond [32, 35]. In addition, galactose- and arabinose-containing procyanidin glycosides related to procyanidin A2 have also been reported for fresh unfermented cacao beans [32] and cacao liquor [36]. HPLC analysis of cacao (chocolate) liquor, the fatty mass that forms upon cooling the liquid obtained by grinding fermented, dried, roasted, skinless cocoa beans, and of cocoa and dark chocolate has shown these products to share similar profiles of monomeric catechins and proanthocyanidins but to differ in the ratio of flavan-3-ols to the total amount of monomeric and oligomeric polyphenols found higher in pure cocoa powder than in cacao liquor or dark chocolate [37, 38]. Furthermore, the procyanidin content of cocoa liquor and chocolate samples has been found to correlate with their antioxidant capacity measured using the oxygen radical absorbance capacity (ORAC) assay as an indicator for potential biological activity [37]. On the average, chocolate and apples are found to have a procyanidin content per serving that is greater (164.7 and 147.1 mg, respectively) than that of red wine and cranberry juice (22.0 and 31.9 mg, respectively) [39].

Additional, although less abundant, polyphenols are the flavonols quercetin, isoquercitrin (quercetin-3-O-glucoside), quercetin-3-O-arabinoside, quercetin-3-O-glucuronide, hyperoside (quercetin-3-O-galactoside), the flavones luteolin, luteolin-7-O-glucoside, apigenin, vitexin (apigenin-8-C-glucoside) and isovitexin (apigenin-6-C-glucoside), and the flavanone naringenin [27, 40], and catechin-C glycoside [41]. The anthocyanin fraction is primarily comprised of cyanidin-3-α-L-arabinoside and cyanidin-3-β-D-galactoside [34].

Cocoa powder is the ground product that remains after a significant portion of the cocoa fat (butter) has been removed from roasted milled cocoa beans by hydraulic pressing. Alternatively, cocoa is prepared from the nibs, the roasted cocoa beans that were deprived of the shell and germ, which are first moistened by steaming and next defatted by passage through a screw-expeller press to yield a solid mass of cocoa [42]. Polyphenols in cocoa comprise 12–18% of its total weight on a dry basis, making it a richer source of flavonoids than other phenolic-rich foods such as onions, red wine, and apples [43]. Addition of a mild alkali to cocoa powder (the Dutching process) yields alkalinized cocoa with superior organoleptic properties but with diminished flavanol content and antioxidant activity in direct proportion to the amount of alkali added and to an increase in the product pH above that of natural cocoa powder (5.3–5.8) [44]. Alkali is known to cause chemically losses of the mean total flavonoid content and, thus, to affect the antioxidant properties and polyphenol bioavailability of cocoa powders. A study by Andrés-Lacueva et al. [27] verified that Dutching of the nibs, liquor, or cocoa powder reduced the content of the flavanol (−)-epicatechin (by 67%) to a greater extent than that of (+)-catechin (by 38%), probably because of its epimerization into (−)-catechin, a less bioavailable form of catechin. A decline was also observed for di-, tri-, and tetrameric procyanidins. In the case of flavonols, quercetin presented the highest loss (86%), whereas quercetin-3-glucuronide, quercetin-3-arabinoside, and isoquercitrin showed a similar decrease (58%, 62%, and 61%, respectively). It was concluded that the large decrease in flavonoid content of natural cocoa powder, together with the observed change in the monomeric flavanol profile that results from the alkalinization treatment, could affect the antioxidant properties and the polyphenol bioavailability of cocoa powder products [37]. Three additional contributing factors to flavan-3-ol decreases in cacao beans are fermentation, roasting, and certain prefermentation treatments. Roasting of the beans at temperatures exceeding 70°C is found to lead not only to appreciable losses of monomeric catechins but also to increases in catechin levels due to the epimerization of (−)-epicatechin to (−)-catechin [45] or to (+)-catechin [46]. Indeed, the total amount of soluble polyphenols in the dried fat-free mass of fresh cocoa beans is 15–20% (which is equivalent to approximately 6% in air-dried cocoa beans containing 54% fat and 6% water), but in fermented beans, it is only about 5%, with 10% or more being regarded as an indication of a poor fermentation [34]. In general, the decrease in total polyphenol content of cocoa beans as a result of fermentation is rather fast since there is a report indicating that cocoa beans fermented for 1–6 days lost 63% of their polyphenol content by day 6 along with 17% drop in antioxidant capacity relative to corresponding values on day 0 [22]. Major losses occur in the monomeric fraction (i.e., (+)-catechin plus (−)-epicatechin), which can drop by >80% of the initial values [45]. Conversely, underfermented and underroasted cacao beans with very high cacao solid percentages (≥90%) contained higher levels of antioxidant phenols than more fermented and more roasted beans but at the cost of an inferior flavor and a greater microbial burden from the fermentation process [25]. On the other hand, a study by Nazaruddin et al. [47] showed that the levels of (−)-epicatechin and (+)-catechin in cacao beans decreased when subjected to different degrees of fermentation but with the losses differing between the two flavanols and proportionally to the degree of fermentation (6–17% and 0.95–1.62%, respectively). The same study also found that prefermentation treatments such as postharvest pod storage, spreading, and pressing also played a decreasing role on catechins and on the astringency of the beans. Hence, fermentation will reduce the level of bitterness and astringency of the cocoa bean by promoting the loss of polyphenols [28]. In this regard, the flavan-3-ols (−)-epicatechin and (+)-catechin and the procyanidins B2, B5, C1, [epicatechin-(4$\beta\rightarrow$8)]3-epicatechin, and [epicatechin-(4$\beta\rightarrow$8)]4-epicatechin are regarded as key contributors to the bitterness and astringency of roasted cacao [48], with the purines caffeine and theobromine, a series of quercetin, naringenin, luteolin, and apigenin glycopyranosides, and a family of amino acid amides also playing a role [49].

A comparative evaluation of the phenolic and flavonoid contents and total antioxidant capacity of beverages prepared from cocoa, black tea, and green tea with red wine has verified cocoa to contain higher levels of total phenolics, expressed as gallic acid equivalents (GAE), and flavonoids, expressed as epicatechin equivalents (ECE), per serving than the other beverages and that the total antioxidant capacity, based on the results of the ABTS (2,2'-azino-bis(3-ethylbenzthiazoline-6-sulfonic acid) and DPPH (2,2-diphenyl-1-picrylhydrazyl) tests, decreased in the order cocoa > red wine > green tea > black tea [18]. Hence, cocoa may offer greater health benefit to humans than teas and red wine [18, 43]. Furthermore, since the type of antioxidant phenol present in each food product varies according to the particular source, being procyanidins in chocolate, various catechins and catechin esters of gallic acid in tea [18], and proanthocyanidins, catechins, flavanones, flavonols, and stilbenes in red wine [24], it is also important to establish the total contents of antioxidant phytocomponents in addition to the individual ones, especially in the case of cocoa and tea [18].

A survey of the oxygen radical absorbance capacity (ORAC), an index of antioxidant activity, and of the total polyphenol, procyanidin, and nonfat cocoa solid (NFCS) contents of top market share cocoa- and chocolate-containing food products sold in the United States showed a strong linear correlation between NFCS and ORAC (R^2=0.9849), total polyphenols, as GAE (R^2=0.9793), and procyanidins (R^2=0.946), respectively, and the greater dependence of the antioxidant capacity on NFCS than on factors such as differences in cocoa bean blends and processing [44]. The same survey determined that antioxidant activities and total polyphenol and procyanidin contents decreased in the order natural cocoas > baking chocolates > dark chocolates > chocolate baking chips > milk chocolate > chocolate syrups. An important contributor to these differences is the cooking process used in the manufacturing of cocoa drinks and baked chocolate-containing products since it is found to negatively affect both the levels of flavan-3-ols and proanthocyanidins and the antioxidant activity, more so at a pH above than below 7.25 [50, 51]. When the procyanidins in cocoa liquors and in chocolate samples were quantified by normal phase high performance liquid chromatography (HPLC) with fluorescence detection, differences in the average content procyanidin oligomers per gram of sample and in the abundance of each oligomer size were detected [37]. In this study, chocolate liquors from different geographical areas were found to differ in their procyanidin content as a function of the extent of their fermentation, with underfermented ones containing substantially more procyanidins than well-fermented ones. Also, a chocolate containing a high concentration of cocoa liquor was richer in procyanidins than dark chocolate, which, in turn, was richer than milk chocolate. When the procyanidin content of the various samples was related to the number of constitutive units of the oligomer, it was apparent that the concentration decreased proportionally to an increase in the number of units. However, in this study, while a high liquor chocolate exhibited measurable amounts of the entire spectrum of oligomer, from dimers up to decamers, dark chocolate failed to demonstrate procyanidins larger than tetramers, and milk chocolate did not show oligomers beyond hexamers. The apparent absence of certain oligomeric sizes in commercial chocolates was ascribed to the inability of the assay method used and to the levels of detected procyanidins falling below its limits of detection [37].

Passage Through and Stability of Chocolate and Cocoa Polyphenols in the Gastrointestinal Tract

Two compartments are involved in the metabolism of flavonoid compounds such as those present in chocolate. One is represented by the small intestine, liver, and kidney, and the other by the colon [52]. The two factors that have received the greatest attention in connection with the health benefits of antioxidant polyphenols in cacao and its products have been the effect of gastrointestinal pH and the role played by the intestinal microflora, since both may contribute to their degradation to products and metabolites demonstrating decreased or no antioxidant activity before they can reach the circulation [53].

The role of pH on the stability of cacao polyphenols has been the subject of multiple evaluations in vitro. Thus, a comparative study of various foods prepared with natural cocoa powder, and including chocolate frosting, hot cocoa drink, chocolate cookies, and chocolate cake, demonstrated changes in (−)-epicatechin content and in antioxidant activity as a function of the product pH and type of leavening agent added to the food product. In general, these parameters were found to decrease proportionally to an increase in pH above 7.25 [50]. Another study investigated the losses in total polyphenols and flavanols, including procyanidins, during the Dutching of natural cocoa powder, and found them to increase linearly with the degree of alkalization in the pH range 6.5–7.61 [54].

In a model study designed to investigate the stability of a series of structurally unrelated polyphenolic compounds, dissolved in buffers in the pH range 3–11, on the basis of changes in the ultraviolet absorption spectra of the phenolic compounds in buffered solutions for periods of up to 72 h verified that while (−)-catechin remained stable at all pH values, phenolic acids like gallic acid or chlorogenic acid were not stable at a high pH [55]. The resistance of (−)-epicatechin and (+)-catechin to acid degradation has also been verified in an environment simulating gastric fluid, with the losses amounting to 8–11% and 7–8%, respectively [56]. Similar results have been observed for (−)-epicatechin present in a freshly prepared green or black tea infusion that was incubated with hydrochloric acid to pH 2.0 at 37°C for 1 h and which found the concentration of (−)-epicatechin to decrease by 7% in green tea but not to change in black tea. In contrast, alkalinization of the brews with sodium bicarbonate to pH 7.5 and incubation at 37°C led to losses in both types of teas that increased with the length of the incubation and reached 34% for green tea and 48% for black tea after 1 h [57]. In contrast, an in vitro investigation by Zhu et al. [58] examining the stability of cocoa monomers (i.e., (−)-epicatechin and (+)-catechin) and dimers (i.e., dimer B2 = epicatechin-(4*β*-8)-epicatechin and dimer B5 = epicatechin -(4*β*-6)-epicatechin) in simulated gastric and intestinal juice and at different pH values verified that the dimers were less stable than the monomers at both acid and alkaline pH and that incubation of dimer B2 and dimer B5 in simulated gastric juice (pH 1.8) or acidic pH resulted in their degradation to (−)-epicatechin and interconversion of one dimer to the other, respectively. On the other hand, incubation of all four compounds in simulated intestinal juice or at alkaline pH led to their complete degradation within several hours. The study concluded that the amount and type of flavanols and procyanidins present in cocoa will determine their stability in the gastrointestinal tract as well as in acidic and alkaline environments. Although confirmatory evidence of the hydrolysis of cocoa procyanidin oligomers to monomers and dimers due to the acidic conditions in the stomach has been achieved by Rios et al. [59] through incubation studies of procyanidin oligomers, ranging in size from dimer to hexamer, in simulated gastric juice (pH 2.0) at 37°C for up to 3.5 h, this conclusion is not universally shared. Indeed, a study in which gastric juice was periodically collected to exhaustion through nasogastric sampling from human subjects that had consumed 500 mL of a cocoa beverage containing 733 mg procyanidin polymers and 351 mg structurally related flavanol monomers showed no evidence of procyanidin hydrolysis in vivo when analyzed by a stability-indicating HPLC method [59]. The apparent discrepancies may stem from differences in proton concentration dictated by the duration of exposure to acidic conditions, being shorter (<50 min) in vivo than in vitro (1–3.5 h), and by the degree of acidity, being lower in vivo (pH 5.4) as a result of the buffering effect of the food bolus on the originally high basal acidity (pH 2) than under in vitro conditions, thus making the use of simulated gastric use an inappropriate model for studying the fate of procyanidins in vivo [59]. Similar conclusions were reached in a study in rats that found that the procyanidin dimer B3 and grape seed procyanidins having a higher degree of polymerization were neither converted to monomers nor gave rise to dimer that could be identified in plasma or urine [60].

In summary, it is apparent that monomeric flavanols are relatively stable in an acid environment such as that of the stomach, in the alkaline surroundings of the intestine, and in buffered media of pH up to 11. In contrast, procyanidins appear to be impervious to acid up to a pH ~ 6.5, beyond which they may start to undergo degradation to monomeric flavanols, with the extent of the degradation increasing proportionally to an increase in pH.

The Interaction of Chocolate and Cocoa Polyphenols with the Intestinal Flora

There is abundant evidence to suggest that gut microbes play an important role in the potential health benefits of dietary polyphenols in humans, especially of those that are not well absorbed in the upper gastrointestinal tract and remain in the colon [61]. In the particular case of cacao and cacao products, it is apparent that, in addition to genetic, microbial, and dietary factors, the bioavailability of the polyphenols may be influenced by food-related factors such as the food processing conditions [27], the food matrix [52], and additional concurrently consumed food product [62], and by physicochemical properties of the polyphenols such as molecular weight, stereochemistry, number of hydrogen bond donor and acceptor groups, lipophilicity, solubility, and pKa [52, 63, 64]. Additionally, intra- and interindividual variations of colonic microbiota among humans may be the result of differences imposed by age, disease, stress, medication, and nutritional status and may directly impact on the type and quantity of metabolic products to be found in biological samples [65, 66].

In cacao, only a small proportion of flavan-3-ols occur as C- [36, 41] and *O*-glycosides [27, 40], with the majority being present unlinked to a carbohydrate moiety. For this reason, the release of the most abundant cacao polyphenols from their matrices following mastication will not require a preliminary deglycosylation step by brush border or microbial enzymes. Furthermore, monomers such as (+)-catechin and certain catechin oligomers such as procyanidins will be expected to be absorbed by intestinal cells by passive diffusion, with metabolic transformations taking place during the course of their uptake into the intestinal epithelium and later in the liver [65, 67]. Consequently, only free catechins will be found in plasma or urine along with conjugates of glucuronic acid and sulfate and with *O*-methylated metabolites arising from hepatic/intestinal metabolism [52]. In contrast, any polyphenol undergoing enterohepatic circulation or remaining unabsorbed from the intestinal tract will pass into the large intestine and become the subject of fecal excretion. Based on the results of studies demonstrating the poor oral bioavailability of cacao polyphenols [52, 64] and taking into account the low recovery of the parent compounds from feces, it has been suggested that polyphenols are extensively metabolized or degraded by the intestinal microflora, especially in the colon [52, 67], to metabolites that after reaching the plasma and becoming circulated are eventually excreted in the urine [53, 66, 68].

Since the pioneering work of Booth et al. [69] in the mid-1950s, describing the formation of 3,4-dihydroxyphenylacetic acid, *meta*-hydroxyphenylacetic acid, and homovanillic acid upon fission of the heterocyclic, oxygen-containing, ring C of flavonols in humans and laboratory animals, numerous studies have clearly established a role for the intestinal microbiota in mediating these and other types of metabolic degradation in mammals [70, 71]. The most compelling evidence of the central role of the intestinal microflora in the metabolic breakdown of flavonoids was to find suppression of ring C fission of (−)-catechin in guinea pigs [72] and of (+)-catechin in the rat [73] by an oral pretreatment with an antibiotic. This conclusion was supported by the results of a parallel study that found (+)-catechin not to generate *meta*-hydroxyhippuric acid when administered by the parenteral route [74].

Current understanding of the involvement of gut microbes in polyphenol conversion within the intestinal tract has been greatly facilitated by the use of model compounds, the identification of specific gut microbes, and in vitro incubation with human fecal or animal cecal suspensions [61, 75]. In the colon, monomeric catechins are known to undergo ring fission and to be converted to phenolic acids and phenyl lactone derivatives. Although more than ten metabolites have been identified in human urine, 3-hydroxyphenylpropionic acid, 5-(3,4-dihydroxyphenyl)-γ-valerolactone, and 5-(3-hydroxyphenyl)-γ-valerolactone appear to be the major ones and to be excreted in the free form or as conjugates of glucuronic acid and, to lesser extent, of sulfate [76]. Further proof of the ability of microbes to catabolize flavan-3-ols has been gathered by incubating (−)-epicatechin with a suspension containing a mixture of intestinal bacteria from human (HIB) or rat (RIB) origin for periods of up to 48 h. Complete degradation was observable by 12 h with HBS and by 24 h with RBS, with the time course of the degradation entailing an initial reductive scission of ring C of the catechin (compound *1*), due to the presence of a free *para*-hydroxyl group (C_4–OH) in the catechin, to form an intermediate exhibiting a phloroglucinol moiety at ring A (compound *2*) which can sequentially become *para*-dehydroxylated in ring B to the 5′-monophenol (compound *3*), then decarboxylated at ring A to yield a (4′5′-dihydroxyphenyl)-3-propan-2-ol which, upon lactonization, converts to 4′,5′-dihydroxyphenyl-γ-valerolactone (compound *4*) [75]. Subsequently, *para*-dehydroxylation of the catecholic lactone or decarboxylation of the phloroglucinol part (ring A) of the *para*-dehydroxylated phloroglucinol intermediate will yield 5′-phenylhydroxy-γ-valerolactone (compound *5*). In turn, microbial hydrolytic fission of *4* and *5* will yield the corresponding 4′,5′-dihydroxylphenylbutyric (compound *6*) and 5′-hydroxyphenyl-butyric (compound *7*) carboxylic acids, respectively, with *6* also converting to *7* upon *para*-dehydroxylation and *7* undergoing 5′-*O*-methylation to form *8*. More drastic ring scission of *4* and *5* may account for the formation of the corresponding 4′,5′- and 5′-hydroxyphenylpropionic acids *9* and *10*, respectively [75].

Several human studies have assessed the role of the intestinal microflora on the formation of low molecular weight phenolic acid metabolites from cocoa polyphenols. In a typical study, human adult volunteers consumed a test meal consisting of 80 g of a flavanol-rich chocolate with bread and water, and urine samples were periodically collected for between 3 and 48 h after the test meal. Subsequent HPLC–electrospray ionization tandem mass spectrometry (HPLC-EI-MS) analysis of the urine samples identified *meta*-hydroxyphenylpropionic acid, ferulic acid, 3,4-dihydroxyphenylacetic acid, *meta*-hydroxyphenylacetic acid, phenylacetic acid, vanillic acid, *meta*-hydroxybenzoic acid, *para*-hydroxybenzoic acid, *para*-hydroxyhippuric acid, and hippuric acid [77]. In another study, use of liquid chromatography (LC)–MS/MS to analyze urine samples from human volunteers and rats orally fed a soluble natural cocoa powder in water (40 g/250 mL and 4.8 g/kg/day, respectively) provided a rather comprehensive view of the types and quantities of phenolic acid metabolites. In human volunteers, caffeic acid, ferulic acid, 3-hydroxyphenylacetic acid, vanillic acid, 3-hydroxybenzoic acid, 4-hydroxyhippuric acid, hippuric acid, (−)-epicatechin, and procyanidin B2 were found to be significantly increased after 24 h of consuming cocoa, with the increases showing large interindividual differences [78]. In rats, however, the pattern of increases in phenolic acid metabolites was slightly different, with the increases including 3,4-dihydroxyphenylpropionic acid, *meta*-coumaric acid, 3-hydroxyphenylacetic acid, protocatechuic acid, vanillic acid, and (−)-epicatechin. In addition, supplementing the rat diet with 10% (w/w) natural cocoa powder also raised the levels of caffeic acid, 3-methoxy-4-hydroxyphenylacetic acid, 3-hydroxybenzoic acid, enterolactone, and procyanidin B2, and made those of *meta*-coumaric acid, caffeic acid, 3-hydroxyphenylacetic acid, protocatechuic acid, vanillic acid, 3-hydroxybenzoic acid, enterolactone, (−)-epicatechin, and procyanidin B2 greater for the 10% w/w natural cocoa diet [78]. In general, these results point to the existence of major differences among phenolic acids bearing hydroxyl groups at both the C3 and C4 positions or only at the C3 position.

Through the use of a pH-controlled, stirred, batch-culture fermentation system reflective of the distal region of the human large intestine, it has been verified that both (−)-catechin and (+)-epicatechin are converted to 5-(3′4′-dihydroxyphenyl)-γ-valerolactone, 5-phenyl-γ-valerolactone, and phenylpropionic and that the formation of these metabolites from (+)-catechin required its initial conversion to (+)-epicatechin [79].

Experimental work relying on the anaerobic incubation of (3*R*)- and (3*S*)-flavan-3-ols and their 3-*O*-gallates with the human intestinal bacterium, *Eubacterium* (*E.*) sp. strain SDG-2, has found this bacterium to demonstrate metabolic stereoselectivity towards each substrate. On the one hand, *E. sp* strain SDG-2 was able to cleave ring C of (3*R*)- and (3*S*)-flavan-3-ols to give 1,3-diphenylpropan-2-ol derivatives, but not that of their 3-*O*-gallates, and on the other, it was shown to *para*-dehydroxylate ring B of the (3*R*)-flavan-3-ols (−)-catechin, (−)-epicatechin, (−)-gallocatechin, and (−)-epigallocatechin, but not of that of the (3*S*)-flavan-3-ols (+)-catechin and (+)-epicatechin [80]. These results suggest that degradation of a flavan-3-ol depends on the presence of a C_4–OH, that 3-O-methylation of (+)-catechin will make this compound resistant to degradation by RIB, and that differences in the rate of degradation by HIB and RIB may result from differences in the type and number of bacteria present in the intestinal tract of each species [75]. The identification of additional metabolic products by human colonic microorganisms have been achieved with the help of an ex vivo experimental model and a stability-indicating HPLC method. In this manner, 3,4,5-trihydroxyphenyl-γ-valerolactone and 3,4-dihydroxyphenyl-γ-valerolactone were identified as metabolites of (+)-catechin and (−)-epicatechin, following a 24-h anaerobic incubation of each flavanol with fresh ileostomy fluid [81]. The increased appearance of hippuric acid in the urine and feces following the ingestion of (+)-catechin by rats has been taken as evidence of subsequent β-oxidation of the phenolic acids generated by microbial metabolism and eventual hepatic conjugation of the resulting benzoic acid with glycine [74, 82].

Further proof of microbial stereoselectivity has been suggested by the work of Aura et al. [65] who detected the transient formation of 3,4-dihydroxyphenylpropionic acid in incubation samples of (+)-catechin, but not of (−)-epicatechin, with fecal suspension from healthy human donors under strict anaerobic conditions. In the same study, although both (+)-catechin and (−)-epicatechin yielded 3-hydroxyphenylpropionic acid and 3-phenylpropionic acid, the concentrations originating from each flavanol were not the same during the first 8 h of incubation. The extents of fermentation of (+)-catechin and (−)-epicatechin were 91.4% and 69.8%, respectively, at the end of 24 h.

An indication on the variability of the microbial metabolic products as a function of the bacterial species and on whether the microbe originates from humans or animals has also been documented [83].

By comparing the urinary metabolic products arising from the biotransformation of a monomeric flavan-3-ol among different animal species in the absence and presence of an antibiotic, it has been possible to conclusively establish the participation of intestinal bacteria at an early stage in the degradative pathway and have, at the same time, revealed the existence of qualitative and quantitative differences in metabolic products among different animal species [83]. For instance, the feeding of (+)-catechin to guinea pigs led to the urinary excretion of detectable amounts of *meta*-hydroxyphenylpropionic acid, a compound that was present in only trace amounts in rats [74]. In rabbits, a similar treatment resulted in the production of vanillic acid and protocatechuic acid [84], products that were not detected in the urine of guinea pigs [83]. By changing the bacterial species, salient difference between the types of metabolites that may originate from the biotransformation of a given flavan-3-ol as a function of the type of microorganism has also been reported. While rat fecal extracts can convert (+)-catechin to *meta*-hydroxyphenylpropionic acid [85], conversion by rabbit intestinal microorganisms was primarily to 5-(3-hydroxyphenyl)-valeric acid and secondarily to 5-(3,4-dihydroxyphenyl)-valeric acid [70]. Cultures of *Lactobacillus hilgardii* isolated from red wines and grown in the presence of (+)-catechin were shown to contain the parent compound along with gallic acid, pyrogallol, catechol, *para*-hydroxybenzoic acid, acetovanillone, and homovanillic acid [86]. A closer look at the role of the stereochemistry of the hydroxyl group at C-2 of the flavan-3-ol structure and of the source of the microbial inoculum in polyphenol microbial metabolism has been obtained by using (+)-catechin and (−)-epicatechin and fecal bacteria originating from two different groups of donors. Moreover, the ring fission site changed by changing the source of the microbial inoculum, but dehydroxylation, was found to proceed in similar manner regardless of the source of the inoculum [66].

Together with monomeric flavan-3-ols like (+)-catechin and (−)-epicatechin, the oligomers and polymers of flavan-3-ols variously designated as procyanidins represent an additional group of polyphenolic compounds that might also contribute to the health benefits of chocolate [87–89]. Procyanidins from fresh cocoa beans are predominantly dimers to decamers, but their content varies widely among commercial cocoas and chocolates since they may precipitate during the fermentation and processing of the beans [90]. Like the catechins, cacao procyanidins appear to be potential candidates for metabolism by the gut microflora because of much lower intestinal absorption than monomeric flavan-3-ols [91], especially when they are larger than trimers [92]. The inverse relationship between the extent of passage through the gut mucosa and the degree of polymerization would predict a high likelihood of microbial metabolism. The correctness of this assumption has been corroborated to a large extent by in vitro fermentation techniques. For example, the incubation of dimeric procyanidins, isolated from a commercial grape seed extract product, with human fecal microbiota for up to 24 h, disclosed the formation of 2-(3,4-dihydroxyphenyl)acetic acid and 5-(3,4-dihydroxyphenyl)-γ-valerolactone as the major metabolites and of 3-hydroxyphenylacetic acid, 4-hydroxyphenylacetic acid, 3-hydroxyphenylpropionic acid, phenylvaleric acids, monohydroxylated phenylvalerolactone, and 1-(3′,4′-dihydroxyphenyl)-3-(2″,4″,6″-trihydroxyphenyl)propan-2-ol as minor products. Also, this study provided further support to the previously held view that the degradation pathway of procyanidins is partly different from that of monomeric flavan-3-ols [91]. The existence of a similar degradative pathway for procyanidin polymers with an average degree of polymerization of six has also been verified in vitro. In this case, incubation of human colonic microflora with nonlabeled and ^{14}C-labeled purified procyanidin polymers under anoxic conditions led to complete degradation of the polymers after 48 h to products such as phenylacetic, phenylpropionic, and phenylvaleric acids, monohydroxylated mainly in the *meta* or *para* position, in yields that were similar to those previously reported for flavanol monomers [92].

Although procyanidins may be hydrolyzed to their constitutive monomers during their passage through the gastrointestinal tract, their catabolism by the gut microflora may not be entirely identical. Indeed, when the action of human fecal microbiota on (−)-epicatechin was compared with that of its dimer pure procyanidin B2 using a static in vitro culture model and extracts from the fermentation broth were analyzed for composition, five phenolic metabolites that were unique to the procyanidin and ten phenolic acids that were common to both substrates were identified. The dominant catabolites, detected at or after a 24-h incubation, were 5-(3′-hydroxyphenyl) valeric acid, 3-(3′-hydroxyphenyl) propionic acid, and phenylacetic acid. Procyanidin B2 was degraded twice as fast as (−)-epicatechin, and not more than ~10% of it was converted to (−)-epicatechin. Additional findings were the isolation of two previously unreported catabolites of procyanidin B2 showing retention of the flavanol ring A and the C4–C8 intermonomer bond, confirmation that microbial catabolism favors the removal of the 4′- rather than of the 3′-hydroxyl group, and occurrence of both α-oxidation and β-oxidation during the breakdown [93]. Somewhat different results were achieved by van't Slot and Humpf [94] using an ex vivo model based on the intestinal microbiota isolated from the cecum of freshly slaughtered pigs and which is rather similar in composition to that of the human intestine. These authors found that incubation of procyanidin B2 with freshly collected cecal inoculum at 37°C for periods of 20 min to 8 h resulted in degradation of the test compound to phenolic carboxylic acids (mainly 4-hydroxybenzoic acid), with lesser amounts of 3,4-dihydroxyphenylpropionic acid and 4-hydroxyphenylacetic acid, and with 3,4-dihydroxyphenylpropionic acid only transiently being detected at 1 h) and phloroglucinol, but not to monomeric flavan-3-ols. The disappearance of the procyanidin after 4–8 h was related to its binding to proteins [94].

Another interesting aspect of the polyphenols–microbiota relationship is the ability of the polyphenols to modulate the composition and activity of the gut microbiota. This ability appears to vary rather widely among different bacteria since experiments with (+)-catechin in rats have shown that this flavanol was able to inhibit the growth of some species, to enhance the growth of others, and to be without effect on a few under in vitro and in vivo conditions [79]. Long-term ingestion of phenolics like those present in cocoa may produce changes in the bacterial population of humans and rodents and, hence, in the levels of metabolites detected in the urine. This is the case of cocoa phenolics such as (−)-epicatechin and (+)-catechin which have shown the ability to significantly repress the growth of *Clostridium difficile* and *Bacteroides spp.*, whereas commensal anaerobes such as *Clostridium* spp. and *Bifidobacterium* spp. were less severely affected [95]. The absence of a common effect on different microbial species was also observed in another study in which an incubation of selected members of the intestinal microbiota with (+)-catechin resulted in a significant increase in the growth of the *Clostridium coccoides–Eubacterium rectale* group, of *Bifidobacterium* spp. and of *Escherichia coli* but in inhibition of the growth of *Clostridium histolyticum* [78, 96]. In addition, a change in the population of bacteria towards tannin-resistant Gram-negative species (i.e., *Enterobacteriaceae* and *Bacteroides*) has been reported after a 3-week administration of condensed tannins to rats [78]. Therefore, the long-term ingestion of cocoa phenolics may have produced changes in the bacterial population of volunteers which may account for the differences in the urine metabolic profile relative to the results reported by Rios et al. [77]. In this context, Gu et al. [97] indicated that while the urinary excretion of 3-hydroxyphenylpropionic acid was high in rats fed a high amount (20–40% of the diet) of sorghum bran over 50 days, the changes in the concentration of hydroxyphenylacetic acid were inversely related to the extent of dietary supplementation with sorghum bran, namely, to be high with a low supplementation and low with a high supplementation.

The Metabolism of Chocolate and Cocoa Polyphenols

The Metabolism of Monomeric Flavanols

Data in support of the biotransformation of flavanols and other flavonoids in the small intestine and colon of the gastrointestinal tract as well as of hepatic metabolism are plentiful. In the oral cavity, monomeric flavanols like (+)-catechin and (−)-epicatechin and procyanidin oligomers (dimers to hexamers) appear to be impervious to the action of saliva, with the same outcome being observed in the esophagus [98]. Studies with a cocoa beverage have shown that in the gastric lumen, both monomeric flavanols and oligomeric procyanidins appear to be rather stable in spite of the existence of in vitro data indicating that at a pH comparable to that of the stomach, flavanol oligomers like those present in cacao bean products may undergo degradation to (−)-epicatechin units or to epicatechin-containing oligomers of a smaller size than those originally present in a cacao-derived food item [98]. The rate and extent of absorption from the digestive tract are found to depend on solute characteristics such as molecular size, stereochemistry, lipophilicity, and binding affinity for food matrix proteins. Additional determining factors may be luminal pH, gastric and intestinal transit time, mucosal membrane permeability, and, in some instances, hepatic first-pass effect. An understanding of the events surrounding the uptake into and diffusion through jejuna and ileal enterocytes has been greatly facilitated by the use of a small intestine perfusion model since it allows for solute biotransformations analogous to those taking place when the flavanols transverses specific segments of the digestive tract to reach the mesenteric circulation. With this experimental model, the serosal side after perfusion of the jejunum with (−)-epicatechin contained the glucuronides at the C5 and C7 positions of the flavanol ring A together with high levels of the 3′-*O*-methylated and 4′-*O*-methylated metabolites as well as the 5′- and 7′-glucuronides of the 3′-*O*-methylated and 3′-*O*-nonmethylated (−)-epicatechin [99, 100]. In this model, ileal metabolism was less and the amount of intact monomeric catechin greater relative to results with the jejunum. Although similar results have been reported for (−)-catechin added to the perfusate of ileal and jejunal preparations, only the glucuronide conjugate and 3′-*O*-methyl metabolite, but no intact parent compound, were detected in the mesenteric circulation [101].

In vivo, (+)-catechin and (−)-epicatechin are directly absorbed in the jejunal part of the small intestine, where they are *O*-methylated, glucuronidated, and *O*-methylated glucuronidated before they are taken to the liver by the portal vein for further metabolism to methyl, glucuronide, and sulfate derivatives [102]. These forms pass into the bile through the enterohepatic circulation to reach the duodenum, from where they may make their way into the colon, where, together with procyanidins that may have escaped small intestinal absorption, will be subjected to metabolic degradation by the intestinal microbiota. Interestingly, however, considerable variation seems to surround the identity and number of the metabolites that have been detected in the plasma, urine, and bile samples. This apparent lack of uniformity have been related to differences in the analytical method used to detect the metabolites, the use or omission of enzymatic sample treatment, and the dose of polyphenol or food product used [103].

The Metabolism of (+)-Catechin

Early efforts to decipher the absorption, metabolism, and excretion of chocolate and cocoa monomeric catechins in mammals were based on the parenteral administration of a single dose of radiolabeled compounds like (+)-[U-^{14}C]-catechin and (+)-[ring A-^{14}C]-catechin, followed by the isolation of urinary metabolites for the measurement of tracer incorporation [83]. In addition, the distribution of the radioactivity allowed the assessment of the preferential entry of the catechin and its metabolites into particular tissues and biological fluids [83, 104]. An alternative approach was to treat a suitable animal species first with a single oral dose of a flavan-3-ol in admixture with a standard animal diet and in the following days to maintain the animal on a standard diet. In human studies, the catechin compounds were administered orally, usually as a capsule, with water and after breakfast. To minimize interference by dietary phenolic acids, the volunteers were placed on a common restricted diet for the duration of the study [105]. In general, the bulk of the early findings reflected the direct involvement of the intestinal microbiota in the generation of an abundance of metabolic products, mainly phenols, phenolic acids, and phenyl lactones, and the utility of in vitro fermentation studies as a confirmatory tool [65, 66].

Over the years, considerable information has accumulated on the mechanisms underlying the biotransformations of monomeric polyphenols in humans and animals, aided immensely by the advent of more powerful, sensitive, and selective analytical instrumentation and techniques. In the particular case of cacao product polyphenolics, this task has been greatly facilitated by the combined use of in vitro and in vivo approaches and recognition of the contribution made by the native intestinal microflora. Earlier studies on the metabolism of both (+)-catechin and (−)-catechin were based on the use of laboratory animals and on comparisons of the metabolic profiles with those derived from humans. For example, the fate of (+)-catechin was studied by Das and Griffiths [83] in the rat and guinea pig by intragastrically administering (+)-[U-^{14}C]-catechin or (+)-[ring *A*-^{14}C]-catechin and collecting respired CO_2 and urine samples for up to 4 days and feces for 2 days. The amount of total radioactivity excreted in a 24-h urine sample after feeding (+)-[U^{14}C]-catechin, as a percentage of the administered dose, was ~52% in the rat and ~63% in the guinea pig, and in the case of [ring *A*-^{14}C]-catechin amounted, respectively, to 54.5% and 55%. Radioautography of fecal extracts from rats receiving (+)-[U-^{14}C]-catechin revealed radioactive areas in positions corresponding to δ-(3-hydroxyphenyl)-γ-valerolactone and δ-(3,4-dihydroxyphenyl)-γ-valerolactone and scintillation counting detected 1.3–1.5% of the dose undergoing fecal elimination. After the administration of [ring A-^{14}C]-catechin to rats, 0.56–0.61% of the dose was excreted in the feces over a 2-day period, and radioautography detected the presence of labeled δ-(3-hydroxyphenyl)-γ-valerolactone and δ-(3,4-dihydroxyphenyl)-γ-valerolactone in fecal extracts. In contrast, feces from guinea pigs showed a low total radioactivity following the administration of (+)-[U-^{14}C]-catechin, namely, 0.07–0.12% of the administered dose, although higher radioactivities were detected after giving [ring A-^{14}C]-catechin. In both instances, no phenyl valerolactones were detected in the feces. It is evident that the major portion of the ^{14}C-related radioactivity administered to both guinea pig and rat is excreted in the urine and breath over the period of examination, with only a small amount appearing in the feces, possibly as result of the short (48 h) collection period, and unaccounted balance possibly present in the intestinal tract and as tissue constituents derived from $^{14}CO_2$.

Furthermore, Das and Sothy [106] examined the biliary and urinary excretion of (+)-catechin in normal male rats after a single intravenous injection of (+)-[U-^{14}C]-catechin. Large amounts of ^{14}C, amounting to 44.5% of the dose, were excreted in the bile as three unconjugated metabolites and two glucuronide conjugates, one of which was a (+)-catechin conjugate. In both the urine and bile, the metabolites showed maximum excretion between 30 and 90 min after dosing with [^{14}C]-catechin. The metabolites, identified as *meta*-hydroxyphenylpropionic acid, *para*-hydroxyphenylpropionic acid, δ-(3-hydroxyphenyl)-γ-valerolactone, and δ-(3,4-dihydroxy- phenyl)- γ-valerolactone and originating from the action of intestinal microorganisms on bile duct-excreted metabolites of (+)-catechin, undergo reabsorption into the portal circulation and later excretion in the urine. These metabolites were not detected in bile from rats having the common bile duct cannulated. From these results, it was concluded that biliary excretion represents an alternative pathway for the excretion of metabolites of intravenously administered (+)-catechin. At the same time, these results confirmed those arrived in a previous study that found that the incubation of rat intestinal contents with (+)-catechin in vitro resulted in the formation of the same phenolic acid and lactone metabolites [107].

An additional study by Das [107] in human volunteers disclosed interspecies metabolic similarities and differences. The oral dosing with (+)-catechin-containing capsules resulted in absorption, rapid metabolism, and excretion of the compound in the urine largely within 24 h. Out of 11 (+)-catechin metabolites detected in the urine, the major metabolites were *meta*-hydroxyphenylpropionic acid, a phenolic acid, and the lactones δ-(3,4-dihydroxyphenyl)-γ-valerolactone and δ-(3-hydroxyphenyl)-γ-valerolactone. The phenolic compounds appeared in the urine as both free and conjugated forms, mainly as glucuronides and less as sulfates. While *meta*-hydroxyphenylpropionic acid has also been found in rat urine [74], *meta*-hydroxybenzoic acid, the major urinary metabolite of (+)-catechin in the guinea pig [72], and vanillic acid and protocatechuic acid detected in the urine of rabbits fed (+)-catechin [84] were not detected in human urine. Several phenyl-γ-valerolactones, earlier reported to be present in rat and guinea pig urine [72], were also found in rabbit urine [84]. Furthermore, several unidentified compounds, totaling six, were absent in the urine of rat and guinea pig. These interspecies differences suggest that in man, the breakdown of (+)-catechin may occur differently from those reported for the guinea pig and rat [72, 74, 83]. The absorption of (+)-catechin in the gastrointestinal tract occurred as early as 6 h after the oral administration of the compound and appeared in the urine together with several unidentified metabolites, both in the free and conjugated forms. The amount excreted in the urine within 24 h was about 7.5% of the administered dose. Unchanged (+)-catechin, totaling up to about 18.6% of the administered dose, appeared in feces collected at 48 h together with *meta*-hydroxyphenylpropionic acid and an unidentified metabolite. The rise in phenolic levels in the blood occurred as early as 6 h after the oral administration of (+)-catechin and paralleled the rise in urinary excretion of phenolic compounds.

To ascertain the metabolizing role of both the small intestine and liver, Donovan et al. [101] examined the absorption of different concentrations of (+)-catechin (ranging from 1 to 100 μM/L) in the rat using an in situ perfusion of the jejunum and ileum technique. In addition to a concentration-dependent absorption of catechin in a 30-min period, the presence of only intact compound in the effluent ruled out intestinal excretion of metabolites as a mechanism of (+)-catechin elimination. Furthermore, the absence of intact (+)-catechin and the presence of the glucuronides of catechin and of 3′-*O*-methyl-(+)-catechin in plasma from mesenteric vein blood pointed to the absorption of (+)-catechin into enterocytes and to its metabolism within. Additional methylation and sulfation of 3′-*O*-methyl catechin for excretion in the bile were observed in the liver. Circulating forms were the glucuronides of catechin and 3′-*O*-methyl catechin. By using an isolated rat liver perfusion system, it was determined that the activity of hepatic sulfotransferases can lead to the formation of three sulfated catechin metabolites, the sulfate–glucuronide and two sulfates [108]. However, sulfation at the intestinal level is also a possibility since both human intestinal cytosolic sulfotransferases and recombinant human sulfotransferases were later shown to sulfate either (+)-catechin or (−)-epicatechin in vitro [109].

After the administration of a single dose of (+)-[U^{14}C]-cyanidanol-3, a commercial product of (+)-catechin, to human volunteers, a mean of 55% of the dose of ^{14}C was found in the urine, with 90% of the urine ^{14}C becoming excreted within 24 h of the administration. Major urinary metabolites were the glucuronides of (+)-catechin and 3′-*O*-methyl-(+)-catechin, and the sulfate of the latter, which collectively accounted for 75% of the urine ^{14}C. Urinary excretion of unchanged (+)-catechin was ≤1.4% of the dose and became apparent between 30 min and 12 h after administration. Metabolites representing products of ring C scission were present in only trace amounts and were identified as 3-hydroxybenzoic acid, 3-hydroxyhippuric acid, and 3-hydroxyphenylpropionic acid. Labeled metabolites remained in the plasma for at least 120 h after administration [110]. The same laboratory investigated the role of the 3-OH group in the metabolic disposal of a catechin by treating rats with 3-palmitoyl-(+)-[U-^{14}C]-catechin by the oral route and examining the urine and feces for radioactivity and metabolites [111]. In a period of 17 days following dosing, 63% of the ^{14}C dose was excreted in the urine, 24% in the feces, and 7% as $^{14}CO_2$, with 3.7% of the dose of ^{14}C persisting in tissues after 28 days. Biliary excretion accounted for 28.2% of the dose at 48 h after dosing, and 3′-*O*-methyl-(+)-catechin glucuronide was identified as the major metabolite in bile. These experiments also demonstrated the dependence of 3-palmitoyl-(+)-catechin on the presence of bile for its intestinal absorption and the resistance of this ester to degradation by intestinal microorganisms in vitro. Subsequent deesterification in plasma and liver yielded free (+)-catechin, which was then converted to the 3′-*O*-methylated product in the liver. As a result, the most abundant urinary metabolites of the (+)-ester were found to be the conjugates of (+)-catechin and 3′-*O*-methyl-(+)-catechin, with smaller amounts of ring scission products and $^{14}CO_2$, in all probability arising from biliary metabolites lacking the ester linkage since they were not excreted by rats with ligated bile ducts [111].

The methylation of flavan-3-ol monomers at the 3-OH position has been established by in vitro and in vivo studies. In the rat, the oral administration of (+)-[U-^{14}C]-catechin led to the detection of a major biliary metabolite, later shown to be the glucuronide of 3′-O-methyl-(+) catechin, by

HPLC–MS [112]. The enzymatic nature of the methylation and the participation of the liver were confirmed by isolating [methyl-^{14}C]-*O*-methyl-(+)-catechin from an incubation medium that contained either a liver homogenate or purified catechol-*O*-methyl-transferase (COMT) plus (+)-catechin and S-adenosyl-L-[methyl-^{14}C]-methionine [112]. A similar conclusion was reached with the help of an extracorporeal pig liver preparation to which (+)-catechin had been added, and detecting the 3′-*O*-methyl derivative in the perfusate [113]. Experiments in which rats were given 3-*O*-methyl-(+)-catechin by the oral or intravenous routes indicated that this compound had been further methylated at the ring B 3′- (or 4′-)OH group to yield 3,3′- (or 4′-)dimethyl-(+)-catechin, which, in turn, was subjected to conjugation with glucuronic acid and the product excreted both in the urine and bile [114]. Although intact 3-*O*-methyl-(+)-catechin was excreted in rat urine following parenteral, but not oral, administration, the amount was not significant. Following the parenteral administration of 3-*O*-[^{14}C] methyl-(+)-catechin to the rat and marmoset, the excretion of radioactivity by both species was similar (approximately 50% in urine and 35% in feces) and different from the mouse, which excreted 79% of radioactivity in the urine and 22% in the feces. In all instances, more than 85% of the ^{14}C in bile was in the form of dimethylcatechin glucuronide, with the amounts of radioactivity being ≥50% in the urine and ≤22% in the feces. These results indicated that biological methylation is the major route of 3-*O*-methyl-(+)-catechin metabolism in all species investigated and that the compound does not undergo the ring fission associated with the parent compound (+)-catechin [114]. Methylation of the 3′-*O*-position of (+)-catechin appears to be a reaction common to man and animals (including the rat, mouse, rabbit, pig, and marmoset), and which may be continued with the formation of the 3′-*O*-methyl-(+)-catechin glucuronide and 3′-*O*-methyl-(+)-catechin sulfate as additional urinary metabolites [115]. Confirmatory evidence has also been derived from a study with human male volunteers that were given a single 2 g oral dose of radiolabeled 3′-*O*-methyl-(+)-catechin and their urine samples analyzed within 120 h of the administration for unchanged compound and metabolites. While the excretion of unchanged compound amounted to <0.8% of the dose, that of the glucuronides of 3,3′-*O*-dimethyl-(+)-catechin was 15.8% and of the glucuronides and sulfates of 3′-*O*-methyl-(+)-catechin 11.4% and 10.6% of the dose, respectively. 3′-*O*-Methyl-(+)-catechin was measurable in plasma for up to 12 h after administration. In comparison with mouse, rat, and marmoset, where the methylated derivatives of 3-*O*-methyl-(+)-catechin accounted for the entirety of the urinary metabolites, in man it only represented about 53%. Additionally, sulfate conjugation of 3′-*O*-methyl-(+)-catechin was observed in man but not in the other species [116]. The existence of common metabolic pathways for (+)-catechin and 3-*O*-methyl-(+)-catechin has been confirmed in human volunteers, separately treated with these compounds, who were found to excrete the glucuronides of 3′-*O*-methyl-(+)-catechin and of 3,3′-*O*-dimethyl-(+)-catechin, respectively, as the major urinary metabolites [117].

In rats fed an animal diet supplemented with 0.25% (+)-catechin for 14 days, the plasma only contained the glucuronides of catechin and 3′-*O*-methyl-(+)-catechin but neither free catechin nor free methylated catechin. About 90–95% of (+)-catechin was methylated in the liver, with about 80% of the original and methylated forms being present in the free forms. In comparison to the plasma levels of metabolites, those in the liver were markedly lower [118].

The contributory role of the liver and biliary circulation to the disposal of (+)-catechin and its 3′-*O*-methyl derivative has been investigated in rats with pharmacologically induced hepatitis. When (+)-[U-^{14}C]catechin was intravenously administered to rats pretreated with D-galactosamine, the biliary and fecal excretion of (+)-catechin metabolites was shown to decrease and the renal excretion to increase relative to normal animals. Although the biliary metabolites were present in similar proportions in D-galactosamine-treated and in control animals, the level of 3′-*O*-methyl (+)-catechin sulfate was markedly decreased and that of the 3′-*O*-methyl-(+)-catechin glucuronide markedly increased [119]. The absence of a significant depressive effect of hepatitis on COMT activity was inferred from the similar overall excretion of 3′-*O*-methyl-(+)-catechin conjugates observed in rats with and without hepatitis. As a result of D-galactosamine-induced hepatitis, glucuronidation of

(+)-catechin metabolites was depressed in liver perfusion experiments but enhanced in liver homogenates. Parallel experiments with lung slices revealed that the lung may represent an extrahepatic site for the metabolism of (+)-catechin in D-galactosamine hepatitis since they were not affected by D-galactosamine as much as the liver [119]. The lack of a significant effect of experimental hepatitis on the overall ability of the rat to metabolize an intravenous dose of 3′-*O*-methyl-[^{14}C]-(+)-catechin by methylation or glucuronidation in spite of a significant increase in metabolite excretion in the urine (by 36%) and significant decline (by 40%) in fecal excretion has also been documented. In addition, in bile duct-cannulated rats, the prior induction of hepatitis lowered the ^{14}C excretion (by ~20%) relative to control values. These data implied that as a result of hepatitis, biliary metabolites reabsorbed from the intestine are preferentially excreted in the urine rather than reexcreted in the bile through an enterohepatic cycle as is the case of normal rats and that the plasma clearance of the metabolites of 3-*O*-methyl-(+)-catechin, but not of the parent compound, is altered [104]. An analogous situation has been verified for the disposition and metabolism of (+)-cyanidanol-3 in human alcoholics with cirrhosis of the liver. Following the oral intake of a 2-g dose of U-^{14}C-(+)-cyanidanol-3, this commercial form of (+)-catechin was detected in the plasma between 0.5 and 24 h after dosing, and the radioactivity associated with its metabolites, measured as ^{14}C levels, was still detectable in the urine at 120 h. In the average, 55% of the dose was excreted in the urine, predominantly as the glucuronide and sulfate conjugates of (+)-cyanidanol-3 and 3′-*O*-methyl-(+)-cyanidanol-3. Overall, however, the metabolic handling of (+)-catechin by the liver of cirrhotic patients was not significantly different from that by normal subjects [120].

Harada et al. [121] have studied the metabolism of orally administered (+)-catechin and (−)-epicatechin in rats. With the help of MS and nuclear magnetic resonance (NMR) analyses, these authors established the site of β-glucuronidation of these flavanols by UDP-(beta)β- glucuronosyl transferase to be the 5-hydroxyl group in ring A. The higher plasma concentration of the 5-*O*-β-glucuronide of (−)-epicatechin than that of (+)-catechin in the plasma was ascribed to the higher rate of intestinal absorption of the former compound than that of the latter. At variance with 3′-*O*-methylation, which has a small reducing effect (~1%) on antioxidant activity, 5-*O*-β-glucuronidation preserved the superoxide anion radical scavenging activity of both (−)-epicatechin and (+)-catechin.

Metabolism of (−)-Epicatechin

The metabolism and plasma antioxidant activity of (−)-epicatechin were investigated in rats by correlating the levels of (−)-epicatechin metabolites with the oxidability of cholesteryl ester (CE) and α-tocopherol depletion. To this effect, rats received an intragastrical dose of (−)-epicatechin (10 or 50 mg/rat), and their plasma was collected at 1 h and 6 h after (−)-epicatechin administration and analyzed by reversed-phase HPLC with electrochemical detection. At 1 h, the plasma contained the glucuronide and glucuronide-sulfate conjugates of both (−)-epicatechin and 3′-*O*-methyl-(−)-epicatechin, the levels of which decreased after 6 h to mostly the 3′-*O*-methylated form. Compared to the control group, the plasma obtained at 1 and 6 h after (−)-epicatechin administration was more resistant to copper sulfate ($CuSO_4$) and 2,2′-azobis(2-amidino-propane) dihydrochloride (AAPH)–induced oxidations on the basis of cholesteryl ester hydroperoxide (CE-OOH) accumulation and decrease in α-tocopherol, which were much lower than in the plasma of control (nontreated) rats due to oxidation. The content of CE-OOH and consumption of α-tocopherol in the plasma from (−)-epicatechin-administered animals were much lower than those expected from the amount of unmetabolized (−)-epicatechin present in the plasma. The 3′-*O*-methylated form was found to be more stable than the free form when plasma from (−)-epicatechin-treated rats was autoxidized at 37°C. These results suggest that (−)-epicatechin metabolites, particularly conjugates in the free form, possess an effective antioxidative activity in blood plasma [122].

The oral absorption and metabolism of (−)-epicatechin have also been examined by Piskula and Terao [123]. After the treatment of rats with (−)-epicatechin (172 μM/kg body weight) and collection of tail blood for 8 h after treatment, HPLC/EI-MS analysis and HPLC with electrochemical detection (EC) permitted the identification and quantitation, respectively, of the various plasma metabolites. Between 0 and 8 h, (−)-epicatechin and six conjugated metabolites, reaching a maximum plasma level at 2 h posttreatment, were identified in the plasma as glucuronides, sulfates, and sulfoglucoronides of both (−)-epicatechin and 3′- (or 4′-)-*O*-methyl-(−)-epicatechin. The plasma levels of (−)-epicatechin were the lowest and the first to be cleared from the plasma after 6 h relative to the metabolites. Among the various metabolites, the levels of the (−)-epicatechin conjugates were always higher than those of the methylated-(−)-epicatechin conjugates, with the peak levels in both instances decreasing in the order sulfoglucuronide > glucuronide >> sulfate > intact flavanol. At 30 min after (−)-epicatechin administration, about 90% of total metabolites in plasma were the glucuronides, about 50% were the sulfates, and about 40% were methylated, with each type increasing thereafter. These results indicated that absorption of (−)-epicatechin into the intestinal mucosa takes place before undergoing immediate glucuronidation by the uridine 5′-diphosphate-glucuronosyltrasferase (UGT) activity of the upper half of the small intestine, cecum, and upper half of the large intestine, followed by delivery by the portal circulation to the liver for sulfation by sulfotransferase. After sulfation and methylation in the liver and, to a lesser extent, in the kidney, the metabolites of (−)-epicatechin are excreted in the bile or circulated in the blood before becoming excreted in the urine. At the same time, they confirmed the previously held view that the level of unmetabolized compound found in the urine is always very low.

In spite of the abundance of information on the metabolism of monomeric catechin in the intestinal tract, this issue is still the subject of controversy. For example, the ability of the intestine to serve as an O-methylating and glucuronidating site for different types of flavonoids has been investigated with model flavonoid compounds that were perfused through a section of jejunum or ileum isolated from the small intestine of rats and analyzing the respective perfusate for metabolites. One obvious advantage of this approach is that it negates the metabolic contribution of the intestinal flora [124]. When (+)-catechin and (−)-epicatechin were the flavonoids circulated across the isolated jejunum in the direction of the serosal side, the emerging perfusate contained the corresponding glucuronides (~45%), O-methylated products (as 3′-*O*-methyl- and 4′-*O*-methyl derivatives) (~30%), and *O*-methylated glucuronides (~20% of total flavanols identified). In contrast, perfusing the same flavanols through the ileum resulted in their appearance on the serosal side unmetabolized and in a higher total percentage of flavanols being transferred than through the jejunum (by ~ fivefold) [99]. On the basis of these results, it is assumed that the jejunum is a major site for *O*-methylation and glucuronidation of flavanol compounds and that the ileum will facilitate the delivery of flavanols entering the portal vein in high concentrations to the liver, where they will be subjected to the activities of hepatic COMT and UGT.

In a study by Harada et al. [121], rats orally intubated with either (+)-catechin or (−)-epicatechin yielded plasma, urine, and bile samples that contained four metabolites, identified as the 5-*O*-β glucuronides of (+)-catechin, (−)-epicatechin, and the 3′-*O*-methylated forms of (+)-catechin and (−)-epicatechin, by MS and NMR analyses, and found to exhibit a high antioxidant activity against the superoxide anion in vitro. Baba et al. [125] evaluated the levels of (−)-epicatechin and its metabolites in the plasma and urine of male volunteers that had consumed either chocolate or cocoa powder. Maximum plasma levels of (−)-epicatechin metabolites were observed at 2 h after chocolate or cocoa consumption, with the sulfate, glucuronide, and sulfoglucuronide appearing as the major plasma and urine metabolites of (−)-epicatechin.

Okushio et al. [126] reported the isolation of (−)-epicatechin along with four unknown metabolites, later characterized as 3′-*O*-methyl-(−)-epicatechin, 4′-*O*-methyl-(−)-epicatechin, (−)-epicatechin-5-*O*-β-glucuronide, and 3′-*O*-methyl-(−)-epicatechin-5-*O*-β-glucuronide by MS and NMR analyses

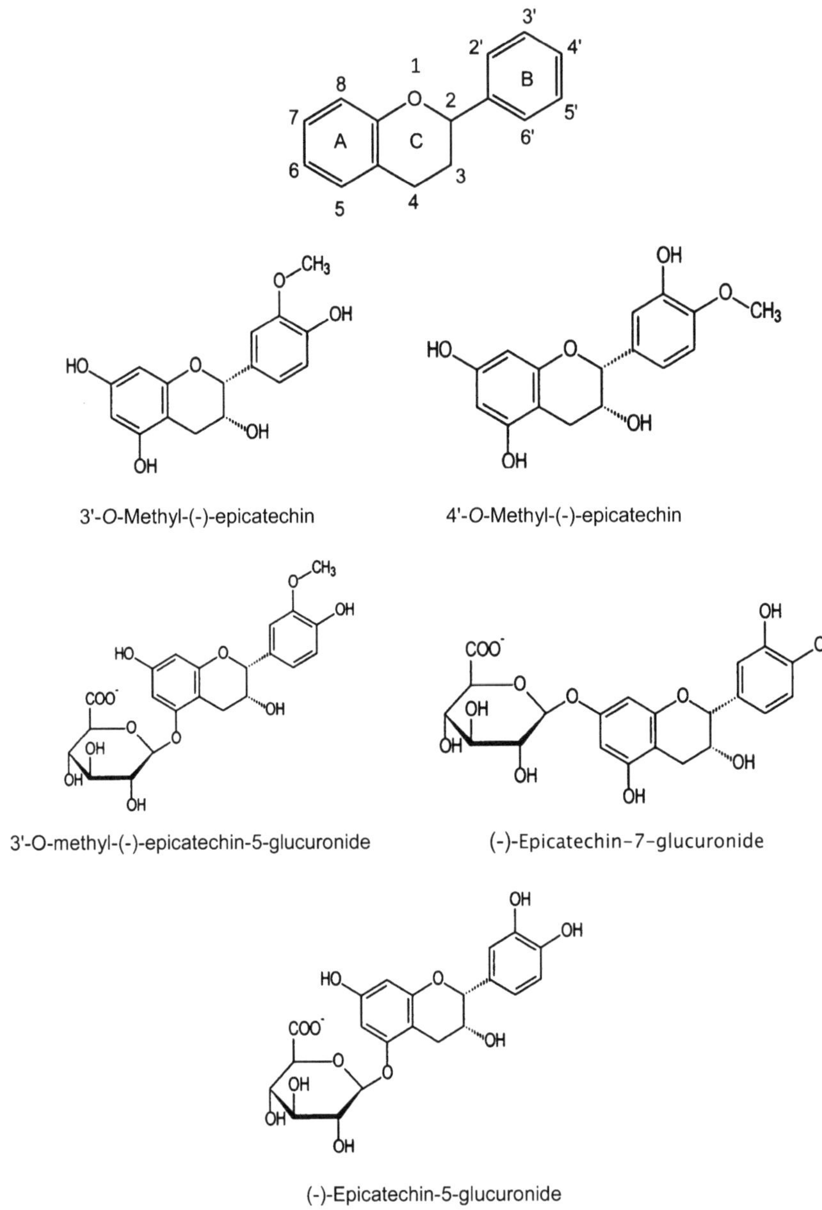

Fig. 17.3 Metabolites of (-)-epicatechin identified in rat urine

from the urine of rats orally treated with (–)-epicatechin. In addition, (–)-epicatechin and its methylated metabolites were also detected in plasma but not in bile (Fig. 17.3).

A comparative study of the metabolism and urinary excretion of (+)-catechin and (–)-epicatechin, orally administered to rats, has shown that the absorbed flavanols were present in the plasma mainly as nonmethylated and 3′-*O*-methylated conjugated metabolites. In animals receiving (–)-epicatechin, the major metabolites were the glucuronide and sulfoglucuronide in the nonmethylated form, and sulfate and glucuronide in the 3′-*O*-methylated form. Excretion of the total amount of metabolites of (–)-epicatechin was greater than the total amount of (+)-catechin metabolites [127]. The 3′-*O*-methylated and 3′-*O*-methylated-glucuronides have also been detected in human plasma [128].

Terao [129] conducted a comparative study of the oral absorption, metabolism, and urinary excretion of a typical flavanol like (−)-epicatechin and of a typical flavonol like quercetin in the rat. The results suggested that after oral administration, the metabolism of both flavonoids begins in the intestinal mucosa where the activity of UGT is at its highest and that they accumulate in the plasma mostly as glucuronide and sulfate conjugates, with small amounts of (−)-epicatechin, but none of quercetin, appearing in the circulation. In turn, the conjugates appeared to exhibit antioxidative activity in the plasma and, hence, the ability to participate in the antioxidant defense of the plasma, a function that has also been described for polyphenolic compounds in red wine [130].

Vaidyanathan and Walle [131] evaluated the absorption of (−)-epicatechin by the intestinal epithelial membrane using a monolayer of human Caco-2 cells and HPLC with diode array detection. In concentrations of 5–50 μM, (−)-epicatechin did not show apical to basolateral absorption but, rather, basolateral to apical efflux. The participation of MRP2, a multispecific organic anion transporter expressed in the apical membrane of Caco-2 cells, was inferred from the 50% reduction in efflux observed in the presence of MK-571, a competitive inhibitor of MRP2. More importantly, the same treatment led to a measurable apical to basolateral absorption of (−)-epicatechin. MK-571 was also found to inhibit the transport of two polar metabolites of (−)-epicatechin, identified as sulfate conjugates and found exclusively on the apical side.

The same authors have also examined the metabolic conjugation of (−)-epicatechin using human liver and intestinal microsomes and cytosol, and recombinant UGT and sulfotransferase (SULT) isoforms and compared the results with those taking place in the rat [132]. Surprisingly, (−)-epicatechin was not glucuronidated by the human liver and small intestinal microsomes, by the human colon microsomes, and by recombinant UGT1A7, which is not present in the liver or intestine. In contrast, (−)-epicatechin was efficiently glucuronidated in the rat liver microsomes with the formation of two glucuronides, and the human liver cytosol efficiently sulfated (−)-epicatechin mainly through the SULT1A1 isoform. For the intestine, both SULT1A1 and SULT1A3 contributed. Other SULT isoforms contributed little. HPLC analysis of the sulfate conjugates showed one major sulfatase-sensitive peak with all tissues. An additional minor sulfatase-resistant peak was formed by the liver and intestinal cytosol as well as with SULT1A1 but not by the Caco-2 cytosol and SULT1A3. In the rat, (−)-epicatechin sulfation was considerably less efficient than in the human liver. These results indicated that sulfation was the major pathway of (−)-epicatechin metabolism in the human liver and intestine and that glucuronidation was unimportant. There was also a large species difference both in glucuronidation and sulfation of (−)-epicatechin between rats and humans [132]. The apparent absorption, blood concentration, maximum plasma concentration, time to maximum plasma concentration, tissue distribution, and fecal, but not the urinary, excretion of a flavanol like (−)-epicatechin were found to be similar to those of the flavonol quercetin in ileostomized rats receiving either 3-[^{3}H]-(2R,3R)]-(−)-epicatechin or [4-^{14}C]-quercetin through a tube inserted in the esophagus [133]. In the study, maximum plasma concentrations and time to maximum were similar for both compounds. Furthermore, the radioactivity from epicatechin was rapidly cleared from the circulation, but increased again from 6 h after administration onward. Furthermore, the amount of radioactivity measured at the end of the ileum 3–7 h after ingestion and the amount of radioactivity excreted in feces at 24 h after ingestion were similar for both (−)-epicatechin and quercetin. More radioactivity was present in the urine during the first 12 h from (−)-epicatechin than from quercetin, but neither compound was found to accumulate in the organs and tissues examined at 24 h after ingestion, except for the gastrointestinal segments [132].

Metabolism of Monomeric Catechins Present in Food Products

Owing to their powerful antioxidant properties, monomeric flavanols and oligomeric procyanidins have been the subject of extensive evaluation in humans and in rodents for absorption from and bioavailability in the digestive tract following their consumption as part of polyphenol-rich food products such as chocolate, cocoa, red wine, green or black tea, and certain fruits. Regardless of the food product, it has become apparent that monomeric flavanol components undergo a ready but limited absorption from the intestinal tract to yield plasma concentrations that are often in the nanomolar or low micromolar range [134], with the bulk becoming metabolized during absorption across the small intestinal mucosa [101], in the liver [101, 135], and in the colon [98] to metabolites that are distributed in the body before being excreted in the urine and bile [121]. A correlation between the dose of catechin consumed and the site of catechin metabolism has been inferred from two key findings in humans, namely, that a large (2,000 mg) dose of (+)-catechin can lead to detectable levels of free catechin in the plasma after 30 min, to traces of methylated catechin in 2 h, and to 40% of the urinary catechin as the methylated product after 8 h [110]. However, all of the catechin was found conjugated, and no free catechin was detected after the consumption of a relatively small initial quantity [136]. These results were taken as a suggestion that the dose determines the primary site of metabolism, with large doses being metabolized primarily in the liver and a small one being metabolized by the intestinal mucosa. In contrast, dimeric and trimeric procyanidins, but not higher oligomers, are detectable in the plasma in their original form along with variable quantities of low molecular weight phenolic acids arising from their metabolic breakdown by the colonic microflora [29].

Baba et al. [137] have found that in human volunteers ingesting chocolate or cocoa, the maximum levels of total epicatechin metabolites in plasma were reached in 2 h, with the sulfate, glucuronide, and sulfoglucuronide conjugates of nonmethylated (−)-epicatechin predominating in the plasma over the methylated forms. The excretion of total epicatechin metabolites within 24 h of chocolate or cocoa intake was in the average about 30% and 25% of the total epicatechin intake. Similar results were observed by these authors in rats fed cocoa powder to determine the plasma antioxidative activity of cocoa polyphenols. Analysis of plasma samples, previously treated with β-glucuronidase and/or sulfatase, by HPLC and HPLC–MS detected several (−)-epicatechin-related compounds identified as free (−)-epicatechin and as glucuronide, sulfate, and sulfoglucuronide conjugates of nonmethylated or methylated epicatechin. The maximum concentration of all (−)-epicatechin metabolites in the plasma occurred at 30–60 min postadministration. In addition, a treatment with cocoa powder significantly reduced the accumulation of lipid peroxides and the depletion of α-tocopherol induced by oxidants like AAPH and $CuSO_4$ when compared to plasma samples collected before administration. These results suggested that (−)-epicatechin present in cocoa powder was absorbed from the digestive tract and subsequently converted to various conjugated forms during its transit in the digestive tract, and that these metabolites became distributed in the plasma and were able to protect plasma components against oxidative activity [125]. In line with this observations, Wang et al. [138] also determined that the consumption of chocolate can have a direct effect on the plasma levels of (−)-epicatechin and on the plasma antioxidative activity against lipid peroxidation. In this case, the ingestion of increasing doses (27, 53, and 80 g) of a procyanidin-rich chocolate, containing 5.3 mg of total procyanidin and 1.3 mg of (−)-epicatechin/g, by human volunteers led to a dose-related increase in plasma (−)-epicatechin levels and in plasma antioxidant capacity and a decrease in plasma TBARS formation and 8-isoprostane levels. Furthermore, the effect of different levels of cocoa powder on the oral absorption and urinary excretion of (−)-epicatechin was evaluated by Baba et al. [139] in rats. Both the sum of plasma (−)-epicatechin metabolites (as glucuronide and/or sulfate in nonmethylate and *O*-methylated forms) at 1 h after administration and the peak plasma level, occurring at 18 h after administration, also increased in a dose-dependent manner. The same dose-related effect was observed on the sum of (−)-epicatechin urine metabolites measured at 18 h after treatment. Furthermore, the sum of (−)-epicatechin metabolites found in the urine was the same from both (−)-epicatechin and cocoa powder. These results were taken as a demonstration of the lack of a significant effect of the different components of cocoa powder on the bioavailability of (−)-epicatechin.

In a study evaluating the urinary metabolites of (–)-epicatechin arising for the oral ingestion of a decaffeinated green tea extract by human subjects or administered intragastrically as a 0.6% solution to rats and mice, Li et al. [140] noted that some interspecies differences existed in the metabolites identified in the urine with the aid of LC/ESI-MS. After green tea administration, the major conjugates of (–)-epicatechin appearing in human urine samples were identified as the monoglucuronides and monosulfates of (-)-epicatechin, as O-methyl-(-)-epicatechin-O-sulfate and as the ring fission metabolite (–)-5-(3′,4′-dihydroxyphenyl)-γ-valerolactone, predominantly in monoglucuronide and monosulfate forms, as additional metabolites. The profile of the urinary metabolites present in the urine of mice was similar to that of humans, but in rats, 2-*O*-sulfate-(–)-epicatechin and 3′-*O*-methyl-(–)-epicatechin-2-*O*-sulfate were not detected in the urine. These results contrast with those of Roura et al. [103] who found that in humans consuming 40 g of cocoa powder as a beverage in 250 mL of whole milk, the metabolism of (–)-epicatechin resulted in the urinary excretion of this flavanol as glucuronide and sulfate conjugates and in the glucuronide as the only form in plasma.

With the help of an HPLC method for the chiral determination of (+)-/(–)-catechin from sulfated and glucuronidated metabolites in human plasma, Ritter et al. [141] were able to compare the bioavailability and metabolism of these two stereoisomers in humans asked to consume beverages made from either an alkalized cocoa powder found containing typical (+)-catechin plus atypical (–)-catechin (most likely arising from the epimerization of (–)-epicatechin by the Dutching process or by the roasting of coco beans) or a nonalkalized cocoa containing only the natural (+)-antipode. The analysis of plasma samples collected before and 2 h after drinking the cocoa beverage, and with or without a preliminary treatment with sulfatase and β-glucuronidase, indicated a significant increase in (+)-/(–)-catechin ratio in the plasma relative to the ratio seen in the drink (0.80 vs. 0.22). A similar effect was noted when the volunteers were asked to consume a drink containing the two enantiomers in milk, in which case the ratio shifted from 0.81 in the drink to 4.05 in the plasma. When comparing the two enantiomers in terms of the concentrations of plasma glucuronides and sulfates, those of the (+)-enantiomers were more abundant than those of the (–)-enantiomers, possibly because of a greater absorption of the former than of the latter catechin. Another noteworthy finding was the detection of procyanidin B2 in the plasma.

The Metabolism of Procyanidins

Procyanidins are one of the most ubiquitously distributed polyphenol compounds in the plant kingdom and a subject of great interest because in submicromolar concentrations they are able to effectively prevent the formation of conjugated dienes in both the induction (preventive antioxidant) and the propagation (chain-breaking antioxidant) phases of lipid peroxidation. Also, by trapping free radicals and by strongly and noncompetitively inhibiting the activity of some enzymes involved in the generation of reactive oxygen species, procyanidins can markedly delay the onset of the breakdown phase of peroxidized lipids, thus inhibiting the formation of degradation products, and in different aqueous lipid dispersions, they can synergistically interact with other natural antioxidants like vitamin E in quenching peroxyl radicals [142].

Procyanidins are oligomeric to polymeric in nature and occur in various chemical types as a result of differences in the type of their constitutive units, the sequence of these units, the positions of the linkages interconnecting the units (C4→C6 or C4→C8 in the type B, with additional C2→OC7 or C2→OC5 bonds in the type A), the number of units (or chain length), and the presence of a substituent (e.g., acyl or glycosyl groups) [143, 144]. Those consisting exclusively of (+)-catechin, (–)-epicatechin, or (–)-epicatechin 3-*O*-gallate units are the most abundant type of procyanidin in plants [145], with less common types being designated as propelargonidin or prodelphinidin if they contain at least one unit of (epi)afzelechin or(epi)gallocatechin, respectively [144], or as prorobinetinidin or profisetinidin if the structure has at least one unit of robinetinidol or fisetinidol, respectively [146]. Because they may arise from the autoxidative or enzymatic condensation of monomeric precursors, procyanidins are also referred to as condensed tannins [34].

Among natural sources of procyanidins, fruits, beans, nuts, cocoa, chocolate, tea, and red wine are recognized as major ones. Chocolate contains higher amounts of procyanidins per unit weight (3.8–4.9 mg/g) than foods such as red wine (close to 0.2 g/L), cranberry juice (0.1 g/L), and apple (0.5–1 mg/g) [48]. A study analyzing the content of procyanidins in cacao liquor samples originating from Africa, Asia, and South America has found it to range from 0.22% to 0.83% and that the levels are inversely correlated with the duration of the cacao fermentation and the formation of volatile flavoring components during cacao roasting [147]. In addition to potent antioxidant activity and health-related benefits [88], procyanidins are also important contributors to the astringency and bitterness of chocolate [48].

The oral absorption of procyanidins is still a subject of debate since there is experimental evidence in support of either absorption or a lack of it. On the one hand, Donovan et al. [60] found no evidence to indicate oral absorption of the procyanidins in a grape seed extract or of dimeric procyanidin B3 after they were fed to rats as part of a meal. In addition, a comparison of the absorption and metabolism of procyanidin B3 to those of monomeric (+)-catechin and (−)-epicatechin and of a grape seed extract containing both monomers and oligomers in rats after a single meal found no evidence of (+)-catechin, (−)-epicatechin, the methylated forms, or of any procyanidin in the plasma. On the other hand, in rats fed the (epi)catechin-supplemented meal, (epi)catechin and 3′-*O*-methyl-(epi)catechin, but not 4′-*O*-methyl-(epi)catechin, were detected in the plasma. At no time in the study were procyanidin dimers (B1, B2, or B3) detected in the plasma of rats fed procyanidins (as B3-containing or grape seed extract-containing meals). In rats fed procyanidin B3 meal, neither catechin nor 3′-*O*-methylcatechin could be detected in the plasma even when plasma was sampled at 9 h and 24 h after contact with intestinal microflora, thus ruling B3 out as a source of catechin in plasma. The plasma of the rats fed the grape seed diets (at 200 mg and 400 mg) contained (+)-catechin, (−)-epicatechin, and their 3′-*O*-methylated conjugates but was not accompanied by additional compounds exhibiting the characteristics of a flavanol.

There are also several lines of evidence linking the oral absorption of procyanidins to their molecular weight. For example, Nakamura and Tonogai [148] used grape seed polyphenol (GSP) and grape seed polyphenol high molecular weight (GSPH) preparations to assess the role of molecular weight of procyanidins in the appearance of metabolic products in the serum and urine. While the GSP was a commercial product for oral use with a declared procyanidin content of >99%, and found to consist of a complex mixture of high molecular weight procyanidins with low molecular weight components such as (+)-catechin, (−)-epicatechin, procyanidins B1, B2, and C1, and gallic acid, the GSPH was obtained by size exclusion fractionation of the GSP product on Sephadex LH-20 to a concentration of procyanidins of 101.6 ± 0.6 w/w %. Oral administration of GSP (0.1–1 g/kg body weight) to rats led to the appearance of only (+)-catechin, (−)-epicatechin, 3′-*O*-methyl-(+)-catechin, and 3′-*O*-methyl-(−)-epicatechin in serum collected periodically over a 20-h period, in concentrations that were dose dependent, with the maximum concentrations of all four metabolites occurring at 3 h after administration. Procyanidin dimers (B1 and B2) and trimers (C1) were not detected at any time. Most of the metabolites were present as conjugates with glucuronic acid or sulfate, with the free forms being present in only trace quantities. The metabolites were cleared from the circulation at 20 h after GSP administration. Excretion of the metabolites in the urine was rapid, with more than 25% appearing in less than 5 h and the bulk (>80%) taking place within 25 h after administration. No metabolites were found in the serum of rats receiving GSPH orally (0.5 g/kg). In the same study, the analysis of the urine from rats given (+)-catechin (0.3 g/kg) showed (+)-catechin and 3′-*O*-methyl-(+)-catechin, which changed to (−)-epicatechin and 3′-*O*-methyl-(−)-epicatechin when the treatment was (−)-epicatechin (0.3 g/kg) and to (+)-catechin, (−)-epicatechin, 3′-*O*-(+)-catechin, and 3′-*O*-(−)-epicatechin when GSP (1 g/kg) was also given. In contrast, after the administration of GSPH, only trace amounts of (+)-catechin, (−)-epicatechin, 3′-*O*-methyl-(+)-catechin, and 3′-*O*-methyl-(−)-

epicatechin were detected in the urine. Taken together, these results suggest that only the monomers of GSPH are absorbed from the intestinal tract and that higher molecular weight procyanidins (MW 578.5–866.8) are more likely to undergo degradation to low molecular weight aromatic compounds by the colonic microflora [92]. Similar results have been reported by Gonthier et al. [59] in a study comparing the intestinal absorption of procyanidin dimer B2, trimer C3, and polymer to that of catechin monomer in rats following a 5-day intake as part of the diet. In rats fed procyanidins, neither parent compound nor catechin derivatives could be detected, but those fed catechin monomer excreted large amounts of catechin and its 3′-*O*-methylated form. Furthermore, 16 metabolites of microbial origin were detected and identified as low molecular weight aromatic acids from all polyphenols. In general, the extent of microbial metabolism decreased proportionally to the degree of complexity of the polyphenol, with the catechin monomer yielding the highest output (~11% of the dose), dimers (~7%) and trimers (~0.7%) yielding intermediate outputs, and the polymer the lowest (≤0.5%). Hence, the degree of procyanidin polymerization has a major impact on their fate in the body which is characterized by a poor absorption through the gut barrier and a limited metabolism by the intestinal microflora as compared to monomeric catechins. In comparison to other flavonoids present in chocolate and other foodstuffs, procyanidins show a poor absorption through the gut barrier as a result of a high molecular weight, with the bulk reaching the colon for degradation by the colonic microflora to various aromatic acids [93].

On the other hand, Sano et al. [149] found that the oral administration of grape seed extract to healthy human volunteers resulted in the detection of procyanidin B1 in the serum at 2 h after intake. In a more recent study by Apeldoorn et al. [91], in situ perfusion of the small intestine of rats with purified procyanidin dimers A1 (epicatechin-(2-*O*-7, 4–8)-catechin), A2 (epicatechin-(2-*O*-7, 4–8)-epicatechin), and B2 (epicatechin-(4–8)-epicatechin); type A trimers, a mixture of A1, B2, and a tetrameric type A; and monomeric epicatechin revealed that the A1 and A2 dimers were absorbed from the small intestine of rats to a greater extent than the B2 dimer. Furthermore, the absorption of type A dimers was shown to be only 5–10% of that of the monomeric (−)-epicatechin and, unlike (−)-epicatechin, not to yield conjugates or methylated metabolites. In contrast, type A trimers were not absorbed to a measurable extent. In addition, the study determined that tetrameric procyanidins enhanced the absorption of B2 but not that of A1. In addition, there is in vitro data confirming the absorption of procyanidins from the gut derived from experiments comparing the absorption of radiolabeled procyanidin dimer and trimer through a cell monolayer derived from the human intestinal cell line Caco-2. In this system, the oligomeric procyanidins were absorbed in contrast to the lack of absorption of procyanidin polymers having an average degree of polymerization of 7 [150], with the magnitude approximating that of (+)-catechin [67]. Further evidence on the ability of procyanidins to reach the circulation has come from work by Baba et al. [151], who reported the oral absorption and urinary excretion of procyanidin B2 along with the degradate (-)-epicatechin and the methylated and/or conjugated metabolites of (-)-epicatechin in the rat.

The passage of procyanidins from the digestive tract into the blood has been assessed in adult human volunteers who were asked to consume a flavanol-rich cocoa beverage (0.375 g/kg in 300 mL of water, average total consumption of 26.4 g, providing 323 mg of monomers and 256 mg of dimers) [152]. Analysis of plasma samples by reversed-phase HPLC with EC detection and LC–MS detected measurable amounts of the dimeric procyanidin B2 together with (−)-epicatechin and (+)-catechin as early as 30 min after cocoa consumption (16±5 nmol/L, 2.61±0.46 μmol/L, and 0.13±0.03 μmol/L, respectively), which reached a maximum by 2 h (41±4 nmol/L, 5.92±0.60 μmol/L, and 0.16±0.03 μmol/L, respectively). Further proof of the oral bioavailability of procyanidin B2 (epicatechin-4β→8-epicatechin), a major component of the polyphenolic fraction of cocoa powder, was obtained by Baba et al. [151] in rats. These authors determined that after the oral administration of procyanidin B2 (50 mg/kg body weight, in water) to rats, LC–MS detected B2, (−)-epicatechin, and 3′-*O*-methyl-(−)-epicatechin in the blood and the same compounds plus conjugates of 3′-*O*-methyl-(−)-epicatechin in the urine. Samples collected from rats receiving only water were devoid of these polyphenols. Exposing the plasma samples from rats on B2 to lipid peroxidation by AAPH or by

$CuSO_4$ found the accumulation of lipid peroxides to be about equal in the former case and significantly lower in the latter case when compared to results for plasma from rats receiving only water.

Pharmacokinetic Aspects of Chocolate and Cocoa Polyphenols

Animal Studies

The oral absorption of (+)-catechin and (−)-epicatechin from a variety of food sources and matrices has been extensively examined both in humans and experimental animals. The bioavailability and pharmacokinetic characteristics of (−)-epicatechin and (+)-catechin were examined by Catterall et al. [153] in the rat by administering the ^{3}H-labeled forms of these flavanols orally and intravenously at dose levels equivalent to those typically consumed by humans in the diet (i.e., 0.58 mg/kg of catechin and 4.5 mg/kg of epicatechin). Both compounds were rapidly absorbed from the gut, with their maximum plasma concentration (t_{max}) occurring at about 1 h postadministration. The blood half-life ($t_{1/2}$) values were 1.10 ± 0.08 h for epicatechin and 1.63 ± 0.22 h for catechin. An extensive biliary excretion was suggested by finding that about one-third of the dose of these compounds was excreted in the urine and two-thirds in the feces following intravenous administration, but when the same doses were delivered orally, only about 5% of the administered doses were recovered in the urine. A poor absorption of monomeric flavanols from the gut was apparent from a comparison of the plasma areas under the serum (or plasma) concentration–time curves (AUCs) following oral and intravenous administrations of each compound and which agreed, in both instances, with an oral bioavailability of less than 5%. The possible degradation of (−)-epicatechin and (+)-catechin by the intestinal microflora was inferred from the observation that tritium was exchanged with water in the blood after oral, but not after intravenous, administration of the flavanols to rat. At a much higher oral dose of (−)-epicatechin (250 mg), Zhu et al. [154] found that in the rat the absolute oral bioavailability of this flavanol was 0.39, the t_{max} to be 2 h, and the urinary and fecal recovery to amount to 4.7% and 11.0% of the administered dose, respectively.

An insight into the plasma pharmacokinetic profiles of (−)-epicatechin metabolites was gained by Piskula and Terao [123] by feeding this flavanol to rats and quantitating the ensuing metabolites in plasma by HPLC–MS over an 8-h period. In the circulation, (−)-epicatechin was found to be accompanied by a slower migrating peak later identified as 3′-*O*-methyl-(−)-epicatechin. With the aid of a combination of enzymatic hydrolysis, up to six metabolites of (−)-epicatechin and methyl-(−)-epicatechin (consisting of the corresponding sulfates, glucuronides, and mixed sulfate–glucuronide conjugates) were quantified in addition to smaller concentrations of the parent compound. Maximum plasma concentrations were observed at 1 h postadministration with (−)-epicatechin and within the first 2 h with all the metabolites, with the levels declining thereafter and dropping sharply after 8 h. Except for glucuronidation, the dominant metabolites in the plasma were those of methyl-(−)-epicatechin. This study provided a chronological view of the metabolism of (−)-epicatechin, which leads to nearly 90% conversion to the glucuronide and to about 50% conversion to the sulfate in the first 30 min postadministration. Within 8 h, the concentration of the sulfate conjugate had slowly risen to 70%. In addition, about 40% of the plasma epicatechin metabolites were methylated in the first 30 min, and their levels continued to increase to 75% after 8 h. Certain metabolic differences appear to exist between (−)-epicatechin and (+)-catechin, which in the rat has been shown to be predominantly converted to the glucuronidated derivatives of both (+)-catechin and 3′O-methyl-(+)-catechin, and with plasma levels of these metabolites starting to decline after 12 h and dropping dramatically at 24 h postadministration [118]. However, Tsang et al. [155] have reported that in rats ingesting 3.6 mg of (−)-epicatechin and 5.8 mg of (+)-catechin as part of a grape seed extract, none of these compounds were detected in the plasma or urine in spite of their presence in the gastrointestinal tract at 1 h postingestion. In contrast, at this time, trace quantities of the glucuronides of (-)-epicatechin and (+)-catechin and of the glucuronides of the methylated derivatives were detected in the duodenum, jejunum and ileum. Furthermore, the glucuronides and glucuronides of the 3′-O-methylated derivatives were present in the plasma and urine, and the sulfates and sulfates of the 3′-*O*-methylated derivatives were detected in the urine over a 24-h period. Overall, only about 36% and 27% of the ingested (−)-epicatechin and (+)-catechin, respectively, were recovered as metabolites in the urine, with the balance most likely undergoing biotransformation to low molecular weight phenolic acids by the intestinal microflora [59, 92] although other factors such as degradation/metabolism in the gastrointestinal tract, transporter-mediated intestinal secretion/efflux, and presystemic gut wall metabolism might also play a contributory role [156]. As demonstrated for (−)-epicatechin present in green tea in a rat model, presystemic extraction is not a major determinant of low oral bioavailability of monomeric flavanols [156].

In a study aimed at determining the level of (−)-epicatechin and its metabolites in rat plasma in terms of suppression of the oxidation of plasma components, Baba et al. [137] fed 1 g/kg of a cocoa powder, containing 7.8 mg of flavanol, to rats as a suspension in water. Analysis of plasma samples disclosed the presence of free (−)-epicatechin and of the glucuronide, sulfate, and sulfoglucuronide conjugates of nonmethylated and methylated (−)-epicatechin. In plasma, (−)-epicatechin was present in much smaller concentrations than its metabolites, with its maximum plasma concentration occurring at 30 min postadministration and that of its metabolites occurring between 30 and 60 min postadministration. Glucuronide conjugates of both nonmethylated and methylated (−)-epicatechin were present in higher concentrations than those of sulfated conjugates. Moreover, administration of cocoa powder significantly reduced the accumulation of lipid peroxides in plasma and significantly reduced the loss of α-tocopherol in plasma exposed to free radical generating systems compared to plasma obtained before administration.

In a subsequent study, Baba et al. [139] compared the levels of (+)-catechin, (−)-epicatechin, and their metabolites in the plasma and urine of rats orally treated with equimolar doses (17.2 mmol/L) of a flavanol present singly or as mixture. After administration of the flavanol preparation as an aqueous solution, the absorbed (+)-catechin and (−)-epicatechin were found to be present in the plasma primarily as metabolites such as nonmethylated or 3′-*O*-methylated conjugates. In rats receiving only (+)-catechin and the mixture of catechins, the primary metabolite of (+)-catechin in plasma was the glucuronide of the nonmethylated form. In contrast, in rats on (−)-epicatechin and on the flavanol mixture, the primary plasma metabolites of (−)-epicatechin were the glucuronide and sulfoglucuronide of the nonmethylated forms and the sulfate of the 3′-*O*-methylated forms. Urinary excretion of the total amount of (−)-epicatechin metabolites in the (−)-epicatechin group was found to be significantly higher than the amount of (+)-catechin metabolites in animals on (+)-catechin. The sum of (+)-catechin metabolites in the urine was significantly lower in rats treated with a flavanol mixture group than in rats only on (+)-catechin, and the sum of (+)-epicatechin metabolites in the treated with a flavanol mixture was also significantly lower than in animals on (−)-epicatechin alone. These results suggest that the bioavailability of (−)-epicatechin is higher than that of (−)-catechin in rats and that, when ingested as a combination, (−)-catechin and (+)-epicatechin might be absorbed competitively in the gastrointestinal tract of rats.

The role of the dose of cocoa on the absorption of urinary excretion of (−)-epicatechin has been evaluated in rats by the oral administration of cocoa powder (150, 750, and 1,500 mg/kg) or (−)-epicatechin (1, 5, and 10 mg/kg) [127]. A dose-dependent increase in the sum of plasma (−)-epicatechin metabolites at 1 h postadministration, the peak plasma concentrations, and the sum of (−)-epicatechin metabolites excreted in the urine within 18 h postadministration of either a cocoa powder or pure (−)-epicatechin. Furthermore, the sum of (−)-epicatechin metabolites excreted in the urine reached the same level in rats receiving either (−)-epicatechin or cocoa powder for equivalent amounts of (−)-epicatechin. Although the increases in plasma (−)-epicatechin with an increase in the dose of (−)-epicatechin administered (as such or as cocoa powder) did not indicate interference with the absorption process, the dose-dependent gradual decrease in the excretion of urinary metabolites within 18 h postadministration suggested decreased bioavailability with increased doses of cocoa powder.

Human Studies

According to Hackett [110], the administration of a single dose of [U-^{14}C]-cyanidanol-3 to human volunteers resulted in a mean urinary excretion of 55% of the dose of ^{14}C, with 90% of the urine ^{14}C becoming excreted within 24 h postadministration. In the urine, three-quarters of the radioactivity was accounted for the glucuronides of (+)-catechin and 3′-*O*-methyl-(+)-catechin, and the sulfate of the latter, with urinary excretion of intact cyanidanol-3 accounting for 0.1–1.4% of the dose, and cyanidanol-3 and metabolites containing the intact flavanol ring system accounting for 90% of the urine ^{14}C. Ring scission was found to represent only minor metabolic pathway that led to the excretion of small amounts of 3-hydroxybenzoic acid, 3-hydroxyhippuric acid, and 3-hydroxyphenylpropionic acid. Unchanged cyanidanol-3, representing less than 10% of the dose, was detected in plasma between 30 min and 12 h after administration, and the radioactivity of the metabolites (as total ^{14}C) was detected for at least 5 days after administration.

Bell et al. [136] determined the changes in plasma (+)-catechin concentrations in a group of male and female human volunteers after the ingestion of a single serving of dealcoholized red wine reconstituted to its original volume either with water (RWW) or with water–alcohol (RWA) and delivering 121,000 nmol of (+)-catechin and 104 mg of total flavonoid. Peak plasma total (+)-catechin, occurring 1 h after red wine ingestion, ranged from 40 to 130 nmol/L after RWA and from 30 to 110 nmol/L after RWW, respectively, which together represented 15- to 65-fold increases over baseline plasma values. On the other hand, the corresponding 8-h AUCs were not significantly different between the two treatment groups and averaged 306.1 ± 29.5 nmol•h/L. Also, the intergroup t_{max} and C_{max} values were not significantly different and averaged 1.44 ± 0.13 h and 76.7 ± 7.5 mmol/L, respectively. In contrast, the $t_{1/2}$ value for the RWA-fed group was 22% shorter (3.17 h) than for the RWW (4.08 h). No significant gender-related differences were noted for the average plasma C_{max}, 8-h AUC and $t_{1/2}$ values. Overall, these results suggested that plasma total (+)-catechin concentrations are responsive to dietary (+)-catechin intake and that the concurrent intake of a moderate amount of alcohol appears not to affect flavonoid absorption and metabolism.

Richelle et al. [157] evaluated the kinetics of appearance of (−)-epicatechin from black chocolate in the plasma of healthy male volunteers. Following the consumption of a single 40 g or 80 g serving of black chocolate, delivering 82 mg and 164 mg of (−)-epicatechin per serving, respectively, together with ad libitum bread intake, with a 1-week interval between doses, the plasma epicatechin level was monitored periodically for up to 8 h postconsumption. Irrespective of the amount of black chocolate consumed, the peak plasma concentration was reached in 2 h with the lower dose (380 nmol/L) and in 2.57 h (700 nmol/L) with the higher dose of dark chocolate. The clearance of the flavanol from the plasma was very fast ($t_{1/2}$ for elimination of 1.9 h for 40 g of chocolate and 2.3 h for 80 g of chocolate), and the mean C_{max} (103 and 196 ng/mL) and AUC (445 and 1,069 ng/mL•h) of the plasma epicatechin kinetics were proportional to the quantity of chocolate ingested.

Rein et al. (2000) [158] used a selective and sensitive HPLC method with coulometric detection to determine the physiological levels of (−)-epicatechin and (−)-epicatechin metabolites in human subjects asked to consume a procyanidin-rich semisweet chocolate. At 2 h after ingestion, the time for t_{max}, there was a 12-fold increase in plasma epicatechin relative to baseline values (from 22 to 257 nmol/L), which was paralleled by a 36% increase in plasma total antioxidant capacity, and a 40%

decrease in plasma lipid peroxidation, assessed on the basis of malondialdehyde formation, all of which reached baseline values by 6 h after ingestion.

In agreement with studies conducted in laboratory animals, Holt et al. [152] determined that the absorption of (−)-epicatechin and (+)-catechin after acute cocoa consumption by human subjects is quite rapid, becoming detectable in the plasma as early as 30 min postingestion and reaching a maximum at 2 h, with the mean (−)-epicatechin level being about 16-fold higher than that of (+)-catechin at this point. Plasma concentrations of these compounds were markedly lower (by ca. sixfold and eightfold, respectively) 6 h after cocoa consumption. In addition, procyanidin dimer B2 was also detected alongside the monomers and found to follow a time course of absorption into and elimination from the plasma similar to that of the monomers. However, since the concentration of procyanidin dimer in the plasma was much lower than that of epicatechin in spite of only a 20% difference in concentration in the cocoa consumed in the study (12.2 mg vs. 9.7 mg per gram of cocoa), it was conjectured that some of procyanidin dimers could have undergone depolymerization to absorbable monomers by the human colonic flora, thus explaining the high ratio of epicatechin to catechin observed in the study.

The existence of a dose–response effect between the amount of chocolate consumed and plasma epicatechin and oxidative damage has been confirmed by Wang et al. [138] in a group of male and female human volunteers. Following the consumption of proportionally increasing servings of procyanidin-rich chocolate (27, 53, and 80 g of a chocolate containing 46 mg of epicatechin and 186 mg of procyanidins/27 g of chocolate), the plasma (−)-epicatechin concentration rose proportionally to the amount of chocolate consumed, with a maximum increase being observed at 2 h postingestion, and the average decrease in plasma epicatechin between the 2- and 6-h point being inversely related to the amount of chocolate consumed.

Tomás-Barberán et al. [159] have assessed the role of the concentration of flavonoid content in cocoa (mainly as flavanol monomers) on flavonoid bioavailability and metabolism in humans. For this purpose, the bioavailability of the flavonoids from two cocoa powders, a conventional cocoa powder and an unfermented, nonroasted, and blanch-treated cocoa powder with an eightfold higher content in (−)-epicatechin and procyanidin B2 than the conventional cocoa, consumed as drinks in semiskimmed milk, was compared in healthy human volunteers using a double-blind crossover design. The content of epicatechin glucuronide, the main metabolite detected in plasma, was fivefold higher upon consumption of flavonoid-enriched cocoa powder as compared to the conventional cocoa powder. A similar trend was observed for the urinary excretion of urinary metabolites, mainly methyl epicatechin sulfate, which ranged from twofold to twelvefold higher depending on the metabolite. As a whole, these results confirm the benefits that can be derived from improved processing of the flavonoid composition of cocoa powders to increase the flavonoid content (mainly flavanol monomers and dimers) and that this enhancement leads to an increase in the cocoa flavonoid metabolites present in plasma and in urine as a consequence of enhanced bioavailability. Taking into account the known poor bioavailability of oligomeric procyanidins, both the higher content in plasma and the higher urine excretion of epicatechin-derived metabolites upon consumption of a flavonoid-enriched cocoa powder appear to be mainly related to the higher content of flavanol monomer than of procyanidin dimer.

The Role of Stereochemical Features

At variance with flavonoids like flavonols and flavones, which are planar molecules, flavan-3-ols and procyanidins are nonplanar by virtue of having a saturated bond between C2 and C3 in the heterocyclic ring C of the flavonoid nucleus. As a result, C2 and C3 are chiral centers, and introduction of a hydroxyl functionality at C3 yields four stereoisomers (Fig. 17.4), of which (+)-catechin and (−)-epicatechin are of common occurrence in plants while (−)-catechin and (+)-epicatechin are quite

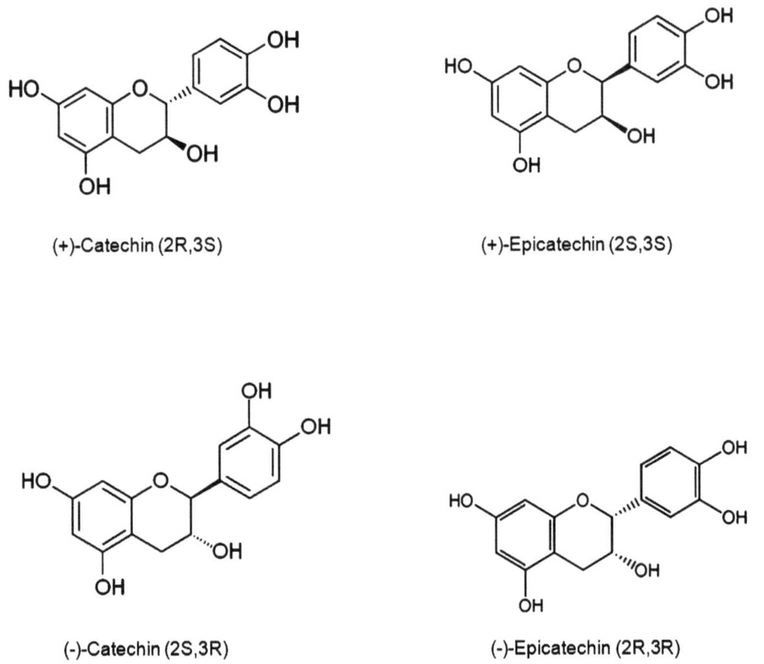

Fig. 17.4 Stereoisomeric forms of catechin and epicatechin present in cacao beans

rare. Furthermore, in oligomeric and polymeric procyanidins, substitution at C4 in ring C will make this position chiral [50]. In cocoa beans and in most commercially available chocolate and cocoa products, the major catechin monomers are (−)-epicatechin and (+)-catechin, with the former compound predominating over the latter. However, a few isolated reports have indicated that the bulk of the catechin fraction is (−)-catechin with (−)-epicatechin being present in lesser concentrations. Such a change has been ascribed to the epimerization of (−)-epicatechin at the C2 position during the manufacturing of the chocolate since this catechin enantiomer is not detected in cacao beans [160]. The same change may take place during the alkalinization treatment used in the manufacturing of cocoa powder products and which has been found to lower total phenolics content as well as the levels of (−)-epicatechin and (+)-catechin by more than 65% and 35%, respectively [27]. In short, (−)-epicatechin may epimerize to (−)-catechin, and (+)-catechin may epimerize to (−)-epicatechin.

A study in rats has disclosed that (−)-epicatechin and (+)-catechin are absorbed from the gastrointestinal tract to different extent when administered as aqueous suspensions and that several metabolites are excreted in the urine proportionally to the amounts absorbed. The existence of a competition for absorption at the gastrointestinal level between the two stereoisomers was inferred from the lower excretion of urinary metabolites observed with their mixture than with (+)-catechin alone [127]. Likewise, Donovan et al. [161] have verified that a difference in intestinal mucosal absorption exists between the (−)- and (+)-enantiomers of catechin using an in situ perfusion technique in the jejunum and ileum of the rat. Upon the addition of equimolar doses of each enantiomer to the perfusing fluid (tested at 10, 30, or 50 μmol/L), the (+)-enantiomer invariably demonstrated a higher absorption than the (−)-enantiomer, with the amount of each enantiomer increasing proportionally to an increase in the amount added to the perfusion fluid. Interestingly, however, the difference in absorption between the two enantiomers became narrower with each dose increment. Moreover, in plasma obtained from the abdominal aorta, the temporal concentrations of (−)-catechin were twofold to eightfold lower than those of the (+)-catechin enantiomer, the form more commonly found in chocolate samples. The same study determined that both (+)- and (−)-catechin were methylated at the 3′-O-position, and that the concentrations of the 3′-*O*-methyl derivative of (−)-catechin were significantly lower than those of (+)-catechin. Enantiomeric 4′-*O*-methylated metabolites were found to be present in very low or in undetectable concentrations in most plasma samples. Moreover, concentrations of (+)- and (−)-catechin in plasma obtained from the mesenteric vein were generally higher than those obtained from aortic plasma, and the trend of the differences in the extent of methylation between the (+)- and (−)-enantiomers of catechin were found to be similar to those observed for aortic plasma.

Ottaviani et al. [162] investigated the influence of the stereochemical configuration of flavanols on their absorption, metabolism, and biological activity. Healthy adult males were asked to consume equal amounts of stereoisomeric (−)-epicatechin, (−)-catechin, (+)-catechin, and (+)-epicatechin (1.5 mg/kg) in a well-defined cocoa-based, low-flavanol dairy-containing drink matrix, and flavanol levels were subsequently determined in plasma and 24-h urine. Based on the amounts of metabolites detected in the plasma and in a 24-h urine, it was apparent that the stereochemical configuration of the various flavanols had a profound influence on their uptake from the gastrointestinal tract, which decreased in the order (−)-epicatechin > (+)-epicatechin > (+)-catechin >> (−)-catechin. An additional difference was the lack of detectable amounts of 4′-*O*-methylated metabolites in addition to non-methylated and 3′-*O*-methylated metabolites seen with all four compounds. Interestingly, when the same stereoisomers were tested at equivalent doses in vivo, only (−)-epicatechin demonstrated significant vasodilatory action, an effect that was independent of antioxidant properties. Overall, these results stress the importance of stereochemical configuration of an ingested flavanol on any health benefits this type of polyphenol may provide when consumed as part of flavanol-containing foods.

The Role of Diet Components

The potential health-related benefits associated with flavanols in dark chocolate and cocoa are closely related with their absorption from the digestive tract, which in turn may be dependent on such factors as their chemical stability in the gastrointestinal tract, metabolism by enterocytes, first-pass metabolism, biotransformation by the activity of the intestinal microflora, and interaction with food components [163]. The role of food on the absorption and pharmacokinetics of cacao products flavanols is of the utmost importance to human health when considering that only a fraction of the polyphenols, fluctuating in the nmol/L range [127, 158], is absorbed and utilized by humans and that food and food components can alter the oral bioavailability of these components as a result of an effect on the motility of gastric secretions and hepatic blood flow, factors which can, in turn, influence the rate and extent of intestinal absorption, presystemic metabolism, and systemic clearance of these important phytochemicals [163]. From studies with drug molecules, a great deal has been learned on the impact that the presence of food in the gastrointestinal tract may have on the bioavailability of polar molecules such as those present in polyphenol-rich food products such as chocolate, tea, and wine. Consequently, the role of the amount of solute ingested, the dose, the role of formulation factors, the size and composition of the ingested food, and the temporal relationship between the ingested food and drug during their transit within the gastrointestinal tract have emerged as important modulating factors to solute oral bioavailability [164].

A review of the literature on food–chocolate flavanols is rather limited and often conflicting. In this regard, one of the most extensively studied contributory factors to a decrease in the bioavailability of flavonoids in general and of monomeric catechins in particular is milk. The possibility that milk might interfere with the oral absorption of antioxidant polyphenols was suggested by Hertog et al. [165], who suggested that the lack of an inverse association between the intake of black tea antioxidant flavonols and risk of ischemic heart disease (IHD) in a population of men from an industrial town in

South Wales, UK, might be related to the abolishment of the plasma antioxidant-raising capacity of the tea flavonols by milk, which is customarily added to tea. This possibility was subsequently investigated by van het Hof et al. [166] in 12 healthy male and female human volunteers, who consumed a single serving of green tea, black tea, or black tea with semiskimmed milk (3 g lyophilized tea solids each). Based on the blood total catechins contents over 8 h, it was determined that the consumption of green tea led to a faster and greater rise in blood total catechins level than an equivalent intake of black tea. More importantly, the addition of milk to black tea did not significantly affect the blood total catechins levels over the entire observation period. Very similar results were reported by Leenen et al. [167] for a group of 21 healthy human volunteers, who were asked to consume a single serving of black tea, green tea, or water with and without milk. Based on the plasma levels of total catechins and values for total antioxidant activity measured using the ferric-reducing antioxidant potential (FRAP) assay, these authors concluded that while both black and green tea were able to significantly increase plasma antioxidant activity proportionally to an increase in plasma total catechins, these results were not different when milk was added to tea. A similar conclusion was reached by Hollman et al. [168], who investigated the absorption of tea flavonols (quercetin glycosides, kaempferol glycosides) in healthy volunteers who consumed black tea with and without milk every 2 h each day for a total of eight cups per day for 3 days. Addition of milk to black tea (15 mL per 135 mL of tea) did not change the AUC of the plasma concentration–time curve of quercetin or kaempferol.

The results of a study by Richelle et al. [169] comparing the antioxidant activity of commonly consumed polyphenol-containing beverages (cocoa, coffee, tea) on a per cup serving basis and in the presence and absence of milk showed that differences in antioxidant activities existed among the various beverages and that milk in the amounts added to the beverages (10% to infused tea bag or to a soluble coffee powder, 66% to a cocoa drink from a reconstituted powder) did not alter the antioxidant activities attained in its absence. To clarify the effects of milk on catechin bioavailability and antioxidant status, Reddy et al. [170] asked 9 apparently healthy adult male volunteers to consume either black tea (prepared by boiling black tea leaf in water, sweetening with sugar, and bringing up to 350 mL with boiling water) or tea with milk (prepared similarly but made up with boiling water to 280 mL and then mixed with sugar and 70 mL of preboiled milk, containing 4% total solid, 1.75% proteins, 3.5%). In the presence of milk, the time course of the plasma total catechin levels peaked as early as 60 min postingestion and then remained rather constant for the next 2.5 h. In the absence of milk, however, the plasma total catechin levels reached a similar value as that seen in the presence of milk at about 90 min, peaked at 120 min postingestion, and yielded a plasma catechin AUC over 180 min that was significantly greater ($p<0.05$) than that in the presence of milk. On the other hand, the changes in plasma antioxidant capacity, as inferred from the FRAP values, reached a maximum at 60 min after consumption of plain black tea and remained fairly constant thereafter. In the presence of milk, the FRAP values rose more slowly and reached a maximum value at 3 h, at which time it was equal to the value seen in the absence of milk.

In a later study by Serafini et al. [171] comparing the total antioxidant capacity (TAC) for samples of dark and milk chocolate, based on the results of FRAP assays, with the corresponding plasma FRAP values and plasma (−)-epicatechin levels derived from the consumption of these chocolates in amounts containing equivalent contents of (−)-epicatechin content (100 g and 200 g, respectively) by healthy fasted human volunteers, showed that dark chocolate consistently yielded higher FRAP values both in vitro and in vivo and was better absorbed from the gastrointestinal tract than milk chocolate (found to contain the equivalent of up to 40 mL of milk/200 g). Surprisingly, the coingestion of the dark chocolate with full-fat milk (100 g/200 mL) resulted in a marked decrease in both the area under the curve (AUC) for a plot of plasma (−)-epicatechin levels against time (4 h) and the plasma TAC but which were still higher than those for milk chocolate. The differences in flavonoid absorption between the two types of chocolate were assumed to be due to the formation of secondary bonds between chocolate flavonoids and milk proteins, which could have negatively affected the bioavailability and, therefore, the antioxidants properties of milk chocolate flavonoids in vivo. This interpretation was questioned by Schroeter et al. [172], who considered that factors such as the chocolate matrix composition, lipid/water content, viscosity, density, and extent of mastication were more important than milk in determining the oral bioavailability of chocolate flavonoids than milk. Moreover, these authors reported that in a human study evaluating the oral bioavailability of (−)-epicatechin from a cocoa made into a beverage either in whole milk (3.25% lipid) or in water (supplemented with carbohydrate and 3.25% lipid as a control), the AUCs for the plots of plasma (−)-epicatechin levels against time from each beverage were not significantly different under isocaloric and isolipidemic conditions. Also, while both cocoa beverages elevated the plasma TAC over baseline values, the difference between their TAC values and between the corresponding AUC values for plasma antioxidant capacity plots against time was not statistically significant. This study concluded that food matrix may play a greater role in determining the extent of oral absorption and biological activity of monomeric flavanols in cocoa products than the presence of milk in the cocoa product. Taking into account the lipid content of the cocoa beverage used in this study and that added as milk to dark chocolate (3% vs. 30%) by Serafini et al. [171], Serafini and Crozier [173] have argued that the lack of effect of milk on cocoa flavanol absorption may have been due to differences in the amount of milk consumed in each case, being higher when it was added to a dark chocolate than to a cocoa beverage. This view is supported by the findings of a study by Lorenz et al. [174] evaluating the protective effects of tea polyphenols on endothelial function in healthy female volunteers. Following the consumption of either 500 mL of freshly brewed black tea, black tea with 10% skim milk, or boiled water as control, flow-mediated dilation (FMD) was measured at 2 h after consumption and compared to values recorded before consumption. Black tea significantly improved FMD compared with water, but its effect was completely blunted by milk. As a confirmation, the same beverages were tested in vitro on isolated rat aortic rings and endothelial cells. Again, while black tea induced vasorelaxation in rat aortic rings and increased the activity of endothelial nitric oxide synthase in endothelial cells, these actions were completely inhibited by the addition of milk to tea. The inhibitory actions of milk were attributed to the formation of complexes between milk proteins and tea catechins.

The modulating influence of milk protein on the uptake and plasma concentrations of cocoa polyphenols was investigated by Keogh et al. [175] in middle-aged men and women who were allowed to consume chocolate drinks prepared in water with added sugar and containing 2 g of polyphenols following an overnight fast. While the milk-free beverage contained powdered chocolate drink, 2.7 g polyphenols, nondairy creamer (NDC), and sugar in 200 mL of water, the milk chocolate drink contained the same ingredients: skimmed milk powder (35% protein, 54% lactose) in place of NDC, cocoa butter, and more sugar in 200 mL of water. This beverage contained 2.45 g of milk proteins. Milk protein did not influence the mean plasma catechins ((+)-catechin, (−)-epicatechin) concentration after ingestion of the chocolate powder, but slightly accelerated their appearance in the blood, which in the case of (−)-epicatechin was significantly different at 1 and 2 h and marginally significant at 4 h after ingestion relative to no milk protein. Milk protein had a small lowering effect on the plasma catechin at 4 h postingestion but not before or after. These results were taken as evidence that neither milk nor milk proteins interfere with the oral bioavailability of monomeric flavanols in a milk chocolate drink and, hence, with the potential health effects associated with chocolate products. A similar conclusion was reached by Roura et al. [176, 177] after determining the effect of milk on the bioavailability of cocoa polyphenols in 21 human volunteers who received either whole milk as a control, cocoa powder dissolved in whole milk, or cocoa powder dissolved in water. LC–MS analysis of the urine for the excretion of (−)-epicatechin metabolites detected one (−)-epicatechin glucuronide and three (−)-epicatechin sulfates from the two cocoa beverages and established that while milk did not have a significant effect on the amount of metabolites excreted in urine, it did, however, alter glucuronide and sulfate excretion rates and the sulfation position. In contrast, Mullen et al. [178] have examined the effects of milk on the bioavailability of cocoa flavan-3-ol metabolites in healthy nonsmoking human volunteers using a protocol that required the volunteers to consume a low flavonoid diet for 2 days before drinking a cocoa beverage, made by dissolving a commercial cocoa powder in

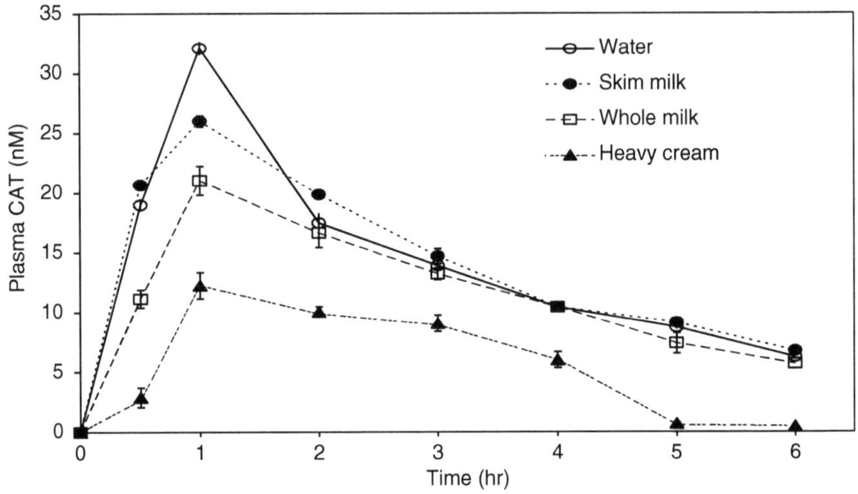

Fig. 17.5 Temporal changes in plasma (+)-catechin levels following its administration to rats at 350 mg/kg as a solution in water or in a dairy product. Each point represents the mean ± S.E.M. for n = 6

water or milk, and containing equimolar (45 μmol) quantities of (−)-epicatechin and (−)-catechin with only tracer amounts of (+)-catechin. Milk was found to neither affect gastric emptying time nor transit time through the small intestine, to have a minor effect on the plasma pharmacokinetics of (epi)catechin-O-sulfate and not to influence that of *O*-methyl-(epi)catechin-*O*-sulfate. On the other hand, milk significantly lowered the urinary excretion of (epi)catechin-*O*-sulfate and (−)-epicatechin-*O*-glucuronide and (epi)catechin-*O*-sulfate and *O*-methyl-(epi)catechin-*O*-sulfate.

To more conclusively single out the role of milk on the bioavailability of chocolate and cocoa flavan-3-ols while circumventing the effects of product matrix, Gossai and Lau-Cam [179] evaluated the effects of type of dairy product (whole milk, skim milk, heavy cream) on the oral absorption of (+)-catechin and (−)-epicatechin in the rat. Each flavanol was administered intragastrically to rats as a solution in water, serving as control, or in a dairy product, at a dose of 350 mg/kg. Plasma levels of each catechin were measured by over a 5-h period as total catechin concentrations after enzymatic hydrolysis of flavanol conjugates. As shown in Figs. 17.5 and 17.6, the highest concentration of a monomeric catechin was achieved from a simple aqueous solution, with the peak concentration being reached at about the same time (1 h) by both catechins when delivered as solutions in water and skim milk. Thereafter, (−)-epicatechin was more rapidly eliminated from the circulation than (+)-catechin. In contrast, whole milk and heavy cream lowered the plasma levels of both (+)-catechin and (−)-epicatechin relative to those from water and skim milk, more so in the case of the former than of the latter compound. However, while the time for maximum plasma concentration remained unchanged in the case of catechin, it was delayed by an additional 2 h in the case of (−)-epicatechin. In looking for an explanation for these contrasting results, the protein and fat contents of the dairy products were measured and found to contain the levels reported in Table 17.1. These values suggested that the fat content rather than the protein content was the major determining factor for the decreases in plasma flavanols seen in the rat. At the same time, these results suggested that proteins do not play a determining role in monomeric catechin absorption from the gastrointestinal tract as fats do since heavy cream,

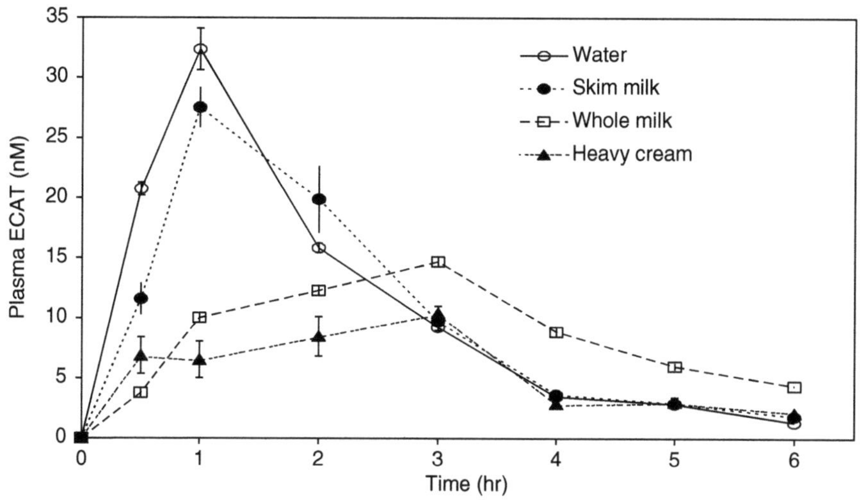

Fig. 17.6 Temporal changes in plasma (−)-epicatechin levels following its administration to rats at 350 mg/kg as a solution in water or in a dairy product. Each point represents the mean ± S.E.M. for n = 6

having the highest fat content and the lowest protein content, was the most inhibiting of plasma catechin levels than either whole milk or skim milk. This correlate can be clearly seen in the plot of AUC values for CAT and ECAT versus fat content in a dairy product shown in Fig. 17.7. Although the binding of low molecular weight polyphenols such as catechins with dietary or endogenous proteins, released from the intestinal wall due to the sloughing off of cells during intestinal digestion, within the gastrointestinal tract is viewed as potential limiting factors to polyphenol absorption in the small intestine [180], such interaction might require the direct binding of the polyphenols to proteins of the intestinal wall during gastrointestinal passage rather than to free dietary proteins.

The interaction of monomeric flavanols in cocoa with specific classes of dietary components has been investigated in healthy normolipidemic fasted subjects by Schramm et al. [163] using a crossover design. In trial 1, the effect of two concentrations of carbohydrate (as sucrose) on the oral absorption of the flavanols from a sugar-free flavanol-rich cocoa in water (providing 1.53 mg of total catechin (catechin + epicatechin) per kg body weight) was evaluated over a period of 3 weeks. Trial 2, carried out as a 4-week crossover study, examined the effect of butter, bread, and meat (as steak). Trial 3, carried out as in trial 1, used bread, grapefruit juice, and whole milk as treatments. In each instance, patients were also asked to drink a volume of water equal to that of a liquid food to serve as a control treatment. Based on the plasma concentrations of epicatechin and catechin measured over 8 h, sugar was shown to increase the absorption of cocoa flavanols and the rate of flavanol elimination in direct proportion to the intake of sugar. In trial 2, bread was the only macronutrient to increase cocoa flavanol absorption significantly relative to water and to a greater extent than butter, with meat exerting no apparent effect. In trial 3, bread, grapefruit juice, and milk increased cocoa flavanol absorption relative to water, with the effect being significant only with bread and grapefruit juice. Combined data from the three trials were consistent with the hypothesis that carbohydrates and carbohydrate-rich foods (bread, grapefruit juice) can increase oral flavanol absorption possibly by exerting a specific

Table 17.1 Protein and fat content of dairy products used in the study

	Protein content (g/240 mL)[a]		Fat content (g/240 mL)
Sample	Declared[b]	Found	Declared
Skim milk	8.00	7.16±1.20	2.00
Whole milk	8.00	7.05±0.96	8.00
Heavy cream	0.00	0.29±0.01	80.00

[a] Values are reported as the mean ± SEM, n = 2
[b] Amount declared by the manufacturer
[c] Amount measured by a colorimetric assay method

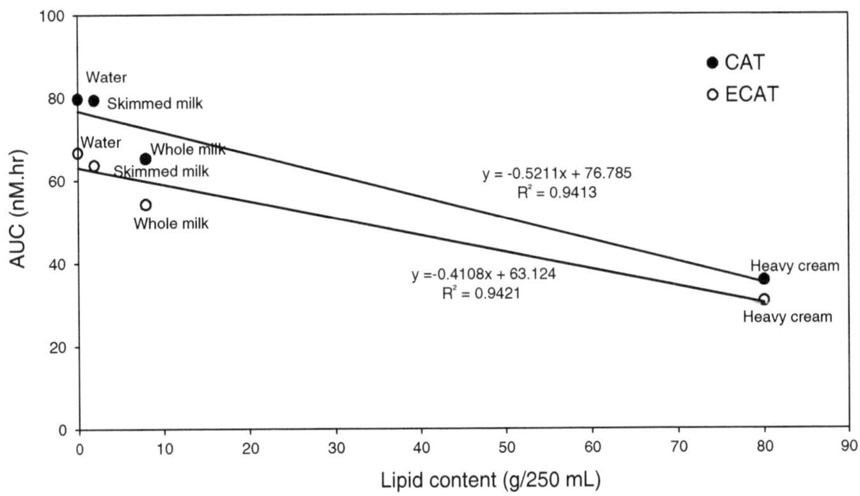

Fig. 17.7 Correlation plots of plasma $AUC_{0-\infty}$ for solutions of (+)-catechin and (−)-epicatechin in a dairy product versus concentration of fat in a dairy product. Each point represents the mean for n = 6 values. For comparative purposes, data from solutions in water are shown alongside

effect on gastrointestinal motility and/or secretion or by stimulating the activity of as yet unidentified flavanol transporter. The contrasting effects of lipid macronutrients noted in this study with those attained in rats with pure monomeric flavanols and dairy products with a high fat content may have stemmed from the closer interaction between an emulsified lipid, as is the case of dairy products, with a flavanol, possibly via micellization, than is possible with a nonemulsified lipid such as butter [26, 179]. On the other hand, the slightly enhancing effect of milk on cocoa flavanol absorption relative to cocoa in water noted in this study is at variance with the lowering effect reported for commercial chocolates in rats [179]. In the absence of information on the fat content of the milk used in the human trial, an alternative possibility for the differing results could be the result of a species difference [181]. The components of a cocoa powder such as fat, protein, carbohydrate, and other polyphenols may not affect the bioavailability of a monomeric flavanol significantly but, instead, slow down the absorption since the plasma levels of the parent molecule and its 3′-*O*-methylated metabolites have been reported to be lower than those derived from the pure flavanol [127].

The Role of Chocolate Type and Chocolate Matrix Characteristics

In vitro digestion models are useful experimental tools to assess the release of polyphenols from food matrices during their transit through the gastrointestinal tract and their transformation as a result of digestion [182]. In addition to providing information of the chemical stability of polyphenols present as purified entities or as components of food items, these models can also help to determine how food matrix components, including carbohydrates, fats, proteins, fiber, and chemical additives, can influence the release of polyphenols from foodstuff and, hence, their digestibility in and bioavailability from the intestinal milieu. Taking advantage of this simple, rapid, and low-cost experimental approach, Ortega et al. [26] have examined the role of matrix fats on the digestibility, stability, and bioaccessibility of polyphenols (procyanidins, phenolic acids, flavones) from two cocoa sources: a cocoa liquor with higher fat (50%) and procyanidin contents and a cocoa powder with lower fat (15%) and procyanidin content. In this study, the cocoa samples were separately added to either a gastric digestion system (made from buffered α-amylase followed by hydrochloric acid to pH 2, and porcine pepsin solution) and duodenal digestion (bicarbonate to pH 6.5, followed by bile salts and pancreatin) and incubated at 37°C for 2 h. Alternatively, the duodenal digestion was carried out using a static or a continuous-flow dialysis setup with the duodenal digestion mixture serving as the receiving solution. Under acidic conditions, cocoa liquor oligomeric procyanidins with a high degree of polymerization were completely hydrolyzed, mainly to dimers and less to monomers, during gastric digestion, with the resulting products remaining stable thereafter. The cocoa powder did not show any dimer. In the duodenal digestion, the procyanidin content of the cocoa liquor was 1.8-fold higher than in the gastric digestion, possibly because of protection of the procyanidin molecules by micelles derived from fat in the cocoa liquor or from bile salts present in the digestion mixture or by becoming incorporated into lipid emulsion droplets formed from the cocoa liquor. Under analogous conditions, the change in procyanidin content for the cocoa powder was only about 1.2-fold higher between duodenal digestion and gastric digestion. The lack of effect of the fat content on the digestibility of the cocoa samples due to either gastric or duodenal digestion was apparent from the increase in the formation of low molecular weight phenolic acids and from the disappearance of ferulic acid from both food samples. During gastric digestion, the flavone glycoside content decreased and that of the aglycones increased, but during duodenal hydrolysis, the levels of flavones, aglycones, and glucosides were all increased. This result may reflect the protective role of matrix fats or added bile acids on the flavone glycosides as was the case of the procyanidins.

Neilson et al. [183] assessed the bioavailability of (−)-epicatechin from a chocolate or cocoa beverage whose matrices varied in macronutrient composition and physical form. The solid confections consisted of different chocolate bars (containing reference dark chocolate [CDK], and a high sucrose [CHS] and high milk protein [CMP]); and liquid confections represented by cocoa beverages (containing sucrose milk protein [BSMP] and nonnutritive sweetener milk protein [BNMP]). All matrices provided the same quantities of (+)-catechin + (−)-epicatechin per serving (40-g bar or 250-mL liquid product). Six human subjects consumed each product in a randomized crossover design, with the serum (−)-epicatechin concentrations being monitored over 6 h postconsumption period. AUCs were similar among chocolate matrices. However, AUCs were higher for BSMP and BNMP (132 and 143 nM/h) than for CMP (101 nM/h), with the same trend applying to the peak serum concentrations (C_{max}) which were also higher for the BSMP and BNMP formulations (43 and 42 nM) than for the CDK and CMP ones (32 and 25 nM). Solid confections showed mean t_{max} values that were higher, although not statistically different, than beverages (1.8–2.3 h vs. 0.9–1.1 h), thus reflecting differences in the shape of the pharmacokinetic curves for beverages and confections. In vitro experiments in which Caco-2 human intestinal epithelial cells were incubated with formulations subjected to a two step digestion, to represent the exposure of enterocytes to concentrations of (+)-catechin and (−)-epicatechin present in the various formulations, yielded total catechin accumulation values that

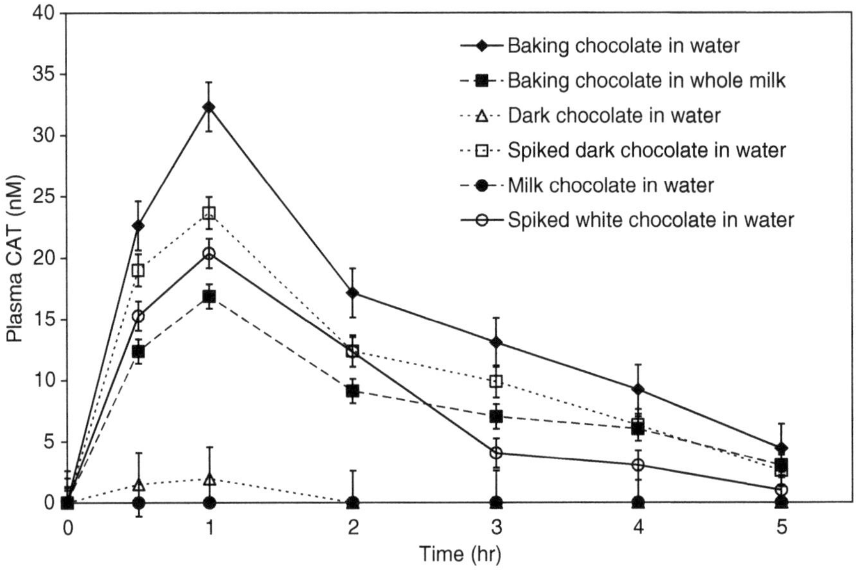

Fig. 17.8 Temporal changes in plasma (+)-catechin levels arising from the oral administration of a dispersion of baking chocolate or of milk chocolate in water, in an amount delivering 350 mg/kg of (+)-catechin, to rats. Each point represents the mean ± S.E.M. for n = 6

did not differ between treatments. These data implied that bioavailability of cocoa flavan-3-ols is likely similar from typical commercial cocoa-based foods and beverages but that the physical form and sucrose content may influence t_{max} and C_{max}.

Using three chocolate confection, identified as dark chocolate reference, high sucrose and high milk protein, and containing equal quantities of the same cocoa powder and equipotent in their contents of monomeric flavanols (~1.60 mg/1.5 g serving as (+)-catechin plus (−)-epicatechin), Neilson et al. [184] investigated the impact of certain chocolate matrix ingredients (i.e., level of cocoa butter, sucrose, soya lecithin, and milk protein) on the oral bioavailability and potential modulating effect on phase II metabolism. Collection of blood for up to 8 h from rats receiving 1.5 g of a confection by intragastric gavage and analysis of the plasma fractions HPLC–MS revealed that the predominant metabolites were *O*-glucuronides and *O*-methyl-*O*-glucuronides. Plasma concentrations of metabolites were generally highest for the high sucrose treatment and lowest for the milk treatment, with the reference dark chocolate treatment generally yielding intermediate concentrations. Milk and sucrose appeared to modulate both metabolism and plasma pharmacokinetics and, to a lesser extent, the overall bioavailability of catechins from chocolate confections.

The role of chocolate matrix composition on the oral absorption and ensuing plasma pharmacokinetics of (+)-catechin and (−)-epicatechin has been investigated in the rat using chocolates with distinct formulation components, namely, baking, dark, milk, and white [179]. Dispersions of each chocolate type both in water and in whole milk, each providing equal amounts (175 mg/kg) of a monomeric catechin, adjusted through spiking with the relevant pure compound if necessary, were administered to rats by intragastric gavage, and plasma monomeric catechins were measured on blood samples collected at 30 min and every 1 h thereafter for up to 6 h posttreatment. A solution of the two catechins in water, in amounts equal to those present in a chocolate dispersion, was also administered to serve as a control (or 100% absorption). As shown in Figs. 17.8 and 17.9, the resulting plasma

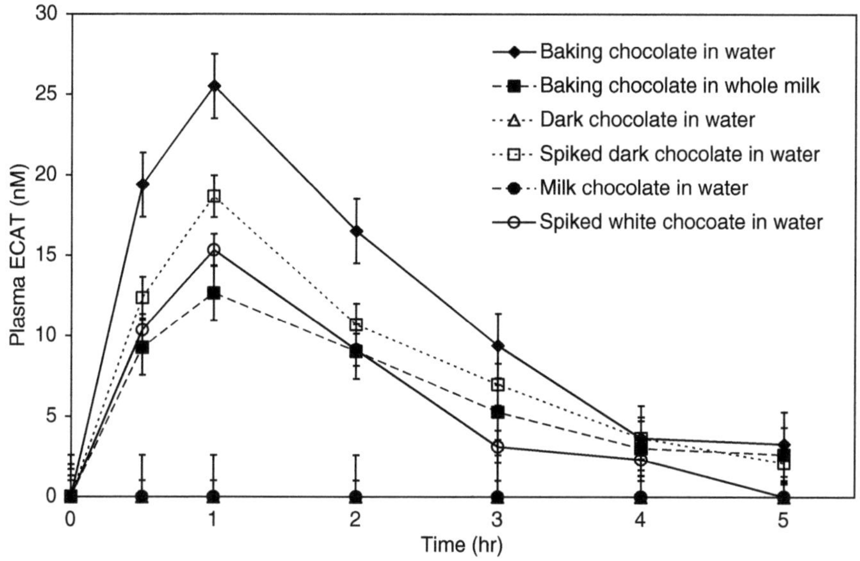

Fig. 17.9 Temporal changes in plasma (–)-epicatechin levels arising from the oral administration of a dispersion of baking chocolate or of milk chocolate in water, in an amount delivering 350 mg/kg of (–)-epicatechin, to rats. Each point represents the mean ± S.E.M. for n = 6

catechin levels varied rather widely among the various types of chocolate, with the differences being attributes to differences in physical properties, matrix components), and matrix characteristics associated with each chocolate [179]. In this connection, oral feeding of an aqueous solution of baking chocolate, exhibiting a friable nature, minimal unctuosity, no added sugar, and a lower content in cocoa butter than the other test chocolates, yielded pharmacokinetic values that approximated those of pure catechins in water most closely than any of the other chocolates (Table 17.2). In the absence of spiking, dark chocolate and white chocolate yielded very low and no levels of the two monomeric catechins. Following spiking, both chocolates yielded discernible plasma levels of the two catechins, but which were 1.4–1.8-fold and 1.4–1.9-fold lower, respectively, than plasma levels from baking chocolate in water. At least two factors may account for these differences: one is the higher contents of cocoa butter, whole milk (in powdered or condensed form), surfactant, sugar, and flavorings present than in baking chocolate [185, 186]. The other factor could be the extent of the physical interactions that may take place between chocolate matrix components during the manufacturing of a chocolate, entailing cycles of melting and solidification of cocoa butter and other fat sources and mixing. Under these conditions, catechin molecules could become incorporated into fat globules which, upon cooling, will create an enclosing polymorphic crystal network that melts at just below body temperature [187] and which may impede the transfer of a catechin into the intestinal mucosa. On the other hand, the extremely low C_{max} and AUC values for (–)-catechin and (+)-epicatechin from a dairy milk chocolate, even after spiking this chocolate with exogenous catechins, could have resulted from the combined effect of a very complex matrix and the presence of both milk and fats, factors which, when coexisting, may jointly contribute to interfering with the release of monomeric catechins from this type of chocolate. To test the role of milk on the absorption of monomeric catechins from a chocolate matrix, baking chocolate, the chocolate demonstrating the highest C_{max} and AUC values for a

Table 17.2 Plasma pharmacokinetic data for (−)-catechin (CAT) and (−)-epicatechin (ECAT), present in or added to, commercial chocolates after oral administration of the chocolate to rats as dispersions in water or in whole milk[a,b,c]

	CAT			ECAT		
Type of chocolate	C_{max} (nM)	t_{max} (h)	AUC (nM.h)	C_{max} (nM)	t_{max} (h)	AUC (nM.h)
Baking in water	$32.32 \pm 1.23^{*}$	1.00	$77.19 \pm 2.03^{*}$	$25.49 \pm 2.17^{*}$	1.00	$60.02 \pm 4.12^{*}$
Baking in milk	$16.86 \pm 0.96^{***}$	1.00	$42.51 \pm 2.62^{***}$	$12.66 \pm 1.03^{***}$	1.00	$32.75 \pm 3.62^{***}$
Dark in water	$2.00 \pm 0.56^{***}$	1.00	$2.22 \pm 1.36^{***}$	$0.00 \pm 0.14^{***}$	1.00	$0.00 \pm 0.35^{***}$
Dark in water, spiked	$23.66 \pm 1.65^{**}$	1.00	$57.12 \pm 2.95^{***}$	$18.66 \pm 1.06^{**}$	1.00	$42.57 \pm 1.09^{**}$
Dairy milk in water	$0.01 \pm 0.12^{***}$	1.00	$0.01 \pm 0.11^{***}$	$0.01 \pm 0.18^{***}$	1.00	$0.01 \pm 0.22^{***}$
Dairy milk, spiked	$0.05 \pm 0.16^{***}$	1.00	$0.07 \pm 0.08^{***}$	$0.02 \pm 0.20^{***}$	1.00	$0.03 \pm 0.12^{***}$
White in water, spiked	$20.37 \pm 3.02^{**}$	1.00	$42.79 \pm 1.09^{***}$	$15.32 \pm 0.99^{***}$	1.00	$31.26 \pm 3.58^{***}$

[a] Values are reported as the mean ± SEM for n = 6 rats
[b] Dark chocolate, dairy milk, and white chocolate were spiked with CAT and ECAT to allow the delivery of 175 mg/kg of each flavanol per serving.
[c] Statistical comparisons were versus the corresponding pure catechin in water: $^{*}p<0.05$, $^{**}p<0.01$, $^{***}p<0.001$

Table 17.3 Protein, fat, and petroleum ether extractable contents of commercial chocolates used in the study

Sample	Declared protein content (g/100 g)	Declared fat content (g/100 g)	Petroleum ether solubles (g/100 g)
Baking chocolate[a]	14.30	50.00	52.63
Dairy milk chocolate[b]	5.00	30.00	31.40
Dark chocolate	5.00	32.50	32.01
White chocolate	0.00	28.60	27.31

[a] Declared by the manufacturer as 7 g of fat and 2 g of protein per 14-g serving
[b] Declared by the manufacturer as 24 g of fat and 3 g of protein per 14-g serving

dispersion in water, was also administered to rats orally as a dispersion in whole milk [175]. From the temporal changes in plasma (+)-catechin and (−)-epicatechin levels (see Figs. 17.7 and 17.8) and pharmacokinetic values (see Table 17.2), it became evident that milk can have a profound negative effect on the bioaccessibility and/or bioavailability of monomeric catechins from the digestive tract and on ensuing pharmacokinetic parameters but the t_{max}. It is plausible that milk, being an emulsion state and having high content in micelles, might have entrapped chocolate catechins in a way that will prevent them from diffusing out of the matrix or from being taken up into enterocytes. Although the same type of experiment was not carried out with other types of chocolate, it is expected that a similar trend will be followed, with the degrees of the interaction being dictated by the matrix characteristics inherent to each type of chocolate. In this context, the role of the fat content on the bioaccessibility of monomeric catechins from a chocolate matrix was investigated by subjecting the various chocolates to an extraction with petroleum ether, and the quantities of solids thus removed were compared with the fat contents declared by each manufacturer [179]. As shown in Table 17.3, three out of the four chocolates tested had declared quantities of total fat that were within 2% of each other (dairy milk 30.0%, dark 32.5%, white 27.3%) and petroleum ether extractables that differed by not more than 5% of each in the most extreme case (dairy 31.4%, dark 32.0%, white 27.3%). A further indication that fats alone are not as important as other features associated with a chocolate is to find that baking chocolate, with the highest declared fat content (50%) and highest content of petroleum ether extractables (52.6%), showed the highest bioavailability of both (+)-catechin and (−)-epicatechin among the chocolate samples examined. The same argument may militate against the possible roles of extensive presystemic first-pass effect and of competitive absorption in the gastrointestinal tract between these two flavanols, which should have affected all the chocolates at least similarly if not equally [127].

Information on the effect of food on the bioavailability of procyanidins is rather scanty. There appears to be only one published study, carried out in rats, which have examined the role of a carbohydrate-rich food on the uptake of procyanidins in vivo [188]. In this instance, Wistar rats were treated with aqueous solutions of either a grape seed extract or of a mixture of a grape seed extract with a carbohydrate-rich food. Under basal conditions, no procyanidins were detected in the plasma. Two hours after the grape seed extract ingestion, several metabolites, such as (−)-catechin glucuronide and (+)-epicatechin glucuronide, were detected, which reached maximum plasma concentrations equal to 6.24 and 9.74 nM, respectively, when the grape seed extract was ingested with carbohydrate-rich food, and 6.32 and 8.71 nM, when grape seed extract was ingested without carbohydrate-rich food. In contrast, the concentration of the free forms of dimers and trimers in the plasma reached a maximum 1 h after administration, but their concentrations were greater after the ingestion of procyanidin-containing extract was consumed without carbohydrate-rich food, reaching a postprandial plasma concentration of 0.57 nM relative to a concentration of 0.12 nM when the grape seed extract was ingested with carbohydrate-rich food. Similarly, the presence of the carbohydrate-rich food reduced the maximum concentration of the trimer in the plasma. Plasma levels of nonconjugated monomers ((+)-catechin, (−)-epicatechin, (−)-epicatechin gallate) were less than 0·03 nM under the two experimental conditions. Among metabolites, the glucuronidated forms of (+)-catechin and (−)-epicatechin were the most abundant, and the methyl-glucuronidated forms were the second most abundant, independently of the presence of the carbohydrate-rich meal. In general, dimeric and trimeric procyanidins were found to be absorbed in vivo, reaching maximum concentrations in the plasma as soon as 1 h postingestion, but their absorption is inhibited by the copresence of carbohydrate-rich food.

References

1. Dauchet L, Amouyel P, Hercberg S, Dallongeville J. Fruit and vegetable consumption and risk of coronary heart disease: a meta-analysis of cohort studies. J Nutr. 2006;136(10):2588–93.
2. Ness AR, Powles JW. Fruit and vegetables, and cardiovascular disease: a review. Int J Epidemiol. 1997; 26(1):1–13.
3. Joshipura KJ, Hu FB, Manson JE, Stampfer MJ, Rimm EB, Speizer FE, et al. The effect of fruit and vegetable intake on risk for coronary heart disease. Ann Intern Med. 2001;134(12):1106–14.
4. Joshipura KJ, Ascherio A, Manson JE, Stampfer MJ, Rimm EB, Speizer FE, et al. Fruit and vegetable intake in relation to risk of ischemic stroke. JAMA. 1999;282(13):1233–9.
5. Sauvaget C, Nagano J, Allen N, Kodama K. Vegetable and fruit intake and stroke mortality in the Hiroshima/Nagasaki life span study. Stroke. 2003;34:2355–60.
6. Utsugi MT, Ohkubo T, Kikuya M, Kurimoto A, Sato RI, Suzuki K, et al. Fruit and vegetable consumption and the risk of hypertension determined by self measurement of blood pressure at home: the Ohasama study. Hypertens Res. 2008;31(7):1435–43.
7. Cho E, Seddon JM, Rosner B, Willett WC, Hankinson SE. Prospective study of intake of fruits, vegetables, vitamins, and carotenoids and risk of age-related maculopathy. Arch Ophthalmol. 2004;122(6):883–92.
8. Watson L, Margetts B, Howarth P, Dorward M, Thompson R, Little P. The association between diet and chronic obstructive pulmonary disease in subjects selected from general practice. Eur Respir J. 2002;20(2):313–8.
9. Tsai CJ, Leitzmann MF, Willett WC, Giovannucci EL. Fruit and vegetable consumption and risk of cholecystectomy in women. Am J Med. 2006;119(9):760–7.
10. Nagano J, Kono S, Preston DL, Moriwaki H, Sharp GB, Koyama K, et al. Bladder-cancer incidence in relation to vegetable and fruit consumption: a prospective study of atomic-bomb survivors. Int J Cancer. 2000;86(1): 132–8.
11. Riboli E, Norat T. Epidemiologic evidence of the protective effect of fruit and vegetables on cancer risk. Am Clin Nutr J. 2003;78(3 Suppl):559S–69.
12. Carter P, Gray LJ, Troughton J, Khunti K, Davies MJ. Fruit and vegetable intake and incidence of type 2 diabetes mellitus: systematic review and meta-analysis. BMJ. 2010;341:c4229.
13. Harding AH, Wareham NJ, Bingham SA, Khaw KT, Luben R, Welch A, et al. Plasma vitamin C level, fruit and vegetable consumption, and the risk of new-onset type 2 diabetes mellitus. Arch Int Med. 2008;168(14): 1493–9.
14. Hamer M, Chida Y. Intake of fruit, vegetables, and antioxidants and risk of type 2 diabetes: systematic review and meta-analysis. J Hypertens. 2007;25(12):2361–9.

15. Nagano J, Kono S, Preston DL, Mabuchi K. A prospective study of green tea consumption and cancer incidence, Hiroshima and Nagasaki (Japan). Cancer Causes Control. 2001;12(6):501–8.
16. Ogunleye AA, Xue F, Michels KB. Green tea consumption and breast cancer risk or recurrence: a meta-analysis. Breast Cancer Res Treat. 2010;119(2):477–84.
17. Arts ICW, Hollman PCH. Polyphenols and disease risk in epidemiologic studies. Am J Clin Nutr. 2005; 81(Suppl):317S–25.
18. Lee KW, Kim YJ, Lee HJ, Lee CY. Cocoa has more phenolic phytochemicals and a higher antioxidant capacity than teas and red wine. J Agric Food Chem. 2003;51(25):7292–5.
19. Bravo L. Polyphenols: chemistry, dietary sources, metabolism, and nutritional significance. Nutr Rev. 1998;56(11):317–33.
20. Halvorsen BL, Carlsen MH, Phillips KM, Bøhn SK, Holte K, Jacobs Jr DR, et al. Content of redox-active compounds (i.e., antioxidants) in foods consumed in the United States. Am J Clin Nutr. 2006;84(1):95–135.
21. Arts IC, van De Putte B, Hollman PC. Catechin contents of foods commonly consumed in The Netherlands. 2. Tea, wine, fruit juices, and chocolate milk. J Agric Food Chem. 2000;48(5):1752–7.
22. Aikpokpodion PE, Dongo LN. Effects of fermentation intensity on polyphenols and antioxidant capacity of cocoa beans. Int J Sustain Crop Prod. 2010;5(4):66–70.
23. Kyi TM, Daud WRW, Mohammad AB, Samsudin MW, Kadhum AAH, Talib MZM. The kinetics of polyphenol degradation during the drying of Malaysian cocoa beans. Int J Food Sci Technol. 2005;40(3):323–31.
24. Lowe B. Experimental cookery, from the chemical and physical standpoint. New York: Wiley; 1937. p. 512–4.
25. McShea A, Ramiro-Puig E, Munro SB, Casadesus G, Castell M, Smith MA. Clinical benefit and preservation of flavonols in dark chocolate manufacturing. Nutr Rev. 2008;66(11):630–41.
26. Ortega N, Reguant J, Romero MP, Macià A, Motilva MJ. Effect of fat content on the digestibility and bioaccessibility of cocoa polyphenol by an in vitro digestion model. J Agr Food Chem. 2009;57(13):5743–9.
27. Andrés-Lacueva C, Monagas M, Khan N, Izquierdo-Pulido M, Urpi-Sarda M, Permanyer J, et al. Flavanol and flavonol contents of cocoa powder products: influence of the manufacturing process. J Agric Food Chem. 2008; 56(9):3111–7.
28. Hii CL, Law CL, Suzannah S, Misnawi, Cloke M. Poyphenols in cocoa (*Theobroma cacao L.*). Asian J Food Agro-Ind. 2009;2(4):702–722.
29. Urpi-Sarda M, Monagas M, Khan N, Llorach R, Lamuela-Raventós RM, Jáuregui O, et al. Targeted metabolic profiling of phenolics in urine and plasma after regular consumption of cocoa by liquid chromatography-tandem mass spectrometry. J Chromatogr A. 2009;1216(43):7258–67.
30. Luna B, Moreno JM, Cruz A, Fernández-González F. Heat-shock and seed germination in a group of Mediterranean plant species growing in a burned area: an approach based on plant functional types. Environ Exp Bot. 2007;60(3):324–33.
31. Othman A, Ismail A, Abdul Ghani N, Adenan I. Antioxidant capacity and phenolic content of cocoa beans. Food Chem. 2007;100(4):1523–30.
32. Porter JF, Parton R, Wardlaw AC. Growth and survival; 57 of Bordetella bronchiseptica in natural waters and in buffered saline without added nutrients. Appl Environ Microbiol. 1991;57(4):1202–6.
33. Kim H, Keeney PG. (-)-Epicatechin content in fermented and unfermented cocoa beans. J Food Sci. 1984;49(4):1090–2.
34. Wollgast J, Anklam E. Review on polyphenols in *Theobroma cacao*: changes in composition during the manufacture of chocolate and methodology for identification and quantification. Food Res Int. 2000;33(6):423–47.
35. Verstraeten SV, Hammerstone JF, Keen CL, Fraga CG, Oteiza PI. Antioxidant and membrane effects of procyanidin dimers and trimers isolated from peanut and cocoa. J Agric Food Chem. 2005;53(12):5041–8.
36. Stark T, Hofmann T. Application of a molecular sensory science approach to alkalized cocoa (*Theobroma cacao*): structure determination and sensory activity of nonenzymatically C-glycosylated flavan-3-ols. J Agric Food Chem. 2006;54(25):9510–21.
37. Adamson GE, Lazarus SA, Mitchell AE, Prior RL, Cao G, Jacobs PH, et al. HPLC method for the quantification of procyanidins in cocoa and chocolate samples and correlation to total antioxidant capacity. J Agric Food Chem. 1999;47(10):4184–8.
38. Natsume M, Osakabe N, Yamagishi M, Takizawa T, Nakamura T, Miyatake H, et al. Analyses of polyphenols in cacao liquor, cocoa, and chocolate by normal-phase and reversed-phase HPLC. Biosci Biotechnol Biochem. 2000;64(12):2581–7.
39. Hammerstone JF, Lazarus SA, Schmitz HH. Procyanidin content and variation in some commonly consumed foods. J Nutr. 2000;130(8S Suppl):2086S–92.
40. Martín MA, Ramos S, Mateos R, Granado Serrano AB, Izquierdo-Pulido M, Bravo L, et al. Protection of human HepG2 cells against oxidative stress by cocoa phenolic extract. J Agric Food Chem. 2008;56(17):7765–72.
41. Hatano T, Miyatake H, Natsume M, Osakabe N, Takizawa T, Ito H, et al. Proanthocyanidin glycosides and related polyphenols from cacao liquor and their antioxidant effects. Phytochemistry. 2002;59(7):749–58.
42. Muselli I. Cocoa study: industry structures and composition. New York/Genoa: UNCTAD Secretariat; 2008.

43. Lamuela-Raventós RM, Romero-Pérez AI, Andrés-Lacueva C, Tornero A. Health effects of cocoa flavonoids. Food Sci Technol Int. 2005;11(3):159–76.
44. Stahl L, Miller KB, Apgar J, Sweigart DS, Stuart DA, McHale N, et al. Preservation of cocoa antioxidant activity, total polyphenols, flavan-3-ols, and procyanidin content in foods prepared with cocoa powder. J Food Sci. 2009;74(6):C456–61.
45. Payne MJ, Hurst WJ, Miller KB, Rank C, Stuart DA. Impact of fermentation, drying, roasting, and Dutch processing on epicatechin and catechin content of cacao beans and cocoa ingredients. J Agric Food Chem. 2011;58(19):10518–27.
46. Caligiani A, Cirlini M, Palla G, Ravaglia R, Arlorio M. Antioxidant and membrane effects of procyanidin dimers and trimers isolated from peanut and cocoa. J Agric Food Chem. 2005;53(12):5041–8.
47. Nazaruddin R, Seng LK, Hassan O, Said M. Effect of pulp preconditioning on the content of polyphenols in cocoa beans (*Theobroma cacao*) during fermentation. Ind Crops Prod. 2006;24(1):87–94.
48. Counet C, Collin S. Effect of the number of flavanol units on the antioxidant activity of procyanidin fractions isolated from chocolate. J Agric Food Chem. 2003;51(23):6816–22.
49. Stark T, Bareuther S, Hofmann T. Sensory-guided decomposition of roasted cocoa nibs (*Theobroma cacao*) and structure determination of taste-active polyphenols. J Agric Food Chem. 2005;53(13):5407–18.
50. Crozier A, Jaganath IB, Clifford MN. Phenols, polyphenols and tannins: an overview. In: Crozier A, Clifford MN, Asahira H, editors. Plant secondary metabolites: occurrence, structure and role in the human diet. Oxford, England: Blackwell; 2006.
51. Dreosti IE. Antioxidant polyphenols in tea, cocoa, and wine. Nutrition. 2000;16(708):692–4.
52. Hollman PC. Absorption, bioavailability, and metabolism of flavonoids. Pharml Biol. 2004;42(S1):74–83.
53. Olthof MR, Hollman PC, Buijsman MN, van Amelsvoort JM, Katan MB. Chlorogenic acid, quercetin-3-rutinoside and black tea phenols are extensively metabolized in humans. J Nutr. 2003;133(6):1806–14.
54. Miller KB, Hurst WJ, Payne MJ, Stuart DA, Apgar J, Sweigart DS, et al. Impact of alkalization on the antioxidant and flavanol content of commercial cocoa powders. J Agric Food Chem. 2008;56(18):8527–33.
55. Friedman M, Jürgens HS. Effect of pH on the stability of plant phenolic compounds. J Agric Food Chem. 2000;48(6):2101–10.
56. Neilson AP, Hopf AS, Cooper BR, Pereira MA, Bomser JA, Ferruzzi MG. Catechin degradation with concurrent formation of homo- and heterocatechin dimers during in vitro digestion. J Agr Food Chem. 2007;55(22): 8941–49.
57. Record IR, Lane JM. Simulated intestinal digestion of green and black teas. Food Chem. 2001;73(4):481–86.
58. Zhu QY, Holt RR, Lazarus SA, Ensunsa JL, Hammerstone JF, Schmitz HH, et al. Stability of the flavan-3-ols epicatechin and catechin and related dimeric procyanidins derived from cocoa. J Agric Food Chem. 2002;50(6):1700–5.
59. Rios LY, Bennett RN, Lazarus SA, Rémésy C, Scalbert A, Williamson G. Cocoa procyanidins are stable during gastric transit in humans. Am J Clin Nutr. 2002;76(5):1106–10.
60. Donovan JL, Manach C, Rios L, Morand C, Scalbert A, Rémésy C. Procyanidins are not bioavailable in rats fed a single meal containing a grapeseed extract or the procyanidin dimer B3. Br J Nutr. 2002;87(4):299–306.
61. Kemperman RA, Bolca S, Roger LC, Vaughan EE. Novel approaches for analysing gut microbes and dietary polyphenols: challenges and opportunities. Microbiology. 2010;156(Pt 11):3224–31.
62. Gu L, House SE, Wu X, Ou B, Prior RL. Procyanidin and catechin contents and antioxidant capacity of cocoa and chocolate products. J Agric Food Chem. 2006;54(11):4057–61.
63. Manach C, Scalbert A, Morand C, Rémésy C, Jiménez L. Polyphenols: food sources and bioavailability. Am J Clin Nutr. 2004;79(5):727–47.
64. Yang CS, Sang S, Lambert JD, Lee MJ. Bioavailability issues in studying the health effects of plant polyphenolic compounds. Mol Nutr Food Res. 2008;52(Suppl 1):S139 51.
65. Aura AM. Microbial metabolism of dietary phenolic compounds in the colon. Phytochem Rev. 2008;7(3):407–29.
66. Aura AM, Mattila I, Seppnen-Lakso T, Miettinen J, Oksman-Caldentey KM, Orešič M. Microbial metabolism of catechin stereoisomers by human faecal microbiota: comparison of targeted analysis and a non-targeted metabolomics method. Phytochem Lett. 2008;1(1):18–22.
67. Scalbert A, Williamson G. Dietary intake and bioavailability of polyphenols. J Nutr. 2000;130(8S Suppl): 2073S–85.
68. Rechner AR, Kuhnle G, Hu H, Roedig-Penman A, van den Braak MH, Moore KP, et al. The metabolism of dietary polyphenols and the relevance to circulating levels of conjugated metabolites. Free Radic Res. 2002;36(11):1229–41.
69. Booth AN, Deeds F, Jones FT, Murray CW. The metabolic fate of rutin and quercetin in the animal body. J Biol Chem. 1956;223(1):251–7.

70. Scheline RR. The metabolism of (+)-catechin to hydroxyphenylvaleric acids by the intestinal microflora. Biochim Biophys Acta. 1970;222(1):228–30.
71. Winter J, Moore LH, Dowell Jr VR, Bokkenheuser VD. C-Ring cleavage of flavonoids by human intestinal bacteria. Appl Environ Microbiol. 1989;55(5):1203–8.
72. Das NP, Griffiths LA. Studies on flavonoid metabolism. Metabolism of (+)-catechin in the guinea pig. Biochem J. 1968;110(3):449–56.
73. Gott DM, Griffiths LA. Effects of antibiotic pretreatments on the metabolism and excretion of [U^{14}C](+)-catechin [(U^{14}C](+)-cyanidanol-3) and its metabolite, 3′-O-methyl-(+)-catechin. Xenobiotica. 1987;17(4):423–34.
74. Griffiths LA. Studies on flavonoid metabolism. Identification of the metabolites of (+)-catechin in rat urine. Biochem J. 1964;92(1):173–9.
75. Meselhy MR, Nakamura N, Hattori M. Biotransformation of (−)-epicatechin 3-O-gallate by human intestinal bacteria. Chem Pharm Bull(Tokyo). 1997;45(5):888–93.
76. Das TK. Rate of adaptation of some of the urea cycle enzymes to a low-protein diet. Proc Nutr Soc. 1971;30(3):79A–80.
77. Rios LY, Gonthier MP, Rémésy C, Mila I, Lapierre C, Lazarus SA, et al. Chocolate intake increases urinary excretion of polyphenol-derived phenolic acids in healthy human subjects. Am J Clin Nutr. 2003; 77(4):912–28.
78. Urpi-Sarda M, Garrido I, Monagas M, Gómez-Cordovés C, Medina-Remón A, Andres-Lacueva C, et al. Profile of plasma and urine metabolites after the intake of almond [*Prunus dulcis* (Mill.) D.A. Webb] polyphenols in humans. J Agric Food Chem. 2009;57(21):10134–42.
79. Tzounis X, Vulevic J, Kuhnle GG, George T, Leonczak J, Gibson GR, et al. Flavanol monomer-induced changes to the human faecal microflora. Br J Nutr. 2008;99(4):782–92.
80. Wang LQ, Meselhy MR, Li Y, Nakamura N, Min BS, Qin GW, et al. The heterocyclic ring fission and dehydroxylation of catechins and related compounds by *Eubacterium* sp. strain SDG-2, a human intestinal bacterium. Chem Pharm Bull(Tokyo). 2001;49(12):1640–3.
81. Schantz M, Erk T, Richling E. Metabolism of green tea catechins by the human small intestine. Biotechnol J. 2010;5(10):1050–9.
82. Meng X, Sang S, Zhu N, Lu H, Sheng S, Lee MJ, et al. Identification and characterization of methylated and ring-fission metabolites of tea catechins formed in humans, mice, and rats. Chem Res Toxicol. 2002; 15(8):1042–50.
83. Das NP, Griffiths LA. Studies on flavonoid metabolism. Metabolism of (+)-[^{14}C] catechin in the rat and guinea pig. Biochem J. 1969;115(4):831–6.
84. Oshima Y, Watanabe H, Isakari S. The mechanisms of catechins metabolism. I. substances in the urine of rabbits administered (+)-catechin. Biochem J. 1958;45(11):861–5.
85. Booth AN, Williams RT. The hydroxylation of catechol acids by intestinal contents. Biochem J. 1963; 88(3):66–7.
86. Alberto MR, Gómez-Cordovés C, Manca de Nadra MC. Metabolism of gallic acid and catechin by Lactobacillus hilgardii from wine. J Agric Food Chem. 2004;52(21):6465–6459.
87. Arteel GE, Schroeder P, Sies H. Reactions of peroxynitrite with cocoa procyanidin oligomers. J Nutr. 2000;130(8 S Suppl):2100S–4.
88. Bladé C, Arola L, Salvadó MJ. Hypolipidemic effects of proanthocyanidins and their underlying biochemical and molecular mechanisms. Mol Nutr Food Res. 2010;54(1):37–59.
89. Tomaru M, Takano H, Osakabe N, Yasuda A, Inoue K, Yanagisawa R, et al. Dietary supplementation with cacao liquor proanthocyanidins prevents elevation of blood glucose levels in diabetic obese mice. Nutrition. 2007;23(4):351–5.
90. Crozier A, Jaganath IB, Clifford MN. Dietary phenolics: chemistry, bioavailability and effects on health. Nat Prod Rep. 2009;26(8):1001–43.
91. Appeldoorn MM, Vincken JP, Gruppen H, Hollman PCH. Procyanidin dimers A1, A2, and B2 are absorbed without conjugation or methylation from the small intestine of rats. J Nutr. 2009;139(8):169–1473.
92. Déprez S, Brezillon C, Rabot S, Philippe C, Mila I, Lapierre C, et al. Polymeric proanthocyanidins are catabolized by human colonic microflora into low-molecular-weight phenolic acids. J Nutr. 2000;130(11):2733–8.
93. Stoupi S, Williamson G, Drynan JW, Barron D, Clifford MN. A comparison of the in vitro biotransformation of (−)-epicatechin and procyanidin B2 by human faecal microbiota. Mol Nutr Food Res. 2010;54(6):747–59.
94. van't Slot G, Humpf HU. Degradation and metabolism of catechin, epigallocatechin-3-gallate (EGCG), and related compounds by the intestinal microbiota in the pig cecum model. J Agric Food Chem. 2009; 57(17):8041–8.

95. Lee KM, Kim WS, Lim J, Nam S, Youn M, Nam SW, et al. Antipathogenic properties of green tea polyphenol epigallocatechin gallate at concentrations below the MIC against enterohemorrhagic Escherichia coli O157:H7. J Food Prot. 2009;72(2):325–31.
96. Smith AH, Mackie RI. Effect of condensed tannins on bacterial diversity and metabolic activity in the rat gastrointestinal tract. Appl Environ Microbiol. 2004;70(2):1104–15.
97. Gu YX, Song YW, Fan LQ, Yuan QS. Antioxidant activity of natural and cultured *Cordyceps* sp. Zhongguo Zhong Yao Za Zhi. 2007;32(11):1028–31.
98. Spencer JP. Metabolism of tea flavonoids in the gastrointestinal tract. J Nutr. 2003;133(10):3255S–61.
99. Kuhnle G, Spencer JP, Schroeter H, Shenoy B, Debnam ES, Srai SK, et al. Epicatechin and catechin are O-methylated and glucuronidated in the small intestine. Biochem Biophys Res Commun. 2000;277(2):507–12.
100. Spencer JPE, Schroeter H, Rechner A, Rice-Evans C. Bioavailability of flavan-3-ols and procyanidins: gastrointestinal tract influences and their relevance to bioactive forms in vivo. Antiox Redox Sig. 2001;3(6): 1023–40.
101. Donovan JL, Crespy V, Manach C, Morand C, Besson C, Scalbert A, et al. Catechin is metabolized by both the small intestine and liver of rats. J Nutr. 2001;131(6):1753–7.
102. Aherne SA, O'Brien NM. Dietary flavonols: chemistry, food content, and metabolism. Nutrition. 2002;18(1): 75–81.
103. Roura E, Andrés-Lacueva C, Jáuregui O, Badia E, Estruch R, Izquierdo-Pulido M, et al. Rapid liquid chromatography tandem mass spectrometry assay to quantify plasma (−)-epicatechin metabolites after ingestion of a standard portion of cocoa beverage in humans. J Agric Food Chem. 2005;53(16):6190–4.
104. Hackett AM, Griffiths LA. The effects of an experimental hepatitis on the metabolic disposition of 3-O-(+)-[14C]methylcatechin in the rat. Drug Metab Dispos. 1983;11(6):602–6.
105. Das NP. Studies on flavonoid metabolism. Absorption and metabolism of (+)-catechin in man. Biochem Pharmacol. 1971;20(12):3435–45.
106. Das NP, Sothy SP. Studies on flavonoid metabolism. Biliary and urinary excretion of metabolites of (+)-(U- ^{14}C) catechin. Biochem J. 1971;125(2):417–23.
107. Das NP. Studies on flavonoid metabolism. Degradation of (+)-catechin by rat intestinal contents. Biochim Biophys Acta. 1969;177(3):668–70.
108. Shali NA, Curtis CG, Powell GM, Roy AB. Sulphation of the flavonoids quercetin and catechin by rat liver. Xenobiotica. 1991;21(7):881–93.
109. Huang C, Chen Y, Zhou T, Chen G. Sulfation of dietary flavonoids by human sulfotransferases. Xenobiotica. 2009;39(4):312–22.
110. Hackett AM, Griffiths LA, Broillet A, Wermeille M. The metabolism and excretion of (+)-[^{14}C]-cyanidanol-3 in man following oral administration. Xenobiotica. 1983;13(5):279–83.
111. Hackett AM, Griffiths LA. The metabolism and excretion of 3-palmitoyl-(+)-catechin in the rat. Xenobiotica. 1982;12(7):447–56.
112. Shaw IC, Griffiths LA. Identification of the major biliary metabolite of (+)-catechin in the rat. Xenobiotica. 1980;10(12):905–11.
113. van der Merwe PJ, Hundt HK. Metabolism of (+)-catechin and some of its C-6 and C-8 substituted derivatives in the isolated perfused pig liver. Xenobiotica. 1984;14(10):795–802.
114. Hackett AM, Griffiths LA. The metabolism and excretion of 3-O-methyl-(+)-catechin in the rat, mouse, and marmoset. Drug Metab Dispos. 1981;9(1):54–9.
115. Hackett AM, Shaw IC, Griffiths LA. 3′-O-methyl-(+)-catechin glucuronide and 3′-O-methyl-(+)-catechin sulphate: new urinary metabolites of (+)-catechin in the rat and the marmoset. Experientia. 1982;38(5):538–40.
116. Hackett AM, Griffiths LA, Wermeille M. The quantitative disposition of 3-O-methyl-(+)-[U-^{14}C]catechin in man following oral administration. Xenobiotica. 1985;15(11):907–14.
117. Wermeille M, Turin E, Griffiths LA. Identification of the major urinary metabolites of (+)-catechin and 3-O-methyl-(+)-catechin in man. Eur J Drug Metab Pharmacokinet. 1983;8(1):77–784.
118. Manach C, Texier O, Morand C, Crespy V, Régérat F, Demigné C, et al. Comparison of the bioavailability of quercetin and catechin in rats. Free Radic Biol Med. 1999;27(11–12):1259–66.
119. Shaw IC, Hackett AM, Griffiths LA. Metabolism and excretion of the liver-protective agent (+)-catechin in experimental hepatitis. Xenobiotica. 1982;12(7):405–16.
120. Smillie MV, Griffiths LA, Male PJ, Wermeille MM. The disposition and metabolism of (+)-cyanidanol-3 in patients with alcoholic cirrhosis. Eur J Clin Pharmacol. 1987;33(3):255–9.
121. Harada M, Kan Y, Naoki H, Fukui Y, Kageyama N, Nakai M, et al. Identification of the major antioxidative metabolites in biological fluids of the rat with ingested (+)-catechin and (−)-epicatechin. Biosci Biotechnol Biochem. 1999;63(6):973–7.

122. Da Silva EL, Piskula M, Terao J. Enhancement of antioxidative ability of rat plasma by oral administration of (−)-epicatechin. Free Radic Biol Med. 1998;24(7–8):1209–16.
123. Piskula MK, Terao J. Accumulation of (−)-epicatechin metabolites in rat plasma after oral administration and distribution of conjugation enzymes in rat tissues. J Nutr. 1998;128(7):1172–8.
124. Spencer JP, Chowrimootoo G, Choudhury R, Debnam ES, Srai SK, Rice-Evans C. The small intestine can both absorb and glucuronidate luminal flavonoids. FEBS Lett. 1999;458(2):224–30.
125. Baba S, Osakabe N, Yasuda A, Natsume M, Takizawa T, Nakamura T, et al. Bioavailability of (−)-epicatechin upon intake of chocolate and cocoa in human volunteers. Free Radic Res. 2000;33(5):635–41.
126. Okushio K, Suzuki M, Matsumoto N, Nanjo F, Hara Y. Identification of (−)-epicatechin metabolites and their metabolic fate in the rat. Drug Metab Dispos. 1999;27(2):309–16.
127. Baba S, Osakabe N, Natsume M, Muto Y, Takizawa T, Terao J. Absorption and urinary excretion of (−)-epicatechin after administration of different levels of cocoa powder or (−)-epicatechin in rats. J Agric Food Chem. 2001;49(12):6050–6.
128. Donovan JL, Luthria DL, Stremple P, Waterhouse AL. Analysis of (+)-catechin, (−)-epicatechin and their 3′- and 4′-O-methylated analogs. A comparison of sensitive methods. J Chromatogr B Biomed Sci Appl. 1999;726(1–2):277–83.
129. Terao J. Dietary flavonoids as antioxidants in vivo: conjugated metabolites of (−)-epicatechin and quercetin participate in antioxidative defense in blood plasma. J Med Invest. 1999;46(3–4):159–68.
130. Yamashita S, Sakane T, Harada M, Sugiura N, Koda H, Kiso Y, et al. Absorption and metabolism of antioxidative polyphenolic compounds in red wine. Ann N Y Acad Sci. 2002;957(1):325–8.
131. Vaidyanathan JB, Walle T. Transport and metabolism of the tea flavonoid (−)-epicatechin by the human intestinal cell line Caco-2. Pharm Res. 2001;18(10):1420–5.
132. Vaidyanathan JB, Walle T. Glucuronidation and sulfation of the tea flavonoid (−)-epicatechin by the human and rat enzymes. Drug Metab Dispos. 2002;30(8):897–903.
133. Abrahamse L, Kloots WJ, van Amelsvoort JM. Absorption, distribution and secretion of epicatechin and quercetin in the rat. Nutr Res. 2005;25(3):305–17.
134. Rimbach G, Melchin M, Moehring J, Wagner AE. Polyphenols from cocoa and vascular health-a critical review. Int J Mol Sci. 2009;10(10):4290–309.
135. Spencer JP, Schroeter H, Shenoy B, Srai SK, Debnam ES, Rice-Evans C. Epicatechin is the primary bioavailable form of the procyanidin dimers B2 and B5 after transfer across the small intestine. Biochem Biophys Res Commun. 2001;285(3):588–93.
136. Bell JRC, Donovan JL, Wong R, Waterhouse AL et al. (+)-Catechin in human plasma after ingestion of a single serving of reconstituted red wine. Am J Clin Nutr. 2000;71(1):103–8.
137. Baba S, Osakabe N, Natsume M, Yasuda A, Takizawa T, Nakamura T, et al. Cocoa powder enhances the level of antioxidative activity in rat plasma. Br J Nutr. 2000;84(5):673–80.
138. Wang JF, Schramm DD, Holt RR, Ensunsa JL, Fraga CG, Schmitz HH, et al. A dose-response effect from chocolate consumption on plasma epicatechin and oxidative damage. J Nutr. 2000;130(8 S Suppl):2115S–9.
139. Baba S, Osakabe N, Natsume M, Muto Y, Takizawa T, Terao J. In vivo comparison of the bioavailability of (+)-catechin, (−)-epicatechin and their mixture in orally administered rats. J Nutr. 2001;131(11):2885–91.
140. Li C, Meng X, Winnik B, Lee MJ, Lu H, Sheng S, et al. Analysis of urinary metabolites of tea catechins by liquid chromatography/electrospray ionization mass spectrometry. Chem Res Toxicol. 2001;14(6):702–7.
141. Ritter C, Zimmermann BF, Galensa R. Chiral separation of (+)/(−)-catechin from sulfated and glucuronidated metabolites in human plasma after cocoa consumption. Anal Bioanal Chem. 2010;397(2):723–30.
142. Bombardelli E, Morazzoni P, Carini M, Aldini G, Maffei FR. Biological activity of procyanidins from *Vitis vinifera* L. Biofactors. 1997;6(4):429–31.
143. Cheynier V. Polyphenols in foods are more complex than often thought. Am J Clin Nutr. 2005; 81(1Suppl):223S–9.
144. Hümmer W, Schreier P. Analysis of proanthocyanidins. Mol Nutr Food Res. 2008;52(12):1381–98.
145. Gabetta B, Fuzzati N, Griffini A, Lolla E, Pace R, Ruffilli T, Peterlongo F. Characterization of proanthocyanidins from grape seeds. Fitoterapia 2000;71(2):162–75.
146. Ferreira D, Marais JP, Slade D. Heterogeneity of the interflavanyl bond in proanthocyanidins from natural sources lacking C-4 (C-ring) deoxy flavonoid nucleophiles. Phytochemistry. 2005;66(18):2216–37.
147. Counet C, Ouwerx C, Rosoux D, Collin S. Relationship between procyanidin and flavor contents of cocoa liquors from different origins. J Agric Food Chem. 2004;52(20):6243–49.
148. Nakamura Y, Tonogai Y. Metabolism of grape seed polyphenol in the rat. J Agric Food Chem. 2003; 51(24):7215–25.
149. Sano A, Yamakoshi J, Tokutake S, Tobe K, Kubota Y, Kikuchi M. Procyanidin B1 is detected in human serum after intake of proanthocyanidin-rich grape seed extract. Biosci Biotechnol Biochem. 2003;67(5):1140–3.
150. Déprez S, Mila I, Scalbert A. Carbon-14 biolabeling of (+)-catechin and proanthocyanidin oligomers in willow tree cuttings. J Agric Food Chem. 1999;47(10):4219–30.

151. Baba S, Osakabe N, Natsume M, Terao J. Absorption and urinary excretion of procyanidin by [epicatechin-(4β-8)-epicatechin] in rats. Free Radic Biol Med. 2002;33(1):142–8.
152. Holt RR, Lazarus SA, Sullards MC, Zhu QY, Schramm DD, Hammerstone JF, et al. Procyanidin dimer B2 [epicatechin-(4β-8)-epicatechin] in human plasma after the consumption of a flavanol-rich cocoa. Am J Clin Nutr. 2002;76(4):798–804.
153. Catterall F, King LJ, Clifford MN, Ioannides C. Bioavailability of dietary doses of ^{3}H-labelled tea antioxidants (+)-catechin and (−)-epicatechin in rat. Xenobiotica. 2003;33(7):743–53.
154. Zhu M, Chen Y, Li RC. Oral absorption and bioavailability of tea catechins. Planta Med. 2000;66(5):444–7.
155. Tsang C, Auger C, Mullen W, Bornet A, Rouanet JM, Crozier A, et al. The absorption, metabolism and excretion of flavan-3-ols and procyanidins following the ingestion of a grape seed extract by rats. Br J Nutr. 2005;94(2):170–81.
156. Cai Y, Anavy ND, Chow HHS. Contribution of presystemic hepatic extraction to the low oral bioavailability of green tea catechins in rats. Drug Metab Dispos. 2002;30(11):1246–9.
157. Richelle M, Tavazzi I, Enslen M, Offord EA. Plasma kinetics in man of epicatechin from black chocolate. Eur J Clin Nutr. 1999;53(1):22–6.
158. Rein D, Lotito S, Holt RR, Keen CL, Schmitz HH, Fraga CG. Epicatechin in human plasma: in vivo determination and effect of chocolate consumption on plasma oxidation status. J Nutr. 2000;130(8 S Suppl):2109S–14.
159. Tomás-Barberán FA, Cienfuegos-Jovellanos E, Marín A, Muguerza B, Gil-Izquierdo A, Cerda B, et al. A new process to develop a cocoa powder with higher flavonoid monomer content and enhanced bioavailability in healthy humans. J Agric Food Chem. 2007;55(10):3926–35.
160. Gotti R, Furlanetto S, Pinzauti S, Cavrini V. Analysis of catechins in *Theobroma cacao* beans by cyclodextrin-modified micellar electrokinetic chromatography. J Chromatogr A. 2006;1112(1–2):345–52.
161. Donovan JL, Crespy V, Oliveira M, Cooper KA, Gibson BB, Williamson G. (+)-Catechin is more bioavailable than (−)-catechin: relevance to the bioavailability of catechin from cocoa. Free Radic Res. 2006; 40(10):1029–34.
162. Ottaviani JI, Momma TY, Heiss C, Kwik-Uribe C, Schroeter H, Keen CL. The stereochemical configuration of flavanols influences the level and metabolism of flavanols in humans and their biological activity in vivo. Free Radic Biol Med. 2011;50(2):237–44.
163. Schramm DD, Karim M, Schrader HR, Holt RR, Kirkpatrick NJ, Polagruto JA, et al. Food effects on the absorption and pharmacokinetics of cocoa flavanols. Life Sci. 2003;73(7):857–69.
164. Charman WN, Porter CJH, Mithani S, Dressman JB. Physicochemical and physiological mechanisms for the effects of food on drug absorption: the role of lipids and pH. J Pharm Sci. 1997;86(3):269–82.
165. Hertog MG, Sweetnam PM, Fehily AM, Elwood PC, Kromhout D. Antioxidant flavonols and ischemic heart disease in a Welsh population of men: the Caerphilly study. Am J Clin Nutr. 1997;65(5):1489–94.
166. van het Hof KH, Kivits GAA, Westrate JA, Tijburg LBM. Bioavailability of catechins from tea: the effect of milk. Eur J Clin Nutr. 1998;52(5):356–9.
167. Leenen R, Roodenburg AJC, Tijburg LBM, Wiseman SA. A single dose of tea with or without milk increases plasma antioxidant activity in humans. Eur J Clin Nutr. 2000;54(1):87–92.
168. Hollman PCH, van het Hof KH, Tijburg LBM, Katan MB. Addition of milk does not affect the absorption of flavonols from tea in man. Free Radic Biol Med. 2001;34(3):297–300.
169. Richelle M, Tavazzi I, Offord E. Comparison of the antioxidant activity of commonly consumed polyphenolic beverages (coffee, cocoa, and tea) prepared per cup serving. J Agric Food Chem. 2001;49(7):3438–42.
170. Reddy VC, Sagar GVV, Sreeramulu D, Venu L, Raghunath M. Addition of milk does not alter the antioxidant activity of black tea. Ann Nutr Metab. 2005;49(3):189–95.
171. Serafini M, Bugianesi R, Maiani G, Valtuena S, De Santis S, Crozier A. Plasma antioxidants from chocolate. Nature. 2003;424(6952):1013.
172. Schroeter H, Holt RR, Orozco TJ, Schmitz HH, Keen CL. Milk and absorption of dietary flavanols. Nature. 2003;426(6968):787–8.
173. Serafini M, Crozier A. Milk and absorption of dietary flavanols. Nature. 2003;426(6968):788.
174. Lorenz M, Jochmann N, von Krosigk A, Martus P, Bauman G, Stangl K, et al. Addition of milk prevents vascular protective effects of tea. Eur Heart J. 2007;28(2):219–23.
175. Keogh JB, McInerney J, Clifton PM. The effect of milk protein on the bioavailability of cocoa polyphenols. J Food Sci. 2007;72(3):S230–3.
176. Roura E, Andrés-Lacueva C, Estruch R, Mata Bilbao ML, Izquierdo-Pulido M, Waterhouse AL, et al. Milk does not affect the bioavailability of cocoa powder flavonoid in healthy human. Ann Nutr Metab. 2007;51(6):493–8.
177. Roura E, Andrés-Lacueva C, Estruch R, Mata Bilbao ML, Izquierdo-Pulido M, Lamuela-Raventós RM. The effect of milk as a food matrix for polyphenols on the excretion profile of cocoa (−)-epicatechin metabolites in healthy human subjects. Br J Nutr. 2008;100(4):846–51.
178. Mullen W, Borges G, Donovan JL, Edwards CA, Serafini M, Lean MEJ, et al. Milk decreases urinary excretion but not plasma pharmacokinetics of cocoa flavan-3-ol metabolites in humans. Am J Clin Nutr. 2009; 89(6):1784–91.

179. Gossai D, Lau-Cam CA. Assessment of the effect of type of dairy product and of chocolate matrix on the oral absorption of monomeric chocolate flavanols in a small animal model. Pharmazie. 2009;64(3):202–9.
180. Carbonaro M, Grant G, Pusztai A. Evaluation of polyphenol bioavailability in rat small intestine. Eur J Nutr. 2001;40(2):84–90.
181. Reinboth M, Wolffram S, Abraham G, Ungemach FFR, Cermak R. Oral bioavailability of quercetin from different quercetin glycosides in dogs. Br J Nutr. 2010;104(2):198–203.
182. Gil-Izquierdo A, Zafrilla P, Tomás-Barberán FA. An in vitro method to simulate phenolic compound release from the food matrix in the gastrointestinal tract. Eur Food Res Technol. 2002;214(2):155–9.
183. Neilson AP, George JC, Mattes RD, Janle EM, Matusheski NV, Ferruzzi MG. Influence of chocolate formulation factors on in vitro bioaccessibility and bioavailability of catechins in humans. J Agric Food Chem. 2009; 57(20):9418–26.
184. Neilson AP, Sapper TN, Janle EM, Rudolph R, Matusheski NV, Ferruzzi MG. Chocolate matrix factors modulate the pharmacokinetic behavior of cocoa flavan-3-ol phase II metabolites following oral consumption by Sprague-Dawley rats. J Agric Food Chem. 2010;58(11):6685–91.
185. Beckett ST. Industrial chocolate manufacture and use. 3rd ed. Oxford, UK: Blackwell; 1999.
186. Fryer P, Pinschower K. The materials science of chocolate. Mater Res Soc Bull. 2000;25(12):25–9.
187. Toro-Vazquez JF, Pérez Martínez D, Dibildox-Alvarado E, Charó-Alonso M, Reyes Hernández J. Rheometry and polymorphism of cocoa butter during crystallization under static and stirring conditions. J Am Oil Chem Soc. 2004;81(2):195–202.
188. Serra A, Macià A, Romero MP, Bladé C, Arola L, Motilva MJ. Bioavailability of procyanidin dimers and trimers and matrix food effects in in vitro and in vivo models. Br J Nutr. 2010;103(7):944–52.

Chapter 18
Biological Activity of Cacao Husk and Mass Lignin-Carbohydrate Complexes

Hiroshi Sakagami and Tomohiko Matsuta

Key Points

- The reports of biological activity of lignin-carbohydrate complex (LCC) have been quite few, as compared with that of lower molecular weight polyphenols, possibly due to the structural complexity.
- Cacao mass and husk LCCs were completely solubilized and sterilized by autoclaving under mild alkaline condition, without losing their biological activities.
- They showed much higher anti-HIV activity, as compared with lower molecular weight polyphenols.
- They showed synergy with vitamin C in cytotoxicity induction and radical scavenging.
- They did not induce NO and cytokine production but enhanced the LPS-induced iNOS protein expression by mouse macrophages, suggesting several new biological activity of LCC distinct from LPS.

Keywords Cacao husk and mass • Lignin-carbohydrate complex • Anti-HIV activity • Vitamin C • LPS contamination • Macrophage • Cytokine • Signaling pathway

Introduction

Cocoa bean, a main raw material of chocolate, has been reported to display antioxidant activity [1], antiarteriosclerosis activity [2], antibacterial activity [3], and antiviral activity [4]. The chemical analysis of the components of cacao such as catechin, epicatechin, proanthocyanidin glycosides, and related polyphenols [5] and lignin as food fibers [6] has been reported.

Lignins are a major class of natural products present in the natural kingdom and are formed by the dehydrogenative polymerization of three monolignols: *p*-coumaryl, *p*-coniferyl, and sinapyl

H. Sakagami, Ph.D. (✉)
Division of Pharmacology, Department of Diagnostic and Therapeutic Sciences, and Meikai Pharmaco-Medical Laboratory, Meikai University School of Dentistry, Sakado, Saitama 350-0283, Japan
e-mail: sakagami@dent.meikai.ac.jp

T. Matsuta, Ph.D.
Meikai Pharmaco-Medical Laboratory, Meikai University School of Dentistry, Sakado, Saitama 350-0283, Japan

R.R. Watson et al. (eds.), *Chocolate in Health and Nutrition*, Nutrition and Health 7,
DOI 10.1007/978-1-61779-803-0_18, © Springer Science+Business Media, LLC 2013

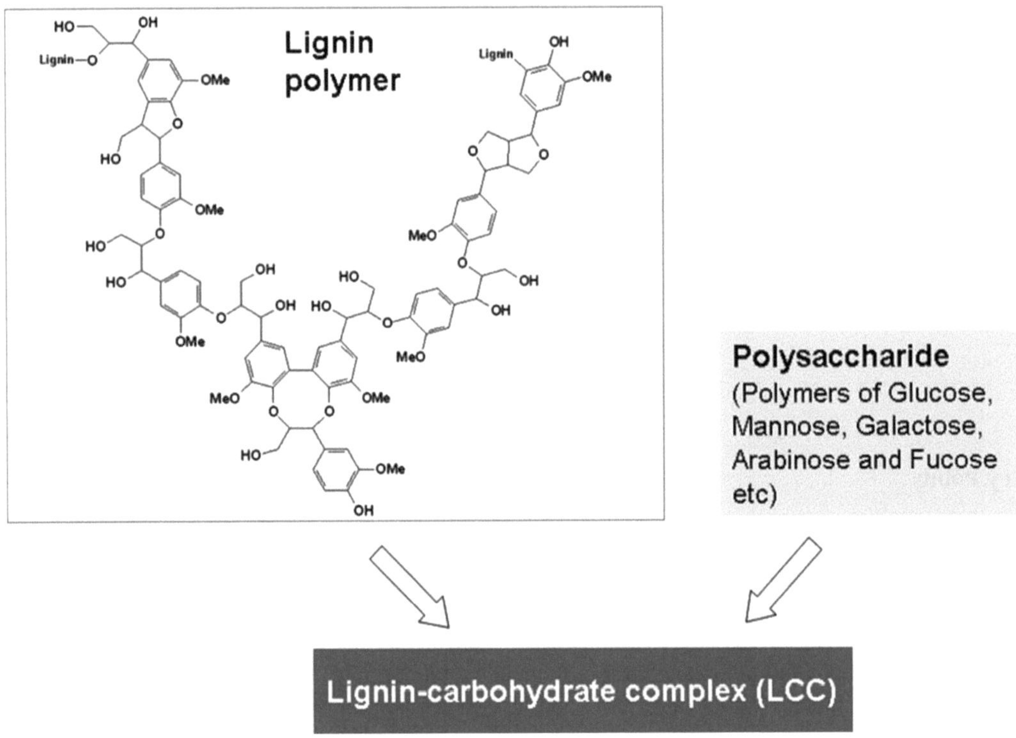

Fig. 18.1 Schematic diagram of LCC formation

alcohols [7]. Some polysaccharides in the cell walls of lignified plants are linked to lignin to form lignin-carbohydrate complex (LCC) (Fig. 18.1). LCC has displayed several unique biological activities such as antihuman immunodeficiency virus (HIV) activity and synergistic actions with vitamin C [8]. However, the physiological role of cacao-derived LCC has not been well characterized.

In order to explore the novel functionality of components of cacao, we have prepared the lignin fractions from cacao husk (the shell of the cacao bean) and cacao mass (paste with cacao husk and germ removed) (Fig. 18.2), using two different methods – either conventional method (Method I) or modified method that improved solubility and sterility (Method II) – and investigated their broad range of biological activities [9, 10].

Modified Method of LCC Preparation

With the increase of molecular weight, the solubility of LCC generally declines, making the sterilization of LCC through Millipore filter difficult. In fact, we experienced the considerable fluctuation of biological activity of LCC samples prepared by Method I, possibly due to its insolubility (Fig. 18.3a). To deal with these problems, we increased the purity of the LCC by repeating the acid-precipitation and solubilization cycle and improving the solubility of LCCs by suspending and sterilizing them in 1.39% $NaHCO_3$ by autoclaving (121°C, 20 min) [9]. This solution is homogenous and thus applicable to any biological assays. We have first confirmed that various biological activities B of LCC prepared by Method I [8] were reproduced by LCC prepared by Method II (see Fig. 18.3b) [9].

Fig. 18.2 Pictures of cacao tree (**a**), cacao beans in the cacao fruit (**b**), cacao husk (**c**), and preparation scheme of cacao mass and husk (**d**)

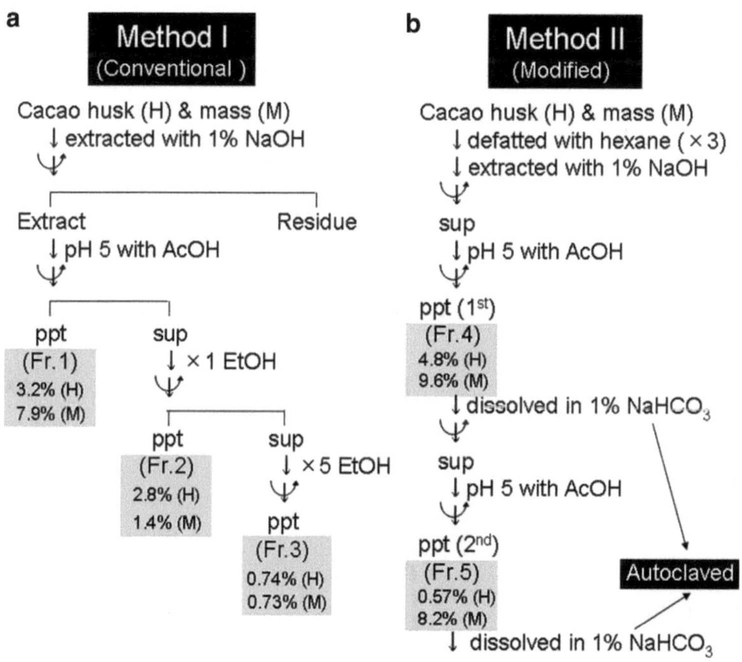

Fig. 18.3 Preparation of cacao husk and cacao mass LCC fractions, prepared by Method I (conventional) (**a**) and Method II (modified) (**b**). All fractions (Frs. 1–5) were dialyzed against excess H_2O and lyophilized. Numbers indicate yield (%). Abbreviations: ppt, precipitate; sup, supernatant (Data cited from References [9] and [10]. With permission)

Yield

LCC Samples Prepared by Method I

The yield of the LCC fraction (prepared by acid precipitation) from the cacao husk by Method I (H-Fr.1) was the greatest (3.2% ± 1.1% (n = 3)), followed by that of the polysaccharide-rich fraction (prepared by 50% ethanol precipitation) (H-Fr.2) (2.8% ± 0.2%) and then that of the low molecular fraction (prepared by 50–83% ethanol precipitation) (H-Fr.3) (0.74% ± 0.15%) (see Fig. 18.3a). The yield of acid-precipitable LCC fraction from the cacao mass (M-Fr.1) was the greatest (7.9% ± 3.4% (n = 3)), followed by that of the polysaccharide-rich fraction (M-Fr.2) (1.4% ± 0.6%) and then that of the low molecular weight fraction (M-Fr.3) (0.73% ± 0.04%) (see Fig. 18.3a). The molecular weight of cacao mass LCC (M-Fr.1) was estimated to be less than 40 kD on gel filtration (Fig. 18.4a–c).

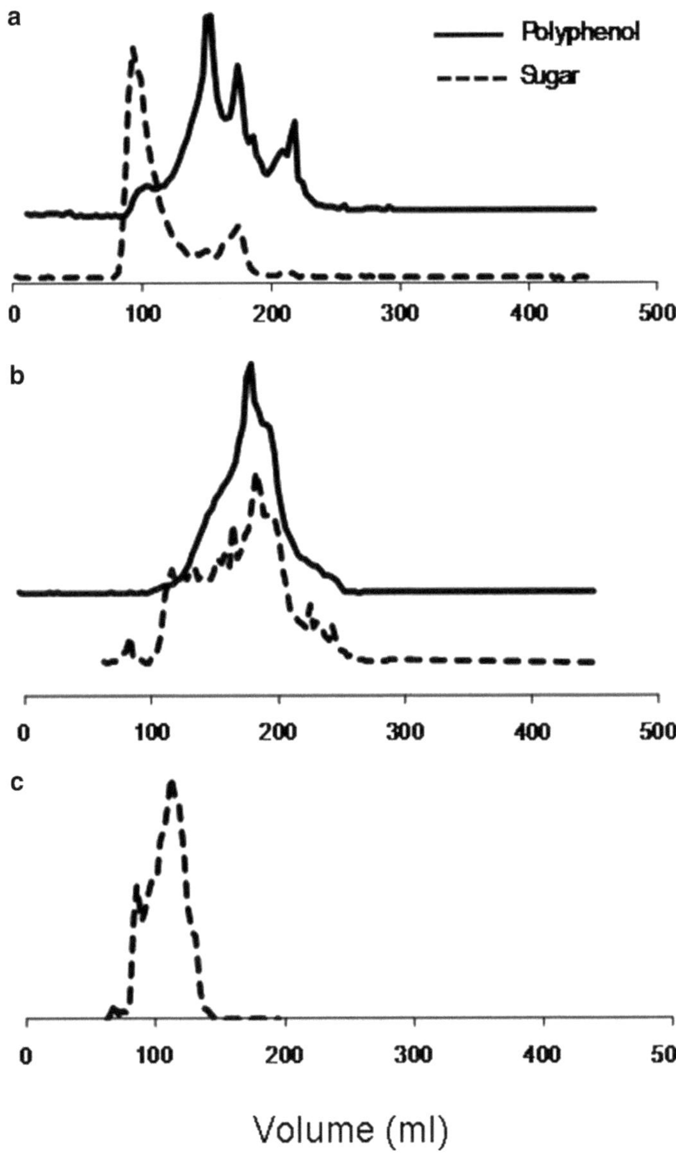

Fig. 18.4 Gel filtration chromatography of cacao mass alkaline extract (**a**), cacao mass LCC (acid precipitation) (**b**), and dextran 40000 (**c**) on TOYOPEARL HW-55F (1.5 i.d. × 100 cm, TOSOH, Tokyo). Sugar (*dotted line*) and polyphenol (*solid line*) content in each fraction were measured by phenol-sulfuric acid method and Folin-Denis method, respectively

LCC Samples Prepared by Method II

The yield of LCC prepared from the cacao mass prepared as the first precipitate by Method II (M-Fr.4) (9.6% ± 1.1%) was almost twice that prepared from the cacao husk (H-Fr.4) (4.8% ± 1.8%) (see Fig. 18.3b). By repeating the alkaline-solubilization and acid-precipitation cycle, the yield of the second precipitate was reduced only by 15% in cacao mass LCC (M-Fr.5) but by 88% in cacao husk LCC (H-Fr.5) (see Fig. 18.3b). Kinetic-chromogenic endotoxin-specific LAL assay demonstrated that M-Fr.5 and H-Fr.5 contained 0.00035 and 0.16 (w/w)% LPS, respectively [10]. To increase the solubility and sterility, samples were dissolved in 1.39% $NaHCO_3$ and then sterilized by autoclave treatment (121°C, 20 min).

Anti-HIV Activity

LCC Sample Prepared by Method I

The anti-HIV activity of cacao husk acid-precipitable LCC fraction (H-Fr.1) (SI = 271) was one order higher than that of the corresponding fractions from the cacao mass (M-F1) (SI = 46) (Table 18.1). The anti-HIV activity of the carbohydrate-rich LCC fraction from the cacao husk (H-Fr.2) (SI = 34,610) was the greatest, one to three orders higher than that of the corresponding fractions from the cacao mass (M-Fr.2) (SI = 3.3) (see Table 18.1). The anti-HIV activity of the low molecular weight fractions from the cacao husk (H-Fr.3) (SI = 115) was one order higher than that of the corresponding fractions from the cacao mass (M-Fr.3) (SI = 17) (see Table 18.1) [9]. The present study demonstrated the potent anti-HIV activity of the cacao husk LCC fractions. Looking on these data, cacao husk seems to be an attractive source for the mass production of anti-HIV substance since its carbohydrate-rich

Table 18.1 Anti-HIV activity of cacao husk mass LCCs[a, b]

Anti-HIV activity ($SI = CC_{50}/EC_{50}$)			Anti-HIV activity ($SI = CC_{50}/EC_{50}$)	
<LCC prepared by Method I>			<Lower molecular weight polyphenols>	
Cacao husk (acid ppt)	H-Fr.1	271 ± 180	Epigallocatechin-3-*O* -gallate	<1.0
Cacao husk (50% ethanol ppt)	H-Fr.2	34,610 ± 56,661	Curcumin	<1.0
Cacao husk (83% ethanol ppt)	H-Fr.3	115 ± 98	Flavonols (3-hydroxyflavones (n = 100))	<1.0
			Isoprenoid-substituted flavonoids	<1.0
Cacao mass (acid ppt)	M-Fr.1	46.0 ± 39.2		
Cacao mass (50% ethanol ppt)	M-Fr.2	3.3 ± 1.6	<Positive control>	
Cacao mass (83% ethanol ppt)	M-Fr.3	17.4 ± 16.4	Dextran sulfate	2,956 ± 1,760
			Curdlan sulfate	11,718 ± 8,426
<LCC prepared by Method II>			AZT	23,261 ± 13,141
Cacao husk LCC (1st ppt)	H-Fr.4	44.5 ± 22.1	ddC	2,974 ± 1,438
Cacao husk LCC (2nd ppt)	H-Fr.5	28.0 ± 16.4		
			LPS	<1.0
Cacao husk LCC (1st ppt)	M-Fr.4	61.3 ± 24.1		
Cacao husk LCC (2nd ppt)	M-Fr.5	39.0 ± 23.6		

[a] Data cited from references [9] and [10]. With permission
[b] Each value represents mean ± SD from four independent experiments

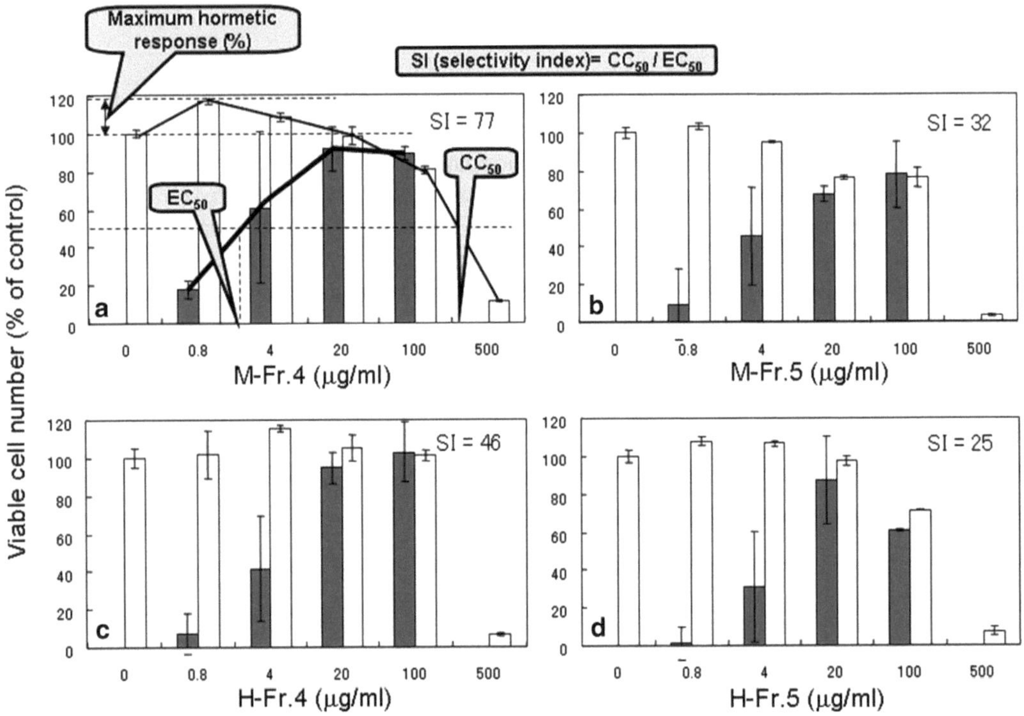

Fig. 18.5 Anti-HIV activity of cacao mass and husk LCCs. HIV-1_{IIIB}-infected (*gray* color) and mock-infected (*white* color) MT-4 cells were incubated for 5 days with the indicated concentrations of cacao husk LCC (H-Fr.4) or cacao mass LCC (M-Fr.4), and the viable cell number was determined by the MTT method and expressed as % of control. Data represent mean ± standard deviation from three determinations. Data derived from four independent experiments with different batches of samples are summarized in Table 12.1 (Data cited from Reference [10]. With permission)

fractions showed surprisingly higher anti-HIV activity than that of tannins (SI = 1–10) [10], flavonoids (SI = 1) [11, 12], and natural lignins from other plant sources and dehydrogenation polymers of phenylpropenoids (so-called synthetic lignin without the sugar moiety) (SI = 10–100) [13] and comparable with that of sulfated polysaccharide (dextran sulfate, curdlan sulfate) and reverse transcriptase inhibitors (AZT, ddC) (see Table 18.1). However, we found that the standard deviation of antiviral activity of polysaccharide-rich LCC fraction (H-Fr.2) (SI = 34,610 ± 56,661) was so large (see Table 18.1). This may be due to the insolubility of the sample that makes the sterilization by filtration through the Millipore filter extremely difficult and thus cause the unexpected bacterial proliferation during the cell culture.

LCC Samples Prepared by Method II

To accurately evaluate the anti-HIV activity of LCCs, it was essential to dissolve completely and sterilize them by autoclave treatment. We found that autoclaved cacao mass LCC (M-Fr.4) showed higher anti-HIV activity (SI = 77) (Fig. 18.5a) than that of autoclaved cacao husk (H-Fr.4)(SI = 46) (see Fig. 18.5c). Superiority of cacao mass over cacao husk was not changed after repeating the alkaline-solubilization and acid-precipitation cycle (compare M-Fr.5 in Fig. 18.5b with H-Fr.5 in Fig. 18.5d). This finding was confirmed by another three independent experiments with different batches of LCC preparations (see Table 18.1). LPS (0.000256–100 μ(mu)g/mL) did not show any anti-HIV activity (SI= ><1.0) (see Table 18.1). We found that lower concentrations (0.8–4 μ(mu)g/mL) of LCCs stimulated the growth of MT-4 cells very slightly (maximum hormetic response = 15.1%) (see Fig. 18.5a).

Anti-influenza Virus Activity

LCC Samples Prepared by Method I

The cacao husk acid-precipitable LCC (H-Fr.1) also potently inhibited the cytopathic effect induced in the MDCK cells by the influenza virus infection. The cacao husk polysaccharide-rich LCC (H-Fr.2) (EC_{50}=0.009 mg/mL; CC_{50}=1.447 mg/mL, SI=155) (Fig. 18.6b) showed higher anti-influenza virus activity than LCC fraction (H-Fr.1) (EC_{50}=0.042 mg/mL; CC_{50}=1.093 mg/mL, SI=26) (Fig. 18.6a and b).

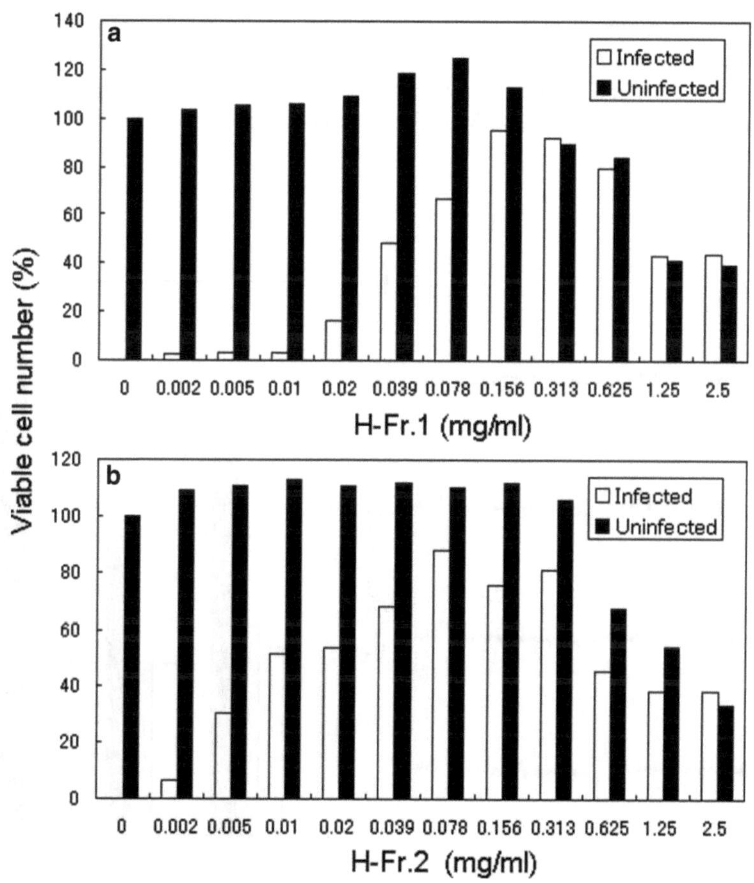

Fig. 18.6 Anti-influenza virus activity of cacao husk LCC fractions. Influenza virus-infected (*white bar*) or mock-infected (*black bar*) MDCK cells were incubated for 3 days with the indicated concentrations of H-Fr.1 (**a**) or H-Fr.2 (**b**), and the viable cell number was determined by MTT method. Each value represents mean from triplicate assays (Data cited from Reference [10]. With permission)

Antibacterial Activity

LCC Samples Prepared by Method I

H-Fr.1, H-Fr.2, M-Fr.1, and M-Fr.2 (1 mg/mL) showed no direct antibacterial activity against *Streptococcus mutans, Actinomyces viscosus, Fusobacterium nucleatum, and Porphyromonas gingivalis* (data not shown).

Synergistic Action of LCC Fractions and Vitamin C

LCC Samples Prepared by Method I

Both acid-precipitable and polysaccharide-rich fractions from cacao husk produced typical broad peak (Fig. 18.7a), and the radical intensity was increased with the increase in pH (Fig. 18.7b). The LCC fractions from the cacao husk (H-Fr.1) reproducibly showed higher radical intensity than the polysaccharide-rich fractions (H-Fr.2) in two independent experiments (Exps.1 and 2).

When the superoxide anion (generated by HX-XOD reaction) was mixed with DMPO, four radical peaks derived from the spin adduct (DMPO-OH) were detected by ESR spectroscopy (data not shown). Sodium ascorbate (vitamin C) (2.5 μ(micro)M) alone did not show any superoxide scavenging

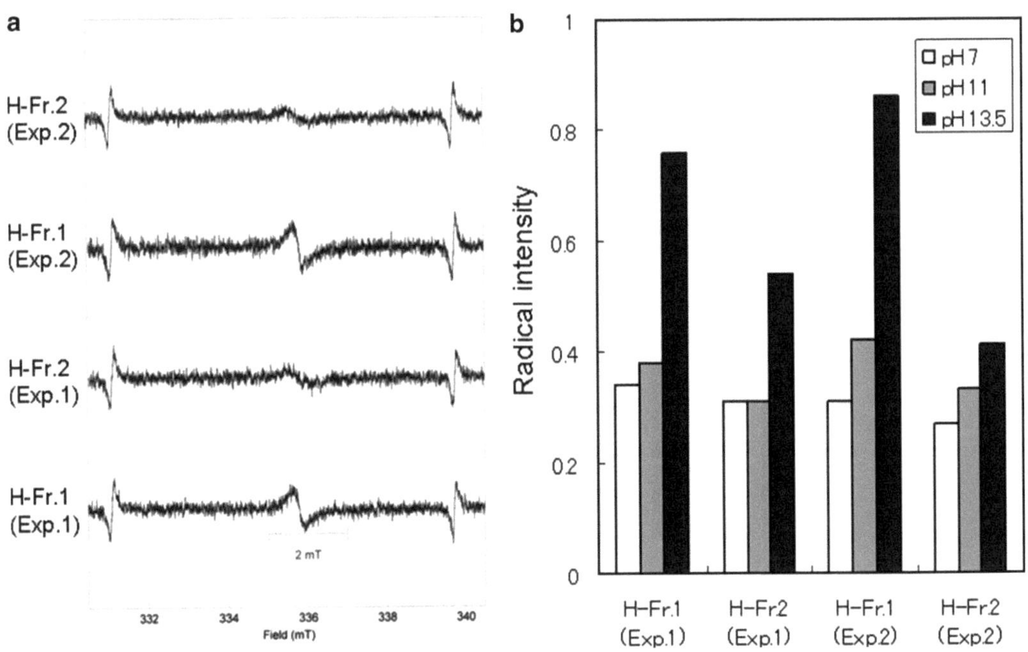

Fig. 18.7 Radical production by cacao husk lignin fractions. (**a**) ESR spectra of cacao husk Fr.1 and Fr.2 (2 mg/mL) prepared by two different experiments (Exp.1 and Exp.2) were measured in 0.1 M KOH (pH 13.5). (**b**) Radical intensity as a function of pH (Data cited from reference [9]. With permission)

Table 18.2 Synergistic superoxide scavenging activity of cacao husk lignin fractions and vitamin C (VC)[a]

	DMPO-OOH radical intensity (% of control)		
Sample	100 μg/mL sample	50 μg/mL sample + 1.25 μM vitamin C	
H-Fr.1 (Exp. 1)	40.8	54.5 < 70.4 [(100 + 40.8)/2]	Synergism
H-Fr.2 (Exp. 1)	91.1	49.1 < 74.6 [(100 + 91.1)/2]	Synergism
H-Fr.1 (Exp. 2)	36.0	46.7 < 68.0 [(100 + 36.0)/2]	Synergism
H-Fr.2 (Exp. 2)	93.6	88.3 < 96.8 [(100 + 93.6)/2]	Synergism
2.5 μM vitamin C	100.0		

[a]Data cited from reference [9]. With permission

Table 18.3 Synergistic hydroxyl radical scavenging activity of cacao husk lignin fractions and vitamin C (VC)[a]

	DMPO-OH radical intensity (% of control)		
Sample	100 μg/mL sample	50 μg/mL sample + 1.25 μM vitamin C	
H-Fr.1	72.4	39.2 < 53.0	Synergism
H-Fr.2	82.0	61.4 > 57.8	
H-Fr.1	70.4	46.2 < 52.0	Synergism
H-Fr.2	91.4	49.1 < 62.5	Synergism
2.5 μM vitamin	33.6		

[a]Data cited from reference [9]. With permission

activity, but in combination with the cacao husk lignin fractions (H-Fr.1, H-Fr.2), the superoxide scavenging activity was synergistically enhanced (Table 18.2).

When the hydroxyl radical (generated by the Fenton reaction) was mixed with DMPO, four radical peaks derived from the spin adduct (DMPO-OH) appeared (data not shown). Vitamin C and the cacao husk lignin fractions (H-Fr.1, H-Fr.2) synergistically scavenged the hydroxyl radical (Table 18.3). Similarly, LCCs from mulberry juice [14] and pine cone extract [8] synergistically enhanced the radical scavenging activity of vitamin C.

We have previously reported that pine cone LCC fraction enhanced both the radical intensity and the cytotoxic activity of vitamin C [15]. The synergism between lignin and vitamin C seems to be a universal phenomenon, suggesting its potential application in cosmetics.

Effect of Cacao Husk on Cigarette Smoke-Induced Cytotoxicity

LCC Samples Prepared by Method I

Cacao husk LCC (H-Fr.1) 100–1,000 μ(micro)g/mL stimulated the proliferation of human normal gingival fibroblast (HGF), but not that of human oral squamous cell carcinoma cell line (HSC-2). Lower concentration of H-Fr.1 (10 μ(micro)g/mL) inhibited the cytotoxicity of cigarette smoke (captured in phosphate buffer, Fig. 18.8a and b) against both HGF (see Fig. 18.8c) and HSC-2 (see Fig. 18.8d) cells. On the other hand, higher concentration (1,000 μ(micro)g/mL) of LCC stimulated the cytotoxicity of cigarette smoke in both cells (see Fig. 18.8c and d) (Sakagami et al., unpublished data).

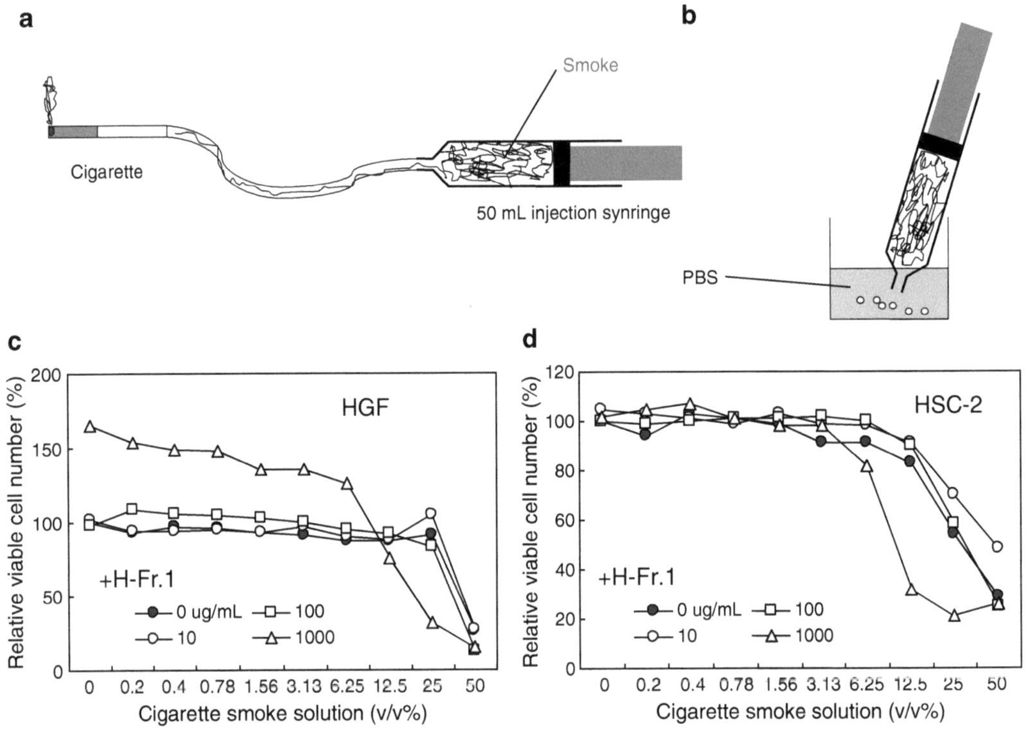

Fig. 18.8 Effect of cacao husk LCC (H-Fr.1) on the cytotoxicity induced by cigarette smoke. Cigarette smoke was corrected by syringe (**a**) and bubbled into phosphate-buffered saline (PBS) (**b**) and then added to normal human gingival fibroblast (HGF) (**c**) and oral squamous carcinoma cells (HSC-2) (**d**)

Macrophage Activation

LCC Samples Prepared by Method II

LPS, above 1 or 10 ng/mL, stimulated the NO production by RAW264.7 and J774.1 cells, respectively (data not shown). Cacao mass and husk LCC samples (7.8–1,000 μ(micro)g/mL), prepared by either single or repeated cycles of alkaline solubilization and acid precipitation, showed similar patterns of NO production by macrophages (Fig. 18.9a–d). However, the apparent NO production observed at LCC concentrations indicated by arrows was due to the action of the contaminating LPS. LPS contamination was more serious in cacao husk LCC. Cacao husk LCC (7.8 μ(micro)g/mL) contained approximately 2.9 ng/mL LPS, a concentration enough to induce NO production. NO determination by Griess method is influenced by colored materials. Both LCC preparations have brownish color and produce pseudopositive reaction, and these values were subtracted from the observed values.

Western blot analysis demonstrated that cacao mass LCC (M-Fr.4) (0.015–50 μ(micro)g/mL) alone did not stimulate iNOS protein expression in RAW264.7 cells but enhanced the LPS-induced iNOS protein expression (Fig. 18.10a). On the other hand, cacao husk LCC (50 μ(micro)g/mL) alone induced iNOS protein expression slightly, possibly due to higher level of LPS contamination (see Fig. 18.10b). But in this case also, combination of cacao husk and LPS induced much higher levels of iNOS protein expression (see Fig. 18.10b). This data suggests that the action point of cacao LCC may be distinct from LPS.

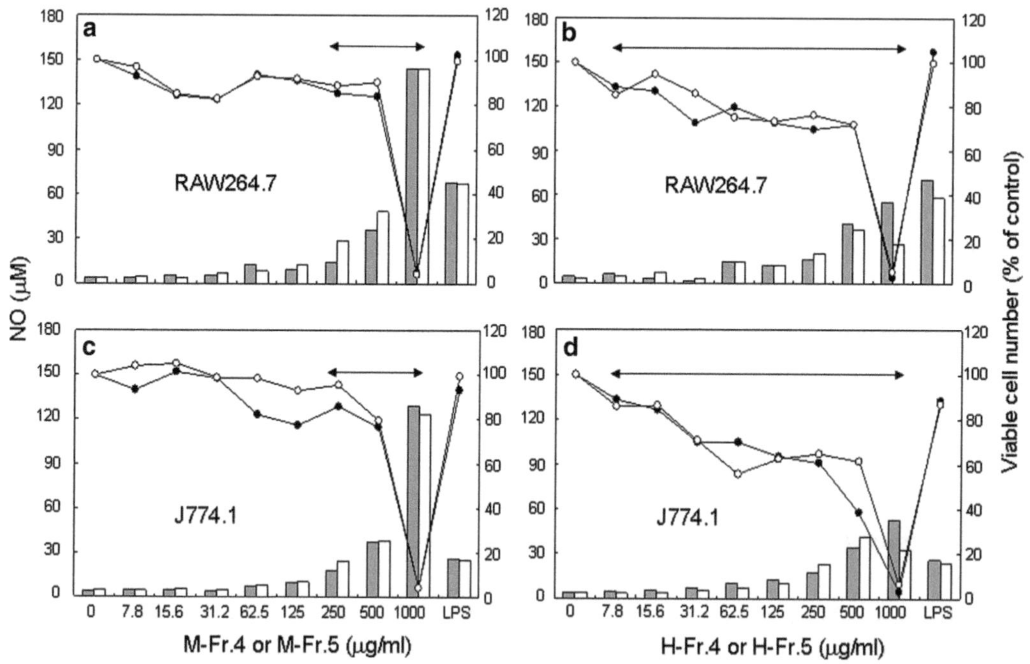

Fig. 18.9 Effect of cacao LCCs on NO production by mouse macrophage-like cells. RAW264.7 (**a**, **b**) and J774.1 cells (**c**, **d**) were incubated for 24 h with the indicated concentrations of cacao mass LCC (M-Fr.4, M-Fr.5) (**a**, **c**) or cacao husk LCC (H-Fr.4, H-Fr.5) (**b**, **d**) (prepared by Method II), and the extracellular NO concentration (*bar*) and relative viable cell number (*circle*) were determined by Griess and MTT method, respectively. *White* and colored symbols represent first (Fr.4) and second precipitation (Fr.5), respectively. *Arrows* indicate the range of LCC concentration, the effect of which on NO production is expected to be significantly affected by LPS contamination. Each value represents mean from duplicate experiments (Data for M-Fr.4 (**a**) and M-Fr.5 (**c**) cited from Reference [10]. With permission)

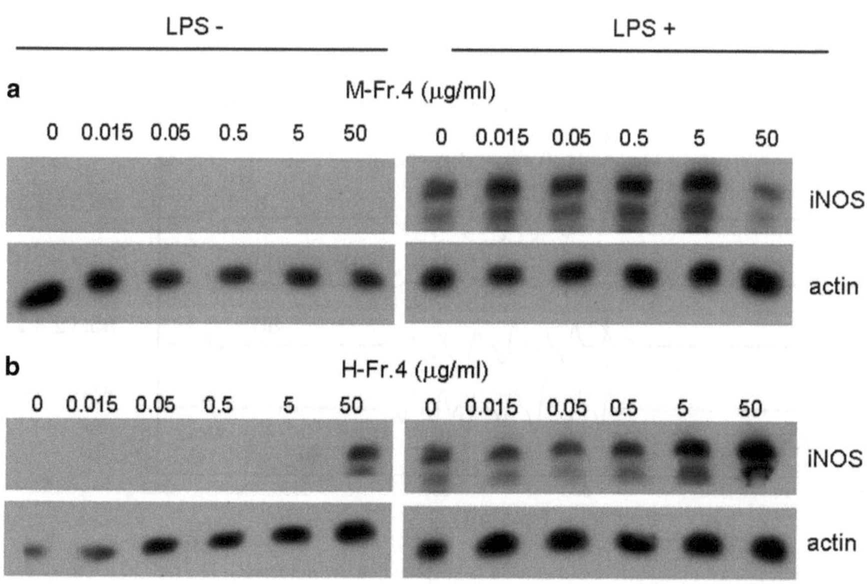

Fig. 18.10 Stimulation of LPS-induced iNOS protein expression by cacao LCCs (prepared by Method II). RAW264.7 cells were incubated for 24 h with the indicated concentrations of cacao mass (M-Fr.4) (**a**) or husk LCCs (H-Fr.4) (**b**) in the presence or absence of LPS (100 ng/mL) and processed for Western blot analysis (Data for M-Fr.4 (**a**) cited from Reference [10]. With permission)

No Radical Scavenging Activity

LCC Samples Prepared by Method II

Cacao mass and husk LCC fractions (M-Fr.4, H-Fr.4) (prepared by Method II) scavenged NO radical, generated from NOC-7 in the presence of carboxy-PTIO (Fig. 18.11). The 50% effective concentration of M-Fr.4 and H-Fr.4 was calculated to be 53 and 172 μ(micro)g/mL, respectively. In contrast, LPS (up to 250 μ(mu)g/mL) did not scavenge superoxide anion, hydroxyl radical, or NO radical. Combination of LPS and vitamin C did not produce synergistic superoxide radical scavenging activity (Sakagami et al., data not shown).

Effect on Cytokine Production

LCC Samples Prepared by Method II

RAW264.7 and J774.1 cells spontaneously produced TNF-α (alpha) into the culture medium (Fig. 18.12a–d). As compared with RAW264.7 cells (see Fig. 18.12a and b), J774.1 cells spontaneously produced one order higher amount of TNF-α (alpha) (see Fig. 18.12c and d), in consistent with previous report [16].

Cacao husk LCC (H-Fr.4, H-Fr.5) (7.8–1,000 μ(micro)g/mL), prepared by either single or two cycles of alkaline extraction and acid precipitation, induced much higher TNF-α (alpha) production

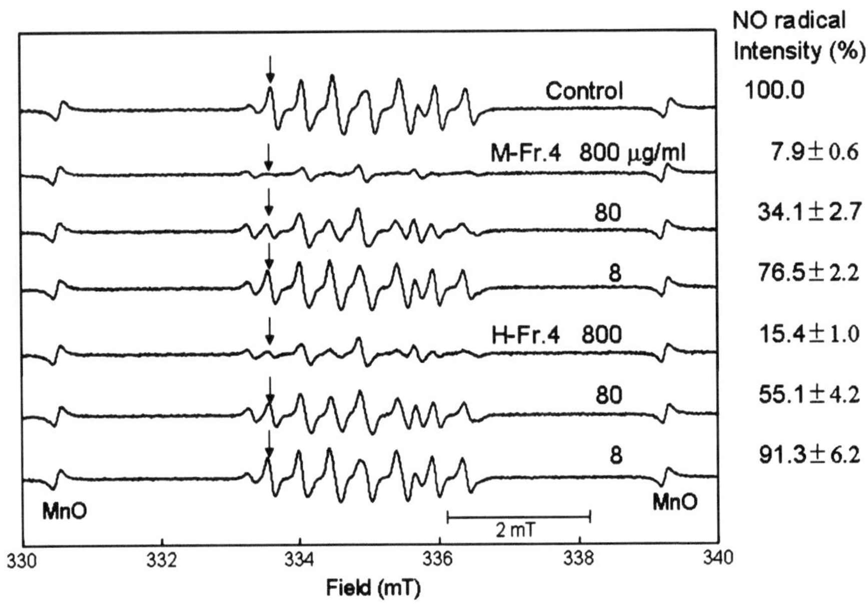

Fig. 18.11 NO radical scavenging activity of cacao LCCs (prepared by Method II). The radical intensity of NO generated from NOC-7 in the presence of carboxy-PTIO without (control) or with 8, 80, or 800 μg/mL of cacao mass (M-Fr.4) or husk LCC (H-Fr.4) was measured with ESR spectroscopy. The NO radical intensity was determined by the second carboxy-PTI peak indicated by *arrows* and expressed as percent of control. Each value represents mean ± SD values from three independent experiments (Data for M-Fr.4, but not for H-Fr.4 cited from reference [9]. With permission)

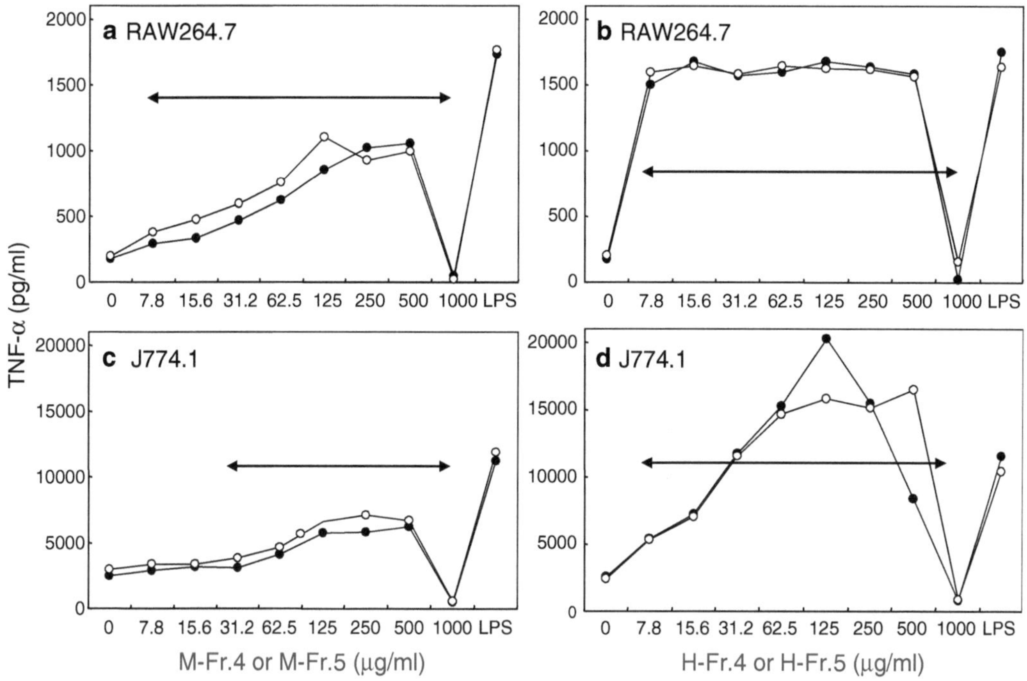

Fig. 18.12 Effect of cacao LCCs on TNF-α (alpha) production by mouse macrophage-like cells. RAW264.7 (**a**, **b**) or J774.1 cells (**c**, **d**) were incubated for 24 h with the indicated concentrations of cacao mass LCC (M-Fr.4 or M-Fr.5) (**a**, **c**) or cacao husk LCC (H-Fr.4 or H-Fr.5) (**b**, **d**), and the extracellular TNF-α concentration (*bar*) was determined by ELISA. *White* and *gray* symbols represent first (M-Fr.4, H-Fr.4) and second (M-Fr.5, H-Fr.5) precipitation samples. *Arrows* indicate the range of LCC concentration, the effect of which on TNF-α (alpha) production is due to contaminating LPS. Each value represents mean from duplicate experiments. Either cacao mass or husk LCC produced only background level of pseudopositive reaction for TNF-α (alpha) determination (<16 and <8 pg/mL TNF-α (alpha), respectively) (data not shown) (Data for M-Fr.4 and M-Fr.5 (**a**, **c**) cited from reference [10]. With permission)

in both RAW264.7 and J774.1 cells (see Fig. 18.12b and d) than cacao mass LCC (M-Fr.4, M-Fr.5) (see Fig. 18.12a and c). However, this apparent TNF-α (alpha) induction by LCCs was due to the action of contaminating LPS. Either cacao mass or husk LCC produced only background level of pseudopositive reaction for TNF-α (alpha) determination, in the absence of cells (<16 and <8 pg/mL TNF-α (alpha), respectively) (data not shown).

LPS (100 ng/mL) significantly ($p < 0.05$) stimulated the IL-1β (beta) (a, b), IFN-α (alpha) (c, d), and IFN-γ (gamma) (e, f) production by RAW264.7 cells (Fig. 18.13). On the other hand, cacao mass LCC (M-Fr.4) did not increase but rather inhibited these cytokine production at higher concentration (a, c, e). Cacao husk LCC (H-Fr.4) at higher concentrations stimulated the production of IL-1β (beta) (b) and IFN-α (alpha) (d), possibly due to the action of contaminating LPS, but not that of IFN-γ (gamma) (f) (see Fig. 18.13).

Effect of LPS Contamination

There is a possibility that cacao mass and husk may have been originally polluted with LPS that is an important component of the outer membrane of Gram-negative bacteria because the bacteria are widely distributed into the natural environment such as water (rivers and sea), air, and soil. In addition,

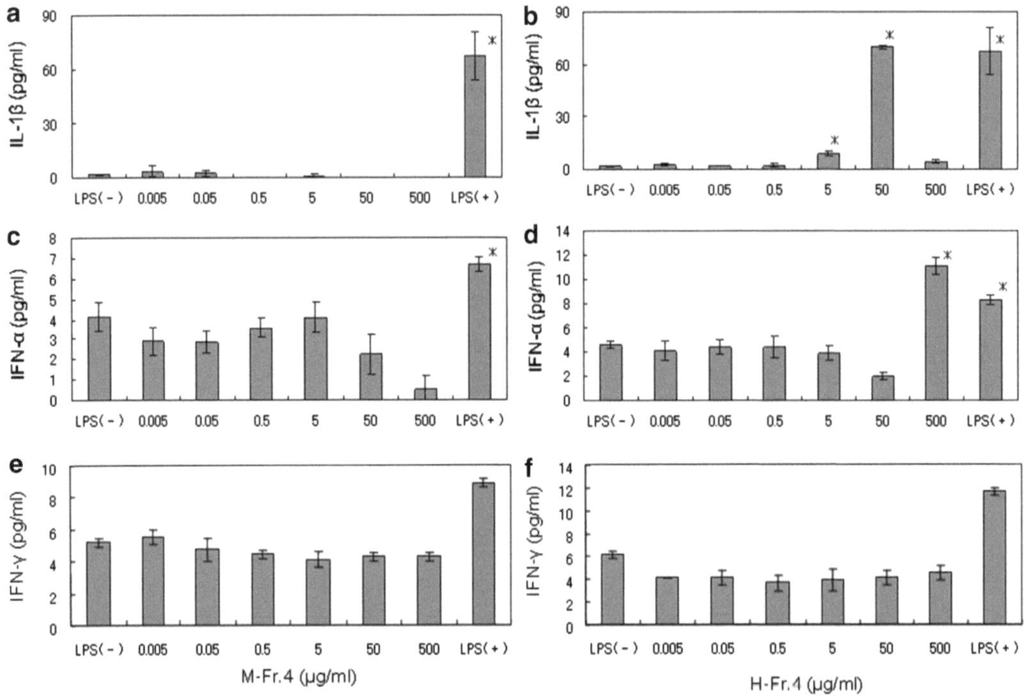

Fig. 18.13 Effect of cacao LCCs on production of other cytokines by RAW264.7 cells. RAW264.7 cells were incubated for 24 h with various concentrations of cacao mass (M-Fr.4) (**a**, **b**, **c**) or cacao husk LCCs (H-Fr.4) (**b**, **d**, **f**), or LPS (100 ng/mL), and extracellular concentration of IL-1β (beta) (**a**, **b**), IFN-α (alpha) (**c**, **d**), and IFN-γ (gamma) (**e**, **f**) was then determined by ELISA. Each value represents mean ± SD of triplicate determinations. Either cacao mass or husk LCC produced only background level of pseudopositive reaction for IL-1β (beta), IFN-α (alpha), and IFN-γ (gamma) determination (data not shown)* <0.05 (Data for M-Fr.4 (**a**, **c**, **e**) cited from reference [10]. With permission)

the contaminated LPS is similarly extracted with alkaline solution and precipitated with acid during the isolation step of LCC. This may explain why cacao husk LCC (outer shell of cacao beans) has been contaminated with LPS more heavily than cacao mass LCC. Higher concentration (100 μ(micro) g/mL) of LPS contamination did not affect the anti-HIV activity, whereas much lower concentration (0.01 ng/mL) of LPS contamination seriously affected the cytokine production in macrophage-like cells. Most of previous studies have not paid attention to such LPS contamination in the LCC preparations. Alkaline extraction step that is necessary for the preparation of LCC has both merit and demerit. Merit is the chemical inactivation of LPS. Demerit is the degradation of LCC into its smaller size. Therefore, the conditions for alkaline extraction should be optimized to maximize the LPS inactivation and minimize the loss of biological activity.

Summary

The present study demonstrated that highly solubilized and sterilized cacao mass LCC, manufactured by autoclave treatment under mild alkaline condition, showed higher anti-HIV activity (SI = 39.0–61.3) than that of LCCs prepared from cacao husk (SI = 28.0–44.5) and other eight plant species (SI = 26.8 ± 30.0) [8]. This was not due to LPS contamination since the EC_{50} of cacao mass is approximately 4 μg/mL (see Fig. 18.2), and LPS (100 μg/mL) showed no anti-HIV activity (see Table 18.1).

The present study confirmed our previous finding that cacao husk LCC stimulated RAW264.7 cells to produce TNF-α (alpha) [9] and further showed that both cacao mass and husk LCCs stimulated both RAW264.7 and J774.1 cells to produce not only TNF-α (alpha) (see Fig. 18.12) but also NO (see Fig. 18.9). Only cacao husk LCC, but not cacao mass LCC, stimulated IL-1β (beta) and IFN-α (alpha) production (see Fig. 18.13b and d). However, this apparent stimulation of cytokine and NO production was mostly due to the action of contaminating LPS.

Our recent DNA microarray analysis demonstrated that relatively higher concentrations of LCC from *Lentinus edodes* mycelia extract (LEM) induced the expression of various immune response-related genes, most of which were overlapped with that induced by LPS [17]. LPS has been reported to induce the production of cytokines by Toll-like receptor (TLR) signaling pathway through TLR4 [18] and activated Janus kinase 2 (JAK2), which compose JAK-STAT (signal transducer and activator of transcription) signaling pathway [19]. Both LEM-LCC and LPS modified the immune response-related gene expression, but LPS more strongly affected the immune response-related gene expression than LEM-LCC [17]. Recently, we found that LCC from LEM, but not LPS, significantly enhanced the expression of dectin-2 gene [20]. It remains to be investigated whether cacao mass LCC induces similar changes in gene expressions and whether lower concentrations of LCCs show different effects, as compared with those observed at higher concentrations.

We have also found that cacao mass (0.015–50 μg/mL) and husk LCCs (0.015–5 μg/mL) alone did not stimulate iNOS protein expression but enhanced the iNOS protein expression triggered by LPS in RAW264.7 cells (see Fig. 18.10). This suggests that the action point of LCCs may differ from that of LPS. We found that cacao husk LCC (50 μg/mL) that contains approximately 80 ng/mL LPS induced only one-fourth of NO production that is expected (see Fig. 18.9b). This reduction of extracellular NO level may be in part due to the NO scavenging activity of cacao husk LCC (see Fig. 18.11). This suggests that cacao husk that contains higher amount of LPS than cacao mass LCC produces NO, but some of which is scavenged by phenylpropenoid portions of LCC.

We have previously reported that protein-bound polysaccharide, PSK, stimulated the differentiating-inducing activity of TNF-α [21] and IFN-γ [22] against human myelogenous leukemic cells toward maturing macrophage-like cells. It remains to be investigated whether cacao LCCs may stimulate the biological activity of TNF-α (alpha) and IFN-γ (gamma).

We have demonstrated that LCCs, present in huge amount in the natural kingdom, showed extremely higher antiviral activity than other lower molecular weight polyphenols such as tannins and flavonoids and synergistic effects with vitamin C. These promising properties of LCC suggest its possible application to various virus-related diseases. In fact, lignin and vitamin C tablet treatment successfully reduced the severity of symptoms and the reduction in the recurrent episodes of HSV-1-infected patients [23].

Acknowledgements This study was supported in part by a Grant-in-Aid from the Ministry of Education, Science, Sports and Culture of Japan (Sakagami, No.19592156).

References

1. Motohashi N, Kawase M, Kurihara T, Shirataki Y, Kamata K, Nakashima H, et al. Relationship between radical intensity and biological activity of cacao husk extracts. Anticancer Res. 1999;19:1125–30.
2. Osakabe N. Polyphenols and atherosclerosis. Vitamin. 2006;80:275–80, (in Japanese).
3. Matsumoto M, Tsuji M, Okuda J, Sasaki H, Nakano K, Osawa K, et al. Inhibitory effects of cacao been husk extract on plaque formation *in vitro* and *in vivo*. Eur J Oral Sci. 2004;112:249–52.
4. Unten S, Ushijima H, Shimizu H, Tsuchie H, Kitamura T, Moritome N, et al. Effect of cacao husk extract on human immunodeficiency virus infection. Lett Appl Microbiol. 1991;14:251–4.
5. Hatano T, Miyatake H, Natsume M, Osakabe N, Takizawa T, Ito H, et al. Proanthocyanidin glycosides and related polyphenols from cacao liquor and their antioxidant effects. Phytochemistry. 1998;59:749–58.

6. Machida N, Suzuki M, Kondo K. Polyphenol. J Pract Pharm. 2006;57(suppl):230–8, (in Japanese).
7. Lewis NG, Yamamoto E. Lignin. Occurrence, biogenesis and biodegradation. Annu Rev Plant Physiol Plant Mol Biol. 1990;41:455–96.
8. Sakagami H, Kushida T, Oizumi T, Makino T. Distribution of lignin carbohydrate complex in plant kingdom and its functionality as alternative medicine. Pharmacol Ther. 2010;128:91–105.
9. Sakagami H, Satoh K, Fukamachi H, Ikarashi T, Simizu A, Yano K, et al. Anti-HIV and vitamin C-synergized radical scavenging activity of cacao husk lignin fractions. In Vivo. 2008;22:327–32.
10. Sakagami H, Kawano M, Thet MM, Hashimoto K, Satoh K, Kanamoto T, et al. Anti-HIV and immunomodulation activities of cacao mass lignin carbohydrate complex. In Vivo. 2011;25:229–36.
11. Fukai T, Sakagami H, Toguchi M, Takayama F, Iwakura I, Atsumi T, et al. Cytotoxic activity of low molecular weight polyphenols against human oral tumor cell lines. Anticancer Res. 2000;20:2525–36.
12. Shirataki Y, Motohashi N, Tani S, Sakagami H, Satoh K, Nakashima H, et al. *In vitro* biological activity of prenylflavanones. Anticancer Res. 2001;21:275–80.
13. Nakashima H, Murakami T, Yamamoto N, Naoe T, Kawazoe Y, Konno K, et al. Lignified materials as medicinal resources. V. Anti-HIV (human immunodeficiency virus) activity of some synthetic lignins. Chem Pharm Bull. 1992;40:2102–5.
14. Sakagami H, Asano K, Satoh K, Takahashi K, Kobayashi M, Koga N, et al. Anti-stress, anti-HIV and vitamin C-synergized radical scavenging activity of mulberry juice fractions. In Vivo. 2007;21:499–506.
15. Satoh K, Ida Y, Ishihara M, Sakagami H. Interaction between sodium ascorbate and polyphenols. Anticancer Res. 1999;19:4177–86.
16. Ito A, Ohsawa F, Natori S. Purification of a cytotoxic protein produced by the murine macrophage-like cell line J774.1 in response to Sarcophaga lectin. J Biochem. 1986;99:9–15.
17. Kawano M, Thet MM, Makino T, Kushida T, Sakagami H. DNA microarray analysis of signaling pathway in macrophages stimulated by lignin-carbohydrate complex from *lentinus edodes* mycelia extract. Anticancer Res. 2010;30:2567–76.
18. Beutler B, Rietschel ET. Innate immune sensing and its roots: the story of endotoxin. Nat Rev Immunol. 2003;3:169–76.
19. Kimura A, Naka T, Muta T, Takeuchi O, Akira S, Kawase I, et al. Suppressor of cytokine signaling-1 selectively inhibits LPS-induced IL-6 production by regulating JAK-STAT. Proc Natl Acad Sci USA. 2005;102:17089–94.
20. Kushida T, Makino T, Tomomura M, Tomomura A, Sakagami H. Enhancement of dectin-2 gene expression by lignin-carbohydrate complex from Lentinus edodes extract (LEM) in mouse macrophage-like cell line. Anticancer Res. 2011;31:1241–8.
21. Sakagami H, Ikeda M, Konno K. Stimulation of tumor necrosis factor-induced human myelogenous leukemic cell differentiation by high molecular weight PSK subfraction. Biochem Biophys Res Commun. 1989;162:597–603.
22. Kim F, Sakagami H, Tanuma S, Konno K. Stimulation of interferon-γ-induced human myelogenous leukemic cell differentiation by high molecular weight PSK subfraction. Anticancer Res. 1990;10:55–8.
23. López BSG, Yamamoto M, Utsumi K, Aratsu C, Sakagami H. Clinical pilot study of lignin-ascorbic acid combination treatment of herpes simplex virus. In Vivo. 2009;23:1011–6.

Part IV
Disease-Related Clinical Studies

Chapter 19
Clinical Benefits of Cocoa: An Overview

Margarida Castell, Francisco Jose Pérez-Cano, and Jean-François Bisson

Key Points

- Cocoa intake enhances antioxidant defenses quickly and over a short period after ingestion.
- Epidemiological studies indicate that cocoa has a cardioprotective effect by improving endothelial function and decreasing platelet aggregation and blood pressure.
- Clinical evidence suggests that cocoa can be a new and interesting food for regulating mood and brain disorders.
- Cocoa flavonoids have in vitro anti-inflammatory effects, and preclinical studies show this potential. The immunomodulatory power of cocoa, demonstrated preclinically, may be beneficial in reducing certain states of autoimmunity and hypersensitivity.
- Although in vitro studies have shown that cocoa flavonoids exert antitumoral effects, further studies are needed.

Keywords Cocoa • Flavonoids • Antioxidant • Cardiovascular diseases • Nervous system • Immune system • Cancer

Introduction

Chocolate was considered medicinal in ancient times [1]. Following on the Olmecs, Mayans, and Aztecs, the Spanish and other Europeans identified many medicinal uses [2]. Chocolate consumption has also long been associated with pleasure. Popular claims confer on chocolate the properties of being a stimulant, relaxant, euphoriant, aphrodisiac, tonic, and antidepressant [3].

M. Castell, Ph.D. (✉) • F.J. Pérez-Cano, Ph.D.
Departament de Fisiologia, Facultat de Farmàcia, Universitat de Barcelona,
Av. Joan XXIII s/n Edifici B, 3a planta, Barcelona 08041, Spain
e-mail: margaridacastell@ub.edu; franciscoperez@ub.edu

J.-F. Bisson, Ph.D.
Department of Cancerology, Human Pathologies & Toxicology, Centre de Recherche ETAP,
13 rue du Bois de la Champelle, Technopôle de Nancy-Brabois, Vandœuvre-lès-Nancy 54500, France
e-mail: jbisson@etap-lab.com

R.R. Watson et al. (eds.), *Chocolate in Health and Nutrition*, Nutrition and Health 7,
DOI 10.1007/978-1-61779-803-0_19, © Springer Science+Business Media, LLC 2013

In recent years, the consumption of cocoa has been associated with healthy effects preventing oxidative stress. Throughout this chapter, and indeed the whole book, the beneficial actions of cocoa products on the cardiovascular system and central nervous system are presented; moreover, experimental studies suggest that cocoa has positive effects on the modulation of cancer development and immune function, including inflammatory process. Here we focus on the recent literature showing the clinical effects of cocoa and, where clinical evidence has not already been reported, on experimental models in vivo.

Evidence of Antioxidant Properties of Cocoa

Many of the beneficial effects of chocolate are associated with the antioxidant effects of the polyphenols contained in cocoa. These polyphenols, mainly flavanols such as catechin, epicatechin, and procyanidins, give chocolate antioxidant activity. In addition, condensed tannins (proanthocyanidins) may also become bioactive antioxidants after colonic fermentation [4, 5]. One serving of dark chocolate is thought to impart a greater antioxidant capacity than the average amount of antioxidants consumed daily in the United States [6]. However, there is some dispute about its antioxidant action when cocoa is consumed with milk: some studies reported a decrease of this activity when added to milk [4], while other studies suggest that milk only lowered the excretion of some urinary metabolites [7].

The antioxidant activity of cocoa flavanols is attributed to their capacity to neutralize free radicals, inhibit the enzymes responsible for reactive oxygen species (ROS) production, chelate metals, and upregulate antioxidant defenses [8]. Quercetin, a minor flavonoid in cocoa, and other nonflavonoid compounds present in cocoa, particularly methylxanthines, also contribute to cocoa's antioxidant activity by neutralizing radicals and chelating metal ions [9]. On the other hand, it has been proposed that ingested flavonoids produce no direct antioxidant effect in vivo but modify protein kinases mediating signal transmission, thus inducing antioxidant gene expression or inhibiting oxidant gene expression [10]. However, despite these antioxidant characteristics, flavonoids in excess or in the presence of redox-active metals can become prooxidants [11].

A large number of in vitro studies have demonstrated the antioxidant properties of cocoa [8, 12]. Going beyond in vitro assays, a smaller number of studies have investigated the effects of cocoa in vivo. Cocoa intake increases total antioxidant capacity (TAC) and decreases lipid oxidation products in plasma [13] and tissues [14], an effect that can be attributed to flavonoid accumulation [15]. Cocoa intake is able to induce antioxidant enzymes such as superoxide dismutase and glutathione peroxidase [16]. Cocoa also improves antioxidant defenses in oxidative stress situations. Thus, a long-term diet supplemented with cocoa fiber reduces lipid peroxidation in hypercholesterolemic rats [17], and cocoa supplementation reduces plasma 8-isoprostane in obese diabetic rats [18].

In addition to the in vivo assays in animals, the antioxidant power of cocoa and chocolate has been assessed in dietary intervention trials in humans [19]. In the plasma of healthy volunteers consuming dark chocolate, increased TAC and decreased presence of lipid oxidation products have been reported [20]. Similar results were found in volunteers who consumed procyanidin-rich chocolate [21]. In these studies, enhancement of TAC was greatest 2 h after chocolate ingestion and returned to basal values 6 h after cocoa intake, probably because of the short plasma half-life of flavonoids and their uptake in cells. A crossover trial on 12 healthy volunteers consuming dark chocolate, dark chocolate with full-fat milk, or milk chocolate showed that 1 h after the ingestion of dark chocolate, plasma TAC values increased by 20%, whereas no changes were reported after ingestion of dark chocolate with milk or milk chocolate [22].

An interesting report shows the action of cocoa on prooxidant situations. The consumption of 100 g of chocolate (0.2% polyphenols) for 2 weeks counteracted oxidative stress in soccer players, as was shown by reductions in plasma malondialdehyde and α-tocopherol increases [23].

There have also been studies that failed to show any antioxidant effects of cocoa, although there were significant health outcomes, as mentioned in the next sections. The lack of such evidence may be due to the fast metabolism of flavonoids or alternatively to the healthy status of the subjects recruited in the study [24].

Cocoa for Cardiovascular Health

Over the last few years, clinical trials have indicated that cocoa-derived products are able to reduce some factors involved in the risk of cardiovascular disease: high blood pressure (BP), excessive platelet activation, and high LDL concentration and oxidation, among others.

Dark chocolate consumption has been associated with lower BP [25–28], although some trials have also reported conflicting results [29, 30]. Recent meta-analyses including 10–13 trials conducted in hypertensive and normotensive individuals in the last 15 years have concluded that dark chocolate is superior to a placebo in reducing systolic hypertension or diastolic prehypertension, and as a result, the authors felt that cocoa products could be recommended as a treatment option for hypertension [31, 32]. However, most of these studies included a relatively small number of subjects, and their results are questionable due to their lack of rigor or parallel bioavailability studies [33, 34]. Davinson et al. [35], using four different doses of cocoa flavanols, report the reduction of the mean arterial BP in 52 postmenopausal women with untreated mild hypertension, although no evidence of dose response was observed. The lowering BP activity of cocoa has also been found in patients with coronary artery disease (CAD) [36]. Blood pressure reduction after the consumption of a high dose of dark chocolate for 3 months in a study performed in 102 patients with prehypertension/stage 1 hypertension and established cardiovascular end-organ damage or diabetes mellitus has been reported [32]. On the other hand, it has been suggested the role of theobromine in decreasing BP. However, Van Den Bogaard et al. [37], in a double-blind placebo trial using flavanol-rich cocoa beverages with natural or added theobromine, concluded that although after 2 h of consumption the central systolic BP was significantly lowered by the theobromine-added beverage, the 24-h ambulatory or central BP was not affected.

The BP-lowering properties of cocoa may be associated in part to modulation of endothelial function [38, 39]. In this regard, a recent study associates the action of flavanols with the optimal nitric oxide (NO) concentrations that produce vasodilatation and therefore a decrease in BP [40]. Cocoa flavanols attenuated an exercise-induced increase in BP in a study of 21 volunteers after a single serving by improving endothelium-dependent flow-mediated dilatation (FMD) [41]. However, other studies have shown an effect on FMD without modulation of BP [42]. Moreover, in recent years, some mechanisms for this effect have been suggested. Heiss et al. [36] recently demonstrated that in patients with CAD, the improvement in endothelial function associated with cocoa flavanols is mediated, after a dietary high-flavanol intervention lasting for 30 days, by an enhancement in the number and function of circulating angiogenic cells. Positive effects on the reduction of platelet aggregation by dark chocolate have been reported as well, but neutral effects have also been found [43].

The saturated fat present in dark chocolate does not adversely affect the blood lipid profile. On the contrary, and due to its richness in polyphenols, both the short- and long-term interventions, performed until now with cocoa, evaluating this aspect seem to significantly reduce serum LDL and total cholesterol (TC) and increase HDL concentrations [44, 45]. In most studies, consumption of cocoa sources induced a reduction in LDL and TC in hypercholesterolemic individuals [43] or in healthy subjects [46].

Moreover, the effects of chronic cocoa consumption on inflammatory cellular and serum biomarkers related to atherosclerosis, an important factor in the development of heart disease, have also been reported [47]. A recent trial involving volunteers with a high risk of cardiovascular diseases taking cocoa with skimmed milk for 4 weeks showed the effect of cocoa on increasing plasma HDL and decreasing plasma oxidized LDL concentrations [48]. Therefore, this action adds new evidence to the beneficial role of cocoa flavonoids in preventing cardiovascular diseases.

Overall, epidemiological studies suggest that cocoa has a clear cardiovascular-protective effect by improving endothelial function and decreasing platelet aggregation and blood pressure. However, more studies focused on establishing optimal doses are required.

Does Cocoa Affect the Nervous System? Clinical Evidence

Scientific proof of the beneficial effects of chocolate and cocoa on the nervous system has emerged over the last decade [49]. Cocoa can have beneficial effects on cognitive function and mood and can also protect nerves from injury and inflammation [50].

Eating chocolate could help to sharpen the mind and give a short-term boost to cognitive skills. It has been reported that the consumption of a cocoa drink boosts blood flow to key areas of the brain for 2–3 h [51]. Increased blood flow to these areas of the brain may help to increase performance in specific tasks and boost general alertness over a short period. These findings raise the prospect of ingredients in chocolate being used to treat vascular impairment, including ischemic cerebrovascular syndromes, dementia, and strokes. The study also suggests that cocoa flavanols could be useful in enhancing brain function in situations where individuals are cognitively impaired such as fatigue, sleep deprivation, or possibly aging. It is suggested that these various independent observations of the effect on blood vessels of drinking flavanol-rich foods could be because of the increase in circulating NO, as mentioned above.

Preliminary or pilot evidence has shown that flavanol-rich cocoa can increase cerebral blood flow both in healthy elderly subjects [52] and in healthy young participants, as measured by functional magnetic resonance imaging (fMRI) in response to a cognitive task (the task-switching paradigm) [53]. It should be noted, however, that in the fMRI study, no significant effects were observed in participants' behavioral reaction times, the effect of switching between two sets of rules, or heart rates after the ingestion of the flavanol-rich cocoa. The authors hypothesized that the fMRI changes may have been related to cognitive changes that were not evident in the behavioral measures used in their project, especially in the young, healthy participants who were probably functioning at a high level of cognitive ability. However, a clinical study [54] failed to support the predicted beneficial effects of short-term consumption of dark chocolate and cocoa on any of the neuropsychological or cardiovascular health-related variables measured. This double-blind, placebo-controlled, fixed-dose, parallel-group clinical trial was performed on 101 healthy volunteer subjects who consumed a 37-g dark chocolate bar and about 240 mL of an artificially sweetened cocoa beverage or similar placebo products everyday for 6 weeks. No significant group-by-trial interactions were found for the neuropsychological (self-report history questionnaire assessing medical and psychiatric histories), hematological (coronary risk panel and C-reactive protein test), or BP variables examined.

It has been reported that, in the elderly, a diet high in some flavonoid-rich foods is associated with better performance in several cognitive abilities in a dose-dependent manner [55]. Other studies indicated that in young adults, in terms of cognitive performance, an acute dose of cocoa flavanols improved spatial memory and performance in some aspects of the choice reaction time task [56, 57]. These effects might be explained by the increased cerebral blood flow caused by cocoa but may also be due to induced retinal blood flow changes.

The effect of chocolate on fatigue appears to have been first described by the Aztec Emperor Montezuma II, who noted, "…the divine drink, which builds up resistance and fights fatigue. A cup of this precious drink [cocoa] permits man to walk for a whole day without food." The Badianus Codex published in 1552 noted the use of cocoa flowers to treat fatigue [2]. According to a recent study [58], consumption of a flavonoid-rich chocolate product eases the symptoms of chronic fatigue. Results of a small, double-blind, randomized, pilot crossover study on ten subjects with chronic fatigue syndrome (CFS) indicated that daily consumption of a high-cocoa liquor/polyphenol-rich chocolate significantly improved symptoms of chronic fatigue after 8 weeks, compared with a cocoa liquor polyphenol-free/low chocolate. Anandamide, previously called arachidonylethanolamide, has a structural similarity with tetrahydrocannibol, and other strongly related compounds that are found in cocoa. There are also compounds like N-acylethanolamines that block the breakdown of anandamide in cocoa. It may be the synergy of these compounds in chocolate that accounts for the results seen in this study. A number of biological systems have been implicated in CFS, and there is mounting evidence that oxidative stress contributes to the disease process and to some of the symptoms of the illness. The benefits of chocolate may be due to the flavonoids protecting cells like neuronal cells from oxidative stress. The brain is more vulnerable to oxidative stress than other organs due to its low-antioxidant protection system and the increased exposure of target molecules to ROS, one of the major damaging agents involved in age-associated decline [59]. Some parameters related to neurotransmission also decline during normal aging [60]. Cellular studies examining the potential mechanisms of neuroprotection by flavonoids have been published, demonstrating that epicatechin prevents neuronal cell death caused by oxidized LDL-induced oxidative stress. Furthermore, it has been shown that these neuroprotective mechanisms involve the modulation of the mitogen-activated protein kinase (MAPK) signaling cascade [61].

Chocolate contains many chemicals that can affect our mood: caffeine, tyramine, flavonoids, phenylethylamine, and others. A theory is that these natural chemicals raise serotonin and endorphin concentrations that make us feel good and have a calming effect on brain function, but this does not happen to everyone [3]. Numerous studies have explored the link between chocolate and mood including depression. Macht et al. [62] gave pieces of chocolate to subjects after viewing film clips that were chosen to induce anger, fear, sadness, and joy. Results indicated that the quality of emotions can affect motivation to eat and therefore responses to consuming chocolate. Later, it was reported that the mood-elevating properties of chocolate can be enhanced with intention [63]. According to another study [64], eating chocolate reduces negative mood compared to drinking water, whereas no or only marginal effects on neutral and positive moods were found.

A study based on a survey using a web-based questionnaire of 2,692 persons who were suffering from clinical depression [65] included 61% individuals classified as "cravers" with the majority being women. This group rated chocolate's capacity to improve their depressed mood as moderate to very important. The study says that, after eating chocolate, the volunteers were more likely to feel significantly less anxious and irritated. It also stated that the craving might predict atypical depression status. It overviewed chocolate's effects on mood state, noting its many psychoactive ingredients, including several biogenic stimulant amines, two analogues of anandamine (producing effects akin to cannabinoid-inducing euphoria), and interactions with several neurotransmitter systems (dopamine, serotonin, and endorphins). Some studies have suggested that carbohydrate craving is more closely linked to the opioid rather than to the serotoninergic system, with endorphins alleviating dysphoria. Thus, chocolate cravings may advance biological mechanisms potentially settling limbic cortex-mediated activation [3].

Phenylethylamine is chemically and pharmacologically related to catecholamines and amphetamine, and its deficit may contribute to a state of depression. The most important methylxanthines found in cocoa are caffeine and theobromine. Like carbohydrates, caffeine could be a self-medication for people who suffer from depressive symptoms. For example, sedation is an important symptom in depression, and methylxanthines can induce a benefit arousal through an interaction with adenosine receptors [66]. Cocoa contains several unsaturated N-acylethanolamines, which are structurally related to anandamide, and high levels of these substances could interact with other active compounds of chocolate and provoke a sensation of well-being [67]. Finally, magnesium, one of the most quantitatively significant minerals in cocoa, is known to be potentially effective for treating depression in relation to the intraneuronal magnesium deficits in depressive patients [68].

Another clinical study showed that depressed moods were significantly related to higher chocolate consumption, with findings being similar in both men and women [69]. Higher Center for Epidemiologic Studies Depression Scale (CES-D) scores within the provisionally depressed range (above the depression screen threshold) were associated with still greater chocolate consumption.

Clinical observations in Parkinson's disease (PD) patients also suggested an increased chocolate consumption [70]. 498 PD patients and their partners were evaluated through a structured self-questionnaire asking for consumption of chocolate and nonchocolate sweets, changes in chocolate consumption during the course of the disease, and depressive symptoms. Consumption of chocolate was significantly higher in PD patients compared to controls. Although reasons for increased chocolate consumption in PD remain elusive, it may hypothetically be a consequence of the high content of various biogenic amines potentially influencing brain monoamine metabolism and/or caffeine analogues with potential antiparkinsonian effects.

Another area of interest of the effects of chocolate on the nervous system is anxiety. A clinical trial performed on 30 human subjects classified into low and high anxiety traits using validated psychological questionnaires showed that daily consumption of 40 g of dark chocolate for up to 14 days reduced the urinary excretion of the stress hormones cortisol and catecholamines [71].

Polyphenols and flavanols, in particular, appear today to offer new and interesting opportunities to regulate mood and brain disorders. Thus, cocoa flavanols could enable us to enjoy the benefits of chocolate as a therapy, without the excessive and potentially adverse effects linked to carbohydrates and lipids. However, further studies are necessary in order to identify the active constituents in the nervous system among the various cocoa polyphenols and to understand their mechanism of action in the brain.

Modulation of Inflammatory Response and Immune Functions by Cocoa

In general, flavonoids are associated with an anti-inflammatory action. In this regard, the flavanols contained in cocoa have also been the object of studies, but more of these have been performed in vitro. The anti-inflammatory role of single cocoa flavonoids (epicatechin, catechin, procyanidins) has been ascertained and is focused on the secretion of inflammatory mediators such as cytokines, NO, and ROS by macrophages and other leukocytes [12]. However, there have been few in vivo and human studies until now.

In an ex vivo approach, it has been demonstrated that a cocoa diet in rats decreases the secretion of tumor necrosis factor alpha (TNF-α), interleukin 6 (IL-6), NO, and ROS from macrophages [72, 73]. More interestingly, there are studies suggesting the anti-inflammatory effect of cocoa on experimental models. The oral administration of a cocoa polyphenolic fraction to mice can inhibit ear edema in a dose-dependent manner [74]. Moreover, rats that received cocoa for a week developed a lower paw edema induced by carrageenan and by bradykinin [73, 75]. On the other hand, the effect of a long cocoa diet on experimental models with chronic inflammation such as adjuvant arthritis and collagen-induced arthritis has been reported. The clinical inflammation in these animals is only mildly modulated by a cocoa diet, although it does decrease the oxidative stress produced by the chronic inflammatory response [76, 77].

With regard to studies on humans, it has been reported that supplementation with cocoa products in healthy humans does not affect inflammation markers [78]; however, a recent cross-sectional analysis showed that the regular intake of dark chocolate by a healthy population in southern Italy is inversely related to serum C-reactive protein concentration [79]. More recently, Monagas et al. [47] reported the effect of cocoa consumption for 4 weeks on some serum inflammatory biomarkers, and they observed the decrease of some adhesion molecules involved in the recruitment of inflammatory cells, thereby suggesting new evidence about the anti-inflammatory potential of cocoa.

With regard to the immune system, the modulatory effect of cocoa flavonoids on lymphocyte cell activation and cytokine secretion has been described in vitro [80]. Regarding in vivo effects, it has been reported that a cocoa-enriched diet is capable of modifying the composition and functionality of several lymphoid tissues, including the gut-associated lymphoid tissue (GALT). In particular, cocoa intake reduced the proportion of Th lymphocytes in the spleen, Peyer's patches, and mesenteric lymph nodes [72, 81]. Moreover, a cocoa diet does not modify the proliferative response or IL-2 secretion in these tissues but does reduce the IL-4 production by splenocytes [72, 82]. Rats fed a cocoa diet showed lower serum IgG, IgM, and IgA concentrations [72], and the diet attenuated antibody response in immunized rats, which mainly affects the Th2-related isotypes [82]. Similarly, a cocoa diet was able to attenuate the specific antibody response in a model of chronic inflammation [76]. Moreover, in the GALT, a cocoa-enriched diet decreased IgA secretion into the gut, and this was accompanied by a reduction in the gene expression of several molecules involved in IgA-secreting cell activation, gut homing, and IgA synthesis [83, 84]. Likewise, cocoa-fed animals showed a modified TLR expression pattern in gut tissues, which may reflect a change in the crosstalk between microbiota and body cells induced by a cocoa diet [83, 84].

In summary, the results obtained from experimental animals suggest the immune modulation capability of cocoa, especially on pathologies where antibodies are harmful mediators such as in autoimmune and allergic diseases. However, no human studies in this area have been performed until now.

Association Between Cocoa Consumption and Cancer

The diet of the Indians from Kuna Island, on the Caribbean coast of Panama, includes a very high intake of flavanol-rich cocoa. As they seem to be protected against the development of cancer, among other diseases, a link between cocoa consumption and cancer has been established [85].

Some studies have demonstrated that cocoa flavonoids and cocoa extracts have biological activities related to antiproliferative and antitumoral effects in vitro and in vivo [86]. Different experimental procedures have shown that either cocoa or cocoa flavonoids exert antioxidant, free-radical scavenging, and metal ion chelating activities, effects on metabolizing enzymes, inhibition of growth factor activities, modulation of MAPK pathways, influence on transcription factors, and effects on proliferation, apoptosis, transformation, migration, and angiogenesis [86]. However, there is a lack of literature regarding epidemiological evidences for the protective effects of cocoa against cancer and overall mortality [87]. There are only a few approaches associating flavonoid intake (including cocoa as a source) with the incidence of different types of cancer [88]. These observational studies, carried out by using the dietary history or food frequency questionnaire methods, consider flavonoid consumption (flavones, flavonols, flavanols, procyanidins, flavanones, and isoflavones) and its association with some types of cancer. The results suggest some association between a flavonoid-rich diet and protection from cancer. Most of the flavonoids considered in these studies are included in cocoa, and sometimes the consumption of chocolate as a source of flavonoids is also considered [88–91].

Summary

Further in vitro and preclinical studies are needed to investigate the mechanisms involved in the actions of cocoa and to justify cocoa's usage as a dietary recommendation for cardiovascular diseases, cognitive function, and perhaps also in immune-mediated diseases and cancer. Specifically, long-term human studies are required to confirm the benefits derived from consumption of cocoa-derived products at its lowest effective intake.

Acknowledgments This study was supported by the Ministerio de Ciencia y Tecnología (AGL2005-02823 and AGL2008-02790) and the Generalitat de Catalunya (SGR2009118).

References

1. Keen CL. Chocolate: food as medicine/medicine as food. J Am Coll Nutr. 2001;20:436S–9.
2. Dillinger TL, Barriga P, Escarcega S, Jimenez M, Salazar Lowe D, Grivetti LE. Food of the gods: cure for humanity? A cultural history of the medicinal and ritual use of chocolate. J Nutr. 2000;130:2057S–72.
3. Parker G, Parker I, Brotchie H. Mood state effects of chocolate. J Affect Disorders. 2006;92:149–59.
4. Tabernero M, Serrano J, Saura-Calixto F. The antioxidant capacity of cocoa products: contribution to the Spanish diet. Int J Food Sci Tech. 2009;41:28–32.
5. Belščak A, Komes D, Horžić D, et al. Comparative study of commercially available cocoa products in terms of their bioactive composition. Food Res Int. 2009;42:707–16.
6. Gu L, House SE, Wu X, Ou B, Prior RL. Procyanidin and catechin contents and antioxidant capacity of cocoa and chocolate products. J Agric Food Chem. 2006;54:4057–61.
7. Mullen W, Borges G, Donovan JL, Edwards CA, Serafini M, Lean ME, et al. Milk decreases urinary excretion but not plasma pharmacokinetics of cocoa flavan-3-ol metabolites in humans. Am J Clin Nutr. 2009;89:1784–91.
8. Ramiro-Puig E, Castell M. Cocoa: antioxidant and immunomodulator. Br J Nutr. 2009;101:931–40.
9. Azam S, Hadi N, Khan UN, Hadi SM. Antioxidant and prooxidant properties of caffeine, theobromine and xanthine. Med Sci Monit. 2003;9:325–30.
10. Stoclet JC, Schini-Kerth V. Dietary flavonoids and human health. Ann Pharm Fr. 2011;69:78–90.
11. Nijveldt RJ, van Nood E, van Hoorn DE, Boelens PG, van Norren K, van Leeuwen PA. Flavonoids: a review of probable mechanisms of action and potential applications. Am J Clin Nutr. 2001;74:418–25.
12. Perez-Cano FJ, Perez-Berezo T, Ramos-Romero S, et al. Is there an anti-inflammatory potential beyond the antioxidant power of cocoa? In: Bishop MR, editor. Chocolate, fast foods and sweeteners: consumption and health. Hauppauge: Nova; 2009. p. 85–104.
13. Lecumberri E, Mateos R, Ramos S, Alía M, Rúperez P, Goya L, et al. Characterization of cocoa fiber and its effect on the antioxidant capacity of serum in rats. Nutr Hosp. 2006;21:622–8.
14. Ramiro-Puig E, Urpi-Sarda M, Perez-Cano FJ, Franch A, Castellote C, Andrés-Lacueva C, et al. Cocoa-enriched diet enhances antioxidant enzyme activity and modulates lymphocyte composition in thymus from young rats. J Agric Food Chem. 2007;55:6431–8.
15. Urpi-Sarda M, Ramiro-Puig E, Khan N, Ramos-Romero S, Llorach R, Castell M, et al. Distribution of epicatechin metabolites in lymphoid tissues and testes of young rats with a cocoa-enriched diet. Br J Nutr. 2010;103:1393–7.
16. Yeh CT, Yen GC. Induction of hepatic antioxidant enzymes by phenolic acids in rats is accompanied by increased levels of multidrug resistance-associated protein 3 mRNA expression. J Nutr. 2006;136:11–5.
17. Lecumberri E, Goya L, Mateos R, Alía M, Ramos S, Izquierdo-Pulido M, et al. A diet rich in dietary fiber from cocoa improves lipid profile and reduces malondialdehyde in hypercholesterolemic rats. Nutrition. 2007;23:332–41.
18. Jalil AM, Ismail A, Pei CP, Hamid M, Kamaruddin SH. Effect of cocoa extract on glucometabolism, oxidative stress, and antioxidant enzymes in obese-diabetic rats. J Agric Food Chem. 2008;56:7877–84.
19. Scheid L, Reusch A, Stehle P, Ellinger S. Antioxidant effects of cocoa and cocoa products ex vivo and in vivo: is there evidence from controlled intervention studies? Curr Opin Clin Nutr Metab Care. 2010;13:737–42.
20. Rein D, Lotito S, Holt RR, Keen CL, Schmitz HH, Fraga CG. Epicatechin in human plasma: in vivo determination and effect of chocolate consumption on plasma oxidation status. J Nutr. 2000;130:2109S–14.
21. Wang JF, Schramm DD, Holt RR, Ensunsa JL, Fraga CG, Schmitz HH, et al. A dose–response effect from chocolate consumption on plasma epicatechin and oxidative damage. J Nutr. 2000;130:2115S–9.
22. Serafini M, Bugianesi R, Maiani G, Valtuena S, De Santis S, Crozier A. Plasma antioxidants from chocolate. Nature. 2003;424:1013.
23. Fraga CG, Actis-Goretta L, Ottaviani JI, Carrasquedo F, Lotito SB, Lazarus S, et al. Regular consumption of a flavanol-rich chocolate can improve oxidant stress in young soccer players. Clin Dev Immunol. 2005;12:11–7.
24. Jalil AMM, Ismail A. Polyphenols in cocoa and cocoa products: Is there a link between antioxidant properties and health? Molecules. 2008;13:2190–219.
25. Hooper L, Kroon PA, Rimm EB, Cohn JS, Harvey I, Le Cornu KA, et al. Flavonoids, flavonoid-rich foods, and cardiovascular risk: a meta-analysis of randomized controlled trials. Am J Clin Nutr. 2008;88:38–50.
26. Corti R, Flammer AJ, Hollenberg NK, Lüscher TF. Cocoa and cardiovascular health. Circulation. 2009;119: 1433–41.

27. Lippi G, Franchini M, Montagnana M, Favaloro EJ, Guidi GC, Targher G. Dark chocolate: consumption for pleasure or therapy? J Thromb Thrombolysis. 2009;28:482–8.
28. Gómez-Juaristi M, González-Torres L, Bravo L, Vaquero MP, Bastida S, Sánchez-Muniz FJ. Beneficial effects of chocolate on cardiovascular health. Nutr Hosp. 2011;26:289–92.
29. Muniyappa R, Hall G, Kolodziej TL, Karne RJ, Crandon SK, Quon MJ. Cocoa consumption for 2 wk enhances insulin-mediated vasodilatation without improving blood pressure or insulin resistance in essential hypertension. Am J Clin Nutr. 2008;88:1685–96.
30. Ried K, Frank OR, Stocks NP. Dark chocolate or tomato extract for prehypertension: a randomised controlled trial. BMC Complement Altern Med. 2009;9:22.
31. Ried K, Sullivan T, Fakler P, Frank OR, Stocks NP. Does chocolate reduce blood pressure? A meta-analysis. BMC Med. 2010;8:39.
32. Desch S, Kobler D, Schmidt J, Sonnabend M, Adams V, Sareban M, et al. Low vs. Higher-dose dark chocolate and blood pressure in cardiovascular high-risk patients. Am J Hypertens. 2010;23:694–700.
33. Egan BM, Lakenx MA, Donovan JL, Woolson RF. Does dark chocolate have a role in the prevention and management of hypertension?: commentary on the evidence. Hypertension. 2010;55:1289–95.
34. Rimbach G, Melchin M, Moehring J, Wagner AE. Polyphenols from cocoa and vascular health – A critical review. Int J Mol Sci. 2009;10:4290–309.
35. Davison K, Berry NM, Misan G, Coat Coates AM, Buckley JD, Howe PR. Dose-related effects of flavanol-rich cocoa on blood pressure. J Hum Hypertens. 2010;24:568–76.
36. Heiss C, Jahn S, Taylor M, Real WM, Angeli FS, Wong ML, et al. Improvement of endothelial function with dietary flavanols is associated with mobilization of circulating angiogenic cells in patients with coronary artery disease. J Am Coll Cardiol. 2010;56:218–24.
37. Van Den Bogaard B, Draijer R, Westerhof BE, van den Meiracker AH, van Montfrans GA, van den Born BJ. Effects on peripheral and central blood pressure of cocoa with natural or high-dose theobromine: a randomized, double-blind crossover trial. Hypertension. 2010;56:839–46.
38. Borchers AT, Keen CL, Hannum SM, et al. Cocoa and chocolate: composition, bioavailability and health implications. J Med Food. 2000;3:77–105.
39. Davison K, Coates AM, Buckley JD, Howe PR. Effect of cocoa flavanols and exercise on cardiometabolic risk factors in overweight and obese subjects. Int J Obes. 2008;32:1289–96.
40. Fraga CG, Litterio MC, Prince PD, Calabró V, Piotrkowski B, Galleano M. Cocoa flavanols: effects on vascular nitric oxide and blood pressure. J Clin Biochem Nutr. 2011;48:63–7.
41. Berry NM, Davison K, Coates AM, Buckley JD, Howe PR. Impact of cocoa flavanol consumption on blood pressure responsiveness to exercise. Br J Nutr. 2010;103:1480–4.
42. Balzer J, Rassaf T, Heiss C, Kleinbongard P, Lauer T, Merx M, et al. Sustained benefits in vascular function through flavanol-containing cocoa in medicated diabetic patients. A double-masked, randomized, controlled trial. J Am Coll Cardiol. 2008;51:2141–9.
43. Faghihzadeh F, Esmaillzadeh A. Impact of cacao consumption on cardiovascular risk factors: review of current evidence. J Isfahan Med Sch. 2010;28:591–605.
44. Pascual V, Valls RM, Sola R. Cacao y chocolate: ¿un placer cardiosaludable? Clin Invest Arterioscler. 2009;21:198–209.
45. Tokede OA, Gaziano JM, Djoussé L. Effects of cocoa products/dark chocolate on serum lipids: a meta-analysis. Eur J Clin Nutr. 2011;65(8):879–86.
46. Baba S, Osakabe N, Kato Y, Natsume M, Yasuda A, Kido T, et al. Continuous intake of polyphenolic compounds containing cocoa powder reduces LDL oxidative susceptibility and has beneficial effects on plasma HDL-cholesterol concentrations in humans. Am J Clin Nutr. 2007;85:709–17.
47. Monagas M, Khan N, Andres-Lacueva C, Casas R, Urpí-Sardà M, Llorach R, et al. Effect of cocoa powder on the modulation of inflammatory biomarkers in patients at high risk of cardiovascular disease. Am J Clin Nutr. 2009;90:1144–50.
48. Khan N, Monagas M, Andres-Lacueva C, Casas R, Urpí-Sardà M, Lamuela-Raventós RM, et al. Regular consumption of cocoa powder with milk increases HDL cholesterol and reduces oxidized LDL levels in subjects at high-risk of cardiovascular disease. Nutr Metab Cardiovasc Dis. 2012. (in press).
49. Cooper KA, Donovan JL, Waterhouse AL, Williamson G. Cocoa and health: a decade of research. Br J Nutr. 2008;99:1–11.
50. Katz DL, Doughty K, Ali A. Cocoa and chocolate in human health and disease. Antioxid Redox Signal. 2011;15(10):2779–811.
51. Macdonald I. Boosting brain power – with chocolate. Am Assoc Adv Sci. 2007. http://www.ultrahealthychocolate.com/medical_studies.html.
52. Sorond FA, Lipsitz LA, Hollenberg NK, Fisher ND. Cerebral blood flow response to flavanol-rich cocoa in healthy elderly humans. Neuropsychiatr Dis Treat. 2008;4:433–40.

53. Francis ST, Head K, Morris PG, Macdonald IA. The effect of flavanol-rich cocoa on the fMRI response to a cognitive task in healthy young people. J Cardiovasc Pharmacol. 2006;47:S215–20.
54. Crews Jr WD, Harrison DW, Wright JW. A double-blind, placebo-controlled, randomized trial of the effects of dark chocolate and cocoa on variables associated with neuropsychological functioning and cardiovascular health: clinical findings from a sample of healthy, cognitively intact older adults. Am J Clin Nutr. 2008;87:872–80.
55. Nurk E, Refsum H, Drevon CA, Tell GS, Nygaard HA, Engedal K, et al. Intake of flavonoid-rich wine, tea, and chocolate by elderly men and women is associated with better cognitive test performance. J Nutr. 2009;139:120–7.
56. Scholey AB, French SJ, Morris PJ, Kennedy DO, Milne AL, Haskell CF, et al. Consumption of cocoa flavanols results in acute improvements in mood and cognitive performance during sustained mental effort. J Psychopharmacol. 2010;24:1505–14.
57. Field DT, Williams CM, Butler LT. Consumption of cocoa flavanols results in an acute improvement in visual and cognitive functions. Physiol Behav. 2011;103:255–60.
58. Sathyapalan T, Beckett S, Rigby AS, Mellor DD, Atkin SL. High cocoa polyphenol rich chocolate may reduce the burden of the symptoms in chronic fatigue syndrome. Nutr J. 2010;9:55–9.
59. Biesalski HK. The role of antioxidants in nutritional support. Nutrition. 2000;16:593–6.
60. Huber A, Stuchbury G, Burkle A, Münch G. Neuroprotective therapies for Alzheimer's disease. Curr Pharm Des. 2006;12:705–17.
61. Abd El Mohsen MM, Kuhnle G, Rechner AR, Schroeter H, Rose S, Jenner P, et al. Uptake and metabolism of epicatechin and its access to the brain after oral ingestion. Free Radic Biol Med. 2002;33:1693–702.
62. Macht M, Roth S, Ellgring H. Chocolate eating in healthy men during experimentally induced sadness and joy. Appetite. 2002;39:147–58.
63. Radin D, Hayssen G, Walsh J. Effects of intentionally enhanced chocolate on mood. Explore. 2007;3:485–92.
64. Macht M, Mueller J. Immediate effects of chocolate on experimentally induced mood states. Appetite. 2007;49:667–74.
65. Parker G, Crawford J. Chocolate craving when depressed: a personality marker. Br J Psychiatry. 2007;191: 351–2.
66. Javelot H, Messaoudi M, Jacquelin C, et al. Antidepressant-like properties of cocoa's polyphenols – The role of flavonoids and flavanols on depression. Agro Food Ind Hi Tech. 2009;20:10–2.
67. di Tomaso E, Beltramo M, Piomelli D. Brain cannabinoids in chocolate. Nature. 1996;382:677–8.
68. Eby GA, Eby KL. Rapid recovery from major depression using magnesium treatment. Med Hypotheses. 2006;67:362–70.
69. Rose N, Koperski S, Golomb BA. Mood food: chocolate and depressive symptoms in a cross-sectional analysis. Arch Intern Med. 2010;170:699–703.
70. Wolz M, Kaminsky A, Löhle M, Storch A, Reichmann H. Chocolate consumption is increased in Parkinson's disease. Results from a self-questionnaire study. J Neurol. 2009;256:488–92.
71. Martin FP, Rezzi S, Peré-Trepat E. Metabolic effects of dark chocolate consumption on energy, gut microbiota, and stress-related metabolism in free-living subjects. J Proteome Res. 2009;8:5568–79.
72. Ramiro-Puig E, Pérez-Cano FJ, Ramírez-Santana C, Castellote C, Izquierdo-Pulido M, Permanyer J, et al. Spleen lymphocyte function modulated by a cocoa-enriched diet. Clin Exp Immunol. 2007;149:535–42.
73. Castell M, Franch A, Ramos-Romero S, et al. Effect of a diet rich in cocoa flavonoids on experimental acute inflammation. In: Keller RB, editor. Flavonoids: biosynthesis, biological effects and dietary sources. Hauppauge: Nova; 2009. p. 213–29.
74. Lee KW, Kundu JK, Kim SO, Chun KS, Lee HJ, Surh YJ. Cocoa polyphenols inhibit phorbol ester-induced superoxide anion formation in cultured HL-60 cells and expression of cyclooxygenase-2 and activation of NF-kappaB and MAPKs in mouse skin in vivo. J Nutr. 2006;136:1150–5.
75. Ramos-Romero S, Ramiro-Puig E, Pérez-Cano FJ, et al. Anti-inflammatory effects of cocoa in rat carrageenin-induced paw oedema. Proc Nutr Soc. 2008;67:E65.
76. Ramos-Romero S, Perez-Cano FJ, Castellote C, Castell M, Franch A. Effect of cocoa-enriched diets on lymphocytes involved in adjuvant arthritis in rats. Br J Nutr. 2012;107:523–32.
77. Ramos-Romero S, Perez-Cano FJ, Perez-Berezo T, Castellote C, Franch A, Castell M. Effect of a cocoa flavonoid-enriched diet on experimental autoimmune arthritis. Br J Nutr. 2012;107:378–87.
78. Mathur S, Devaraj S, Grundy SM, Jialal I. Cocoa products decrease low density lipoprotein oxidative susceptibility but do not affect biomarkers of inflammation in humans. J Nutr. 2002;132:3663–7.
79. di Giuseppe R, di Castelnuovo A, Centritto F, Zito F, De Curtis A, Costanzo S, et al. Regular consumption of dark chocolate is associated with low serum concentrations of C reactive protein in a healthy Italian population. J Nutr. 2008;138:1939–45.
80. Perez-Cano FJ, Franch A, Perez-Berezo T, et al. The effects of flavonoids on the immune system. In: Preedy V, Watson R, editors. Bioactive foods and chronic disease states. New York: Elsevier; 2011.

81. Ramiro-Puig E, Pérez-Cano FJ, Ramos-Romero S, Pérez-Berezo T, Castellote C, Permanyer J, et al. Intestinal immune system of young rats influenced by cocoa-enriched diet. J Nutr Biochem. 2008;19:555–65.
82. Perez-Berezo T, Ramiro-Puig E, Perez-Cano FJ, Castellote C, Permanyer J, Franch A, et al. Influence of a cocoa-enriched diet on specific immune response in ovalbumin-sensitized rats. Mol Nutr Food Res. 2009;53:389–97.
83. Perez-Berezo T, Franch A, Ramos-Romero S, Castellote C, Pérez-Cano FJ, Castell M. Cocoa-enriched diets modulate intestinal and systemic humoral immune response in young adult rats. Mol Nutr Food Res. 2011;55:S56–66.
84. Perez-Berezo T, Franch A, Castellote C, Castell M, Pérez-Cano FJ. Mechanisms involved in down-regulation of intestinal IgA in rats by high cocoa intake. J Nutr Biochem. 2012. (in press).
85. Hollenberg NK, Fisher NDL, McCullough ML. Flavanols, the Kuna, cocoa consumption, and nitric oxide. J Am Soc Hypertens. 2009;3:105–12.
86. Perez-Cano FJ, Franch A, Castellote C, et al. Cocoa and the immune system and proliferative disorders. In: Watson RR, Zibadi S, Preedy VR, editors. Dietary components and immune function. Nutrition and health. Part 5. New York: Humana Press; 2010. p. 469–96.
87. Maskarinec G. Cancer protective properties of cocoa: a review of the epidemiologic evidence. Nutr Cancer. 2009;61:573–9.
88. Theodoratou E, Kyle J, Cetnarskyj R, Farrington SM, Tenesa A, Barnetson R, et al. Dietary flavonoids and the risk of colorectal cancer. Cancer Epidemiol Biomarkers Prev. 2007;16:684–93.
89. Le ML, Murphy SP, Hankin JH, Wilkens LR, Kolonel LN, et al. Intake of flavonoids and lung cancer. J Natl Cancer Inst. 2000;92:154–60.
90. Knekt P, Kumpulainen J, Järvinen R, Rissanen H, Heliövaara M, Reunanen A, et al. Flavonoid intake and risk of chronic diseases. Am J Clin Nutr. 2002;76:560–8.
91. Peterson J, Lagiou P, Samoli E, Lagiou A, Katsouyanni K, La Vecchia C, et al. Flavonoid intake and breast cancer risk: a case-control study in Greece. Br J Cancer. 2003;89:1255–9.

Chapter 20
Endothelial Control of Vascular Tone by Chocolate and Other Polyphenols

Cyril Auger, Noureddine Idris-Khodja, and Valérie B. Schini-Kerth

Key Points

- Endothelial cells play a key role in the control of vascular tone via the release of potent vasodilators such as nitric oxide (NOS) and endothelium-derived hyperpolarizing factor (EDHF).
- Major cardiovascular diseases are characterized by an endothelial dysfunction as indicated by blunted endothelium-dependent relaxations and often also by the development of endothelium-dependent contractile responses involving cyclooxygenase-derived metabolites of arachidonic acid acting on TP receptors to contract the vascular smooth muscle.
- Polyphenol-rich sources and, in particular, those containing high levels of flavanols and anthocyanins are potent activators of the endothelial formation of NO and EDHF.
- Flavanol-rich sources such as cocoa/chocolate have been shown to improve the endothelial function in major cardiovascular diseases such as hypertension, coronary artery diseases, and diabetes.

Keywords Endothelial function • Nitric oxide • Polyphenol-rich sources • Cocoa/chocolate • Cardiovascular diseases

Introduction

Endothelial cells, a monolayer covering the luminal surface of all blood vessels, have a strategic localization separating the flowing blood from the underlying thrombogenic vascular wall. The endothelium is an important autocrine, paracrine, and endocrine organ with a weight of approximately 1.8 kg and a surface of approximately 700 m^2. Besides being a semipermeable barrier, endothelial cells have a key role in maintaining vascular homeostasis. They contribute to maintain blood fluidity and control

C. Auger, Ph.D. • N. Idris-Khodja, Pharm D, Ph.D.
UMR CNRS 7213 – Laboratory of Biophotonics and Pharmacology,
Strasbourg University, 74, route du Rhin, Illkirch 67401, France
e-mail: cyril.auger@unistra.fr; noureddine.idris-khodja@unistra.fr

V.B. Schini-Kerth, Ph.D. (✉)
UMR CNRS 7213–Laboratory of Biophotonics and Pharmacology, Strasbourg University,
74, route du rhin, Illkirch 67401, France
e-mail: valerie.schini-kerth@unistra.fr

R.R. Watson et al. (eds.), *Chocolate in Health and Nutrition*, Nutrition and Health 7,
DOI 10.1007/978-1-61779-803-0_20, © Springer Science+Business Media, LLC 2013

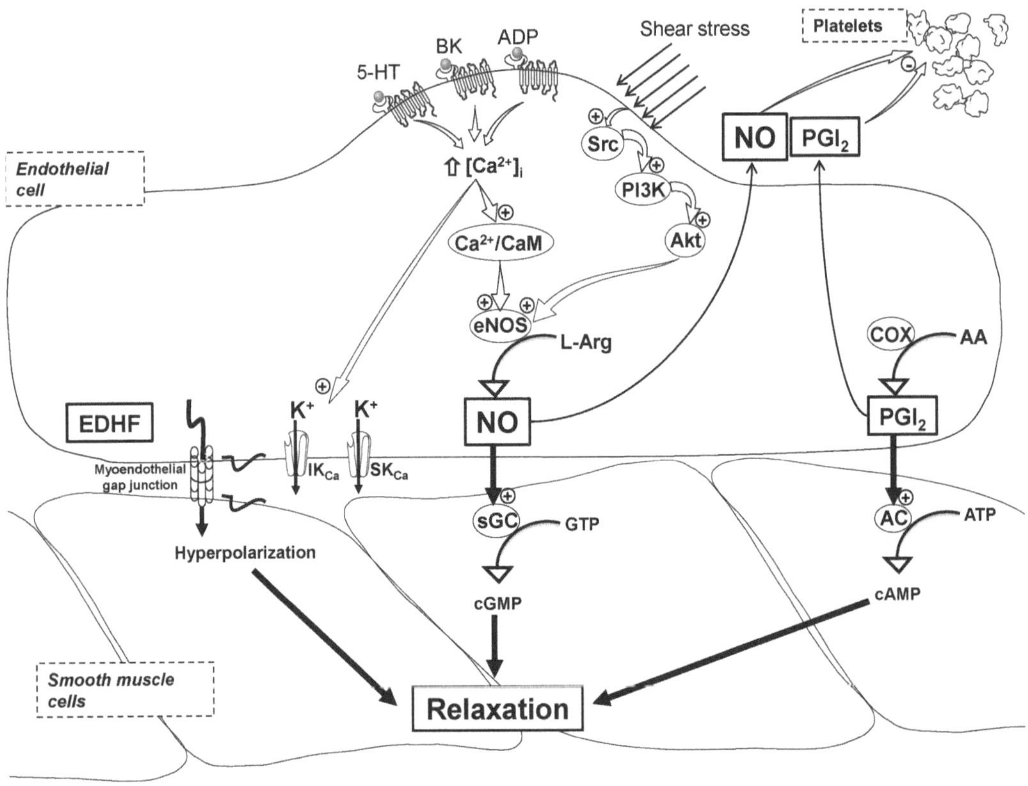

Fig. 20.1 Endothelial cells play an important role in the control of vascular homeostasis, mostly by releasing several potent endothelium-derived relaxing factors. Endothelial nitric oxide synthase (eNOS) generates nitric oxide (NO) from L-arginine (L-Arg) after its activation by shear stress via the Src/PI3-kinase/Akt pathway or by a variety of agonists (5-HT, serotonin; BK, bradykinin; ADP, adenosine diphosphate) via the calcium/calmodulin pathway (Ca^{2+}/CaM). NO relaxes vascular smooth muscle cells via activation of soluble guanylyl cyclase (sGC). Endothelium-derived hyperpolarizing factor (EDHF) involves the activation of endothelial SK_{Ca} and IK_{Ca} channels (small and intermediate conductance Ca^{2+}-activated K^+ channels, respectively), inducing hyperpolarization ( ) of the endothelium, which is, thereafter, transmitted, in part, to the underlying vascular smooth cells via myoendothelial gap junctions with subsequent relaxation. Prostacyclin (PGI_2) is also a vasodilator and also inhibits platelet activation. Abbreviations: AA, arachidonic acid; AC, adenylyl cyclase; ATP, adenosine triphosphate; cAMP, cyclic adenosine monophosphate; cGMP, cyclic guanosine monophosphate; GTP, guanosine triphosphate; PI3K, phosphatidylinositol 3-kinase; sGC, soluble guanylyl cyclase; Src, Src family kinase

vascular tone mostly via the formation and/or release of several potent vasodilators such as nitric oxide (NO), endothelium-derived hyperpolarizing factor (EDHF), and prostacyclin (PGI_2) (Fig. 20.1). Nitric oxide, a diffusible gas, is generated from L-arginine by a constitutively expressed enzyme termed endothelial NO synthase (eNOS) in response to a great variety of physiological agonists. The endothelial formation of NO is constantly adjusted by a variety of local stimulators, such as shear stress induced by flowing blood on the endothelial surface, platelet-derived products such as adenosine diphosphate and serotonin, products derived from the coagulation cascade such as thrombin, and local autacoids such as bradykinin (see Fig. 20.1). Nitric oxide will diffuse toward the underlying smooth muscle to enhance the activity of soluble guanylyl cyclase, leading to the relaxation of the vascular smooth muscle via the cyclic GMP pathway (see Fig. 20.1). In addition, NO helps to maintain blood fluidity by preventing platelet adhesion and activation and also by inhibiting the expression of tissue factor, the physiological activator of blood coagulation. In addition, NO also has antiatherosclerotic properties by inhibiting the expression of several proatherosclerotic factors, such as adhesion molecules, growth factors, and proangiogenic factors. In some blood vessels, like in the coronary circulation and the microcirculation, EDHF and NO contribute to regulation of vascular tone (see Fig. 20.1). Although the nature of EDHF can vary depending on the type of blood vessel and the species, EDHF is often characterized by the hyperpolarization of endothelial cells, which is transmitted to the underlying vascular smooth muscle via myoendothelial gap junctions. The hyperpolarization of the vascular smooth muscle will, in turn, decrease vascular tone subsequent to the inactivation of voltage-operated calcium channels. PGI_2 produced by the arachidonic acid cascade can also induce relaxation of some blood vessels by enhancing the activity of adenylyl cyclase, leading to subsequent activation of the cyclic AMP relaxing pathway in the vascular smooth muscle (see Fig. 20.1).

Endothelial dysfunction is a hallmark of major types of cardiovascular diseases, such as hypertension, coronary artery diseases, and diabetes. The endothelial dysfunction is characterized by blunted endothelium-dependent relaxations affecting both the NO and the EDHF component and often also by the development of endothelium-dependent contractile responses involving vasoconstrictor prostanoids, which act on TP receptors to contract the vascular smooth muscle [1, 2]. Since the endothelial dysfunction appears before structural changes in the arterial wall, it has been suggested to be an important early event in the initiation and development of the pathology. Oxidative stress and, in particular, NADPH oxidase-derived superoxide anions have been suggested to have a major role in endothelial dysfunction in part by chemically reacting with NO, thereby decreasing its bioavailability and also by decreasing the EDHF component [3, 4].

It is now well established that diet, one of the most important lifestyle factors, can strongly influence the incidence of cardiovascular diseases. Indeed, clinical studies have indicated that diets rich in fruits and vegetables [5], beverages such as red wine [6, 7] and green tea [8], and the regular consumption of cocoa and chocolate [9] are associated with a reduced risk of cardiovascular diseases. The protective effect has been attributed, at least in part, to their high polyphenol content, which are molecules with one or more hydroxyl group on at least one aromatic ring. Polyphenols can be classified into two groups, the flavonoids and the nonflavonoids. The beneficial effect of polyphenol-rich sources on the cardiovascular system appears to be mediated by two major flavonoid classes: the flavanols, such as catechin and epicatechin, and their oligomers, the procyanidins and the anthocyanins, such as cyanidin and petunidin, which give the red-purple color to fruits such as grapes. Polyphenols might protect the cardiovascular system by different mechanisms. In particular, they have been shown to prevent oxidation of low-density lipoproteins [10], platelet adhesion and aggregation [11, 12], and smooth muscle cell proliferation and migration [13, 14]. They have also been shown to have antiatherosclerotic [15, 16] and antithrombotic properties [12]. In addition, the beneficial effect may also be due to their ability to affect directly the endothelial function by increasing the formation of the vasoprotective factors, NO and EDHF. For review, see reference [17].

This article reviews experimental and clinical evidence suggesting that cocoa, a rich source of polyphenols and flavonoids, improves the endothelial function in health and diseases, and potential biologic mechanisms.

Polyphenol-Rich Sources Cause Endothelium-Dependent Relaxations: Role of NO and EDHF

Although in vitro and animal studies on the vasorelaxant effects of cocoa/chocolate are relatively scarce, the ability of several polyphenols and polyphenol-rich sources to induce vasorelaxation has been observed in numerous studies for more than two decades.

Indeed, in 1993, Fitzpatrick et al. reported that grape-derived products, namely, wines and grape juices, induce endothelium-dependent relaxations in rat aortic rings. For this purpose, rings of 3–4 mm from rat aorta were suspended in organ chambers and maintained at 37°C in a physiological salt solution to determine changes in isometric tension. After precontraction of rings, the addition of increasing concentrations of grape-derived products induced potent relaxations in rings with an intact endothelium, whereas only minor effects were observed in those without endothelium [18]. Moreover, the endothelium-dependent relaxation was inhibited by competitive inhibitors of eNOS, indicating the involvement of NO. The characterization of the active compounds has indicated that the vasorelaxing effects induced by either a grape seed or a pine extract are due, at least in part, to procyanidins [19–21]. Moreover, a procyanidin-rich Crataegus extract also induced potent NO-mediated relaxations in porcine coronary artery rings [22].

Similarly to grape seed and pine extract, a red wine extract induced full relaxation in intact aortic rings, whereas only small relaxations were observed in rings without endothelium at a thousand times higher concentration [23]. The red wine extract also stimulated the endothelial formation of NO, which was determined using electron paramagnetic spectroscopy using an NO-specific spin trap [23]. The ability of polyphenol-rich products to induce endothelium-dependent NO-mediated relaxations has since been reported in several conductance and resistance arteries from animals, such as the rat and rabbit aorta [18, 23–25], the rat mesenteric artery [26], the rat cerebral arterioles [27, 28], and porcine retinal arterioles [29] and coronary artery [30]. Moreover, NO-mediated relaxations to polyphenol-rich products have also been observed in isolated human coronary and internal mammary artery [31, 32] as well as in human saphenous vein [32]. Noteworthy, a study reported that procyanidin-enriched fractions from a cocoa extract were able to induce relaxations in rabbit aortic rings subsequent to the activation of eNOS [33].

However, although NO-mediated relaxations are often an important component of the endothelium-dependent relaxation induced by polyphenols, they may also involve an EDHF component, especially in resistance arteries and the coronary circulation. The first study demonstrating the involvement of EDHF in the polyphenol-induced relaxation was performed using porcine coronary artery rings and a red wine extract [34]. These authors showed that red wine extract is able to induce in a concentration-dependent manner an endothelium-dependent EDHF-mediated relaxation associated with hyperpolarization of the vascular smooth muscle. Moreover, the polyphenol-induced EDHF-mediated relaxation has also been observed in porcine coronary artery rings using the procyanidin-rich Concord grape juice [35], as well as in the mesenteric arterial bed using several polyphenol-rich natural sources [26, 36–38].

Interestingly, while polyphenols have been described as antioxidant in vivo [10], the endothelium-dependent relaxation appears to involve an intracellular redox-sensitive mechanism. Indeed, the endothelium-dependent NO- and EDHF-mediated relaxations to red wine extract in porcine coronary artery rings were significantly reduced by membrane-permeant analogues of both superoxide dismutase and catalase [34, 39]. Moreover, the native enzymes (superoxide dismutase and catalase), which are unable to cross the cell membrane, did not affect these relaxations. Similar observations were made with other polyphenol-rich sources such as grape juice [35], a grape skin extract [40], a procyanidin-rich Crataegus extract [22], and isolated polyphenols such as epigallocatechin gallate [41]. Taken together, these results suggest that the intracellular generation of reactive oxygen species mediates the stimulatory effect of polyphenols on the endothelial formation of NO- and EDHF-mediated responses (Fig. 20.2).

Further characterization of the polyphenol-induced signaling pathway leading to the formation of NO and activation of EDHF-mediated responses has been undertaken in cultured endothelial cells. A recent study indicated that (−)epicatechin, the major monomeric flavonoid present in cocoa [42], activates eNOS by phosphorylation of serine 633 and 1177, two positive regulatory sites, as well as by dephosphorylation of threonine 495, a negative regulatory site, in human coronary artery endothelial cells [43]. Moreover, several studies using cultured endothelial cells demonstrated that the polyphenol-induced prooxidant response triggers the Src/PI3-kinase/Akt pathway, leading to the activation of both NO and EDHF pathways (see Fig. 20.2) [30, 35, 39, 43, 44].

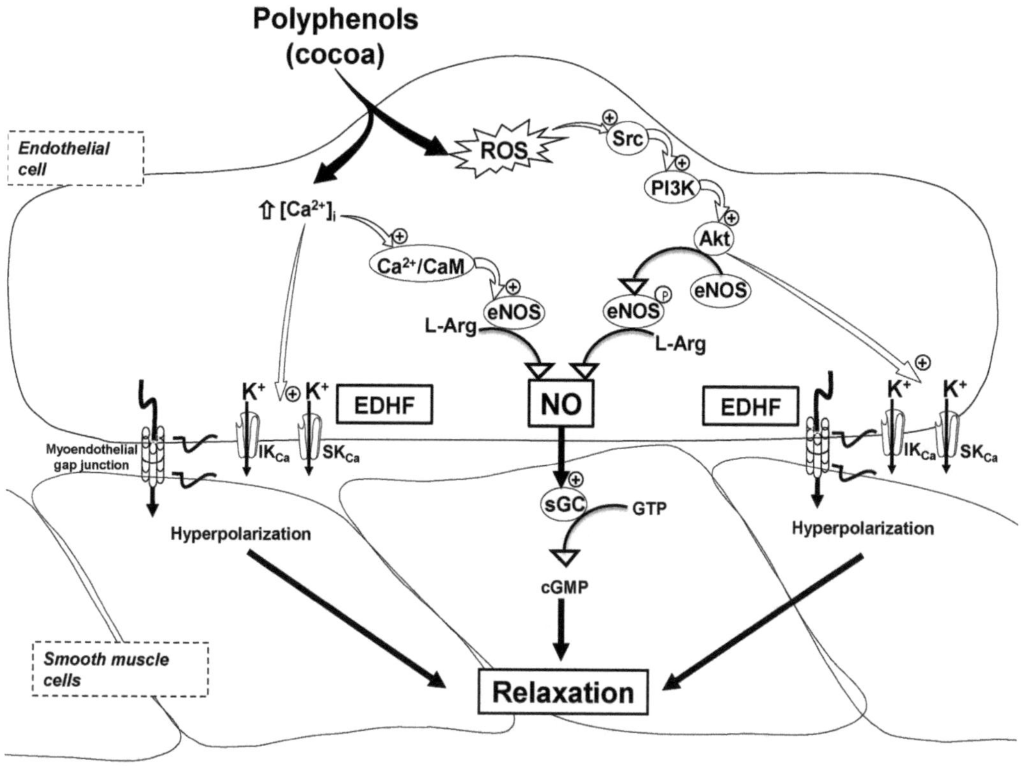

Fig. 20.2 Polyphenols and polyphenol-rich sources induce endothelium-dependent NO- and EDHF-mediated relaxations. Polyphenols are potent inducers of the endothelial formation of nitric oxide (NO) and endothelium-derived hyperpolarizing factor (EDHF) via a redox-sensitive mechanism. Abbreviations: AA, arachidonic acid; ATP, adenosine triphosphate; cAMP, cyclic adenosine monophosphate; cGMP, cyclic guanosine monophosphate; COX, cyclooxygenase; eNOS, endothelial NO synthase; GTP, guanosine triphosphate; IK_{Ca}, intermediate conductance calcium-activated potassium channels; L-Arg, L-arginine; PI3K, phosphatidylinositol 3-kinase; sGC, soluble guanylyl cyclase; SK_{Ca}, small conductance calcium-activated potassium channels; Src, Src family kinase

In addition to the redox-sensitive PI3-kinase/Akt-dependent pathways, NO and EDHF can also be activated by an increase in the calcium signaling pathway in endothelial cells; both of these pathways act most likely in a synergistic manner (see Fig. 20.2). Indeed, the epicatechin-induced activation of eNOS is partially reduced in the absence of extracellular calcium [45]. Interestingly, while the absence of calcium reduces the dephosphorylation of threonine 495, suggesting it is a calcium-dependent event, the phosphorylation of the activator sites serine 633 and serine 1177 was not altered, suggesting that they are not regulated by the calcium signaling pathway [45]. Moreover, the formation of NO in cultured endothelial cell induced by red wine extract is associated with an increase in the intracellular concentration of calcium [46].

Polyphenol-Rich Sources Improve Endothelial Function in Experimental Animals

Recent studies indicate that cocoa and other polyphenol-rich sources may exert beneficial effect on the endothelial function in vivo. Indeed, several studies have shown that oral supplementation with polyphenol-rich sources improves the endothelial function in various animal models of cardiovascular pathologies associated with endothelial dysfunction such as hypertension, atherosclerosis, and diabetes.

The study by Cienfuegos-Jovellanos et al. reported that a single oral administration of a polyphenol-enriched cocoa powder significantly decreased blood pressure of spontaneously hypertensive rats, a genetic model of hypertension [47]. Moreover, the highest dose of cocoa powder (300 mg/kg body weight) was able to induce an antihypertensive effect similar to 50 mg/kg of captopril, a well-known antihypertensive drug, while being without effect in control normotensive rats [47]. Similarly, ingestion of a soluble cocoa fiber product partially inhibited the development of hypertension in spontaneously hypertensive rats [48]. Blood pressure reduction in spontaneously hypertensive rats has also been reported after ingestion of other polyphenol-rich sources such as blueberry [49], green and black tea extracts [50], grape seed extract [51], and red wine [52]. Moreover, the reduction in blood pressure induced by oral supplementation of red wine was associated with an improved endothelial function [52]. In addition, red wine products have also been shown to improve the endothelial function and reduce blood pressure in several other models of hypertension such as the N^G-nitro-L-arginine-induced hypertension [53, 54], the angiotensin II-induced hypertension [55], and the DOCA salt-induced hypertension [56].

Among other cardiovascular diseases associated with an endothelial dysfunction, atherosclerosis is of particular interest due to its involvement in human cardiovascular death toll. Dietary supplementation with either cocoa liquor polyphenols or cocoa powder for 6 months significantly reduced the extent of atherosclerotic lesions in the aorta of Kurosawa and Kusanagi-hypercholesterolemic rabbits, a spontaneously hypercholesterolemic model [57, 58].

Interestingly, cocoa may also improve cardiovascular health and endothelial function by preventing the associated risk factors such as diabetes and the metabolic syndrome. Indeed, supplementation of the diet with cocoa significantly reduced the blood glucose level in both diabetic rats and mice [59, 60]. Similarly, ingestion of proanthocyanidin-rich grape extracts significantly reduced blood pressure in a model of insulin resistance, the fructose-fed rat [61].

Finally, recent data suggest that the polyphenol-rich cocoa may also prevent endothelial dysfunction associated with aging. Indeed, oral administration of a polyphenol-rich red wine extract improved aging-related endothelial dysfunction as well as the exercise capacity of middle-aged rats [62].

Cocoa/Chocolate and Endothelial Function in Humans in Health and Disease

A certain number of clinical studies have evaluated the effect of intake of cocoa-derived products on the endothelial function in healthy humans but also in patients presenting risk factors for cardiovascular diseases such as hypertension, diabetes, smoking, and coronary artery disease. The noninvasive measurement of the endothelial function is often assessed in humans using brachial artery flow-mediated vasodilatation. For this purpose, high-frequency ultrasound assessments of changes in brachial artery diameter are determined at baseline and after a 5-min arterial occlusion induced by an inflated blood pressure cuff. Following release of the inflated cuff, blood flow resumes, triggering the activation of endothelial cells by shear stress, which leads to the PI3-kinase/Akt-dependent activation of eNOS, and ultimately the NO-mediated vasodilatation. In addition, the endothelial function can also be evaluated by peripheral vasodilatation in response to ischemia as assessed by pulse-wave amplitude on the finger. Endothelial dysfunction is indicated by reduced flow-mediated vasodilatation, which is usually associated with the presence of risk factors and increases with the number of risk factors.

In healthy subjects, daily intake of a flavanol-rich cocoa drink (821 mg total flavanols per day) for 4 days has been shown to cause peripheral vasodilatation and an improvement in the vasodilator response to ischemia as assessed by pulse-wave amplitude on the finger [63]. N^G-nitro-L-arginine methyl ester, an NO synthase inhibitor, prevented the cocoa drink-induced vasodilatation demonstrating the involvement of NO [63]. Intake of a high-flavanol cocoa drink (917 mg total flavanols) but not of a low-flavanol (37 mg total flavanols) cocoa drink also increased at 1–3 h brachial artery flow-mediated vasodilatation in healthy subjects; this effect was paralleled by an augmentation of nitroso species (RXNO, an indicator of the formation of NO metabolites) and flavanols/metabolites in plasma [64]. Moreover, oral ingestion of pure epicatechin (1 and 2 mg/kg) resulted in a similar time-dependent increase in flow-mediated vasodilatation as the high-flavanol cocoa drink [64]. Daily intake of high-flavonoid (213 mg procyanidins, 46 mg epicatechin) dark chocolate bar (46 g) by healthy subjects improved endothelium-dependent flow-mediated dilation of the brachial artery as compared to a low-flavonoid chocolate bar, and this effect is associated with an increased plasma epicatechin concentration [65]. Moreover, acute ingestion of drinks containing 2–26 g of cocoa dose-dependently (from 9.3 to 146.0 mg of flavanols) increased brachial artery flow-mediated vasodilatation in healthy older humans (63 ± 2 years) [66]. A flavonoid-rich dark chocolate (cacao polyphenols 550 mg/day) improved also the coronary circulation in healthy adults as assessed by noninvasive transthoracic Doppler echocardiography [67]. Altogether, these observations indicate that flavanol-rich cocoa products induce NO-mediated endothelium-dependent dilatation in several vascular beds in healthy humans and that this effect is mediated, at least in part, by epicatechin.

In patients presenting with at least one cardiovascular risk factor such as coronary artery disease, hypertension, diabetes, or smoking, intake of 100 ml of a high-flavanol cocoa drink containing 176 mg of flavanols but not a low-flavanol cocoa drink (less than 10 mg of flavanols) reversed endothelial dysfunction as assessed by flow-mediated dilation of the brachial artery [68]. The improved vasodilatation was paralleled with an increased plasma NO bioavailability [68]. In hypertensive patients with impaired glucose tolerance and in patients with untreated essential hypertension, intake of a 100 g of a flavanol-rich dark chocolate bar, but not a flavanol-free white chocolate bar, improved flow-mediated dilation of the brachial artery, insulin sensitivity, and blood pressure [69, 70]. Intake of 100 g of a flavanol-rich dark chocolate bar also reduced systolic and diastolic blood pressure in elderly subjects with isolated hypertension [71]. Moreover, daily intake of 6.3 g of dark chocolate containing 30 mg of polyphenols in subjects with untreated upper-range prehypertension or stage 1 hypertension without concomitant risk factors reduced both systolic and diastolic blood pressure accompanied by an increased NO bioavailability and appearance of cocoa polyphenols in plasma [9]. Although a beneficial effect of cocoa/chocolate on blood pressure has been observed in several studies, others did not observe a beneficial effect on blood pressure. Indeed, intake of a flavanol-rich cocoa drink (approximately 900 mg flavanols per day) in subjects with essential hypertension enhanced insulin-mediated vasodilatation without affecting blood pressure and insulin resistance [72]. Moreover, daily intake of 50 g of dark chocolate with 70% cocoa containing 750 mg polyphenols in prehypertensive healthy adults did not improve blood pressure [73]. Besides hypertension, intake of flavanol-rich cocoa/chocolate has also been shown to improve endothelial dysfunction in smokers [74, 75], diabetic patients [76], and heart transplant recipients [77], whereas no such effect was observed in a study with patients with established coronary artery diseases [78].

Summary

Numerous experimental and clinical studies have indicated that flavanol-rich cocoa and chocolate are able to improve the endothelial function in health and disease mostly by stimulating the formation of NO, which, in turn, enhances the protective effect of endothelial cells on the vascular function. The beneficial effect of cocoa-derived products is often mimicked by flavanols, suggesting that they are active ingredients. Further studies are needed to better characterize the active flavanols/metabolites and their bioavailability, the dose-effect relationship, and the molecular targets/mechanisms underlying the vasoprotective effect of cocoa.

References

1. Feletou M, Vanhoutte PM. Endothelial dysfunction: a multifaceted disorder (the Wiggers Award Lecture). Am J Physiol Heart Circ Physiol. 2006;291(3):H985–1002.
2. Tang EH, Vanhoutte PM. Endothelial dysfunction: a strategic target in the treatment of hypertension? Pflugers Arch. 2010;459(6):995–1004.
3. Griendling KK, Sorescu D, Ushio-Fukai M. NAD(P)H oxidase: role in cardiovascular biology and disease. Circ Res. 2000;86(5):494–501.
4. Harrison DG, Cai H, Landmesser U, Griendling KK. Interactions of angiotensin II with NAD(P)H oxidase, oxidant stress and cardiovascular disease. J Renin Angiotensin Aldosterone Syst. 2003;4(2):51–61.
5. Mezzano D, Leighton F, Martinez C, Marshall G, Cuevas A, Castillo O, et al. Complementary effects of Mediterranean diet and moderate red wine intake on haemostatic cardiovascular risk factors. Eur J Clin Nutr. 2001;55(6):444–51.
6. Naissides M, Pal S, Mamo JC, James AP, Dhaliwal S. The effect of chronic consumption of red wine polyphenols on vascular function in postmenopausal women. Eur J Clin Nutr. 2006;60(6):740–5.
7. Marfella R, Cacciapuoti F, Siniscalchi M, Sasso FC, Marchese F, Cinone F, et al. Effect of moderate red wine intake on cardiac prognosis after recent acute myocardial infarction of subjects with Type 2 diabetes mellitus. Diabet Med. 2006;23(9):974–81.
8. Duffy SJ, Keaney Jr JF, Holbrook M, Gokce N, Swerdloff PL, Frei B, et al. Short- and long-term black tea consumption reverses endothelial dysfunction in patients with coronary artery disease. Circulation. 2001; 104(2):151–6.
9. Taubert D, Roesen R, Lehmann C, Jung N, Schomig E. Effects of low habitual cocoa intake on blood pressure and bioactive nitric oxide: a randomized controlled trial. JAMA. 2007;298(1):49–60.
10. Frankel EN, Kanner J, German JB, Parks E, Kinsella JE. Inhibition of oxidation of human low-density lipoprotein by phenolic substances in red wine. Lancet. 1993;341(8843):454–7.
11. Freedman JE, Parker III C, Li L, Perlman JA, Frei B, Ivanov V, et al. Select flavonoids and whole juice from purple grapes inhibit platelet function and enhance nitric oxide release. Circulation. 2001;103(23):2792–8.
12. Wollny T, Aiello L, Di TD, Bellavia V, Rotilio D, Donati MB, et al. Modulation of haemostatic function and prevention of experimental thrombosis by red wine in rats: a role for increased nitric oxide production. Br J Pharmacol. 1999;127(3):747–55.
13. Iijima K, Yoshizumi M, Hashimoto M, Akishita M, Kozaki K, Ako J, et al. Red wine polyphenols inhibit vascular smooth muscle cell migration through two distinct signaling pathways. Circulation. 2002;105(20):2404–10.
14. Iijima K, Yoshizumi M, Hashimoto M, Kim S, Eto M, Ako J, et al. Red wine polyphenols inhibit proliferation of vascular smooth muscle cells and downregulate expression of cyclin A gene. Circulation. 2000;101(7):805–11.
15. Auger C, Caporiccio B, Landrault N, Teissedre PL, Laurent C, Cros G, et al. Red wine phenolic compounds reduce plasma lipids and apolipoprotein B and prevent early aortic atherosclerosis in hypercholesterolemic golden Syrian hamsters (Mesocricetus auratus). J Nutr. 2002;132(6):1207–13.
16. Auger C, Gerain P, Laurent-Bichon F, Portet K, Bornet A, Caporiccio B, et al. Phenolics from commercialized grape extracts prevent early atherosclerotic lesions in hamsters by mechanisms other than antioxidant effect. J Agric Food Chem. 2004;52(16):5297–302.
17. Schini-Kerth VB, Auger C, Etienne-Selloum N, Chataigneau T. Polyphenol-induced endothelium-dependent relaxations role of NO and EDHF. Adv Pharmacol. 2010;60:133–75.
18. Fitzpatrick DF, Hirschfield SL, Coffey RG. Endothelium-dependent vasorelaxing activity of wine and other grape products. Am J Physiol Heart Circ Physiol. 1993;265(2 Pt 2):H774–8.
19. Fitzpatrick DF, Bing B, Maggi DA, Fleming RC, O'Malley RM. Vasodilating procyanidins derived from grape seeds. Ann NY Acad Sci. 2002;957:78–89.
20. Fitzpatrick DF, Fleming RC, Bing B, Maggi DA, O'Malley RM. Isolation and characterization of endothelium-dependent vasorelaxing compounds from grape seeds. J Agric Food Chem. 2000;48(12):6384–90.
21. Fitzpatrick DF, Bing B, Rohdewald P. Endothelium-dependent vascular effects of Pycnogenol. J Cardiovasc Pharmacol. 1998;32(4):509–15.
22. Anselm E, Socorro VF, Dal-Ros S, Schott C, Bronner C, Schini-Kerth VB. Crataegus special extract WS 1442 causes endothelium-dependent relaxation via a redox-sensitive Src- and Akt-dependent activation of endothelial NO synthase but not via activation of estrogen receptors. J Cardiovasc Pharmacol. 2009;53(3):253–60.
23. Andriambeloson E, Kleschyov AL, Muller B, Beretz A, Stoclet JC, Andriantsitohaina R. Nitric oxide production and endothelium-dependent vasorelaxation induced by wine polyphenols in rat aorta. Br J Pharmacol. 1997;120(6):1053–8.
24. Andriambeloson E, Magnier C, Haan-Archipoff G, Lobstein A, Anton R, Beretz A, et al. Natural dietary polyphenolic compounds cause endothelium-dependent vasorelaxation in rat thoracic aorta. J Nutr. 1998;128(12): 2324–33.

25. Cishek MB, Galloway MT, Karim M, German JB, Kappagoda CT. Effect of red wine on endothelium-dependent relaxation in rabbits. Clin Sci (Lond). 1997;93(6):507–11.
26. Dal-Ros S, Bronner C, Schott C, Kane MO, Chataigneau M, Schini-Kerth VB, et al. Angiotensin II-induced hypertension is associated with a selective inhibition of endothelium-derived hyperpolarizing factor-mediated responses in the rat mesenteric artery. J Pharmacol Exp Ther. 2009;328(2):478–86.
27. Soares De Moura R, Costa Viana FS, Souza MA, Kovary K, Guedes DC, Oliveira EP, et al. Antihypertensive, vasodilator and antioxidant effects of a vinifera grape skin extract. J Pharm Pharmacol. 2002;54(11):1515–20.
28. Chan SL, Tabellion A, Bagrel D, Perrin-Sarrado C, Capdeville-Atkinson C, Atkinson J. Impact of chronic treatment with red wine polyphenols (RWP) on cerebral arterioles in the spontaneous hypertensive rat. J Cardiovasc Pharmacol. 2008;51(3):304–10.
29. Nagaoka T, Hein TW, Yoshida A, Kuo L. Resveratrol, a component of red wine, elicits dilation of isolated porcine retinal arterioles: role of nitric oxide and potassium channels. Invest Ophthalmol Vis Sci. 2007;48(9):4232–9.
30. Ndiaye M, Chataigneau M, Lobysheva I, Chataigneau T, Schini-Kerth VB. Red wine polyphenol-induced, endothelium-dependent NO-mediated relaxation is due to the redox-sensitive PI3-kinase/Akt-dependent phosphorylation of endothelial NO-synthase in the isolated porcine coronary artery. FASEB J. 2005;19(3):455–7.
31. Flesch M, Schwarz A, Bohm M. Effects of red and white wine on endothelium-dependent vasorelaxation of rat aorta and human coronary arteries. Am J Physiol. 1998;275(4 Pt 2):H1183–90.
32. Rakici O, Kiziltepe U, Coskun B, Aslamaci S, Akar F. Effects of resveratrol on vascular tone and endothelial function of human saphenous vein and internal mammary artery. Int J Cardiol. 2005;105(2):209–15.
33. Karim M, McCormick K, Kappagoda CT. Effects of cocoa extracts on endothelium-dependent relaxation. J Nutr. 2000;130(8S Suppl):2105S–8.
34. Ndiaye M, Chataigneau T, Andriantsitohaina R, Stoclet JC, Schini-Kerth VB. Red wine polyphenols cause endothelium-dependent EDHF-mediated relaxations in porcine coronary arteries via a redox-sensitive mechanism. Biochem Biophys Res Commun. 2003;310(2):371–7.
35. Anselm E, Chataigneau M, Ndiaye M, Chataigneau T, Schini-Kerth VB. Grape juice causes endothelium-dependent relaxation via a redox-sensitive Src- and Akt-dependent activation of eNOS. Cardiovasc Res. 2007; 73(2):404–13.
36. Kwan CY, Zhang WB, Deyama T, Nishibe S. Endothelium-dependent vascular relaxation induced by Eucommia ulmoides Oliv. bark extract is mediated by NO and EDHF in small vessels. Naunyn Schmiedebergs Arch Pharmacol. 2004;369(2):206–11.
37. Kwan CY, Zhang WB, Sim SM, Deyama T, Nishibe S. Vascular effects of Siberian ginseng (Eleutherococcus senticosus): endothelium-dependent NO- and EDHF-mediated relaxation depending on vessel size. Naunyn Schmiedebergs Arch Pharmacol. 2004;369(5):473–80.
38. Rocha AP, Carvalho LC, Sousa MA, Madeira SV, Sousa PJ, Tano T, et al. Endothelium-dependent vasodilator effect of Euterpe oleracea Mart. (Acai) extracts in mesenteric vascular bed of the rat. Vascul Pharmacol. 2007;46(2):97–104.
39. Ndiaye M, Chataigneau T, Chataigneau M, Schini-Kerth VB. Red wine polyphenols induce EDHF-mediated relaxations in porcine coronary arteries through the redox-sensitive activation of the PI3-kinase/Akt pathway. Br J Pharmacol. 2004;142(7):1131–6.
40. Madeira SV, Auger C, Anselm E, Chataigneau M, Chataigneau T, Soares de Moura R, et al. eNOS activation induced by a polyphenol-rich grape skin extract in porcine coronary arteries. J Vasc Res. 2009;46(5):406–16.
41. Auger C, Kim JH, Chabert P, Chaabi M, Anselm E, Lanciaux X, et al. The EGCg-induced redox-sensitive activation of endothelial nitric oxide synthase and relaxation are critically dependent on hydroxyl moieties. Biochem Biophys Res Commun. 2010;393(1):162–7.
42. Natsume M, Osakabe N, Yamagishi M, Takizawa T, Nakamura T, Miyatake H, et al. Analyses of polyphenols in cacao liquor, cocoa, and chocolate by normal-phase and reversed-phase HPLC. Biosci Biotechnol Biochem. 2000;64(12):2581–7.
43. Ramirez-Sanchez I, Maya L, Ceballos G, Villarreal F. (–)-epicatechin activation of endothelial cell endothelial nitric oxide synthase, nitric oxide, and related signaling pathways. Hypertension. 2010;55(6):1398–405.
44. Kim JA, Formoso G, Li Y, Potenza MA, Marasciulo FL, Montagnani M, et al. Epigallocatechin gallate, a green tea polyphenol, mediates NO-dependent vasodilation using signaling pathways in vascular endothelium requiring reactive oxygen species and Fyn. J Biol Chem. 2007;282(18):13736–45.
45. Ramirez-Sanchez I, Maya L, Ceballos G, Villarreal F. (–)-Epicatechin induces calcium and translocation independent eNOS activation in arterial endothelial cells. Am J Physiol Cell Physiol. 2011;300(4):C880–7.
46. Martin S, Andriambeloson E, Takeda K, Andriantsitohaina R. Red wine polyphenols increase calcium in bovine aortic endothelial cells: a basis to elucidate signalling pathways leading to nitric oxide production. Br J Pharmacol. 2002;135(6):1579–87.
47. Cienfuegos-Jovellanos E, Quinones MM, Muguerza B, Moulay L, Miguel M, Aleixandre A. Antihypertensive effect of a polyphenol-rich cocoa powder industrially processed to preserve the original flavonoids of the cocoa beans. J Agric Food Chem. 2009;57(14):6156–62.

48. Sanchez D, Quinones M, Moulay L, Muguerza B, Miguel M, Aleixandre A. Changes in arterial blood pressure of a soluble cocoa fiber product in spontaneously hypertensive rats. J Agric Food Chem. 2010;58(3):1493–501.
49. Shaughnessy KS, Boswall IA, Scanlan AP, Gottschall-Pass KT, Sweeney MI. Diets containing blueberry extract lower blood pressure in spontaneously hypertensive stroke-prone rats. Nutr Res. 2009;29(2):130–8.
50. Negishi H, Xu JW, Ikeda K, Njelekela M, Nara Y, Yamori Y. Black and green tea polyphenols attenuate blood pressure increases in stroke-prone spontaneously hypertensive rats. J Nutr. 2004;134(1):38–42.
51. Peng N, Clark JT, Prasain J, Kim H, White CR, Wyss JM. Antihypertensive and cognitive effects of grape polyphenols in estrogen-depleted, female, spontaneously hypertensive rats. Am J Physiol Regul Integr Comp Physiol. 2005;289(3):R771–5.
52. Lopez-Sepulveda R, Jimenez R, Romero M, Zarzuelo MJ, Sánchez M, Gómez-Guzmán M, et al. Wine polyphenols improve endothelial function in large vessels of female spontaneously hypertensive rats. Hypertension. 2008;51(4):1088–95.
53. Bernatova I, Pechanova O, Babal P, Kysela S, Stvrtina S, Andriantsitohaina R. Wine polyphenols improve cardiovascular remodeling and vascular function in NO-deficient hypertension. Am J Physiol Heart Circ Physiol. 2002;282(3):H942–8.
54. Pechanova O, Bernatova I, Babal P, Martínez MC, Kyselá S, Stvrtina S, et al. Red wine polyphenols prevent cardiovascular alterations in L-NAME-induced hypertension. J Hypertens. 2004;22(8):1551–9.
55. Sarr M, Chataigneau M, Martins S, Schott C, El Bedoui J, Oak MH, et al. Red wine polyphenols prevent angiotensin II-induced hypertension and endothelial dysfunction in rats: role of NADPH oxidase. Cardiovasc Res. 2006;71(4):794–802.
56. Jimenez R, Lopez-Sepulveda R, Kadmiri M, Romero M, Vera R, Sánchez M, et al. Polyphenols restore endothelial function in DOCA-salt hypertension: role of endothelin-1 and NADPH oxidase. Free Radic Biol Med. 2007;43(3):462–73.
57. Kurosawa T, Itoh F, Nozaki A, Nakano Y, Katsuda S, Osakabe N, et al. Suppressive effect of cocoa powder on atherosclerosis in Kurosawa and Kusanagi-hypercholesterolemic rabbits. J Atheroscler Thromb. 2005;12(1): 20 8.
58. Kurosawa T, Itoh F, Nozaki A, Nakano Y, Katsuda S, Osakabe N, et al. Suppressive effects of cacao liquor polyphenols (CLP) on LDL oxidation and the development of atherosclerosis in Kurosawa and Kusanagi-hypercholesterolemic rabbits. Atherosclerosis. 2005;179(2):237–46.
59. Tomaru M, Takano H, Osakabe N, Yasuda A, Inoue K, Yanagisawa R, et al. Dietary supplementation with cacao liquor proanthocyanidins prevents elevation of blood glucose levels in diabetic obese mice. Nutrition. 2007; 23(4):351–5.
60. Ruzaidi A, Amin I, Nawalyah AG, Hamid M, Faizul HA. The effect of Malaysian cocoa extract on glucose levels and lipid profiles in diabetic rats. J Ethnopharmacol. 2005;98(1–2):55–60.
61. Al-Awwadi NA, Araiz C, Bornet A, Delbosc S, Cristol JP, Linck N, et al. Extracts enriched in different polyphenolic families normalize increased cardiac NADPH oxidase expression while having differential effects on insulin resistance, hypertension, and cardiac hypertrophy in high-fructose-fed rats. J Agric Food Chem. 2005;53(1): 151–7.
62. Dal-Ros S, Zoll J, Lang AL, Auger C, Keller N, Bronner C, Geny B, Schini-Kerth VB, et al. Chronic intake of red wine polyphenols by young rats prevents aging-induced endothelial dysfunction and decline in physical performance: role of NADPH oxidase. Biochem Biophys Res Commun. 2011;404(2):743–9.
63. Fisher ND, Hughes M, Gerhard-Herman M, Hollenberg NK. Flavanol-rich cocoa induces nitric-oxide-dependent vasodilation in healthy humans. J Hypertens. 2003;21(12):2281–6.
64. Schroeter H, Heiss C, Balzer J, Kleinbongard P, Keen CL, Hollenberg NK, et al. (−)-Epicatechin mediates beneficial effects of flavanol-rich cocoa on vascular function in humans. Proc Natl Acad Sci USA. 2006;103(4):1024–9.
65. Engler MB, Engler MM, Chen CY, Malloy MJ, Browne A, Chiu EY, et al. Flavonoid-rich dark chocolate improves endothelial function and increases plasma epicatechin concentrations in healthy adults. J Am Coll Nutr. 2004;23(3):197–204.
66. Monahan KD, Feehan RP, Kunselman AR, Preston AG, Miller DL, Lott ME. Dose-dependent increases in flow-mediated dilation following acute cocoa ingestion in healthy older adults. J Appl Physiol. 2011;111(6):1568–74.
67. Shiina Y, Funabashi N, Lee K, Murayama T, Nakamura K, Wakatsuki Y, et al. Acute effect of oral flavonoid-rich dark chocolate intake on coronary circulation, as compared with non-flavonoid white chocolate, by transthoracic Doppler echocardiography in healthy adults. Int J Cardiol. 2009;131(3):424–9.
68. Heiss C, Dejam A, Kleinbongard P, Schewe T, Sies H, Kelm M. Vascular effects of cocoa rich in flavan-3-ols. JAMA. 2003;290(8):1030–1.
69. Grassi D, Desideri G, Necozione S, Lippi C, Casale R, Properzi G, et al. Blood pressure is reduced and insulin sensitivity increased in glucose-intolerant, hypertensive subjects after 15 days of consuming high-polyphenol dark chocolate. J Nutr. 2008;138(9):1671–6.

70. Grassi D, Necozione S, Lippi C, Croce G, Valeri L, Pasqualetti P, et al. Cocoa reduces blood pressure and insulin resistance and improves endothelium-dependent vasodilation in hypertensives. Hypertension. 2005;46(2): 398–405.
71. Taubert D, Berkels R, Roesen R, Klaus W. Chocolate and blood pressure in elderly individuals with isolated systolic hypertension. JAMA. 2003;290(8):1029–30.
72. Muniyappa R, Hall G, Kolodziej TL, Karne RJ, Crandon SK, Quon MJ. Cocoa consumption for 2 wk enhances insulin-mediated vasodilatation without improving blood pressure or insulin resistance in essential hypertension. Am J Clin Nutr. 2008;88(6):1685–96.
73. Ried K, Frank OR, Stocks NP. Dark chocolate or tomato extract for prehypertension: a randomised controlled trial. BMC Complement Altern Med. 2009;9:22.
74. Hermann F, Spieker LE, Ruschitzka F, Sudano I, Hermann M, Binggeli C, et al. Dark chocolate improves endothelial and platelet function. Heart. 2006;92(1):119–20.
75. Heiss C, Finis D, Kleinbongard P, Hoffmann A, Rassaf T, Kelm M, et al. Sustained increase in flow-mediated dilation after daily intake of high-flavanol cocoa drink over 1 week. J Cardiovasc Pharmacol. 2007;49(2):74–80.
76. Balzer J, Rassaf T, Heiss C, Kleinbongard P, Lauer T, Merx M, et al. Sustained benefits in vascular function through flavanol-containing cocoa in medicated diabetic patients a double-masked, randomized, controlled trial. J Am Coll Cardiol. 2008;51(22):2141–9.
77. Flammer AJ, Hermann F, Sudano I, Spieker L, Hermann M, Cooper KA, et al. Dark chocolate improves coronary vasomotion and reduces platelet reactivity. Circulation. 2007;116(21):2376–82.
78. Farouque HM, Leung M, Hope SA, Baldi M, Schechter C, Cameron JD, et al. Acute and chronic effects of flavanol-rich cocoa on vascular function in subjects with coronary artery disease: a randomized double-blind placebo-controlled study. Clin Sci (Lond). 2006;111(1):71–80.

Chapter 21
Chocolate Flavonoids in the Prevention of Arterial Disease

Nancy J. Correa-Matos and Catherine Christie

Key Points

- Chocolate, which provides a concentrated source of energy because of its high-fat content, belongs to a class of polyphenols known as flavonoids, which contain such celebrated foods as red grapes, tea, soy, and garlic.
- Dark chocolate contains the highest amount of flavonoid-rich cocoa.
- Both short- and long-term ingestion of chocolate products result in an increase in serum antioxidant capacity and a decrease in LDL oxidation, both linked to reduced heart disease risk.
- Health benefits from chocolate consumption related to cardiovascular disease include the health neutrality of its saturated fatty acid (stearic acid) and protective effects of its flavonols, including flavanols and procyanidins.
- Health benefits from chocolate consumption related to cardiovascular disease prevention include anti-inflammatory functions, which prevent development of fatty streaks in the beginning stages of the atherosclerotic process.

Keywords Chocolate flavonoids • Heart disease prevention • Increased antioxidant capacity • Decreased LDL oxidation • Increased chocolate consumption worldwide

Introduction

Cardiovascular disease is the leading cause of death worldwide, claiming millions of deaths in both industrialized and developing nations. With over 17 million deaths a year, the World Health Organization (WHO) has labeled diseases of the heart a pandemic that recognizes no border [1]. Of the 13 million heart attack and stroke deaths each year, most of the deaths come from developing nations. Massive public health campaigns representing collaboration between the WHO and member

N.J. Correa-Matos, Ph.D., RD, LND, L/N (✉)
Department of Nutrition and Dietetics, Brooks College of Health,
University of North Florida, 1 UNF Drive, Jacksonville, FL 32256-2673, USA
e-mail: ncorrea@unf.edu

C. Christie, Ph.D., RD, LD/N, FADA
Brooks College of Health, University of North Florida, 1 UNF Drive, Jacksonville, FL 32256-2673, USA
e-mail: c.christie@unf.edu

R.R. Watson et al. (eds.), *Chocolate in Health and Nutrition*, Nutrition and Health 7,
DOI 10.1007/978-1-61779-803-0_21, © Springer Science+Business Media, LLC 2013

health agencies have begun to make small strides in certain regions through education, preventive services, and treatment [1].

While cardiovascular disease rates are decreasing slightly, environmental factors, such as nutrition, physical activity, and smoking, play a crucial role in the development of this chronic and deadly disease. Studies have demonstrated a link between excessive intake of saturated fat consumption and heart disease, indicating a critical role of diet in heart disease prevention [2]. High intake of saturated fat, along with genetic disposition, may be a primary risk factor for dyslipidemia and elevated low-density lipoprotein cholesterol. Long-term consequences of avoiding preventative measures may result in clogging of the arteries or oxidative damage. Oxidative damage in the pathophysiology of cardiovascular disease is a multifactorial process involving low-density lipoprotein (LDL) cholesterol, inflammation, and vasoconstriction, which results in the formation of atherosclerotic plaques and thrombosis.

For years, public health officials and nutrition experts have emphasized fruit and vegetable consumption for its role in displacing fat consumption and increasing antioxidant and phytochemical content. Diets rich in plant-based foods contain significant amounts of fiber, water, vitamins, and minerals and are low in sodium, saturated fats, and trans fats.

With recent advances in food analysis technology, chocolate, previously considered to play a minute role in human nutrition, has recently been making headlines with regard to its possible cardioprotective characteristics. Chocolate, which provides a concentrated source of energy because of its high-fat content, belongs to a class of polyphenols known as flavonoids, which contain such celebrated foods as red grapes, tea, soy, and garlic. Cocoa contains many oligomeric polymers of the flavonoids, which include flavonol, procyanidin, and anthocyanidin, and the phenolic compounds caffeic acid, chlorogenic acid, p-coumaric acid, phenylacetic acid, and phloretic acid [3]. While some flavonoid contents are lost in the fermentation and production of chocolate, its potential for supplying polyphenol-rich health benefits should not be disregarded. Dark chocolate contains the highest amount of flavonoid-rich cocoa. However, when milk is added to chocolate and produces one of the most popular forms of chocolate, the milk contributes small amounts of protein, vitamins, and minerals, and substantial amounts of added saturated fat. The primary fat found in chocolate is cocoa butter, which contains the saturated fatty acid, stearic acid. This fatty acid, the flavonoid content of chocolate, and subsequent effects on lipoprotein oxidation, inflammation biomarkers, platelet aggregation, vasodilation, as well as cardiovascular health, in general, have become frequent topics of current research [4].

With chocolate's reputation for overconsumption, high calories, and added sugars, cocoa and chocolate products are still under scrutiny by many health professionals for reported health claims. This chapter provides a background into the historical origins and uses of chocolate as well as its production and composition. It also provides a review of the rapidly increasing research literature available about the role of chocolate in cardiovascular health and the potential for this world-renowned and highly desirable sweet to aid in the prevention of diseases of the heart.

Background

The cacao tree is a delicate tropical plant that grows 15–25 ft tall in the shade of trees of other crops in hot, rainforest climates. In fact, cacao can only be cultivated within a 20° span of the equator. Before being introduced to Spain and Europe during the colonization period, the cacao tree, or cocoa, originated from the Central or South American region [5]. The first European to discover chocolate was Christopher Columbus in 1502; however, it was a second discovery by Hernan Cortes in 1528 that resulted in chocolate being introduced to the European world. Today, cocoa is grown in Central and South America, the Caribbean, and in West African countries such as Ivory Coast and Ghana, which produce most of the world's supply [5].

The words cocoa and chocolate originated from the Mayan and Aztec terms cacao and cacahuatl and have their origins in reported divine discoveries by Gods in the Central and South American mountains [6]. The scientific name, *Theobroma cacao,* means "food of the Gods," appropriately referenced by the ancient cultures that first cultivated and enjoyed the cocoa beans, which are actually seeds. Cocoa grows in oval-shaped pods ranging an average of 10 in. and is orange to orange-yellow color when mature. The insides of the pods are filled with a cream-colored pulp holding anywhere from 20 to 50 cocoa beans. Cocoa has two main species, Criollo and Forastero [5]. While Forastero, found mainly in Brazil and Africa, is more prevalent because of its higher yield and easy cultivation, Criollo, found in Central America and the Caribbean, is better known for its exceptional quality and grade.

History

The existing views and uses of chocolate have changed considerably from the early days of its discovery. The Mesoamerican cultures viewed chocolate as magical and mystical. It played an important role in the diet for individuals of royalty, wealth, and power [7]. Chocolate had many uses, ranging from medication to currency, and was originally served as a frothy drink to warriors, merchants, and individuals of nobility. Far from the cocoa beverage known today, Mayan and Aztec cocoa beverage consisted of a frothy drink served cold, spicy, and bitter, with honey, corn, or seasonings such as pepper occasionally being added [8].

Most of the present varieties of chocolate have until recently been considered to have little nutritional value. However, in ancient times, the unprocessed cocoa was valued and used for its healing properties for many health problems or complications. Cocoa was considered to reduce intestinal and nervous distress, inflammation, infection, fever, and gout; have anticancer effects; stimulate the kidney; prolong longevity; and treat tuberculosis. Of significance to this chapter is the fact that chocolate was viewed as being of value for strengthening the heart and for treatment of angina. The medical concept of using chocolate for various ailments continued until the twentieth century, when an impartial and more scientific review of these beliefs began to develop [6].

Since that time, the uses and cultivation of chocolate have continued to evolve and have extended globally. When cocoa was introduced to the Spanish, sugar and other flavors such as vanilla were added and served steaming hot. Following the traditions of the Aztecs and Mayans, it was reserved for aristocrats, royalty, and the most noble of men. Kept secret from the rest of Europe for almost a hundred years, drinking chocolate as both a fashionable and medicinal drink quickly spread across Europe's elite after Spain built its first cocoa processing plant in 1580 [5].

Processing

For hundreds of years, the production of chocolate was unchanged and too expensive for consumption by the masses because of the high costs associated with cocoa grinding. However, during the era of the Industrial Revolution, many changes to the preparation methods of chocolate occurred that resulted in the varieties and uses of chocolate that we know today [7].

Today, the production and manufacturing of chocolate has become a precise and detailed science. After roasting and removal of the dried bean fragments, the concentrated pieces of cocoa are then grinded to liquefy the cocoa butter, creating nonalcoholic chocolate liquor [5]. At this stage, the chocolate liquor can be used to make either cocoa powder or chocolate using different processes. Which one is produced depends on whether the cocoa butter is removed from the liquor during processing. Flavonoid-rich chocolate liquor is a bitter liquid used in many delicious baking recipes. Another common baking ingredient is creamy cocoa butter, the vegetable fat from the cocoa bean extracted from the chocolate liquor. Whether being consumed as sweet white, milk, or dark chocolate or used as a soft creamy mixture for coating, dipping, or glazing, chocolate is available in many sweet or bitter varieties. Dark chocolate, available as a dark, semisweet, or bittersweet treat, is the darkest of eating chocolate and is high in chocolate liquor with minimal added sweeteners. Milk chocolate is made using minimal amounts of chocolate liquor and offers the least amount of flavonoids. Subsequently, the addition of an alkaline during the production of dark and milk chocolate contributes to a reduction of flavonoid content.

Chocolate Composition

Chocolate contains important nutrients that can contribute to the prevention of cardiovascular diseases: lipids, minerals, and polyphenols. From the cocoa plant to the commercially available chocolate, the benefits of chocolate outweigh the problems of excess calorie consumption.

Lipids

Cocoa butter is the source of fat in chocolate. Cocoa butter contains saturated fatty acids, stearic acid (18:0, 35%) and palmitic acid (16:0, 25%), and the unsaturated fatty acid oleic acid (18:1, 35%) [4, 9]. While saturated fatty acid consumption is associated with increased cardiovascular disease risk, stearic acid does not raise plasma LDL cholesterol and has been considered cholesterol neutral [10, 11]. Stearic acid is metabolically different from the other long-chain saturated fatty acids. Several animal and human studies have reported a lower absorption rate for stearic acid as compared to other fatty acids [12, 13]. Another significant factor in its relative absorption rate may be where the stearic acid appears on the triglyceride molecule [14, 15]. Recently, a study found that the lipid composition in chocolate might prevent LDL oxidation while modulate the rate of cholesterol absorption [16]. However, more research is needed to establish the exact mechanisms for the distinct neutral cholesterolemic effects of chocolate lipids.

Minerals

Cocoa beans contain several minerals, and the amount present in processed chocolate depends greatly on the amount of cocoa bean solids present. Dark chocolate has a higher mineral content than milk chocolate [4]. Adequate amounts of calcium, iron, magnesium, phosphorus, potassium, sodium, and zinc can be found in 41 g or a 14.5-ounce serving of cocoa, dark chocolate, and milk chocolate [17]. Calcium, potassium, and magnesium consumption have been related to the prevention of cardiovascular disease owing to their role in reducing blood pressure [18]. In fact, magnesium is required for normal heart rhythms, muscle, and nerve function. Magnesium deficiency is associated with elevated C-reactive protein, blood pressure, and cholesterol as well as arrhythmias and tachycardia [19]. An ounce of dark chocolate contains about 35 mg of magnesium, and a six-ounce bar will provide about half the DRI (320–400 mg). Milk chocolate only contains about half of the magnesium of dark chocolate and owing to the sugar content may result in increased urinary magnesium losses [20].

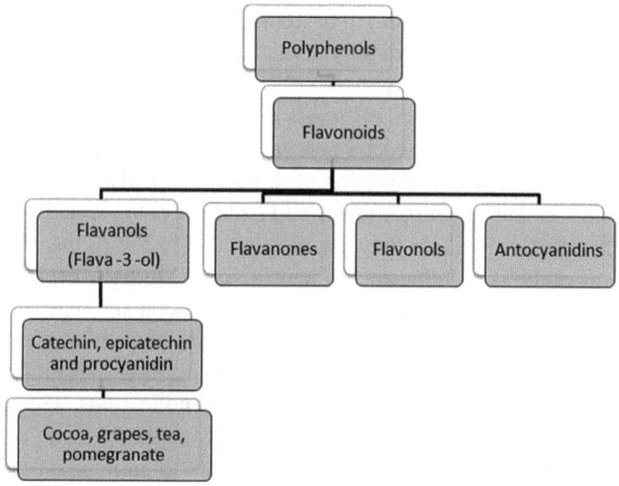

Fig. 21.1 Classification of polyphenols associated with prevention of cardiovascular diseases

Flavonoids

Cocoa and chocolate products, and particularly dark chocolate, contain polyphenol compounds that may provide cardiovascular protection (Fig. 21.1). While phenols consist of one aromatic ring containing at least one hydroxyl group, polyphenols consist of more than one aromatic ring with each ring containing at least one hydroxyl group. Flavonoids, a main class of polyphenols found in cocoa and chocolate, have a C6-C3-C6 backbone structure. The flavonoid group procyanidin is composed of flavan-3-ol monomers and their respective oligomeric chains of catechins and epicatechins, commonly bonded through a $4 \rightarrow 6$ or $4 \rightarrow 8$ linkage or doubly linked with a second interflavonoid bond formed by C-O oxidative coupling at the $2 \rightarrow O^7$ positions [21, 22]. While polyphenols and simple flavonoids are ubiquitous in nature and have been the past focus of food composition research, the use of normal phase HPLC has allowed for separation and quantification of larger oligomers found in cocoa and chocolate [23, 24].

Cocoa and chocolate contain larger oligomeric procyanidins that may contribute to more protective health benefits. Procyanidin oligomers have been identified in raw cocoa and in dark chocolate [24]. Chocolate contains more total procyanidin content by weight compared to wine, cranberry juice, and many varieties of apples. Over 10% of the weight of cocoa powder is flavonoids. However, when compared by serving size, the procyanidin content in chocolate and apples is equivocal. Apples offer more procyanidins per kilocalorie on average than chocolate, 1.3 and 0.9 mg/kcal, respectively [21]. High-cocoa liquor chocolate contains a substantial level of procyanidins, up to decamers, and because of the partial alkalization, dark chocolate contains a lower level. Additionally, the procyanidin content in cocoa is closely related to its oxygen radical absorbance capacity (ORAC) values, which indicates that procyanidins are the primary contributors to cocoa antioxidant capacity [23].

Bioavailability of Active Components

Under the acidic conditions of stimulated gastric juice, oligomeric procyanidins are hydrolyzed to monomer and dimeric units so that large quantities of epicatechins may be released and absorbed into the small intestine [25]. However, in vivo investigation of gastric contents every 10 min after cocoa beverage ingestion until the stomach was emptied showed no changes in the HPLC profile of procyanidins, and no depolymerization of cocoa procyanidins occurred in the stomach [26] keeping its stability in the gastric tissue [27]. The discrepancies may be explained by the increased duration of exposure and increased pH levels in the in vitro study. Received into the small intestine intact, the high molecular weight of cocoa procyanidins is too high for absorption. The major compound present in the portal vein after ingestion of dimer units is likely to be epicatechin, which is conjugated to only a small extent and not subject to O-methylation. Although epicatechin dimers are not extensively absorbed, it is likely they are bioavailable in the forms of epicatechin monomers [28]. The dimeric procyanidins (−)-epicatechin and (+)-catechin have been detected in the plasma of human subjects within 30 min of ingesting flavanol-rich cocoa and reached maximum concentrations by the second hour [29]. A recent investigation estimating the amounts of phenolic acids formed by the microflora and excreted in the urine of humans subjects after consumption of polyphenol-rich chocolate resulted in the excretion of the following six phenolic acids: *m*-hydroxyphenylpropionic acid, ferulic acid, 3,4-dihydroxyphenylacetic acid, *m*-hydroxyphenylacetic acid, vanillic acid, and *m*-hydroxybenzoic acid. These phenolic acids also contribute to antioxidant protection and suggest that the antioxidant effects of chocolate may not be solely expressed by the absorption of catechin molecules but also by the absorption of microbial phenolic acid metabolites [30].

Role of Chocolate Components in Cardiovascular Disease

Antioxidant Function

The structural characteristics of chocolate flavonoids allow for their potent antioxidant capacity owing to their ability to donate hydrogen ions in the process of scavenging free radicals and chelating metal ions [3, 31, 32]. The flavonoid content of various foods (Table 21.1) and beverages has been quantified by using high-performance liquid chromatography (HPLC) and then further correlated to their antioxidant activity by using oxygen radical absorbance capacity (ORAC) assay [21, 23, 24]. These methods are used as measuring devices and indicators of potential health benefits of the procyanidin content present in foods. As previously described, chocolate was found to have greater procyanidin content by

Table 21.1 Catechin/epicatechin concentrations found in food[a]

Source	Flavanol content, mg/kg or mg/L
Chocolate	460–610
Beans	350–550
Apricots	100–250
Cherries	50–220
Peaches	50–140
Blackberries	130
Apples	20–120
Green tea	100–800
Black tea	60–500
Red wine	80–300
Cider	40

[a]Reprinted with permission from Corti R, Flammer AJ, Hollenberg NK, Lüscher TF. Cocoa and cardiovascular health. Circulation. 2009;119(10):1433–1441

weight and higher ORAC values when compared to other antioxidant-rich foods and beverages such as wine, cranberry juice, and apples [21, 23, 24]. However, other studies have shown that this antioxidant and free scavenger protection is transient and that the protection in serum does not last more than 2 h, which is less than compared to vitamin E and vitamin C [33].

Both short- and long-term ingestion of chocolate products results in an increase in serum antioxidant capacity and a decrease in LDL oxidation [34–36]. LDL oxidation plays a significant role in the development of atherosclerotic lesions in cardiovascular disease (CVD). The potential for chocolate to reduce the risk of CVD by preventing LDL oxidation has been verified by both in vitro and in vivo studies via modulation of the enzymes involved in oxidation, 5-lipoxygenase and cyclooxygenases [37, 38]. However, a diet plentiful in fruits, vegetables, and other antioxidant-rich foods is not to be neglected [34–36].

Circulating plasma lipoproteins are susceptible to oxidation. Once oxidized, lipoproteins can accumulate and adhere to arterial walls. This process signals the influx of monocytes to the injured site. When the monocytes enter the endothelium, they convert into macrophages. Oxidized LDL is picked up by the macrophages and converted into foam cells, which in turn stimulate the release of immune mediating cytokines, leukocytes, and additional monocytes, which leads to plaque formation and accumulation [39].

Anti-inflammatory Function

Inflammation is the body's first immune response to damage, infection, or irritation. Inflammation is affected by many cellular components in the blood such as transcription factors and lipoxygenases. During the atherogenic process, the inflammatory response plays a major role in CVD progression [40]. The inflammatory response is stimulated by a cascade of events within the endothelial and subendothelial areas of the arteries. Endothelial dysfunction activates proinflammatory enzymes, which produce free radicals, thus contributing to injury [31, 32]. The inflammatory response promotes the development of fatty streaks seen in the beginning stages of the atherosclerotic process, which may lead to more severe arterial lesions distinctive of coronary heart disease. Thus, it is important to limit the chronic inflammatory response. The biomarker serum C-reactive protein has been used as a measure of inflammation.

Cyclooxygenase Inactivation

The inflammatory response within the body starts with an injury or stimulus. The result of the injury or other stimulus is the release of phospholipase $A_{2,}$ which stimulates the release of arachidonic acid. Arachidonic acid can then either be metabolized by cyclooxygenase 1 and 2 (COX-1 COX-2) or by the 5-lipoxygenase pathway. The COX pathway is responsible for the production of prostaglandins. COX-1 is responsible for baseline levels of prostaglandins, and COX-2 produces prostaglandins through activation (inflammatory response). COX-1, which is found in platelets, activates thromboxane A2 (TXA2), causes blood vessels to constrict, and promotes platelet aggregation. COX-2, by contrast, is expressed in blood vessels and activates prostaglandin I2 (PGI2), which dilates blood vessels and prevents the activation of platelets. Imbalance on the ratio TXA2: PG is related with the formation of a thrombus and infarction [41, 42]. The flavanols in chocolate seem to inhibit the activity of cyclooxygenase and to increase PGI2, preventing further platelet aggregation and inducing vasodilation [42], protecting the heart from damage.

Modulation of NF-kappa B Pathway

One component in chocolate that has an effect on anti-inflammatory response is epigallocatechin gallate (EGCG) (a plant flavonol). Epigallocatechin gallate is a flavonol found in dark chocolate that has been shown to inhibit NF-κB activation in human cell lines [3] by keeping the IκB/NF-κB complex. This complex prevents the translocation of NF-κB into the nucleus and further transcription of mRNA for proinflammatory cytokines, interleukin-1 beta, IL-2, and TNF-alpha, among others [43]. Chronic inflammation secondary to the activation of these cytokines leads to permanent damage to the muscle of the heart or heart failure [44].

Nitric Oxide Pathway Induction

Oligomeric flavonoids found in cocoa and chocolate may also be an important dietary source of protection against oxidative stress, which may occur in states of inflammation. One form of oxidative stress is associated with increased production of nitric oxide and peroxynitrite, which leads to tissue damage. When (−)-epicatechin and respective procyanidin oligomers isolated from the seeds of *Theobroma cacao* were examined for their ability to protect against peroxynitrite-dependent oxidation and nitration reactions, the tetrameric oligomers were more efficient than the monomeric epicatechin [45]. However, studies suggest that the intermediary procyanidins do not directly react with peroxynitrite but most likely with reactive oxidizing/nitrating intermediaries [45, 46]. Nitric oxide (NO) is a precursor to the prooxidant and nitrating intermediaries and can be affected by chocolate consumption. NO levels were shown to increase in individuals with diminished endothelial function who consumed a high-level flavonol cocoa drink; however, subjects with normal endothelial function did not receive the same beneficial effect. Therefore, the flavonols found in chocolate protect against peroxynitrate and could play an important anti-inflammatory role, which may have more than one mechanism of action related to oxidative stress [47].

Another function of cocoa flavanols is to aid in the function of NO in increasing the transcription of the phase II antioxidant enzymes involved in endothelial normal function [48] and preventing platelet aggregation [49]. This benefit is warranted by different mechanisms: inhibition of oxidative processes via blocking phospholipase activity and further peroxide production, inhibiting platelets aggregation receptors and lipooxygenase and cyclooxygenase pathways [33].

Control of Hypertension

The benefits of chocolate increasing endothelial NO, and thus reducing oxidation, have been shown to reduce blood pressure. A study conducted in hypertensive patients receiving 6 g of dark chocolate for 18 weeks found a significant reduction in blood pressure and increased level of NO [50] when compared to white chocolate. Although studies looking for the modulation of blood pressure via inhibition of angiotensin-converting enzymes in vitro have evidenced the effect of chocolate lowering blood pressure, more research is needed to see effects in humans [50]. The amount of stearic fatty acids in cocoa has been associated also with lower diastolic pressure in humans [51], but the exact mechanism must be elucidated.

Reduction of Atherosclerosis

Risk factor for the formation of the atherogenic plaque is the oxidation of LDL cholesterol and lipid. Several studies have shown that the consumption of chocolate increased high-density lipoproteins (HDL; good cholesterol), lowered triglycerides [11], inhibited LDL oxidation, and reduced total cholesterol [16]. The consumption of nonfermentable fiber (cellulose and bran) in unprocessed cocoa can also contribute to reduce total fat and cholesterol absorption [52]. The mechanism by which the flavonoid catechin (found mostly in dark chocolate) reduces serum cholesterol is via a reduction in its absorption when using the rodent model [53].

Platelet Modulation

The suppression of platelet aggregation is thought to play a role in decreasing the thrombolytic activity of blood platelets, thus potentially reducing cardiovascular damage. Research conducted both in vitro and in vivo has demonstrated chocolate's ability to suppress platelet activation [54–56]. The effects of cocoa beverages provided to subjects showed platelet suppression in both nonstimulated and stimulated platelets, while also decreasing platelet activation marker expression [54, 55].

Whole chocolate's effect on platelet activity has also shown beneficial results. Researchers found that after consumption of dark chocolate, platelet aggregation was inhibited whereas no effect was seen in those consuming white or milk chocolate [56]. This effect is thought to be derived from the flavonoids effect on altering eicosanoid synthesis. A recent study tested the efficacy of a high-flavonol chocolate bar, containing 148 mg of flavonols, to a control group who consumed a chocolate bar containing 3.33 mg of flavonols [57]. The consumption of this high-procyanidin chocolate resulted in an increase in plasma prostacyclin concentrations and a decrease in both the plasma leukotriene concentration and the plasma leukotriene prostacyclin ratio. This ability of flavonols to modulate eicosanoid synthesis may be due to the flavonol's capacity to affect enzymes that synthesize or degrade eicosanoids. Flavonols can also modulate enzymes in the leukotriene-synthesizing process by inhibiting 5-lipoxygenase. Inhibition of 5-lipoxygenase issues a cascading effect, which in turn inhibits production of leukotrienes. One of these leukotrienes is leukotriene A4 (LTA4), and LTA4 is then converted to leukotriene B4 (LTB4). LTB4's function is platelet aggregation. Flavonols inhibition of 5-lipoxygenase through this process ultimately leads to a decrease in platelet aggregation.

Improving Insulin Resistance

Flavanols have been shown to improve insulin sensitivity in nondiabetic obese patients with hyperglycemia receiving 100 g of dark chocolate for 15 days [58]. Most of the benefits are associated with the anti-inflammatory properties and the benefits to the endothelial tissue.

How Much Chocolate?

Table 21.2 shows evidence on the cardioprotective role of chocolate. However, it is important to be aware that the chocolate candy can be high in calories. Selection of lower-calorie chocolate is recommended. Moreover, higher benefits are conferred by the dark chocolate due to the high amount of flavonoids, such as epicatechin, among others [9].

Summary

Regular consumption of chocolate and its flavonols is increasing worldwide. Chocolate and the flavonols in chocolate are not simply preferred foods due to their flavor and texture but have many potentially beneficial and multifaceted effects on organs and tissues. Health benefits from chocolate

Table 21.2 Evidence of the cardioprotective effects of chocolate

Heart benefit	Dose	Model	Author
Vasodilation via NO synthesis	176–185 mg of flavonols in 2 h	Humans	Heiss et al. 2005 [62]
Reduction in blood pressure	6 g of dark chocolate containing 30 mg of polyphenols/day for 18 weeks	Humans	Taubert et al. 2007 [50]
Reduction in inflammation (lowered C-reactive protein), increase in HDL	700 mg of dark chocolate rich in flavonoids/day for 1 week	Humans	Hamed et al. 2008 [61]
Reduction of 5% of LDL cholesterol	81–163 mg of cocoa powder with epicatechin for 4 weeks	Mild hypercholesterolemic patients	Baba et al. 2007 [59]
Reduction in total cholesterol	37 g of dark chocolate bar for 6 weeks	Humans	Crews et al. 2008 [60]
Increase in serum antioxidant capacity and a decrease in LDL oxidation	22 g of cocoa powder, 16 g of dark chocolate	Humans	Wan et al. 2002 [35]; Wang et al. 2000 [36]

consumption related to CVD include the health neutrality of its saturated fatty acid (stearic acid) and protective effects of its flavonols, including flavanols and procyanidins, mostly found in dark chocolate. These include antioxidant capacity and ability to prevent LDL oxidation and reduce development of atherosclerotic lesions, anti-inflammatory functions which prevent development of fatty streaks in the beginning stages of the atherosclerotic process, and suppression of platelet aggregation through several mechanisms potentially reducing cardiovascular damage. Much of this research, while promising, has been conducted through short-term trials of chocolate consumption. Long-term trials are needed to confirm these beneficial health effects and CVD risk reduction before chocolate can be recommended as a healthful food. The current research suggests that chocolate can be included as part of a healthy diet with potential health benefits and without adverse CVD consequences, especially within a calorie-controlled diet that does not promote obesity.

References

1. World Health Organization. The Atlas of heart disease and stroke. Available at: http://www.who.int/cardiovascular_diseases/resources/atlas/en/. Accessed 1 Mar 2006.
2. Eyre H, Kahn R, Robertson RM. Preventing, cancer, cardiovascular disease, and diabetes: a common agenda for the American Cancer Society, the American Diabetes Association, and the American Heart Association. Diabetes Care. 2004;27:1812–24.
3. Borchers AT, Keen CL, Hannum SM, Gershwin ME. Cocoa and chocolate: composition, bioavailability, and health implications. J Med Food. 2000;3(2):77–104.
4. Steinberg FM, Bearden MM, Keen CL. Cocoa and chocolate flavonoids: implications for cardiovascular health. J Am Diet Assoc. 2003;103:215–23.
5. Chocolate Manufacturers Association. The story of chocolate. Available at: www.chocolateusa.org. Accessed 23 Feb 2006.
6. Dillinger TL, Barriga P, Escarcega S, Jimenez M, Lowe DS, Grivetti LE. Food for the Gods: cure for humanity? A cultural history of the medicinal and ritual uses of chocolate. J Nutr. 2000;30:2057S–72.
7. Morton M, Morton F. Chocolate: an illustrated history. New York: Crown Publishers; 1986.
8. Telly C. Chocolate: its quality and flavor (which is the world's best chocolate). In: Szogyi A, editors. Chocolate: food of the gods. Westport: Greenwood Press; 1997:165–166; USDA (2005) National Nutrient Database for Standard Reference, Release 18. Available at: http://www.ars.usda.gov/ba/bhnrc/ndl. Accessed 23 Jan 2006.

9. Corti R, Flammer AJ, Hollenberg NK, Lüscher TF. Cocoa and cardiovascular health. Circulation. 2009;119(10): 1433–41.
10. Kris-Etherton PM, Derr JA, Mitchell DC. The role of fatty acid saturation on plasma lipids, lipoproteins and apolipoproteins: effects of whole food diets high in cocoa butter, olive oil, soybean oil, dairy butter and milk chocolate on the plasma lipids of young men. Metabolism. 1993;42:130–4.
11. Kris-Etherton PM, Derr JA, Mustad VA, Seligson FH, Pearson TA. Effects of a milk chocolate bar a day substituted for a high carbohydrate snack increases high-density lipoprotein cholesterol in young men on a NCEP/AHA step one diet. Am J Clin Nutr. 1994;60:1037S–42.
12. Jones AE, Stolinski M, Smith RD, Murphy JL, Wootten SA. Effects of fatty acid chain length and saturation on the gastrointestinal handling and metabolic disposal of dietary fatty acids in women. Br J Nutr. 1999;81:37–43.
13. Baer DJ, Judd JT, Kris-Etherton PM, Zhao G, Emken EA. Stearic acid absorption and its metabolizable energy value are minimally lower than those of other fatty acids in healthy men fed mixed diets. J Nutr. 2003;133: 4129–34.
14. Nestel PJ, Pomeroy S, Kay S, Sasahara T, Yamashita T. Effect of a stearic acid-rich structured triacylglycerol on plasma lipid concentrations. Am J Clin Nutr. 1998;68:1196–201.
15. Brink EJ, Haddeman E, de Fouw NJ, Westrate JA. Positional distribution of stearic acid-rich, structured triacylglycerol and dietary calcium concentration determines the apparent absorption of these fatty acids in rats. J Nutr. 1995; 125:2379–2387.
16. Mursu J, Voutilainen S, Nurmi T, Rissanen TH, Virtanen JK, Kaikkonen J, et al. Dark chocolate consumption increases HDL cholesterol concentration and chocolate fatty acids may inhibit lipid peroxidation in healthy humans. Free Radic Biol Med. 2004;37(9):1351–9.
17. USDA. 2005. National nutrient database for standard reference, Release 18. Available at: http://www.ars.usda.gov/ba/bhnrc/ndl. Accessed 23 Jan 2006.
18. Executive Summary of the Third Report of the National Cholesterol Education Program (NCEP). Expert panel on detection, evaluation and treatment of high blood cholesterol in adults (Adult Treatment Panel III). JAMA. 2001; 285:2486–2498.
19. King DE, Mainous 3rd AG, Geesey ME, Woolson RF. Dietary magnesium and C-reactive protein levels. J Am Coll Nutr. 2005;24(3):166–71.
20. Barbagallo M, Dominguez LJ, Galioto A, Ferlisi A, Cani C, Malfa L, et al. Role of magnesium in insulin action, diabetes and cardio-metabolic syndrome X. Mol Aspects Med. 2003;24(1–3):39–52.
21. Hammerstone JF, Lazarus SA, Schmitz HH. Procyanidin content and variation in some commonly consumed foods. J Nutr. 2000;130:2086S–91.
22. Prior RL, Lazarus SA, Cao G, Muccitelli H, Hammerstone JF. Identification of procyanidins and anthocyanins in blueberries and cranberries (*Vaccinium* sp) using high-performance liquid chromatography/mass spectrometry. J Agric Food Chem. 2001;49:1270–6.
23. Adamson GE, Lazarus SA, Mitchell AE, Prior RL, Cao G, Jacobs PH, et al. HPLC method for the quantification of procyanidins in cocoa and chocolate samples and correlation to total antioxidant capacity. J Agric Food Chem. 1999;47:4184–8.
24. Hammerstone JF, Lazarus SA, Mitchell AE, Rucker R, Schmidt HH. Identification of procyanidins in cocoa (Thembroma cacao) and chocolate using high-performance liquid chromatography/mass spectrometry. J Agric Food Chem. 1999;47:490–6.
25. Spencer JPE, Chaudry F, Pannala AS, Srai SK, Debnam E, Rice-Evans C. Decomposition of cocoa procyanidins in the gastric milieu. Biochem Biophys Res Commun. 2000;272:236–41.
26. Rios LY, Bennett RN, Lazarus SA, Remesy C, Scalbert A, Williamson G. Cocoa procyanidins are stable during gastric transit in humans. Am J Clin Nutr. 2002;76:1106–10.
27. Mullen W, Borges G, Donovan JL, et al. Milk decreases urinary excretion but not plasma pharmacokinetics of cocoa flavan-3-ol metabolites in humans. Am J Clin Nutr. 2009;89:1784–91.
28. Spencer JP, Schroeter H, Shenoy B, Srai SKS, Debnam ES, Rice-Evans C. Epicatechin is the primary bioavailable form of the procyanidin dimers B2 and B5 after transfer across the small intestine. Biochem Biophys Res Commun. 2001;285:88–593.
29. Holt RR, Lazarus SA, Sullards MC, Zhu QY, Schramm DD, Hammerstone JF, et al. Procyanidin dimer B2 [epicatechin-(4 β)-epicatechin] in human plasma after the consumption of a flavanol-rich cocoa. Am J Clin Nutr. 2002;76:798–804.
30. Rios LY, Gonthier MP, Remesy C, Mila I, Lapierre C, Lazarus SA, et al. Chocolate intake increases urinary excretion of polyphenol-derived phenolic acids in healthy human subjects. Am J Clin Nutr. 2003;77:912–8.
31. Keen CL, Holt RR, Oteiza PI, Fraga CG, Schmitz HH. Cocoa antioxidants and cardiovascular health. Am J Clin Nutr. 2005;81(1):298s–303.
32. Ding EL, Hutfless SM, Ding X, Girotra S. Chocolate and prevention of cardiovascular disease: a systematic review. Nutr Metab (Lond). 2006; 3:2. (BioMed Central ISSN 1743–7075).

33. Galleano M, Oteiza PI, Fraga CG. Cocoa, chocolate, and cardiovascular disease. J Cardiovasc Pharmacol. 2009; 54(6):483–90.
34. Mathur S, Devaraj S, Grundy SM, Jialal I. Cocoa products decrease low density lipoprotein susceptibility but do not affect biomarkers of inflammation in humans. J Nutr. 2002;132:3663–7.
35. Wan Y, Vinson JA, Etherton TD, Proch J, Lazarus SA, Kris-Etherton PM. Effects of cocoa powder and dark chocolate on LDL oxidative susceptibility and prostaglandin concentrations in humans. Am J Clin Nutr. 2002;74:596–602.
36. Wang JF, Schramm DD, Holt RR, Ensunsa JL, et al. A dose–response effect from chocolate consumption on plasma epicatechin and oxidative damage. J Nutr. 2000;130(8s):2115s–9s.
37. Zhang WY, Liu HQ, Xie KQ, et al. Procyanidin dimer B2 [epicatechin-(4beta-8)-epicatechin] suppresses the expression of cyclooxygenase-2 in endotoxin-treated monocytic cells. Biochem Biophys Res Commun. 2006;345:508–15.
38. Lee KW, Kundu JK, Kim SO, et al. Cocoa polyphenols inhibit phorbol ester-induced superoxide anion formation in cultured HL-60 cells and expression of cyclooxygenase-2 and activation of NF-kappaB and MAPKs in mouse skin in vivo. J Nutr. 2006;136:1150–5.
39. Rader DJ. Inhibition of cholesteryl ester transfer protein activity: a new therapeutic approach to raising high-density lipoprotein. Curr Atheroscler Rep. 2004;6(5):398–405.
40. Libby P, Ridker PM, Maseri A. Inflammation and atherosclerosis. Circulation. 2002;106(18):135–6.
41. Selmi C, Cocchi CA, Lanfredini M, Keen CL, Gershwin ME. Chocolate at heart: the anti-inflammatory impact of cocoa flavanols. Mol Nutr Food Res. 2008;52(11):1340–8.
42. Grosser T. The pharmacology of selective inhibition of COX-2. Thromb Haemost. 2006;96(4):393–400.
43. Bremner P, Heinrich M. Natural products as targeted modulators of the nuclear factor-kappaB pathway. J Pharm Pharmacol. 2002;54(4):453–72.
44. Sharma R, Coats AJ, Anker SD. The role of inflammatory mediators in chronic heart failure: cytokines, nitric oxide, and endothelin-1. Int J Cardiol. 2000;72(2):175–86.
45. Arteel GE, Sies H. Protection against peroxynitrite by cocoa polyphenol oligomers. FEBS Lett. 1999;462: 167–70.
46. Arteel GE, Schroeder P, Sies H. Reactions of peroxynitrite with cocoa procyanidin oligomers. J Nutr. 2000;130: 2100S–4.
47. Sies H, Schewe T, Heiss C, Kelm M. Cocoa polyphenols and inflammatory mediators. Am J Clin Nutr. 2005;8: 304S–12.
48. Mann GE, Rowlands DJ, Li FY, de Winter P, Siow RC. Activation of endothelial nitric oxide synthase by dietary isoflavones: role of NO in Nrf2-mediated antioxidant gene expression. Cardiovasc Res. 2007;75(2):261–74.
49. Flammer AJ, Hermann F, Sudano I, et al. Dark chocolate improves coronary vasomotion and reduces platelet reactivity. Circulation. 2007;116:2376–82.
50. Taubert D, Roesen R, Lehmann C, Jung N, Schömig E. Effects of low habitual cocoa intake on blood pressure and bioactive nitric oxide: a randomized controlled trial. JAMA. 2007;298(1):49–60.
51. Simon JA, Fong J, Bernert Jr JT. Serum fatty acids and blood pressure. Hypertension. 1996;27:303–7.
52. Jenkins DJ, Kendall CW, Vuksan V, Vidgen E, Wong E, Augustin LS, et al. Effect of cocoa bran on low-density lipoprotein oxidation and fecal bulking. Arch Intern Med. 2000;160(15):2374–9.
53. Raederstorff DG, Schlachter MF, Elste V, Weber P. Effect of EGCG on lipid absorption and plasma lipid levels in rats. J Nutr Biochem. 2003;14(6):326–32.
54. Rein D, Paglieroni TG, Pearson DA, Wun T, Schmitz HH, Gosselin R, et al. Cocoa and wine polyphenols modulate platelet activation and function. J Nutr. 2000;130:2120s–6.
55. Rein D, Paglieroni TG, Wun T, Pearson DA, Schmitz HH, Gosselin R, et al. Cocoa inhibits platelet activation and function. Am J Clin Nutr. 2000;72:30–5.
56. Innes AJ, Kennedy G, McLaren M, Bancroft AJ, Belch J. Dark chocolate inhibits platelet aggregation in healthy volunteers. Platelets. 2003;14(5):325–7.
57. Schramm DD, Wang JF, Holt RR, Ensunsa JL, Gonsalves JL, Lazarus SA, et al. Chocolate procyanidins decrease the leuotriene-prostacyclin ratio in human and human aortic endothelial cells. Am J Clin Nutr. 2001;73:36–40.
58. Grassi D, Lippi C, Necozione S, Desideri G, Ferri C. Short-term administration of dark chocolate is followed by a significant increase in insulin sensitivity and a decrease in blood pressure in healthy persons. Am J Clin Nutr. 2005;81(3):611–4.
59. Baba S, Natsume M, Yasuda A, et al. Plasma LDL and HDL cholesterol and oxidized LDL concentrations are altered in normo- and hypercholesterolemic humans after intake of different levels of cocoa powder. J Nutr. 2007;137:1436–41.
60. Crews WD Jr, Harrison DW, Wright JW. A double-blind, placebo-controlled, randomized trial of the effects of dark chocolate and cocoa on variables associated with neuropsychological functioning and cardiovascular health: clinical findings from a sample of healthy, cognitively intact older adults. Am J Clin Nutr. 2008;87(4):872–80.

61. Hamed MS, Gambert S, Bliden KP, Bailon O, Singla A, Antonino MJ, Hamed F, Tantry US, Gurbel PA. Dark chocolate effect on platelet activity, C-reactive protein and lipid profile: a pilot study. South Med J. 2008;101(12):1203–8.
62. Heiss C, Kleinbongard P, Dejam A, Perré S, Schroeter H, Sies H, Kelm M. Acute consumption of flavanol-rich cocoa and the reversal of endothelial dysfunction in smokers. J Am Coll Cardiol. 2005;46(7):1276–83.

Chapter 22
Chocolate and Coronary Heart Disease

Imre Janszky and Kenneth J. Mukamal

Key Points

- No study has found chocolate clearly harmful in relation to clinical cardiovascular events. However, no study has examined or was designed to examine excessive consumption explicitly.
- In long-term observational studies, there is suggestive but not definitive evidence that chocolate consumption – at least in moderation – might inversely be associated with coronary heart disease and other cardiovascular events.
- Laboratory and short-term randomized studies indicate that cardiovascular protection from chocolate consumption is biologically plausible, and there are several potential mechanisms to explain the observed inverse association as causal effects of chocolate.
- However, owing to several potential methodological weaknesses in the long-term observational studies, we cannot yet make a firm conclusion on the long-term cardiovascular effects of chocolate.
- We need more long-term studies with a rigorous design; a long-term randomized trial focusing on clinical events is especially missing.

Keywords Chocolate • Coronary heart disease • Cardiovascular diseases • Flavonoid antioxidants • Stearic acid

Introduction

Cardiovascular disease is the leading cause of death worldwide. According to WHO estimates, 17 million people die as a result of cardiovascular disorders every year, which accounts for nearly 30% of all deaths. The most common cardiovascular disorder is coronary heart disease (CHD), which occurs when the supply of blood to heart muscle cells is hampered due to the narrowing of the coronary vessels and is responsible for over seven million deaths every year [1].

I. Janszky, M.D., Ph.D. (✉)
Public Health Sciences, Karolinska Insitutet, Stockholm, Sweden
e-mail: Imre.Janszky@ki.se

K.J. Mukamal, M.D., MPH
Department of Medicine, Beth Israel Deaconess Medical Center, Brookline, MA, USA
e-mail: kmukamal@bidmc.harvard.edu

R.R. Watson et al. (eds.), *Chocolate in Health and Nutrition*, Nutrition and Health 7,
DOI 10.1007/978-1-61779-803-0_22, © Springer Science+Business Media, LLC 2013

Along with physical activity and tobacco use, diet is among the most important modifiable lifestyle factors affecting CHD risk. Since chocolate is widely consumed and is a rich source of several biologically active compounds that can potentially affect the occurrence of CHD, there is great interest and indeed a pressing need to understand the potential influence of chocolate consumption on CHD risk.

The randomized controlled trial is the most powerful research design to establish causality in medicine. However, no randomized controlled trial has yet been conducted on chocolate consumption and occurrence of CHD. Such a trial would necessarily be large, long-term, and consequently very costly. Apart from that, it is questionable whether such trial could achieve successful blinding, that is, to make participants unaware of whether they consume chocolate or some kind of a placebo [2]. As such, the feasibility of testing chocolate in a true trial of CHD events is at best uncertain.

Several experimental studies have been conducted to explore the potential cardiovascular effects of chocolate and its components. However, these trials have all been short-term and focused upon physiological outcomes like blood pressure, endothelial function, or lipid levels, rather than the occurrence of CHD itself. Effects observed in short-term physiological trials cannot necessarily be extrapolated to long-term effects and to clinical endpoints. Thus, in absence of large-scale long-term randomized trials, we have to rely on prospective observational studies – in tandem with feeding studies – to evaluate chocolate's possible effects on clinical CHD events.

In this chapter, we review the current epidemiological evidence regarding the association of chocolate with CHD incidence and mortality, review the methodological challenges inherent in these observational studies, and review potential biological mechanisms that may play important roles in the etiology of observed associations.

Chocolate Consumption and CHD Risk: Long-Term Observational Studies

Although a number of other study designs have been used, here we review eight relevant long-term prospective observational studies which can shed light on possible long-term cardiovascular effects of chocolate consumption. These prospective studies all measured chocolate consumption prior to the occurrence of cardiovascular events, eliminating a potential source of bias if patients are (possibly inappropriately) counseled to avoid chocolate during cardiac rehabilitation, and represent the strongest available information we have to date. Table 22.1 provides a short summary of design and main findings of these studies.

In the Harvard Alumni Study [3], candy consumption was associated with a 27% lower all-cause mortality after adjustment for potential confounding factors. The lowest mortality was found among moderate users of candy. Those who consumed candies one to three times a month had 36% lower mortality compared to those who never or almost never used candies. However, in this study, chocolate and nonchocolate candies were not differentiated, and all-cause mortality was the outcome. While this is undoubtedly important, it does not directly inform the relationship of chocolate consumption with CHD, although it is likely that a considerable part of all-cause mortality was due to CHD.

The next relevant publication on the topic came from the Nurses' Health Study [4] in which no evidence was found for a major adverse or beneficial effect of chocolate on CHD incidence.

Two relevant publications derive from the Iowa Women's Health Study. At 13 years of follow-up, catechin intake from selected dietary sources was investigated. Catechin consumption from chocolate had a dose-dependent inverse association with CHD mortality [5]. When the same cohort was analyzed again slightly differently after 16 years of follow-up, there was some evidence that chocolate consumption was inversely associated with stroke incidence [6].

In the Zutphen Elderly Study, intake of 24 cocoa-containing food items was repeatedly assessed by a cross-check dietary history method [7]. The intake of cocoa from individual foods was summed to yield actual cocoa in grams per day for each subject. There was an inverse dose-dependent association

Table 22.1 Summary of prospective cohort studies on habitual chocolate intake and clinical cardiovascular events

Reference	Study subjects	Follow-up time (years)	Exposure	Main results	Multivariable adjustments
Lee and Paffenbarger (1998) [3]	Harvard Alumni Study 7,841 Men free of cardiovascular disease and cancer	5	Chocolate and nonchocolate candy users where compared to never/ almost never users	RR for all-cause mortality = 0.73 (95% CI, 0.60–0.89)	Age, BMI, energy expenditure, smoking, intake of alcohol, red meat, and vegetables, and vitamin or mineral supplement use
Hu et al. (2000) [4]	Nurses' Health Study	14	Number of servings of 1 oz (=28 g) chocolate	<1 Servings of chocolate per month, RR for CHD incidence = 1.15 (95% CI: 0.96–1.37)	Coronary risk factors and intake of meats and dairy products
	80,032 Middle-aged women with no known cardiovascular disease, cancer, hypercholesterolemia, or diabetes		Reference group = almost never	1 Serving per week, RR = 1.08 (95% CI, 0.88–1.32) 3–4 Servings peer week, RR = 1.11 (95% CI, 0.92–1.34)	
Arts et al. (2001) [5]	Iowa Women's Health Study	13	Tertiles of catechin consumption from chocolate	Middle tertile, RR for CHD mortality = 0.94 (95% CI, 0.78–1.13)	Age, marital status, education, estrogen replacement therapy, smoking, waist-hip ratio, physical activity, hypertension, diabetes, intake of total energy, lipids, alcohol, whole grains, vitamins, and intake of other sources of catechins like different fruits, wine, and tea
Mink et al. (2007) [6]	34,492 Postmenopausal women free of cardiovascular diseases	16	Reference = lowest tertiles	Highest tertile, RR = 0.88 (95% CI, 0.71–1.08)	
			Chocolate consumers were compared to nonconsumers	RR for CHD mortality = 0.98 (95% CI, 0.88–1.10) RR for stroke mortality = 0.85 (95% CI, 0.70–1.03)	Age, energy intake, marital status, education, blood pressure, diabetes, BMI, waist-to-hip ratio, physical activity, smoking, and estrogen use
Buijsse et al. (2006) [7]	Zutphen Elderly Study	15	Tertiles of cocoa intake	Middle tertile, RR for myocardial infarction = 0.70 (95% CI, 0.47–1.05)	Age, BMI, physical activity, smoking, diet prescription, medication, intake of alcohol, vegetables, fruit, meat, dairy fat, nuts and seeds, sugar confectionary other than chocolate, cookie and savory foods, coffee, and total calories
	470 Elderly Dutch men free of chronic diseases and diabetes		Reference group = the lowest tertile	Highest tertile, RR = 0.50 (95% CI, 0.32–0.78)	

(continued)

Table 22.1 (continued)

Reference	Study subjects	Follow-up time (years)	Exposure	Main results	Multivariable adjustments
Janszky et al. (2009) [8]	Stockholm Heart Epidemiology Program 1,169 Nondiabetic myocardial infarction survivors	8.5	Number of usual portions (50 g) of chocolate Reference group = nonusers	>1 Per month, RR for cardiac mortality = 0.73 (0.41–1.31), up to once per week, RR = 0.56 (0.32–0.99), twice or more per week, RR = 0.34 (0.17–0.70)	Age, sex, smoking, obesity, physical inactivity, education and intake of alcohol, coffee, and nonchocolate sweets
Mostofsky et al. (2010) [9]	Swedish Mammography Cohort 31,826 Women without a history of diabetes or heart failure or myocardial infarction	9	Number of servings of chocolate (19-30 g per serving) Reference group = no chocolate intake	1–3 Servings of chocolate per month, RR for heart failure = 0.74 (0.58–0.95) 1–2 Servings per week, RR = 0.68 (0.50–0.93) 3–6 Servings per week, RR = 1.09 (0.74–1.62) ≥1 Servings per day, RR = 1.23 (0.73–2.08)	Age, education, marital status, postmenopausal hormone use, smoking, BMI, physical activity, history of hypertension and hypercholesterolemia, family history of myocardial infarction, total energy intake, and intake of alcohol
Buijsse et al. (2010) [10]	Potsdam Arm of the European Prospective Investigation 19,357 Participants free of stroke and myocardial infarction	8	Top quartile of chocolate consumption was compared to the bottom quartile	RR for stroke = 0.52 (0.30–0.89) RR for myocardial infarction = 0.73 (0.47–1.15) RR for combination of myocardial infarction and stroke = 0.61 (0.44–0.87)	Age, sex, employment status, education, anthropometrics, smoking status, physical activity, diabetes, and intake of total energy, alcohol, fruit, vegetables, red and processed meat, dairy, coffee, tea, and cereal fiber
Lewis et al. (2010) [15]	1,216 Elderly Australian women	9.5	Those who consumed chocolate at least once a week were compared to those consuming it less than once a week	RR for all atherosclerotic vascular disease events = 0.76 (0.60–0.97) RR for CHD events = 0.65 (0.46–0.94) RR for heart failure = 0.41 (0.22–0.76)	Age, body mass index, socioeconomic status, and energy intake at baseline

RR relative risk, *CHD* coronary heart disease

between cocoa intake and cardiovascular mortality. Men in the highest tertile of cocoa intake had half of the cardiovascular mortality of those in the lowest tertile in the multiadjusted models. This method may have been particularly likely to reduce misclassification and identify true associations, a potentially important limitation discussed below.

In a secondary prevention context, we found a strong inverse, dose-dependent association between chocolate intake and cardiac mortality in Swedish survivors of a first myocardial infarction [8]. Chocolate consumption generally had a similarly inverse but weaker association with total mortality and nonfatal outcomes, like recurrent acute myocardial infarction, congestive heart failure, or stroke. In another Swedish study, moderate consumption of chocolate had a U-shaped association with hospitalization and death due to heart failure [9].

Two other studies from other nations found potential benefits related to chocolate. Chocolate consumption had an inverse dose-dependent association with cardiovascular events in a population-based German study [10]. In an Australian study including elderly women, there was also a relatively strong inverse association with all atherosclerotic vascular disease events, with CHD events and especially with heart failure.

Finally, some other investigations are worthy of mention, although these did not have a prospective cohort design. In an Italian case–control study including 760 cases of a first myocardial infarction and 682 controls, chocolate consumption had a strong inverse association with risk of acute myocardial infarction (AMI) [11]. After controlling for age, sex, education, diabetes, obesity, hypertension, hyperlipidemia, cholesterol level, body mass index, cigarette smoking, alcohol and coffee drinking, and family history of AMI, the relative risk for an increment of one serving of chocolate per day was 0.81 (95% confidence interval [CI], 0.69–0.96).

An intriguing ecologic study in Panama compared the cause-specific death rates on San Blas islands and the mainland. The rate of deaths due to cardiovascular causes among the island-dwellers was nearly one-tenth of that of mainland Panama. Given their shared genetic heritage, this protection appeared to be environmental and not genetic. The authors suggested that the very high intake of unprocessed cocoa by indigenous dwellers of the San Blas islands is one of the explanations for their finding [12].

In cross-sectional analyses of the National Heart, Lung, and Blood Institute Family Heart Study, there was a strong dose-dependent inverse association between chocolate consumption and prevalence of CHD. Compared to subjects who did not report any chocolate intake, the odds ratio for CHD was 0.43 (95% CI, 0.28–0.67) for the highest category of intake, that is, over 5 times/week when adjusting for age, sex, family CHD risk group, energy intake, education, nonchocolate candy intake, linolenic acid intake, smoking, alcohol intake, exercise, and fruit and vegetables [13, 14]. Moreover, in the same study, there was a similar dose-dependent inverse association between frequency of chocolate consumption and prevalence of calcified atherosclerotic plaques in the coronary arteries [13, 14]. A similar finding was observed in the aforementioned Australian study [15] where apart from the prospective association of chocolate consumption and vascular events, chocolate was also related to presence of carotid focal plaques. Women who consumed chocolate at least weekly had a lower prevalence of carotid atherosclerotic plaques compared with women who consumed chocolate less than weekly. However, the authors found no association with common carotid artery intima–media thickness.

Methodological Limitations of the Prospective Studies on Chocolate and CHD

The aforementioned prospective studies had several methodological limitations that are important to consider when attempting to determine the true long-term cardiovascular effects of chocolate or giving public health recommendations on chocolate consumption. Some of these limitations are inherent in any observational studies, but others are specific to the specific study question.

As with any observational study, unevenly distributed characteristics associated with chocolate intake habits and with risk for cardiovascular outcomes could have lead to over- or underestimation of the true effect of chocolate consumption; that is, any observational study is susceptible to confounding, and this may be particularly true for a behavior like chocolate consumption that certainly clusters with other others. However, all these aforementioned studies extensively controlled for potential confounding factors. Although one can never exclude the possibility of residual or uncontrolled confounding (i.e., the possibility that any important factor was not included or not well measured), it is important to remember that any remaining confounder would need to be strongly associated with both chocolate use and cardiovascular events to be able to alter the results considerably. Furthermore, any unintentionally missing factor would need to be unrelated to the factors included in the multivariable models to cause a misleading result.

Prospective cohort studies can provide seriously biased results if a large proportion of participants are lost from follow-up. Generally, the follow-up in these studies was rather complete, often backed up by regional or national mortality registries or sometimes by nationwide registration of all hospitalization events [8, 9]. The quality of information on outcomes was also adequate on a general level. Outcomes were ascertained either by reliable health registries, by review of death certificates, or by review and verification of medical records.

However, the quality of information on the exposure, that is, chocolate consumption, was generally far from ideal. All studies necessarily depended on self-report of chocolate intake. Although in some studies efforts were made to improve the quality of the reported information on chocolate consumption by relying on validated questionnaires, using repeated exposure assessments or by testing reproducibility, the real consumption could have potentially been considerably misclassified.

Misclassifying chocolate intake could have a variety of consequences. If the misclassification was not related to the risk of CHD (i.e., it mainly introduced noise), it typically would be expected to lead to an underestimation of the true effects; that is, the true effect of chocolate could be even stronger than these studies suggest. However, for example, if subjects with an unhealthy lifestyle tended to report lower chocolate intake because of guilt, while individuals with healthier lifestyles did not share this guilt, this could have caused an overestimation of the effects of chocolate in these studies [16]. Unfortunately, it is not simple to determine which of these types of misclassification might be more likely.

The great heterogeneity of chocolate and cocoa products further complicates this issue. Chocolate can be dark, milk, or white, based upon its cocoa content. None of the aforementioned prospective studies differentiated between these basic types of chocolate, and the Harvard Alumni did not even differentiate between chocolate or nonchocolate candy. Due to differences in type and origin of cocoa beans and differences in their processing and in preparation of chocolate, there can be a considerable variation concerning the chemical composition or the bioavailability of the components even within dark or milk chocolates [2, 17]. Moreover, components of chocolate other than cocoa can also play an important role in determining the cardiovascular effect of these products. Milk chocolate contains more fat and sugar than dark and is less rich in cocoa solids [18]. The milk content may also decrease the absorption of cocoa-related antioxidants in the gastrointestinal tract [19], although this effect remains uncertain [2, 20]. However, the uptake of cocoa antioxidants and therefore their bioavailability in humans could be increased significantly by concurrent consumption of carbohydrates [21]. Thus, the differential effects of dark and milk chocolate is an unresolved issue.

The type of chocolate consumed is likely to be important in evaluating the plausibility of these results. Short-term human trials that have shown vascular benefit from chocolate typically investigated the effects of dark chocolate or cocoa per se. Fraga et al. observed lower oxidative stress, improved lipid profile, and decrease in blood pressure in a short-term experimental trial using milk chocolate [22], but there is generally far less evidence supporting a cardiovascular benefit of milk or white chocolate than for cocoa or dark chocolate.

The prospective observational studies on habitual chocolate intake and clinical cardiovascular events that we have summarized above typically assumed that the members of their cohorts consumed milk rather than dark chocolate, generally only based on the greater use of milk chocolate in most countries. However, this is obviously an imperfect assumption, and it is made more complex by variability in the types of chocolate themselves. For example, the cocoa content of chocolate products and the definition of dark and milk chocolate vary by country. In the United States, milk chocolate has to contain a minimum of 10% of cocoa solids, dark chocolate 15%. The corresponding proportions in European Union are 25% and 35% [23]. This considerable difference in the proportion of cocoa solids might partly explain why chocolate appeared to be somewhat more effective in European cohorts than in the United States.

A systematic error could also arise in prospective observational studies on cardiovascular effects of chocolate from the dynamic relationship between ill-health and chocolate intake. Some of those abstaining from chocolate could have actually stopped consuming chocolate due to illness or due to medical advice. Thus, if individuals with preexisting disease or obesity tend to stop eating chocolate and are also at high risk for CHD, this could have explained the observed protective effect of chocolate consumption. Diabetes would appear to be the most likely medical reason that patients would be counseled to remove chocolate from their diets. To address this issue, several studies excluded participants with diabetes or statistically controlled for diabetes. Apart from that, several studies also examined the role of nonchocolate sweets. Those patients who stop chocolate consumption due to health problems would likely also reduce intake of other sweets. No consistent protective effects were observed in these studies for sweets other than chocolate, tending to support the likelihood of benefit for chocolate specifically.

Possible Biological Mechanisms

The results of the observational studies on habitual chocolate intake are generally compatible with a moderate-to-strong protective cardiovascular effect. Is protection by chocolate biologically plausible? Short-term experimental studies have helped to clarify this question by investigating several biological mechanisms by which chocolate may influence cardiovascular risk.

Cocoa and chocolate is a rich source of biologically active compounds. Cocoa and several chocolate products contain high-quantity and high-quality flavonoid antioxidants often exceeding that of

Table 22.2 Potential cardioprotective effects of flavonoids in chocolate

Endothelial nitric oxide	↑
Endothelial function	↑
Blood pressure	↓
Proinflammatory cytokine expression	↓
Glucose tolerance	↑
Platelet aggregation	↓
Angiotensin converting enzyme activity	↓
HDL	↑
LDL	↓
LDL oxidation	↓

other flavonoid-rich foods like black or green tea or red wine [24, 25]. As summarized in Table 22.2, several studies have demonstrated potential cardioprotective effects of intake of chocolate or cocoa products that are attributed to flavonoids. The effects on endothelium and platelets are mediated by the increased bioactive nitric oxide availability due to stimulation of nitric oxide synthesis by flavonoids [2, 16, 21, 26]. These powerful mechanisms alone provide very suggestive evidence that chocolate has beneficial vascular effects.

Fat in chocolate is especially high in stearic acid (around 30% of all fatty acids) [27]. The effect of stearic acid on cardiovascular health remains controversial, but a recent systematic review concluded that stearic acid has beneficial or neutral effects on blood pressure and clotting parameters and does not adversely affect traditional lipid risk factors [24]. This contrasts with other saturated fats and suggests that the benefit of chocolates antioxidants may not be outweighed by the fat content of cocoa mass.

Chocolate also contains several minerals, like magnesium, zinc, selenium, calcium, potassium, and copper, in relatively high amounts that might impact cardiovascular health [26].

Generally, the prospective cohort studies on habitual intake and clinical cardiovascular events were limited in their ability to test whether these plausible biological mechanisms are indeed responsible for the observed protective effect of chocolate. In the Potsdam Arm of the European Prospective Investigation, blood pressure explained 16%, 10%, and 12% of the lower risk for myocardial infarction, for stroke, and for their combination in association with chocolate consumption [10]. In the Stockholm Heart Epidemiology Program, we found no evidence that blood pressure, lipids, inflammatory markers, coagulatory factors, homocysteine, or markers of glucose tolerance mediated the observed inverse association between chocolate consumption and cardiac mortality [8]. Similarly, blood pressure was not found to be a mediator in the Zutphen Elderly Study [7]. However, none of these studies had widespread measures of true endothelial function or platelet reactivity that might respond directly to chocolate intake; hence, this remains an important open question.

Summary

In sum, at the present time, there is strongly suggestive but not definitive evidence that chocolate consumption is inversely associated with coronary heart disease and other cardiovascular events. This is consistent with the findings from several short-term randomized studies which found beneficial effects on cardiovascular risk markers and with the fact that chocolate is an exceptionally rich source of several potentially cardioprotective bioactive substances. Cardiovascular protection from chocolate consumption is thus biologically plausible, and there are several potential mechanisms to explain the observed inverse association. However, due to potential methodological weaknesses in the long-term observational studies, we cannot yet make a definitive conclusion on long-term cardiovascular effects of chocolate. Thus, while generally we see no reason to discourage chocolate intake, and it remains a reasonable part of a healthy diet when consumed in moderation, we cannot yet recommend that individuals adopt chocolate intake specifically to improve their cardiovascular health. We need more long-term studies and especially long-term randomized trials focusing on clinical events before chocolate intake can be recommended specifically for prevention of cardiovascular disease.

References

1. World Health Organization. The atlas of heart disease and stroke. Geneva: WHO; 2011.
2. Cooper KA, Donovan JL, Waterhouse AL, Williamson G. Cocoa and health: a decade of research. Br J Nutr. 2008;99(1):1–11.
3. Lee IM, Paffenbarger Jr RS. Life is sweet: candy consumption and longevity. BMJ. 1998;317(7174):1683–4.

4. Hu FB, Stampfer MJ, Willett WC. Reply to PM Kris-Etherton et al. Am J Clin Nutr. 2007;72(4):1059a–60.
5. Arts IC, Jacobs Jr DR, Harnack LJ, Gross M, Folsom AR. Dietary catechins in relation to coronary heart disease death among postmenopausal women. Epidemiology. 2001;12(6):668–75.
6. Mink PJ, Scrafford CG, Barraj LM, Harnack L, Hong CP, Nettleton JA, et al. Flavonoid intake and cardiovascular disease mortality: a prospective study in postmenopausal women. Am J Clin Nutr. 2007;85(3):895–909.
7. Buijsse B, Feskens EJ, Kok FJ, Kromhout D. Cocoa intake, blood pressure, and cardiovascular mortality: the Zutphen elderly study. Arch Intern Med. 2006;166(4):411–7.
8. Janszky I, Mukamal KJ, Ljung R, Ahnve S, Ahlbom A, Hallqvist J. Chocolate consumption and mortality following a first acute myocardial infarction: the Stockholm Heart Epidemiology Program. J Intern Med. 2009;266(3): 248–57.
9. Mostofsky E, Levitan EB, Wolk A, Mittleman MA. Chocolate intake and incidence of heart failure: a population-based prospective study of middle-aged and elderly women. Circ Heart Fail. 2010;3(5):612–6.
10. Buijsse B, Weikert C, Drogan D, Bergmann M, Boeing H. Chocolate consumption in relation to blood pressure and risk of cardiovascular disease in German adults. Eur Heart J. 2010;31(13):1616–23.
11. Gallus S, Tavani A, La Vecchia C. Response to chocolate, well-being and health among elderly men: chocolate and acute myocardial infarction in a case–control study from Italy. Eur J Clin Nutr. 2009;63(4):588–9.
12. Bayard V, Chamorro F, Motta J, Hollenberg NK. Does flavanol intake influence mortality from nitric oxide-dependent processes? Ischemic heart disease, stroke, diabetes mellitus, and cancer in Panama. Int J Med Sci. 2007;4(1):53–8.
13. Djousse L, Hopkins PN, Arnett DK, Pankow JS, Borecki I, North KE, et al. Chocolate consumption is inversely associated with calcified atherosclerotic plaque in the coronary arteries: the NHLBI Family Heart Study. Clin Nutr. 2011;30(1):38–43.
14. Djousse L, Hopkins PN, North KE, Pankow JS, Arnett DK, Ellison RC. Chocolate consumption is inversely associated with prevalent coronary heart disease: the National Heart, Lung, and Blood Institute Family Heart Study. Clin Nutr. 2011;30(2):182–7.
15. Lewis JR, Prince RL, Zhu K, Devine A, Thompson PL, Hodgson JM. Habitual chocolate intake and vascular disease: a prospective study of clinical outcomes in older women. Arch Intern Med. 2010;170(20):1857–8.
16. Heiss C, Keen CL, Kelm M. Flavanols and cardiovascular disease prevention. Eur Heart J. 2010;31(21):2583–92.
17. Visioli F, Bernaert H, Corti R, Ferri C, Heptinstall S, Molinari E, et al. Chocolate, lifestyle, and health. Crit Rev Food Sci Nutr. 2009;49(4):299–312.
18. Dreosti IE. Antioxidant polyphenols in tea, cocoa, and wine. Nutrition. 2000;16(7–8):692–4.
19. Serafini M, Bugianesi R, Maiani G, Valtuena S, De Santis S, Crozier A. Plasma antioxidants from chocolate. Nature. 2003;424(6952):1013.
20. Schroeter H, Holt RR, Orozco TJ, Schmitz HH, Keen CL. Nutrition: milk and absorption of dietary flavanols. Nature. 2003;426(6968):787–8. discussion 788.
21. Rimbach G, Melchin M, Moehring J, Wagner AE. Polyphenols from cocoa and vascular health-a critical review. Int J Mol Sci. 2009;10(10):4290–309.
22. Fraga CG, Actis-Goretta L, Ottaviani JI, Carrasquedo F, Lotito SB, Lazarus S, et al. Regular consumption of a flavanol-rich chocolate can improve oxidant stress in young soccer players. Clin Dev Immunol. 2005;12(1):11–7.
23. McKevith B. New chocolate regulations. Nutr Bull. 2003;28:197–8.
24. Ding EL, Hutfless SM, Ding X, Girotra S. Chocolate and prevention of cardiovascular disease: a systematic review. Nutr Metab Lond. 2006;3:2.
25. Vinson JA, Proch J, Zubik L. Phenol antioxidant quantity and quality in foods: cocoa, dark chocolate, and milk chocolate. J Agric Food Chem. 1999;47(12):4821–4.
26. Steinberg FM, Bearden MM, Keen CL. Cocoa and chocolate flavonoids: implications for cardiovascular health. J Am Diet Assoc. 2003;103(2):215–23.
27. Connor WE. Harbingers of coronary heart disease: dietary saturated fatty acids and cholesterol. Is chocolate benign because of its stearic acid content? Am J Clin Nutr. 1999;70(6):951–2.

Chapter 23
Dark Chocolate and (Pre-)Hypertension

Karin Ried

Key Points

- Flavanol-rich chocolate and cocoa products may have a small but significant effect on blood pressure.
- The blood pressure–reducing effect of cocoa is more pronounced in systolic hypertension than systolic prehypertension and normotension, but is independent of diastolic status.
- Even a small daily dosage of cocoa flavanols (e.g., 6 g of dark chocolate) appears to have a beneficial effect on blood pressure.
- Higher sugar content in chocolate appears to attenuate the blood pressure–lowering effect of cocoa, particularly in overweight and obese individuals.
- Our analysis indicates cocoa to be more effective in reducing blood pressure in younger individuals in the short term compared with older individuals.

Keywords Chocolate • Cocoa • Flavanols • Blood pressure • Hypertension • Meta-analysis

The Kuna Indians

The interest in the effect of cocoa on blood pressure started with the discovery that an island population in Central America, the Kuna Indians, had a distinctively low rate of hypertension coupled with a consistent healthy low blood pressure unaffected by age [1, 2].

The majority of the Kuna Indians live on the San Blas Island off Panama (population approximately 35,000); those Kuna Indians who migrated to the mainland manifest a higher prevalence of hypertension as well as an age-dependent rise of blood pressure, implying that lifestyle factors such as diet, rather than genetics, play a protective role [3]. Island-dwelling Kuna Indians consume about 3–4 cups of cocoa drinks on average per day, while the mainland-dwelling Kuna Indians consume up

K. Ried, Ph.D., M.Sc., GDPH, Cert. Integrative Medicine (✉)
Discipline of General Practice, The University of Adelaide, Adelaide, South Australia, Australia
e-mail: karin.ried@adelaide.edu.au

R.R. Watson et al. (eds.), *Chocolate in Health and Nutrition*, Nutrition and Health 7,
DOI 10.1007/978-1-61779-803-0_23, © Springer Science+Business Media, LLC 2013

to 10 times less cocoa [3, 4]. Average high salt intake was not associated with the differences in blood pressure [3]. Mean blood pressure of the island-dwelling adult Kuna Indians hovers around 110 mmHg systolic and 70 mmHg diastolic, while on the mainland observed age-related blood pressure rise and hypertension prevalence is comparable to that of Western populations [1].

High blood pressure is a critically important risk factor for cardiovascular disease and attributes to 37% of cardiovascular deaths in Western populations [5]. The association between cardiovascular risk and blood pressure levels are continuous [6], with the risk of ischemic heart disease and stroke halved for every 20 mmHg reduction in systolic blood pressure (SBP) and 10 mmHg diastolic blood pressure (DBP) [7]. Even small reductions in blood pressure therefore can substantially reduce cardiovascular events at a population level.

Cocoa Consumption Today: Modern Chocolate

Cocoa is extracted from cacao beans, the fatty seeds of the *Theobroma cacao* tree. Cocoa is rich in flavanols, particularly epicatechin, catechin, and procyanidins, proposed to be responsible for the blood pressure–lowering effect [8, 9]. Flavanols are also found in other plant-derived produce, including beans, apricots, blackberries, apples, and tea leaves, albeit in a lower concentration than in cocoa products (460–610 mg/kg of flavanol monomers; 4–5 g/kg of flavanol polymers) [10, 11]. Flavanol intake is, however, also dependent on serving size, and flavanol content depends on the processing of the cacao beans and raw cocoa.

Traditionally, cocoa was consumed as a cold unsweetened drink of raw dried cacao powder, often mixed with starch and spices by the native Indians, but this was considered bitter and unpalatable for the early European explorers, including Christopher Columbus in 1502 and Hernánd Cortés in 1519. The Spanish brought cocoa to Europe, to which sugar was added and the drink was heated [12, 13]. Subsequent roasting (up to 120°C), mixing (conching), alkalizing (dutching), and adding sugar, milk, vanilla, and lecithin emulsifiers make chocolate as we know it today [14]. Various chocolate manufacturers have fine-tuned the processing leading to different flavors and smoothness of chocolates, but also to altered cocoa and flavanol content in various cocoa products.

Dark chocolate contains larger amounts of cocoa (50–85%) than milk chocolate (20–30%). Different processing procedures however influence the flavanol content of the cocoa in the chocolate; a 70% cocoa-containing chocolate bar from one company therefore might not contain the same amount of flavanols and flavanol composition as a 70% chocolate bar from another company. Content and composition of flavanols depend on the variety and ripeness of cocoa beans used as well as the manufacturing steps.

Fresh and fermented cocoa beans contain about 10% of flavanols (100 mg/g), the cocoa powder consumed by the Kuna Indians contains about 3.6% of flavanols, and cocoa-rich dark chocolate on the market contains about 0.5% of flavanols [15, 16]. Moreover, heavy dutching, the alkalizing of chocolate to pH 7–8, can reduce the flavanol content to less than 10 mg/100 g (0.001%).

Furthermore, research suggests that the monomeric portion of cocoa flavanols, epicatechin and catechin, and to a lesser extent the polymeric flavanols, the procyanidins, are linked to blood pressure and vasoactive effects [4]. Modern processing of cacao reduces the monomeric flavanol content and influences the epicatechin/catechin ratio [17].

Fresh and fermented cocoa beans contain between 2.5 and 16.5 mg of epicatechin per gram depending on the variety, the growing region, and the harvesting practices [18, 19], whereas processed cocoa retains only 2–18% of the original epicatechin, due to roasting and dutching [17].

Due to the large variation in flavanol content in chocolate and cocoa products, it is critical to compare the dosages of flavanols rather than simply the amounts of chocolate or administered cocoa products in clinical trials investigating the effect of cocoa on blood pressure.

Blood Pressure–Lowering Mechanisms of Cocoa

The blood pressure–lowering properties of cocoa have been linked to the formation of endothelial nitric oxide (NO), which promotes vasodilation and consequently lowers blood pressure.

Increased NO production might be triggered by upregulation of NO synthase through the insulin-mediated signaling pathway [20]. Insulin sensitivity was shown to be improved after cocoa intake in a number of trials [21–24], but not in one [25]. Secondly, cocoa flavanols have been shown to inhibit angiotensin-converting enzyme (ACE) activity, and hence reduce blood pressure [26, 27]. Thirdly, there is also evidence to suggest cocoa flavanols to act as powerful antioxidants within the cardiovascular system, upregulating NO-synthase activity and hence reducing blood pressure [28, 29].

The Effect of Chocolate on Blood Pressure

Meta-analyses of Intervention Trials

In the last decade, several clinical trials have investigated the effect of chocolate and cocoa products on blood pressure. Our team has conducted a Cochrane systematic review [30] and updated previous meta-analyses [31–33], including published studies until 2010.

Methods

Search Strategy and Selection of Trials

We searched the Medline and Cochrane databases for randomized controlled trials investigating the effect of chocolate or cocoa products on blood pressure published between 1955 and 2010. In addition, we searched reference lists of published trials and reviews.

We included trials in the meta-analysis if the control group received a placebo or a low flavanol–containing cocoa product (powder as drink, bar or tablet), the duration of the intervention was at least 2 weeks, and data needed for meta-analysis were published or were available from authors (mean systolic or diastolic blood pressure [SBP/DBP] and standard deviation [SD]).

The quality of the studies was assessed based on reported randomization, blinding, assessment of blood pressure, loss to follow-up, industry funding source, and assessment of compliance and dietary chocolate intake of participants.

Analysis

We conducted meta-analysis using the Cochrane Program Review Manager version 5.1 [34] and the Stata program version 10 [35]. Owing to high heterogeneity between trials, we used a random effects model. The pooled effect size or overall weighted mean difference was obtained using the mean (SD) SBP/DBP of the treatment and control groups at the end of the intervention and the participant numbers in each of the groups. We adjusted for baseline differences in SBP/DBP between intervention and control groups by incorporating these in the meta-analysis model.

We explored whether other factors could have influenced meta-analysis results, by testing the following variables in meta-regression analysis using the Stata program version 10: dosage of flavanol content (total or monomers) in chocolate or cocoa products, blinding, baseline blood pressure, theobromine content, sugar content, body mass index (BMI), duration, and age.

Characteristics of Included Studies

A total of 21 studies involving 947 participants were included in the meta-analysis (Table 23.1) [21–23, 25, 36–50]. Ten studies used commercially available chocolate, and 11 studies used flavanol-rich cocoa powder (tablet, bar, or mixed with water or milk) and compared the effect to a control group, which either took flavanol-free placebo (white chocolate, milk, or placebo pill) or low-flavanol powder, respectively. The active intervention group received either dark chocolate of 6–100 g (6 g equals one piece of a 100 g dark chocolate bar) containing 50–90% cocoa, milk-chocolate-based confectionary (105 g of <10% cocoa [39]), or flavanol-enriched cocoa powder, containing a dosage of 30–1,080 mg (mean = 545.5 mg) of flavanols per day. Trials ran between 2 and 18 weeks.

Meta-analysis Results

Pooled meta-analyses results including all 21 studies revealed a significant blood pressure–reducing effect of cocoa-rich products compared with control (Fig. 23.1):

Mean change SBP (95%CI): −2.65 (−4.54, −0.75) mmHg, $p < 00001$, n = 21
Mean change DBP (95%CI): −2.05 (−3.29, −0.81) mmHg, $p < 0.0001$, n = 20 available for DBP

Albeit significant, the reduction in *systolic* blood pressure weakened with increasing number of studies compared to previous meta-analyses by:

1. Ried (2010) [33] (15 trials): mean change SBP (95%CI): −3.16 (−5.08, −1.23) mmHg, $p = 0.001$
2. Desch (2010) [32] (10 trials): mean change SBP (95%CI): −4.52 (−5.87, −3.16) mmHg, $p < 001$
3. Taubert (2007) [31](5 trials): mean change SBP (95%CI): −4.7 (−7.6, −1.8) mmHg, $p = 0.002$

Overall reduction in *diastolic* blood pressure in our updated meta-analysis is also a little smaller than reported in earlier meta-analyses:

1. Ried (2010) [33] (15 trials): mean change DBP (95%CI): −2.02 (−3.35, 0.69) mmHg, $p = 0.003$
2. Desch (2010) [32] (10 trials): mean change DBP (95%CI): −2.5 (−3.9, 1.2) mmHg, $p < 0.001$
3. Taubert (2007) [31](5 trials): mean change DBP (95% CI): −2.8 (−4.8, −0.8) mmHg, $p = 0.006$

This trend toward a null effect raises the question: What was done differently in later trials leading to a less pronounced effect of cocoa on blood pressure when all available data are summarized?

Early trials reporting a significant blood pressure–reducing effect of chocolate and included in the first meta-analysis by Taubert (2007) [31] had similar study designs, all comparing dark versus white chocolate over a period of 2 weeks [21, 36, 39]. The majority of the later trials used specifically formulated cocoa powder, flavanol-enriched for the treatment group and a low-flavanol product for the comparison group, and/or later trials observed participants over longer periods of time.

To assess whether the observed blood pressure–lowering effect of cocoa was associated with factors such as flavanol dosage, choice of control, or duration of intervention, we conducted subgroup meta-analyses as well as meta-regression analyses.

Dosage: Chocolate and Cocoa Products

Ten studies used commercially available chocolate products, including Ritter Sport (Halbbitter, 50% cocoa, Germany, n = 5), Cuorenero (90% cocoa, Italy, n = 1), Haighs (70% cocoa, Australia, n = 1), Meiji (80% cocoa, Japan, n = 1), Galaxy/Dove/Mars (50% cocoa, USA, n = 1), and M&M Dove/Mars

Table 23.1 Characteristics of included studies[a]

Study	Study design	Participants cocoa/control	Dosage, cocoa[b] brand	Cocoa product	Dosage total flavanols (mg)	Dosage epi-/catechin (mg)	Theobromine (mg)	Duration (weeks)	Baseline SBP	Baseline DBP
Taubert (2003) [36]	C	13/13	100 g Ritter Sport Halbbitter, D	50% DC/WC	500/0	89/0	560/0	2	153.3/153.6	84.5/84.2
Murphy (2003) [37]	P	13/15	6×234 mg Cocoapro Mars, US	HiFl/LoFl tablets	234/6.4	–	–	4	118/116	78/76
Engler (2004) [38]	P	11/10	46 g Dove Mars, US	HiFl/LoFl DC	213/0	46/0		2	121/112.8	68.1/66.1
Fraga (2005) [39]	C	14/14	105 g M&M Mars, US	16% MC/WC	168/0	39/0	179/0	2	123/123	72/71
Grassi (2005a) [21]	C	15/15	100 g Ritter Sport, D	50% DC/WC	500/0	89/0	560/0	2	112.9/113.2	74/73.8
Grassi (2005b) [21]	C	20/20	100 g Ritter Sport, D	50% DC/WC	500/0	89/0	560/0	2	141.3/141.1	92.4/91.8
Taubert (2007) [40]	P	22/22	6.3 g Ritter Sport, D	50% DC/WC	30/0	7.1/0	26.4/0	18	147.7/147.5	86.4/86.7
Crews (2008) [41]	P	45/45	37 g bar + 12 g powder, Hershey, US	HiFl/LoFl bar and drink	755/41	–	–	6	126.8/128.6	74.2/75
Grassi (2008) [22]	C	19/19	100 g Cuorenero, Ital	90% DC/WC	1,080/0	150/0	1,700/0	2	141.1/140.9	91.2/91.1
Muniyappa (2008) [25]	C	20/20	62 g Cocoapro Mars, US	HiFl/LoFl drink	902/28	236/10	674/654	2	141/141	91/91
Davison (2008a) [23]	P	12/11	Mars, US	HiFl/LoFl drink	902/36	–	337/327	12	124124	76/77
Davison (2008b) [23]	P	13/13	Mars, US	HiFl/LoFl drink	902/36		337/327	12	126/121	78/74
Al-Faris (2008) [42]	P	30/29	100 g Galaxy/Dove Mars, US	50% DC/WC	500/0	10.6/0	–	2	115.9/115.2	73/72.8
Shiina (2009) [43]	P	20/19	45 g Meiji, Jpn	80% DC/WC	550/0	–	–	2	116.4/121.6	64.7/72.2
Ried (2009) [44]	P	11/10	50 g Haighs, AUS	70% DC/placebo pill	750/0	–	–	8	135/135.7	83.6/77.8
Monagas (2009) [45]	C	42/42	40 g Nutrexpa, Spain	HiFl/0 in milk	495/0	56.5/0	440/0	4	138/138	84/84
Bogaard (2010) [46]	C	41/41	3.6 g Aticoa, Barry Callebaut, Belg	HiFl/LoFl drink	529/0	38/0	543 (TEC+NTC)/0	3	141.7/141.7	84.2/84.2
Heiss (2010) [47]	C	16/16	Cocoapro Mars, US	HiFl/LoFl drink in milk or water	750/18	130/6	186/192	4	132/131	–
Desch (2010) [48]	P	43/48	25 g/6 g Ritter Sport, D	50% DC/DC	125/30	29.5/7.1	110/26.4	12	135.7/134.3	79.3/75.8
Davison (2010) [49]	p	13/14	232 g Mars, US	HiFl/LoFl drink	1,052/33	266/12	460.5/402.2	6	143/145.4	83.2/88
Njike (2011) [50]	C	38/38	22 g Hershey, US	HiFl/LoFl drink	805/9	69/0	436/0	6	123.3/123.6	68.5/67.3

Abbreviations: – not given, *DC* dark chocolate, *WC* white chocolate, *MC* milk chocolate, *HiFl* high-flavanol cocoa powder, *LoFl* low-flavanol cocoa powder, *TEC* theobromine-enriched chocolate, *NTC* natural dose theobromine chocolate, *P* parallel trial, *C* crossover trial, *AUS* Australia, *Belg* Belgium, *D* Germany, *Ital* Italy, *Jpn* Japan, *US* United States

[a]Trial groups included in meta-analysis are the following: Ried (2009) (phase 1), Bogaard (2010) (TEC+NTC groups average), and Njike (2010) (sugar+sugarless groups average). All dosages are given per day

[b]Cocoa-rich treatment group

a **SBP all trials (n=21)**

Study or Subgroup	Mean Difference	SE	Cocoa Total	Control Total	Weight	Mean Difference IV, Random, 95% CI
Taubert 03	-5.1	0.73	13	13	7.0%	-5.10 [-6.53, -3.67]
Murphy 03	-1	4	13	15	3.2%	-1.00 [-8.84, 6.84]
Engler 04	1.8	4.43	11	10	2.9%	1.80 [-6.88, 10.48]
Fraga 05	-4	1.6	14	14	6.1%	-4.00 [-7.14, -0.86]
Grassi 05a	-6.5	1.49	15	15	6.2%	-6.50 [-9.42, -3.58]
Grassi 05b	-11.3	0.95	20	20	6.8%	-11.30 [-13.16, -9.44]
Taubert 07	-2.8	2.28	22	22	5.2%	-2.80 [-7.27, 1.67]
Crews 08	-0.53	2.64	45	45	4.7%	-0.53 [-5.70, 4.64]
Grassi 08	-3.7	0.7	19	19	7.0%	-3.70 [-5.07, -2.33]
Muniyappa 08	-1	1.6	20	20	6.1%	-1.00 [-4.14, 2.14]
Davison 08a	-6.1	3.46	12	11	3.8%	-6.10 [-12.88, 0.68]
Davison 08b	1.6	4.5	13	13	2.8%	1.60 [-7.22, 10.42]
Al-Faris 08	-7.1	2.19	30	29	5.3%	-7.10 [-11.39, -2.81]
Shiina 09	0.6	3.82	20	19	3.4%	0.60 [-6.89, 8.09]
Ried 09	2.9	6.55	11	10	1.7%	2.90 [-9.94, 15.74]
Monagas 09	3	2.72	42	42	4.6%	3.00 [-2.33, 8.33]
Bogaard 10	0.25	1.54	41	41	6.1%	0.25 [-2.77, 3.27]
Heiss 10	-5	3.23	16	16	4.0%	-5.00 [-11.33, 1.33]
Desch 10	-0.3	2.43	43	48	5.0%	-0.30 [-5.06, 4.46]
Davison 10	-2	5.22	13	14	2.3%	-2.00 [-12.23, 8.23]
Njike 11	3.2	1.72	39	39	5.9%	3.20 [-0.17, 6.57]
Total (95% CI)			**472**	**475**	**100.0%**	**-2.65 [-4.54, -0.75]**

Heterogeneity: Tau² = 12.88; Chi² = 112.12, df = 20 (P < 0.00001); I² = 82%
Test for overall effect: Z = 2.74 (P = 0.006)

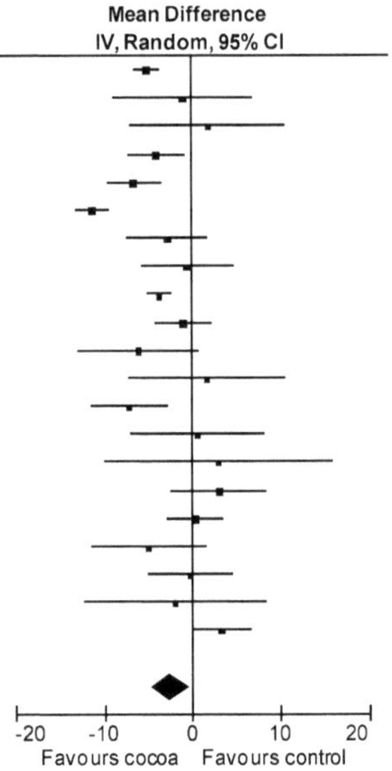

b **DBP all trials (n=20)**

Study or Subgroup	Mean Difference	SE	Cocoa Total	Control Total	Weight	Mean Difference IV, Random, 95% CI
Taubert 03	-1.9	0.99	13	13	8.4%	-1.90 [-3.84, 0.04]
Murphy 03	-1	3.39	13	15	3.2%	-1.00 [-7.64, 5.64]
Engler 04	1	2.76	11	10	4.1%	1.00 [-4.41, 6.41]
Fraga 05	-4	1.6	14	14	6.8%	-4.00 [-7.14, -0.86]
Grassi 05a	-3.9	1.03	15	15	8.3%	-3.90 [-5.92, -1.88]
Grassi 05b	-7.6	0.94	20	20	8.6%	-7.60 [-9.44, -5.76]
Taubert 07	-1.9	1.15	22	22	0.0%	-1.90 [-4.15, 0.35]
Crews 08	0.07	1.6	45	45	6.8%	0.07 [-3.07, 3.21]
Grassi 08	-3.7	0.78	19	19	8.9%	-3.70 [-5.23, -2.17]
Muniyappa 08	1	1.6	20	20	6.8%	1.00 [-2.14, 4.14]
Davison 08a	-4.6	2.3	12	11	0.0%	-4.60 [-9.11, -0.09]
Davison 08b	-0.3	2.88	13	13	0.0%	-0.30 [-5.94, 5.34]
Al-Faris 08	-5.4	1.41	30	29	7.3%	-5.40 [-8.16, -2.64]
Shiina 09	1.4	3.54	20	19	3.0%	1.40 [-5.54, 8.34]
Ried 09	1.4	4.62	11	10	2.0%	1.40 [-7.66, 10.46]
Monagas 09	1	1.6	42	42	6.8%	1.00 [-2.14, 4.14]
Bogaard 10	-0.8	0.93	41	41	8.6%	-0.80 [-2.62, 1.02]
Desch 10	0.5	1.62	43	48	0.0%	0.50 [-2.68, 3.68]
Davison 10	-2.1	3.26	13	14	3.4%	-2.10 [-8.49, 4.29]
Njike 11	-1.25	1.44	38	39	7.2%	-1.25 [-4.07, 1.57]
Total (95% CI)			**365**	**365**	**100.0%**	**-2.14 [-3.58, -0.70]**

Heterogeneity: Tau² = 5.42; Chi² = 58.06, df = 15 (P < 0.00001); I² = 74%
Test for overall effect: Z = 2.91 (P = 0.004)

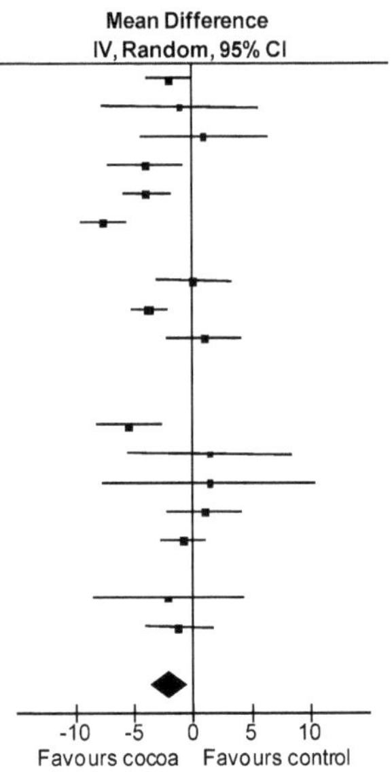

Fig. 23.1 Meta-analysis of trials on the effect of cocoa products on blood pressure. This figure is copublished, with permission, and based on the Cochrane Review by Ried K, Sullivan TR, Fakler P, Frank OR, Stocks NP. Effect of chocolate on blood pressure. Cochrane Database Syst Rev. 2011; In press (likely Issue 12). Cochrane Reviews are regularly updated as new evidence emerges and in response to feedback. The Cochrane Library should be consulted for the most recent version of the review

(<10% cocoa, USA, N = 1). Eleven trials were provided with especially formulated flavanol-enriched/low-flavanol cocoa powder by Mars, USA (n = 7), Hershey, US (n = 2), Nutrexpa, Spain (n = 1), and Barry Callebaut, Belgium (n = 1) (see Table 23.1).

Dosage: Total Flavanols

Cocoa content and processing of the trial product determined flavanol dosage. In the majority of trials (n = 10), the active group consumed 500 to 750 mg of flavanols per day, compared with less than 250 mg per day in five studies or more than 750 mg per day in six studies. The control group received either a flavanol-free product (n = 12) or a low flavanol–containing cocoa powder (n = 9). Flavanol dosage of low-flavanol products ranged between 6.4 and 41 mg per day (see Table 23.1).

Total flavanol content in the active treatment group was not significantly associated with blood pressure change. However, if we distinguished between studies which had a true control group with no flavanol intake versus control groups consuming low-flavanol products, we found a significant difference in effect size between these groups.

Meta-analysis of trials with true control groups revealed a significant blood pressure–reducing effect:

Mean change SBP (95% CI): −3.70 (−6.02, −1.36) mmHg, $p = 0.002$, n = 12
Mean change DBP (95% CI): −2.71 (4.26, −1.5) mmHg, $p < 0.001$, n = 11

In contrast, there was no significant difference between treatment and control groups of trials using a low flavanol–containing product in the control group:

Mean change SBP (95%CI): −0.54 (−2.46, 1.39) mmHg, $p = 0.59$, n = 9
Mean change DBP (95%): −0.55 (−1.90, 0.79) mmHg, $p = 0.42$, n = 8

Five of the low-flavanol *control* groups received higher dosages of flavanols than the *active* intervention group in the trial by Taubert (2007) using a flavanol-free control group [31].

It is plausible that any differences found between the treatment and control groups were weakened in trials using high-flavanol versus low-flavanol products, as the low-flavanol product may also have an effect on blood pressure. In fact, trials studying acute effects reported a sustained accumulative effect on vascular function when flavanols were taken daily over 1 or 3 weeks [51, 52]. In addition, one trial included in our review provided a small amount of daily dark versus white chocolate (6.3 g) over a 4-month period and reported a cumulative effect on blood pressure at 6, 12–18 weeks, respectively [40]: mean SBP (SD) change of −0.5 (1.6), −2.0 (1.7), −2.8 (1.6) mmHg and mean DBP (SD) change of −0.4 (1.5), −1.6 (1.0), −1.9 (1.5) mmHg.

Blinding

The 11 trials using low-flavanol cocoa products as control aimed to facilitate "blinding" or "masking" of participants to minimize any expectation bias or placebo effect. In addition, other energy and nutrient components in the high- versus low-flavanol trial products were matched to eliminate other potential confounding factors. However, the masked control products used in 9 of the 11 trials contained small but potentially sufficient amounts of flavanols to influence a vascular response and blood pressure change. Only two trials used a blinded design with flavanol-free control groups [38, 46]. Sensitivity analysis of blinded trials versus unblinded (open-label) trials confirmed the results of our subgroup meta-analyses by flavanol content of the control group (flavanol-free versus low-flavanol) as described above.

Dosage of Flavanol Monomers: Epicatechin + Catechin

Fifteen out of the 21 trials measured the proportion of monomeric flavanols. Dosage of epicatechin and catechin ranged between 7 and 236 mg per day in the intervention group, and 7–12 mg in four control groups. We did not find an association between monomeric flavanol dosage and blood pressure change using meta-regression analysis.

Baseline Blood Pressure: Hypertensive, Prehypertensive, Normotensive

Our previous meta-analysis revealed a difference in effect of cocoa products on blood pressure dependent on hypertension status at baseline. While blood pressure was significantly lowered in people with systolic hypertension (≥140 mmHg) or diastolic prehypertension (≥80 mmHg), no significant effect of cocoa on normal blood pressure (<140/80 mmHg) was evident [33].

The inclusion of six additional trials in this meta-analysis confirmed our previous findings.

A significant blood pressure–lowering effect was evident in the hypertensive subgroup (>140 mmHg) but not for SBP <140 mmHg at baseline: Hypertensive subgroup: mean SBP change (95%CI): −3.99 (−7.02, −0.97) mmHg, $p=0.01$; normotensive subgroup: mean SBP change (95%CI): −1.78 (−4.07, 0.52) mmHg, $p=0.13$. However, a significant blood pressure–lowering effect was evident independent of diastolic baseline blood pressure.

Theobromine

It has been suggested that theobromine found in cocoa might be indicative for vasoactivity and thus blood pressure reduction of cocoa products [53]. Theobromine is the bitter alkaloid of the cacao plant and also found in other plants, such as tea and the cola nut. Other similar compounds, the methylxanthines, include caffeine in coffee.

The theobromine content of cacao products reported in a selection of studies included in our meta-analysis ranged between 400 and 1,980 mg/100 g, with daily dosage intake of participants between 26 and 1,700 mg of theobromine.

It is questionable whether the chocolate and cocoa products are palatable if large amounts of the bitter theobromine are included. In addition, very high dosages of theobromine might have side effects in humans. While some animals, such as dogs, might succumb to theobromine poisoning from as little as 50 g of chocolate for a smaller dog and 400 g for an average-sized dog due to slow metabolism of theobromine [54], it is estimated that a 60-kg human would need to consume about 4.5 kg of dark chocolate with natural-containing theobromine to be poisoned [55]. However, theobromine-enriched cocoa powder containing about 30 times more theobromine than commercial chocolate has been used in small quantities in research [46]. Adverse effects of such theobromine-enriched cocoa might be expected by intake of only 150 g.

Our meta-regression analysis indicated theobromine not to be associated with the effect of cocoa on blood pressure.

Sugar

Commercially available milk chocolate contains about 50% of sucrose (range 45–65 g/100 g), and dark chocolate about 30% sugar (range 13–45%). Chocolate and cocoa products tested in the studies included in our review contained between 0% and 80% of sugar. One study compared the effect of

80% sugared with sugar-free cocoa powder on blood pressure and endothelial function [50]. A larger beneficial effect on endothelial function was observed with sugar-free cocoa powder compared with sugared cocoa powder, which might be explained by reduced micro- and macrocirculation in the arteries due to acute hyperglycemia [56].

Our meta-analysis of 21 trials indicated a larger beneficial effect on blood pressure if test products contained low amounts of sugar (<10 g/day). The beneficial effect of cocoa products with low sugar content was even more pronounced in overweight and obese individuals (BMI > 25):

Mean SBP (95%CI) change: −2.62 (−5.13, 0.11) mmHg, $p = 0.04$, <10 g sugar in cocoa/day, BMI > 25
Mean DBP (95CI) change: −2.34 (−4.19, −0.50) mmHg, $p = 0.01$, <10 g sugar in cocoa/day, BMI > 25
Mean SBP (95%CI) change: −7.53 (−5.37, 2.32) mmHg, $p = 0.44$, >10 g sugar in cocoa/day, BMI > 25
Mean DBP (95%CI) change: −1.08 (−4.11, 1.94) mmHg, $p = 0.48$, >10 g sugar in cocoa/day, BMI > 25

While we did not identify body mass index (BMI) to be a predicting variable for the effect of cocoa on blood pressure, if viewed in combination with sugar intake, the differences in effect could be explained by an increased insulin resistance often associated with higher BMI. Skeletal muscle fat disrupts the insulin-mediated signaling pathway, and higher levels of insulin can cause elevation of blood pressure [20].

Duration

To be included in our meta-analysis, the studies needed to have run for a minimum of 2 weeks duration. The majority of included studies (n = 9 out of 21) chose this relatively short time frame, a further eight trials studied participants for 3–8 weeks, one study with two trial arms ran for 12 weeks, and one followed participants up until 18 weeks. However, the longest study gave participants the smallest dose of chocolate (6 g of 50% cocoa chocolate containing 30 mg of flavanols) [40].

Pooled meta-analysis of trials of 2 weeks duration revealed a significant blood pressure–lowering effect:

Mean SBP change (95%CI): −4.81 (−7.21, −2.41) mmHg, $p < 0.001$, n = 9
Mean DBP change (95% CI): −3.19 (−5.00, −1.38) mmHg, $p = 0.0006$, n = 9

In contrast, pooled meta-analysis of longer-term trials (>2 weeks) obliterated the effect of cocoa on blood pressure:

Mean SBP change (95%CI): −0.14 (−1.76, 1.48) mmHg, $p = 0.87$, n = 12
Mean DBP change (95% CI): −0.85 (−1.82, 0.12) mmHg, $p = 0.09$, n = 11

It is important to note that seven out of the nine trials (78%) of 2 weeks duration also had a flavanol-free control group; therefore, this subgroup analysis by duration might be confounded by the level of cocoa flavanol used in the control groups.

Furthermore, several 2-week trials asked participants to consume a large amount of chocolate per day (1 bar = 100 g). This might be unpractical in the long term and might cause some unwanted side effects such as weight gain. Therefore, while shorter-term trials appear more effective, meta-analysis results by duration should be interpreted cautiously.

Age

Our meta-analysis indicated that the effect of cocoa on blood pressure was attenuated with increasing age. While cocoa had a significant blood pressure–lowering effect in trials with younger participants (mean age < 50 years, mean ± SE = 35.7 ± 9 years), we did not find a significant difference between treatment and control groups of trials with older participants (mean age > 50 years, mean ± SE = 60.6 ± 9 year). This effect was more pronounced if we compared trials of participants with a mean aged of less than 60 years versus more than 60 years:

Mean SBP change (95% CI): –4.5 (7.41, –1.73) mmHg, $p = 0.002$, n = 10, <50 years
Mean DBP change (95% CI): –3.85 (–5.45, –2.26) mmHg, $p < 0.001$, n = 9, <50 years
Mean SBP change (95% CI): –0.91 (–3.21, 1.39) mmHg, $p = 0.44$, n = 11 >50 years
Mean DBP change (95% CI): –0.79 (–1.66, 0.08) mmHg, $p = 0.07$, n = 10, >50 years

The age-related effect seen in these subsets of studies might be associated with the structural and biochemical changes in the arterial wall apparent with aging [57] and subsequent vascular reactivity to stimuli. Age-related changes include arterial stiffening in association with decrease of elastin and increase of collagen and glycosaminoglycans [57]. In addition, endothelin-1, a potent vasoconstrictor protein, is elevated in older adults [58], and endothelial oxidative stress compromising NO availability is more pronounced in the elderly [59]. Cocoa flavanols have been shown to reduce vascular resistance, arterial stiffness, and endothelin-1 and are potent scavengers of free radicals [4, 29, 60], leading to improved vascular function.

In the short-term studies included in our review, the effect of cocoa on blood pressure might be more pronounced in younger individuals due to the age-related decrease of vascular reactivity to physiological stimuli such as cocoa flavanols.

Other Studies Excluded from Meta-analysis

Our search identified seven additional studies, reported in ten articles, which investigated the effect of chocolate or cocoa products on blood pressure. We excluded these seven studies in our meta-analysis because studies investigated the acute effects 2 h after ingestion of cocoa (n = 3) [24, 61, 62], data required for meta-analysis were not available (n = 2) [52, 63], or studies administered a high dosage of flavanol-rich cocoa to both active and control groups (n = 2) [64–66].

Acute beneficial effects of flavanol-rich cocoa included improved flow-mediated dilation (FMD) [24, 62], improved coronary vascular function, and decreased platelet adhesion [61].

Summary

In conclusion, our meta-analysis of 21 short-term intervention studies involving 947 participants suggests flavanol-rich chocolate and cocoa products to have a small but significant blood pressure–reducing effect (mean change SBP ± SE: –2.7 ±1.9 mmHg, $p < 0.0001$; mean change DBP ± SE: –2.1 ± 1.2 mmHg, $p < 0.0001$). This effect was more pronounced in systolic hypertensives compared with systolic normotensives but was independent of diastolic status.

Relatively small dosages of cocoa flavanols taken daily for an extended period appear to have beneficial effects on vascular function, including blood pressure; however, flavanol content in commercially available chocolates varies greatly. In addition to the cocoa flavanol content, less sugar appears to improve the blood pressure–lowering effect, particularly in overweight and obese individuals, potentially due to compromised insulin sensitivity.

Vascular responsiveness to cocoa appears to decrease with age. We found cocoa to reduce blood pressure more readily in younger people (<50 years, mean ± SE = 35.7 ± 9 years) compared with older people (>50 years, mean ± SE = 60.6 ± 9 years), with the effect being more pronounced for systolic blood pressure.

Our meta-analysis did not suggest the blood pressure–lowering properties of cocoa to be associated with theobromine content.

References

1. Hollenberg N. Vascular action of cocoa flavanols in humans: the roots of the story. J Cardiovasc Pharmacol. 2006;47(Suppl 2):S99–102, discussion S119–121.
2. Kean BH. The BP of the Kuna Indians. Am J Trop Med Hyg. 1944;24:341.
3. McCullough ML, Chevaux K, Jackson L, Preston M, Martinez G, Schmitz HH, et al. Hypertension, the Kuna, and the epidemiology of flavanols. J Cardiovasc Pharmacol. 2006;47(Suppl 2):S103–9, discussion 119-121.
4. Schroeter H, Heiss C, Balzer J, Kleinbongard P, Keen CL, Hollenberg NK, et al. (−)-Epicatechin mediates beneficial effects of flavanol-rich cocoa on vascular function in humans. Proc Natl Acad Sci USA. 2006;103:1024–9.
5. Martiniuk AL, Lee CM, Lawes CM, Ueshima H, Suh I, Lam TH, et al. Hypertension: its prevalence and population-attributable fraction for mortality from cardiovascular disease in the Asia-Pacific region. J Hypertens. 2007;25:73–9.
6. McInnes GT. Lowering blood pressure for cardiovascular risk reduction. J Hypertens. 2005;23(Suppl):S3–8.
7. Lewington S, Clarke R, Qizilbash N, Peto R, Collins R. Prospective studies collaboration. Age-specific relevance of usual blood pressure to vascular mortality: a meta-analysis of individual data for one million adults in 61 prospective studies. Lancet. 2002;360:1903–13.
8. Corti R, Flammer AJ, Hollenberg NK, Lüscher TF. Cocoa and cardiovascular health. Circulation. 2009;119:1433–41.
9. Heiss C, Kelm M. Chocolate consumption, blood pressure, and cardiovascular risk. Eur Heart J. 2010;31:1554–6.
10. Fernandez-Murga L, Tarín JJ, García-Perez MA, Cano A. The impact of chocolate on cardiovascular health. Maturitas. 2011;69:312–21.
11. Hammerstone JF, Lazarus SA, Schmitz HH. Procyanidin content and variation in some commonly consumed foods. J Nutr. 2000;130:2086S–92.
12. Dillinger TL, Barriga P, Escárcega S, Jimenez M, Salazar Lowe D, Grivetti LE. Food of the gods: cure for humanity? A cultural history of the medicinal and ritual use of chocolate. J Nutr. 2000;130:2057–72.
13. Lippi D. Chocolate and medicine: dangerous liaisons? Nutrition. 2009;25:1100–3.
14. Beckett ST. The science of chocolate. 2nd ed. Cambridge: RSC Publishing; 2008.
15. Chaitman BR, Schmitz HH, Keen CL. Cocoa flavanols and cardiovascular health. Business Briefing: US Cardiology 2006. www.touchbriefings.com/pdf/1601/Chaitman.pdf. 23–6.
16. Chevaux KA, Jackson L, Villar ME, et al. Proximate, mineral and procyanidin content of certain foods and beverages consumed by the Kuna Amerinds of Panama. J Food Comp Anal. 2001;14:553–63.
17. Payne MJ, Hurst WJ, Miller KB, Rank C, Stuart DA. Impact of fermentation, drying, roasting, and Dutch processing on epicatechin and catechin content of cacao beans and cocoa ingredients. J Agric Food Chem. 2010;58:10518–27.
18. Wollgast J, Anklam E. Review on polyphenols in Theobroma cacao: changes in composition during manufacture of chocolate and methodology for identification and quantification. Food Res Int. 2000;33:423–47.
19. Kim HC, Keeney PG. (-)Epicatechin content in fermented and unfermented cocoa beans. J Food Sci. 1984;49:1090–2.
20. Addison S, Stas S, Hayden MR, Sowers JR. Insulin resistance and blood pressure. Curr Hypertens Rep. 2008; 10:319–25.
21. Grassi D, Necozione S, Lippi C, Croce G, Valeri L, Pasqualetti P, et al. Cocoa reduces blood pressure and insulin resistance and improves endothelium-dependent vasodilation in hypertensives. Hypertension. 2005;46:398–405.
22. Grassi D, Desideri G, Necozione S, Lippi C, Casale R, Properzi G, et al. Blood pressure is reduced and insulin sensitivity increased in glucose-intolerant, hypertensive subjects after 15 days of consuming high-polyphenol dark chocolate. J Nutr. 2008;138:1671–6.
23. Davison K, Coates AM, Buckley JD, Howe PR. Effect of cocoa flavanols and exercise on cardiometabolic risk factors in overweight and obese subjects. Int J Obes (Lond). 2008;32:1289–96.
24. Faridi Z, Njike VY, Dutta S, Ali A, Katz DL. Acute dark chocolate and cocoa ingestion and endothelial function: a randomized controlled crossover trial. Am J Clin Nutr. 2008;88:58–63.
25. Muniyappa R, Hall G, Kolodziej TL, Karne RJ, Crandon SK, Quon MJ. Cocoa consumption for 2 wk enhances insulin-mediated vasodilatation without improving blood pressure or insulin resistance in essential hypertension. Am J Clin Nutr. 2008;88:1685–96.
26. Actis-Goretta L, Ottaviani JI, Fraga CG. Inhibition of angiotensin converting enzyme activity by flavanol-rich foods. J Agric Food Chem. 2006;54:229–34.
27. Persson IA, Persson K, Hägg S, Andersson RG. Effects of cocoa extract and dark chocolate on angiotensin-converting enzyme and nitric oxide in human endothelial cells and healthy volunteers – a nutrigenomics perspective. J Cardiovasc Pharmacol. 2011;57:44–50.
28. Keen CL, Holt RR, Oteiza PI, Fraga CG, Schmitz HH. Cocoa antioxidants and cardiovascular health. Am J Clin Nutr. 2005;81:298S–303.
29. Serafini M, Bugianesi R, Maiani G, Valtuena S, De Santis S, Crozier A. Plasma antioxidants from chocolate. Nature. 2003;424:1013.

30. Ried K, Sullivan TR, Fakler P, Frank OR, Stocks NP. Effect of chocolate on blood pressure. *Cochrane Database of Systematic Reviews* 2010, Issue 12. Art. No.: CD008893.
31. Taubert D, Roesen R, Schömig E. Effect of cocoa and tea intake on blood pressure: a meta-analysis. Arch Intern Med. 2007;167:626–34.
32. Desch S, Schmidt J, Kobler D, Sonnabend M, Eitel I, Sareban M, et al. Effect of cocoa products on blood pressure: systematic review and meta-analysis. Am J Hypertens. 2010;23:97–103.
33. Ried K, Sullivan T, Fakler P, Frank OR, Stocks NP. Does chocolate reduce blood pressure? A meta-analysis. BMC Med. 2010;8:39.
34. Review Manager (RevMan) [Computer program]. Version 5.1 Copenhagen: The Nordic Cochrane Centre, The Cochrane Collaboration. 2011. http://www.cc-ims.net/RevMan. Released 22 Mar 2011.
35. StataCorp. Stata statistical software: release 10. College Station: StataCorp LP; 2007.
36. Taubert D, Berkels R, Roesen R, Klaus W. Chocolate and blood pressure in elderly individuals with isolated systolic hypertension. JAMA. 2003;290:1029–30.
37. Murphy KJ, Chronopoulos AK, Singh I, Francis MA, Moriarty H, Pike MJ, et al. Dietary flavanols and procyanidin oligomers from cocoa (*Theobroma cacao*) inhibit platelet function. Am J Clin Nutr. 2003;77:1466–73.
38. Engler MB, Engler MM, Chen CY, Malloy MJ, Browne A, Chiu EY, et al. Flavonoid-rich dark chocolate improves endothelial function and increases plasma epicatechin concentrations in healthy adults. J Am Coll Nutr. 2004;23:197–204.
39. Fraga CG, Actis-Goretta L, Ottaviani JI, Carrasquedo F, Lotito SB, Lazarus S, et al. Regular consumption of a flavanol-rich chocolate can improve oxidant stress in young soccer players. Clin Dev Immunol. 2005;12:11–7.
40. Taubert D, Roesen R, Lehmann C, Jung N, Schömig E. Effects of low habitual cocoa intake on blood pressure and bioactive nitric oxide: a randomized controlled trial. JAMA. 2007;298:49–60.
41. Crews Jr WD, Harrison DW, Wright JW. A double-blind, placebo-controlled, randomized trial of the effects of dark chocolate and cocoa on variables associated with neuropsychological functioning and cardiovascular health: clinical findings from a sample of healthy, cognitively intact older adults. Am J Clin Nutr. 2008;87:872–80.
42. Al-Faris NA. Short-term consumption of a dark chocolate containing flavanols is followed by a significant decrease in normotensive population. Pak J Nutr. 2008;7:773–81.
43. Shiina Y, Funabashi N, Lee K, Murayama T, Nakamura K, Wakatsuki Y, et al. Acute effect of oral flavonoid-rich dark chocolate intake on coronary circulation, as compared with non-flavonoid white chocolate, by transthoracic Doppler echocardiography in healthy adults. Int J Cardiol. 2009;131:424–9.
44. Ried K, Frank OR, Stocks NP. Dark chocolate or tomato extract for prehypertension: a randomised controlled trial. BMC Complement Altern Med. 2009;9:22.
45. Monagas M, Khan N, Andres-Lacueva C, Casas R, Urpí-Sardà M, Llorach R, et al. Effect of cocoa powder on the modulation of inflammatory biomarkers in patients at high risk of cardiovascular disease. Am J Clin Nutr. 2009;90:1144–50.
46. van den Bogaard B, Draijer R, Westerhof BE, van den Meiracker AH, van Montfrans GA, van den Born BJ. Effects on peripheral and central blood pressure of cocoa with natural or high-dose theobromine: a randomized, double-blind crossover trial. Hypertension. 2010;56:839–46.
47. Heiss C, Jahn S, Taylor M, Real WM, Angeli FS, Wong ML, et al. Improvement of endothelial function with dietary flavanols is associated with mobilization of circulating angiogenic cells in patients with coronary artery disease. J Am Coll Cardiol. 2010;56:218–24.
48. Desch S, Kobler D, Schmidt J, Sonnabend M, Adams V, Sareban M, et al. Low vs. higher-dose dark chocolate and blood pressure in cardiovascular high-risk patients. Am J Hypertens. 2010;23:694–700.
49. Davison K, Berry NM, Misan G, Coates AM, Buckley JD, Howe PR. Dose-related effects of flavanol-rich cocoa on blood pressure. J Hum Hypertens. 2010;24:568–76.
50. Njike VY, Faridi Z, Shuval K, Dutta S, Kay CD, West SG, et al. Effects of sugar-sweetened and sugar-free cocoa on endothelial function in overweight adults. Int J Cardiol. 2011;149:83–8.
51. Heiss C, Finis D, Kleinbongard P, Hoffmann A, Rassaf T, Kelm M, et al. Sustained increase in flow-mediated dilation after daily intake of high-flavanol cocoa drink over 1 week. J Cardiovasc Pharmacol. 2007;49:74–80.
52. Balzer J, Rassaf T, Heiss C, Kleinbongard P, Lauer T, Merx M, et al. Sustained benefits in vascular function through flavanol-containing cocoa in medicated diabetic patients a double-masked, randomized, controlled trial. J Am Coll Cardiol. 2008;51:2141–9.
53. Kelly CJ. Effects of theobromine should be considered in future studies. Am J Clin Nutr. 2005;82:486–7, author reply 487–488.
54. Strachan ER, Bennett A. Theobromine poisoning in dogs. Vet Rec. 1994;134:284.
55. Rusconi M, Conti A. *Theobroma cacao* L., the food of the gods: a scientific approach beyond myths and claims. Pharmacol Res. 2010;61:5–13.
56. Akbari CM, Saouaf R, Barnhill DF, Newman PA, LoGerfo FW, Veves A. Endothelium-dependent vasodilatation is impaired in both microcirculation and macrocirculation during acute hyperglycemia. J Vasc Surg. 1998; 28:687–94.

57. O'Rourke M. Arterial stiffness, systolic blood pressure, and logical treatment of arterial hypertension. Hypertension. 1990;15:339–47.
58. Donato AJ, Gano LB, Eskurza I, Silver AE, Gates PE, Jablonski K, et al. Vascular endothelial dysfunction with aging: endothelin-1 and endothelial nitric oxide synthase. Am J Physiol Heart Circ Physiol. 2009;297:H425–32.
59. Taddei S, Virdis A, Ghiadoni L, Salvetti G, Bernini G, Magagna A, et al. Age-related reduction of NO availability and oxidative stress in humans. Hypertension. 2001;38:274–9.
60. Loke WM, Hodgson JM, Proudfoot JM, McKinley AJ, Puddey IB, Croft KD. Pure dietary flavonoids quercetin and (−)-epicatechin augment nitric oxide products and reduce endothelin-1 acutely in healthy men. Am J Clin Nutr. 2008;88:1018–25.
61. Flammer AJ, Hermann F, Sudano I, Spieker L, Hermann M, Cooper KA, et al. Dark chocolate improves coronary vasomotion and reduces platelet reactivity. Circulation. 2007;116:2376–82.
62. Berry NM, Davison K, Coates AM, Buckley JD, Howe PR. Impact of cocoa flavanol consumption on blood pressure responsiveness to exercise. Br J Nutr. 2010;103:1480–4.
63. Farouque HM, Leung M, Hope SA, Baldi M, Schechter C, Cameron JD, et al. Acute and chronic effects of flavanol-rich cocoa on vascular function in subjects with coronary artery disease: a randomized double-blind placebo-controlled study. Clin Sci (Lond). 2006;111:71–80.
64. Almoosawi S, Fyfe L, Ho C, Al-Dujaili E. The effect of polyphenol-rich dark chocolate on fasting capillary whole blood glucose, total cholesterol, blood pressure and glucocorticoids in healthy overweight and obese subjects. Br J Nutr. 2010;103:842–50.
65. Allen RR, Carson L, Kwik-Uribe C, Evans EM, Erdman Jr JW. Daily consumption of a dark chocolate containing flavanols and added sterol esters affects cardiovascular risk factors in a normotensive population with elevated cholesterol. J Nutr. 2008;138:725–31.
66. Erdman Jr JW, Carson L, Kwik-Uribe C, Evans EM, Allen RR. Effects of cocoa flavanols on risk factors for cardiovascular disease. Asia Pac J Clin Nutr. 2008;17(Suppl 1):284–7.

Chapter 24
Polyphenol-Rich Cocoa and Chocolate: Potential Role in the Prevention of Diabetes

Suzana Almoosawi and Emad Al-Dujaili

Key Points

- Diabetes is one of the world's leading causes of mortality. Failure to control diabetes could lead to the development of peripheral vascular and cardiovascular complications.
- In vitro and animal studies suggest that polyphenols in cocoa and chocolate have the potential to protect against diabetes and diabetes-related complications via a variety of molecular and physiological mechanisms. However, data from human and epidemiological studies remain limited.
- Further research is required before a definite role for polyphenol-rich cocoa or chocolate in preventing and managing diabetes-related complications is established.

Keywords Cocoa • Chocolate • Polyphenols • Insulin resistance • Diabetes • Obesity

Introduction

Diabetes is one of the leading causes of mortality in both developed and developing countries. Worldwide, diabetes affects over 220 million people, with projected prevalence and related deaths expected to double between 2005 and 2030 [1].

Diabetes mellitus is generally classified as type 1 and type 2. Type 1 diabetes is characterized by an early onset in childhood and arises as a result of failure of pancreatic β(beta)-cells to secrete insulin [1]. In type 2 diabetes, the function of pancreatic β(beta)-cells remains normal; however, the ability of skeletal muscle and other insulin-target tissues to utilize glucose is compromised [1]. Failure to control diabetes results in hyperglycemia, which if left untreated leads to the development of complications such as retinopathy, neuropathy, cardiovascular disease, and kidney failure. To date, the cause of type

S. Almoosawi, Ph.D. (✉)
Population Nutrition Research Group, MRC Human Nutrition Research,
Elsie Widdowson Laboratory, Cambridge, UK
e-mail: suzana.almoosawi@mrc-hnr.cam.ac.uk

E. Al-Dujaili, B.Sc., Ph.D.
Department of Dietetics, Nutrition and Biological Sciences, Queen Margaret University, Edinburgh, Scotland
e-mail: ealdujaili@qmu.ac.uk

R.R. Watson et al. (eds.), *Chocolate in Health and Nutrition*, Nutrition and Health 7,
DOI 10.1007/978-1-61779-803-0_24, © Springer Science+Business Media, LLC 2013

1 diabetes remains largely unknown; hence, it is not possible yet to design dietary recommendations to prevent its occurrence. By contrast, type 2 diabetes is viewed as a lifestyle disease brought about by a combination of a sedentary lifestyle and excess adiposity [1]. In fact, current recommendations for the treatment and prevention of type 2 diabetes include adoption of a healthy lifestyle characterized by increased physical activity, maintenance of a healthy body weight, reducing adiposity, and adopting a healthy, balanced diet [2]. Since type 2 diabetes accounts for 90% of all diabetes cases [1], understanding its etiology and the role of diet and nutrition in preventing or reducing its occurrence might be central to devising effective health-promoting strategies.

Polyphenols are one of the most widespread groups of antioxidants found in our diet. One of the richest sources of polyphenols is cocoa [3]. Mounting evidence suggests a protective role for polyphenols against chronic disease such as cardiovascular disease, hypertension, and diabetes [4, 5]. This protective role is thought to arise as a result of the antioxidant properties of polyphenols and their ability to alter the activity of various enzymes involved in metabolism, increase nitric oxide bioavailability, and reduce inflammation [4, 5].

The following critical review summarizes the current evidence from epidemiological studies, in vitro and animal studies and randomized clinical trials on the protective role of polyphenol-rich cocoa and its related products, chocolate, against diabetes. Potential mechanisms by which cocoa and chocolate polyphenols can protect against diabetes and its related complications are also discussed.

Epidemiological Studies

Evidence of the inverse association between cocoa consumption and chronic diseases such as diabetes and hypertension originates from studies on the Kuna Indians, a native population of San Blas Islands in Panama [6, 7]. The Kuna Indians have the highest flavonoid intake of any community, consuming on average 900 mg of polyphenols per day [6]. This high polyphenol intake is thought to be obtained from commonly ingested cocoa beverages. Studies comparing dietary behaviour of traditional Kuna Indians with Indian migrants residing in Panama City have shown that island-dwelling Kunas consume 5–6 cups of cocoa-containing beverages per day [6]. The procyanidin content of these cocoa beverages has been estimated to be as high as 1,880 mg procyanidins per 4–8 oz [8]. This led several researchers to hypothesize that polyphenol-rich cocoa may be protective against chronic diseases. This hypothesis is supported by the lower prevalence of chronic diseases like stroke, hypertension, and diabetes in island-dwelling Kunas compared to other Panamanians who consume commercially available polyphenol-poor cocoa products [9]. In fact, it has been observed that deaths due to diabetes mellitus are up to four times lower among island-dwelling Kunas compared to city dwellers [6].

Findings of the beneficial effects of polyphenol-rich cocoa as well as polyphenol-rich chocolate have also been observed in several other large Western cohorts, including the Dutch Zutphen Elderly cohort [10], the Potsdam European Prospective Investigation into Cancer cohort [11], and the Iowa's Women's Health study [12]. More recently, Cassidy and colleagues [13] reported a 5–7% reduction in risk of hypertension in men and women aged ≤60 years associated with consumption of catechin and epicatechin, respectively. This study used data from three large cohorts with an average follow-up of 14 years and included 87,242 women from the Nurses' Health Study (NHS) II, 46,672 women from the NHS I, and 23,043 men from the Health Professionals Follow-Up Study.

Most of the aforementioned observational studies used robust statistical analysis adjusting for important dietary and lifestyle confounders, including total energy intake, age, and body mass index. Nevertheless, findings from epidemiological studies should be interpreted with caution as they are subject to several limitations. First, it is difficult to obtain an estimate of the polyphenol content of cocoa and chocolate products on population level, and most epidemiological studies also do not differentiate between the types of cocoa and chocolate products consumed (white, milk, or dark chocolate). Invariably, most Western populations are more likely to consume commercially available cocoa or milk chocolate that is poor in polyphenols. This means that findings from observational studies may underestimate the true extent of the beneficial properties of polyphenol-rich cocoa and chocolate. As a result, more stringent animal and human studies conducted under controlled conditions are required to confirm the findings of epidemiological studies. Second, most studies employed single nutrient or food analyses while adjusting for some nutrient and food intake. Such analysis does not take into account the fact that chocolate, like all other foods, is not eaten in isolation. Recently, Hamer and Mishra proposed the use of dietary patterns for studying the interactive effects of polyphenols and habitual diet [14]. This approach takes into account potential synergetic effects between polyphenol intake and other dietary constituents [14]. This approach also recognizes that foods are consumed in combination and that intake of foods is often highly correlated [14]. However, chocolate consumption is likely to form part of unhealthy dietary and lifestyle habits, as evident by the Seguimiento Universidad de Navarra Study, a study that included 5,880 participants, and that showed a strong link between chocolate consumption and snacking [15]. Thus, to overcome the limitations of single nutrient and food analyses, future epidemiological studies should investigate the relation between cocoa or chocolate consumption and chronic disease and mortality after adjusting for overall dietary pattern. This will permit estimating the independent effect of cocoa or chocolate consumption on health. Finally, it is important to highlight that most observational studies have focused on cardiovascular disease outcomes and hypertension, and only limited evidence exists in relation to the association between cocoa or chocolate consumption and diabetes. The few studies that have examined the association between chocolate consumption and insulin resistance found that chocolate either protects against diabetes [6], is only weakly associated with diabetes risk [16], or adversely affects glucose tolerance and body mass index [17]. Such inconsistencies in findings could be attributed to variations in the definition of cocoa and chocolate, with some studies placing sugary confectionaries and other polyphenol-poor cocoa and chocolate confectionaries in the same category as polyphenol-rich cocoa and chocolate. The latter is further complicated by the difficulty arising in estimating polyphenol intake from cocoa and chocolate. As such, evidence from epidemiological studies in relation to diabetes remains limited, and further research is warranted in this field.

In Vitro and Animal Studies

Most of our knowledge of the hypoglycemic effects of polyphenol-rich cocoa and chocolate comes from mechanistic studies on cell lines and animals. Although most of these studies have been criticized for using pharmacological doses of cocoa polyphenols, they provide key insight into the potential mechanisms by which cocoa polyphenols may protect against diabetes. Accordingly, cocoa polyphenols have been shown to act on several stages of glucose metabolism. In the gastrointestinal tract, cocoa polyphenols have been reported to downregulate the enzymatic activity of salivary and pancreatic α-amylase, with larger polymers, such as procyanidins, being more effective than simpler phenols [18–20]. Cocoa polyphenols have also been shown to reduce intestinal glucose absorption by directly interacting with glucose transporters [21, 22]. More specifically, epicatechin gallate and quercetin-3-O-glucoside have been demonstrated to inhibit facilitated intestinal glucose transporter, GLUT2 [22] and sodium-dependent glucose transporters, namely, GLUT1 and sodium-dependent glucose transporter-1 [21]. These changes in carbohydrate breakdown and glucose absorption could be hypothesized to account for the improved glycemic response seen in animals following consumption of cocoa.

The beneficial effects of cocoa and its polyphenols on glycemic control have been demonstrated in several animal models of diabetes. In a model of type 2 diabetes, the diet of db/db diabetic obese mice was enriched with different doses of cacao liquor procyanidins for 3 weeks. Cacao procyanidins produced a dose-dependent reduction in blood glucose and fructosamine levels [23]. The authors concluded that dietary intake of cacao beans could be "beneficial in preventing the onset of type 2 diabetes" [23]. Similarly, in a series of studies, Amin and colleagues examined the effect of cocoa extract on glucose tolerance in normoglycemic and hyperglycemic rats [24]. Using three concentrations of cocoa extract, they observed that glucose tolerance was improved in hyperglycemic rats following acute administration of a diet enriched with 3% cocoa extract. By contrast, normoglycemic rats showed a reduction in glucose concentrations 1 and 1.5 h after oral glucose tolerance tests after ingestion of 0.5%, 1.0%, and 3.0% cocoa extract. However, none of these effects were sustained after 2 weeks of ingestion. In a later study, Amin and colleagues induced diabetes in 90 male Sprague–Dawley rats and extended the cocoa treatment period to 4 weeks [25]. Glucose was monitored at the beginning and at the end of intervention. The rats were divided into two groups: normoglycemic and hyperglycemic group. Each group was further subdivided into four groups: a control, 1% cocoa extract, 2% cocoa extract, and 3% cocoa extract. Cocoa extract was administered by direct gastric intubation. In normoglycemic rats, no significant changes in glucose levels were observed between baseline and week 4. In diabetic rats, plasma glucose levels decreased by 47% and 57% after feeding with 1% and 3% cocoa extract, respectively. Amin and colleagues attributed these effects to the ability of cocoa to enhance basal insulin secretion, as demonstrated by some of their unpublished in vitro work on pancreatic cell lines [24, 26]. Indeed, cocoa has a protective effect against cytokine-induced β(beta)-cell damage [27]. Pro-inflammatory factors have long been recognized to play a role in the pathogenesis of diabetes. One such pro-inflammatory factor is interleukin-1β(beta). Interleukin-1β(beta) is activated by nuclear factor-κ(kappa)B [27]. Its activation results in stimulation of inducible nitric oxide synthase (iNOS) expression, leading to nitric oxide overproduction and subsequently β(beta)-cell damage. Epicatechin has been shown to downregulate nuclear factor-κB [27]. This downregulation protects against interleukin-1β(beta)-induced β(beta)-cell damage, which leads to a reduction in nitrite levels and improvement in pancreatic insulin release [27]. Administration of (−)-epicatechin in doses of 30 mg/kg twice daily for 4–5 days to alloxan-induced diabetic rats has also been reported to normalize blood glucose concentration via regeneration of necrotic pancreatic β(beta)-cells [28–31], a finding that has not been observed by later studies [32–34]. Nevertheless, more recent evidence suggests that administration of (−)-epicatechin to rats can preserve pancreatic β(beta)-cell morphology against damage induced by streptozotocin, leading to improved pancreatic insulin secretion [35].

In addition to the above pathways, multiple other mechanisms can explain the protective effects of cocoa polyphenols against diabetes. (−)-Epicatechin has been reported to possess insulin-like activity, including stimulation of oxygen uptake into insulin-target tissues (i.e., adipose tissue), inhibition of lipolysis, and increase of glucose uptake and glycogenesis [36]. (−)-Epicatechin has also been shown to reduce lipid peroxidation and improve antioxidant enzyme activity in rats, effects that are accompanied by a reduction in blood glucose levels [37, 38]. Cocoa supplementation can reportedly reduce circulating plasma free fatty acid [38], increase high-density lipoprotein cholesterol concentration, and reduce low-density lipoprotein cholesterol and triglyceride levels [25]. Both changes in oxidative stress and free fatty acids play a key role in the pathogenesis of diabetes and may influence the development of diabetes-related complications [39, 40]. As such, polyphenol-rich cocoa and is main product, chocolate, can be argued to have the potential to protect against diabetes and its related risk factors and complications via various mechanisms, as outlined in Fig. 24.1. This could partially supported by some of the few studies that have shown that consumption of cocoa-derived procyanidins by streptozotocin rats inhibits diabetes-induced cataract formation by reducing oxidative stress [41].

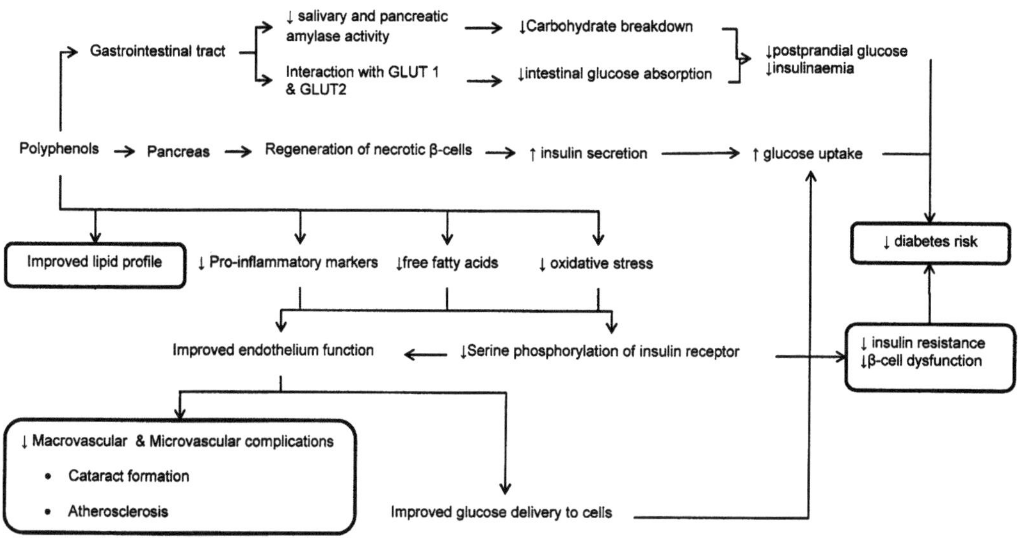

Fig. 24.1 Potential mechanisms by which cocoa and chocolate polyphenols protect against diabetes and its associated macrovascular and microvascular complications

Limitations

One of the major criticisms of animal studies investigating the hypoglycemic properties of cocoa extract is the use of obese type 2 diabetes genetic models and streptozotocin-induced diabetic rat models (see reference [42]). In the first model, animals are genetically modified to develop a phenotype similar in characteristic to type 2 diabetes with hyperglycemia, hyperinsulinemia, and obesity. In the second model, streptozotocin produces inflammatory lesions in pancreatic β(beta)-cell function, which leads to progressive decline in β(beta)-cell function and eventually diabetes in a similar manner to the autoimmune response seen in type 1 diabetes [43]. These models have been criticized because they do not mimic the natural disease progression and the metabolic characteristics of human type 2 diabetes [42]. Thus, Jalil et al. proposed the use of rats fed a high-fat diet and injected with a low dose of streptozotocin for studying the hypoglycemic effects of cocoa polyphenols [42]. In this model of type 2 diabetes, insulin resistance precedes hyperglycemia and occurs as a consequence of obesity, which then leads to type 2 diabetes [42]. Jalil et al. observed that cocoa extract supplemented with polyphenols (2.17 mg epicatechin, 1.52 mg catechin, 0.25 mg procyanidin dimer, and 0.13 mg trimer per 1 g cocoa extract) and methylxanthines (3.55 mg caffeine and 2.22 mg theobromine per 1 g cocoa extract) acutely reduced postprandial glucose levels as assessed by the oral glucose tolerance test [42]. However, no significant improvements in plasma glucose level, insulin level, and insulin sensitivity after 4-week consumption of the cocoa extract were observed [42]. As a result, Jalil et al. argued that cocoa may not directly affect insulin secretion. Alternatively, differences in composition and dosage of cocoa extracts and the use of different diabetes models, reflecting differences in etiology of disease, may have yielded inconsistent results [42].

Role of Other Cocoa and Chocolate Constituents

Cocoa and dark chocolate contain high levels of minerals and methylxanthines (caffeine and theobromine) [44]. These compounds can potentially influence glucose metabolism. Magnesium acts as a cofactor at several stages of carbohydrate metabolism affecting glucose transport, insulin secretion, and insulin activity [45]. In prospective cohorts, magnesium intake has also been inversely related to type 2 diabetes incidence [45]. Caffeine, on the other hand, has been shown to reduce insulin sensitivity in healthy individuals [46] and impair glucose metabolism in type 2 diabetics [47, 48]. Thus, it could be speculated that caffeine may have counteracted the hypoglycemic properties of cocoa polyphenols in Jalil's 4-week trial leading to the null findings. Theobromine is present in dark chocolate in concentrations as high as 237–519 mg per 50 g [49]. However, its role in glucose metabolism remains largely unknown. Thus, future studies should consider providing data on the composition of cocoa extract to enable comparison between studies. Potential interactions between the different cocoa constituents also need to be investigated. Further research using physiological doses would also be recommended.

Human Studies

In 2005, Grassi and colleagues hypothesized that polyphenol-rich dark chocolate may improve insulin sensitivity [50], a hypothesis that stemmed from earlier studies that showed that cocoa and dark chocolate polyphenols reduce blood pressure by increasing insulin-dependent nitric oxide bioavailability [51–54]. This led to the series of pioneering studies which showed that consumption of polyphenol-rich dark chocolate improves insulin resistance, insulin sensitivity, and fasting glucose levels in healthy [50], hypertensive [55], and glucose-intolerant hypertensive adults [56]. Since then, a number of human interventions have been completed that have looked at the effect of polyphenol-rich cocoa and dark chocolate on glucose metabolism in overweight and obese [57, 58], insulin-resistant [59], and diabetic adults [60] (Table 24.1). Currently, more clinical trials are underway to investigate the effect of polyphenol-rich cocoa on endothelial dysfunction, kidney function, and cerebral blood flow in diabetic patients (see clinicaltrials.gov).

Early Human Studies

In their first trial, Grassi and colleagues compared the effect of 100 g of polyphenol-rich dark chocolate versus the effect of polyphenol-deficient white chocolate on several parameters of glucose metabolism [50]. In this crossover randomized-controlled trial, healthy normotensive volunteers were given 100 g of dark chocolate with a 500-mg polyphenol content or 90 g of white chocolate for 15 days and then crossed over to the other treatment separated by a 1-week washout period [50]. They found that polyphenol-rich dark chocolate improved homeostasis model assessment of insulin resistance (HOMA-IR) and quantitative insulin sensitivity check index (QUICKI) compared to polyphenol-deficient white chocolate. In a later study, Grassi et al. reported significant improvements in HOMA-IR, QUICKI, and oral glucose tolerance test in patients with essential hypertension at the end of 15 days intervention with 100 g of polyphenol-rich dark chocolate [55]. In their latest intervention, Grassi et al. showed improvements in HOMA-IR, QUICKI, insulin sensitivity index (ISI), and β(beta)-cell function as calculated by corrected insulin response (CIR120) in hypertensive subjects with impaired glucose tolerance [56]. These changes in insulin sensitivity and β(beta)-cell function were correlated

Table 24.1 Effect of polyphenol-rich cocoa and chocolate on different measures of glucose metabolism

Author	Study design	Population	Polyphenol dose	Placebo	Measurement	Change
Mathur (2002) [69]	Nonrandomized trial, subjects studied at the end of 6 weeks DC and at the end of 6-week follow-up period	N = 25 healthy: 12 females, 13 males	36.9 g chocolate and 30.95 g cocoa drink, 651 mg procyanidins	Habitual diet	FG	No
Grassi (2005) [50]	Randomized crossover, 15 days	N = 15 healthy: 7 males, 8 females	100 g chocolate, 500 mg polyphenols, 88 mg flavanols	90 g white chocolate	OGTT, HOMA-IR, QUICKI, FI	Yes
Grassi (2005) [55]	Randomized crossover, 15 days	N = 20 never-treated, grade I patients with essential hypertension: 10 males, 10 females	100 g chocolate, 500 mg polyphenols, 88 mg flavanols	90 g white chocolate	OGTT, HOMA-IR, QUICKI, ISI, FI	Yes
Stote (2007) [59]	Randomized, parallel-group, 5 days	N = 20 subjects: 12 insulin-resistant, 1 diabetic, 7 healthy; 10 males, 10 females	Three cocoa procyanidin doses: 900, 400, and 200 mg procyanidins taken twice daily	Control beverage, 22 mg procyanidins	OGTT, HOMA-IR	No
Taubert (2007) [62]	Randomized, parallel control, 18 weeks	N = 44 upper-range prehypertension or stage 1 hypertension: 24 females, 20 males	6.3 g chocolate, 30 mg polyphenols	30 g white chocolate	FG	No
Balzer (2008) [60]	Randomized, parallel control, 30 days	N = 41 type 2 diabetics: 29 females, 12 males	18 g cocoa, 321 mg flavanols taken thrice daily (total 963 mg flavanols)	18 g cocoa, 25 mg flavanols taken thrice daily (total 75 mg flavanols)	FG, HbA_{1C}	Yes
Davison (2008) [57]	Randomized, parallel control, 12 weeks	N = 49 BMI ≥25 kg/m^2: 31 females, 18 males	Cocoa drink, 902 mg flavanols	Low-flavanol cocoa drink	HOMA-IR	No
Grassi (2008) [56]	Randomized crossover, 15 days	N = 19 hypertensives with impaired glucose tolerance: 11 males, 8 females	100 g chocolate, 500 mg polyphenols	90 g white chocolate	OGTT, HOMA-IR, QUICKI, ISI, CIR_{120}, FI	Yes
Almoosawi (2010) [58]	Randomized crossover, 2 weeks	N = 14 BMI ≥25 kg/m^2	20 g chocolate, 500 mg polyphenols, 1,000 mg polyphenols	No control	FG	Yes
Almoosawi (2011) [65, 70]	Randomized crossover, 4 weeks	N = 42: 21 BMI < 25 kg/m^2, 21 BMI ≥25 kg/m^2	20 g chocolate, 500 mg polyphenols	20 g chocolate, negligible polyphenol content	FG, HOMA-IR, QUICKI, FI	Yes

FG fasting glucose, *OGTT* oral glucose tolerance test, *HOMA-IR* homeostasis model assessment of insulin resistance, *QUICKI* quantitative insulin sensitivity check index, *FI* fasting insulin, HbA_{1C}, glycosylated hemoglobin, *ISI* insulin sensitivity index, CIR_{120} corrected insulin response

with improvement in flow-mediated dilation and blood pressure, thereby providing support for the basis of the hypothesis that led Grassi and colleagues to investigate the beneficial properties of polyphenol-rich dark chocolate on glucoregulatory biomarkers.

Limitations of Early Human Studies

As is the case with most studies, there were several limitations to the studies conducted by Grassi and colleagues. First, Grassi and colleagues used large amounts of chocolate providing 480 kcal. Inevitably, such large intake is not sustainable in the long term and could lead to weight gain, as demonstrated by Desch et al. who found that consumption of 25 g of polyphenol-rich dark chocolate by overweight and obese individuals over 3 months increases body weight [61]. This suggests that any beneficial effect of chocolate consumption could be outweighed by the detrimental effects of weight gain. Second, Grassi et al. used white chocolate as control, which does not permit effective blinding of volunteers. White chocolate also differs from dark chocolate in its macronutrient and fatty acid composition because it does not contain any methylxanthines or magnesium. This means that any observed changes in glucose or insulin levels seen in Grassi et al.'s trials cannot be attributed directly to polyphenol intake. Regardless of these limitations, Grassi et al.'s findings triggered novel interventions that examined the effect of cocoa or lower dark chocolate intake on glucose metabolism. These studies, as outlined below, produced conflicting findings.

Evidence from Human Studies in Insulin-Resistant and Diabetic Patients

In a short-term study conducted by Stote and colleagues, twice-daily administration of different doses of procyanidin-rich cocoa (22–900 mg procyanidins) to insulin-resistant men and women for 5 days did not change fasting glucose levels, HOMA-IR, and ISI [59]. By contrast, in overweight and obese subjects, Davison et al. found that 12 weeks of twice-daily ingestion of a cocoa beverage (902 mg flavanols) improved insulin resistance in addition to reducing blood pressure and ameliorating flow-mediated dilation [57]. However, this effect was not observed with habitual dark chocolate intake [62], whereby ingestion of 6.3 g of dark chocolate (30 mg of polyphenols) for 18 weeks did not produce any significant changes in glucose or insulin levels [62].

In another study conducted on type 2 diabetic patients undergoing oral or insulin antidiabetic therapy, Balzer et al. demonstrated that consumption of cocoa (963 mg and 75 mg flavanols) for 30 days improved endothelium function and glycemic control, as indicated by the reduction in glycosylated hemoglobin levels [60]. The latter effect occurred despite the absence of any effect on fasting glucose, suggesting that cocoa intake might be more essential in terms of improving long-term glycemic control and possibly reducing vascular complications in diabetes [60]. Alternatively, the use of a cocoa beverage containing 27 g of carbohydrates, of which 15 g was sugar, may have attenuated the hypoglycemic effect of cocoa polyphenols. This implies that future studies should perhaps use sugar-free cocoa in investigating the impact of cocoa consumption on diabetics. The fact that patients were receiving antidiabetic therapy may also suggest that patients were at a more advanced stage of disease, which potentially implies that timing of intervention may determine the extent of the impact of cocoa consumption on glucose metabolism.

Together, these studies may suggest that a longer duration and a higher dose of polyphenols could be required to achieve a marked reduction in glucose levels. It can also be postulated that improvement in glycemic regulation can occur without a detectable reduction in fasting glucose. Alternatively, polyphenol intake may be more important in protecting against diabetes-related complication as opposed to glycemic control. In this respect, a recent study conducted on type 2 diabetics found that 4-week consumption of 15 g chocolate with 85% cocoa content (6.6 mg epicatechin) improved atherosclerotic profile by increasing HDL cholesterol and improving the cholesterol to high-density cholesterol ratio [63].

Evidence from Human Studies in Overweight and Obese Subjects

In a randomized crossover study conducted by our group, overweight and obese subjects were given 20 g of dark chocolate containing 500 or 1,000 mg polyphenols for 2 weeks, separated by a 1-week washout period [58]. Both polyphenol doses were found to produce similar reductions in fasting capillary glucose levels and blood pressure [58]. We concluded that increasing polyphenol intake above 500 mg/day may not necessarily provide additional health benefits. Indeed, one study has found that high doses of polyphenols produce pro-oxidant effects [64], and, as outlined earlier, findings from studies that used high doses of polyphenols from cocoa are conflicting [57, 59, 60]. Based on our preliminary results, we decided to examine the hypoglycemic properties of dark chocolate containing 500 mg of polyphenols in overweight and obese women. We selected a placebo dark chocolate that was matched for macronutrient and methylxanthine composition. Accordingly, overweight and obese women were found to respond more effectively to the hypoglycemic and hypotensive properties of polyphenol-rich dark chocolate compared to women with a normal-range body mass index [65]. In fact, 4-week consumption of dark chocolate with 500 mg of polyphenols produced a body mass index–dependent improvement in fasting glucose, insulin sensitivity, and blood pressure. To our knowledge, this was the first study that investigated differences in response between individuals with normal body mass index and those with a body mass index of 25 kg/m^2 or greater. Our finding may imply that individuals with a high cardiometabolic risk may benefit more from polyphenol intake than healthy individuals. This could be partially reinforced by the greater reduction in blood pressure seen following consumption of polyphenol-rich dark chocolate by hypertensives [55] compared to normotensive adults [50].

Overweight and obese individuals were more adversely affected by the ingestion of the placebo chocolate used in our study (polyphenol-poor dark chocolate), as indicated by the significant rise in fasting insulin, HOMA-IR following placebo [65]. This finding has not been reported by earlier studies [50, 55, 56] but is consistent with the findings of Holt et al., who reported that chocolate confectionery (Mars Bar) and chocolate-flavored breakfast cereal (Coco Pops) have an insulin index 50–60% higher than that predicted by their glycolic index [66]. Similarly, Brand-Miller et al. have observed that commercially available cocoa powder can produce hyperinsulinemia by increasing postprandial insulin secretion [67]. This hyperinsulinemic effect appears to be higher than the one induced by products with a similar fat content, like potato chips and croissants [67]. Brand-Miller et al. argued that stearic acid in cocoa butter or amino acids in cocoa might account for the enhanced insulin response of cocoa products [67]. However, to date, no studies have attempted to identify the agent responsible for the hyperinsulinemic effects of polyphenol-poor cocoa or chocolate. Therefore, it remains unclear whether it is the high fat content or other potential constituents that account for the hyperinsulinemic effects of polyphenol-poor cocoa or chocolate. The combined evidence from the aforementioned studies may also suggest that the health benefits of polyphenol-rich cocoa and chocolate consumption should be placed into two categories: (1) health benefits provided over and above usual intake, such as the glucose and blood pressure-lowering effects, and that can be directly attributed to polyphenols and (2) indirect health effects, such as preventing hyperinsulinemia induced by polyphenol-poor cocoa or chocolate. The latter effects do not provide additional health benefits; rather, they counteract the deleterious effects of polyphenol-poor placebo. This hypothesis warrants further investigation since it can improve our understanding of the role of polyphenols in health.

This is particularly important since in addition to raising insulin and promoting insulin resistance, we observed a marked rise in oxidative stress and salivary cortisol in women with normal-range BMI and in overweight and obese women following a 4-week ingestion of placebo. Polyphenol-rich dark chocolate, on the other hand, improved antioxidant status and did not affect cortisol levels. Cocoa butter induces oxidative stress both in vitro and in vivo. High-fat feeding has also been shown to stimulate basal and stress-induced hypothalamic activity in animal studies. This evidence reinforces once again the possibility that in the absence of polyphenols, high-fat products such as chocolate may adversely affect metabolism. This led us to hypothesize that commercially available cocoa and its products that are deficient in polyphenols could have deleterious effects on health, both through their energy density and ability to raise insulin and cortisol levels.

Summary

Evidence from in vitro and animal studies suggests a potential role for polyphenol-rich cocoa and chocolate in protecting against insulin resistance and diabetes-related complications. However, data from epidemiological and human studies remain minor and conflicting. Further research is needed to identify the polyphenol dose, treatment duration, and stage of disease intervention required to achieve diabetes-protective effects. Future studies should also adopt more robust study designs ensuring effective blinding of volunteers. The latter could be achieved by using a placebo product that is matched for appearance, taste, texture, and macro- and micronutrient composition.

Finally, it remains to be stated that because of its high caloric and fat value, recommending chocolate as a health food in the prevention and management of diabetes-related complication may be counterintuitive. Currently, most of the evidence of the hypoglycemic and insulin-sensitizing effects of chocolate polyphenols comes from studies that use large amounts of chocolate, sometimes up to 100 g/day. Since individuals with type 2 diabetes are likely to be obese, including chocolate as part of their diet might lead to weight gain that is likely to counteract any beneficial effect of polyphenols. From a public perspective, the absence of labels on cocoa and chocolate products that provide information on the polyphenol content of these products can also potentially be misleading since neither the darkness of chocolate nor its cocoa content serve as guides of polyphenol content [68]. Thus, caution should be used in interpreting the findings from current studies. Nevertheless, with recent advances in the processing method of cocoa beans, it may be possible in future to use the natural cocoa beans and the derived powder as part of a healthy balanced diet in the prevention and management of chronic diseases. This is particularly true in relation to cocoa since it possesses all the physical and physiological advantages of chocolate in terms of flavor and polyphenols without providing excess calories. Until then, we need to continue to explore the sweet secrets of *Theobroma cacao*, the food of the Gods.

References

1. WHO. Diabetes. Page. http://www.who.int/mediacentre/factsheets/fs312/en/. Updated 2011. Accessed 20 April 2011
2. Dyson PA, Kelly T, Deakin T, Duncan A, Frost G, Harrison Z, et al. Diabetes UK evidence-based nutrition guidelines for the prevention and management of diabetes. Diabet Med. 2011;28(11):1282–8. doi:10.1111/j.1464-5491.2011.03371.x.
3. Lee KW, Kim YJ, Lee HJ, Lee CY. Cocoa has more phenolic phytochemicals and a higher antioxidant capacity than teas and red wine. J Agric Food Chem. 2003;51:7292–5.
4. Corti R, Flammer AJ, Hollenberg NK, Luscher TF. Cocoa and cardiovascular health. Circulation. 2009;119:1433–41.
5. Keen CL, Holt RR, Oteiza PI, Fraga CG, Schmitz HH. Cocoa antioxidants and cardiovascular health. Am J Clin Nutr. 2005;81:298S–303.

6. Bayard V, Chamorro F, Motta J, Hollenberg NK. Does flavanol intake influence mortality from nitric oxide-dependent processes? Ischemic heart disease, stroke, diabetes mellitus, and cancer in Panama. Int J Med Sci. 2007;4:53–8.
7. McCullough ML, Chevaux K, Jackson L, Preston M, Martinez G, Schmitz HH, et al. Hypertension, the Kuna, and the epidemiology of flavanols. J Cardiovasc Pharmacol. 2006;47(Suppl 2):S103–9. discussion 119–121.
8. Chevaux KA, Jackson L, Villar ME, et al. Proximate, mineral and procyanidin content of certain foods and beverages consumed by the Kuna Amerinds of Panama. J Food Comp Anal. 2001;14:553–64.
9. Hollenberg NK, Fisher ND, McCullough ML. Flavanols, the Kuna, cocoa consumption, and nitric oxide. J Am Soc Hypertens. 2009;3:105–12.
10. Buijsse B, Feskens EJ, Kok FJ, Kromhout D. Cocoa intake, blood pressure, and cardiovascular mortality: the Zutphen Elderly Study. Arch Intern Med. 2006;166:411–7.
11. Buijsse B, Weikert C, Drogan D, Bergmann M, Boeing H. Chocolate consumption in relation to blood pressure and risk of cardiovascular disease in German adults. Eur Heart J. 2010;31:1616–23.
12. Mink PJ, Scrafford CG, Barraj LM, Harnack L, Hong CP, Nettleton JA, et al. Flavonoid intake and cardiovascular disease mortality: a prospective study in postmenopausal women. Am J Clin Nutr. 2007;85:895–909.
13. Cassidy A, O'Reilly EJ, Kay C, Sampson L, Franz M, Forman JP, et al. Habitual intake of flavonoid subclasses and incident hypertension in adults. Am J Clin Nutr. 2011;93:338–47.
14. Hamer M, Mishra G. Role of functional foods in primary prevention: cranberry extracts and cholesterol lowering. Clin Lipidol. 2009;4:141–3.
15. Alonso A, de la Fuente C, Beunza JJ, Sanchez-Villegas A, Martinez-Gonzalez MA. Chocolate consumption and incidence of hypertension. Hypertension. 2005;46:e21–2. author reply e22.
16. Oba S, Nagata C, Nakamura K, Fujii K, Kawachi T, Takatsuka N, et al. Consumption of coffee, green tea, oolong tea, black tea, chocolate snacks and the caffeine content in relation to risk of diabetes in Japanese men and women. Br J Nutr. 2010;103:453–9.
17. Williams DE, Prevost AT, Whichelow MJ, Cox BD, Day NE, Wareham NJ. A cross-sectional study of dietary patterns with glucose intolerance and other features of the metabolic syndrome. Br J Nutr. 2000;83:257–66.
18. Quesada C, Bartolomé B, Nieto O, Gómez-Cordovés C, Hernández T, Estrella I. Phenolic inhibitors of α-amylase and trypsin enzymes by extracts from pears, lentils, and cocoa. J Food Prot. 1996;59:185–92.
19. McDougall GJ, Shapiro F, Dobson P, Smith P, Blake A, Stewart D. Different polyphenolic components of soft fruits inhibit alpha-amylase and alpha-glucosidase. J Agric Food Chem. 2005;53:2760–6.
20. McDougall GJ, Stewart D. The inhibitory effects of berry polyphenols on digestive enzymes. Biofactors. 2005;23:189–95.
21. Cermak R, Landgraf S, Wolffram S. Quercetin glucosides inhibit glucose uptake into brush-border-membrane vesicles of porcine jejunum. Br J Nutr. 2004;91:849–55.
22. Chen CH, Hsu HJ, Huang YJ, Lin CJ. Interaction of flavonoids and intestinal facilitated glucose transporters. Planta Med. 2007;73:348–54.
23. Tomaru M, Takano H, Osakabe N, Yasuda A, Inoue K, Yanagisawa R, et al. Dietary supplementation with cacao liquor proanthocyanidins prevents elevation of blood glucose levels in diabetic obese mice. Nutrition. 2007;23:351–5.
24. Amin I, Faizul HA, Azli R. Effect of cocoa powder extract on plasma glucose levels in hyperglycaemic rats. Nutr Food Sci. 2004;34:116–21.
25. Ruzaidi A, Amin I, Nawalyah AG, Hamid M, Faizul HA. The effect of Malaysian cocoa extract on glucose levels and lipid profiles in diabetic rats. J Ethnopharmacol. 2005;98:55–60.
26. Ruzaidi R, Amin I, Nawalyah AG, et al. Hypoglycaemic properties of Malaysian cocoa (*Theobroma Cacao*) polyphenols extract. http://eprints.utm.my/946/. Updated 10 June 2010. Accessed 10 March 2011.
27. Kim MJ, Ryu GR, Kang JH, Sim SS, Min DS, Rhie DJ, et al. Inhibitory effects of epicatechin on interleukin-1beta-induced inducible nitric oxide synthase expression in RINm5F cells and rat pancreatic islets by down-regulation of NF-kappaB activation. Biochem Pharmacol. 2004;68:1775–85.
28. Chakravarthy BK, Gupta S, Gode KD. Antidiabetic effect of (−)-epicatechin. Lancet. 1982;2:272–3.
29. Chakravarthy BK, Gupta S, Gambhir SS, Gode KD. Pancreatic beta-cell regeneration in rats by (−)-epicatechin. Lancet. 1981;2:759–60.
30. Chakravarthy BK, Gupta S, Gambhir SS, Gode KD. The prophylactic action of (−)-epicatechin against alloxan induced diabetes in rats. Life Sci. 1981;29:2043–7.
31. Chakravarthy BK, Gupta S, Gode KD. Functional beta cell regeneration in the islets of pancreas in alloxan induced diabetic rats by (−)-epicatechin. Life Sci. 1982;31:2693–7.
32. Sheehan EW, Stiff DD, Duah F, Slatkin DJ, Schiff Jr PL, Zemaitis MA. The lack of effectiveness of (−)-epicatechin against alloxan induced diabetes in Wistar rats. Life Sci. 1983;33:593–7.
33. Ryle PR, Barker J, Gaines PA, Thomson AD, Chakraborty J. Alloxan-induced diabetes in the rat – protective action of (−) epicatechin? Life Sci. 1984;34:591–5.

34. Bone AJ, Hii CS, Brown D, Smith W, Howell SL. Assessment of the antidiabetic activity of epicatechin in streptozotocin-diabetic and spontaneously diabetic BB/E rats. Biosci Rep. 1985;5:215–21.
35. Kim MJ, Ryu GR, Chung JS, Sim SS, Min DS, Rhie DJ, et al. Protective effects of epicatechin against the toxic effects of streptozotocin on rat pancreatic islets: in vivo and in vitro. Pancreas. 2003;26:292–9.
36. Ahmad F, Khalid P, Khan MM, Rastogi AK, Kidwai JR. Insulin like activity in (–) epicatechin. Acta Diabetol Lat. 1989;26:291–300.
37. Quine SD, Raghu PS. Effects of (–)-epicatechin, a flavonoid on lipid peroxidation and antioxidants in streptozotocin-induced diabetic liver, kidney and heart. Pharmacol Rep. 2005;57:610–5.
38. Jalil AM, Ismail A, Pei CP, Hamid M, Kamaruddin SH. Effects of cocoa extract on glucometabolism, oxidative stress, and antioxidant enzymes in obese-diabetic (Ob-db) rats. J Agric Food Chem. 2008;56:7877–84.
39. Evans JL, Goldfine ID, Maddux BA, Grodsky GM. Oxidative stress and stress-activated signaling pathways: a unifying hypothesis of type 2 diabetes. Endocr Rev. 2002;23:599–622.
40. Evans JL, Goldfine ID, Maddux BA, Grodsky GM. Are oxidative stress-activated signaling pathways mediators of insulin resistance and beta-cell dysfunction? Diabetes. 2003;52:1–8.
41. Osakabe N, Yamagishi M, Natsume M, Yasuda A, Osawa T. Ingestion of proanthocyanidins derived from cacao inhibits diabetes-induced cataract formation in rats. Exp Biol Med (Maywood). 2004;229:33–9.
42. Jalil AB, Ismail A, Chong PP, Hamid M, Hasbullah S, Kamaruddin S. Effects of cocoa extract containing polyphenols and methylxanthines on biochemical parameters of obese-diabetic rats. J Sci Food Agric. 2008;89:130–7.
43. Rees DA, Alcolado JC. Animal models of diabetes mellitus. Diabet Med. 2005;22:359–70.
44. Kelly CJ. Effects of theobromine should be considered in future studies. Am J Clin Nutr. 2005;82:486–7. author reply 487–488.
45. Larsson SC, Wolk A. Magnesium intake and risk of type 2 diabetes: a meta-analysis. J Intern Med. 2007; 262:208–14.
46. Graham TE, Sathasivam P, Rowland M, Marko N, Greer F, Battram D. Caffeine ingestion elevates plasma insulin response in humans during an oral glucose tolerance test. Can J Physiol Pharmacol. 2001;79:559–65.
47. Lane JD, Barkauskas CE, Surwit RS, Feinglos MN. Caffeine impairs glucose metabolism in type 2 diabetes. Diabetes Care. 2004;27:2047–8.
48. Lane JD, Feinglos MN, Surwit RS. Caffeine increases ambulatory glucose and postprandial responses in coffee drinkers with type 2 diabetes. Diabetes Care. 2008;31:221–2.
49. Smit HJ, Gaffan EA, Rogers PJ. Methylxanthines are the psycho-pharmacologically active constituents of chocolate. Psychopharmacology (Berl). 2004;176:412–9.
50. Grassi D, Lippi C, Necozione S, Desideri G, Ferri C. Short-term administration of dark chocolate is followed by a significant increase in insulin sensitivity and a decrease in blood pressure in healthy persons. Am J Clin Nutr. 2005;81:611–4.
51. Karim M, McCormick K, Kappagoda CT. Effects of cocoa extracts on endothelium-dependent relaxation. J Nutr. 2000;130:2105S–8.
52. Fisher ND, Hughes M, Gerhard-Herman M, Hollenberg NK. Flavanol-rich cocoa induces nitric-oxide-dependent vasodilation in healthy humans. J Hypertens. 2003;21:2281–6.
53. Engler MB, Engler MM, Chen CY, Malloy MJ, Browne A, Chiu EY, et al. Flavonoid-rich dark chocolate improves endothelial function and increases plasma epicatechin concentrations in healthy adults. J Am Coll Nutr. 2004;23:197–204.
54. Taubert D, Berkels R, Roesen R, Klaus W. Chocolate and blood pressure in elderly individuals with isolated systolic hypertension. JAMA. 2003;290:1029–30.
55. Grassi D, Necozione S, Lippi C, Croce G, Valeri L, Pasqualetti P, et al. Cocoa reduces blood pressure and insulin resistance and improves endothelium-dependent vasodilation in hypertensives. Hypertension. 2005;46:398–405.
56. Grassi D, Desideri G, Necozione S, Lippi C, Casale R, Properzi G, et al. Blood pressure is reduced and insulin sensitivity increased in glucose-intolerant, hypertensive subjects after 15 days of consuming high-polyphenol dark chocolate. J Nutr. 2008;138:1671–6.
57. Davison K, Coates AM, Buckley JD, Howe PR. Effect of cocoa flavanols and exercise on cardiometabolic risk factors in overweight and obese subjects. Int J Obes Lond. 2008;32:1289–96.
58. Almoosawi S, Fyfe L, Ho C, Al-Dujaili E. The effect of polyphenol-rich dark chocolate on fasting capillary whole blood glucose, total cholesterol, blood pressure and glucocorticoids in healthy overweight and obese subjects. Br J Nutr. 2010;103:842–50.
59. Stote KS, Clevidence BA, Baer DJ. Effect of cocoa and green tea consumption on glucoregulatory biomarkers in insulin-resistant men and women [abstract]. Federation of American Societies for Experimental Biology, 28 April–2 May, Washington, DC; 2007.
60. Balzer J, Rassaf T, Heiss C, Kleinbongard P, Lauer T, Merx M, et al. Sustained benefits in vascular function through flavanol-containing cocoa in medicated diabetic patients a double-masked, randomized, controlled trial. J Am Coll Cardiol. 2008;51:2141–9.

61. Desch S, Kobler D, Schmidt J, Sonnabend M, Adams V, Sareban M, et al. Low vs. higher-dose dark chocolate and blood pressure in cardiovascular high-risk patients. Am J Hypertens. 2010;23:694–700.
62. Taubert D, Roesen R, Lehmann C, Jung N, Schomig E. Effects of low habitual cocoa intake on blood pressure and bioactive nitric oxide: a randomized controlled trial. JAMA. 2007;298:49–60.
63. Mellor DD, Sathyapalan T, Kilpatrick ES, Beckett S, Atkin SL. High-cocoa polyphenol-rich chocolate improves HDL cholesterol in type 2 diabetes patients. Diabet Med. 2010;27:1318–21.
64. Sakano K, Mizutani M, Murata M, Oikawa S, Hiraku Y, Kawanishi S. Procyanidin B2 has anti- and pro-oxidant effects on metal-mediated DNA damage. Free Radic Biol Med. 2005;39:1041–9.
65. Almoosawi S, Fyfe L, Al-Dujaili EAS. Polyphenol rich dark chocolate: effect on glucoregulatory biomarkers, blood pressure, lipids and glucocorticoids in healthy and obese subjects. Proceedings of 4th International Conference on Polyphenols in Health; Harrogate; 2008.
66. Holt SH, Miller JC, Petocz P. An insulin index of foods: the insulin demand generated by 1000-kJ portions of common foods. Am J Clin Nutr. 1997;66:1264–76.
67. Brand-Miller J, Holt SH, de Jong V, Petocz P. Cocoa powder increases postprandial insulinemia in lean young adults. J Nutr. 2003;133:3149–52.
68. Hollenberg NK, Fisher ND. Is it the dark in dark chocolate? Circulation. 2007;116:2360–2.
69. Mathur S, Devaraj S, Grundy SM, Jialal I. Cocoa products decrease low density lipoprotein oxidative susceptibility but do not affect biomarkers of inflammation in humans. J Nutr. 2002;132:3663–7.
70. Almoosawi S. Polyphneols in green coffee bean and chocolate: effect on glucoregulatory biomarkers blood pressure and lipid profile in obesity. Saarbrücken: Verlag/Dr. Müller; 2011.

Chapter 25
Polyphenol-Rich Dark Chocolate in Treatment of Diabetes Mellitus Risk Factors

Emad Al-Dujaili, Catherine Tsang, and Suzana Almoosawi

Key Points

- Type 2 diabetes is generally characterized by hyperglycemia, insulin resistance (reduced insulin sensitivity), and obesity, which is associated with not only diabetes but also hyperlipidemia and hypertension.
- Evidence from the past 10 years demonstrates that moderate intake of dark chocolate may exert protective effects against the development of type 2 diabetes risk factors; however, large-scale clinical trials are required.
- Several mechanisms have been proposed to explain the chocolate's positive effects, not only on insulin sensitivity and vascular endothelial function but also its metabolic (fat and carbohydrate), antihypertensive, antithrombotic, and anti-inflammatory effects.
- Polyphenol-rich chocolate may potentially be used as a nutraceutical medication to help treat diabetes, strokes, and vascular dementia; thus, future studies should provide information on polyphenol content and the flavanol plasma levels achieved.

Keywords Type 2 diabetes • Risk factors • Chocolate • Polyphenols • Antioxidants • Treatment

Introduction

Diabetes mellitus, and in particular type 2, is generally characterized by hyperglycemia, insulin resistance (reduced insulin sensitivity), and obesity. Obesity is associated with not only diabetes but also hyperlipidemia and hypertension. The coexistence of these diseases is known as metabolic syndrome,

E. Al-Dujaili, B.Sc., Ph.D. (✉)
Department of Dietetics, Nutrition and Biological Sciences, Queen Margaret University, Edinburgh, Scotland
e-mail: ealdujaili@qmu.ac.uk

C. Tsang, B.Sc. (Hons), MMedSci, Ph.D.
School of Contemporary Sciences, University of Abertay, Dundee, Scotland

S. Almoosawi, Ph.D.
Population Nutrition Research Group, MRC Human Nutrition Research, Elsie Widdowson Laboratory, Cambridge, UK

R.R. Watson et al. (eds.), *Chocolate in Health and Nutrition*, Nutrition and Health 7,
DOI 10.1007/978-1-61779-803-0_25, © Springer Science+Business Media, LLC 2013

a high-risk factor for cardiovascular disease. Insulin resistance is considered a key feature of these diseases and is defined as a state requiring more insulin in order to obtain the biological effects required with a lower insulin level in the normal state. This metabolic abnormality is induced by obesity, especially increased visceral fat, via the enhancement of inflammation and hypoadiponectinemia. Adipose tissues have now become increasingly recognized as an active endocrine organ that produces and secretes a wide array of bioactive substances, including inflammatory mediators. The glucocorticoid cortisol plays a critical role in controlling short-term and long-term glucose metabolism; it also has a potential role in regulating other inflammatory cytokines and adipokines (see Chap. 24).

Cocoa beans are now widely known to provide a rich source of polyphenolic antioxidants, specifically, "flavanols." Their antioxidant concentration per mass is believed to be about 10%, which is the highest concentration level of antioxidants in any known food, and their importance stems mainly from their antioxidant properties and ability to regulate various biological/biochemical processes [1]. Polyphenols have been implicated in the prevention of a number of oxidative-related diseases, including cardiovascular disease (CVD), hypertension, and diabetes. They are often categorized into four groups depending on the number of phenol rings contained in their structure: phenolic acids, flavonoids, stilbenes, and lignans [1, 2]. So far, thousands of polyphenols have been identified in the diet [3], and the average dietary intake of these plant products is estimated to be in the range of 1 g per day [2, 4]. Nevertheless, the maximum plasma concentration of polyphenols rarely exceeds 1 μM following ingestion of 10–100 mg of a single phenolic compound. Cocoa, chocolate, and especially dark chocolate contain high levels of low-molecular-weight flavonoids such as (−)-epicatechin and (+)-catechin [5] as well as dimer to hexamer procyanidins, these procyanidins being characteristic of cocoa and chocolate products [6, 7]. Interestingly, cocoa has also been shown to contain more polyphenols and flavonoids per serving than red wine, green tea, or black tea. In fact, flavanols are believed to be responsible for the healthy effect of chocolate on the cardiovascular system. However, similar to many natural foods, the antioxidant values of the raw cocoa fall markedly during the chocolate-making process. Therefore, by consuming high-quality cocoa products that contain bioactive flavanols, insulin function could be improved, blood sugar is better controlled, and possibly, the incidence of prediabetes is lowered. The molecules in cocoa credited for the "healthy heart" effects of certain cocoas and dark chocolate may well be used to help treat diabetes, stroke, and vascular dementia and could soon be available to nutraceutical companies and industry for production and development into new medications in the near future. In this chapter, a summarized overview describing the effect of chocolate in the treatment of diabetes risk factors will be presented and whether chocolate can be consumed as potential adjunct therapy by type 2 diabetic patients.

Definition of Healthy Chocolate

It is a common misunderstanding that any dark chocolate contains high levels of polyphenols. In fact, there is no relationship between the darkness of chocolate and polyphenol content. In other words, not every dark chocolate contains high polyphenols. So we have to always define the healthy chocolate as the polyphenol-rich chocolate and not dark chocolate [8]. The use of the term "dark chocolate" is misleading; there is nothing about the color of the chocolate that will tell you the flavanol content. A key step in the manufacturing chain of edible chocolate, where significant loss of flavanols occurs, after fermentation, is an alkalization step called dutching. The Dutchman van Houten discovered 200 years ago that adding alkali-potash to cocoa nibs would enhance the taste, texture, and appearance of the cocoa. Dutched cocoa bitterness was therefore eliminated, together with most of the active flavanols. The underreported effect of alkalization is, in fact, darkening of cocoa, so that a very dark chocolate might be essentially devoid of flavanols. Commercially, the so-called dark chocolates mainly refer to the percentage of nonfat cocoa present in the product in relation to other constituents such as milk, fat, and sugar that are used commercially to reflect chocolate quality [9]. However, it is accepted that the flavanols and procyanidins contained mostly in commercial dark chocolate and not white chocolate that are most likely to be associated with the observed health benefits. Therefore, monitoring the flavanol content of chocolate as the key determinant of quality in commercially available chocolate products would be essential for consumer assurance.

The main healthy chemicals in chocolate are believed to be epicatechin and its esters, as tests showed that pure epicatechin consumed by people had much the same effect as flavanol-rich cocoa. Studies have shown that epicatechin is, at least in part, responsible for the beneficial vascular effects that are observed after the consumption of flavanol-rich chocolate [10–12]. This chapter will summarize the results of several studies for the potential benefits of cocoa flavanols for treatment of serious risk factors that are mostly implicated in the onset of diabetes mellitus.

Chocolate and Carbohydrate Metabolism in DM

Cocoa and polyphenol-rich chocolate (PRC) act on different multiple sites of carbohydrate metabolism. Cocoa polyphenols have been shown to downregulate (alpha) α-amylase, an enzyme involved in the breakdown of carbohydrate [13] and that proanthocyanidins can in vitro inhibit salivary and pancreatic α-amylase activity [14]. Intestinal glucose transporter, GLUT2, and sodium-dependent glucose transporters, namely, GLUT1, have been shown to be inhibited by polyphenols via direct interactions between epicatechin gallate or quercetin-3-O-glucoside and the respective transporters, which results in slower carbohydrate breakdown and delayed absorption, hence improved glycemic response [15, 16]. Consumption of a cocoa extract has been shown to reduce blood glucose in diabetic rats [17] and in diabetic obese mice [18]. In humans, PRC consumption was found to improve insulin resistance, insulin sensitivity, and fasting glucose levels in healthy [19], hypertensive [20], glucose-intolerant hypertensive [21], and obese subjects [22]. Most recently, Grassi et al. [21] reported improvement in HOMA-IR, QUICKI, and insulin-sensitivity index as well as (beta) β-cell function in hypertensive subjects with impaired glucose tolerance following 15 days of 100 g DC and that the changes were correlated with improvement in flow-mediated dilation, thereby confirming a previous observation that an association exists between insulin resistance and endothelium. Also, Muniyappa et al. [23] have reported that cocoa consumption for 2 weeks enhances insulin-mediated vasodilation without improving blood pressure or insulin resistance in essential hypertension.

Some researchers, on the other hand, failed to show any significant improvement in glucose, HOMA-IR and insulin sensitivity following 5 days of daily consumption of procyanidin-rich cocoa beverage by insulin-resistant men and women [24, 25]. This may suggest that a longer duration and a higher dose of polyphenols are required to achieve a significant reduction in glucose levels. It can also be postulated that improvement in glycemic regulation can occur without detectable reduction in fasting glucose. Balzer et al. [26] investigated the effect of regular cocoa intake on diabetics and showed a significant reduction of hemoglobin A_{1c}, implying improved glycemic control despite not having any effect on fasting glucose levels.

A number of antiobesity agents, such as orlistat, are potent competitive inhibitors of pancreatic lipase. These agents interfere with the hydrolysis and absorption of dietary carbohydrates and lipids. Pancreatic α-amylase, lipase, and phospholipase A_2, which are delivered into the intestinal lumen as constituents of pancreatic juices, are the major enzymes involved in the hydrolysis of dietary starch and fat. It is increasingly recognized that polyphenols can influence carbohydrate and lipid metabolism by affecting the activity of these digestive enzymes. Inhibition of α-amylase in vitro by cocoa phenolic extracts has been reported, with procyanidins showing a greater inhibitory activity [27].

Chocolate and Blood Pressure

The first observations of an association between cocoa consumption and blood pressure were reported from the Kuna population of Panama where Kuna Indians were found to have low rates of CVD and hypertension despite consuming large quantities of salt [28]. This cocoa consumption was found to be associated with a threefold higher urinary nitrate/nitrite excretion in Kuna Indians compared to mainland dwellers, who did not consume flavonoid-rich cocoa [29]. Activation of nitric oxide (NO) system was suggested to be the underlying mechanism by which cocoa consumption reduced blood pressure [29, 30]. Reductions in both systolic blood pressure (SBP) and diastolic blood pressure (DBP) were reported by short-term randomized controlled trials [31, 32]. In their latest study, Grassi et al. [21] also reported a reduction in 24-h SBP and DBP in glucose-intolerant hypertensive patients. The effect of solid versus liquid cocoa and sugared versus sugar-free cocoa has been compared, and it was found that acute consumption of dark chocolate (DC) and sugar-free cocoa by overweight and obese individuals has reduced SBP and DBP by 3.2 or 1.2 mmHg and 1.4 or 1.2 mmHg, respectively [32]. In contrast, sugared cocoa has not been shown to reduce blood pressure.

Long-term (12 weeks) randomized controlled trials have also reported improvement in blood pressure following consumption of flavonoid-rich cocoa containing 902 mg flavanols [22, 25]. This effect was correlated with an increase in S-nitrosoglutathione, thereby confirming previous observational studies of the relation between cocoa consumption, NO system, and blood pressure. It is worth mentioning that Taubert et al. [25] stated that although the reduction in blood pressure was significant, none of their participants attained lower-range prehypertension (130/85) or an optimal blood pressure (120/80), suggesting that statistical significance does not necessarily imply clinical relevance. Similar reductions in blood pressure following cocoa ingestion have been correlated with a 50% reduction in cardiovascular and all-cause mortality [33]. This suggests that including cocoa or PRC in the diet of hypertensive subjects or those at risk of hypertension (overweight and obese) may be more beneficial in the long-term reduction of hypertension-related complications rather than in controlling hypertension itself. This is also particularly important for diabetics, where cocoa and PRC might be beneficial in preventing diabetes-associated complications rather than in improving glycemic control. Before recommending DC and cocoa to patients who are on antihypertensive treatment, potential effects of the chocolate phenols should be evaluated. Taubert et al. [25] stated that "the cumulative phenol dose may determine the magnitude of transcriptional NO synthase activation and subsequent fall in blood pressure." Angiotensin II is produced through the conversion of angiotensin I by angiotensin converting enzyme. Procyanidins (dimer to hexamers) and epigallocatechin have been shown to be effective inhibitors of angiotensin converting enzyme [34]. Chocolate extracts have also been shown to inhibit this enzyme in rat kidney membrane, the effect being related to the concentration of flavanols within the chocolate extract [35]. Cocoa flavonoids were shown to attenuate angiotensin II-induced activation of MAPK-dependent pathway [36] that can lead to decreased endothelin-1 and reactive oxygen species (ROS) production and subsequently improved NO bioavailability. However, in vivo inhibition of angiotensin converting enzyme by chocolate or cocoa has not yet been reported.

It is worth mentioning, however, that other studies have failed to report any significant changes in blood pressure following 2 weeks of 46 g DC (213 mg procyanidins) ingestion [37–39]. In healthy subjects, another study [40] attributed the lack of effect on blood pressure (BP) to normal baseline blood pressure levels. However, Grassi et al. [20] subjects also had a mean SBP of 113.9 mmHg, but a significant 6.4-mmHg reduction in SBP levels was observed. The differences in these studies could be due to several factors: individual baseline differences in BP, duration of intervention, time point at which blood pressure was measured, dose of flavonoids used, and the instruments used to monitor blood pressure.

Chocolate and Vascular Health

Dark chocolate has been shown to be protective against heart attacks, as it can lower their incidence [41–43]. It also lowers the incidence of death by coronary heart disease [44, 45]. Epidemiological data suggest that intake of flavanol-rich cocoa was inversely associated with 15-year cardiovascular and all-cause mortality in older males [46–48]. Moreover, (−)-epicatechin and its metabolite, epicatechin-7-*O*-glucuronide, have been identified as independent predictors of some of the vascular effects associated with the consumption of a flavanol-rich beverage [10, 49, 50]. The most widely reported and consistent finding was the inverse association between chocolate flavonoid intake and coronary heart disease mortality in men and women by several studies over many countries [33, 51–56]. Chocolate flavanoids have been shown to protect the arteries in a dose-response bioavailability manner in humans [57, 58]. There are several mechanisms by which flavanoids may be protective against cardiovascular disease: antioxidant, antiplatelet, anti-inflammatory effects, as well as increasing HDL, lowering blood pressure, and improving endothelial function [10, 59]. Moreover, Balzer et al. [26] showed that in both acute (single-dose) ingestion of cocoa, containing increasing concentrations of flavanols (75, 371, and 963 mg), and 30-day intervention studies, there was an improvement in vascular function in medicated diabetic patients as assessed by an increase in FMD (flow-mediated dilation). They concluded that diets rich in flavanols reverse vascular dysfunction in diabetes, highlighting therapeutic potentials in cardiovascular disease.

Cocoa flavanols were found to have an aspirin-like effect (reduction in platelet aggregation). Dark chocolate induces coronary vasodilation, improves coronary vascular function, and decreases platelet adhesion 2 h after consumption [60]. These beneficial effects were paralleled by a significant reduction of serum oxidative stress and were positively correlated with changes in serum epicatechin concentration. Several studies (as reviewed in [61]) focused on the in vivo effects of food sources containing (−)-epicatechin and (1)-catechin, as well as procyanidin oligomers and polymers on platelet function. Platelet function was measured using the platelet function analyzer (PFA-100) [62–66]. This device assesses shear stress-induced platelet aggregation and found that acute intake of flavanols produced 3–11% inhibition of collagen-epinephrine-induced closure time per 100 mg flavanols consumed 2–6 h after ingestion. Other studies also found a significant inhibition of platelet aggregation and activation upon acute or chronic intake of dietary flavanols from cocoa [38, 60, 67–69]. Such effects may be nutritionally relevant, as 100 mg flavanols can be obtained from 11 g of dark chocolate with a cocoa content of 70% w/w [70, 71], from 52 g of milk chocolate [72, 73], or from 50 to 100 mL of a cocoa drink containing 8% w/v pure cocoa [66, 70, 71].

Chocolate and Endothelium Function

Numerous studies have indicated that flavanols may exert significant vascular protection not only from their antioxidant properties but also increased nitric oxide availability. Nitric oxide is a critical component in healthy blood flow and blood pressure control, two very important factors in controlling diabetes. Studies have also shown that antioxidants in cocoa help the body process NO, thus preventing fatty substances in the blood from oxidizing and narrowing the arteries, leading to high blood pressure, another serious diabetes risk factor. Chocolate flavonoids have been shown to improve NO bioavailability through a variety of mechanisms. Quercetin increases endothelial nitric oxide synthase (NOS) activity, while epicatechin decreases superoxide levels by directly scavenging it [74]. The dimer procyanidin B2 and (−)-epicatechin glucuronide could both scavenge superoxide and inhibit NADPH oxidase [75]. Polyphenols have also been shown to prevent eNOS uncoupling due to decreased BH4, an eNOS cofactor [76]. Under pathological conditions, BH4 levels are reduced and dihydrobiopterin (BH2) levels are increased; eNOS becomes dysfunctional and produces superoxide rather than NO. As for NADPH oxidase, its inhibition is of particular relevance to diabetes since increased glucose levels have been reported to increase NADPH oxidase activity, leading to increased ROS production [77, 78]. Flavonoids in chocolate could also improve NO bioavailability by modulating the activity of arginase, an enzyme in the urea cycle. Here, (−)-epicatechin and its metabolites have been shown to lower arginase-2 mRNA expression and activity in human endothelial cells in vitro [79].

In vivo studies have now been published to investigate effects of cocoa and PRC on endothelium function. In a short-term study [39], consumption of 100 mL of a flavanol-rich cocoa (containing 176 mg flavan-3-ols) drink by patients with at least one cardiovascular risk factor (hypertension, hyperlipidemia, or diabetes) for 2 consecutive days increased flow-mediated dilation maximally after 2 h of consumption. Another study produced vasodilation and improved vasodilator response to ischemia in finger arteries after 5 days [37]. Similar observations were reported in smokers following consumption of 100 mL of a cocoa drink [59]. Acute consumption of 100 g of chocolate has also been shown to increase both resting and hyperemic brachial artery diameter in healthy subjects [80]. Schroeter et al. [10] identified (−)-epicatechin and its metabolite epicatechin-7-O-glucuronide as independent predictors of vascular effects of flavanol-rich cocoa and their effect on NO. Chocolate containing 500 mg polyphenols for 2 weeks in patients with essential hypertension [20, 80] produced a similar effect and improvements in flow-mediated dilation. This suggested that the PRC effect on endothelium function could be sustained with regular consumption of flavonoid-rich products and that desensitization does not seem to occur. This hypothesis was substantiated by other studies in hypercholesterolemic women [66, 81]. Recently, sugar-free cocoa has been shown to ameliorate flow-mediated dilation to a better extent than sugared cocoa [32]. This finding is particularly relevant to diabetic patients, as increased glucose levels promote ROS production and reduced NO bioavailability. Reducing the sugar content of the cocoa beverage could thus augment the improvement in endothelium function seen after acute and chronic consumption of cocoa in medicated diabetic patients.

Chocolate, Atherosclerosis, and Lipid Profile

Polyphenol-rich chocolate improves blood lipid profile by increasing HDL (good) cholesterol and lowering LDL (bad) cholesterol and its susceptibility to oxidation, apparently by increasing plasma antioxidant capacity and decreasing ROS formation [82–84]. Consumption of 22 g cocoa or 16 g PRC, containing 446 mg procyanidins, for 4 weeks by healthy subjects has been shown to increase total antioxidant capacity and lead to an 8% increase in LDL oxidation lag time [84]. Prolongation of LDL oxidation lag time and improvements in lipid profile have also been reported by other studies [21, 85, 86] with reported reduction in LDL cholesterol of 11–15% [26, 87]. Jalil et al. [27] attributed these effects to the potential ability of polyphenols to inhibit pancreatic lipase, promote cholesterol excretion in feces, and reduce hepatic lipid accumulation through activation of adenosine monophosphate (AMP)-activated protein kinase, attenuation of hepatic secretion of apolipoprotein-B100, and stimulation of hepatic LDL receptors expressions. However, Vinson et al. [88] demonstrated that while cocoa polyphenols prolong lag time of LDL + VLDL oxidation, cocoa butter decreases it, resulting in increased LDL oxidation. This was thought to be due to the high content of fatty acids in cocoa butter, rendering it atherogenic. Recently, Mellor et al. [89] have reported that high-cocoa polyphenol-rich chocolate improves HDL cholesterol in type 2 diabetes patients. The study has been well planned and controlled (double-blind crossover); however, the number of participants was too small ($n = 12$). Two groups of type 2 diabetic patients on stable medication were randomly selected. One group was given 45 g of 85% dark chocolate for 8 weeks, and the second group was given a placebo that was the same color but contained no cocoa solids. None of the participants gained weight, despite eating 45 g of chocolate per day, and the HDL cholesterol increased significantly with dark chocolate compared to the placebo. There was also a decrease in the total cholesterol/HDL ratio in the dark chocolate group. The researchers concluded that consumption of dark chocolate in moderation is not harmful to diabetics and may actually offer some benefits.

Consumption of cocoa results in reduced atherosclerotic lesion size in animal models of atherosclerosis and is reported to reduce wall thickness and cholesterol content of aorta, although the change in the latter was not significant [90]. Acute reduction in F2-isoprostane levels, a measure of lipid peroxidation, has also been reported following ingestion of cocoa [91]. Another study [86] found a negative correlation between plasma concentrations of HDL, but not LDL cholesterol, and oxidized LDL, suggesting that HDL could influence LDL oxidation through inhibition of monocyte chemotaxis, breakdown of lipid peroxide via paraoxonase, reverse cholesterol transporter via lecithin-cholesterol transferase, and inhibition of vascular endothelium activation via apolipoprotein A1. It is important to note also that some studies failed to report any significant change in lipid profile following chocolate intake [19, 25, 83, 92].

Cocoa polyphenols could prevent atherosclerosis development without significantly altering lipid profile. Kurosawa et al. [90] demonstrated that cocoa increased resistance against LDL oxidation and decreased TBARS (an oxidative stress marker) significantly at the end of the first 2 and 3 months of the 6-month trial without significantly altering plasma cholesterol, TG, or phospholipids. This has led to decreased foam cell formation and reduction in the area of atherosclerotic lesion following cocoa consumption. Cocoa is also rich in fiber. Cocoa fiber obtained from cocoa husks was found to counteract the effects of lipid-rich diet by preventing fat-induced decline in HDL cholesterol and fat-induced rise in total cholesterol (TC), LDL cholesterol, and lipid peroxidation [93]. Therefore, polyphenol-rich chocolate seems to be effective in improving the atherosclerotic cholesterol profile in patients with diabetes by increasing HDL cholesterol and improving the cholesterol/HDL ratio without affecting weight.

Chocolate and Glucocorticoid Hormones

Cortisol can play a critical role in controlling short-term and long-term glucose metabolism as well as regulating other inflammatory cytokines and adipokines [94]. In vitro, cortisol has been shown to reduce adipose tissue IL-6 release [95], diminish adiponectin gene expression [96] and secretion, and in conjunction with insulin, stimulate leptin release from adipocytes [97]. Similar changes in circulating adiponectin levels have also been reported in vivo with endogenous cortisol overproduction or exogenous cortisol administration [94]. Cortisol is also known to inhibit TNF-α release from macrophages [98], which is overexpressed in adipose tissue of obese insulin-resistant subjects and is associated, alongside other inflammatory mediators, with enhanced adipose tissue inflammation [99].

Polyphenols could influence cortisol and corticosteroid metabolism in several ways. For instance, ingestion of licorice and grapefruit juice polyphenols has been reported to increase plasma cortisol causing severe hypokalemia and hypertension [100]. Moderate intake of licorice was found to increase salivary cortisol and reduce cortisone [101] without influencing blood pressure. Similarly, drinking grapefruit juice, rich in the flavonoids naringenin, quercetin, and hesperetin, has been shown to decrease urinary cortisone/cortisol ratio [102, 103]. These effects occur through inhibition of 11 (beta)β hydroxysteroid dehydrogenase (11βHSD2) [104, 105]. In contrast, extracts of freshly ground roasted *Coffea arabica* have been shown to inhibit 11βHSD1 activity and to prevent glucocorticoid-receptor translocation, glucocorticoid-induced expression of PEP-CK, and gluconeogenesis [106], an antidiabetic action of coffee. Recently, the composition of dietary macronutrients has been shown to produce both acute and chronic effects on cortisol metabolism [107, 108]. In fact, consumption of dietary fat or high-fat low-carbohydrate diets has been found to replicate the changes in cortisol metabolism and 11βHSD1 activity that are seen in obesity and the metabolic syndrome [109–111].

Chocolate and Inflammatory Markers

Type 2 diabetes mellitus (DM) and insulin resistance are now well known to be influenced by inflammation that is provoked by immune cells called macrophages, which lead to disturbance of various pro-inflammatory markers (e.g., C-reactive protein and TNF-α) [99]. Chocolate has anti-inflammatory effects by inhibiting the lipoxygenase pathway, and cocoa polyphenols have been shown to decrease inflammation via several mechanisms [112, 113], namely, inhibition of mitogen-induced activation of T cells, polyclonal activation of B cells, reduced expression of interleukin-2 messenger RNA, and reduced secretion of IL-2 by T cells; all of these are inflammatory markers. Other studies have also found that chocolate procyanidins can modulate a variety of other cytokines (e.g., IL-5, TNF-α, TGF-β), reducing their inflammatory effects [114–118]. Recently, a significant association between inflammation and moderate consumption of cocoa products was found in a study comparing subjects that ate chocolate regularly in the form of dark chocolate (n = 824) with subjects that did not eat chocolate for at least 1 year (n = 1,317). Serum C-reactive protein concentration in the subgroup that had up to one serving (20 g of cocoa) every 3 days was significantly lower than in both nonconsumers and subjects having higher consumption [119].

Summary

Diabetes type 2 is generally characterized by hyperglycemia, insulin resistance (reduced insulin sensitivity), and obesity. Obesity is associated with not only diabetes but also hyperlipidemia and hypertension. Coexistence of these diseases is known as metabolic syndrome, a high-risk factor for cardiovascular disease. Insulin resistance is considered a key feature of these diseases and is defined as a state requiring more insulin in order to obtain the biological effects achieved with a lower insulin level in the normal state. This metabolic abnormality is induced by obesity, especially increased visceral fat, via the enhancement of inflammation and hypoadiponectinemia. Adiponectin is an adipocytokine specifically and abundantly expressed in adipose tissue that directly sensitizes to insulin, and its level is inversely correlated to the percentage of body fat in adults. Treatment with thiazolidine derivatives, one type of antidiabetic drug, ameliorates insulin resistance and increases serum adiponectin level in type 2 diabetic patients with hypoadiponectinemia. Therefore, improvement of insulin resistance and adiponectin levels is expected to be an effective therapeutic strategy for the improvement and/or prevention of diabetes as well as metabolic syndrome.

Polyphenol-rich chocolate has the potential to be included in the diet of individuals with non-insulin-dependent type 2 diabetes as part of a balanced approach to diet and lifestyle, with a potential reduction in cardiovascular risk and without detrimental effects on weight or glycemic control. Mellor et al. [89] reported an improvement in atherosclerotic cholesterol profile in patients with type 2 diabetes by increasing HDL cholesterol and improving the cholesterol/HDL ratio following PRC intake. The mechanism of action remains unclear; however, it has been suggested that phenolics may have either an insulin-sensitizing effect [19] or there is reduced inflammation independent of insulin action [82] through reduced lipid peroxidation.

Recent studies have investigated the potential of flavanol-rich cocoa and chocolate to influence vascular health. Epicatechin and its metabolite, epicatechin-7-*O*-glucuronide, have been identified as independent predictors of some of the vascular effects associated with the consumption of a flavanol-rich chocolate. In addition, polyphenol-rich dark chocolate not only reduced blood pressure and improved lipid profile and endothelial function but also ameliorated insulin sensitivity and (beta)β-cell function [21]. Therefore, targeted dietary and nutrition supplements that can influence disease risk factors would probably be beneficial in the prevention and treatment of type 2 DM. However, knowing the scarcity of studies on diabetics and that type 2 diabetics tend to be obese, caution must be considered when recommending cocoa or chocolate to such patients.

Evidence suggesting a potential role of diet in regulating glucocorticoid metabolism and the ability of polyphenols to both adversely and positively affect cortisol metabolism, investigating the effect of chocolate polyphenols on cortisol metabolism in overweight and obese individuals, and assessing the association between cortisol, glucose, insulin, and blood pressure and lipid profile might be essential in elucidating the mechanisms by which these phenolic compounds influence the above metabolic parameters. It has also been observed that the hypotensive effect of cocoa might be mediated through increasing nitric oxide levels. Cocoa flavanols have been shown to increase nitric oxide production by cultured human vascular endothelial cells and improve endothelium-dependent vasorelaxation in brachial arteries of healthy humans [92]. The renin-angiotensin-aldosterone system is known to be an important regulator of blood pressure homeostasis. In vivo inhibition of angiotensin converting enzyme in humans following ingestion of polyphenol-rich pomegranate juice has been reported, providing further evidence for the need to investigate potential modulation of angiotensin converting enzyme activity by polyphenols.

Finally, over the past 10 years, studies have demonstrated that moderate PRC or dark chocolate consumption may exert protective effects against the development of type 2 DM risk factors and cardiovascular disease. Several mechanisms have been proposed to explain this positive influence not only on insulin sensitivity and vascular endothelial function but also metabolic (fat and carbohydrate), antihypertensive, antithrombotic, and anti-inflammatory effects, including changes in oxidant defense mechanisms, cytokine production, and platelet function. Nevertheless, due to several limitations of the data available, future studies should provide information about type of chocolate used, its polyphenol content, especially flavanols, and the flavanol plasma levels achieved. Furthermore, they have to consider other potentially active substances present in cocoa (e.g., potassium, magnesium, and theobromine) that were not discussed in this chapter.

References

1. Manach C, Scalbert A, Morand C, et al. Polyphenols: food sources and bioavailability. Am J Clin Nutr. 2004;79(5):727–47.
2. Scalbert A, Williamson G. Dietary intake and bioavailability of polyphenols. J Nutr. 2000;130(8S):2073S–85.
3. de Lorgeril M, Salen P. The Mediterranean-style diet for the prevention of cardiovascular diseases. Public Health Nutr. 2007;9(1A):118–23.
4. Saura-Calixto F, Goñi I. Definition of the Mediterranean diet based on bioactive compounds. Crit Rev Food Sci Nutr. 2009;49(2):145–52.
5. Osakabe N, Yamagishi M, Sanbongi C, et al. The antioxidative substances in cacao liquor. J Nutr Sci Vitaminol. 1998;44(2):313–21, (Tokyo).
6. Adamson GE, Lazarus SA, Mitchell AE, et al. HPLC method for the quantification of procyanidins in cocoa and chocolate samples and correlation to total antioxidant capacity. J Agric Food Chem. 1999;47(10):4184–8.
7. Natsume M, Osakabe N, Yamagishi M, et al. Analyses of polyphenols in cacao liquor, cocoa, and chocolate by normal-phase and reversed-phase HPLC. Biosci Biotechnol Biochem. 2000;64(12):2581–7.
8. Hollenberg NK, Fisher NDL. Is it the dark in dark chocolate? Circulation. 2007;116(21):2360–2.
9. Langer S, Marshall LJ, Day AJ, et al. fLavanols and methylxanthines in commercially available dark chocolate: a study of the correlation with nonfat cocoa solids. J Agric Food Chem. 2011;59(15):8435–41,PMID: 21699218.
10. Schroeter H, Heiss C, Balzer J, et al. (–)-Epicatechin mediates beneficial effects of flavanol-rich cocoa on vascular function in humans. Proc Natl Acad Sci USA. 2006;103(4):1024–9.
11. Williamson G, Manach C. Bioavailability and bioefficacy of polyphenols in humans. II. Review of 93 intervention studies. Am J Clin Nutr. 2005;81:243S–55.
12. Loke WM, Hodgson JM, Proudfoot JM, et al. Pure dietary flavonoids quercetin and (–)-epicatechin augment nitric oxide products and reduce endothelin-1 acutely in healthy men. Am J Clin Nutr. 2008;88:1018–25.
13. Quesada C, Bartolomé B, Nieto O, et al. Phenolic inhibitors of α -amylase and trypsin enzymes by extracts from pears, lentils and cocoa. J Food Prot. 1996;59:185–92.

14. McDougall GJ, Shapiro F, Dobson P, et al. Different polyphenolic components of soft fruits inhibit alpha-amylase and alpha-glucosidase. J Agric Food Chem. 2005;53(7):2760–6.
15. Cermak R, Landgraf S, Wolffram S. Quercetin glucosides inhibit glucose uptake into brush-border-membrane vesicles of porcine jejunum. Br J Nutr. 2004;91(6):849–55.
16. Chen CH, Hsu HJ, Huang YJ, et al. Interaction of flavonoids and intestinal facilitated glucose transporters. Planta Med. 2007;73(4):348–54.
17. Ruzaidi A, Amin I, Nawalyah AG, et al. The effect of Malaysian cocoa extract on glucose levels and lipid profiles in diabetic rats. J Ethnopharmacol. 2005;98(1–2):55–60.
18. Tomaru M, Takano H, Osakabe N, et al. Dietary supplementation with cacao liquor proanthocyanidins prevents elevation of blood glucose levels in diabetic obese mice. Nutrition. 2007;23(4):351–5.
19. Grassi D, Lippi C, Necozione S, et al. Short-term administration of dark chocolate is followed by a significant increase in insulin sensitivity and a decrease in blood pressure in healthy persons. Am J Clin Nutr. 2005;81:611–4.
20. Grassi D, Necozione S, Lippi C, et al. Cocoa reduces blood pressure and Insulin resistance and improves endothelium-dependent vasodilation in hypertensive. Hypertension. 2005;46:398–405.
21. Grassi D, Desideri G, Necozione S, et al. Blood pressure is reduced and insulin sensitivity increased in glucose-intolerant, hypertensive subjects after 15 days of consuming high-polyphenol dark chocolate. J Nutr. 2008;138:1671–6.
22. Davison K, Coates AM, Buckley JD, et al. Effect of cocoa flavanols and exercise on cardiometabolic risk factors in overweight and obese subjects. Int J Obes (Lond). 2008;32:1289–96.
23. Muniyappa R, Hall G, Kolodziej TL, et al. Cocoa consumption for 2 wk enhances insulin-mediated vasodilatation without improving blood pressure or insulin resistance in essential hypertension. Am J Clin Nutr. 2008;88(6):1685–96.
24. Stote KM, Clevidence BA, Baer DJ. Effect of cocoa and green tea consumption on glucoregulatory biomarkers in insulin resistant men and women. FASEB J. 2007;21:847–917.
25. Taubert D, Roesen R, Lehmann C, et al. Effects of low habitual cocoa intake on blood pressure and bioactive nitric oxide. JAMA. 2007;298:49–60.
26. Balzer J, Rassaf T, Heiss C, et al. Sustained benefits in vascular function through flavanol-containing cocoa in medicated diabetic patients a double-blind, randomized, controlled trail. J Am Coll Cardiol. 2008;51:2141–9.
27. Jalil AM, Ismail A, Pei CP, et al. Effect of cocoa extract on glucometabolism, oxidative stress and antioxidant enzymes in obese-diabetic (Ob-db) rats. J Agric Food Chem. 2008;56:7877–84.
28. McCullough ML, Chevaux K, Jackson L, et al. Hypertension, the Kuna, and the epidemiology of flavanols. J Cardiovasc Pharmacol. 2006;47:119–21.
29. Hollenberg NK. Vascular action of cocoa flavanols in humans: the roots of the story. J Cardiovasc Pharmacol. 2006;47(Suppl 2):S99–102.
30. Bayard V, Chamorro F, Motta J, et al. Does flavanol intake influence mortality from nitric oxide-dependent processes? Ischemic heart disease, stroke, diabetes mellitus, and cancer in Panama. Int J Med Sci. 2007;4:53–8.
31. Taubert D, Berkels R, Roesen R, et al. Chocolate and blood pressure in elderly individuals with isolated hypertension. JAMA. 2003;290:1029–30.
32. Faridi Z, Nijke VY, Dutta S, et al. Acute dark chocolate and cocoa ingestion and endothelium function: a randomized controlled crossover trial. Am J Clin Nutr. 2008;88:58–63.
33. Buijsse B, Feskens EJ, Kok FJ, et al. Cocoa intake, blood pressure, and cardiovascular mortality: the Zutphen elderly study. Arch Intern Med. 2006;166:411–7.
34. Actis-Goretta L, Ottaviani JI, Keen CL, et al. Inhibition of angiotensin converting enzyme (ACE) activity by flavan-3-ols and procyanidins. FEBS Lett. 2003;555(3):597–600.
35. Actis-Goretta L, Ottaviani JI, Fraga CG. Inhibition of angiotensin converting enzyme activity by flavanol-rich foods. J Agric Food Chem. 2006;54(1):229–34.
36. Lee KW, Kundu JK, Kim SO, et al. Cocoa polyphenols inhibit phorbol ester-induced superoxide anion formation in cultured HL-60 cells and expression of cyclooxygenase-2 and activation of NF-kappaB and MAPKs in mouse skin in vivo. J Nutr. 2006;136(5):1150–5.
37. Fisher ND, Hughes M, Gerhard-Herman M, et al. Flavanol-rich cocoa induces nitric-oxide-dependent vasodilation in healthy humans. J Hypertens. 2003;21:2281–6.
38. Murphy KJ, Chronopoulos AK, Singh I, et al. Dietary flavanols and procyanidin oligomers from cocoa (*Theobroma cacao*) inhibit platelet function. Am J Clin Nutr. 2003;77:146–1473.
39. Heiss C, Dejam A, Kleinbongard P, et al. Vascular effects of cocoa rich in flavan-3-ols. JAMA. 2003; 290(8):1030–1.
40. Crews Jr WD, Harrison DW, Wright JW. A double-blind, placebo-controlled, randomized trial of the effects of dark chocolate and cocoa on variables associated with neuropsychological functioning and cardiovascular health: clinical findings from a sample of healthy, cognitively intact older adults. Am J Clin Nutr. 2008;87(4):872–80.
41. Kris-Etherton PM, Keen CL. Evidence that the antioxidant flavonoids in tea and cocoa are beneficial for cardiovascular health. Curr Opin Lipidol. 2002;13(1):41–9.

42. Steinberg FM, Bearden MM, Keen CL. Cocoa and chocolate flavonoids: implications for cardiovascular health. J Am Diet Assoc. 2003;103(2):215–23.
43. Keli SO, Hertog MG, Feskens EJ, et al. Dietary flavonoids, antioxidant vitamins, and incidence of stroke: the Zutphen study. Arch Intern Med. 1996;156(6):637–42.
44. Hertog MG, Kromhout D, Aravanis C, et al. Flavonoid intake and long-term risk of coronary heart disease and cancer in the seven countries study. Arch Intern Med. 1995;155(4):381–6.
45. Rimm EB, Katan MB, Ascherio A, et al. Relation between intake of flavonoids and risk for coronary heart disease in male health professionals. Ann Intern Med. 1996;125(5):384–9.
46. Geleijnse JM, Launer LJ, Van der Kuip DA, et al. Inverse association of tea and flavonoid intakes with incident myocardial infarction: the Rotterdam study. Am J Clin Nutr. 2002;75(5):880–6.
47. Knekt P, Jarvinen R, Reunanen A, et al. Flavonoid intake and coronary mortality in Finland: a cohort study. BMJ. 1996;312(7029):478–81.
48. Hertog MG, Feskens EJ, Hollman PC, et al. Dietary antioxidant flavonoids and risk of coronary heart disease: the Zutphen elderly study. Lancet. 1993;342(8878):1007–11.
49. Arts IC, Jacobs DRJ, Harnack LJ, et al. Dietary catechins in relation to coronary heart disease death among post-menopausal women. Epidemiology. 2001;12(6):668–75.
50. Panneerselvam M, Tsutsumi YM, Bonds JA, et al. Dark chocolate receptors: epicatechin-induced cardiac protection is dependent on {delta}-opioid receptor stimulation. Am J Physiol Heart Circ Physiol. 2010;299(5): H1604–9.
51. Knekt P, Kumpulainen J, Jarvinen R, et al. Flavonoid intake and risk of chronic diseases. Am J Clin Nutr. 2002;76(3):560–8.
52. Hirvonen T, Pietinen P, Virtanen M, et al. Intake of flavonols and flavones and risk of coronary heart disease in male smokers. Epidemiology. 2001;12(1):62–7.
53. Yochum L, Kushi LH, Meyer K, et al. Dietary flavonoid intake and risk of cardiovascular disease in postmenopausal women. Am J Epidemiol. 1999;149(10):943–9.
54. Mink PJ, Scrafford CG, Barraj LM, et al. Flavonoid intake and cardiovascular disease mortality: a prospective study in postmenopausal women. Am J Clin Nutr. 2007;85:895–909.
55. Campia U, Panza JA. Flavanol-rich cocoa: a promising new dietary intervention to reduce cardiovascular risk in type 2 diabetes. J Am Coll Cardiol. 2008;51(22):2150–2.
56. Corti R, Flammer AJ, Hollenberg NK, et al. Cocoa and cardiovascular health. Circulation. 2009; 119(10):1433–41.
57. Richelle M, Tavazzi I, Enslen M, et al. Plasma kinetics in man of epicatechin from black chocolate. Eur J Clin Nutr. 1999;53(1):22–6.
58. Wang JF, Schramm DD, Holt RR, et al. A dose-response effect from chocolate consumption on plasma epicatechin and oxidative damage. J Nutr. 2000;130(8S Suppl):2115S–9.
59. Heiss C, Kleinbongard P, Dejam A, et al. Acute consumption of flavanol-rich cocoa and the reversal of endothelial dysfunction in smokers. J Am Coll Cardiol. 2005;46(7):1276–83.
60. Flammer AJ, Hermann F, Sudano I, et al. Dark chocolate improves coronary vasomotion and reduces platelet reactivity. Circulation. 2007;116:2376–82.
61. Ostertag LM, O'Kennedy N, Kroon PA, et al. Impact of dietary polyphenols on human platelet function – a critical review of controlled dietary intervention studies. Mol Nutr Food Res. 2010;54:60–81.
62. Bordeaux B, Yanek LR, Moy TF, et al. Casual chocolate consumption and inhibition of platelet function. Prev Cardiol. 2007;10:175–80.
63. Polagruto JA, Gross HB, Kamangar F, et al. Platelet reactivity in male smokers following the acute consumption of a flavanol-rich grapeseed extract. J Med Food. 2007;10:725–30.
64. Rein D, Paglieroni TG, Wun T, et al. Cocoa inhibits platelet activation and function. Am J Clin Nutr. 2000; 72:30–5.
65. Shenoy SF, Keen CL, Kalgaonkar S, et al. Effects of grape seed extract consumption on platelet function in post-menopausal women. Thromb Res. 2007;121:431–2.
66. Wang-Polagruto JF, Villablanca AC, Polagruto JA, et al. Chronic consumption of flavanol-rich cocoa improves endothelial function and decreases vascular cell adhesion molecule in hypercholesterolemic postmenopausal women. J Cardiovasc Pharmacol. 2006;47:S177–86.
67. Heptinstall S, May J, Fox S, et al. Cocoa flavanols and platelet and leukocyte function: recent *in vitro* and *ex vivo* studies in healthy adults. J Cardiovasc Pharmacol. 2006;47:S197–205.
68. Hermann F, Spieker LE, Ruschitzka F, et al. Dark chocolate improves endothelial and platelet function. Heart. 2006;92:119–20.
69. Innes AJ, Kennedy G, McLaren M, et al. Dark chocolate inhibits platelet aggregation in healthy volunteers. Platelets. 2003;14:325–7.
70. Nutrient Data Laboratory, Beltsville Human Nutrition Research Center, Agricultural Research Service, US Department of Agriculture (USDA), Beltsville; 2004.

71. Nutrient Data Laboratory, Food Composition Laboratory, Beltsville Human Nutrition Research Center, Agricultural Research Service, US Department of Agriculture (USDA), Beltsville; 2007.
72. Kyle JAM, Duthie GG. In: Andersen OM, Markham KR, editors. Flavonoids: chemistry, biochemistry, and applications. Boca Raton: CRC Press/Taylor and Francis; 2006. p. 219–62.
73. Rasmussen SE, Frederiksen H, Krogholm KS, et al. Dietary proanthocyanidins: occurrence, dietary intake, bioavailability, and protection against cardiovascular disease. Mol Nutr Food Res. 2005;49:159–74.
74. Sánchez M, Galisteo M, Vera R, et al. Quercetin downregulates NADPH oxidase, increases eNOS activity and prevents endothelial dysfunction in spontaneously hypertensive rats. J Hypertens. 2006;24(1):75–84.
75. Steffen Y, Gruber C, Schewe T, et al. Mono-O-methylated flavanols and other flavonoids as inhibitors of endothelial NADPH oxidase. Arch Biochem Biophys. 2008;469(2):209–19.
76. Kawashima S. The two faces of endothelial nitric oxide synthase in the pathophysiology of atherosclerosis. Endothelium. 2004;11(2):99–107.
77. Akbari CM, Saouaf R, Barnhill DF, et al. Endothelium-dependent vasodilatation is impaired in both microcirculation and macrocirculation during acute hyperglycemia. J Vasc Surg. 1998;28(4):687–94.
78. Title LM, Cummings PM, Giddens K, et al. Oral glucose loading acutely attenuates endothelium-dependent vasodilation in healthy adults without diabetes: an effect prevented by vitamins C and E. J Am Coll Cardiol. 2000;36(7):2185–91.
79. Schnorr O, Brossette T, Momma TY, et al. Cocoa flavanols lower vascular arginase activity in human endothelial cells in vitro and in erythrocytes in vivo. Arch Biochem Biophys. 2008;476(2):211–5.
80. Vlachopoulos C, Aznaouridis K, Alexopoulos N, et al. Effect of dark chocolate on arterial function in healthy individuals. Am J Hypertens. 2005;18(6):785–91.
81. Heiss C, Finis D, Kleinbongard P, et al. Sustained increase in flow-mediated dilation after daily intake of high-flavanol cocoa drink over 1 week. J Cardiovasc Pharmacol. 2007;49(2):74–80.
82. Mursu J, Voutilainen S, Nurmi T, et al. Dark chocolate consumption increases HDL cholesterol concentration and chocolate fatty acids may inhibit lipid peroxidation in healthy humans. Free Radic Biol Med. 2004;37(9):1351–9.
83. Mathur S, Devaraj S, Grundy SM, et al. Cocoa products decrease low density lipoprotein oxidative susceptibility but do not affect biomarkers of inflammation in humans. J Nutr. 2002;132(12):3663–7.
84. Wan Y, Vinson JA, Etherton TD, et al. Effects of cocoa powder and dark chocolate on LDL oxidative susceptibility and prostaglandin concentrations in humans. Am J Clin Nutr. 2001;74(5):596–602.
85. Osakabe N, Baba S, Yasuda A, et al. Daily cocoa intake reduces the susceptibility of low-density lipoprotein to oxidation as demonstrated in healthy human volunteers. Free Radic Res. 2001;34(1):93–9.
86. Baba S, Osakabe N, Kato Y, et al. Continuous intake of polyphenolic compounds containing cocoa powder reduces LDL oxidative susceptibility and has beneficial effects on plasma HDL-cholesterol concentrations in humans. Am J Clin Nutr. 2007;85(3):709–17.
87. Fraga CG, Actis-Goretta L, Ottaviani JI, et al. Regular consumption of a flavanol-rich chocolate can improve oxidant stress in young soccer player. Clin Dev Immunol. 2005;12:11–7.
88. Vinson JA, Proch J, Bose P, et al. Chocolate is a powerful ex vivo and in vivo antioxidant, an antiatherosclerotic agent in an animal model, and a significant contributor to antioxidants in the European and American diets. J Agric Food Chem. 2006;54(21):8071–6.
89. Mellor DD, Sathayapalan T, Kilpatrick ES, et al. High-cocoa polyphenol-rich chocolate improves HDL cholesterol in type-2 diabetes patients. Diabet Med. 2010;27(11):1318–21.
90. Kurosawa T, Itoh F, Nozaki A, et al. Suppressive effect of cocoa powder on atherosclerosis in Kurosawa and Kusanagi-hypercholesterolemic rabbits. J Atheroscler Thromb. 2005;12(1):20–8.
91. Wiswedel I, Hirsch D, Kropf S, et al. Flavanol-rich cocoa drink lowers plasma F(2)-isoprostane concentrations in humans. Free Radic Biol Med. 2004;37(3):411–21.
92. Engler MB, Engler MM, Chen CY, et al. Flavonoid-rich dark chocolate improves endothelial function and increases plasma epicatechin concentrations in healthy adults. J Am Coll Nutr. 2004;23(3):197–204.
93. Ramos S, Moulay L, Granado-Serrano AB, et al. Hypolipidemic effect in cholesterol-fed rats of a soluble fiber-rich product obtained from cocoa husks. J Agric Food Chem. 2008;56(16):6985–93.
94. Fallo F, Scarda A, Sonino N, et al. Effect of glucocorticoids on adiponectin: a study in healthy subjects and in Cushing's syndrome. Eur J Endocrinol. 2004;150(3):339–44.
95. Fried SK, Bunkin DA, Greenberg AS. Omental and subcutaneous adipose tissues of obese subjects release interleukin-6: depot differences and regulation by glucocorticoid. J Clin Endocrinol Metab. 1998;83:847–50.
96. Degawa-Yamauchi M, Moss KA, Bovenkerk JE. Regulation of adiponectin expression in human adipocytes: effects of adiposity, glucocorticoids, and tumor necrosis factor alpha. Obes Res. 2005;13(4):662–9.
97. Kanu A, Fain JN, Bahouth SW, et al. Regulation of leptin release by insulin, glucocorticoids, G(i)-coupled receptor agonists, and pertussis toxin in adipocytes and adipose tissue explants from obese humans in primary culture. Metabolism. 2003;52(1):60–6.

98. Swantek JL, Cobb MH, Geppert TD. Jun N-terminal kinase/stress-activated protein kinase (JNK/SAPK) is required for lipopolysaccharide stimulation of tumor necrosis factor alpha (TNF-alpha) translation: glucocorticoids inhibit TNF-alpha translation by blocking JNK/SAPK. Mol Cell Biol. 1997;17(11):6274–82.
99. Wellen KE, Hotamisligil GS. Obesity-induced inflammatory changes in adipose tissue. J Clin Invest. 2003;112(12):1785–8.
100. Sardi A, Geda C, Nerici L, et al. Rhabdomyolysis and arterial hypertension caused by apparent excess of mineralocorticoids: a case report. Ann Ital Med Int. 2002;17(2):126–9.
101. Al-Dujaili EAS, Kenyon CJ, Nicol MR, et al. Liquorice and glycyrrhetinic acid increase DHEA and deoxycorticosterone levels in vivo and in vitro by inhibiting adrenal SULT2A1 activity. Mol Cell Endocrinol. 2011;336:102–9.
102. Almoosawi S, Fyfe L, Ho C, et al. The effect of polyphenol-rich dark chocolate on fasting capillary whole blood glucose, total cholesterol, blood pressure and glucocorticoids in healthy overweight and obese subjects. Br J Nutr. 2009;103(6):842–50.
103. Lee YS, Lorenzo BJ, Koufis T, et al. Grapefruit and its flavonoids inhibit 11beta-hydroxysteroid dehydrogenase. Clin Pharmacol Ther. 1996;59:62–71.
104. Guo J, Reidenberg MM. Inhibition of 11beta-hydroxysteroid dehydrogenase by bioflavonoids and their interaction with furosemide and gossypol. J Lab Clin Med. 1998;132:32–8.
105. Song D, Lorenzo B, Reidenberg MM. Inhibition of 11-beta-hydroxysteroid dehydrogenase by gossypol and bioflavonoids. J Lab Clin Med. 1992;120:792–7.
106. Atanasov AG, Dzyakanchuk AA, Schweizer RA, et al. Coffee inhibits the reactivation of glucocorticoids by 11beta-hydroxysteroid dehydrogenase type 1: a glucocorticoid connection in the anti-diabetic action of coffee? FEBS Lett. 2006;580(17):4081–5.
107. Basu R, Singh R, Basu A, et al. Effect of nutrient ingestion on total-body and splanchnic cortisol production in humans. Diabetes. 2006;55:667–74.
108. Wake DJ, Homer NZ, Andrew R, et al. Acute in vivo regulation of 11ß-hydroxysteroid dehydrogenase type 1 activity by insulin and intralipid infusions in humans. J Clin Endocrinol Metab. 2006;91:4682–8.
109. Stimson RH, Johnstone AM, Homer NZ, et al. Dietary macronutrient content alters cortisol metabolism independently of body weight changes in obese men. J Clin Endocrinol Metab. 2007;92(11):4480–4.
110. Morton NM, Ramage L, Seckl JR. Down-regulation of adipose 11ß-hydroxysteroid dehydrogenase type 1 by high-fat feeding in mice: a potential adaptive mechanism counteracting metabolic disease. Endocrinology. 2004;145:2707–12.
111. Drake AJ, Livingstone DE, Andrew R, et al. Reduced adipose glucocorticoid reactivation and increased hepatic glucocorticoid clearance as an early adaptation to high-fat feeding in Wistar rats. Endocrinology. 2005; 146:913–9.
112. Sies H, Schewe T, Heiss C, Kelm M. Cocoa polyphenols and inflammatory mediators. Am J Clin Nutr. 2005;81(1):304S–12.
113. Sanbongi C, Suzuki N, Sakane T. Polyphenols in chocolate, which have antioxidant activity, modulate immune functions in humans in vitro. Cell Immunol. 1997;177(2):129–36.
114. Mao TK, Powell J, Van de Water J, et al. The effect of cocoa procyanidins on the transcription and secretion of interleukin 1 beta in peripheral blood mononuclear cells. Life Sci. 2000;66(15):1377–86.
115. Mao T, Van De Water J, Keen CL, et al. Cocoa procyanidins and human cytokine transcription and secretion. J Nutr. 2000;130(8S Suppl):2093S–9.
116. Mao TK, Van de Water J, Keen CL, et al. Effect of cocoa flavanols and their related oligomers on the secretion of interleukin-5 in peripheral blood mononuclear cells. J Med Food. 2002;5(1):17–22.
117. Mao TK, van de Water J, Keen CL, et al. Modulation of TNF-alpha secretion in peripheral blood mononuclear cells by cocoa flavanols and procyanidins. Dev Immunol. 2002;9(3):135–41.
118. Mao TK, Van De Water J, Keen CL, et al. Cocoa flavonols and procyanidins promote transforming growth factor-beta1 homeostasis in peripheral blood mononuclear cells. Exp Biol Med (Maywood). 2003;228(1):93–9.
119. di Giuseppe R, Di Castelnuovo A, Centritto F. Regular consumption of dark chocolate is associated with low serum concentrations of C-reactive protein in a healthy italian population. J Nutr. 2008;138:1939–45.

Chapter 26
Chocolate and Cancer Prevention?

Gertraud Maskarinec

Key Points

- Given the many ways that flavanols interact with cells and tissues, catechins and procyanidins in cocoa products may contribute to cancer prevention.
- Experimental studies in cell lines and animal models indicate that cocoa products and their active ingredients influence pathways with anticarcinogenic properties. They have the ability to lower oxidative stress and chronic inflammation, inhibit cell proliferation, and modify gut microbiota.
- Doses of catechins and procyanidins in experimental settings are often much higher than can be achieved in humans.
- Human intervention studies have reported favorable changes in biomarkers assessing antioxidant status, inflammatory markers, and intestinal microbiota.
- The sparse epidemiologic literature offers weak support for a reduction in cancer incidence and mortality related to high flavanol intake from all sources, but the evidence for cocoa products is inconclusive. Colorectal cancer has received the most attention of all cancer sites.
- Dietary assessment methods and nutritional databases face many challenges in accurately measuring intake of cocoa products and their flavanol content.
- For future epidemiologic studies related to cocoa and chocolate products, valid dietary assessment methods that capture the large variety of chocolate and cocoa products as well as other flavanol-containing foods are needed.
- The relative intake of flavanols from cocoa and chocolate products may be too small to detect their distinct cancer-protective effect in population-based studies with diverse nutritional habits and a wide range of risk factors for cancer.

Keywords Cancer • Prevention • Flavanols • Catechins • Procyanidins • Cocoa • Chronic inflammation • Oxidative stress • Gut microbiota • Epidemiology • Intervention studies • Dietary assessment

G. Maskarinec, M.D., Ph.D. (✉)
University of Hawaii Cancer Center, Honolulu, HI, USA
e-mail: Gertraud@crch.hawaii.edu

R.R. Watson et al. (eds.), *Chocolate in Health and Nutrition*, Nutrition and Health 7,
DOI 10.1007/978-1-61779-803-0_26, © Springer Science+Business Media, LLC 2013

Introduction

Chocolate has properties of food and drugs, as can be seen from its medicinal use by Olmecs, Mayans, Aztecs, and Europeans [1, 2]. In recent times, studies of cocoa products have reported beneficial effects in terms of risk for cardiovascular disease, particularly hypertension and hyperlipidemia [3–5]. As described elsewhere in this book, cocoa contains large amounts of polyphenolic compounds, flavanols (also called flavan-3-ols), primarily catechins, as well as procyanidins, polymeric condensation products (dimers to decamers) of catechins. These are formed during fermentation [6], an oxidative process that causes the dark color and determines the flavor of cocoa products [7]. In natural cocoa powders, the respective concentrations of catechins and procyanidins were reported as 2.9 and 41 mg/g, respectively [8, 9]. Although experimental studies using procyanidins from cocoa are limited, abundant research on these compounds from other sources (e.g., grapes and red wine) has identified biologic mechanisms that may lead to beneficial health effects. Other constituents of cocoa products (e.g., theobromine) have also been researched [10], but to a lesser degree, and will not be considered in this chapter.

One of the functions of flavanols is protection against oxidative stress due to their reducing ability [11]; a particularly strong antioxidant capacity (AOC) has been demonstrated for procyanidins [8, 12]. Although there is growing epidemiologic evidence for potential cancer-protective effects of other flavonoids, such as found in green tea and soy beans [13–16], similar investigations for cocoa products are more limited. This chapter will take three approaches to consider the case for a cancer-protective effect of cocoa products. First, potential biologic mechanisms, how polyphenolic compounds contained in cocoa and chocolate may protect against cancer, will be presented. Second, the limited epidemiologic literature examining the association between chocolate intake and cancer risk will be summarized. This will include analogous investigations of other foods that are rich in flavanols and procyanidins. Third, dietary assessment methods will be described, and methodological issues that need to be addressed to perform better epidemiologic studies in the future will be discussed.

Potential Cancer-Protective Mechanisms

As abundant research on polyphenolic compounds from grapes and red wine indicates, polyphenols exert many biologic actions that may lower cancer risk. To name a few, stimulation of the immune system, modulation of enzyme activity related to detoxification, regulation of hormone metabolism, cell cycle arrest in tumor cells, apoptosis, and reduction in cell proliferation may all contribute to beneficial health effects [17–20]. For this chapter, the focus will be on the four more commonly researched mechanisms of action in cocoa products: antioxidant activity, anti-inflammatory effects, inhibition of cell proliferation, and direct effects in the gut [2, 19, 20]. The following sections present examples of experimental studies and short-term trials with cocoa products in human studies that illustrate the four mechanisms.

Antioxidative Activity

Due to their chemical structure, polyphenols are able to capture reactive oxygen species (ROS) that are produced as part of inflammatory and endogenous processes and play a role in carcinogenesis [21, 22]. A strong correlation between procyanidin content of cocoa products and its AOC was demonstrated using oxygen radical absorbance capacity (ORAC) assays [8]. For example, natural cocoa powders contained the highest levels of AOC, unsweetened chocolates and chocolate liquor were intermediate, while milk chocolates had the lowest concentration of AOC. One serving of cocoa (5 g) or chocolate (15 or 40 g, depending upon the type of chocolate) provides AOC amounts that exceed the amount in many commonly consumed foods [8].

Several experimental studies have looked specifically at the antioxidative actions of compounds derived from cocoa. Phenolics from cacao liquor extracts showed inhibitory effects on lipid peroxidation in microsomes and on the autoxidation of linoleic acid. These effects were attributed to the radical-scavenging activity in the peroxidation chain reactions [23]. In a functional genome analysis of Caco-2 cells, a human colon adenocarcinoma cell line, treatment with polyphenolic cocoa extract modified the expression of several genes involved in the cellular response to oxidative stress [24]. The ability of dimer and trimer procyanidins isolated from cocoa to inhibit lipid peroxidation was demonstrated through their interaction with membrane phospholipids in liposomes [25]. It appears that the procyanidins are able to provide protection against oxidation through this mechanism. In a different research approach, fecal free radical production was lowered as a result of administering chocolate that had a high procyanidin content [26]. In a murine lymphoma model, the albumin fraction of semi-fermented dry cacao showed free radical scavenging capacity [27]. The authors suggested that cacao seeds may be considered a source of potential antitumor peptides. In a human feeding study, dose–response increases in the plasma AOC and decreases in plasma lipid oxidation products were observed 2 h after the ingestion of a procyanidin-rich chocolate containing 5.3 mg total procyanidin/g, suggesting that epicatechin may contribute to the ability of plasma to scavenge free radicals and to inhibit lipid peroxidation [28].

As a result of the relatively low levels of flavanols in body fluids, as compared to concentrations in experimental studies, it is more difficult to demonstrate antioxidant effects in human interventions. The nutritional interventions varied between chocolate (6–105 g/day), cocoa drinks (100–300 mL per day), and flavanol tablets (234 mg per day) and lasted between 1 day and 18 weeks. Oxidative stress was assessed either as the concentration of lipid peroxidation products: malondialdehyde (MDA), an end product of peroxidation [22], plasma thiobarbituric acid-reactive substances (TBARS), and F_2 isoprostanes [29]; as AOC using ORAC, total peroxyl radical-trapping antioxidant potential (TRAP), trolox equivalents (TEAC), or ferric-reducing ability (FRAP) [21]; or as a susceptibility of body fluids to ex vivo oxidation [21] (e.g., resistance of low-density lipoprotein [LDL] to oxidation or increase in lag time) [30]. In eleven studies that assessed AOC, five used the TEAC, three the TRAP, two the ORAC, and one the FRAP method. Four studies showed an increase in AOC [28, 31–33], but seven did not [34–40]. TBARS decreased in two interventions [28, 31], but not in another one [35]. MDA decreased in two studies [37, 39] and not in another one [40]. Isoprostanes were reduced only in one study [37], but not in six other investigations [28, 33, 34, 36, 38, 41]. The LDL oxidation lag increased in two studies [32, 34] and decreased in one study [38], but no change was observed in another trial [36]. As a recent review of 19 trials concluded [42], total plasma AOC, as well as most markers of oxidative stress, was not influenced by short-term trials with cocoa products. It appeared that low-density lipoprotein oxidation ex vivo and in vivo decreased, but this may have been due to a direct beneficial effect of cocoa flavanols on lipid and protein oxidation.

Anti-inflammatory Effects

Given the increasing evidence for a role of chronic inflammation in cancer development [43–48], in particular for colorectal cancer [49], polyphenolic agents may be cancer protective through that route. Markers of inflammation, such as C-reactive protein (CRP), interleukins (IL), tumor necrosis factor (TNF)-α (alpha), and cytotoxic activity of peripheral-blood lymphocytes, have been used in experimental studies and associated with cancer risk in prospective studies [48–50]. In a cross-sectional design within a large Italian cohort, the levels of CRP were compared between 1,317 subjects with no chocolate intake and 824 subjects who ate dark chocolate regularly with a mean of 5.7 g (range 0.7–20 g) per day [51]. Even after adjustment for confounders, serum CRP concentrations remained significantly lower in chocolate consumers (1.10 mg/L) than nonconsumers (1.32 mg/L). In a J-shaped distribution, the lowest CRP concentrations were observed in consumers of up to 1 serving (20 g) of dark chocolate every 3 days. In the same report, chocolate consumers also had a significantly lower mean body mass index than nonconsumers.

Nutritional interventions with inflammatory biomarkers as end point reveal conflicting results. In a trial of 42 patients at high risk of cardiovascular disease, the results suggested a positive influence of cocoa polyphenols on the modulation of inflammatory mediators, including IL-6 and CRP, P-selectin, and intercellular adhesion molecule-1 (ICAM-I) [52]. Three studies measured CRP and observed no change during the interventions [34, 53, 54]. Neither were the levels of the cytokines IL6, IL1β, and TNF-α modified during 6 weeks of intervention with cocoa and dark chocolate [34]. An intervention with 40-g cocoa powder in 500 mL of skim milk for 4 weeks positively influenced a combination of circulating inflammatory markers among 42 volunteers [52]. A recent trial in healthy volunteers tested the effect of cocoa-derived flavanols in a randomized crossover study with 494-mg cocoa flavanols per day during the high-cocoa flavanol period and 23-mg cocoa flavanols in the low-cocoa drink [55]. Approximately 20% of the flavanols were monomers; the rest were dimers to decamers. During the high-cocoa flavanol, plasma catechin levels increased to 73 nmol/L, while CRP decreased significantly from 0.27 to 0.19 mg/mL.

Antiproliferative Action

Accumulating research favors more anticarcinogenic functions of polyphenols in general [17, 56, 57] and cocoa ingredients in particular [58–60]. For example, in experiments with liver cells, phenolic-rich extracts of unroasted and roasted cocoa prevented drug-induced cytotoxicity possibly by inducing autophagy [61]. Potential beneficial effects for cells overexpressing Erb-B2 such as in certain neoplasias were suggested in an experiment, in which procyanidins from cocoa lowered the expression of this gene [62]. A pentameric procyanidin from cocoa inhibited the growth of human breast cancer cells [58]. Similarly, cocoa polyphenols extracts had an antiproliferative effect on prostate cancer cell growth, but not on normal cells [63]. Lower pancreatic carcinogenesis was demonstrated in rats with cacao liquor that inhibited in vitro mutagenicity of a carcinogen, as well as rat pancreatic carcinogenesis in the initiation stage [60]. The same investigators also published results showing that cocoa liquor exerts chemopreventive effects in the lung without any promoting influence in other major organs [59].

Direct Effects in the Gut

Because procyanidins do not appear to be degraded in the stomach, they may also interact directly with the gut mucosa or the gut microbiota [2, 55, 64]. For example, the gut microbiota in rats was modified by the administration of procyanidins demonstrating the possibility that these compounds alter the composition of intestinal bacteria [65]. In a study with healthy volunteers described above [55], a change in gut microbiota composition was observed in addition to the increase in plasma catechin levels and the decrease in CRP during the high-cocoa flavanol period. The bifidobacterial and lactobacilli populations increased, while the clostridia counts decreased. These observations are important in the context of recent evidence that different bifidobacteria and other gut microbiota may play a role in the development of gastrointestinal and other cancers [66, 67] (e.g., by affecting the renewal of gut epithelial cells or by altering cancer risk in other tissues through pathways such as nutrient harvesting, metabolism of xenobiotics, and the immune system) [68].

Epidemiologic Studies

Estimation of Intake

Several newly developed databases [9, 69, 70] have made it possible to assess regular intake of flavanols across individuals and populations. The consumption of cocoa products varies widely across populations. In the European Prospective Investigation into Cancer and Nutrition (EPIC), the mean intake of chocolate was reported as 6–7 g per day, with the highest intake of 10 g (for women) and 12 g (for men) in Great Britain and the lowest intake in Greece and Italy [71]. A cross-sectional investigation in the Netherlands showed that chocolate consumption contributed 20% of daily catechin intake in children (5 mg), 6% in adults (3 mg), and 3% in the elderly (2 mg) [72]. For the Spanish diet, the mean daily intake of cocoa products was estimated at 8.6 g, accounting for 10% of the total AOC of daily intake [73]. In an Italian population, the investigators estimated that 13–14% of procyanidins were obtained from chocolate products [74]. Based on national survey data in the United States, the mean intake of procyanidins was estimated at 58 mg per day or more with 18% of total procyanidin intake from chocolate; apples and grapes were the other major sources [75]. Among 1,500 healthy women in the Long Island Study [76], daily flavanol intake was estimated at 174 mg with tea, chocolate, cherries, and apples as the major sources. However, the high intake has to be considered in light of the relatively low bioavailability of these compounds; catechins and especially their polymers are poorly absorbed and rapidly eliminated [77, 78]. In comparison to genistein, the flavonoid with the highest bioavailability, only 2–8% of monomeric flavanols are bioavailable [79].

Comparison of Populations

After noting, in 1944, the low blood pressure among Kuna Indians in Panama who consume 900 mg of flavanols in cocoa drinks every day [80], causes of deaths for Kuna Indians living in the San Blas islands were compared with the population in mainland Panama who did not consume cocoa drinks. Lower mortality rates for cancer and other chronic diseases were found among islanders than in mainland Panama, but possible confounding by other lifestyle factors was not considered [81]. The potential adverse effects of cocoa on testicular cancer were investigated in an ecologic study. The research was based on the hypothesis that chocolate may disturb thc hormonal equilibrium in utero and predispose the fetus to hypospadia, a risk factor for testicular cancer [82]. Using data on per capita consumption of cocoa and the prevalence of hypospadia for each country, the researchers found a greater than twofold increase in cocoa intake during 1961 and 2008 and a strong correlation with hypospadias in 20 countries ($r=0.76$) and with testicular cancer incidence for 18 countries ($r=0.86$). However, no information on confounders or on individual-level associations was available.

Case-Control Studies

A number of case-control studies specifically investigated chocolate and cocoa intake in relation to cancer [15] In addition, some studies explored flavanol-containing foods as a group and showed reduced risks for several cancers with higher catechin and procyanidin intake [15, 83, 84], but they did not always specify the nutritional source for these compounds. In a 2005 review of seven prospective and four case-control studies on catechins [15], suggestive findings were only observed for lung and colorectal cancer in prospective studies, but no significant associations were observed for the case-control studies. In the small body of literature, adenomas and colorectal cancer are the most commonly studied outcomes. A study in North Carolina observed a nonsignificantly lower prevalence of adenomatous polyps that showed a dose–response relation associated with the consumption of chocolate candy [85]. A protective effect of flavanols not specifically from cocoa products was observed in a Scottish case-control study [86]. Colorectal cancer risk was reduced by 26% for epicatechin ($p = 0.02$) and by 22% for procyanidins ($p = 0.03$). A high chocolate dietary pattern, identified through cluster analysis in a French study, showed no significant association with any stage of colorectal disease ranging from polyps to adenomas and colorectal cancer [87]. In an earlier report from the same study, chocolate was identified as a risk factor for colorectal cancer [88]. No effect of a chocolate–cereal dietary pattern on breast and colorectal cancer was found in a Canadian study [89]. In a series of case-control studies from Italy that investigated a variety of flavonoids [74], procyanidins were associated with a lower colorectal cancer risk; interestingly, the protection was stronger for procyanidins with a higher degree of polymerization. The OR for the highest versus the lowest quintile of intake was 0.82 for monomers and dimers combined, 0.88 for monomers, 0.75 for dimers, 0.74 for all polymers with three or more mers, 0.84 for trimers, 0.80 for 4–6 mers, 0.79 for 7–10 mers, 0.69 for more than 10 mers, and 0.74 for total procyanidins [90]. The protective effect was also stronger for rectal than colon cancer. The authors proposed that, given their poor absorption rate of procyanidins, they may exert a local effect in the colon and rectum.

A few studies have been published on other cancer sites. An investigation of pancreatic cancer reported an elevated risk for chocolate candy, in particular among men with an OR of 2.4 for those who reported at least one serving per day [91]. The authors suggested that the saturated fat and/or sugar content of the candies may be responsible for this association. In the Italian net of studies, stomach cancer risk was lower with higher intake of procyanidins [92]; the risk estimates were 0.44 for monomers and dimers and 0.36 for trimers and higher. Higher epicatechin, anthocyanidin, and procyanidin intakes were protective against non-Hodgkin lymphoma (NHL) in a cancer registry-based case-control study [83]. In one Italian study, no significant relation of ovarian cancer with flavanols was detected [93].

Cohort Studies

Four prospective cohorts included questions on cocoa and chocolate as part of their nutritional assessment procedures and reported mortality and/or cancer outcomes: the Zutphen Elderly Study from the Netherlands, [94] the Iowa Women's Study [95], the Harvard Alumni Study [96], and the Leisure World Cohort Study [97].

The Zutphen Elderly Study is a prospective cohort study of 806 men ages 65–84 years at baseline [98]. Based on the measurement of 6 catechins in 120 foods, the mean catechin intake at cohort entry was estimated at 72 ± 47.8 mg. Chocolate contributed 3% of the total catechin intake, tea 87%, and apples 8%. After 15 years of follow-up, the adjusted relative risk for overall mortality among men in the highest tertile (>2.5 g per day) was 0.53 when compared to the lowest tertile (<0.5 g per day) of cocoa intake [99]. After 10 years of follow-up, 96 incident epithelial cancers were recorded, including 42 cases of lung cancer [94]. Overall catechin intake was not associated with epithelial cancer or lung cancer after adjustment for confounders. However, catechins from chocolate were nonsignificantly inversely associated with lung cancer incidence and all epithelial cancers; the respective risk estimates were 0.76 (95% CI 0.29–2.02) and 0.89 (95% CI 0.45–1.80).

In the Iowa Women's Study, mean intake of catechins was 25±32 mg/day at baseline [95]. Tea contributed 56% of catechins, apples and pears 26%, but the percentage derived from chocolate was not reported separately [100]. The association with cardiovascular mortality was close to significance ($p=0.06$) with a risk estimate of 0.92 (95% CI 0.84–1.00) for chocolate consumers compared to non-consumers. Although catechin intake from tea was weakly protective against rectal cancer among women in this cohort, no separate risk estimates were shown for chocolate [95]. A more recent analysis of the Iowa Women's Study found no association of chocolate consumption with non-Hodgkin lymphoma although total procyanidin intake was protective with a 30% lower for the highest category [101]. For specific foods, the lower risk was observed for all fruits and vegetables and for apple juice/cider.

Mortality, not cancer incidence, was studied in the other two investigations [96, 97]. In the Harvard Alumni Study, subjects who reported consuming candy 1–3 times per month had a 27% lower risk of mortality [96]. In the Leisure World Cohort Study [97], the subjects with the most frequent chocolate consumption did not experience a reduction in mortality, but those who reported occasional chocolate intake had a statistically significant lower mortality by 6%.

Methodological Issues

Dietary Assessment

For good quality studies in nutritional epidemiology, an accurate assessment of dietary intake is essential. In comparison to other foods, measuring chocolate consumption poses additional challenges, for example, the sporadic and seasonal intake of cocoa products. The major issue, however, is the wide variety of chocolate recipes and ingredients. It is particularly important to distinguish dark chocolate from milk chocolate because the content of catechins and procyanidins is highest in natural cocoa powders (2.9–3.5 and 32–49 mg/g), followed by unsweetened chocolate (1.5–2.8 and 9–25 mg/g), dark chocolate (0.8–1.6 and 9–10 mg/g), and milk chocolate (0.23–0.32 and 2.2–3.1 mg/g) [8]. Furthermore, nutritional flavonoid databases do not include values for the large variety of cocoa beans from various geographic areas processed with different methods; however, catechin and procyanidin contents vary due to the strong influence of fermentation on flavanol content [102].

In many of the existing food frequency questionnaires (FFQs), only one or two questions are directed toward chocolate products because the usual goal is to capture the entire diet [51, 95]. Some studies simply asked about chocolate intake as part of a lifestyle questionnaire [96, 97]. Examples of FFQs that collected information on cocoa products are described in more detail to illustrate how the wide variation in assessment methods may influence the exposure information used to estimate the association with disease outcomes. Some of the investigations reported reliability measures that were comparable to those for other foods [91, 95].

1. In a cross-sectional study on meal patterns, it appears that candies and chocolates were part of the same question [103].
2. For a set of Italian case-control studies [74], one question on chocolate was asked in a validated FFQ with 78 items.
3. The Iowa Women's Study used a validated FFQ with 127 items [104] that was based on a questionnaire developed at Harvard and contained two items related to chocolate: chocolate bars or pieces (e.g., Hersheys, M&M's) and candy bars (e.g., Snickers, Milky Way, Reese's) plus items for candy without chocolate and brownies. A validation study reported correlations of 0.45–0.83 between the FFQ and 28-day food records for chocolate-containing foods [95].

4. In a similar approach, a San Francisco study on pancreatic cancer [91] applied a 131-item FFQ, also modified from a questionnaire originally developed at Harvard. Two questions asked about pure chocolate candy bars or packets of candy (e.g., M&M's and other mixed candy bars, such as Snickers/Milky Way/Reese's). This FFQ reported a correlation coefficient of 0.41 for chocolate, as measured by dietary record and questionnaire.
5. The EPIC study in Italy asked participants about intake of a standard dose of 20 g of chocolate considering type (dark, milk, nut, or any type) and frequency of intake [51], while the EPIC questionnaire in Germany inquired how frequently 50-g chocolate bars were consumed [105].
6. The Spanish EPIC questionnaire collected nutritional data using a dietary history with around 600 food items [106]. The list included 14 items with cocoa or chocolate: bun stuffed with chocolate, croissant with chocolate, chocolate biscuits, chocolate cakes, chocolate nut spread, instant cocoa drink, chocolate for cooking/melting, milk chocolate, white chocolate, dark chocolate, chocolate truffles, chocolate ice cream, chocolate mousse, and chocolate milkshake.
7. The Dutch Zutphen cohort study estimated total intake of cocoa based on 24 cocoa-containing foods collected by a dietary history method [94, 99]. The reliability of intake was shown by repeated assessments; baseline consumption of confectionary and cocoa was significantly associated with intake after 5 and 10 years.

Nutritional Databases

Information on the content of catechins and procyanidins can be obtained from different databases that were developed during the last 10 years and are now available on the Internet [9, 69, 70]. The database provided by the US Department of Agriculture provides information for flavanols in six categories of products [69] derived from six different reports [107–112], but does not include more recent analyses [8]. It appears that the estimates for catechin content are in fairly good agreement as long as the type of chocolate is known [107]. Measuring procyanidins poses challenges over measuring catechins because of their oligomer structure [75]. A separate USDA database has combined levels of procyanidins in a number of foods, primarily cocoa products, fruits, and vegetables [70]. Baking chocolate and cocoa power had by far the highest concentration of total procyanidins with respective values of 1.7 and 14 mg/g [70]. More recently, a French database called Phenol-Explorer has compiled values for a large range of polyphenols in foods [9].

Biomarkers for Exposure Assessment

As a tool to assess flavanol exposure more directly and without bias due to self-reports, epicatechin has been measured in body fluids. Intake of cocoa products and epicatechin measured in serum shows an excellent dose–response relation. On average, serum levels increase by 200 nmol/L for 100 mg catechins from cocoa products when measured 2 h after intake, the peak time [77]. Individual studies have reported serum levels of 1.5 μmol/L [113], 5 μmol/L [114], and 6 μmol/L [115] after respective intakes of approximately 100, 200, and 300 mg catechins from chocolate products. However, in contrast to isoflavones, for which serum levels and urinary excretion are strongly correlated with intake [116, 117], exposure to flavanols cannot be assessed as accurately through a biologic biomarker because of the lower bioavailability and the short half-life of catechins [77]. Therefore, measuring catechins in serum relies on short-term intake and does not provide information on long-term exposure [77]. Only in the case of regular and frequent consumption, catechin levels in blood and urine represent actual exposure, as shown by an analysis from the Shanghai Women's Health Study that indicated a good correlation between urinary metabolites of catechins with self-reported green tea intake [118].

Summary

It is obvious that the question mark in the title of this article cannot be resolved at this time. From a mechanistic point of view, the flavanols contained in cocoa products appear to have the ability to lower cancer risk. However, the amounts consumed in comparison to fruits, vegetables, tea, and other polyphenol-containing foods are modest in most populations and may be difficult to distinguish from one another. So far, the evidence from epidemiologic investigations is sparse and inconclusive. The weak support for a reduction in cancer incidence comes primarily from investigations that looked at exposure to flavanols from all sources, not just cocoa products [15, 74, 94, 101]. Some studies indicate adverse effects on cancer risk [88, 91], but exposure misclassification and confounding with other lifestyle factors may be responsible for spurious associations.

Although several short-term trials have reported favorable modifications in biomarkers assessing antioxidant or anti-inflammatory effects, the small changes may be more relevant to cardiovascular disease outcomes than cancer risk. Yet, even in the context of cardiovascular disease, the relevance of changes in these biomarkers is not fully understood. As a thorough evaluation of the existing evidence concluded, the hypothesis that polyphenol-rich foods exert beneficial health effects through antioxidant activity has not been established [79]. Nevertheless, additional nutritional interventions may be able to elucidate the effects of flavanols contained in cocoa products on specific markers to examine their possible mechanisms of action. The ideal trials will have a well-defined exposure, placebo drinks or foods, good compliance measures, and valid outcome biomarkers or preferably a panel of biomarkers [4, 119].

The high concentration of procyanidins with their strong antioxidant potential is a distinctive property of cocoa products; chocolate and apples contain more procyanidins per serving than red wine and cranberry juice [6], and the AOC of cocoa products is higher than for many other foods [120]. The presence of these compounds may be a good reason to continue research into the cancer-protective effects of cocoa [70], although their ability to be effective as antioxidants in humans may be limited by their low bioavailability [77, 79]. For some individuals, cocoa products constitute a larger proportion of their diet than green tea, wine, or soy beans [12, 72, 73], all of which are under investigation for protection against chronic conditions. For better research in the future, improved methods to obtain valid estimates of catechin and procyanidin intake from the large variety of foods including cocoa products are necessary so that the effects of sugar, fat, and energy content of chocolate products can be separated from the effects of flavanols and from those of other flavanol-containing foods. This requires improvements of FFQ and nutrient databases or the use of alternate dietary assessment methods such as dietary records. Nevertheless, the relative intake of flavanols from cocoa and chocolate products may be too small to detect their separate effect in population-based studies with a wide range of nutritional, behavioral, and genetic risk factors for cancer.

References

1. Bruinsma K, Taren DL. Chocolate: food or drug? J Am Diet Assoc. 1999;99:1249–56.
2. Keen CL. Chocolate: food as medicine/medicine as food. J Am Coll Nutr. 2001;20:436S–9.
3. Taubert D, Roesen R, Schomig E. Effect of cocoa and tea intake on blood pressure: a meta-analysis. Arch Intern Med. 2007;167:626–34.
4. Cooper KA, Donovan JL, Waterhouse AL, Williamson G. Cocoa and health: a decade of research. Br J Nutr. 2008;99:1–11.
5. Jia L, Liu X, Bai YY, Li SH, Sun K, He C, et al. Short-term effect of cocoa product consumption on lipid profile: a meta-analysis of randomized controlled trials. Am J Clin Nutr. 2010;92:218–25.
6. Hammerstone JF, Lazarus SA, Schmitz HH. Procyanidin content and variation in some commonly consumed foods. J Nutr. 2000;130:2086S–92S.
7. Schwan RF, Wheals AE. The microbiology of cocoa fermentation and its role in chocolate quality. Crit Rev Food Sci Nutr. 2004;44:205–21.

8. Gu L, House SE, Wu X, Ou B, Prior RL. Procyanidin and catechin contents and antioxidant capacity of cocoa and chocolate products. J Agric Food Chem. 2006;54:4057–61.
9. Perez-Jimenez J, Neveu V, Vos F, Scalbert A. Identification of the 100 richest dietary sources of polyphenols: an application of the phenol-explorer database. Eur J Clin Nutr. 2010;64(Suppl 3):S112–20.
10. Knight IE. Chocolate and cocoa. Malden: Blackwell Sciences; 1999.
11. Scalbert A, Williamson G. Dietary intake and bioavailability of polyphenols. J Nutr. 2000;130:2073S–85S.
12. Prior RL, Gu L. Occurrence and biological significance of proanthocyanidins in the American diet. Phytochemistry. 2005;66:2264–80.
13. Adlercreutz H, Mazur W. Phyto-oestrogens and Western diseases. Ann Med. 1997;29:95–120.
14. Dreosti IE. Antioxidant polyphenols in tea, cocoa, and wine. Nutrition. 2000;16:692–4.
15. Arts IC, Hollman PC. Polyphenols and disease risk in epidemiologic studies. Am J Clin Nutr. 2005;81:317S–25S.
16. Yang CS, Sang S, Lambert JD, Hou Z, Ju J, Lu G. Possible mechanisms of the cancer-preventive activities of green tea. Mol Nutr Food Res. 2006;50:170–5.
17. Scalbert A, Manach C, Morand C, Remesy C, Jimenez L. Dietary polyphenols and the prevention of diseases. Crit Rev Food Sci Nutr. 2005;45:287–306.
18. Fernandez-Panchon MS, Villano D, Troncoso AM, Garcia-Parrilla MC. Antioxidant activity of phenolic compounds: from in vitro results to in vivo evidence. Crit Rev Food Sci Nutr. 2008;48:649–71.
19. Aron PM, Kennedy JA. Flavan-3-ols: nature, occurrence and biological activity. Mol Nutr Food Res. 2008;52:79–104.
20. Nandakumar V, Singh T, Katiyar SK. Multi-targeted prevention and therapy of cancer by proanthocyanidins. Cancer Lett. 2008;269:378–87.
21. Dotan Y, Lichtenberg D, Pinchuk I. Lipid peroxidation cannot be used as a universal criterion of oxidative stress. Prog Lipid Res. 2004;43:200–27.
22. Mayne ST. Antioxidant nutrients and chronic disease: use of biomarkers of exposure and oxidative stress status in epidemiologic research. J Nutr. 2003;133(Suppl 3):933S–40S.
23. Hatano T, Miyatake H, Natsume M, Osakabe N, Takizawa T, Ito H, et al. Proanthocyanidin glycosides and related polyphenols from cacao liquor and their antioxidant effects. Phytochemistry. 2002;59:749–58.
24. Noe V, Penuelas S, Lamuela-Raventos RM, Permanyer J, Ciudad CJ, Izquierdo-Pulido M. Epicatechin and a cocoa polyphenolic extract modulate gene expression in human Caco-2 cells. J Nutr. 2004;134:2509–16.
25. Verstraeten SV, Hammerstone JF, Keen CL, Fraga CG, Oteiza PI. Antioxidant and membrane effects of procyanidin dimers and trimers isolated from peanut and cocoa. J Agric Food Chem. 2005;53:5041–8.
26. Record IR, McInerney JK, Noakes M, Bird AR. Chocolate consumption, fecal water antioxidant activity, and hydroxyl radical production. Nutr Cancer. 2003;47:131–5.
27. Preza AM, Jaramillo ME, Puebla AM, Mateos JC, Hernandez R, Lugo E. Antitumor activity against murine lymphoma L5178Y model of proteins from cacao (Theobroma cacao L.) seeds in relation with in vitro antioxidant activity. BMC Complement Altern Med. 2010;10:61.
28. Wang JF, Schramm DD, Holt RR, Ensunsa JL, Fraga CG, Schmitz HH, et al. A dose–response effect from chocolate consumption on plasma epicatechin and oxidative damage. J Nutr. 2000;130:2115S–9S.
29. Tsimikas S. Measures of oxidative stress. Clin Lab Med. 2006;26:571–90.
30. Sies H, Schewe T, Heiss C, Kelm M. Cocoa polyphenols and inflammatory mediators. Am J Clin Nutr. 2005;81:304S–12.
31. Rein D, Lotito S, Holt RR, Keen CL, Schmitz HH, Fraga CG. Epicatechin in human plasma: in vivo determination and effect of chocolate consumption on plasma oxidation status. J Nutr. 2000;130:2109S–14.
32. Wan Y, Vinson JA, Etherton TD, Proch J, Lazarus SA, Kris-Etherton PM. Effects of cocoa powder and dark chocolate on LDL oxidative susceptibility and prostaglandin concentrations in humans. Am J Clin Nutr. 2001;74:596–602.
33. Serafini M, Bugianesi R, Maiani G, Valtuena S, De SS, Crozier A. Plasma antioxidants from chocolate. Nature. 2003;424:1013.
34. Mathur S, Devaraj S, Grundy SM, Jialal I. Cocoa products decrease low density lipoprotein oxidative susceptibility but do not affect biomarkers of inflammation in humans. J Nutr. 2002;132:3663–7.
35. Murphy KJ, Chronopoulos AK, Singh I, Francis MA, Moriarty H, Pike MJ, et al. Dietary flavanols and procyanidin oligomers from cocoa (Theobroma cacao) inhibit platelet function. Am J Clin Nutr. 2003;77:1466–73.
36. Engler MB, Engler MM, Chen CY, Malloy MJ, Browne A, Chiu EY, et al. Flavonoid-rich dark chocolate improves endothelial function and increases plasma epicatechin concentrations in healthy adults. J Am Coll Nutr. 2004;23:197–204.
37. Wiswedel I, Hirsch D, Kropf S, Gruening M, Pfister E, Schewe T, et al. Flavanol-rich cocoa drink lowers plasma F(2)-isoprostane concentrations in humans. Free Radic Biol Med. 2004;37:411–21.
38. Mursu J, Voutilainen S, Nurmi T, Rissanen TH, Virtanen JK, Kaikkonen J, et al. Dark chocolate consumption increases HDL cholesterol concentration and chocolate fatty acids may inhibit lipid peroxidation in healthy humans. Free Radic Biol Med. 2004;37:1351–9.

39. Fraga CG, Actis-Goretta L, Ottaviani JI, Carrasquedo F, Lotito SB, Lazarus S, et al. Regular consumption of a flavanol-rich chocolate can improve oxidant stress in young soccer players. Clin Dev Immunol. 2005;12:11–7.
40. Heiss C, Finis D, Kleinbongard P, Hoffmann A, Rassaf T, Kelm M, et al. Sustained increase in flow-mediated dilation after daily intake of high-flavanol cocoa drink over 1 week. J Cardiovasc Pharmacol. 2007;49:74–80.
41. Taubert D, Roesen R, Lehmann C, Jung N, Schomig E. Effects of low habitual cocoa intake on blood pressure and bioactive nitric oxide: a randomized controlled trial. JAMA. 2007;298:49–60.
42. Scheid L, Reusch A, Stehle P, Ellinger S. Antioxidant effects of cocoa and cocoa products ex vivo and in vivo: is there evidence from controlled intervention studies? Curr Opin Clin Nutr Metab Care. 2010;13:737–42.
43. Aggarwal BB, Shishodia S, Sandur SK, Pandey MK, Sethi G. Inflammation and cancer: how hot is the link? Biochem Pharmacol. 2006;72:1605–21.
44. Balkwill F, Mantovani A. Inflammation and cancer: back to Virchow? Lancet. 2001;357:539–45.
45. Coussens LM, Werb Z. Inflammation and cancer. Nature. 2002;420:860–7.
46. Siemes C, Visser LE, Coebergh JW, Splinter TA, Witteman JC, Uitterlinden AG, et al. C-reactive protein levels, variation in the C-reactive protein gene, and cancer risk: the Rotterdam Study. J Clin Oncol. 2006;24:5216–22.
47. Trichopoulos D, Psaltopoulou T, Orfanos P, Trichopoulou A, Boffetta P. Plasma C-reactive protein and risk of cancer: a prospective study from Greece. Cancer Epidemiol Biomarkers Prev. 2006;15:381–4.
48. Il'yasova D, Colbert LH, Harris TB, Newman AB, Bauer DC, Satterfield S, et al. Circulating levels of inflammatory markers and cancer risk in the health aging and body composition cohort. Cancer Epidemiol Biomarkers Prev. 2005;14:2413–8.
49. Tsilidis KK, Branchini C, Guallar E, Helzlsouer KJ, Erlinger TP, Platz EA. C-reactive protein and colorectal cancer risk: a systematic review of prospective studies. Int J Cancer. 2008;123:1133–40.
50. Selmi C, Mao TK, Keen CL, Schmitz HH, Eric GM. The anti-inflammatory properties of cocoa flavanols. J Cardiovasc Pharmacol. 2006;47(Suppl 2):S163–71.
51. di Giuseppe R, Di Castelnuovo A, Centritto F, Zito F, De Curtis A, Costanzo S, et al. Regular consumption of dark chocolate is associated with low serum concentrations of C-reactive protein in a healthy Italian population. J Nutr. 2008;138:1939–45.
52. Monagas M, Khan N, Andres-Lacueva C, Casas R, Urpí-Sardà M, Llorach R, et al. Effect of cocoa powder on the modulation of inflammatory biomarkers in patients at high risk of cardiovascular disease. Am J Clin Nutr. 2009;90:1144–50.
53. Crews Jr WD, Harrison DW, Wright JW. A double-blind, placebo-controlled, randomized trial of the effects of dark chocolate and cocoa on variables associated with neuropsychological functioning and cardiovascular health: clinical findings from a sample of healthy, cognitively intact older adults. Am J Clin Nutr. 2008;87:872–80.
54. Kurlandsky SB, Stote KS. Cardioprotective effects of chocolate and almond consumption in healthy women. Nutr Res. 2006;26:509–16.
55. Tzounis X, Rodriguez-Mateos A, Vulevic J, Gibson GR, Kwik-Uribe C, Spencer JP. Prebiotic evaluation of cocoa-derived flavanols in healthy humans by using a randomized, controlled, double-blind, crossover intervention study. Am J Clin Nutr. 2011;93:62–72.
56. Yang CS, Landau JM, Huang MT, Newmark HL. Inhibition of carcinogenesis by dietary polyphenolic compounds. Annu Rev Nutr. 2001;21:381–406.
57. Gaikwad NW, Yang L, Muti P, Meza JL, Pruthi S, Ingle JN, et al. The molecular etiology of breast cancer: evidence from biomarkers of risk. Int J Cancer. 2008;122:1949–57.
58. Ramljak D, Romanczyk LJ, Metheny-Barlow LJ, Thompson N, Knezevic V, Galperin M, et al. Pentameric procyanidin from Theobroma cacao selectively inhibits growth of human breast cancer cells. Mol Cancer Ther. 2005;4:537–46.
59. Yamagishi M, Natsume M, Osakabe N, et al. Chemoprevention of lung carcinogenesis by cacao liquor proanthocyanidins in a male rat multi-organ carcinogenesis model. Cancer Lett. 2003;191:49–57.
60. Yamagishi M, Natsume M, Osakabe N, Nakamura H, Furukawa F, Imazawa T, et al. Effects of cacao liquor proanthocyanidins on PhIP-induced mutagenesis in vitro, and in vivo mammary and pancreatic tumorigenesis in female Sprague–Dawley rats. Cancer Lett. 2002;185:123–30.
61. Arlorio M, Bottini C, Travaglia F, Locatelli M, Bordiga M, Coïsson JD, et al. Protective activity of Theobroma cacao L. phenolic extract on AML12 and MLP29 liver cells by preventing apoptosis and inducing autophagy. J Agric Food Chem. 2009;57:10612–8.
62. Kenny TP, Keen CL, Jones P, Kung HJ, Schmitz HH, Gershwin ME. Pentameric procyanidins isolated from Theobroma cacao seeds selectively downregulate ErbB2 in human aortic endothelial cells. Exp Biol Med (Maywood). 2004;229:255–63.
63. Jourdain C, Tenca G, Deguercy A, Troplin P, Poelman D. In-vitro effects of polyphenols from cocoa and beta-sitosterol on the growth of human prostate cancer and normal cells. Eur J Cancer Prev. 2006;15:353–61.
64. Rios LY, Bennett RN, Lazarus SA, Remesy C, Scalbert A, Williamson G. Cocoa procyanidins are stable during gastric transit in humans. Am J Clin Nutr. 2002;76:1106–10.
65. Smith AH, Mackie RI. Effect of condensed tannins on bacterial diversity and metabolic activity in the rat gastrointestinal tract. Appl Environ Microbiol. 2004;70:1104–15.

66. Lampe JW. The human microbiome project: getting to the guts of the matter in cancer epidemiology. Cancer Epidemiol Biomarkers Prev. 2008;17:2523–4.
67. Roberfroid M, Gibson GR, Hoyles L, McCartney AL, Rastall R, Rowland I, et al. Prebiotic effects: metabolic and health benefits. Br J Nutr. 2010;104(Suppl 2):S1–63.
68. Turnbaugh PJ, Ley RE, Hamady M, Fraser-Liggett CM, Knight R, Gordon JI. The human microbiome project. Nature. 2007;449:804–10.
69. Nutrient Laboratory ARS. USDA database for the flavonoid content of selected foods, release 2.1. U.S. Department of Agriculture, 2003; http://www.nal.usda.gov/fnic/foodcomp/Data/Flav/flav.html. Accessed on 19 July 2011.
70. Nutrient Laboratory ARS. USDA database for the proanthocyanidin content of selected foods. U.S. Department of Agriculture, 2004; http://www.nal.usda.gov/fnic/foodcomp/Data/PA/PA.html. Accessed on 19 July 2011.
71. Wirfalt E, McTaggart A, Pala V, Gullberg B, Frasca G, Panico S, et al. Food sources of carbohydrates in a European cohort of adults. Public Health Nutr. 2002;5:1197–215.
72. Arts IC, Hollman PC, Feskens EJ, Bueno de Mesquita HB, Kromhout D. Catechin intake and associated dietary and lifestyle factors in a representative sample of Dutch men and women. Eur J Clin Nutr. 2001;55:76–81.
73. Tabernero M, Serrano J, Saura-Calixto F. The antioxidant capacity of coca products: contribution to the Spanish diet. Int J Food Sci Technol. 2006;41(Suppl 1):28–32.
74. Rossi M, Bosetti C, Negri E, Lagiou P, La VC. Flavonoids, proanthocyanidins, and cancer risk: a network of case-control studies from Italy. Nutr Cancer. 2010;62:871–7.
75. Gu L, Kelm MA, Hammerstone JF, Beecher G, Holden J, Haytowitz D, et al. Concentrations of proanthocyanidins in common foods and estimations of normal consumption. J Nutr. 2004;134:613–7.
76. Fink BN, Steck SE, Wolff MS, Kabat GC, Gammon MD. Construction of a flavonoid database for assessing intake in a population-based sample of women on Long Island, New York. Nutr Cancer. 2006;56:57–66.
77. Manach C, Williamson G, Morand C, Scalbert A, Remesy C. Bioavailability and bioefficacy of polyphenols in humans. I. Review of 97 bioavailability studies. Am J Clin Nutr. 2005;81:230S–42.
78. Serra A, Macia A, Romero MP, Valls J, Bladé C, Arola L, et al. Bioavailability of procyanidin dimers and trimers and matrix food effects in in vitro and in vivo models. Br J Nutr. 2010;103:944–52.
79. Hollman PC, Cassidy A, Comte B, Heinonen M, Richelle M, Richling E, et al. The biological relevance of direct antioxidant effects of polyphenols for cardiovascular health in humans is not established. J Nutr. 2011;141:989S–1009S.
80. Kean BH. The blood pressure of the Cuna Indians. Am J Trop Med. 1944;24:341–3.
81. Bayard V, Chamorro F, Motta J, Hollenberg NK. Does flavanol intake influence mortality from nitric oxide-dependent processes? Ischemic heart disease, stroke, diabetes mellitus, and cancer in Panama. Int J Med Sci. 2007;4:53–8.
82. Giannandrea F. Correlation analysis of cocoa consumption data with worldwide incidence rates of testicular cancer and hypospadias. Int J Environ Res Public Health. 2009;6:568–78.
83. Frankenfeld CL, Cerhan JR, Cozen W, Davis S, Schenk M, Morton LM, et al. Dietary flavonoid intake and non-Hodgkin lymphoma risk. Am J Clin Nutr. 2008;87:1439–45.
84. Arts IC. A review of the epidemiological evidence on tea, flavonoids, and lung cancer. J Nutr. 2008; 138:1561S–6.
85. McKelvey W, Greenland S, Sandler RS. A second look at the relation between colorectal adenomas and consumption of foods containing partially hydrogenated oils. Epidemiology. 2000;11:469–73.
86. Theodoratou E, Kyle J, Cetnarskyj R, Farrington SM, Tenesa A, Barnetson R, et al. Dietary flavonoids and the risk of colorectal cancer. Cancer Epidemiol Biomarkers Prev. 2007;16:684–93.
87. Rouillier P, Senesse P, Cottet V, Valleau A, Faivre J, Boutron-Ruault MC. Dietary patterns and the adenomacarcinoma sequence of colorectal cancer. Eur J Nutr. 2005;44:311–8.
88. Boutron-Ruault MC, Senesse P, Faivre J, Chatelain N, Belghiti C, Meance S. Foods as risk factors for colorectal cancer: a case-control study in Burgundy (France). Eur J Cancer Prev. 1999;8:229–35.
89. Nkondjock A, Ghadirian P. Associated nutritional risk of breast and colon cancers: a population-based case-control study in Montreal, Canada. Cancer Lett. 2005;223:85–91.
90. Rossi M, Negri E, Parpinel M, Lagiou P, Bosetti C, Talamini R, et al. Proanthocyanidins and the risk of colorectal cancer in Italy. Cancer Causes Control. 2010;21:243–50.
91. Chan JM, Wang F, Holly EA. Sweets, sweetened beverages, and risk of pancreatic cancer in a large population-based case-control study. Cancer Causes Control. 2009;20:835–46.
92. Rossi M, Rosato V, Bosetti C, Lagiou P, Parpinel M, Bertuccio P, et al. Flavonoids, proanthocyanidins, and the risk of stomach cancer. Cancer Causes Control. 2010;21:1597–604.
93. Rossi M, Negri E, Lagiou P, Talamini R, Dal Maso L, Montella M, et al. Flavonoids and ovarian cancer risk: a case-control study in Italy. Int J Cancer. 2008;123:895–8.
94. Arts IC, Hollman PC, Bueno de Mesquita HB, Feskens EJ, Kromhout D. Dietary catechins and epithelial cancer incidence: the Zutphen elderly study. Int J Cancer. 2001;92:298–302.

95. Arts IC, Jacobs Jr DR, Gross M, Harnack LJ, Folsom AR. Dietary catechins and cancer incidence among post-menopausal women: the Iowa Women's Health Study (United States). Cancer Causes Control. 2002;13:373–82.
96. Lee IM, Paffenbarger Jr RS. Life is sweet: candy consumption and longevity. BMJ. 1998;317:1683–4.
97. Paganini-Hill A, Kawas CH, Corrada MM. Non-alcoholic beverage and caffeine consumption and mortality: the Leisure World Cohort Study. Prev Med. 2007;44:305–10.
98. Arts IC, Hollman PC, Feskens EJ, Bueno de Mesquita HB, Kromhout D. Catechin intake might explain the inverse relation between tea consumption and ischemic heart disease: the Zutphen Elderly Study. Am J Clin Nutr. 2001;74:227–32.
99. Buijsse B, Feskens EJ, Kok FJ, Kromhout D. Cocoa intake, blood pressure, and cardiovascular mortality: the Zutphen Elderly Study. Arch Intern Med. 2006;166:411–7.
100. Mink PJ, Scrafford CG, Barraj LM, Harnack L, Hong CP, Nettleton JA, et al. Flavonoid intake and cardiovascular disease mortality: a prospective study in postmenopausal women. Am J Clin Nutr. 2007;85:895–909.
101. Thompson CA, Habermann TM, Wang AH, Vierkant RA, Folsom AR, Ross JA, et al. Antioxidant intake from fruits, vegetables and other sources and risk of non-Hodgkin's lymphoma: the Iowa Women's Health Study. Int J Cancer. 2010;126:992–1003.
102. Camu N, Gonzalez A, De Winter T, Van Schoor A, De Bruyne K, Vandamme P, et al. Influence of turning and environmental contamination on the dynamics of populations of lactic acid and acetic acid bacteria involved in spontaneous cocoa bean heap fermentation in Ghana. Appl Environ Microbiol. 2008;74:86–98.
103. Berteus FH, Torgerson JS, Sjostrom L, Lindroos AK. Snacking frequency in relation to energy intake and food choices in obese men and women compared to a reference population. Int J Obes Lond. 2005;29:711–9.
104. Munger RG, Folsom AR, Kushi LH, Kaye SA, Sellers TA. Dietary assessment of older Iowa women with a food frequency questionnaire: nutrient intake, reproducibility, and comparison with 24-hour dietary recall interviews. Am J Epidemiol. 1992;136:192–200.
105. Buijsse B, Weikert C, Drogan D, Bergmann M, Boeing H. Chocolate consumption in relation to blood pressure and risk of cardiovascular disease in German adults. Eur Heart J. 2010;31:1616–23.
106. Zamora-Ros R, Andres-Lacueva C, Lamuela-Raventos RM, Berenguer T, Jakszyn P, Barricarte A, et al. Estimation of dietary sources and flavonoid intake in a Spanish adult population (EPIC-Spain). J Am Diet Assoc. 2010;110:390–8.
107. Arts IC, van De PB, Hollman PC. Catechin contents of foods commonly consumed in The Netherlands. 1. Fruits, vegetables, staple foods, and processed foods. J Agric Food Chem. 2000;48:1746–51.
108. Arts IC, van De PB, Hollman PC. Catechin contents of foods commonly consumed in The Netherlands. 2. Tea, wine, fruit juices, and chocolate milk. J Agric Food Chem. 2000;48:1752–7.
109. de Pascual-Teresa S, Santos-Buelga C, Rivas-Gonzalo JC. Quantitative analysis of flavan-3-ols in Spanish foodstuffs and beverages. J Agric Food Chem. 2000;48:5331–7.
110. Sakakibara H, Honda Y, Nakagawa S, Ashida H, Kanazawa K. Simultaneous determination of all polyphenols in vegetables, fruits, and teas. J Agric Food Chem. 2003;51:571–81.
111. Hertog MGL, Hollman PC, van De Putte B. Content of potentially anticarcinogenic flavonoids of tea infusions, wines, and fruit juices. J Agric Food Chem. 1993;41:1242–6.
112. Bonvehi JS, Coll F. Evaluation of bitterness and astringency of polyphenolic compounds in cocoa powder. Food Chem. 1997;60:365–70.
113. Schramm DD, Karim M, Schrader HR, Holt RR, Kirkpatrick NJ, Polagruto JA, et al. Food effects on the absorption and pharmacokinetics of cocoa flavanols. Life Sci. 2003;73:857–69.
114. Baba S, Osakabe N, Yasuda A, Natsume M, Takizawa T, Nakamura T, et al. Bioavailability of (−)-epicatechin upon intake of chocolate and cocoa in human volunteers. Free Radic Res. 2000;33:635–41.
115. Holt RR, Lazarus SA, Sullards MC, Zhu QY, Schramm DD, Hammerstone JF, et al. Procyanidin dimer B2 [epicatechin-(4beta-8)-epicatechin] in human plasma after the consumption of a flavanol-rich cocoa. Am J Clin Nutr. 2002;76:798–804.
116. Maskarinec G, Singh S, Meng L, Franke AA. Dietary soy intake and urinary isoflavone excretion among women from a multiethnic population. Cancer Epidemiol Biomarkers Prev. 1998;7:613–9.
117. Franke AA, Hebshi SM, Pagano I, Kono N, Mack WJ, Hodis HN. Urine accurately reflects circulating isoflavonoids and ascertains compliance during soy intervention. Cancer Epidemiol Biomarkers Prev. 2010;19:1775–83.
118. Luo J, Gao YT, Chow WH, Shu XO, Li H, Yang G, et al. Urinary polyphenols and breast cancer risk: results from the Shanghai Women's Health Study. Breast Cancer Res Treat. 2010;120:693–702.
119. Wong E, Freiberg M, Tracy R, Kuller L. Epidemiology of cytokines: the Women on the Move through Activity and Nutrition (WOMAN) Study. Am J Epidemiol. 2008;168:443–53.
120. Halvorsen BL, Carlsen MH, Phillips KM, Bøhn SK, Holte K, Jacobs Jr DR, et al. Content of redox-active compounds (ie, antioxidants) in foods consumed in the United States. Am J Clin Nutr. 2006;84:95–135.

Chapter 27
The Effects of Cocoa- and Chocolate-Related Products on Neurocognitive Functioning

W. David Crews Jr., David W. Harrison, Kim P. Gregory, Bon Kim, and Allison B. Darling

Key Points

- To date, there appear to be only nine published studies in the scientific literature that have examined the effects of cocoa- and chocolate-related products on neurocognitive functioning.
- The majority (i.e., seven of nine) of the published studies that have examined the effects of cocoa- and chocolate-related products on neurocognitive processes have found that the compounds are associated with significantly improved and/or preserved aspects of cognitive functioning.
- Increasing evidence suggests that dietary flavonoids and methylxanthines, such as those commonly found in cocoa and chocolate products, may possess neuroprotective, neuroenhancing, and neurostimulating effects.

Keywords Cocoa • Chocolate • Neurocognitive • Neuropsychological • Cognitive • Memory • Brain

W.D. Crews Jr., Ph.D. (✉)
Behavioral Neuroscience Laboratory, Department of Psychology,
Virginia Polytechnic Institute & State University, Blacksburg, VA, USA
e-mail: wdcrewsjr@aol.com

D.W. Harrison, Ph.D.
Behavioral Neuroscience Laboratory, Department of Psychology,
Virginia Polytechnic Institute & State University, Blacksburg, VA, USA
e-mail: dwh@vt.edu

K.P. Gregory, M.A.
Department of Early Childhood Education, Virginia Western Community College, Roanoke, VA, USA
e-mail: kimpgregory@gmail.com

B. Kim, B.S.
Department of Psychology, Liberty University, Lynchburg, VA, USA
e-mail: bkim@liberty.edu

A.B. Darling, B.S.
Behavioral Neuroscience Laboratory, Department of Psychology, Virginia Polytechnic Institute and State University, Blacksburg, VA, USA
e-mail: allied13@vt.edu

R.R. Watson et al. (eds.), *Chocolate in Health and Nutrition*, Nutrition and Health 7,
DOI 10.1007/978-1-61779-803-0_27, © Springer Science+Business Media, LLC 2013

Introduction

In recent years, there has been increased interest in the potential health-related benefits of antioxidant and phytochemical-rich [1, 2] cocoa and chocolate, especially in relation to how the products may promote cardiovascular and cerebrovascular health and provide corresponding protective effects [3–38]. In particular, cocoa- and chocolate-related products have been shown to decrease/inhibit both low-density lipoprotein (LDL) oxidation and total cholesterol levels [5–12] as well as platelet activation/function [5, 13–17], enhance serum lipid profiles [5, 11, 18, 19], favorably modify eicosanoid synthesis [20], lower blood pressure (BP) [5, 18, 21–26], promote endothelium-dependent relaxation/dilation [5, 27–30], increase cerebral blood flow [31–34], and inhibit free radical–induced erythrocyte hemolysis [35]. Studies have also suggested that cocoa flavanols/procyanidins may possess immunoregulatory effects, help to modulate immune responses, and reduce inflammation [5, 36–38].

Similarly, there has been growing interest in the possible neurocognitive and neuroprotective effects of cocoa- and chocolate-related compounds. The purpose of this chapter is to provide a review of the published scientific literature (through May 2011) that has examined the efficacy of cocoa- and chocolate-related products on the neurocognitive processes and functioning of both humans and laboratory animals. This chapter comprises the following sections: “Human Studies,” “Animal Studies,” “Proposed Mechanisms of Action,” “Summary,” and “Directions for Future Research.” It should be noted, however, that this chapter will not include studies that have examined neurocognitive processes as a function of cocoa- and/or chocolate-related fluctuations in mood/affect, product likings, cravings, “addictions,” or satiety, changes evoked by the chemosensory aspects of cocoa and/or chocolate, or studies that have not involved the actual consumption of cocoa- and/or chocolate-related products.

Human Studies

To date, there appear to have been a total of seven published studies that have examined the effects of cocoa- and/or chocolate-related products on human neurocognitive functioning.

Smit and his associates [39] were among the first to investigate the psychopharmacological activity of cocoa powder and methylxanthines found in a 50-g dark chocolate bar via two double-blind, placebo-controlled, within-subjects, repeated measures, design studies. In the first investigation, 20 participants (mean age = 32.6 ± 11.2 years) completed a test battery consisting of a 25-item mood questionnaire and tasks involving simple reaction time, rapid visual information processing, and manual dexterity (i.e., tapping) prior to, and at 1- and 2-h intervals after, receiving two active treatments consisting of identical amounts of methylxanthines contained in a 50-g bar of dark chocolate (i.e., 11.6 g of Cadbury’s Bournville cocoa powder or 250 mg theobromine + 19 mg of caffeine) or placebo. Findings from this study demonstrated that the cocoa powder and methylxanthines mixture exhibited very similar outcome profiles, with strong treatment effects found for an energetic arousal mood construct and on the simple reaction time task, versus placebo. On the rapid visual information processing task, significantly improved performances were noted with consumption of the methylxanthines, as compared to placebo, while a nonsignificant marginal improvement was found after the ingestion of the cocoa powder. No significant effects were found for either treatment on the tapping task.

In the second study [39], 22 participants (mean age = 35.4 ± 14.2) consumed visually identical 60-g portions of chocolate containing either no methylxanthines (similar to the amounts found in white chocolate), a low quantity of methylxanthines (i.e., 8 mg caffeine + 100 mg theobromine; similar to the amounts found in milk chocolate), a high quantity of methylxanthines (i.e., 20 mg caffeine + 250 mg theobromine; similar to the amounts found in dark chocolate), or a 60-ml water placebo. A task schedule similar to the first study was utilized; however, treatments were administered and consumed over 30 min in four equal portions prior to beginning the next task in the treatment battery. The findings from this study appeared to confirm the results of the first study, as a significant overall treatment effect was found for the energetic arousal mood construct, while high-methylxanthine chocolate significantly decreased reaction time on the simple reaction time task, versus the no-methylxanthines chocolate, and both low- and high-methylxanthine chocolate groups significantly improved their performances on the rapid visual information processing task, as compared to the no-methylxanthines chocolate group. Similar to the first study, no significant effects were found on the tapping task. Taken together, Smit and his colleagues [39] concluded that the cocoa content in dark chocolate has notable psychopharmacological activity (e.g., psychostimulant effects), which can be attributed to the presence of caffeine and/or theobromine.

Francis and colleagues [33] examined the effect of flavanol-rich cocoa on the functional magnetic resonance imaging (fMRI) blood oxygenation level–dependent (BOLD) response to a cognitive task (a "task-switching" paradigm using both "switch" and "non-switch" conditions) in 16 healthy, young participants (18–30 years of age). Participants consumed a high-flavanol cocoa drink (172 mg of flavanols per drink) for 5 days prior to one fMRI session and the administration of the switching task and a low flavanol cocoa beverage (13 mg of flavanols per drink) for 5 days prior to another fMRI session and switching task administration in a double-blind, counterbalanced fashion. Despite a significantly increased BOLD response during the cognitive task-switching paradigm after consumption of the high, versus low, flavanol-rich cocoa drink, no significant effects were observed in participants' behavioral reaction times, switch costs, or heart rates across treatments. The authors [33] hypothesized, however, that the BOLD changes may have been related to cognitive changes (e.g., strategy formation, cognitive effort) that were not exhibited on the standard behavioral measures that were utilized in the study, especially in young, healthy participants who were likely already functioning at a relatively high level of cognitive ability.

Our research group [40] conducted the first-known clinical trial of the short-term efficacy of dark chocolate and cocoa on variables associated with neuropsychological functioning and cardiovascular health in older adults. A sample of 101 healthy, cognitively intact adults, 60 years of age and older, were examined via a 6-week, randomized, double-blind, placebo-controlled, parallel-group design. Participants were randomly assigned to receive a 37-g dark chocolate bar (approximately 11 g of natural cocoa and 397.30 mg of total proanthocyanins) and one 8-oz (237 mL) cup of an artificially sweetened cocoa beverage (approximately 11 g of natural cocoa and 357.41 mg of total proanthocyanins) per day or similarly matched placebo products. To examine neurocognitive functioning, participants were administered a series of neuropsychological tests assessing short-term verbal and visual memory, cognitive processing speed, sequencing and flexibility, response speed and inhibition, sustained attention and concentration, visuomotor persistence, and mental energy at pretreatment baseline and again after 6 weeks of treatment. Cardiovascular health-related variables included measurements of blood pressure, pulse rate, cholesterol, triacylglycerols, C-reactive protein, and body mass that were obtained at pretreatment baseline and after 6 weeks of treatment. A midpoint assessment (after 3 weeks of treatment) was also conducted for the blood pressure, pulse rate, and mental energy variables. In an effort to ensure sustained dosing, participants were requested to consume either one chocolate bar or 8 oz of the cocoa beverage, or a similar placebo product, 2 h before their scheduled assessments. The results of the clinical trial failed to support the predicted effects of short-term (6 weeks) dark chocolate and cocoa consumption on any of the neuropsychological or cardiovascular health-related variables included in the study. Consumption of dark chocolate and cocoa was, however, associated with significantly higher pulse rates at the 3- and 6-week assessments. It was speculated that a diversity of factors may have contributed to the overall absence of positive findings, including the fact that participants appeared very healthy and cognitively intact, possessed relatively high mean levels of education (more than 15 years) and likely more test-taking experience that enabled them to optimize their neuropsychological performances, that the sample size many not have been of sufficient magnitude to detect significant group differences, that the treatment duration was too short or quantity of chocolate/cocoa consumed was too low to demonstrate the hypothesized treatment effects, and/or that control of every aspect of participants' diets was not attempted.

In a population-based, cross-sectional study of 2,031 participants (70–74 years of age), Nurk and colleagues [41] investigated the relationships between habitual intake of flavonoid-rich chocolate, wine, and tea during the previous year and performance on a diversity of cognitive tests. Participants completed a cognitive battery consisting of six tests measuring episodic memory, cognitive processing speed and sequencing abilities, perceptual speed, visuospatial skills, global cognition, and access to semantic memory, as well as a take-home food frequency questionnaire that assessed their habitual food consumption. Over the 1-year period, participants' mean intakes for chocolate, wine, and tea were 3.8 g/day, 22 mL/day, and 222 mL/day, respectively. The results from the study indicated that participants who consumed any of the three flavonoid-rich products, as compared to nonconsumers, exhibited significantly better mean cognitive test scores and a lower prevalence of poor cognitive performances, with the best test scores and lowest chance of poor test performances observed in participants who consumed all three compounds. Multivariate models that included adjustments for sex, education, smoking status, vitamin supplement usage, history of cardiovascular disease, diabetes, and total energy intake indicated that chocolate consumers, versus nonconsumers, obtained significantly better test scores on all measures, except on the test of visuospatial skills. Dose-dependent associations between the food products and cognition were also observed, with the maximum effect for chocolate noted at an intake of approximately 10 g/day. The authors noted that the true effects of chocolate might be even stronger than observed because the types of chocolate consumed by participants were not specified and because not all types of chocolate contain equal amounts of flavonoids. Nurk and associates [41] also acknowledged that while the utilization of a large population-based sample and six different cognitive tests were study strengths, weaknesses included use of a cross-sectional design and a volunteer sample (i.e., possible recruitment bias), potential error in the estimates of nutrients, and related use of food-based analysis of population data. Furthermore, it did not appear that participants' intake of cocoa, versus chocolate, was assessed or reported in the study.

In another randomized, double-blind, placebo-controlled, balanced, three-period crossover trial, Scholey and his colleagues [42] examined the acute cognitive and subjective effects of cocoa flavanols consumption during sustained mental demand. Thirty young, healthy volunteer adults (mean age = 21.9 years) completed six cycles of the 10-min Cognitive Demand Battery (consisting of the Serial Threes subtraction task, the Serial Sevens subtraction task, the Bakan Rapid Visual Information Processing task, and a self-rated mental fatigue visual analogue scale) 90 min after the ingestion of drinks containing either 520 or 994 mg of cocoa flavanols or a nutrient-matched, low cocoa flavanols (46 mg of cocoa flavanols) control beverage, with a 3-day washout period between drinks. Participants also completed the State-Trait Anxiety Inventory state scale before and after consumption of each beverage. Findings from this study indicated that, as compared to the control beverage, intake of drinks containing 520 and 994 mg of cocoa flavanols significantly enhanced Serial Threes performances (purported to measure attention, concentration, and procedural learning), while consumption of the 994-mg cocoa flavanols drink resulted in significantly faster Bakan Rapid Visual Information Processing task performances but more errors on the Serial Sevens task (purported to measure not only attention and concentration but also relying more on working memory and executive functioning). Only the 520-mg cocoa flavanol beverage significantly improved ratings of mental fatigue. No significant treatment effects were noted for any of the beverages on the number of correct Serial Sevens completed, on the Bakan Rapid Visual Information Processing task accuracy, or the State-Trait Anxiety Inventory's state anxiety scale. According to the authors [42], the most robust and largest number of effects on mood and cognition were evident for the 520-mg cocoa flavanols drink and occurred 130 min post-consumption, which appeared to correspond to peak cocoa flavanol and epicatechin levels reported in previous studies.

Recently, Field and his associates [43] utilized a randomized, single-blinded, order counterbalanced, crossover design to investigate the effects of an acute dose of cocoa flavanols on visual and cognitive functioning in 30 healthy adults (18–25 years of age). Participants were administered a series of visual (i.e., contrast sensitivity, motion coherence threshold, and motion integration time threshold) and cognitive (i.e., visual spatial working memory for location and choice reaction time tasks) tests 2 h after the consumption of either 35 g of CHOXI+ dark chocolate (containing 773 mg of cocoa flavanols, 38 mg of caffeine, and 222 mg of theobromine) or 35 g of white chocolate (containing only trace amounts of cocoa flavanols, caffeine, and theobromine). The two treatments were scheduled 1 week apart, and the experimenters remained blind as to the types of chocolates consumed by participants. Results from the visual tests indicated that, as compared to the white chocolate control condition, the high cocoa flavanol treatment significantly improved visual contrast sensitivity and reduced the time required to detect motion direction; however, no significant effects were found on the minimum proportion of coherent motion that could be detected. For the cognitive tests, the high cocoa flavanols treatment, as compared to the control, significantly enhanced visual spatial working memory and performance on some aspects of the choice reaction time task (i.e., during the predictable phase).

Animal Studies

To date, there appear to have been only two published studies that have examined the effects of cocoa-related products on laboratory animals' neurocognitive functioning.

Rozan and colleagues [44] utilized 64 male Wistar/Han IGS rats to investigate the preventative effects of short-term administration of a cocoa polyphenolic extract (Acticoa powder) on free radicals produced by leucocytes after heat exposure and the extract's protective effects on subsequent cognitive impairments. Rats were randomly divided into the following four groups depending on the treatment and exposure/no exposure to heat: vehicle rats without heat exposure (controls), vehicle rats with heat exposure, vitamin E (200 mg/kg^{-1}/day of α-tocopherol, which was used as an antioxidant reference) with heat exposure, and Acticoa powder (22.9 mg/kg^{-1}/day, 36.9% total polyphenols) with heat exposure. The Acticoa powder, vitamin E, and vehicle were administered to the rats orally for 14 days prior to 40°C heat, with 50% relative humidity exposure for 2 h on day 15. Blood samples were subsequently obtained from the rats 1 day post-heat exposure to measure free radical production by leucocytes, followed by a light extinction test (performed on day 18, 3 days after heat exposure) to assess memory and learning processes and the Morris water maze test (performed on days 20 and 21, 5 and 6 days after heat exposure) to assess spatial learning and related long-term memory. The experimenters remained blinded to the treatment variables during all cognitive testing. The results revealed that on the day after heat exposure, free radical production by leukocytes was significantly reduced in the Acticoa powder- and vitamin E-treated rats, as compared to controls. Furthermore, the control rats, and those treated with Acticoa powder or vitamin E, unlike the vehicle rats exposed to heat, demonstrated significant lever discrimination on the light extinction test by pressing more often on the active, versus inactive, lever. On the Morris water maze test, escape latencies decreased significantly over trials in the control and Acticoa powder- and vitamin E-treated rats, whereas the group of vehicle rats exposed to heat did not. The authors [44] noted that the daily oral administration of Acticoa powder or vitamin E protected heat-exposed rats from cognitive impairments (i.e., preserved short- and long-term memory processes), potentially by counteracting the overproduction of free radicals and that the cognitive outcomes may be related to preservation of brain function as a result of reduced heat exposure, inflammatory aggression, or enhanced brain plasticity.

Similarly, Bisson and associates [45] examined the effects of long-term administration of the cocoa polyphenolic extract, Acticoa, on cognitive functions, urinary dopamine levels, and lifespan in 18 aged male Wistar-Unilever rats. The laboratory animals were randomly assigned to orally receive either Acticoa powder (24 mg/kg/day, 34.9 g of total polyphenols/100 g) or a control vehicle (spring water) for 12 months from the age of 15–27 months and tested via light extinction and water maze paradigms both prior to receiving the treatments (i.e., at 9, 13, and 15 months of age) and again during treatment (i.e., at 17, 21, and 25 months of age). Urinary dopamine levels and lifespans were also evaluated during the study, and all of the cognitive tests were recorded by experimenters who remained blind to the treatments administered. On the light extinction test, rats in the Acticoa treatment group displayed significantly more total lever pressing activity at 17, 21, and 25 months of age and also significant discrimination between active and inactive levers at 17 and 21 months of age, as compared to controls. Results for the water maze paradigm demonstrated that while performance declined in the control group at 21 and 25 months of age, for the Acticoa powder treatment group, water maze performance remained stable. Escape latencies were also lower for the rats treated with Acticoa powder, versus controls, at 21 and 25 months of age. Furthermore, the Acticoa powder was noted to preserve high levels of urinary free dopamine, a neurotransmitter associated with such functions as smooth controlled movements, attention, efficient memory, and problem-solving, and to prolong the lifespans of rats treated with Acticoa powder, versus controls, by about 11% over the 27-month test period. According to the study's authors [45], these findings were indicative of the beneficial effects of the cocoa polyphenolic extract on spatial memory and short- and long-term learning and that Acticoa powder may be useful in decreasing age-related brain impairments and cognitive deficits.

Proposed Mechanisms of Action

While a detailed discussion of the proposed neuroprotective, neuroenhancing, and neurostimulating mechanisms that may be associated with cocoa- and chocolate-related products is beyond the scope of this chapter, a brief overview of these mechanisms follows.

Over the years, an array of potentially pharmacologically active substances have been identified in cocoa and chocolate (see Smit [46] for an overview); however, most reportedly do not by themselves result in significant effects in humans, secondary to such factors as their extremely low concentrations in cocoa-related products and their inabilities to reach and/or cross the blood–brain barrier [46]. In contrast, the methylxanthines in cocoa and chocolate, and in particular, caffeine and theobromine, have been suggested to be responsible for at least a portion of the products' (psycho)pharmacological activity, although potential interactive and synergistic effects among the many constituents of cocoa and chocolate also remain possible [39, 46]. For example, in the two previously reviewed studies by Smit and associates [39], cocoa powder and a methylxanthine mixture (i.e., caffeine and theobromine) were found to exhibit very similar outcome profiles, with strong treatment effects found for an energetic arousal mood construct and on a simple reaction time task. In light of these results, the authors [39] concluded that the cocoa content in dark chocolate has notable psychopharmacological activity (e.g., psychostimulant effects) that can be attributed to the presence of caffeine and/or theobromine.

In addition to methylxanthines, cocoa and cocoa-related products are rich in polyphenolic compounds and, in particular, flavonoids [47–49]. In recent years, an ever-increasing body of evidence has indicated that plant-derived, dietary flavonoids, such as those commonly found in notable quantities in cocoa and chocolate, may possess neuroprotective, neuroenhancing, and neurostimulating effects, including an ability to protect neurons from injury resulting from oxidative stress and neurotoxins, the potential to suppress or inhibit neuroinflammatory processes, the promotion and enhancement of existing neuronal and neurocognitive functioning, and the ability to stimulate cerebral blood flow and induce neuronal regeneration and neurogenesis [49–57].

While the bioactivity of flavonoids has historically been attributed to their antioxidant properties, via either their abilities to scavenge reactive species or through their impact on the intracellular redox status, their ability to act as antioxidants in vivo, especially in the brain where their concentrations are very low, is purportedly limited [49–55].

Although the precise mechanisms/processes via which flavonoids may act within the brain remain unresolved at present, a number of other mechanisms have been proposed by Spencer and his colleagues [49–55], including (1) their apparent abilities to modulate and interact with a number of critical neuronal protein and lipid kinase signaling pathways in the brain, resulting in an inhibition of apoptosis stimulated by neurotoxic species and the promotion of neuronal survival and synaptic plasticity, and (2) their noted capabilities to produce beneficial effects in the cerebral and peripheral vascular systems and enhance blood flow, which may result in angiogenesis, neurogenesis (e.g., in the hippocampus), and changes in neuronal morphology.

Lastly, it has been hypothesized that the antioxidant and health-related benefits of the cocoa-related polyphenols and flavonoids may be either enhanced or reduced, either directly or indirectly, by such factors as bioavailability, antioxidant status, and the status (e.g., health) of the participants being studied, as well as by the presence of methylxanthines, peptides, and minerals [48].

Summary

Taken together, there appear to have been nine published studies [33, 39–45] (through May 2011) in the scientific literature that have examined the effects of cocoa- and chocolate-related products on neurocognitive functioning. Seven of these studies (one publication [39] included two studies) involved human participants [33, 39–43], while two other investigations employed laboratory animals [44, 45].

Regarding the studies involving human participants, the sample sizes utilized in five of the investigations [33, 39, 42, 43] were relatively small (i.e., $n \leq 30$), while a medium-sized sample (i.e., $n = 101$) was examined in one clinical trial [40], and a large sample was utilized in a population-based study [41]. With the exception of two studies that exclusively employed older (i.e., 60 years of age or older) adults [40, 41], the remaining projects generally examined younger participants whose ages or mean ages fell between 18 and 35.4 years [33, 39, 42, 43]. Four of the investigations examined "healthy" individuals [33, 40, 42, 43], while the health statuses of the participants in the remaining three studies were not specified [39, 41].

Five of the human investigations employed double-blind procedures [33, 39, 40, 42], with within-subjects/crossover methodology being utilized in four of these studies [33, 39, 42, 43] and a parallel-group design used in the other [40]. Among the two remaining human studies, one used a single-blinded, crossover design [43], while the population-based study utilized cross-sectional methods [41]. Within these studies, an array of high and low flavonoid/flavanol-rich cocoa- and chocolate-related products and dosages were employed [33, 39, 40, 42, 43], as well as methylxanthine mixtures in two studies [39]. The specific types of cocoa- and chocolate-related products consumed in the population-based study [41] were not precisely defined. The duration of the studies' treatments/assessments ranged from approximately 3 h or less in four investigations [39, 42, 43], to 5 days [33] and 6 weeks [40] in two additional investigations, to a cross-sectional, retrospective, self-report of chocolate consumption over a 1-year period in the population-based study [41].

A wide array of outcome measures, assessing a diversity of neurocognitive and neuropsychological processes were administered in the human studies. These ranged from simple reaction time tasks, to objective, standardized tests (e.g., portions of the Wechsler Memory Scale-III [40, 58], Trail Making Test [40, 59]) frequently utilized in clinical practice, to subjective, self-report questionnaires (e.g., a mood questionnaire [39]). Overall, five out of seven of the studies [39, 41–43] that examined the effects of cocoa- and chocolate-related products in humans found that the products were associated with significant enhancement/improvement of certain aspects of individuals' neurocognitive functioning, and, in particular, on tasks assessing simple reaction time [39], rapid visual information processing [39, 42], energetic arousal [39], episodic memory [41], cognitive processing speed and sequential abilities [41], perceptual speed [41], global cognition [41], access to semantic memory [41], Serial Threes subtractions [42], visual contrast sensitivity [43], detection of motion direction [43], visual spatial working memory [43], choice reaction time [43], and mental fatigue [42].

Regarding the two investigations involving laboratory animals, both studies utilized strains of male Wistar rats, with total sample sizes of 64 [44] and 18 [45] animals. Aged rats were examined in one study [45], while the ages of the animals in the other study [44] were not specified. Both of the studies' methodologies [44, 45] utilized random assignment of the rats to parallel treatment and control groups. The experimenters who conducted the neurocognitive testing in both investigations [44, 45] also remained blind to the treatments that had been administered to the rats. Furthermore, both of the animal studies utilized similar doses of a cocoa polyphenolic extract (i.e., Acticoa powder), whereas the treatment duration was only 14 days in one investigation [44] but 12 months in the other [45]. Similar outcome measures were employed across the two studies, and both investigations found that the Acticoa powder was associated with significantly improved or preserved neurocognitive functioning in the rats, and, in particular, on light extinction and water maze tasks assessing short- and long-term learning and spatial memory [44, 45].

Taken together, the majority (i.e., seven out of nine) of the studies [39, 41–45] published to date (through May 2011) that have examined the effects of cocoa- and chocolate-related products on the neurocognitive processes of humans and laboratory animals have found that the compounds are associated with significantly improved and/or preserved aspects of cognitive functioning. While the precise mechanisms via which cocoa- and chocolate-related products may act within the brain remain unresolved at present, in recent years, an ever-increasing body of evidence has indicated that plant-derived, dietary flavonoids, such as those commonly found in notable quantities in cocoa and chocolate, may possess neuroprotective, neuroenhancing, and neurostimulating effects [49–57]. Furthermore, the methylxanthines in cocoa and chocolate, and in particular, caffeine and theobromine, have also been suggested to be responsible for at least a portion of the products' (psycho)pharmacological activity [39, 46]. It has also been hypothesized that the antioxidant and health-related benefits of the cocoa-related polyphenols and flavonoids may be either enhanced or reduced by such factors as bioavailability, antioxidant status, and the status (e.g., health) of the participants being studied, as well as by the presence of methylxanthines, peptides, and minerals [48].

Directions for Future Research

While the majority of studies cited here provide evidence of the efficacy of cocoa- and chocolate-related products to enhance and/or preserve the neurocognitive functioning of humans and laboratory animals, based on the diversity of samples, methodologies, cocoa- and chocolate-related products, doses, treatments, and outcome variables utilized in these studies, as well as the null results found in two of the investigations, future research is required to more precisely identify and define the true effects of cocoa and chocolate on neurocognitive functioning.

Well-designed, larger-scale, prospective studies are needed that not only replicate the findings of the previous investigations but that also extend the scientific literature and better elucidate the efficacy of cocoa- and chocolate-related products to have neuroprotective, neuroenhancing, and neurostimulating effects. In particular, future research is required with diverse populations (e.g., of varying ages, genders, health and cognitive statuses) that examine different types and formulas of cocoa- and chocolate-related products (with various polyphenol/flavonoid and methylxanthine contents), dosage regimens, and treatment/assessment durations (e.g., acute, short-term, and long-term timeframes). These investigations should also precisely identify their samples' and treatment products' characteristics, as well as their specific inclusionary/exclusionary criteria. Furthermore, outcome measures should be carefully selected to ensure that they have been documented to be reliable, valid, and sensitive measures of particular neurocognitive processes and that they decrease the possibility of practice effects over successive administrations (e.g., via use of alternate forms) or ceiling effects (e.g., via use of measures requiring sustained effort leading to a progressive performance decline).

Finally, future research is needed to better identify and clarify the precise mechanisms and modes of action that may underlie the proposed neuroprotective, neuroenhancing, and neurostimulating effects of cocoa- and chocolate-related products, as well as the potential synergistic interactions between the polyphenols/flavonoids and methylxanthines in the compounds.

References

1. Adamson GE, Lazarus SA, Mitchell AE, Prior RL, Cao G, Jacobs PH, et al. HPLC method for the quantification of procyanidins in cocoa and chocolate samples and correlation to total antioxidant capacity. J Agric Food Chem. 1999;47(10):4184–8.
2. Hammerstone JF, Mitchell AE, Lazarus SA, Schmitz H, Keen CL, Rucker RB. Identification of procyanidins in cocoa (Theobroma cacao) using high performance liquid chromatography (HPLC)/mass spectrometry (MS). J Agric Food Chem. 1999;47:490–6.
3. Steinberg FM, Bearden MM, Keen CL. Cocoa and chocolate flavonoids: implications for cardiovascular health. J Am Diet Assoc. 2003;103(2):215–23.
4. Kris-Etherton PM, Keen CL. Evidence that the antioxidant flavonoids in tea and cocoa are beneficial for cardiovascular health. Curr Opin Lipidol. 2002;13(1):41–9.
5. Ding EL, Hutfless SM, Ding X, Girotra S. Chocolate and prevention of cardiovascular disease: a systematic review. Nutr Metab. 2006;3(2):1–12.
6. Kondo K, Hirano R, Matsumoto A, Igarashi O, Itakura H. Inhibition of LDL oxidation by cocoa. Lancet. 1996;348(9040):1514.
7. Lotito SB, Actis-Goretta L, Renart ML, et al. Influence of oligomer chain length on the antioxidant activity of procyanidins. Biochem Biophys Res Commun. 2000;276:945–51.
8. Mathur S, Devaraj S, Grundy SM, Jialal I. Cocoa products decrease low density lipoprotein oxidative susceptibility but do not affect biomarkers of inflammation in humans. J Nutr. 2002;132(12):3663–7.
9. Osakabe N, Yasuda A, Natsume M, Takizawa T, Terao J, Kondo K. Catechins and their oligomers linked by C4 C8 bonds are major cacao polyphenols and protect low-density lipoprotein from oxidation in vitro. Exp Biol Med. 2002;227(1):51–6.
10. Tokede OA, Gaziano JM, Djousse L. Effects of cocoa products/dark chocolate on serum lipids: a meta-analysis. Eur J Clin Nutr 2011; E-pub. Available at: http://www.nature.com/ejcn/journal/vaop/ncurrent/full/ejcn201164a.html. Accessed 24 May 2011.
11. Wan Y, Vinson JA, Etherton TD, Proch J, Lazarus SA, Kris-Etherton PM. Effects of cocoa powder and dark chocolate on LDL oxidative susceptibility and prostaglandin concentrations in humans. Am J Clin Nutr. 2001;74: 596–602.
12. Waterhouse AL, Shirley JR, Donovan JL. Antioxidants in chocolate. Lancet. 1996;348(9030):834.
13. Holt RR, Schramm DD, Keen CL, Lazarus SA, Schmitz HH. Chocolate consumption and platelet function. JAMA. 2002;287(17):2212–3.
14. Murphy KJ, Chronopoulos AK, Singh I, Francis MA, Moriarty H, Pike MJ, et al. Dietary flavanols and procyanidin oligomers from cocoa (Theobroma cacao) inhibit platelet function. Am J Clin Nutr. 2003;77:1466–73.
15. Pearson DA, Paglieroni TG, Rein D, Wun T, Schramm DD, Wang JF, et al. The effects of flavanol-rich cocoa and aspirin on ex vivo platelet function. Thromb Res. 2002;106:191–7.
16. Rein D, Paglieroni TG, Pearson DA, Wun T, Schmitz HH, Gosselin R, et al. Cocoa and wine polyphenols modulate platelet activation and function. J Nutr. 2000;130:2120S–6.
17. Rein D, Paglieroni TG, Wun T, Pearson DA, Schmitz HH, Gosselin R, et al. Cocoa inhibits platelet activation and function. Am J Clin Nutr. 2000;72:30–5.
18. Grassi D, Necozione S, Lippi C, Croce G, Valeri L, Pasqualetti P, et al. Cocoa reduces blood pressure and insulin resistance and improves endothelium-dependent vasodilation in hypertensives. Hypertension. 2005;46:398–405.
19. Mursu J, Voutilainen S, Nurmi T, Rissanen TH, Virtanen JK, Kaikkonen J, et al. Dark chocolate consumption increases HDL cholesterol concentration and chocolate fatty acids may inhibit lipid perioxidation in healthy humans. Free Radic Biol Med. 2004;37(9):1351–9.
20. Schramm DD, Wang JF, Holt RR, Ensunsa JL, Gonsalves JL, Lazarus SA, et al. Chocolate procyanidins decrease the leukotriene-prostacyclin ratio in humans and human aortic endothelial cells. Am J Clin Nutr. 2001;73:36–40.

21. Desch S, Schmidt J, Kobler D, Sonnabend M, Eitel I, Sareban M, et al. Effect of cocoa products on blood pressure: systematic review and meta-analysis. Am J Hypertens. 2010;23(1):97–103.
22. Grassi D, Lippi C, Necozione S, Desideri G, Ferri C. Short-term administration of dark chocolate is followed by a significant increase in insulin sensitivity and a decrease in blood pressure in healthy persons. Am J Clin Nutr. 2005;81:611–4.
23. Ried K, Sullivan T, Fakler P, Frank OR, Stocks NP. Does chocolate reduce blood pressure? A meta-analysis. BMC Med 2010; 8(39): Available at: http://www.biomedcentral.com/1741-7015/8/39. Accessed 24 May 2011.
24. Taubert D, Berkels R, Roesen R, Klaus W. Chocolate and blood pressure in elderly individuals with isolated systolic hypertension. JAMA. 2003;290(8):1029–30.
25. Taubert D, Roesen R, Lehmann C, Jung N, Schomig E. Effects of low habitual cocoa intake on blood pressure and bioactive nitric oxide: a randomized controlled trial. JAMA. 2007;298(1):49–60.
26. Taubert D, Roesen R, Schomig E. Effect of cocoa and tea on blood pressure: a meta-analysis. Arch Intern Med. 2007;167(7):626–34.
27. Faridi Z, Njike VY, Dutta S, Ali A, Katz DL. Acute dark chocolate and cocoa ingestion and endothelial function: a randomized controlled crossover trial. Am J Clin Nutr. 2008;88(1):58–63.
28. Fisher NDL, Hughes M, Gerhard-Herman M, Hollenberg NK. Flavanol-rich cocoa induces nitric-oxide-dependent vasodilation in healthy humans. J Hypertens. 2003;21:2281–6.
29. Heiss C, Dejam A, Kleinbongard P, Schewe T, Sies H, Kelm M. Vascular effects of cocoa rich in flavan-3-ols. JAMA. 2003;290(8):1030–1.
30. Karim M, McCormick K, Kappagoda CT. Effects of cocoa extracts on endothelium-dependent relaxation. J Nutr. 2000;130:2105S–8.
31. Dinges DF. Cocoa flavanols, cerebral blood flow, cognition, and health: going forward. J Cardiovasc Pharmacol. 2006;47(Suppl. 2):S221–3.
32. Fisher NDL, Sorond FA, Hollenberg NK. Cocoa flavanols and brain perfusion. J Cardiovasc Pharmacol. 2006;47(Suppl 2):S210–4.
33. Francis ST, Head K, Morris PG, MacDonald IA. The effect of flavanol-rich cocoa on the fMRI response to a cognitive task in healthy young people. J Cardiovasc Pharmacol. 2006;47(Suppl 2):S215–20.
34. Sorond FA, Lipsitz LA, Hollenberg NK, Fisher NDL. Cerebral blood flow response to flavanol-rich cocoa in healthy elderly humans. Neuropsychiatr Dis Treat. 2008;4(2):433–40.
35. Zhu QY, Holt RR, Lazarus SA, Orozco TJ, Keen CL. Inhibitory effects of cocoa flavanols and procyanidin oligomers on free radical-induced erythrocyte hemolysis. Exp Biol Med. 2002;227:321–9.
36. Mao TK, Powell J, Van de Water J, Keen CL, Schmitz HH, Hammerstone JF, et al. The effect of cocoa procyanidins on the transcription and secretion of interleukin 1β in peripheral blood mononuclear cells. Life Sci. 2000;66(15):1377–86.
37. Mao T, Van de Water J, Keen CL, Schmitz HH, Gershwin ME. Cocoa procyanidins and human cytokine transcription and secretion. J Nutr. 2000;130:2093S–9.
38. Sanbongi C, Suzuki N, Sakane T. Polyphenols in chocolate, which have antioxidant activity, modulate immune functions in humans in vitro. Cell Immunol. 1997;177:129–36.
39. Smit HJ, Gaffan EA, Rogers PJ. Methylxanthines are the psycho-pharmacologically active constituents of chocolate. Psychopharmacology. 2004;176:412–9.
40. Crews WD Jr, Harrison DW, Wright JW. A double-blind, placebo-controlled, randomized trial of the effects of dark chocolate and cocoa on variables associated with neuropsychological functioning and cardiovascular health: clinical findings from a sample of healthy, cognitively intact older adults. Am J Clin Nutr. 2008;87:872–80.
41. Nurk E, Refsum H, Drevon CA, Tell GS, Nygaard HA, Engedal K, et al. Intake of flavonoid-rich wine, tea, and chocolate by elderly men and women is associated with better cognitive test performance. J Nutr. 2009;139: 120–7.
42. Scholey AB, French SJ, Morris PJ, Kennedy DO, Milne AL, Haskell CF. Consumption of cocoa flavanols results in acute improvements in mood and cognitive performance during sustained mental effort. J Psychopharmacol. 2010;24(10):1505–14.
43. Field DT, Williams CM, Butler LT. Consumption of cocoa flavanols results in an acute improvement in visual and cognitive functions. Physiol Behav. 2011;103:255–60.
44. Rozan P, Hidalgo S, Nejdi A, Bisson JF, Lalonde R, Messaoudi M. Preventive antioxidant effects of cocoa polyphenolic extract on free radical production and cognitive performances after heat exposure in Wistar rats. J Food Sci. 2007;72(3):S203–6.
45. Bisson JF, Nejdi A, Rozan P, Hidalgo S, Lalonde R, Messaoudi M. Effects of long-term administration of a cocoa polyphenolic extract (Acticoa powder) on cognitive performances in aged rats. Br J Nutr. 2008;100:94–101.
46. Smit HJ. Theobromine and the pharmacology of cocoa. In: Fredholm BB, editor. Methylxanthines, handbook of experimental pharmacology 200. Berlin: Springer; 2011. p. 201–34.
47. Keen CL. Chocolate: food as medicine/medicine as food. J Am Coll Nutr. 2001;20(5):436S–9.

48. Jalil AMM, Ismail A. Polyphenols in cocoa and cocoa products: is there a link between antioxidant properties and health? Molecules. 2008;13:2190–219.
49. Spencer JPE. Food for thought: the role of dietary flavonoids in enhancing human memory, learning and neuro-cognitive performance. Proc Nutr Soc. 2008;67:238–52.
50. Spencer JPE. Flavonoids: modulators of brain function? Br J Nutr. 2008;99:ES60–77.
51. Vauzour D, Vafeiadou K, Rodriguez-Mateos A, Rendeiro C, Spencer JPE. The neuroprotective potential of flavonoids: a multiplicity of effects. Genes Nutr. 2008;3:115–26.
52. Spencer JPE. Flavonoids and brain health: multiple effects underpinned by common mechanisms. Genes Nutr. 2009;4:243–50.
53. Spencer JPE, Vauzour D, Rendeiro C. Flavonoids and cognition: the molecular mechanisms underlying their behavioural effects. Arch Biochem Biophys. 2009;492:1–9.
54. Spencer JPE. The impact of fruit flavonoids on memory and cognition. Br J Nutr. 2010;104:S40–7.
55. Spencer JPE. The impact of flavonoids on memory: physiological and molecular considerations. Chem Soc Rev. 2009;38:1152–61.
56. Patel AK, Rogers JT, Huang X. Flavanols, mild cognitive impairment, and Alzheimer's dementia. Int J Clin Exp Med. 2008;1:181–91.
57. Shah ZA, Li R, Ahmad AS, Kensler TW, Yamamoto M, Biswal S, et al. The flavanol (–)-epicatechin prevents stroke damage through the Nrf2/HO1 pathway. J Cereb Blood Flow Metab. 2010;30:1951–61.
58. Wechsler D. WMS-III administration and scoring manual. San Antonio: The Psychological Corporation; 1997.
59. Reitan RM. Trail making test manual for administration and scoring. Tucson: Reitan Neuropsychology Laboratory; 1992.

Chapter 28
Flavonoid-Rich Chocolate and Cognitive Improvement

Eha Nurk

Key Points

- There is some evidence that flavanols, but also methylxanthine compounds, in cocoa may have a positive effect on cognitive functioning.
- The results from clinical trials indicate that cocoa flavanols are beneficial in improving the vascular function and increasing cerebral blood flow and may therefore have potential for treatment of vascular impairment, including dementias and strokes, but also potential for maintaining an adequate cognitive functioning.
- Long-term administration of a cocoa polyphenolic extract may be beneficial in retarding age-related brain impairments, including cognitive deficits in normal aging, as demonstrated in animal studies.
- The efficacy of flavanol-rich cocoa on cognitive performance has been shown in a couple of human trials, but there are also a few experiments that yielded no improvement, presumably due to the ceiling effect in cognitive testing.
- Based on population-based cross-sectional study chocolate intake up to 10 g/day is associated with better performance in several cognitive abilities in a dose-dependent manner.
- Studies on cell cultures have shown that neuroprotective actions of cocoa flavonoids involve a number of effects within the brain, including a potential to protect neurons against injury induced by neurotoxins and an ability to suppress neuroinflammation, and that cocoa may therefore have a potential to promote memory, learning, and cognitive function.

Keywords Alzheimer's disease • Cerebral blood flow • Chocolate • Cocoa • Cognitive functioning • Dementia • Flavanols

In 1662 Henry Stubbe (1632–1676), an English physician, writer, and scholar, wrote the first book on chocolate, "The Indian Nectar, or a discourse concerning Chocolata" [1]. In that book, he writes: "The Nature of the Cacao-nut, and the other Ingredients of that Composition, is examined, and stated according to the Judgment and Experience of the Indians, and Spanish Writers, who lived in the

E. Nurk, M.D., M.Phil., Ph.D. (✉)
Department of Surveillance and Evaluation, National Institute for Health Development, Tallinn, Estonia

Department of Nutrition, Institute of Basic Medical Sciences, University of Oslo, Oslo, Norway
e-mail: eha.nurk@medisin.uio.no; eha.nurk@tai.ee

R.R. Watson et al. (eds.), *Chocolate in Health and Nutrition*, Nutrition and Health 7,
DOI 10.1007/978-1-61779-803-0_28, © Springer Science+Business Media, LLC 2013

Indies, and others; with fundry additional Observations made in England: The ways of compounding and preparing Chocolata are enquired into; its Effects, as to its alimental and Venereal quality, as well as Medicinal (especially in Hypochondriacal Melancholy) are fully debated" [1]. However, instructions for the medicinal use of cacao have been found already in documents from early Colonial Era in the Badianus Codex (1552) and the Florentine Codex (1590) [2], but chocolate as the brain-strengthening remedy was first mentioned by Stubbe [1, 2].

In 1996, 444 years after the first factual documentation of cocoa as an herbal remedy, Waterhouse and colleagues [3] described an in vitro experiment stating that the action and content of polyphenols from cocoa could be considered as a dietary source of antioxidants. In their study [3], polyphenols were extracted from commercial cocoa and chocolate, and the polyphenol content and antioxidant activity against low-density lipoprotein oxidation was measured. They found a potent inhibition by the cocoa polyphenols: at 5 μmol/L total polyphenols (expressed as gallic acid equivalents), cocoa phenols inhibited 75% low-density lipoprotein oxidation while pure catechin inhibited oxidation 87%. Those findings were again to open up a whole new area of chocolate/cocoa and health. Since then, there has been increased interest in the potential health-related benefits of antioxidant- and phytochemical-rich dark chocolate and cocoa. Among other benefits, there is hope that cocoa flavanols may increase blood flow to the brain. As when the flow of blood to the brain slows over time, the result may be structural damage and dementia. Therefore, long-term improvements in brain blood flow could impact cognitive behavior, offering future potential for debilitating brain conditions, including cognitive impairment and dementia.

Although chocolate is akin to cocoa, they are two different terms and are not interchangeable. In this chapter, the terms are used alternately, according to each particular study under consideration. During the recent years, cocoa products and moderate chocolate consumption have been promoted to be part of healthy diet due to the health benefits ascribed to selected cocoa components. Specifically, cocoa as a plant and chocolate as food contain a series of chemicals that can interact with cell and tissue components, providing protection against the development and amelioration of different pathological conditions [4]. In the cocoa bean, there are found more than 200 compounds that are thought to be beneficial for the human body [5]. The major interest has been focused on polyphenols that are so abundant in cocoa. Single servings of cocoa and cocoa products contain more phenolic antioxidants than most foods [6, 7]. As cocoa is predominantly consumed as chocolate, the high polyphenol content of cocoa renders chocolate particular interest of the nutritional viewpoints, and the evidence that chocolate has nutritional benefits is mounting. Cocoa, in its unprocessed state, is particularly rich in flavanols, which are one subclass of polyphenols that are present in plants as nonconjugated molecules, including (−)-epicatechin and (+)-catechin, and as flavanol-based oligomers of these molecules, known as procyanidins [8, 9].

The majority of relevant health effects of cocoa and chocolate have been related to cardiovascular disease. A growing number of clinical studies indicate that regular ingestion of flavanol-rich cocoa exerts a range of effects potentially favorable to vascular health, for example, reducing elevated blood pressure, increasing insulin sensitivity, decreasing inflammation, improving endothelium dependent relaxation, contributing to vasodilatation to improve blood flow, suppressing platelet aggregation, exerting antioxidant protections against free radicals, inhibiting low-density lipoprotein oxidation, modulating immune function of the proinflammatory cytokines, and positively influencing production of eicosanoids to foster their role in cardiovascular health [7, 9–13]. Vascular health, in turn, is closely related to neurocognitive disorders. There are at least three important aspects related to vascular diseases and dementia or Alzheimer's disease: shared risk factors, clinical manifestations, and overlapping neuropathologic findings [13]. Specifically, increasing age, apolipoprotein E ε4 genotype, hypertension, arteriosclerosis, diabetes mellitus, smoking, and atrial fibrillation are common risk factors for both dementia and vascular diseases [14]. Cerebral infarction may often be related to clinical manifestations of pathologically diagnosed Alzheimer's disease, as during the first year after cerebral infarction, the incidence of dementia is nine times greater than expected [13, 15], and after the first year, there is a 50% increase in the likelihood of developing Alzheimer's disease [13, 15, 16]. Moreover, stroke shortens the duration of preclinical Alzheimer's disease [13, 16]. In relation to neuropathologic findings, there are neurofibrillary plaques and tangles in vascular dementia and vascular pathology in Alzheimer's disease [13, 17]. In addition to the commonly seen vascular lesions such as cerebral amyloid angiopathy, microvascular degeneration, and periventricular white matter lesions, there are numerous structural and functional cerebromicrovascular abnormalities in Alzheimer's disease, including impairment of blood-brain barrier, decreased microvascular density, vascular distortions, functional alteration and derangement of cerebral endothelial function, and arteriolar changes such as lipohyalinosis [13]. Thus, tight relationship between vascular diseases and dementia or Alzheimer's disease makes the research on cocoa flavanols even more promising, suggesting that cocoa flavanols may provide a dietary approach to maintaining cardiovascular function and health and pointing towards new possibilities for cocoa flavanol-based interventions for vascular complications associated with cognitive functioning. So far, the number of studies related to effects of cocoa and chocolate on cognition has been limited.

The combinations of antioxidants or anti-inflammatory polyphenolic compounds found in fruits and vegetables have shown efficacy in aging – preventing the onset of age-related deficits, enhancing cognitive abilities, and even reversing cognitive deficits [18, 19]. The antioxidant properties of polyphenols derived from cocoa products and their beneficial effects on cognition have been demonstrated in a couple of animal studies. Rozan et al. [20] evaluated the antioxidant activity of Acticoa powder, a cocoa polyphenolic extract obtained using the ACTICOA (Barry Callebaut, Zurich, Switzerland) process from cocoa beans, in plasma. They found that daily oral administration for 14 consecutive days of ACTICOA powder protected Wistar/Han IGS rats from cognitive impairments after heat exposure by counteracting the overproduction of free radicals. In another study, administration of a cocoa polyphenolic extract (ACTICOA powder) to old Wistar-Unilever rats during 1 year was beneficial in retarding age-related brain impairments, including cognitive deficits in normal aging and perhaps neurodegenerative diseases, as ACTICOA powder improved cognitive performances in light extinction and water maze paradigms, increased lifespan, and preserved high urinary free dopamine levels [11]. In the latter study, both long- and short-term memory processes were improved by the cocoa polyphenolic extract, as evaluated by month-to-month or trial-to-trial performances demonstrating beneficial effects of ACTICOA powder on spatial learning [11].

In human studies, the favorable effects of cocoa and chocolate on cognition have yielded mixed results, showing consistently that cocoa flavanols are beneficial in improving vascular function and increasing cerebral blood flow and therefore may have potential for treatment of vascular impairment, including dementias and strokes. The effect on cognitive behavior is less evident.

In a pilot study among 27 healthy elderly subjects, the flavanol-rich cocoa intake (920 mL of cocoa daily in four equal doses including 821 mg total flavanols) over a 5-day period induced vasodilatation [13]. Because ingestion of a control drink, flavanol-poor cocoa, led to vasodilator responses that were significantly smaller, these experiments implicate flavanols as the primary polyphenols responsible for cocoa-induced vasodilatation. Fisher et al. [13] also demonstrated that in healthy elderly subjects, ingestion of 900 mg of flavanols daily for 1 week increased cerebral blood flow as measured by transcranial Doppler ultrasound. In another study, Sorond et al. [21] used transcranial Doppler ultrasound to measure mean blood velocity in the middle cerebral artery in 13 healthy elderly volunteers (59–83 years old) in response to the regular intake of flavanol-rich cocoa. They found that 2 weeks of regular flavanol-rich cocoa intake, providing 900 mg of cocoa flavanols daily, resulted in a significant increase in peak cerebral blood flow response in the middle cerebral artery. One more human study confirmed the beneficial role of cocoa flavanols on improving vascular function and increasing cerebral blood flow. Among 16 young female subjects (18–30 years old), the relationship between cerebral blood flow and a single acute dose (450 mg flavanols) of flavanol-rich cocoa was evaluated by functional magnetic resonance imaging [22]. The study demonstrated that flavanol-rich cocoa can increase the cerebral blood flow to gray matter. Although blood oxygenation level-dependent response to a cognitive task switching paradigm was significantly increased in the presence of flavanol-rich cocoa in that pilot study, there was no effect on behavioral responses such as reaction time, switch cost, or error rate for the letter-digit pair switching task [22]. The reason why no improvement in behavioral response was seen may be that the study participants, the healthy young subjects, were already performing at a high level of cognitive ability, which would be very difficult to improve upon [22].

Similarly, the findings from a double-blind, placebo-controlled, randomized, clinical trial among individuals of 60 years and older failed to support the predicted beneficial effects of short-term (6 weeks) consumption of dark chocolate (37.0 g; containing 60% cacao, ~11 g natural cocoa, and 397 mg total proanthocyanins/g) and cocoa (237 mL cup of the cocoa beverage; dry weight 12 g, of which ~11 g was natural cocoa containing 357 mg total proanthocyanins/g) on any of the neuropsychological, hematologic, or physiologic variables included in the investigation [23]. The authors hypothesize that the relatively high mean levels of education (above 15 years) of subjects in the study may have negatively affected the results, as higher levels of education compared to lower levels have been associated with greater cognitive reserve and enhanced neuropsychological test performances. Furthermore, persons with higher educational levels typically have significantly more test-taking experiences than persons with less education do [23].

In contrast, in the randomized, controlled, double-blinded, balanced three-period crossover trial, Scholey et al. [24] demonstrated that the acute consumption of cocoa flavanols can improve performance and reduce "mental fatigue" during highly effortful cognitive processing in healthy young participants. The cognitive assessment included a Cognitive Demand Battery comprising two serial subtraction tasks (Serial Threes and Serial Sevens), the Bakan Rapid Visual Information Processing task, and a paper-and-pencil mental fatigue scale. Compared to control drink, the cognitive improvements were evident following consumption of both 520 mg cocoa flavanols and 994 mg cocoa flavanols, whereas the 520 mg cocoa flavanols drink appeared to be more beneficial. Similarly, in connection with subjective outcomes, mental fatigue ratings were also lowest following consumption of 520 mg cocoa flavanols, but there were no effects after consumption of 994 mg cocoa flavanols. Also, there were no significant treatment effects on Rapid Visual Information Processing task.

Consistently to report by Scholey et al., improvements in cognitive performance due to cocoa flavanols were demonstrated in a trial by Field et al. [25] employing a randomized, single-blinded, order counterbalanced, crossover design in which 30 healthy young adults (18–25 years old) consumed both dark chocolate containing 720 mg cocoa flavanols and a matched quantity of white chocolate, with a 1 week interval between testing sessions. In the latter study, cocoa flavanols improved spatial memory and performance on some aspects of the choice reaction time task. In addition, they reported acute effects of cocoa flavanols on the efficiency of visual function. The acute effects in these two trials [24, 25] can be explained by increased cerebral blood flow caused by cocoa flavanols, and because the improvements were seen in several different tests, these results are indicative of a general mechanism underlying the improvements observed, such as increased cerebral blood flow may boost a motivation or attentiveness on the tasks as suggested by Field et al. [25].

Moreover, the results of these recent trials [24, 25] support earlier positive associations from the population-based Hordaland Health Study of 2,031 elderly (70–74 years old) individuals [26]. However, the study design was cross-sectional, and therefore interpretation as long-term associations should be considered with caution. In the Hordaland, study participants were asked to report their habitual chocolate intake during the past year and completed also a cognitive test battery including the Kendrick Object Learning Test (episodic memory), part A in Trail Making Test (sensorimotor speed), short version of Mini-Mental State Examination (global cognition), modified versions of the Digit Symbol Test (perceptual speed and executive function), Block Design (visuospatial skills), and Controlled Oral Word Association Test (semantic memory). Chocolate users compared to nonusers performed better in all six different cognitive tests and had significantly reduced risk for poor test performance in most tests, whereas the mean intake of chocolate among users was as little as ~8 g/day. Moreover, there was a dose-response relationship: the cognitive performance improved with increasing chocolate intake, and a maximum beneficial effect on cognitive performance was gained at a mean intake of chocolate of ~10 g/day [26]. These findings are in accord with an earlier study that even very modest consumption of chocolate may significantly contribute to total polyphenol intake [27] and may thus be beneficial for cognition. Furthermore, the real effect of polyphenols in chocolate may be even stronger because not all chocolates are equally good sources of flavanols (the health benefits ascribed to chocolate relate nearly exclusively to the dark, bittersweet-tasting chocolate and to products with a cocoa content of 60% or more), and the type of chocolate consumed was not specified in the Hordaland study.

In general, cocoa containing high flavanol content is relatively difficult to obtain due to common postharvest handling and processing procedures used in the food industry that can dramatically reduce the flavanol content of chocolate and cocoa [7, 9]. The more processed the chocolate, the more flavonoids (and all of their related health benefits) are lost. In addition, milk, cream, sugar, or other additives added to chocolate dilute the cocoa concentration and reduce the flavanol content per weight. Thus, commercially available cocoa-based products, what one can buy in stores, are routinely very low and/or inconsistent in flavanol content [28]. Although the unprocessed cocoa is the best in regard to flavonoids, certain cocoas can be manufactured to be extraordinarily rich in flavanols. Therefore, chocolate is relatively variable with some products containing essentially no flavonoids (0.09 mg procyanidin/g), whereas others are high in flavonoids (4 mg procyanidin/g) [29]. On that account, approximate estimates of the amounts of flavonoid-rich chocolate needed to elicit acute and chronic effects are 38 and 125 g, respectively [29].

Beyond the flavanols, there are methylxanthine compounds in cocoa, caffeine, and theobromine that may influence cognitive functioning. In one study by Smit et al. [30], two active treatments containing identical amounts of methylxanthines (encapsulated cocoa powder equivalent to a 50 g chocolate bar and isolated caffeine and theobromine equivalent to a 50 g chocolate bar) against placebo were used. In another study [30], the treatments were 60 g portions of chocolate prepared to look identical but contain differing amounts of methylxanthines: zero, low (8 mg caffeine + 100 mg theobromine), and high methylxanthines (20 mg caffeine + 250 mg theobromine). In the first study, cocoa powder and isolated caffeine and theobromine both significantly decreased reaction time and increased energetic arousal compared to placebo. Isolated caffeine and theobromine also significantly increased scores on a rapid visual information processing task and increased hedonic tone (a measure of positive mood) compared to placebo, whereas cocoa powder was associated with a nonsignificant increase. The chocolate treatments of varying methylxanthine levels in the second study produced similar results. High methylxanthines chocolate significantly decreased reaction time while both low and high methylxanthines chocolates increased rapid visual information processing task scores relative to placebo.

Studies on cell cultures have suggested that oxidative stress induced by reactive oxygen species is strongly associated with the pathogenesis of neurodegenerative disorders such as Alzheimer's disease and that apoptosis is a crucial pathway in neuronal cell death. Cocoa procyanidin fraction and procyanidin B2 prevent the H_2O_2-induced apoptosis in rat pheochromocytoma (PC12) cells from oxidative stress through blocking the phosphorylation of c-Jun N-terminal protein kinase and p-38 mitogen-activated protein kinase [31]. Also, an aldehylic product of membrane lipid peroxidation – 4-hydroxynonenal – is reported to be elevated in the brains of Alzheimer's disease patients, and it mediates the induction of neuronal apoptosis in the presence of oxidative stress [32]. In another study, Cho et al. [33] showed that cocoa procyanidin fraction and procyanidin B2-protected PC12 cells against 4-hydroxynonenal-induced apoptosis by blocking mitogen-activated protein kinase kinase 4 activity as well as accumulation of reactive oxygen species. In addition, Ramiro-Puig et al. [34] have shown the potential neuroprotective action of cocoa flavonoids by reducing reactive oxygen species production and modulating mitogen-activated protein kinase activation: cells incubated with cocoa extract or (−)-epicatechin reduced reactive oxygen species production in a dose-dependent manner, reaching 35% inhibition. Moreover, c-Jun N-terminal protein kinase and p-38 involved in apoptosis were down-modulated by cocoa extract and (−)-epicatechin with p-38 inhibition reaching up to 70% [34]. Additional health benefits of the major cocoa flavonoids, epicatechin and catechin, on the neurotoxicity induced by amyloid β protein in rat PC12 cells have been demonstrated by Heo and Lee [35]. Thus, neuroprotective actions of cocoa flavonoids involve a number of effects within the brain, including a potential to protect neurons against injury induced by neurotoxins, an ability to suppress neuroinflammation, and may therefore have a potential to promote memory, learning, and cognitive function.

In addition to potential beneficial direct effect of chocolate on cognition, chocolate may have an indirect negative impact as food cravings may impair cognitive performance. Chocolate is the most commonly and intensely craved food in Western cultures [36] as chocolate fulfills two innate preferences: one for sweet taste, and one for creamy texture, making chocolate one of the most palatable foods available [37]. In a study among 96 female undergraduate students, Tiggemann et al. [38] examined the impact of chocolate craving on the three specific components of working memory – the phonological loop, the visuospatial sketch pad, and the central executive. They demonstrated that chocolate cravings (under manipulated craving conditions) selectively disrupt performance on visuospatial tasks. According to the working memory model, this occurs because food cravings compete for limited visuospatial working memory resources [38].

Finally, with regard to natural cocoa ingestion, there is an intriguing phenomenon, the Kuna Indians, native to islands off the coast of Panama, drink several servings of unprocessed cocoa a day and appear to be the only known society which salts its food and yet is virtually free from hypertension [39]. Although the prevalence of senile dementia among the Kuna has not been formally assessed, other Third World cultures in which hypertension and stroke are quite rare are characterized by a near absence of dementia [40]. As there seems to be a correlation between excellent cerebrovascular health and freedom of dementia, one may hypothesize that there is also a strong association between cocoa drink usage and cognitive health [40].

In conclusion, beneficial effects of cocoa or chocolate on cognitive functioning have been reported in a couple of animal studies [11, 20], while similar effects have not been found in all human trials [22, 23]. However, the most recent trials [24, 25] and a population-based cross-sectional study [26] support the positive findings from animal trials. There is evidence that cocoa flavanols may increase blood flow to the brain [13, 21, 22], and long-term improvements in brain blood flow could impact cognitive behavior (as long-term administration of cocoa flavanols to aging rats was associated with preservation of youthful cognitive performance [11]), offering future potential for impairing brain conditions leading dementia and stroke. The considerations of limited human studies also suggest that regular consumption of cocoa flavanols might have important potential for promoting retention of cognitive function as people age. The effects of cocoa flavanols on cognitive functioning may act or interact through a number of signaling pathways, for example, protection from neurodegradation, increased perfusion, decreased neuroinflammation, modulation of neuronal function, and neuronal apoptosis, but the specific mechanisms by which cocoa flavonoids improve cognitive function remain to be clarified and are still the subject of ongoing research. There is need for further carefully designed studies that consider possible ceiling effects of cognitive testing, encompass participants with different educational levels, and involve higher number of subjects, as the sample sizes in conducted trials have been relatively small increasing variance and risk of type two error, biasing towards nil. Finally, the fact that chocolate is considered as the third highest daily source of antioxidants for US consumers, preceded by coffee and tea [6], makes it an even more fascinating subject to study its health benefits.

References

1. Stubbe H. The Indian Nectar, or a discourse concerning Chocolata. London: J. C. for Andrew Crook at the Sign of the Green Dragon in St. Paul's Church-yard; 1662.
2. Dillinger TL, Barriga P, Escarcega S, et al. Food of the gods: cure for humanity? A cultural history of the medicinal and ritual use of chocolate. J Nutr. 2000;130(8S Suppl):2057S–72.

3. Waterhouse AL, Shirley JR, Donovan JL. Antioxidants in chocolate. Lancet. 1996;348(9030):834.
4. Galleano M, Oteiza PI, Fraga CG. Cocoa, chocolate, and cardiovascular disease. J Cardiovasc Pharmacol. 2009;54(6):483–90.
5. Visioli F, Bernaert H, Corti R, et al. Chocolate, lifestyle, and health. Crit Rev Food Sci Nutr. 2009;49(4):299–312.
6. Vinson JA, Proch J, Bose P, et al. Chocolate is a powerful ex vivo and in vivo antioxidant, an antiatherosclerotic agent in an animal model, and a significant contributor to antioxidants in the European and American Diets. J Agric Food Chem. 2006;54(21):8071–6.
7. Katz DL, Doughty K, Ali A. Cocoa and chocolate in human health and disease. Antioxid Redox Signal. 2011;15:2779–881.
8. Fraga CG. Cocoa, diabetes, and hypertension: should we eat more chocolate? Am J Clin Nutr. 2005;81(3):541–2.
9. Engler MB, Engler MM. The emerging role of flavonoid-rich cocoa and chocolate in cardiovascular health and disease. Nutr Rev. 2006;64(3):109–18.
10. Alspach G. The truth is often bittersweet ...: chocolate does a heart good. Crit Care Nurse. 2007;27(1):11–5.
11. Bisson JF, Nejdi A, Rozan P, et al. Effects of long-term administration of a cocoa polyphenolic extract (Acticoa powder) on cognitive performances in aged rats. Br J Nutr. 2008;100(1):94–101.
12. Fisher ND, Hollenberg NK. Aging and vascular responses to flavanol-rich cocoa. J Hypertens. 2006;24(8):1575–80.
13. Fisher NDL, Sorond FA, Hollenberg NK. Cocoa flavanols and brain perfusion. J Cardiovasc Pharmacol. 2006;47(Suppl 2):S210–4.
14. Breteler MM. Vascular risk factors for Alzheimer's disease: an epidemiologic perspective. Neurobiol Aging. 2000;21(2):153–60.
15. Kokmen E, Whisnant JP, O'Fallon WM, et al. Dementia after ischemic stroke: a population-based study in Rochester, Minnesota (1960–1984). Neurology. 1996;46(1):154–9.
16. Kalaria RN. The role of cerebral ischemia in Alzheimer's disease. Neurobiol Aging. 2000;21(2):321–30.
17. Kalaria RN, Ballard C. Overlap between pathology of Alzheimer disease and vascular dementia. Alzheimer Dis Assoc Disord. 1999;13(Suppl 3):S115–23.
18. Joseph JA, Shukitt-Hale B, Casadesus G. Reversing the deleterious effects of aging on neuronal communication and behavior: beneficial properties of fruit polyphenolic compounds. Am J Clin Nutr. 2005;81(1 Suppl):313S–6.
19. Brown LA, Riby LM, Reay JL. Supplementing cognitive aging: a selective review of the effects of ginkgo biloba and a number of everyday nutritional substances. Exp Aging Res. 2010;36(1):105–22.
20. Rozan P, Hidalgo S, Nejdi A, et al. Preventive antioxidant effects of cocoa polyphenolic extract on free radical production and cognitive performances after heat exposure in Wistar rats. J Food Sci. 2007;72(3):S203–6.
21. Sorond FA, Lipsitz LA, Hollenberg NK, et al. Cerebral blood flow response to flavanol-rich cocoa in healthy elderly humans. Neuropsychiatr Dis Treat. 2008;4(2):433–40.
22. Francis ST, Head K, Morris PG, et al. The effect of flavanol-rich cocoa on the fMRI response to a cognitive task in healthy young people. J Cardiovasc Pharmacol. 2006;47(Suppl 2):S215–20.
23. Crews Jr WD, Harrison DW, Wright JW. A double-blind, placebo-controlled, randomized trial of the effects of dark chocolate and cocoa on variables associated with neuropsychological functioning and cardiovascular health: clinical findings from a sample of healthy, cognitively intact older adults. Am J Clin Nutr. 2008;87(4):872–80.
24. Scholey AB, French SJ, Morris PJ, et al. Consumption of cocoa flavanols results in acute improvements in mood and cognitive performance during sustained mental effort. J Psychopharmacol. 2010;24(10):1505–14.
25. Field DT, Williams CM, Butler LT. Consumption of cocoa flavanols results in an acute improvement in visual and cognitive functions. Physiol Behav. 2011;103(3–4):255–60.
26. Nurk E, Refsum H, Drevon CA, et al. Intake of flavonoid-rich wine, tea, and chocolate by elderly men and women is associated with better cognitive test performance. J Nutr. 2009;139(1):120–7.
27. Scalbert A, Williamson G. Dietary intake and bioavailability of polyphenols. J Nutr. 2000;130(8S Suppl):2073S–85.
28. Hollenberg NK. Vascular action of cocoa flavanols in humans: the roots of the story. J Cardiovasc Pharmacol. 2006;47(Suppl 2):S99–102.
29. Kris-Etherton PM, Keen CL. Evidence that the antioxidant flavonoids in tea and cocoa are beneficial for cardiovascular health. Curr Opin Lipidol. 2002;13(1):41–9.
30. Smit HJ, Gaffan EA, Rogers PJ. Methylxanthines are the psycho-pharmacologically active constituents of chocolate. Psychopharmacology (Berl). 2004;176(3–4):412–9.
31. Cho ES, Lee KW, Lee HJ. Cocoa procyanidins protect PC12 cells from hydrogen-peroxide-induced apoptosis by inhibiting activation of p38 MAPK and JNK. Mutat Res. 2008;640(1–2):123–30.
32. Poli G, Schaur RJ. 4-Hydroxynonenal in the pathomechanisms of oxidative stress. IUBMB Life. 2000;50(4–5):315–21.
33. Cho ES, Jang YJ, Kang NJ, et al. Cocoa procyanidins attenuate 4-hydroxynonenal-induced apoptosis of PC12 cells by directly inhibiting mitogen-activated protein kinase kinase 4 activity. Free Radic Biol Med. 2009;46(10):1319–27.
34. Ramiro-Puig E, Casadesus G, Lee HG, et al. Neuroprotective effect of cocoa flavonoids on in vitro oxidative stress. Eur J Nutr. 2009;48(1):54–61.

35. Heo HJ, Lee CY. Epicatechin and catechin in cocoa inhibit amyloid beta protein induced apoptosis. J Agric Food Chem. 2005;53(5):1445–8.
36. Hetherington MM, MacDiarmid JI. "Chocolate addiction": a preliminary study of its description and its relationship to problem eating. Appetite. 1993;21(3):233–46.
37. Drewnowski A, Krahn DD, Demitrack MA, et al. Taste responses and preferences for sweet high-fat foods: evidence for opioid involvement. Physiol Behav. 1992;51(2):371–9.
38. Tiggemann M, Kemps E, Parnell J. The selective impact of chocolate craving on visuospatial working memory. Appetite. 2010;55(1):44–8.
39. Hollenberg NK, Martinez G, McCullough M, et al. Aging, acculturation, salt intake, and hypertension in the Kuna of Panama. Hypertension. 1997;29(1 Pt 2):171–6.
40. McCarty MF. Toward prevention of Alzheimer's disease – potential nutraceutical strategies for suppressing the production of amyloid beta peptides. Med Hypotheses. 2006;67(4):682–97.

Chapter 29
Cocoa and Chocolate: Chemistry, Biochemistry, and Beneficial Brain Effects

Maria Patrizia Carrieri and Joe A. Vinson

Key Points

- Epicatechin metabolites are the proposed brain bioactive compounds from cocoa and dark chocolate.
- Cerebral blood flow increase is the major mechanism for cognitive effects mediated by cocoa and dark chocolate.
- Dark chocolate and cocoa can affect mood, anxiety, depression, and chronic fatigue syndrome.

Keywords Cerebral blood flow • Epicatechin • Nitric oxide • Vasodilation • Metabolites • Anxiety • Fatigue • Chronic fatigue syndrome • Cognitive function • Aging • Cardiovascular • Cerebrovascular • Mood • Depression

Introduction

> Anything good in life is either immoral, illegal or fattening.
>
> – Oscar Wilde (1854–1900)

The famous Italian dessert, named "tiramisu," meaning "pick me up," in the sense of regaining strength, reenergizing, or waking-up, probably owes its name to its ingredients, especially coffee and chocolate, whose positive effects on cognitive functions were already known. Another expression, "Seven days without chocolate makes one weak," reinforces the common belief that chocolate can have an impact on mood and fatigue.

M.P. Carrieri, Ph.D.
INSERM, U912 (SESSTIM); Aix Marseille Univ, IRD, UMR-S912; ORS PACA, Observatoire Régional de la Santé Provence Alpes Côte d'Azur, Marseille, France
e-mail: pmcarrieri@aol.com

J.A. Vinson, Ph.D. (✉)
Department of Chemistry, University of Scranton, Scranton, PA, USA
e-mail: joe.vinson@scranton.edu

R.R. Watson et al. (eds.), *Chocolate in Health and Nutrition*, Nutrition and Health 7,
DOI 10.1007/978-1-61779-803-0_29, © Springer Science+Business Media, LLC 2013

Chocolate consumption is now regarded by two opposing viewpoints. On the one hand, the increasing epidemic of obesity [1] and associated comorbidities [2] in high-resource settings has led to chocolate and especially milk chocolate–based snacks (such as chocolate bars sold from vending machines) being banned as "unhealthy food" and excluded from most low-calorie and low-carbohydrate diets. Moreover, chocolate has started being labeled as a food that potentially leads to addictive behaviors and craving risks for regular users [3, 4].

On the other hand, recent research has completely enlightened the multiple health benefits associated with dark chocolate consumption, mainly attributable to its antioxidant properties. Polyphenols, the most frequently occurring natural antioxidants, found in cocoa, certain fruits, and common beverages like wine, tea, and coffee have been shown to have positive cardiovascular effects (vasoactive and antithrombotic activity) [5] and to play a role in the prevention and treatment of several diseases associated with oxidative stress, such as cancer and inflammation [6, 7]. They may help to prevent neurodegenerative diseases and diabetes mellitus [8]. Like caffeine, chocolate consumption also seems to slow progression of liver fibrosis [9].

Chemistry and Biochemistry of Cocoa

Most animal and human studies have been carried out with cocoa and dark chocolate because milk chocolate contains a lower amount of cocoa solids and white chocolate has only fat and sugar and a little flavoring. It is important to know that chemically cocoa and commercial dark chocolate containing a large percentage of cocoa solids (35–85%) are extremely complex mixtures. They contain hundreds of compounds, as do all plant-derived foods and beverages. However, psychoactive components of cocoa and chocolate, such as anandamide and phenethylamine which are often touted in the popular press, are present in parts per billion to parts per million, which are not high enough doses for cocoa or chocolate to affect the brain. Theobromine, chemically similar to caffeine, is the major stimulant in cocoa and chocolate and is present at concentrations of several hundred milligrams per serving [10]. The structures of these two stimulants are similar and shown in Figs. 29.1 and 29.2.

A recent human study with 700 mg of theobromine found that it decreased both blood pressure and decreased self-reported calmness. It had no effect on psychomotor performance [11]. The authors conclude that theobromine has no CNS-mediated effects.

Chemically the two important polyphenol compounds present in cocoa and in dark chocolate and to a lesser extent in milk chocolate are (–)-catechin and (+)-epicatechin, whose structures are shown in Figs. 29.3 and 29.4. These flavanols are hypothesized to be the major cocoa bioactives in the body, including the brain.

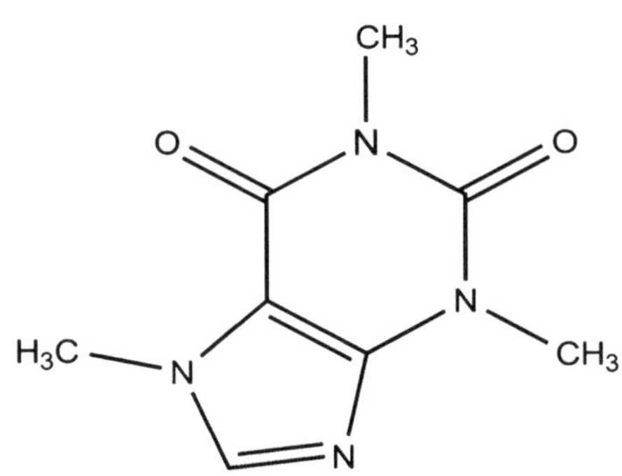

Fig. 29.1 Caffeine

Fig. 29.2 Theobromine

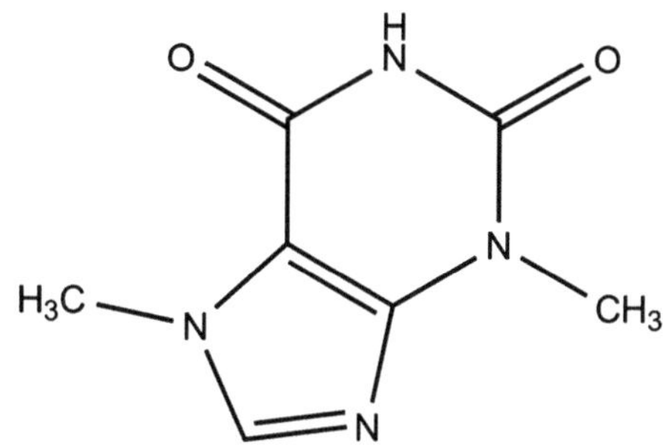

Fig. 29.3 (–)-catechin

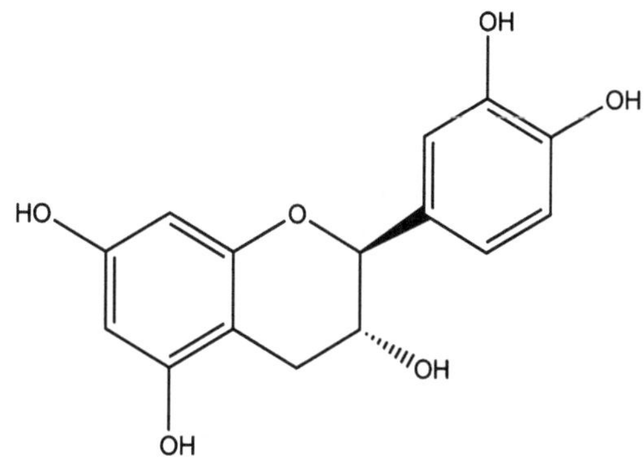

Fig. 29.4 (+)-epicatechin

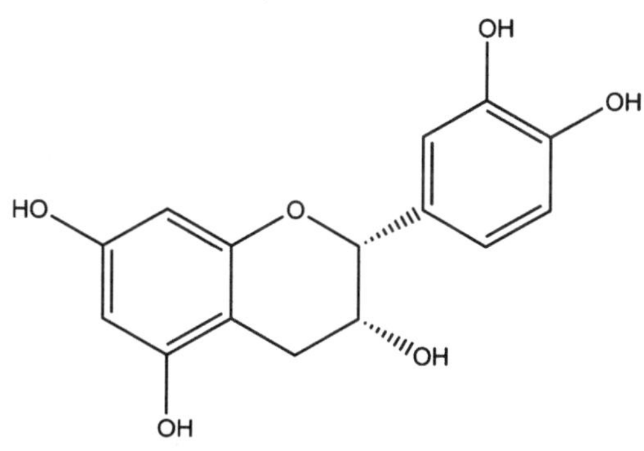

It is apparent that they are stereoisomers of each other. In contrast to fruits, cocoa has (–)-catechin, which is formed from the (+)-enantiomer by processing. These flavanols are excellent antioxidants and have almost identical antioxidant properties. Oligomers (procyanidins B1, B2, B5, and C1) and polymers account for 90% or greater of total polyphenols in cocoa and chocolate and of monomers such as catechin and epicatechin 5–10% [12]. Large molecules with molecular masses >500 amu cannot be absorbed by the body so monomers are the compounds in cocoa and chocolate that may potentially be present in plasma and thus in the brain. However, (–)-catechin is poorly bioavailable and has not been detected by LC-MS-MS in plasma after animal [13] and also after human consumption [14].

Health effects derived from cocoa polyphenols depend on their bioavailability. In terms of flavanols, bioavailability is influenced by their degree of polymerization [15]. Monomers are well absorbed in the small intestine. As the result of phase II enzymes, (–)-epicatechin is converted into glucuronidated and sulfated metabolites as well as into methylated metabolites, which in turn may also be glucuronidated and sulfated. It is also possible that the conjugated metabolites may be deconjugated back to original aglycones such as epicatechin in tissues and organs [16]. However, the absorption of dimeric procyanidins in humans seems to be very limited [17], and polymeric procyanidins are not well absorbed in their native form. These polyphenols reach the colon, where they are biotransformed by the intestinal microbiota into hydroxyphenylvalerolactones and a series of phenolic acids, [14] which may be further absorbed and consequently exert multiple biological activities.

Blood–Brain Barrier Cell Studies

The first barrier to absorption of polyphenols is the intestines, where endothelial cells uptake the compounds and where some metabolism takes place. However, the blood–brain barrier (BBB) is a formidable one. Currently, there is substantial evidence to show that the cerebral endothelium is the main cellular structure responsible for the BBB in mammals [18]. Among the elements that distinguish brain endothelium from intestinal endothelium are complex tight junctions and the presence of specific transporters, including efflux carriers. These properties make the BBB a regulatory interface that selectively limits passage of most small polar molecules and macromolecules from the cerebrovascular circulation to the brain. The BBB exerts tight control over transendothelial molecular traffic and contributes to the regulation of brain extracellular fluid [19]. Therefore, hydrophilic compounds cannot cross the BBB in the absence of specific mechanisms such as membrane transporters or endocytosis. RBE-4 cells (rat capillary cerebroendothelial cells) were found to transport quercetin (a lipophilic polyphenol), catechin (a more hydrophilic polyphenol), and cyanidin-3-glucoside (a cationic hydrophilic polyphenol). The efficiency of transport (apical to basolateral side) ranged from 3% to 23% in a time-dependent manner. This same group investigated the transport of (+)-catechin (not found in processed cocoa) and (−)-epicatechin in both RBE4 and an immortalized human cerebral endothelial cell line recently developed as a model of the human BBB [19]. Transport efficiencies were similar for the two cell lines and varied between 8% and 28% and were time dependent. Both aglycones and glucuronides were found in the cell media after incubation, indicating that the cells had glucuronidating enzymes present. The transport efficiency of epicatechin was much higher than catechin, indicating their stereochemistry greatly affected transport.

There is plenty of evidence that polyphenols are neuroprotective whether from cocoa, tea, red wine, and grape seed extract (all high polyphenol sources). This is thought to be due to the protection of neuronal cells from oxidative stress through induction of antioxidant defenses, modulation of signaling cascades, apoptosis, and/or by the synthesis/degradation of the amyloid (beta)β-peptide [20]. For example, one epidemiological study showed that consumption of tea, red wine, and chocolate in moderate amounts slows the cognitive decline during normal human aging [21].

Aging and atherosclerosis are major risk factors for stroke and have been associated with a decrease in nitric oxide (NO) production and an increase in reactive oxygen species generation [22]. The following study was predicated on the hypothesis that severe dyslipidemia and associated oxidative stress could accelerate age-related decline in cerebrovascular endothelial function and cerebrovascular blood flow (CBF). Catechin treatment (30 mg/kg/day) in the drinking water for 3 months improved cerebrovascular flow–mediated dilation (endothelial function) and learning abilities in atherosclerotic mice by decreasing cerebral superoxide and increasing CBF during stimulation [23]. This is a very high dose of catechin and is not feasible in terms of giving humans cocoa.

An in vivo study with rat aortal rings showed that only (−)-epicatechin among the catechin and epicatechin stereoisomers was capable of provoking an arterial dilation [24]. Thus, only (−)-epicatechin causes vasodilation presumably by an NO-dependent pathway. This result should be transferable to the cerebrovascular system.

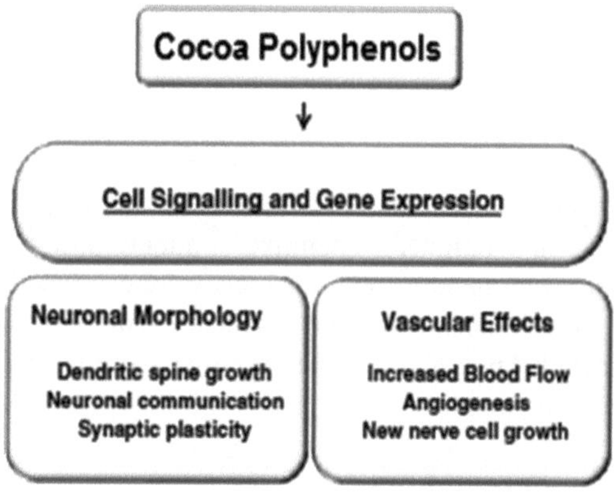

Fig. 29.5 Proposed brain mechanism of cocoa polyphenols

Figure 29.5 illustrates the mechanisms by which polyphenols affect brain function, as proposed by Spencer [25]. We have revised it to make it specific to cocoa.

However, research on the possible impact of chocolate compounds on mood is complex for many reasons. First, many experimental studies either use chocolate bars which contain a considerable quantity of sugar or carbohydrates or are based on the use of chocolate as a palatable food (i.e., subjects in experiments choose the type of chocolate they consider more palatable). Moreover, in most experimental studies, only short-term effects on mood are evaluated. Whether relief after craving and "addiction-like" behaviors are due more to other compounds present in chocolate bars (like carbohydrates and sugar) than to the components of cocoa itself remains a controversial issue.

In their review examining the mood state effects of chocolate, Parker and coauthors [26] assessed chocolate effects with respect to two different phenomena: food craving and emotional eating. The authors suggest that chocolate craving is mainly determined by a desire for hedonic reward and that dopamine is the main neurotransmitter released after eating chocolate in the aim of an enjoyable sensory experience. By contrast, emotional eating is more characterized by carbohydrate craving and is motivated by the desire for the calming effects of endorphins to alleviate dysphoria or other negative states. However, according to the authors, the benefits from chocolate consumption in subjects using it as self-medication are ephemeral, as chocolate is more likely to protract rather than discontinue a dysphoric mood. Furthermore, the authors also underline that any attempt to study the effects on mood following ingestion of chocolate as a palatable food containing carbohydrates may be confounded by several factors, such as "time in relation to eating, whether subjects are cravers or noncravers, and the actual mood state (e.g., depression, anxiety, ennui)."

In this chapter, we took into account all these limitations in assessing the impact of chocolate consumption on mood states. For this reason, we mainly focused on studies including subjects who were "noncravers" and those not subject to emotional eating. We also assessed the short- and long-term effects that cocoa and its derivatives may have in reducing the burden of anxiety symptoms, depression, fatigue as well as improving cognitive performance and other brain responses to stress. Animal models and experimental and observational studies, both in healthy individuals and in individuals with morbidities involving mood disorders or which affect cognitive performance, were included. This chapter also aims to elucidate the possible causal path that may lead to mood and cognitive performance improvements after chocolate/cocoa intake by an analysis of the chemistry properties of cocoa's compounds and their effects on brain functions.

Methods

We searched BIOSIS (ISI-CE BIOSIS previews), PubMed, and Embase for relevant publications written in English, French, and Spanish between 2000 and 2011. Our search had several keywords: "cocoa," "chocolate," "flavonoids," "anxiety," "depression," "depressive symptoms," "mood," "dysphoria," "fatigue," "cognitive," "attention deficit," and "craving." We searched reference lists from retrieved manuscripts. Abstracts were examined to assess eligibility for inclusion.

For biochemistry and chemistry, PubMed was searched between 2000 and 2011, and keywords included polyphenols, theobromine, caffeine, cerebral blood flow, catechin, epicatechin, and endothelial function.

Chocolate, Depression, and Anxiety

Personality Traits, Chocolate Craving, and Mood Disorder Relief

As depression and anxiety disorders can increase chocolate and/or food craving [27] and both of these are considered markers of specific psychological traits, the relief of depression and anxiety-like symptoms in individuals experiencing chocolate craving or chocolate emotional eating may be mediated by the specific psychiatric traits of these individuals. These results may make it difficult to interpret the role that regular chocolate consumption may play on mood disorders.

An extensive body of literature exists on the issue of chocolate craving and the causes of craving. This includes the psychiatric profiles of individuals experiencing relief from craving and symptoms after chocolate consumption. While a literature analysis on chocolate craving is not the objective of the current review and will be considered in another chapter in this volume (Chap. 34), it is, however, important to detail some specific results that may be helpful in understanding that the response to chocolate consumption may depend on the psychological traits of individuals.

A study conducted in 2001 analyzed attitudes [28] toward chocolate (craving and consumption) in a general population using a factor analysis. The following two dimensions were identified: the first included both "obsessional thoughts about chocolate consumption" and one's "weakness" for chocolate while under emotional stress. The second dimension included both "guilt and negative feelings after consumption" and "guilt or worry about body shape." Other studies that aimed to make the link between attitudes toward chocolate and psychiatric traits showed that individuals who reported eating chocolate for "self-medication" were more likely to present hysteroid dysphoria-like symptoms (i.e., depressive symptoms associated with feelings of rejection) [29], and those characterized by considerable levels of chocolate craving reported chocolate consumption when feeling bored, anxious, and having a dysphoric mood [30]. Although a desire for chocolate was shown to be associated with depression, another study found no association between chocolate craving and suicidal thoughts [31].

An observational interventional study was conducted to understand whether chocolate craving could be satisfied by chocolate consumption. Three different modes of satisfying chocolate craving (chocolate, white chocolate, and cocoa powder) [4] were offered to volunteers who were subject to chocolate craving. Individuals experiencing chocolate craving could use one of the three modes and record on a scale whether or not their craving had been satisfied. Results showed that the critical factors satisfying chocolate craving were the taste and mouthfeel and not the quantity of active compounds present in the consumed chocolate. Although the study did not randomize subjects to three arms and imposed an evaluation of craving relief following an intentional choice of the subject, this study confirms that chocolate craving is probably the expression of a specific eating disorder whose relief may be attributable more to sensation satisfaction than to the effect of chocolate compounds.

As already mentioned in the introduction, while those who define themselves as "chocolate addicts" or "chocolate cravers" do not generally seem to be subject to classical eating disorders, they are more prone to suffer from depression and anxiety [32]. On the other hand, chocolate craving also occurs more frequently in individuals experiencing emotional dysregulation, or in those with anxiety and anger, which are common in persons with hysteroid dysphoria [26, 29].

The fact that chocolate craving (and consumption) is more frequently found in individuals with mood disorders explains why any investigation of the association between chocolate use and depression in cross-sectional observational studies may fail to disentangle the causes of chocolate consumption from depression and anxiety symptoms' relief. This is also why some reviews cannot conclude whether chocolate consumption improves mood disorders [33].

Definition: Depression and Anxiety

Clinical depression is a common mental disorder characterized by depressed mood for weeks or months (as opposed to a transitory "blue" mood); it is often accompanied by loss of interest or pleasure, feelings of guilt or low self-esteem, hopelessness, disturbed sleep or appetite, low energy, and poor concentration. Persistent somatic symptoms – such as headaches, digestive disorders, and chronic pain – may be signs of clinical depression.

According to the WHO report in 2007 [34], it is estimated that depression affects about 121 million people worldwide and was "the leading cause of disability as measured by YLDs and the 4th leading contributor to the global burden of disease (DALYs) in 2000. By the year 2020, depression is projected to reach 2nd place of the ranking of DALYs calculated for all ages, both sexes."

While the definition of major depression is based on a combination of symptoms lasting for at least 2 weeks, there is a considerable variety of depression subtypes (unipolar depression, biological depression, manic depression, seasonal affective disorder, dysthymia, etc.).

Generalized anxiety disorders can also coexist with clinical depression and other mental disorders and are characterized by excessive longstanding worry about health, job responsibilities, finances, the safety of one's children, or even being late for appointments. These feelings of worry and dread are difficult to control and can be accompanied by physical symptoms such as headaches, pain from muscle tension, frequent urination, difficulty swallowing, a "lump in the throat," or exaggerated startled responses.

Anxiety and depression are major public health problems that are reaching epidemic levels in many medium- and high-income countries. In 2002, depression accounted for 4.5% of the worldwide total burden of disease (in terms of disability-adjusted life years). It is also responsible for the greatest proportion of burden attributable to nonfatal health outcomes, accounting for almost 12% of total years lived with disability worldwide [34]. Treatment of depression and anxiety is tailored to the individual, the severity and cause of the depressive episode, as well as anxiety disorders all being taken into account.

Although the mechanism provoking depression and anxiety has still not been clearly elucidated, the main trigger is known to be exposure to chronic stress. Recent research now suggests that depressive episodes can occur as "a normally self-limiting but highly error-prone process of recuperation from stress-triggered neuronal microdamage" [35].

Depression and anxiety are disorders or co-occurring disorders with multidimensional clinical manifestations. This is why evaluating the role that chocolate/cocoa intake may have on "depression" and anxiety manifestations or on chronic stress consequences is not an easy task. Such evaluation may require a classification of the type of depressive and anxiety symptoms on which chocolate intake may be effective and of the functions that may be affected by such disorders.

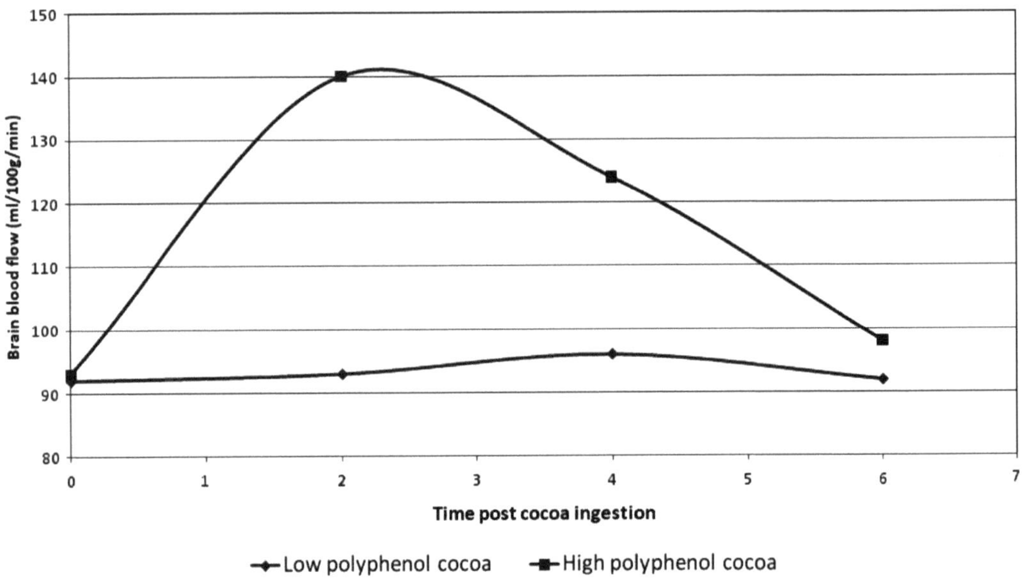

Fig. 29.6 Time course of the average CBF across gray matter after consumption of a single dose of cocoa drink low (13 mg) and high (172 mg) in polyphenol

Biochemistry Issues

There have been numerous articles indicating that cerebrovascular blood flow (CBF) is reduced in those with depression and Alzheimer's. In a 2010 article, subjects with major depression (unipolar) were measured for CBF before and after remission. A significant increase in CBF was seen during remission [36]. In a 2011 study, major depressive disorder (unipolar or bipolar) patients with homogenous demographic and clinical features were subjected to repetitive transcranial magnetic stimulation, and basal CBF was measured. Nonresponders had significantly lower CBF in several brain areas compared with positive responders [37]. We propose that an increase in CBF following cocoa consumption is the major mechanism for cocoa's beneficial effect of depression.

Another landmark experiment by an international group showed that in healthy male adults, ingestion of high polyphenol cocoa was associated with acute elevations of circulating NO_x species and an enhanced flow-mediated dilation response of arteries (endothelial function improvement) and an increased microcirculation [38]. Epicatechin and plasma concentrations of its metabolites significantly correlated with a vascular effect. Neither catechin nor its metabolites correlated, indicating the lack of bioactivity of this epimer. An in vivo study with rat aortal rings showed that only (−)-epicatechin was capable of provoking arterial dilation among the catechin and epicatechin stereoisomers [24]. Thus, only (−)-epicatechin causes vasodilation presumably by an NO-dependent pathway.

The conclusion of both animal and human studies is that (−)-epicatechin and its metabolites mediate the vascular effects of cocoa. This has been amply demonstrated in studies with respect to the cardiovascular effects of cocoa and chocolate. Our hypothesis is that (−)-epicatechin also mediates the cerebrovascular effect of both.

An intervention study with young subjects in 2006 showed that polyphenol-rich cocoa increased CBF by 50% and that this elevated CBF returned to baseline 6 h postconsumption [39]. These results are illustrated in Fig. 29.6. This curve indicated a long-lasting acute effect of cocoa on blood flow.

Aging is associated with frontal subcortical microangiopathy and executive cognitive dysfunction, suggesting that elderly individuals may have impaired metabolic activation of CBF to the frontal lobes. Aging is marked by reductions in CBF [40] and in the speed of cognitive processing [41].

As a result of earlier studies indicating that coronary artery blood flow and endothelial function (NO dependent) were increased by cocoa, other groups began investigating the effect of cocoa on the human brain. Blood flow has a fundamental role in brain function. Neuroimaging techniques, such as functional magnetic resonance imaging (fMRI), which are based on increased blood flow accompanying increased neural activity in the brain, have widely demonstrated the relationships of cerebrovascular responses and cerebrovascular health to human cognitive and emotional functions. In 2006, a study was published demonstrating that CBF, as measured by Doppler ultrasound, was increased in healthy elderly subjects after consuming 900 mg of cocoa polyphenols [42]. In a larger study, the same research team gave cocoa polyphenols to 34 elderly subjects and found that the increase in CBF postcocoa consumption was larger after 2 weeks of supplementation than after 1 week [43]. It is important to keep in mind that this acute increase in CBF after cocoa occurred within a background of decreased CBF arising from the caffeine and theobromine stimulants present in cocoa.

Another aspect of brain health is the issue of stress. Stress, of course, is common in everyday life and is believed to affect happiness, health, and cognition. A study in 2005 with young adults found that CBF increased with mild to moderate stress induced by a mental arithmetic task [44]. The second utilized 30 healthy adults in a double-blind study who consumed 520, 994 mg polyphenols, or a control beverage during an hour of very difficult cognitive tests [45]. Consumption of cocoa flavanols resulted in acute improvements in mood and cognitive performance during sustained mental effort. Only the 520-mg beverage resulted in an improvement of mood, as measured by a mental fatigue scale. Both doses improved one of the measures of cognitive function. The study's authors speculated that the mechanisms were related to the proven effects of cocoa polyphenols on endothelial function and blood flow, that is, nitric oxide activation.

A second study utilized the new burgeoning field of metabolomics [46]. Thirty subjects consumed 40 g of dark chocolate (74% cocoa solids) daily for 14 days. Urine and plasma samples were collected three times during the trial. This is the first and only study that used dark chocolate, the other human studies having used a cocoa beverage. Concentrations of compounds changed by the chocolate were measured by NMR and MS. Human subjects with a higher anxiety trait showed a distinct metabolic profile indicative of a different energy homeostasis and gut microbial activities. Dark chocolate reduced the urinary excretion of the stress hormone cortisol and catecholamines and partially normalized stress-related differences in energy metabolism. Moreover, the anxiety trait-related metabolic differences observed in urine (e.g., levels of hippurate, *p*-cresol sulfate, glycine, citrate, *trans*-aconitate, proline, DOPA, and (beta)β-alanine) tended to be normalized toward the levels observed in low anxiety trait subjects, whereas metabolic signatures in blood plasma were maintained over the duration of the clinical trial. The study provides strong evidence that the daily consumption of 40 g of dark chocolate (equivalent to eating one chocolate bar) over a period of 2 weeks is sufficient to modify the metabolism of free-living and healthy human subjects, by changing both host and gut microbial metabolism.

Effect of Cocoa-Containing Food on Anxiety and Depression

In order to better explore the role that chocolate intake can play on anxiety or depression, we selected only studies whose focus was NOT on subjects reporting suffering from chocolate craving or eating disorders.

Short-Term Effects of Cocoa or Chocolate Intake on Mood

It has already been shown that craving for the rewards given by chocolate seems to be intensified when a depressive mood is induced in both animal and human models [27]. To evaluate whether chocolate as a palatable food might modify induced mood states, several experimental studies were performed on human subjects, all based on the principle of inducing emotions through watching video clips and evaluating self-administration of chocolate or other palatable foods.

Macht's 2002 [47] study reported results on 24 healthy men who experienced anger, fear, sadness, and joy induced through film clips. Individuals abstained from eating chocolate 2 or 8 h before the film clip. After watching the clips, they received pieces of chocolate. Hours of deprivation were correlated with motivation and the most affective responses to eating chocolate. Sadness was found to decrease appetite. Experiencing joy increased appetite and was found to be associated with a tendency to eat more chocolate, a more pleasant taste and more stimulating than for sadness. The co-existence of other research that shows increased eating response to sadness confirms the existence of two different types of "emotional eating": emotion-congruent modulation of eating and eating to regulate emotions.

Another study conducted more recently by Macht and coauthors [48] tested two hypotheses: (1) chocolate eating can immediately affect negative, but not positive or neutral mood, and (2) that this effect is due to palatability. To test the first hypothesis, they conducted an experiment including 48 healthy women and men of normal weight and assessed the effects of eating a piece of chocolate and drinking water on negative, positive, and neutral mood states induced by watching film clips. Their results showed that eating chocolate reduced negative mood compared with drinking water, whereas no major effects were found on neutral and positive moods. To test the second hypothesis, the researchers conducted a second experiment including 113 healthy women and men of normal weight and compared effects of eating palatable and unpalatable chocolate on negative mood. They also measured the duration of chocolate-induced mood change. This experiment showed that negative mood improved after eating palatable chocolate in contrast to unpalatable chocolate or eating nothing at all. However, this effect was short-lived in both experiments, that is, it disappeared after 3 min. These studies simply show that the benefits of chocolate as a palatable food on mood-induced states have short-term effectiveness, and this effectiveness derives mainly from chocolate's palatability.

Long-Term Effects of Chocolate on Mood

Some evidence about the antidepressant effects of cocoa polyphenolic extract is provided by an experiment conducted on rats [49] undergoing a forced swimming test at different cocoa polyphenolic extract exposures. The outcome measured was the duration of immobility after the swimming test. The purpose of this study was to assess the effects of treatment with cocoa polyphenolic extract on locomotor activity in an open field test and on immobility in the forced swimming test in rats, which was a useful experimental method for screening for a wide range of antidepressants (including tricyclics and monoamine oxidase inhibitors).

Cocoa polyphenolic extract was found to be associated with a significant reduction in the duration of rat immobility (depression-related) for both doses of 24 and 48 mg/kg/14 days, confirming the hypothesis that cocoa polyphenolic extract has antidepressant-like effects.

Another study conducted on rats was designed to better investigate the anxiolytic effects of short- and long-term administration of cacao mass [50]. The authors used the elevated T-maze test, which is a model of anxiety for animals. Two different experiments were conducted: during the first, rats were administered cacao mass (100 mg/100 g body weight) and immediately performed the elevated T-maze test. This experiment showed that cacao mass administration significantly abolished delayed avoidance latency, but did not change escape latency, that is, it reduced conditional fear-related behavior, but did not modify the concentration of brain monoamines or emotion-related neurotransmitters, such as norepinephrine, serotonin, and dopamine, in the rat brain. In the second experiment, rats were fed using a cacao mass–containing diet for 2 weeks before performing the elevated T-maze test. Unlike for short-term administration, chronic consumption of cacao mass tended to increase avoidance latency and did not change escape latency. It also enhanced brain serotonin concentrations and serotonin turnover. This result suggests that chronic consumption of cacao did not affect fear-related behaviors, but was involved in brain monoamine metabolism. The two experiments put together suggest that short-term cacao mass consumption may have an anxiolytic effect, although this does not seem to be true for chronic consumption.

One study [51] conducted in pregnant women suggested that chocolate consumption can produce multiple effects at environmental and psychological levels. The study aimed to assess whether chocolate consumption and stress experienced during pregnancy could predict infant temper as rated by their mothers at 6 months of age. The study collected chocolate consumption and intensity of stress during pregnancy in 305 pregnant healthy mothers. The results highlighted that mothers reporting chocolate intake during pregnancy were more likely to positively rate the temperament of their infants, while prenatal stress was associated with lower rating, especially in mothers reporting no chocolate consumption during pregnancy.

Some research has evaluated the effect of chocolate consumption on mood in the elderly. Strandberg and colleagues [52] conducted a study in older men (born between 1919 and 1934) who had been followed up since 1960. In 2002–2003, a mailed questionnaire was used to assess the health status of all those still alive. The response rate was 69% (N = 860), but only individuals reporting a preference for either chocolate or candy were selected. Individuals preferring chocolate to candy exhibited significantly lower levels of depression (Zung's Scale), fewer feelings of loneliness, more feelings of happiness, and plans for the future.

A French intervention study [53] conducted in 209 older individuals tested the effect of prolonged dark chocolate administration (2 months, 30 g/day). Among those screened with depression by the mini Geriatric Depression Scale (GDS), one quarter no longer had depressive symptoms after 2 months. Interestingly, after 1 year from the beginning of the study, 67% patients of those who had recovered again presented with depressive symptoms.

A cross-sectional analysis [54] conducted on a sample of 931 adults from San Diego, California, without diabetes or any known coronary artery disease and who did not use antidepressants, was conducted to study the relationship between chocolate consumption and depressive symptoms, as measured by Center for Epidemiological Studies Depression Scale (CES-D). The CES-D score was dichotomized to classify patients with a positive depression screen result (CES-D score > or = 16) and those with probable major depression (CES-D score > or = 22). Individuals were provided with chocolate servings each week, while chocolate consumption frequency and rate data were gathered, based on the Fred Hutchinson Food Frequency Questionnaire. Chocolate consumption was compared for those with lower versus higher CES-D scores. This study highlighted a significant association between chocolate number of servings/month and possible depression as individuals with probable major depression reported a higher number of servings/month (11.8 servings/month), and there was a significantly linear "dose–response" relationship. However, the authors remain cautious about a possible causal relationship because, as previously mentioned, this cross- sectionally assessed relationship may be confounded by many factors (e.g., time in relation to eating, whether subjects were cravers or noncravers, and the current mood state).

In 2007, WHO's World Health Survey collected data on health and health-related outcomes and their determinants in samples of adults aged 18 years and older [55]. This survey showed that respondents with depression and other chronic conditions had much lower mean health scores when compared with respondents who had only one chronic condition. For this reason, it may be particularly meaningful to assess the possible effect of cocoa intake on anxiety and/or depression in patients with chronic conditions. However, studies focusing on anxiety or depression in patients already affected by a chronic disease are sparse. Carrieri et al. [56] reported results about the association between chocolate intake and depressive symptoms from a 3-year follow-up of HIV–HCV infected patients enrolled in the ANRS Hepavih CO-13 cohort. Depressive symptoms in HIV-infected patients are very frequent [57] and have been found to be associated with HIV clinical progression, independently of adherence to treatment [58]. In this study, individuals who reported no chocolate consumption in the previous 6 months had almost a twofold risk of presenting with probable (CES-D score >16) or major depression (CES-D >22). This relationship was confirmed even when using French sex-specific CES-D cut-off points for probable depression (17 for men and 23 for women). It is possible that the effect was observed more easily in a French population because in Europe, dark chocolate must contain at least 35% chocolate liquor, and dark chocolate consumption is more frequent in France.

To summarize, despite some evidence from animal models when we focus on acute effects just after a short period of administration, the possible effects of exposure to cocoa derivatives on anxiety and depressive symptoms seem negligible. The positive impact of more prolonged periods of administration on mood disorders becomes more evident, but these results are not completely consistent across the different studies. More research needs to be conducted on different populations and different types of mood disorders to confirm these observations.

Cognitive Functions and Fatigue

Cognitive Functions

What are commonly named "cognitive functions" include four types of different processes [59]: (1) receptive functions allowing acquisition, management, classification, and integration of information; (2) memory and learning allowing the stocking and recalling of information received; (3) thinking and reasoning concerning the mental organization and reorganization of information; (4) expressive functions allowing communication and action.

Possible Mechanisms

One animal study showed that cocoa polyphenols improved cognitive function [60]. The most bioavailable polyphenol in cocoa (−)-epicatechin was then tested in a proof-of-concept experiment to see if it improved cognitive function similar to the effect of cocoa on mice and also to determine a mechanism [61]. Epicatechin was mixed with the diet at 500 or 5 mg/kg body weight and given to mice for 20 days. It was shown to improve memory and upregulated genes which are associated with learning and downregulated markers of neurodegeneration. Angiogenesis and neuronal spine density were increased and associated with spatial memory. A daily dose for 13 days of 3 or 30 mg/g of epicatechin caused a dose–response increase in brain metabolites of epicatechin. Most importantly, injection of a mixture of the epicatechin metabolites improved spatial memory and thus convincingly demonstrated the cause and effect of the metabolites on cognitive brain function.

Chronic fatigue syndrome (CFS) is a medically unexplained illness. Initially it was ascribed to muscle defects but later was shown to be an impaired neuropsychological function [62, 63]. CBF was found to be lower in every region of the brain measured in CFS subjects [62]. Not surprisingly after 8 weeks of intervention, polyphenol-rich dark chocolate (with 85% cocoa solids) was found to significantly improve the standard fatigue score and anxiety and depression score in comparison to an isocaloric low polyphenol chocolate [64].

A proof-of-concept paper, albeit in mice, investigated the effect of (−)-epicatechin on fatigue; (−)-epicatechin enhances fatigue resistance and oxidative capacity in mouse muscle [65]. Epicatechin alone or epicatechin plus exercise significantly improved mice treadmill performance by 50% versus water alone or water plus exercise. Also, enhanced in situ muscle fatigue resistance was observed with epicatechin. One very important finding was that there were significant increases in capillarity for epicatechin versus administering water only. Exercise further increased skeletal muscle capillarity when coupled with epicatechin. Structural and metabolic changes in both skeletal and cardiac muscles resulted in greater endurance capacity. The increased capillarity we believe is paralleled by an increase in neurons, as outlined in the previous mechanism figure.

Studies have been conducted to assess whether administering cocoa or chocolate could result in acute improvement of cognitive performance. Usually such studies mainly focus on understanding to what extent prolonged exposures to cocoa can slow the deterioration of cognitive performance associated with aging. Models based on rats can easily provide evidence about the possible effects of prolonged administration of cocoa extracts on cognitive functions. The possible role of (−)-epicatechin in reducing neurodegenerative disorders such as Parkinson's and Alzheimer's diseases has also been highlighted [66, 67].

Bisson and coauthors [60] conducted a randomized experiment on Wistar-Unilever rats. One arm received a cocoa polyphenolic extract (ACTICOA powder, Barry Callebaut, Zurich, Switzerland), for 1 year (rats were between 15 and 27 months old), orally delivered at the dose of 24 mg/kg/day. The other arm was the control group. The objective of the study was to verify whether cocoa polyphenolic extract could affect the onset of age-related cognitive deficits, urinary free dopamine levels, and life span. Age-related cognitive deficits were assessed with light extinction and water maze tests, in which rats were released from the same starting position. The results showed that ACTICOA powder significantly improved cognitive performances in light extinction and water maze paradigms, increased life span, and preserved high urinary free dopamine levels. In other words, short- and long-term memory processes as well as spatial learning were all improved by the cocoa polyphenolic extract.

A similar randomized experiment with shorter exposure to cocoa polyphenolic extract (14 days) was performed on in Wistar rats [68] that were randomly assigned to four groups (control group, control group with heat exposure [HE], ACTICOA powder with HE, vitamin E supplementation with HE) with the aim of evaluating the antioxidant activity of ACTICOA powder (AP) and cognitive performance after HE. Both AP and vitamin E supplementation were effective in reducing the overproduction of free radicals by leucocytes. Heat exposure caused cognitive impairment (measured by light extinction and water maze tests) in the control group after HE, but the daily oral administration of AP or vitamin E protected rats from cognitive impairments after HE by counteracting the overproduction of free radicals.

Scholey and coauthors explored whether cocoa flavonols (CF) could improve cognitive performance during mental efforts in healthy humans [45]. They conducted randomized controlled three-period crossover trials on 30 healthy adults. The three periods corresponded to administration of drinks containing 520-, 994-mg CF, or a control period. The subjects underwent repeated 10-min battery comprising of two serial subtraction tasks (serial threes and serial sevens), a rapid visual processing task, and a mental fatigue scale. Consumption of both 520- and 994-mg CF significantly improved serial threes performance, while the 994-mg CF beverage significantly speeded the rapid visual processing task. However, increases in self-reported "mental fatigue" were significantly attenuated by the consumption of the 520-mg CF beverage only. This study confirms that the acute consumption of CF can improve performance and reduce "mental fatigue" during high-effort cognitive processing in healthy young participants.

Another more recent randomized crossover trial [69] assessing the impact of cocoa flavonols on acute improvement in visual and cognitive functions was conducted in 30 healthy adults who consumed either dark chocolate containing 720-mg CF or white chocolate at different time periods, with a 1-week interval between testing sessions. The following several functions were tested: visual contrast sensitivity, motion sensitivity, and direction and cognitive performance. Assessment used visual spatial working memory for location tasks and a choice reaction time task designed to engage processes of sustained attention and inhibition. Compared to "control" periods, cocoa flavonols significantly improved visual contrast sensitivity and reduced the time required to detect motion direction. It also enhanced spatial memory and performance, but had no effect on motion sensitivity. These acute effects can be explained by increased cerebral blood flow caused by administering cocoa flavonols. The authors also suggest that the chronic effects of cocoa flavonol supplementation on visual function may be larger than the acute effects reported here owing to the neuroprotective properties of dietary flavonoids, along with their effectiveness at suppressing inflammatory processes, and enhancing endothelial function – which, on the whole, are likely to be beneficial for the retina and visual cortex.

These two studies, conducted on healthy adults, provide consistent results about the acute effects of cocoa flavonol intake on specific cognitive functions and partly confirm what has already been observed in animal models. The question now is whether these properties of cocoa flavonols on cognitive functions can also be observed in the elderly. The two studies available examining this issue do not provide consistent results, and it is not clear whether this is due to their design or to the methods used to estimate cognitive functions.

The first, conducted by Nurk and colleagues [21], is an observational cross-sectional study assessing the relationship between intake of flavonoids (chocolate, wine, and tea) and cognitive performance in a large study sample of (N = 2031), cognitive performance was measured by a cognitive test battery including the Kendrick Object Learning Test, the Trail Making Test part A (TMT-A), modified versions of the Digit–Symbol Test, Block Design, Mini-Mental State Examination, and Controlled Oral Word Association Test. Participants who consumed chocolate, wine, or tea exhibited significantly better test scores and lower prevalence of poor cognitive performance than those who did not. Individuals consuming all three presented the best cognitive score. A dose–response relationship was also observed between the number of flavonoids consumed (chocolate wine and tea) and the lowest risks for poor test performance. The association between intake of these flavonoids and cognition performance was dose dependent, but the effect was modest for chocolate. This study suggests that for the elderly, a diet high in certain flavonoid-rich foods can potentially improve cognitive performance in a dose-dependent manner.

However, these results were not confirmed by a trial conducted in the elderly whose objective was to assess the short-term (6 weeks) effects of dark chocolate and cocoa on neuropsychological functioning in healthy older adults. The trial randomized 101 individuals to either a placebo arm or to a treatment arm receiving a 37-g dark chocolate bar and 237 mL of an artificially sweetened cocoa beverage or similar placebo products each day for 6 weeks. Neuropsychological functioning was assessed through the Selective Reminding Test, the Wechsler Memory Scale-III Faces I and Faces II subtests, the Trail Making Test, the Stroop Color–Word Test, the Wechsler Adult Intelligence Scale-III Digit Symbol–Coding subtest, and the Activation–Deactivation Adjective Check List. The authors did not find any difference in neuropsychological variables examined. This study based on a randomized trial failed to demonstrate any acute impact of cocoa intake on neuropsychological functioning. Further research is still needed to evaluate which cognitive functions may be improved by prolonged cocoa administration in the elderly.

Fatigue

The medical definition of fatigue is physical and/or mental exhaustion that can be triggered by stress, medication, overwork, physical/mental illness, or disease. The last part of this definition implies that fatigue can also occur within a framework of clinical manifestations of mental disorders like depression or other diseases implying neurocognitive impairment. Chronic fatigue syndrome (CFS) is a debilitating condition that may frequently start in adults after acute infectious disease episodes (e.g., chickenpox or mononucleosis) as a post-viral fatigue. Risk factors describing why a post-viral fatigue degenerates into a chronic condition are still unknown [70].

However, CFS can also occur without no exposure to infectious disease episodes, and it may be multifactorial [71], possibly related to chronic debilitating diseases or chronic exposure to specific treatments [72]. It is worth noting that an imbalance in various neurotransmitters including serotonin was detected in patients affected by CFS [73]. It is estimated that prevalence rates range from 0.2% to 2.6% [74]. Whatever the cause, CFS impairs patients' daily lives and is associated with poor quality of life [74].

The first to describe the possible positive effects of chocolate on fatigue was the Aztec Emperor Montezuma (who reigned between 1502 and 1520). Antifatigue effects were also described in the Badanius codex (1552) [75].

One study providing some evidence of the antifatigue effects of chocolate consumption was conducted by Sathyapalan and colleagues [64]. The authors included ten consecutive patients with CFS and severe fatigue (scores > 10) on the Chalder Fatigue Scale [76]. Subjects with comorbid psychiatric disorders (according to the DSM IV) and those taking more than 10 g of chocolate per day were excluded. Using a crossover design, the authors compared consumption of 1 bar/day (15 g) polyphenol-rich chocolate containing 85% cocoa solids versus cocoa liquor–free chocolate. The Chalder fatigue score improved significantly after 8 weeks of consumption of active high cocoa polyphenol chocolate but decreased significantly during the period of cocoa liquor–free chocolate. The authors discuss that the results can be attributed to the high polyphenol content within active chocolate and possible synergies among components like anandamide and N-acylethanolamines.

Possible Mechanisms

Chronic fatigue syndrome (CFS) is a medically unexplained illness. Initially it was ascribed to muscle defects, but this was shown to be an impaired neuropsychological function [62] with 18 measurable and significant deficits. CBF was found to be lower in every region of the brain measured in CFS subjects [62]. Not surprisingly, polyphenol-rich dark chocolate (85% cocoa solids) was found after 8 weeks of intervention to significantly improve a standard fatigue score and anxiety and depression score versus an isocaloric low polyphenol chocolate [64].

A proof-of-concept paper, albeit in mice, investigated the effect of (−)-epicatechin on fatigue; (−)-epicatechin enhances fatigue resistance and oxidative capacity in mouse muscle [65]. Epicatechin alone or epicatechin plus exercise significantly improved treadmill performance (50%) versus water alone or water plus exercise. Also, an enhanced in situ muscle fatigue resistance was observed with epicatechin. One very important finding was that there were significant increases in capillarity versus water and exercise further increased skeletal muscle capillarity when coupled with epicatechin. Structural and metabolic changes in both skeletal and cardiac muscles resulted in greater endurance capacity. The increased capillarity we believe is paralleled by an increase in neurons, as outlined in the mechanism scheme.

Summary

This analysis of the literature mainly focusing on both chemistry and biochemistry properties of cocoa compounds and the effects of cocoa intake on mood disorders and cognitive functions clearly shows that despite the short-term effects seen mostly on cognitive functions, there is some evidence to show that sustained administration of cocoa derivatives can also reduce the intensity of anxiety and depressive symptoms as well as significantly improving cognitive functions, especially in the elderly. The evidence is stronger on cognitive performance probably because even short-term effects are marked. However, more experimental and observational studies excluding individuals suffering from chocolate craving and eating disorders are needed to confirm these results. Research in this field should target patients with both of these chronic conditions as well as mood or cognitive disorders, arising either from the chronicity of some diseases (like HIV, hepatitis, or cancer) or as treatment-related side effects. In particular, it is crucial to study whether combining treatment for depression or anxiety with specific interventions like cocoa and other flavonoid administration could help to minimize the toxicity burden of certain treatments. One important example is provided by hepatitis C treatment based on PEG-interferon, ribavirin, and other new forthcoming drugs whose severe associated mood disorders (depression, anger, anxiety) are major causes of treatment refusal, nonadherence, and treatment interruption.

It is already known that some patients undergoing chemotherapy suffer from mood alterations and neurocognitive impairment symptoms. However, as the main clinical worry is survival, associated side effects are often neglected. Nevertheless, they seriously impair patients' quality of life. Further research is also needed to better understand the effects of each chocolate compound and their magnitude on each mood disorder whether or not produced by chronic stress.

We believe that a plethora of evidence in the cardiovascular field for the mechanisms of cocoa polyphenols can be carried over into the area of the brain. Indeed, in several of the studies mentioned here, the authors made a similar hypothesis. The increase in CBF in gray matter suggests that cocoa polyphenols have potential in terms of the prevention and treatment of cerebrovascular deficits such as stroke, Alzheimer's, and dementia. We look forward to future investigations into chocolate and the brain.

Acknowledgments M. P. Carrieri would like to thank Jude Sweeney for the English revision and editing of the sections she wrote in this manuscript.

References

1. Rokholm B, Baker JL, Sorensen TI. The levelling off of the obesity epidemic since the year 1999 – a review of evidence and perspectives. Obes Rev. 2010;11(12):835–46.
2. Schelbert KB. Comorbidities of obesity. Prim Care. 2009;36(2):271–85.
3. Hetherington MM, MacDiarmid JI. "Chocolate addiction": a preliminary study of its description and its relationship to problem eating. Appetite. 1993;21(3):233–46.
4. Rozin P, Levine E, Stoess C. Chocolate craving and liking. Appetite. 1991;17(3):199–212.
5. Vinson JA, Proch J, Bose P, et al. Chocolate is a powerful ex vivo and in vivo antioxidant, an antiatherosclerotic agent in an animal model, and a significant contributor to antioxidants in the European and American diets. J Agric Food Chem. 2006;54(21):8071–6.
6. Maskarinec G. Cancer protective properties of cocoa: a review of the epidemiologic evidence. Nutr Cancer. 2009;61(5):573–9.
7. Selmi C, Cocchi CA, Lanfredini M, et al. Chocolate at heart: the anti-inflammatory impact of cocoa flavanols. Mol Nutr Food Res. 2008;52(11):1340–8.
8. Scalbert A, Manach C, Morand C, et al. Dietary polyphenols and the prevention of diseases. Crit Rev Food Sci Nutr. 2005;45(4):287–306.
9. De Gottardi A, Berzigotti A, Seijo S, et al. Dark chocolate attenuates the post-prandial increase in HVPG in patients with cirrhosis and portal hypertension. J Hepatol. 2010;52(Suppl 1):S9.
10. Smit HJ. Theobromine and the pharmacology of cocoa. In: Fredholm BB, editor. Methylxanthines. Handbook of experimental pharmacology, vol. 200. Berlin: Springer; 2001.
11. Mitchell ES, Slettenaar M, vd Meer N, et al. Differential contributions of theobromine and caffeine on mood, psychomotor performance and blood pressure. Physiol Behav. 2011;104(5):816–22.
12. Gu L, House SE, Wu X, Ou B, et al. Procyanidin and catechin contents and antioxidant capacity of cocoa and chocolate products. J Agric Food Chem. 2006;54(11):4057–61.
13. Baba S, Osakabe N, Natsume M, et al. In vivo comparison of the bioavailability of (+)-catechin, (−)-epicatechin and their mixture in orally administered rats. J Nutr. 2001;131(11):2885–91.
14. Urpi-Sarda M, Monagas M, Khan N, et al. Targeted metabolic profiling of phenolics in urine and plasma after regular consumption of cocoa by liquid chromatography-tandem mass spectrometry. J Chromatogr A. 2009;1216(43):7258–67.

15. Manach C, Scalbert A, Morand C, et al. Polyphenols: food sources and bioavailability. Am J Clin Nutr. 2004;79(5):727–47.
16. O'Leary KA, Day AJ, Needs PW, Sly WS, O'Brien NM, Williamson G. Flavonoid glucuronides are substrates for human liver beta-glucuronidase. FEBS Lett. 2001;503(1):103–6.
17. Holt RR, Lazarus SA, Sullards MC, et al. Procyanidin dimer B2 [epicatechin-(4beta-8)-epicatechin] in human plasma after the consumption of a flavanol-rich cocoa. Am J Clin Nutr. 2002;76(4):798–804.
18. Cserr HF, Bundgaard M. Blood-brain interfaces in vertebrates: a comparative approach. Am J Physiol. 1984;246(3 Pt 2):R277–88.
19. Faria A, Pestana D, Teixeira D, et al. Insights into the putative catechin and epicatechin transport across blood-brain barrier. Food Funct. 2011;2(1):39–44, Epub 17 Nov 2010.
20. Stevenson DE, Hurst RD. Polyphenolic phytochemicals – just antioxidants or much more? Cell Mol Life Sci. 2007;64(22):2900–16.
21. Nurk E, Refsum H, Drevon CA, et al. Intake of flavonoid-rich wine, tea, and chocolate by elderly men and women is associated with better cognitive test performance. J Nutr. 2009;139(1):120–7.
22. Miller AA, De Silva TM, Judkins CP, et al. Augmented superoxide production by Nox2-containing NADPH oxidase causes cerebral artery dysfunction during hypercholesterolemia. Stroke. 2010;41(4):784–9.
23. Drouin A, Bolduc V, Thorin-Trescases N, et al. Catechin treatment improves cerebrovascular flow-mediated dilation and learning abilities in atherosclerotic mice. Am J Physiol Heart Circ Physiol. 2011;300(3):H1032–43.
24. Ottaviani JI, Momma TY, Heiss C, et al. The stereochemical configuration of flavanols influences the level and metabolism of flavanols in humans and their biological activity in vivo. Free Radic Biol Med. 2011;50(2):237–44.
25. Spencer JP. Flavonoids and brain health: multiple effects underpinned by common mechanisms. Genes Nutr. 2009;4(4):243–50.
26. Parker G, Crawford J. Chocolate craving when depressed: a personality marker. Br J Psychiatry. 2007;191: 351–2.
27. Willner P, Benton D, Brown E, et al. "Depression" increases "craving" for sweet rewards in animal and human models of depression and craving. Psychopharmacology (Berl). 1998;136(3):272–83.
28. Cramer KM, Hartleib M. The attitudes to chocolate questionnaire: a psychometric evaluation. Pers Individ Differ. 2001;31(3):931–42.
29. Schuman M, Gitlin MJ, Fairbanks L. Sweets, chocolate, and atypical depressive traits. J Nerv Ment Dis. 1987; 175(8):491–5.
30. Hill AJ, Weaver CF, Blundell JE. Food craving, dietary restraint and mood. Appetite. 1991;17(3):187–97.
31. Lester D, Bernard D. Liking for chocolate, depression, and suicidal preoccupation. Psychol Rep. 1991;69(2):570.
32. Dallard I, Cathebras P, Sauron C, et al. Is cocoa a psychotropic drug? Psychopathologic study of a population of subjects self-identified as chocolate addicts. Encéphale. 2001;27(2):181–6.
33. Parker G, Parker I, Brotchie H. Mood state effects of chocolate. J Affect Disord. 2006;92(2–3):149–59.
34. http://www.who.int/whosis/whostat2007_10highlights.pdf
35. Wager-Smith K, Markou A. Depression: a repair response to stress-induced neuronal microdamage that can grade into a chronic neuroinflammatory condition? Neurosci Biobehav Rev. 2011;35(3):742–64.
36. Vakilian A, Iranmanesh F. Assessment of cerebrovascular reactivity during major depression and after remission of disease. Ann Indian Acad Neurol. 2010;13(1):52–6.
37. Richieri R, Boyer L, Farisse J, et al. Predictive value of brain perfusion SPECT for rTMS response in pharmacoresistant depression. Eur J Nucl Med Mol Imaging. 2011;38(9):1715–22.
38. Schroeter H, Heiss C, Balzer J, et al. (–)-Epicatechin mediates beneficial effects of flavanol-rich cocoa on vascular function in humans. Proc Natl Acad Sci USA. 2006;103(4):1024–9.
39. Francis ST, Head K, Morris PG, et al. The effect of flavanol-rich cocoa on the fMRI response to a cognitive task in healthy young people. J Cardiovasc Pharmacol. 2006;47(Suppl 2):S215–20.
40. Larsson A, Skoog I, Aevarsson, et al. Regional cerebral blood flow in normal individuals aged 40, 75 and 88 years studied by 99Tc(m)-d, l-HMPAO SPET. Nucl Med Commun. 2001;22(7):741–6.
41. Krieg Jr EF, Chrislip DW, Letz RE, et al. Neurobehavioral test performance in the third National Health and Nutrition Examination Survey. Neurotoxicol Teratol. 2001;23(6):569–89.
42. Fisher ND, Sorond FA, Hollenberg NK. Cocoa flavanols and brain perfusion. J Cardiovasc Pharmacol. 2006;47(Suppl 2):S210–4.
43. Sorond FA, Lipsitz LA, Hollenberg NK, et al. Cerebral blood flow response to flavanol-rich cocoa in healthy elderly humans. Neuropsychiatr Dis Treat. 2008;4(2):433–40.
44. Wang J, Rao H, Wetmore GS, et al. Perfusion functional MRI reveals cerebral blood flow pattern under psychological stress. Proc Natl Acad Sci USA. 2005;102(49):17804–9.
45. Scholey AB, French SJ, Morris PJ, et al. Consumption of cocoa flavanols results in acute improvements in mood and cognitive performance during sustained mental effort. J Psychopharmacol. 2010;24(10):1505–14.
46. Martin FP, Rezzi S, Pere-Trepat E, et al. Metabolic effects of dark chocolate consumption on energy, gut microbiota, and stress-related metabolism in free-living subjects. J Proteome Res. 2009;8(12):5568–79.

47. Macht M, Roth S, Ellgring H. Chocolate eating in healthy men during experimentally induced sadness and joy. Appetite. 2002;39(2):147–58.
48. Macht M, Mueller J. Immediate effects of chocolate on experimentally induced mood states. Appetite. 2007;49(3):667–74.
49. Messaoudi M, Bisson JF, Nejdi A, et al. Antidepressant-like effects of a cocoa polyphenolic extract in Wistar-Unilever rats. Nutr Neurosci. 2008;11(6):269–76.
50. Yamada T, Yamada Y, Okano Y, et al. Anxiolytic effects of short- and long-term administration of cacao mass on rat elevated T-maze test. J Nutr Biochem. 2009;20(12):948–55.
51. Raikkonen K, Pesonen AK, Jarvenpaa AL, et al. Sweet babies: chocolate consumption during pregnancy and infant temperament at six months. Early Hum Dev. 2004;76(2):139–45.
52. Strandberg TE, Strandberg AY, Pitkala K, et al. Chocolate, well-being and health among elderly men. Eur J Clin Nutr. 2008;62(2):247–53.
53. Golman T, Buzzi J, Carre S. Efficacité du chocolat noir (70%) contre la dépression des patients agés résidant en E.H.P.A.D=Efficacy of dark chocolate (70%) on depression in patients living in a home for the dependent elderly (EHPAD). La Revue francophone de gériatrie et de gérontologie. 2007;XIV(133):136–9.
54. Rose N, Koperski S, Golomb BA. Mood food: chocolate and depressive symptoms in a cross-sectional analysis. Arch Intern Med. 2010;170(8):699–703.
55. WHO Survey Data Centre.: World Health Survey. Geneva, World Health Organization. http://surveydata.who.int/ (2007).
56. Carrieri MP, Cohen J, Winnock M, et al. Chocolate intake, depression, and clinical progression in HIV-HCV coinfected patients: still more questions than answers. Arch Intern Med. 2010;170(17):1607, author reply 1608–1609.
57. Preau M, Bonnet A, Bouhnik AD, et al. Anhedonia and depressive symptomatology among HIV-infected patients with highly active antiretroviral therapies (ANRS-EN12-VESPA). Encéphale. 2008;34(4):385–93.
58. Bouhnik AD, Preau M, Vincent E, et al. Depression and clinical progression in HIV-infected drug users treated with highly active antiretroviral therapy. Antivir Ther. 2005;10(1):53–61.
59. Terminologie de neuropsychologie et de neurologie du comportement. Recherche et réd. Louise Bérubé., c1991., 176 p. Reproduit avec la permission de Les Éditions de la Chenelière Inc., p. 7.
60. Bisson JF, Nejdi A, Rozan P, et al. Effects of long-term administration of a cocoa polyphenolic extract (Acticoa powder) on cognitive performances in aged rats. Br J Nutr. 2008;100(1):94–101.
61. van Praag H, Lucero MJ, Yeo GW, et al. Plant-derived flavanol (−)epicatechin enhances angiogenesis and retention of spatial memory in mice. J Neurosci. 2007;27(22):5869–78.
62. Busichio K, Tiersky LA, Deluca J, et al. Neuropsychological deficits in patients with chronic fatigue syndrome. J Int Neuropsychol Soc. 2004;10(2):278–85.
63. Ciccone DS, Busichio K, Vickroy M, et al. Psychiatric morbidity in the chronic fatigue syndrome: are patients with personality disorder more physically impaired? J Psychosom Res. 2003;54(5):445–52.
64. Sathyapalan T, Beckett S, Rigby AS, et al. High cocoa polyphenol rich chocolate may reduce the burden of the symptoms in chronic fatigue syndrome. Nutr J. 2010;9:55.
65. Nogueira L, Ramirez-Sanchez I, Perkins GA, et al. (−)-Epicatechin enhances fatigue resistance and oxidative capacity in mouse muscle. J Physiol. 2011;589:4015–631.
66. Pan T, Jankovic J, Le W. Potential therapeutic properties of green tea polyphenols in Parkinson's disease. Drugs Aging. 2003;20(10):711–21.
67. Schroeter H, Williams RJ, Matin R, et al. Phenolic antioxidants attenuate neuronal cell death following uptake of oxidized low-density lipoprotein. Free Radic Biol Med. 2000;29(12):1222–33.
68. Rozan P, Hidalgo S, Nejdi A, et al. Preventive antioxidant effects of cocoa polyphenolic extract on free radical production and cognitive performances after heat exposure in Wistar rats. J Food Sci. 2007;72(3):S203–6.
69. Field DT, Williams CM, Butler LT. Consumption of cocoa flavanols results in an acute improvement in visual and cognitive functions. Physiol Behav. 2011;103(3–4):255–60.
70. Jason LA, Porter N, Brown M, et al. CFS: A review of epidemiology and natural history studies. Bull IACFS ME. 2009;17(3):88–106.
71. Dinos S, Khoshaba B, Ashby D, et al. A systematic review of chronic fatigue, its syndromes and ethnicity: prevalence, severity, co-morbidity and coping. Int J Epidemiol. 2009;38(6):1554–70.
72. Griffith JP, Zarrouf FA. A systematic review of chronic fatigue syndrome: don't assume it's depression. Prim Care Companion J Clin Psychiatry. 2008;10(2):120–8.
73. Ortega-Hernandez OD, Cuccia M, Bozzini S, et al. Autoantibodies, polymorphisms in the serotonin pathway, and human leukocyte antigen class II alleles in chronic fatigue syndrome: are they associated with age at onset and specific symptoms? Ann N Y Acad Sci. 2009;1173:589–99.
74. Reid S, Chalder T, Cleare A, et al. Chronic fatigue syndrome. BMJ. 2000;320(7230):292–6.
75. Dillinger TL, Barriga P, Escarcega S, et al. Food of the gods: cure for humanity? A cultural history of the medicinal and ritual use of chocolate. J Nutr. 2000;130(8S Suppl):2057S–72.
76. Chalder T, Berelowitz G, Pawlikowska T, et al. Development of a fatigue scale. J Psychosom Res. 1993;37(2):147–53.

Part V
Non–Disease-Related Clinical Studies

Chapter 30
Chocolate: Mood

Marlene M. Millen and Beatrice A. Golomb

Key Points

- Chocolate and mood are strongly intertwined in the popular imagination. The strong body of chocolate lore is *not* matched by a strong body of evidence.
- Theoretically, chocolate bears many substances and properties that could influence mood, favorably and adversely. The balance of favorable to adverse effects could differ by individual and timing after chocolate consumption.
- Favorable effects on mood might be mediated by methylxanthines (caffeine and theobromine), catechins, biogenic amines like serotonin and ethylamine, and cannabinoid receptor agonists like anandamine. Antioxidant functions and cell energy support through mitochondrial biogenesis (mediated by epicatechin) might support positive mood.
- Detrimental effects on mood might be mediated by methylxanthine excess or withdrawal, effects of copper, or effects of trans-fatty acids (where these have polluted the chocolate product).
- Empirically, few studies in humans have examined the relation of chocolate to mood. Based on the scant extant literature, mood associations appear to be favorable immediately after eating chocolate that is perceived as palatable. In studies not focused upon the immediate effect, greater chocolate consumption has been linked to lower mood, but it is not clear which (if either) drives which.

Keywords Chocolate • Mood • Depression • Methylxanthines • Biogenic amines • Anandamine • Catechins • Antioxidant • Mmm

M.M. Millen, M.D.
Department of Internal Medicine, San Diego School of Medicine,
University of California, San Diego, CA, USA
e-mail: mmillen@ucsd.edu

B.A. Golomb, M.D., Ph.D. (✉)
Department of Medicine and Department of Family and Preventive Medicine,
San Diego School of Medicine, University of California, La Jolla, CA, USA
e-mail: bgolomb@ucsd.edu

R.R. Watson et al. (eds.), *Chocolate in Health and Nutrition*, Nutrition and Health 7,
DOI 10.1007/978-1-61779-803-0_30,

Introduction

Chocolate is cheaper than therapy and you don't need an appointment.

– Catherine Aitken

Effects of chocolate on mood appear to be strongly established in the popular imagination: Chocolate is viewed as a food that improves mood. How strong is the basis of support for this? A range of mechanisms by which chocolate may influence mood have been identified, both favorable and adverse.

In this chapter, we will review the mechanisms by which chocolate might influence mood, in the shorter and longer term. Additionally, the empirical evidence on the association of chocolate to mood in humans – surprisingly limited – will be reviewed.

Method

Articles describing or discussing potential mechanism were identified using a PubMed search on date May 5, 2011, with English language articles including the title word "chocolate" and keywords "mechanism" and "mood." Additional targeted searches focused on specific mechanisms.

Articles bearing original level data on human associations of chocolate to mood were identified using a PubMed search on date May 5, 2011, for English language articles, using keywords "chocolate," "cocoa," "mood," "anxiety," and "depression." These were supplemented by articles based on author familiarity.

Results

Potential mechanisms of chocolate's effects on mood, established in animal studies, are listed in Table 30.1. This table includes both substances and actions.

Empirical studies in humans, associating chocolate with mood, include those listed in Table 30.2.

Discussion

Despite much lore and public interest in chocolate and mood, we identified only eight articles with original data examining this relation in humans.

Two studies were experimental, but focused on very short-term effects [1, 2]. One of these examined mood triggered by an emotion-inducing film clip [2]. Elicited mood in response to a stranger's observed plight (in the case of film-clip-triggered emotion) might plausibly be labeled empathy, and it might be worthwhile to ask whether it is desirable to blunt empathic modulation of emotional tone. The relation of this type of exogenous/empathic mood to depression requires elucidation. However, in general, immediate effects of palatable chocolate on mood were favorable, consistent with popular perception, though the effect dissipated quickly. Palatability of the chocolate related to the mood benefits [2]. One could conjecture common chemical underpinnings for effects on palatability and mood, among a range of potential explanations.

The other studies assessed how mood related to chocolate consumption, as assessed by some versions of chocolate consumption recall (i.e., these examined intermediate-term effects; chocolate was

Table 30.1 Mechanisms relevant to mood effects of chocolate

Mechanism	Putative direction	Reason	Comments
Antioxidation (polyphenols)	Favorable	Oxidative stress has been linked to major depression [19] and is thought to play a role in its pathophysiology [20]. Chocolate is rich in polyphenol antioxidants [21–25], which may protect against oxidative injury	
Blood-brain barrier (BBB) defense	Favorable	Antioxidants (see above) defend the BBB, further protecting the brain from toxic insults (themselves often promoting oxidative stress) [26–29]	Theobromine itself crosses the BBB [30]
Biogenic amines Dimethylamine Ethylamine Histamine Isobutylamine Normetanephrine Octopamine Phenylethylamine Serotonin Synephrine Trimethylamine Tryptamine Tyramine	Favorable or adverse	Biogenic amines in chocolate have potential drug-like neurophysiological effects [31, 32] Several target the serotonin and adrenergic systems that are targeted by antidepressant medications Cocoa increased serotonin turnover in rats [33]. Its effects on amines were presumed to relate to effects on anxiety	In the Yamada study, cocoa was anxiolytic in the short term, but not with chronic use [33]
Sympathomimetic effects from biogenic amines	Favorable or adverse	Biogenic amines in chocolate (e.g., PEA, tyramine) have sympathomimetic/stimulant effects [34] Adrenergic as well as serotonergic agents are major constituents of pharmacological antidepressants See also, below, methylxanthines and methylxanthine withdrawal	See also methylxanthines
Methylxanthines, caffeine, and theobromine, as stimulants	Favorable or adverse	Chocolate contains caffeine and theobromine (both methylxanthines), which may lead to monoamine/adrenergic activation, which can have antidepressant effects [32, 35, 36] However, excess caffeine may also produce dysphoric and anxiogenic effects [37, 38] See also sleep, methylxanthine withdrawal	See also sleep

(continued)

Table 30.1 (continued)

Mechanism	Putative direction	Reason	Comments
Methylxanthine withdrawal	Adverse	Withdrawal from methylxanthines (the literature focuses more on caffeine) is linked to dysphoric mood [39–42]	
Magnesium repletion	Favorable	Chocolate is rich in magnesium and could help correct magnesium deficiency if present. Favorable effects on cell function may translate to benefits to psychological well-being and mood [32]	
Tetrahydro-beta-carbolines	Unclear	Tetrahydro-beta-carbolines have been designated as potential neuroalkaloids in chocolate, but mood/behavior ramifications remain unclear [43]	
Epicatechin/mitochondrial density, capillarity	Favorable	Epicatechin, derived from chocolate, increased mitochondrial biogenesis as well as capillarity in a study in rats [44]. These effects, which likely undergirded benefits to exercise capacity that were observed, [44] benefit cell energy, which is relevant to mood [7] Moreover, improved exercise capacity may increase exercise, with potential mood benefits [45, 46]	
Trans fat: proinflammatory, prooxidant	Adverse	Some chocolate bars and chocolate food products bear trans-hydrogenated fats. Trans fats inhibit delta-6 desaturase and thus long-chain omega-3 production [47]. Long-chain omega-3s show favorable mood and behavior associations in many studies [48–50]	Omega-3 [51]: Our data show an association of trans fats to aggression (and also depression, unpublished) [13], and other data show a relation to depression [12]

Sleep disruption	Adverse	Methylxanthines are linked to sleep disruption particularly in sensitive individuals [38, 52, 53]	Short-term sleep deprivation has antidepressant effect, but ongoing adverse or inadequate sleep may be prodepressant [54, 55]
Cannabinoid agonist fatty acids	Uncertain	Fatty acids in chocolate may stimulate the endogenous cannabinoid system, bind/serve as agonists for the cannabinoid receptor Anandamine N-oleoylethanolamine N-linoleoylethanolamine 1. Bind/serve as agonist for cannabinoid receptors [56, 57] 2. May be associated with reinforcing action of chocolate [58]	
Microflora contamination	Potentially adverse	Potential for abreaction to microorganisms or toxins they elaborate (e.g., fungal biotoxins) [59, 60]	
Copper balance	Adverse	Copper is extensively used as a fungicide in cocoa crops, not excluding crops that are labeled organic [14]. Chocolate contains measurable copper, more so for dark chocolate [14, 61] High copper and high copper-zinc ratio are linked to depression [15–18]. (We speculate the Cu/Zn SOD and antioxidant defense is involved in this)	
Sensory	Favorable	Hedonic (presumed) [32]. Cannot exclude neurochemical/chemical effects yielding positive hedonic experience, which is then associatively attributed to sensory aspects. (Indeed, according to one analysis, epicatechin content dominated sensory quality prediction for cocoa [62])	

Table 30.2 Chocolate and mood in humans: empirical evidence

Citation	Sample size	Design	Sample characteristics	Chocolate assessment	Mood outcome assessment	Result	Direction/comment
Barkeling et al. 2002 [3]	362 women	Cross-sectional	Nondiabetic Swedish women age 34–64	Past 2 months chocolate vs. depressive sx now	Self-rating scale for affective syndromes (CPRS S-A)	Among assessed sweet foods, only chocolate was related to mood: ($r=0.17$, $P<0.01$, more chocolate associated with more adverse mood)	
Wolz et al. 2009 [63]	274 PD; 234 controls	Cross-sectional	Parkinson disease patients mean age, 66.2; sex, 34.7% males	Chocolate recall	Beck Depression Inventory	Higher chocolate intake in patient with PD. No association with depression scores.	This article may indirectly bear on chocolate consumption and mood. It shows increased chocolate (but not other sweets) is consumed in PD, a condition mediated by oxidative stress driving mitochondrial damage in a vicious cycle. (Among those with PD, chocolate consumption was not linked to mood.) It *suggests* but does not conclusively demonstrate adaptive consumption of chocolate (which boosts antioxidant defenses and possibly mitochondrial function, in a condition in which oxidative stress and mitochondrial dysfunction play central pathogenic roles). This may be relevant to the Rose 2010 article linking adverse mood (since depressed mood may also have a mitochondrial/oxidative basis) to chocolate consumption
Strandberg et al. 2008 [64]	1,367	Cross-sectional	Elderly males average age 76	Intermediate term: compared elderly who preferred chocolate to other sweets (excluded those who ate no sweets)	Zung depression score	Positive mood associations with feeling of loneliness ($P=0.01$), feeling of happiness ($P=0.01$), having plans for the future ($P=0.0002$), and the Zung depression score ($P=0.02$)	Favorable, association (elderly males)
Rose et al. 2010 [4]	931	Cross-sectional	Male and female adults, 20–85; not using antidepressants and without diabetes or CAD	Intermediate term: 4-day food record and self-rated chocolate times/week	CES-D	Increase in chocolate consumption associated with higher CES-D depression scores, including more positive depression screens and probable major depression	Unfavorable association (broadly sampled adults)

Carrieri et al. 2010 [5]	574	Cohort	HCV-HIV patients	6-month recall of chocolate intake	CES-D	Individuals who did not consume chocolate were more likely to present probable major depression (odds ratio, 1.8, [95% confidence interval, 1.2–2.7]; $P=.006$). However, abstainer effects are a consideration [7]	Favorable association (HCV/HIV patients)
Macht et al. 2007 [2]	48	Experimental (chocolate vs. water)* on film-stimulated emotion	Normal weight men and women	Momentary mood. Gave chocolate vs. water after viewing film clips to induce mood effect	25-Point scale of momentary mood	Chocolate (vs. water) improved self-rated mood after sad film ($p=0.009$), but not neutral film. Borderline after positive film ($p=0.07$)	Favorable, experimental
Macht et al. 2007 [2]	113	Experimental study (palatable vs. unpalatable chocolate vs. nothing) on film-stimulated emotion*	Normal weight men and women	Momentary mood: measured baseline before film clip, 1, 2, 3 min after chocolate ingestion	25-Point scale of momentary mood	Palatable chocolate had short-lived benefit to experimentally induced negative mood state vs. unpalatable or nothing. Effect disappeared after 3 min	Palatability: subjects rated one type of milk and three types of "plain" chocolate (70%, 85%, and 99% cocoa) for palatability 1/3 Given the chocolate they rated most palatable; 1/3 given the chocolate they rated least palatable; 1/3 given nothing Film clip to induce sad mood
Macht 2006 [1]	37	Experimental	Healthy female psychology class students age 19–30 in Germany	Experimentally administered – Ritter Sport chocolate bar (50 g and 270 kcal) vs. an apple (170 g and 90 kcal) or nothing	Mood was rated on a bipolar scale from 0 ("extremely bad") to 10 ("extremely good"). Assessed at times 5, 30, 60, and 90 min after eating	Mood rated significantly higher after eating chocolate bar vs. apple ($P<0.001$). (Mood in turn rated higher after eating apple than nothing)	Favorable, experimental, but short time frame for effect

Abbreviations: *CAD* coronary artery disease, *CES-D* Center for Epidemiological Studies Depression scale, *BDI* Beck Depression Inventory, *HCV* hepatitis C virus, *HIV* human immunodeficiency virus, *PD* Parkinson disease

*Subjects first rated which of six "Ritter Sport" chocolate types (Alfred Ritter GmbH) they preferred and later received one piece of chocolate (5 g) immediately after viewing a film clip (four pieces in all). Preferences: 12: milk chocolate (MC) + hazelnuts and raisins ("Rum-Traube-Nuss," 6 per deprivation condition, i.e., 6:6), 9 MC + cocoa cream paste ("Nugat," 4:5), 7 MC + yogurt filling ("Olympia," 4:3); 4 MC + desiccated coconut ("Cocos," 2:2); 9 white chocolate + hazelnuts ("Weisse Voll-nuss," 4:5); 7 plain chocolate ("Halbbitter," 4 :3)

not necessarily eaten shortly prior to survey administration). Of these, one was primarily an examination of chocolate in Parkinson disease, and one was published only as a letter to the editor and expressed findings in coinfection with HIV-HCV. Both the latter study and one study focused on elderly men bore favorable associations, in these specialized samples. Two others were in generally healthy groups: One was in women [3] and one in a relatively broad samples of adults [4]. In both of these studies, more unfavorable mood scores were associated with higher chocolate consumption, specifically. Thus, in the former study, this was true for chocolate and not other sweets [3]. In our own (the latter) study, chocolate showed the association; calories, carbohydrates, caffeine, and fat did not [4].

In the HIV-HCV sample, the study examined abstinence from chocolate versus consumption of chocolate [5]. There are issues specific to the specialized population and to the study design. First, chocolate, like coffee, may benefit the underlying liver disease in HCV, with implications to mood sequelae of illness. Benefits, if present, may therefore be illness specific. Second, abstainer effects may be present, whereby persons who do not consume, or tolerate, products like chocolate may fail to tolerate them due to other adverse circumstances (such as impaired ability to metabolize or detoxify certain classes of substance) that may also bear on the risk of depression [6].

The most general samples identified a significant adverse association of chocolate to mood in adult women and also men. However, even in these relatively broader samples in the other studies cited, directionality and causality remain undefined by cross-sectional association. It cannot be ascertained from a cross-sectional design to what degree low mood motivated chocolate consumption, chocolate consumption fostered low mood, both (e.g., through short-term favorable and intermediate-term adverse mechanisms), or neither (e.g., common factors fostering both low mood and chocolate consumption, among other possibilities) [4].

We add the speculation that both favorable and adverse associations may be causally valid with the same (nonimmediate) time course, in different individuals, perhaps even for related reasons. As one possibility, suppose persons in need of antioxidant defense and/or mitochondrial support for mood support are adaptively driven to consume chocolate more frequently. This is one potential interpretation of the Parkinson disease data. Depression with somatic symptoms, at least, has been strongly linked to mitochondrial dysfunction [7], which is strongly linked to Parkinson disease, and could account for why those with Parkinson disease eat more chocolate. Additionally, a range of oxidative stressors have been linked to adverse mood and behavior (copper, discussed here, represents one example). Notably, our study found chocolate to be linked to aggression, as well as depression [8]. Then, in a generally healthy population, more frequent chocolate consumption could statistically reflect persons with greater underlying pathology – that is, greater need for the adaptation. In contrast, in a population selected for a condition linked to oxidative stress and/or mitochondrial dysfunction (elderly have greater mitochondrial dysfunction [9, 10], and HIV medications are mitochondrial toxins [11]), then *failure* to consume chocolate may reflect failure to benefit from a beneficial adaptation, providing favorable associations to mood. Clearly, experimental studies examining mood as a function of chocolate consumption are needed to assess this and other hypotheses and to determine the causal direction (if any) or, perhaps, causal directions.

Indeed, a range of mechanisms were identified that could produce favorable or adverse effects of chocolate on mood or potentially both with differing time course. *Benefits* may arise via effects on neurochemistry (serotonin, phenylethylamine, cannabinoid receptor effects), cell energy and antioxidation (mitochondrial density benefits of epicatechin), and phosphodiesterase inhibition via theobromine. *Adverse* effects on mood may arise by the same phosphodiesterase inhibition, which may impair sleep in sensitive individuals (sleep impairment and deprivation are linked to depression, not just as effects of depression), or indeed by withdrawal effects that arise following the favorable effects. (Care must be taken in stipulating assessed time course of association.) Chocolate products sometimes contain trans-hydrogenated fats, which have been linked – as has chocolate – to depression [4, 12] and aggression [8, 13]. Chocolate can be high in copper, in part through use of copper as a fungicide in cacao crops [14], and high copper/zinc ratios are linked adversely to mood and behavior [15–18].

The question remains unanswered: What does chocolate do to mood? But what does emerge, in reviewing the extant literature, is a richer (no pun intended) picture of the implications of chocolate to brain function and to mood – with potential for modulation by individual differences based on biological factors in the individual, and differences in the chocolate products.

Implications: As chocolate is increasingly recommended for cardiovascular benefits and continues to be enjoyed for itself,[1] there is a need for more empirical research to better define the causal effect, or effects, of chocolate on mood and attending to different time courses of effect, impact of different composition of the chocolate product, and effect modifiers in the individual.

Acknowledgments We wish to thank Sabrina Koperski for editorial and administrative assistance. Golomb wishes to thank her coauthor for having fitting initials: MMM.

References

1. Macht M, Dettmer D. Everyday mood and emotions after eating a chocolate bar or an apple. Appetite. 2006; 46:332–6.
2. Macht M, Mueller J. Immediate effects of chocolate on experimentally induced mood states. Appetite. 2007; 49:667–74.
3. Barkeling B, Linne Y, Lindroos AK, Birkhed D, Rooth P, Rossner S. Intake of sweet foods and counts of cariogenic microorganisms in relation to body mass index and psychometric variables in women. Int J Obes Relat Metab Disord. 2002;26:1239–44.
4. Rose N, Koperski S, Golomb BA. Mood food: chocolate and depressive symptoms in a cross-sectional analysis. Arch Intern Med. 2010;170:699–703.
5. Carrieri MP, Cohen J, Winnock M, Salmon D. Chocolate intake, depression, and clinical progression in HIV-HCV coinfected patients: still more questions than answers. Arch Intern Med. 2010;170:1607, author reply 8–9.
6. Golomb BA, Koperski S, Rose N. Chocolate consumption and effects on serotonin synthesis – reply. Arch Intern Med. 2010;170:1608–9.
7. Gardner A, Boles RG. Symptoms of somatization as a rapid screening tool for mitochondrial dysfunction in depression. Biopsychosoc Med. 2008;2:7.
8. Golomb BA, Koperski S, White HL. Chocolate consumption is linked to aggression. Circulation. 2011; 173(Suppl):P53.
9. Smigrodzki RM, Khan SM. Mitochondrial microheteroplasmy and a theory of aging and age-related disease. Rejuvenation Res. 2005;8:172–98.
10. Kujoth GC, Hiona A, Pugh TD, Someya S, Panzer K, Wohlgemuth SE, et al. Mitochondrial DNA mutations, oxidative stress, and apoptosis in mammalian aging. Science. 2005;309:481–4.
11. Herman JS, Easterbrook PJ. The metabolic toxicities of antiretroviral therapy. Int J STD AIDS. 2001;12:555–62, quiz 63–4.
12. Sanchez-Villegas A, Verberne L, De Irala J, Ruíz-Canela M, Toledo E, Serra-Majem L, et al. Dietary fat intake and the risk of depression: the SUN Project. PLoS One. 2011;6:e16268.
13. Golomb BA, Evans MA, Dimsdale JE. Trans fat consumption linked to aggression. Circulation. 2008;117:e237.
14. Dos Santos WN, da Silva EG, Fernandes MS, Araujo RG, Costa AC, Vale MG, et al. Determination of copper in powdered chocolate samples by slurry-sampling flame atomic-absorption spectrometry. Anal Bioanal Chem. 2005;382:1099–102.
15. Narang RL, Gupta KR, Narang AP, Singh R. Levels of copper and zinc in depression. Indian J Physiol Pharmacol. 1991;35:272–4.
16. Manser WW, Khan MA, Hasan KZ. Trace element studies on Karachi population. Part IV: blood copper, zinc, magnesium and lead levels in psychiatric patients with depression, mental retardation and seizure disorders. J Pak Med Assoc. 1989;39:269–74.
17. Crayton JW, Walsh WJ. Elevated serum copper levels in women with a history of post-partum depression. J Trace Elem Med Biol. 2007;21:17–21.

[1] "Candy doesn't have to have a point. That's why it's candy." – Charlie, from *Charlie and the Chocolate Factory.*

18. Schlegel-Zawadzka M, Zieba A, Dudek D, Zak-Knapik J, Nowak G. Is serum copper a "trait marker" of unipolar depression? A preliminary clinical study. Pol J Pharmacol. 1999;51:535–8.
19. Khanzode SD, Dakhale GN, Khanzode SS, Saoji A, Palasodkar R. Oxidative damage and major depression: the potential antioxidant action of selective serotonin re-uptake inhibitors. Redox Rep. 2003;8:365–70.
20. Bilici M, Efe H, Koroglu MA, Uydu HA, Bekaroglu M, Deger O. Antioxidative enzyme activities and lipid peroxidation in major depression: alterations by antidepressant treatments. J Affect Disord. 2001;64:43–51.
21. Miller KB, Stuart DA, Smith NL, Lee CY, McHale NL, Flanagan JA, et al. Antioxidant activity and polyphenol and procyanidin contents of selected commercially available cocoa-containing and chocolate products in the United States. J Agric Food Chem. 2006;54:4062–8.
22. Vinson JA, Proch J, Bose P, Muchler S, Taffera P, Shuta D, et al. Chocolate is a powerful ex vivo and in vivo antioxidant, an antiatherosclerotic agent in an animal model, and a significant contributor to antioxidants in the European and American Diets. J Agric Food Chem. 2006;54:8071–6.
23. Gu L, House SE, Wu X, Ou B, Prior RL. Procyanidin and catechin contents and antioxidant capacity of cocoa and chocolate products. J Agric Food Chem. 2006;54:4057–61.
24. Serafini M, Bugianesi R, Maiani G, Valtuena S, De Santis S, Crozier A. Plasma antioxidants from chocolate. Nature. 2003;424:1013.
25. Spadafranca A, Martinez Conesa C, Sirini S, Testolin G. Effect of dark chocolate on plasma epicatechin levels, DNA resistance to oxidative stress and total antioxidant activity in healthy subjects. Br J Nutr. 2010;103: 1008–14.
26. Lehner C, Gehwolf R, Tempfer H, Krizbai I, Hennig B, Bauer HC, et al. Oxidative stress and blood-brain barrier dysfunction under particular consideration of matrix metalloproteinases. Antioxid Redox Signal. 2011;15: 1305–23.
27. Barichello T, Lemos JC, Generoso JS, Cipriano AL, Milioli GL, Marcelino DM, et al. Oxidative stress, cytokine/chemokine and disruption of blood-brain barrier in neonate rats after meningitis by *Streptococcus agalactiae*. Neurochem Res. 2011;36(10):1922–30.
28. Lukic-Panin V, Deguchi K, Yamashita T, Shang J, Zhang X, Tian F, et al. Free radical scavenger edaravone administration protects against tissue plasminogen activator induced oxidative stress and blood brain barrier damage. Curr Neurovasc Res. 2010;7:319–29.
29. Lochhead JJ, McCaffrey G, Quigley CE, Lochhead JJ, McCaffrey G, Quigley CE, et al. Oxidative stress increases blood-brain barrier permeability and induces alterations in occludin during hypoxia-reoxygenation. J Cereb Blood Flow Metab. 2010;30:1625–36.
30. Eteng MU, Eyong EU, Akpanyung EO, Agiang MA, Aremu CY. Recent advances in caffeine and theobromine toxicities: a review. Plant Foods Hum Nutr. 1997;51:231–43.
31. Pastore P, Favaro G, Badocco D, Tapparo A, Cavalli S, Saccani G. Determination of biogenic amines in chocolate by ion chromatographic separation and pulsed integrated amperometric detection with implemented wave-form at Au disposable electrode. J Chromatogr A. 2005;1098:111–5.
32. Bruinsma K, Taren DL. Chocolate: food or drug? J Am Diet Assoc. 1999;99:1249–56.
33. Yamada T, Yamada Y, Okano Y, Terashima T, Yokogoshi H. Anxiolytic effects of short- and long-term administration of cacao mass on rat elevated T-maze test. J Nutr Biochem. 2009;20:948–55.
34. Broadley KJ, Akhtar Anwar M, Herbert AA, Fehler M, Jones EM, Davies WE, et al. Effects of dietary amines on the gut and its vasculature. Br J Nutr. 2009;101:1645–52.
35. Smit HJ. Theobromine and the pharmacology of cocoa. Handb Exp Pharmacol. 2011;200:201–34.
36. Hadfield MG. Caffeine and the olfactory bulb. Mol Neurobiol. 1997;15:31–9.
37. Richardson NJ, Rogers PJ, Elliman NA, O'Dell RJ. Mood and performance effects of caffeine in relation to acute and chronic caffeine deprivation. Pharmacol Biochem Behav. 1995;52:313–20.
38. Smith A. Effects of caffeine on human behavior. Food Chem Toxicol. 2002;40:1243–55.
39. Ozsungur S, Brenner D, El-Sohemy A. Fourteen well-described caffeine withdrawal symptoms factor into three clusters. Psychopharmacology (Berl). 2009;201:541–8.
40. Ratliff-Crain J, O'Keeffe MK, Baum A. Cardiovascular reactivity, mood, and task performance in deprived and nondeprived coffee drinkers. Health Psychol. 1989;8:427–47.
41. Lane JD, Phillips-Bute BG. Caffeine deprivation affects vigilance performance and mood. Physiol Behav. 1998;65:171–5.
42. Phillips-Bute BG, Lane JD. Caffeine withdrawal symptoms following brief caffeine deprivation. Physiol Behav. 1997;63:35–9.
43. Herraiz T. Tetrahydro-beta-carbolines, potential neuroactive alkaloids, in chocolate and cocoa. J Agric Food Chem. 2000;48:4900–4.
44. Villareal F. Chocolate: an exercise mimetic? In: Southwest chapter American College of Sports Medicine 2010 annual meeting, San Diego: American College of Sports Medicine; 2010.

45. Blumenthal J, Babyak M, Moore K, Craighead WE, Herman S, Khatri P, et al. Effects of exercise training on older patients with major depression. Arch Intern Med. 1999;159:2349–56.
46. Craft LL, Landers DM. The effect of exercise on clinical depression and depression resulting from mental illness: a meta-analysis. J Sport Exerc Psychol. 1998;20:339–57.
47. Larque E, Perez-Llamas F, Puerta V, Girón MD, Suárez MD, Zamora S, et al. Dietary trans fatty acids affect docosahexaenoic acid concentrations in plasma and liver but not brain of pregnant and fetal rats. Pediatr Res. 2000;47:278–83.
48. Hibbeln JR, Ferguson TA, Blasbalg TL. Omega-3 fatty acid deficiencies in neurodevelopment, aggression and autonomic dysregulation: opportunities for intervention. Int Rev Psychiatry. 2006;18:107–18.
49. Hibbeln JR, Lewis M, Johnson J, Lin YH, Umhau JC. Low omega-3 fatty acid status and increased risk of suicide death among active duty US military: A case control comparison. ISSFAL International Congress Maastricht Netherlands; June 2010.
50. Hibbeln JR, Salem Jr N. Dietary polyunsaturated fatty acids and depression: when cholesterol does not satisfy. Am J Clin Nutr. 1995;62:1–9.
51. Kurata N, Privett OS. Effects of dietary trans acids on the biosynthesis of arachidonic acid in rat liver microsomes. Lipids. 1980;15:1029–36.
52. Vorona RD, Ware JC. Exacerbation of REM sleep behavior disorder by chocolate ingestion: a case report. Sleep Med. 2002;3:365–7.
53. Okudaira N, Kripke DF, Mullaney DJ. Theophylline delays human sleep phase. Life Sci. 1984;34:933–8.
54. Flemons WW, Tsai W. Quality of life consequences of sleep-disordered breathing. J Allergy Clin Immunol. 1997;99:S750–6.
55. Guilleminault C, Lin CM, Goncalves MA, Ramos E. A prospective study of nocturia and the quality of life of elderly patients with obstructive sleep apnea or sleep onset insomnia. J Psychosom Res. 2004;56:511–5.
56. di Tomaso E, Beltramo M, Piomelli D. Brain cannabinoids in chocolate. Nature. 1996;382:677–8.
57. James JS. Marijuana and chocolate. AIDS Treat News. 1996;257:3–4.
58. Maccioni P, Pes D, Carai MA, Gessa GL, Colombo G. Suppression by the cannabinoid CB1 receptor antagonist, rimonabant, of the reinforcing and motivational properties of a chocolate-flavoured beverage in rats. Behav Pharmacol. 2008;19:197–209.
59. Kuiper-Goodman T. Mycotoxins: risk assessment and legislation. Toxicol Lett. 1995;82–83:853–9.
60. Warzecha A, Jakubowska B, Siwicka H. Microflora of cocoa pulp and its effect on contamination of ready-made confectionery. Rocz Panstw Zakl Hig. 1985;36:154–63.
61. Joo S-J, Betts NM. Copper intakes and consumption patterns of chocolate foods as sources of copper for individuals in the 1987–1988 Nationwide Food Consumption Survey. Nutr Res. 1996;16:41–52.
62. Wood JE, Allaway D, Boult E, Scott IM. Operationally realistic validation for prediction of cocoa sensory qualities by high-throughput mass spectrometry. Anal Chem. 2010;82:6048–55.
63. Wolz M, Kaminsky A, Lohle M, Koch R, Storch A, Reichmann H. Chocolate consumption is increased in Parkinson's disease. Results from a self-questionnaire study. J Neurol. 2009;256:488–92.
64. Strandberg TE, Strandberg AY, Pitkala K, Salomaa VV, Tilvis RS, Miettinen TA. Chocolate, well-being and health among elderly men. Eur J Clin Nutr. 2008;62:247–53.

Chapter 31
Chocolate: Psychopharmacological Aspects, Mood, and Addiction

Nuno Rodrigues-Silva

Key Points

- Chocolate has specific orosensory qualities that provide strong hedonic effects, which in turn relate to changes in the opioidergic and dopaminergic systems. This may induce soothing effects, which are especially relevant to people with certain types of mood depression (e.g., atypical depression).
- Chocolate consumption may be a personality marker of high neuroticism and represent a strategy to cope with the propensity to experience intense negative emotions.
- Regarding the psychoactive substances present in chocolate, at least methylxanthines (theobromine and caffeine) have significant psychopharmacological effects. Methylxanthines have an impact on mood and improve performance in specific tasks. Methylxanthines may therefore play the role of chocolate as an "energizer."
- Chocolate addiction should be regarded as an inaccurate and misleading term. "Moreishness" may be a better term. It translates as the ambivalence associated with the desire to eat more chocolate (due to high palatability) and self-imposed dietary restraint (due to personal and cultural beliefs), which limits chocolate intake before satiation is attained, making the "desire for chocolate" more salient (craving).
- The orosensory qualities of chocolate are uncontested determinants for chocolate liking, but methylxanthines also contribute to an acquired taste for chocolate liking.
- PEA and anandamide are unlikely to exert significant biological effects. Also, an increase in brain serotonin related to carbohydrates after chocolate consumption is unlikely to occur.
- Anandamide-related compounds, salsolinol, THBCs, and magnesium cannot be discarded as effective psychoactive agents (see Table 31.1).

Keywords Chocolate craving • Moreishness • Emotional eating • Chocolate psychoactivity • (Beta) β-phenylethylamine • Anandamide • Methylxanthines

N. Rodrigues-Silva, M.D. (✉)
Department of Psychiatry and Mental Health, Vila Nova de Gaia/Espinho Hospital Center, Rua Conceição Fernandes, 4434-502 Vila Nova de Gaia, Portugal
e-mail: ruxnux@gmail.com

R.R. Watson et al. (eds.), *Chocolate in Health and Nutrition*, Nutrition and Health 7,
DOI 10.1007/978-1-61779-803-0_31, © Springer Science+Business Media, LLC 2013

Table 31.1 Chocolate types: Differential properties

	Dark chocolate	Milk chocolate	White chocolate
Micronutrients[a]	+++	++	+
Sugar/fat[b]	+/++	+++	+++

[a]These refer to potentially psychoactive constituents in chocolate via direct or indirect mechanisms and discussed in the text below

[b]Sugar and fat relate to the orosensory properties of chocolate and its hedonic effects

Introduction

In 1753, Carl Linnaeus named it Theobroma cacao, "food of the gods." Chocolate's medicinal applications can be traced back to the Aztecs, who used it to promote strength before military and sexual conquests, among many others [1]. The Badianus Codex (1552) noted the use of cacao flowers to treat fatigue, and subsequent sixteenth- to early twentieth-century manuscripts consistently suggested that chocolate stimulated the nervous systems of apathetic, exhausted, or feeble patients [2]. Until the eighteenth century, chocolate had been consumed as a medical beverage. From this time on, chocolate became closely linked with milk, and Van Houten refined chocolate into a more digestible form, mixing the beans with potash. This powder soon became known as cocoa and was promoted as "the food prescribed by doctors." Subsequently, there was a shift to industrial marketing campaigns about the benefits of chocolate as a nutritious food [1]. By the middle end of the twentieth century, the widespread supposed benefits of chocolate became the interest of scientific research. Psychoactive components in chocolate have been proposed to explain its effects on mood, cognition, and behavior. β-phenylethylamine and other biogenic amines, anandamide, and related derivatives were the most proclaimed, owing to their interesting pharmacodynamic properties. Many conclusions were hastily drawn. This jump to conclusions happened as pharmacokinetic properties were not taken into account, and the mere presence of a psychoactive substance was confounded with effective biological actions after chocolate ingestion. Theobromine, a rather specific methylxanthine to chocolate, was discarded as not possessing significant pharmacological activity, and less interest was drawn to it, compared to caffeine. However, recent research is refocusing on theobromine and caffeine as suitable pharmacologic agents to explain at least part of the psychoactive effects of chocolate and its additive or synergistic interactions with each other.

In addition to its rich composition, chocolate also has highly palatable characteristics, from flavor to texture, related to its sweet and fat composition, that elicit strong hedonic effects. Much research also focuses on these qualities to explain chocolate liking and emotional eating.

The main objective of this chapter is to provide an overview of the accumulated research on chocolate's psychopharmacological properties, its effects on mood, and a discussion regarding chocolate craving and addiction.

Orosensory Properties and Hedonic Effects of Chocolate

Chocolate fulfills two innate preferences: one for sweet taste and one for creamy texture, making chocolate one of the most palatable foods available [3]. Each food has its own characteristic orosensory properties, and as such, chocolate has specific orosensory properties. However, its specific composition makes it, probably, the most palatable food in the world. Palatable foods overturn regulatory mechanisms of appetite control and also stimulate neural systems implicated in drug dependence. Some researchers see foods high in fats and sweet as similar to drugs of abuse [3]. Ingestion of food, particularly sweets, can induce the release of endogenous opiates [4]. However, palatability and not merely the presence of sweetness seems to be the most important factor in inducing neural changes in the opioid structures [5]. Opioid antagonists are stated to block the hedonically driven component of overeating [3]. Besides endorphins, other substances, like dopamine, critically contribute to hedonic responses [6]. It has been suggested that endorphins enhance dopaminergic activity in the mesolimbic pathways to alter the reward value of food [7].

Hedonic effects have been postulated to be major determinants of mood changes immediately after chocolate ingestion. It has been found that the positive changes in mood after chocolate consumption were more pronounced a short time (5 and 30 min) after ingestion, sustaining the point that early mechanisms, such as orosensory stimulation, contribute to a greater extent than psychoactive substances contained in chocolate [8], which would take longer to exert its effects. However, one may argue that rather than orosensory stimulation, memory and reward anticipation of chocolate effects could also induce earlier than expected positive emotions due to psychoactive substances present in chocolate. To ascertain if anticipation of nutrient-derived changes or psychopharmacologic substances occurs immediately after chocolate consumption, it should be evaluated if regular chocolate consumption relates to stronger conditioned responses.

Chocolate Craving and Addiction

Why do people crave chocolate? This is an intensely debated question. Probably the solution relies in multiple answers, depending on the specific subject and context. One view regards craving for chocolate and carbohydrates in general as a homeostatic response to nutrient deficiency (e.g., magnesium deficiency).

Although psychoactive substances present in chocolate have been speculated to explain chocolate craving, most of the time with insufficient evidence, the orosensory effects of chocolate and the corresponding strong hedonic effects elicited have been proved by solid evidence to be the hallmark of chocolate craving and even chocolate addiction ("chocoholism") [9]. Many experts claim that the orosensory properties of chocolate and the desire for sensory gratification for chocolate are sufficient to explain the motivation to eat chocolate [7].

The theory of emotional eating states that certain individuals eat to cope with stress and negative emotional experiences. It has been noticed that during negative mood states, preferences for "junk food" arise; during positive mood states, there is a preference for healthy food [10]. It was argued that food rich in carbohydrates elicits nutrient-dependent changes in neurotransmitter systems, mainly through an increase in serotonin synthesis [11], and subsequently can have an impact on mood. Additionally, hedonic effects provided by highly palatable food have a neurochemical impact on opioid and dopaminergic systems, providing a soothing strategy to counteract negative mood states [12].

Besides the hedonic effects of chocolate and nutrient-dependent changes in neurotransmitters systems, evidence supports that at least some psychoactive constituents present in chocolate may contribute to chocolate liking. In a well-designed study, it was shown that methylxanthines (theobromine and caffeine) in amounts found in a 50-g bar probably contribute to chocolate liking. This effect was highly significant and reinforced with repeated exposition [13]. In this way, methylxanthines present in chocolate are probably related to an acquired taste for chocolate, besides its innate appeal (fat and sugar).

One may argue that if psychoactive substances were involved in chocolate craving, then chocolate-containing cocoa powder would satisfy craving and black chocolate would be the most preferred and "addictive." However, milk chocolate is the most preferred, compared to dark and white chocolate [9]. On the other hand, why is milk chocolate preferred to white chocolate? If the orosensory properties of chocolate, like sweetness and texture produced by fat, are the factors involved in chocolate liking, why do people not prefer white chocolate as much as milk chocolate, as the orosensory properties are similar? One plausible explanation for this is that milk chocolate concentrates all the important constituents and characteristics related to chocolate liking separately present in white and black chocolate. It is a highly palatable food due to its appealing orosensory properties (sugar provides sweetness, and fat provides texture, derived from cocoa butter, a major component of white chocolate), and it contains effective psychoactive substances, theobromine and caffeine (and all the other psychoactive compounds with possible biological actions) derived from cocoa solids, the major constituent of dark chocolate (Table 31.1).

Does chocolate addiction exist? Although apparent similarities exist between drug use and chocoholism, like mood effects, external cue control of appetite, and reinforcement, the majority of cases reported as "chocoholism" should not be viewed as addictive. A different perspective on this has been proposed, giving a prominent role to psychological processes of ambivalence and attribution operating together with normal appetite control mechanisms, hedonic characteristics of chocolate, and socially and culturally determined perceptions of the appropriate intakes [14]. Besides a wide overlap of the brain circuits underlying the rewarding effects of drugs and food, "chocoholism" should not be viewed as an addiction because it lacks crucial processes central to drug addiction, like neuroadaptive changes and withdrawal effects. Moreover, most so-called chocoholics eat moderate amounts of chocolate, as the study showing highest average amounts of chocolate consumption reported to be 12.5 bars per week, a rather moderate amount. However, very rare cases show incredibly high amounts of chocolate intake (70 bars per week), and a morbid process should be involved in this cases [15].

Why do so many people claim to be chocolate addicts? On the one hand, chocolate consumption is thought to be pleasurable and to increase energy; on the other hand, guilt is elicited by virtue of its culturally bounded connotations, such as its link to obesity and "sweet tooth," providing the ground for ambivalence towards chocolate consumption ("nice but naughty"). Moment to moment, eating is regulated by the orosensory effects of food, which provide positive feedback, and the post-ingestive effects of food, which contribute to negative feedback. When this balance changes in favor of the post-ingestive effects, satiety occurs. The more palatable the food is, higher post-ingestive effects will be needed to induce satiety (i.e., a greater amount of food) [7]. This is the case for chocolate, a highly palatable food that consequently requires high intake amounts to attain satiety. However, this is not a staple component of the diet but something that "should be eaten with restraint," and "I would like to eat more but I can't" settles in. This will not stop the desire to eat chocolate, and it will make it even more salient, a craving-like feeling builds up. So, when one attempts to explain one's difficult-to-resist relationship with chocolate, one erroneously states it as an addiction. According to this view, chocolate is the most craved food because it relates to the highest positive difference between palatability and eating restraint, as the eater limits chocolate consumption long before the appetite for it has been satiated. This concept is best reflected by the term "moreishness."

Psychoactive Constituents in Chocolate

Chocolate is abundant in psychoactive compounds, which favored rich speculation. In this regard, research focus has changed over the years, from appealing and exotic hypotheses regarding anandamide and 2-phenylethylamine to methylxanthines.

Several compounds have been identified in chocolate as potentially exerting psychoactive effects (Table 31.2). Nevertheless, the mere presence of psychoactive substances does not mean psychoactivity. This and sparse research on these compounds led to much more speculation than solid evidence, as, based on research results, at present one can only, with few exceptions, hypothesize but not firmly conclude. Now, it would be very important for research to "fill the gaps" and ascertain whether each

Table 31.2 Potential psychoactive substances present in chocolate

	PEA	Anandamide	Anandamide-related compounds	Theobromine	Caffeine	Carbohydrates	Salsolinol	Beta-carbolines	Magnesium
Psychoactive effects	–	–	+/–	+	+	–	–	+/–	+/–
Possible interactions[a]	SAL/β-carbolides (+)			Caffeine (+)	Theobromine (+)	Proteins (–)			
Mechanism	MAO-B inhibition, dopamine, 5-HT	Cannabinoid receptors	Anandamide (indirectly)	Adenosine antagonism	Adenosine antagonism	5-HT			Mg2+ depletion, dopamine

–, not likely to possess biological activity after oral consumption of chocolate, although with psychoactive properties; –/+, not possible to conclude if significant biological activity after oral consumption of chocolate; +, likely to induce psychoactive effects after oral consumption of chocolate

[a]Possible interactions with other chocolate constituents that may increase (+) or decrease (–) its biologic effects

compound present in chocolate with psychoactive properties has significant biological effects after oral ingestion, as consumed in everyday life.

Biogenic Amines: 2-Phenylethylamine

2-Phenylethylamine (PEA) is an endogenous trace amine present in the human brain [16]. It is a common constituent of several foods like chocolate, cheese, and wines. Chocolate is the food with highest reported amounts of PEA, containing on average 3.5–8.02 μg/g [17], even though this is a smaller amount than previously reported by the British Food Manufacturing Industries Research Foundation [18]. PEA is called the "endogenous amphetamine" because research has focused attention on its amphetamine-like actions. A great deal of interest regarding PEA was generated when G-protein-coupled receptors specific to PEA were discovered [19].

PEA occurs in trace amounts in the brain tissue and has a heterogeneous distribution, with the highest concentrations in mesolimbic structures, where monoamines are also abundant. PEA is synthesized by the decarboxylation of phenylalanine in dopaminergic neurons and rapidly metabolized by monoamine oxidase type B (MAO-B), an enzyme that possesses high affinity for PEA. The uptake and release of noradrenaline (NA), dopamine (DA), and serotonin (5-HT) is significantly induced by PEA [20], and it can also interact directly with receptors of several of the monoamines, such as direct DA receptor agonism [21], and could be a modulator of monoamines neurons. However, all the actions of PEA mentioned above occur at concentrations at least 100 times higher than occur endogenously, except in situations where PEA levels are very high, like after MAO-B inhibition.

Early research found low levels of PEA in depression and high levels in mania [22] and also a relation to antipsychotic effects [20]. Speculation regarding PEA and psychiatric disorders was greatly promoted by Liebowitz and Klein [23], who named an affective disorder as "hysteroid dysphoria" characterized by an attention-seeking behavior, depressed mood reactive to feelings of rejection, and frequently associated to sweet craving, especially chocolate, a source of PEA [24]. Also, PEA was related to feelings of euphoria as experienced in sexual activity, mainly in animal studies where PEA was injected in the brain [25].

A recent study showed that acute intravenous administration of PEA induced a significant increase in striatal 5-HT level and chronic administration led to the opposite, as well as several behavioral changes. However, when PEA was orally administered, there was no difference [19]. The ingestion of PEA in chocolate, although present in relatively high concentrations, seems irrelevant in healthy people since biogenic amines are rapidly and efficiently metabolized by fairly abundant MAO-B enzymes in the mucosa of the small intestine, in the liver, and in the kidney [26].

Speculation regarding PEA-mediated chocolate effects is not supported by research, and intravenous or intracerebral injection and oral administration are quite different things. As even researchers jumped to conclusions regarding PEA effects and food containing PEA, also popular beliefs regarding PEA rapidly grew, relating it to love, sex, and romance and claiming chocolate as a "sex substitute"[27].

To summarize, PEA is not a probable candidate to explain, even partially, the psychoactive or any other effects of chocolate in humans. A few exceptions should be considered: individuals with altered gastrointestinal anatomy (e.g., gastric ulcers) or physiology; those on MAO-B inhibitor medication [28]; also some individuals with a constitutional deficiency in MAO-B activity, as some people may experience PEA-induced migraine associated with decreased MAO-B activity [18].

Tyramine is also present in chocolate (3.8 μg/g) in milk chocolate [17], but in much lower amounts than other foods like cheese (0–624 μg/g) [29]. Tyramine effects vary with the menstrual cycle, with sensitivity higher in the perimenstrual period [9], but it is unlikely to play any role as a psychoactive agent as present in chocolate in daily regular amounts.

Anandamide and Related Compounds

Anandamide is thought to be an endogenous neurotransmitter or neuromodulator acting on cannabinoid receptors, mimicking the effects of plant-derived cannabinoid drugs [30].

In 1996, a scientific correspondence in *Nature* brought chocolate and cannabinoids-like compounds to the scene. The authors showed that chocolate contains anandamide and two related compounds, the latter in higher concentrations than anandamide. White chocolate did not contain any of these compounds [31]. This study provided evidence for the presence of new compounds possibly explaining chocolate effects in the brain and behavior, via direct (anandamide) or indirect (two related compounds) effects in the cannabinoid system. The two related compounds mentioned are N-acylethanolamines, N-oleoylethanolamine and N-linoleoylethanolamine and do not activate the cannabinoid receptors, but they have been found to inhibit significantly anandamide hydrolysis in rat brain cells, indirectly increasing anandamide levels [32]. However, the question remained whether the concentrations of these compounds would be sufficient to produce significant effects in vivo.

This research opened ground to new claims regarding chocolate. A short time after, this was even used as an argument in a Belgium court. The defendant's lawyer, to clear the accused of dealing and smoking marijuana after a positive cannabinoids urine immunoassay screening test, claimed that the client ate a large amount of chocolate, which would increase, directly and/or indirectly, endogenous cannabinoids (anandamide), leading to a false-positive cannabinoid (plant-derived) test [33]. Obviously, the lawyer's lack of preparation in the field led him to a completely misleading interpretation. Anandamide and Δ^9-THC (and metabolites), although with similar actions and effects, could not be assumed to cross-react. Anyway, laboratory confirmation was performed, and no cross-reactivity was found.

Anandamide is present in chocolate in minute amounts, approximately 0.05 μg/g, and its bioavailability is about 5%. Regardless of the bioavailability, if we assume anandamide as bioavailable and potent as THC, the necessary chocolate amount to be ingested, to get the minimum blood level necessary for a person weighting 70 kg to experience a "high," would be 25 kg [27]. This makes anandamide as present in chocolate very improbable to be directly related to the psychoactive effects of chocolate.

Remarkably, the two N-acylethanolamines related to anandamide are present in chocolate in much higher amounts, possibly in a magnitude of 10^3–10^4 times [31], meaning that a 25-g chocolate bar may have more than 1 mg of this compound. The concentration of N-linoleoylethanolamine that inhibits anandamide hydrolysis by 50% is ≈5 μM [31]. This makes it more reasonable for these compounds to increase indirectly anandamide's concentration in the brain. Nonetheless, for now, this can be only speculation, and further research needs to elucidate the pharmacokinetics of these compounds after oral consumption of chocolate.

Salsolinol and Tetrahydro-β-Carbolines

Salsolinol (SAL) and tetrahydro-β-carbolines (THBCs) are morphine-like alkaloids, produced endogenously as a condensation product of acetaldehyde with dopamine or between acetaldehyde and indoleamines [34], respectively, and have been speculated to play a role in the etiology of alcohol dependence [35]. THBCs are present in chocolate in variable amounts depending on chocolate type (1.4, 5.5, and 3.3 mg/kg for white, milk, and dark chocolate, respectively) [36]. SAL is present in chocolate in higher amounts than THBCs and also differs in white, milk, and dark chocolate (5, 20, and 25 mg/kg, respectively) [37]. Based in the possible, but still controversial, role in alcohol dependence, SAL has been hypothesized as a potential agent involved in the psychoactive effects of chocolate. SAL has been shown to have reinforcing properties when administered in the central tegmental area, mediated by the activation of DA neurons and local 5-HT3 receptors [38]. SAL is a dopaminergic compound, which binds to dopamine receptors, mainly the D3 receptor, and it also inhibits the formation of cAMP, the release of β-endorphin and ACTH by the hypophysis [37].

A critical step to hypothesize chocolate as having those effects would be SAL and/or THBCs blood levels to rise after oral intake of chocolate. The literature remains scarce regarding this, but a recent study shows that the consumption of bananas, which, depending on the maturation phase, contain SAL in amounts as high as approximately 4 mg/kg (sum of both enantiomers), significantly raised plasma SAL levels after oral ingestion. The study was performed in humans, but a parallel study in rats also showed a drastic increase in SAL blood levels, but did not relate to increased levels in the brain, precluding the link necessary to establish the plausibility for psychoactive effects induced by SAL from a dietary source [35]. Although alcohol has been reported to increase SAL and THB in certain areas of the brain [39], this is likely to be derived from in situ synthesis [40]. This study supported the idea that peripheral SAL may not significantly contribute to the brain SAL concentration. Nevertheless, the study included a small sample, 11 people and 5 rats, for both the control and experimental groups and reported that amounts of SAL in chocolate are considerably higher than for bananas.

Besides this, THBCs also act as mild MAO inhibitors that could lead to increased PEA levels through inhibition of metabolization by MAO-B, amplifying PEA's effects [27].

Methylxanthines in Chocolate: Theobromine and Caffeine

Methylxanthines are among the most widely consumed psychoactive substances in the world [41]. Although caffeine can be found in several products like coffee, tea, and guarana, theobromine is consumed only in cocoa. Contrary to other compounds present in chocolate with potential psychoactive effects, methylxanthines have been studied more extensively (mainly caffeine) regarding its pharmacodynamic and pharmacokinetic properties.

Theobromine was put aside since early research which regarded it devoid of significant effects in the CNS [42]. However, research on theobromine is reemerging, and a reappraisal of the issue may reveal new and exciting insights. Chocolate represents the richest food source of theobromine and the main source of theobromine in our diet. A 50-g dark chocolate bar contains 17–36 mg of caffeine and 237–519 mg of theobromine [43, 44].

Theobromine slowly increases after oral administration, with peak concentration about 3 h after. Interestingly, it has been found that theobromine absorption from chocolate was faster and produced higher maximum plasma concentrations compared to theobromine in capsules, peaking after 2 h. In contrast, caffeine is rapidly absorbed, and peak plasma concentration occurs in about 30 minutes, although there is conflicting evidence in this point [45]; however, if ingested in chocolate, the absorption is delayed and peaks approximately 1.5–2 h after, and it produces lower maximum plasma concentration [46]. Theobromine attained roughly four times higher concentrations than the maximum plasma levels for caffeine [47]. The clearance rate in acutely administered theobromine is half of that for caffeine (1.2 mL/min/kg for theobromine) and, unexpectedly, reduces upon chronic administration (1.2 to 0.75 mL/min/kg), although variability between subjects is high [27].

The median discriminative threshold dose of caffeine in capsules is 18 mg [47]. As peak plasmatic levels of caffeine after chocolate consumption are approximately 75% of those after caffeine capsules ingestion, the threshold for the psychoactive effects of caffeine present in chocolate should be higher than 18 mg. Nevertheless, a chocolate bar normally contains 36 mg of caffeine [46] and therefore should contain the necessary dose to produce subjective effects. For theobromine, the median discriminative threshold dose has been found at 560 mg, but a 50-g chocolate bar contains approximately 237–519 mg (and less in milk chocolate). Nevertheless, two points should be made on this, the discriminative dose for theobromine was shown to be quite variable and may be much lower (100 mg) for sensitive individuals [47]. Also, discriminative doses have been evaluated for theobromine and caffeine separately; however, the overlap in their mechanisms of action may entail important additive or even synergistic effects, resulting in lower discriminative doses for each of these compounds, which should be a subject for further study.

Regarding its pharmacodynamic properties, caffeine and theobromine competitively antagonize adenosine effects at both the A_1 and A_{2A} receptors [48]. High doses of caffeine can also activate phosphodiesterase, but this is usually not the case with normal human caffeine consumption [49]. In general, caffeine's potency is considerably higher than theobromine, up to a 10:1 relation. However, specific affinities to adenosine receptors A_1 and A_{2A} differ for caffeine and theobromine. Compared to caffeine, theobromine is at least ten times less active at the A_{2A} receptor but only two to three times less potent at the A_1 receptor [50]. On the contrary, theobromine is present in much higher concentrations in chocolate, a 10:1 ratio on average [27], but peak plasma concentrations of theobromine compared to caffeine after chocolate consumption, due to slower absorption and less bioavailability, diminish this difference to a 4:1 ratio. Taking this into account, we can roughly say that it is reasonable to assume that theobromine is at least (or even more) as potent at the A_1 receptors as caffeine [49].

Adenosine has been postulated to be a homeostatic regulatory factor to match the rate of energy consumption to substrate supply. Adenosine exerts quite complex actions in the SNC, but in general, adenosine has an inhibitory effect on transmitter release, and there is some evidence to suggest that the release of excitatory neurotransmitters is more strongly inhibited than inhibitory neurotransmitters [51]. Adenosine A_1 receptors are found in almost all brain areas, with highest levels in the hippocampus, cerebral and cerebellar cortex, and some thalamic nuclei. Only moderate levels are found in the dorsal striatum and nucleus accumbens. Adenosine A_{2A} is located in dopamine-rich regions of the brain and, more specifically, is colocalized with some dopamine D_2 receptors [49]. Methylxanthines increase the turnover rate of monoamines; the rate of firing of noradrenergic neurons in the locus coeruleus increases as well as the mesocortical cholinergic neurons. The evidence is strong relating A_1 receptors and dopamine D_1 receptors, also affecting glutamate and GABA systems [49].

It has been shown that caffeine consumption (at 75 and 150 mg) improved mood and reduced tenseness, but this might be no more than withdrawal reversal, and controversy still runs on this issue. Psychomotor and cognitive performances are well-studied topics, including mental arithmetic, learning and information processing, reaction times, hand steadiness, and tapping rate. In general, there is a positive trend in performance during these tasks [52] and systematic analysis of the different components of such tasks suggests that caffeine may improve sensory input and motor output without altering central processing functions [53]. Of note, during long-term administration of caffeine, there is an effect inversion regarding some effects related to adaptive changes in the brain that can be beneficial rather than detrimental, e.g., while acute doses increase seizure susceptibility, long-term use leads to decreased seizure susceptibility [49]. There is accumulating evidence that caffeine use may reduce suicidal tendencies, perhaps acting as antidepressant [49].

It has been stated that methylxanthines are the psychopharmacological active constituents of chocolate. In a well-designed study, encapsulated cocoa powder representative of a 50-g bar of dark chocolate was compared to placebo to evaluate its effects in mood and subjective alertness and performance in simple reaction time and rapid visual information processing tasks. Cocoa powder in capsules was used to distinguish the effects of active compounds in chocolate from its orosensory properties. Strong effects were found for the mood construct energetic arousal and the simple reaction time task (but not for the rapid visual information processing task). Remarkably, in a subsequent study, the administration of capsules containing the same amount of caffeine and theobromine as in the cocoa powder capsules showed superimposable effects [54]. This provided clear evidence that a 50-g bar of dark chocolate has measurable psychoactive effects and that methylxanthines contribute to it. Nevertheless, we cannot assume that all the psychopharmacological effects of chocolate are due to methylxanthines because performance in other domains were not tested, and so one cannot disclose nonstudied psychoactive effects of other compounds present in chocolate.

In brief, methylxanthines are present in chocolate in biological relevant amounts and are well absorbed in the bloodstream, reaching the CNS to act as nonselective adenosine receptors antagonists, which will act on multiple neurotransmitter systems like the dopaminergic and exert effects in mood, cognition, and behavior.

Methylxanthines have been used for a long time to promote wakefulness. Caffeine and theobromine act as adenosine receptors as antagonists, which is believed to be the cause of its counteracting effects on sleep [55].

The waking-maintaining cell groups in the brainstem consist of serotonin, noradrenaline, and acetylcholine neurons. Additionally, in the hypothalamus/basal forebrain, histamine, orexin, and acetylcholine neurons also serve this function. Adenosine receptors are present in all of these neurons, making them targets for caffeine and theobromine, as antagonists at these receptor sites [55]. The cholinergic basal forebrain is probably one major area for mediating the sleep-inducing effects of adenosine by inhibition of these wake-promoting neurons via the A_1 receptor [48]. In the meanwhile, this view has been challenged in A1-receptor knockout mice, in which caffeine did not affect vigilance, but did reduce sleep in A_{2A}-receptor knockout mice, regarding A_{2A} receptors as major players in the vigilance state [56].

Homeostatic sleep drive is an important feature of sleep regulation, which increases the drive for sleep as the day goes on, presumably due to fatigue, diminishing at night with rest. Homeostatic sleep drive is often described as sleep propensity or "sleep pressure" and it is necessary for initiation of sleep. Adenosine is linked to homeostatic drive and appears to accumulate as this drive increases during the day and to diminish with rest [57].

It is well known that caffeine ingested at bedtime affects sleep negatively. Generally, more than 200 mg of caffeine is needed to affect sleep significantly. The most prominent effects are shortened total sleep time, prolonged sleep latency, increases of the initial light sleep EEG stages, and decreases of the later deep sleep EEG stages [49].

Extensive research of caffeine effects in sleep can be found in the literature, but theobromine has been scarcely studied in this domain.

Nutrient-Dependent Chocolate Effects on Neurotransmitters

Magnesium

Chocolate has one of the highest magnesium levels reported of all foods, approximately 100 mg/g [58], except white chocolate, which contains magnesium in much lower amounts, about 12 times lower than milk chocolate [59]. Magnesium deficiency results in selective depletion of dopamine in the CNS, a major neurotransmitter involved in euphoria, satisfaction, and addiction [60]. Additionally, magnesium deficiency is related to anxiety [61], and its administration has been related to reduced hyperexcitability in children [62] and attenuated posttraumatic depression/anxiety in rats [63]. It has been speculated that chocolate consumption could be induced in part by magnesium deficiency [64], and in some cases, doses of magnesium relieved chocolate craving [65]. Furthermore, magnesium deficiency has been associated to the premenstrual syndrome [66], which in turn has been related to increased craving for chocolate [67]. However, this craving was not specific for chocolate, and it was more generally related to sweet foods, most of them with insignificant levels of magnesium. Moreover, many other foods high in magnesium are not craved during the premenstrual syndrome.

Stress-induced magnesium deficiency is another hypothesis linking chocolate craving to magnesium as glucocorticoids and mineralocorticoids increase renal excretion and reduce the intestinal absorption of magnesium [68, 69]. However, these are only disparate unrelated findings, and using the scarce evidence available, there is no sufficient evidence to suggest that magnesium is implicated in craving for chocolate and its psychoactive effects.

Carbohydrates and Serotonin

Chocolate is a prototypic sweet food and, pertaining to a general effect as a carbohydrates-rich food, could lead to an increase in serotonin synthesis [12].

The ingestion of carbohydrates stimulates the release of insulin which, along with its anabolic effects, promotes amino acids in the blood to enter muscle cells, except for tryptophan [70]. This will cause a relative increase of tryptophan over other amino acids, which compete for passing the blood–brain barrier. As a consequence, more tryptophan enters the brain, and there is an increase in the synthesis of serotonin [71], a neurotransmitter postulated to have a major role in mood disorders and a target of many psychopharmaceuticals.

However, these effects only occur if the ingested high-carbohydrate food is also low in protein, as demonstrated by a study showing that the ingestion of a mixture of carbohydrates and as low as 6% protein reversed the increase in serotonin synthesis [72]. This probably occurs because in general food protein has relatively less tryptophan than other amino acids, neutralizing the effect of relatively high blood tryptophan concentrations induced by insulin secretion [73].

This last consideration is important as chocolate's protein content is generally greater than 6%, which possibly precludes any effect on serotonin synthesis via the mechanism described above.

Chocolate, Mood, and Emotions

It is widely claimed that chocolate is a stimulant, euphoriant, tonic, and antidepressant. However, this is a widely debated matter. Many studies reported that chocolate has an impact in the emotional state, although it is unclear how these changes are mediated. Chocolate may relate to mood in several and very different ways. Besides positive emotional changes, also negative emotional changes are related to chocolate eating, normally elicited by guilt, associated with the desire for something that is culturally bounded with negative connotations related to body shape.

A large, well-designed, cross-sectional study, conducted in the United States, found that people with higher depression ratings consume more chocolate, extending both to men and women. Several nutrients that could confound these results, such as caffeine, fat, carbohydrate, and energy intake, were not related to mood symptoms, suggesting a specific role for chocolate [74]. However, whether causality exists between the two variables cannot be ascertained by this study type.

In line with Liebowitz' "hysteroid dysphoria,"[23] Parker et al. stated that chocolate craving in depressed people was an efficient discriminator of atypical depression, identifying two-thirds of people with three or more symptoms of atypical depression. Interestingly, chocolate craving was related to the personality high-order construct neuroticism, specifically irritability and rejection sensitivity. Also, it was associated to the self-perception of chocolate as reducing anxiety and irritability and to the increased use of other self-soothing strategies [12, 75]. Neuroticism, along with its higher tendency to experience negative feelings, has been related to higher reactivity of the limbic system. Chocolate craving in depression could be related to homeostatic strategies to reestablish emotional stability.

Chocolate may be consumed to cope with negative mood as verified in a study, in which a negative induced mood state was reversed after chocolate eating. However, this effect was only verified after eating palatable chocolate, but not unpalatable chocolate. Nevertheless, this study reported only the very immediate effects of chocolate and provided more evidence for the importance of the hedonic effects of eating chocolate [76].

When ingested in response to a dysphoric state as an emotional eating strategy, it has been stated that chocolate may provide transient relief but prolong rather that shorten the dysphoric mood [7]. In fact, several studies showed that resisting craving produced a more positive mood [77].

However, data shown above should be cautiously interpreted, and chocolate should not be prematurely discarded of antidepressant activity. Chocolate's mood effects need to be further explored as confusion reigns in this domain. Many conclusions about chocolate's effects in mood have been extrapolated from observational studies and from studies focusing on high-energy or highly palatable foods. Quite different results have been found which may be related to a heterogeneous population consuming chocolate due to quite different purposes. Potential confounders are many and must be ascertained. Distinguishing between craver and noncraver populations and the subjects' actual mood state should be defined – whether mood varies in time with chocolate ingestion, are effects acute and/or chronic, how long do the effects last, and differences after short-term and long-term administration. The latter seems relevant as research did not explore in a clinical trial setting the long-term effects of chocolate administration on mood, as the lack of an effect in the short-term does not imply the same in the long-term, being the case for antidepressant drugs taking at least 4 to 6 weeks to produce significant effects in mood.

Summary

Pharmacological effects of chocolate have been deducted merely acknowledging the presence of pharmacologically active compounds, potentially leading to a dramatic overestimation of their effects, regardless of the pharmacokinetic properties of its components and its minute quantities present in chocolate compared to biologically effective amounts, as was the case with PEA and anandamide.

A food like chocolate contains multiple constituents, and, therefore, different effects can be related to different compounds, and we should also account for possible additive or synergistic interactions between constituents.

Accumulating evidence points to complex effects related to chocolate eating. Both palatability and chocolate psychoactive constituents seem to account for chocolate's effects. It may be stated that hedonic effects relate to immediate effects and psychoactive constituents (methylxanthines) have a more retarded effect. In line with this, eating chocolate may be associated with quite different motives and purposes due to its conjunction of disparate and heterogeneous properties and mechanisms, providing several possibilities for its use.

Future research still has a long way to run to clearly establish the effects of chocolate in mood, cognition, and behavior. As these are the oldest claimed effects of chocolate and although extensive research has been performed in this regard, uncertainty is the current state.

Although initially underestimated, methylxanthines are candidates as major determinants for the psychoactive effects of chocolate. Nevertheless, it needs to be evaluated to which domains of psychomotor and cognitive functioning do this effects extend and also to distinguish the role of theobromine and caffeine and the possible addictive or synergistic interactions of this substances, which act on the same receptors types. Additionally, research in methylxanthines as present in chocolate should be a major influence and a model to research in other psychoactive substances, therefore providing well-designed and consistent research, hopefully diminishing speculation and the establishment of well-grounded evidence.

References

1. Wilson P. Centuries of seeking chocolate's medicinal benefits. Lancet. 2010;376(9736):158–9.
2. Dillinger T, Barriga P, Escárcega S, et al. Food of the gods: cure for humanity? A cultural history of the medicinal and ritual use of chocolate. J Nutr. 2010;130(8S):2057S–72.
3. Drewnowski A, Bellisle F. Is sweetness addictive? Nutr Bull. 2007;32:52–60.
4. Grigson P. Like drugs for chocolate: separate rewards modulated by common mechanism. Physiol Behav. 2002;76:389–95.
5. Levine A. The animal model in food intake regulation: examples from the opioid literature. Physiol Behav. 2006;89:92–6.
6. Epstein L, Leddy J. Food reinforcement. Appetite. 2006;46:22–5.
7. Parker G, Parker I, Brotchie H. Mood state effects of chocolate. J Affect Disord. 2006;92:149–56.
8. Dettmer D, Macht M. Everyday mood and emotions after eating a chocolate bar or an apple. Appetite. 2006;46:332–6.
9. Michener W. Rozin P; pharmacological versus sensory factors in the satiation of chocolate craving. Physiol Behav. 1994;56:419–22.
10. Lymn B. The nutritional values and food group characteristics of foods preferred during various emotions. J Psychol. 1982;112:121–7.
11. Silva N. Chocolate consumption and effects on serotonin synthesis. Arch Intern Med. 2010;170(17):1608.
12. Parker G, Crawford J. Chocolate craving when depressed: a personality marker. Br J Psychiatry. 2007;191:351–2.
13. Smit H, Blackburn R. Reinforcing effects of caffeine and theobromine as found in chocolate. Psychopharmacology. 2005;181:101–6.
14. Rogers P, Smit H. Food craving and food "addiction": a critical review of the evidence from a biopsychosocial perspective. Pharmacol Biochem Behav. 2000;66:3–14.
15. Hetherington M, Macdairmid J. "Chocolate addiction": a preliminary study of its description and its relationship to problem eating. Appetite. 1993;21:233–46.
16. Reynolds G, Riederer P, Sandler M, et al. Amphetamine and 2-phenylethylamine in post-mortem Parkinsonian brain after (−)deprenyl administration. J Neural Transm. 1978;43:271–7.
17. Hurst W, Toomey P. High-performance liquid chromatographic determination of four biogenic amines in chocolate. Analyst. 1981;106:394–402.
18. Sandler M, Youdim M, Hanington E. A phenylethylamine oxidising defect in migraine. Nature. 1974;250(464): 335–7.
19. Sengupta T, Mohanakumar K. 2-Phenylethylamine, a constituent of chocolate and wine, causes mitochondrial complex-I inhibition, generation of hydroxyl radicals and depletion of striatal biogenic amines leading to psychomotor dysfunctions in Balb/c mice. Neurochem Int. 2010;57:637–46.
20. Paterson I, Juorio A, Boulton A. 2-Phenylethylamine: a modulator of catecholamine transmission in the mammalian central nervous system? J Neurochem. 1990;55(6):1827–37.
21. Antelman S, Edwards D, Lin M. Phenylethylamine: evidence for a direct, postsynaptic dopamine-receptor stimulatory action. Brain Res. 1977;127:317–22.
22. Davis B, Boulton A. The trace amines and their acidic metabolites in depression an overview. Prog Neuropsychopharmacol Biol Psych. 1994;18:17–45.
23. Liebowitz M, Klein D. Hysteroid dysphoria. Psychiatr Clin North Am. 1979;2:555–75.
24. Moller S. Serotonin, carbohydrates, and atypical depression. Pharmacol Toxicol. 1992;71(Suppl 1):61–71.
25. Kohl J, Francoeur R. The scent of eros. New York: Continuum; 1995.
26. Askar A, Morad M. Lebensmittelvergiftung. I. Toxine in Naturlichen Lebensmitteln. Alimento. 1980;19:59–66.
27. Smit H. Theobromine and the pharmacology of cocoa. In: Fredholm B, editor. Methylxanthines. Handbook of experimental pharmacology. Heidelberg: Springer; 2011. p. 201–34.
28. ten Brink B, Damink C, Joosten H, et al. Occurrence and formation of biologically active amines in foods. Int J Food Microbiol. 1990;11(73):84.
29. Devane W. Isolation and structure of a brain constituent that binds to the cannabinoid receptor. Science. 1992; 258:1946–9.
30. Di Tomaso E, Beltramo M, Piomelli D. Brain cannabinoids in chocolate. Nature. 1996;382:677–8.
31. Desarnaud F, Cadas H, Piomelli D. Anandamide amidohydrolase activity in rat brain microsomes. J Biol Chem. 1995;270(11):6030–5.
32. Tytgat J, Boven M, Daenens P. Cannabinoid mimics in chocolate utilized as an argument in court. Int J Legal Med. 2000;113:137–9.
33. Quertemont E, Tambour S, Tirelli E. The role of acetaldehyde in the neurobehavioral effects of ethanol: a comprehensive review of animal studies. Prog Neurobiol. 2005;75:247–74.

34. Lee J, Vijay A, Hamakazi K, et al. A critical evaluation of influence of ethanol and diet on salsolinol enantiomers in humans and rats. Alcohol Clin Exp Res. 2010;34(2):242–50.
35. Herraiz T, Chaparro C. Tetrahydro-ß-carbolines, potential neuroactive alkaloids, in chocolate and cocoa. J Agric Food Chem. 2000;48(10):4900–4.
36. Melzig M, Putscher I, Heklein P. In vitro pharmacological activity of the tetrahydroisoquinoline salsolinol present in products from Theobroma cacao L. like cocoa and chocolate. J Ethnopharmacol. 2000;73:153–9.
37. Rodd Z, Oster S, Ding Z, et al. The reinforcing properties of salsolinol in the ventral tegmental area: evidence for regional heterogeneity and the involvement of serotonin and dopamine. Alcohol Clin Exp Res. 2008;32(2): 230–9.
38. Rojkovicova T, Mechref Y, Starkey J, et al. Quantitative chiral analysis of salsolinol in different brain regions of rats genetically predisposed to alcoholism. J Chromatogr B Analyt Technol Biomed Life Sci. 2008;863:206–14.
39. Gilbert R. Caffeine consumption. In: Spiller G, editor. The methylxanthine beverages and foods: chemistry, consumption, and health effects. New York: Liss; 1984. p. 185–213.
40. Origitano T, Hannigan J, Collins MA. Brain Res. 1981;224(2):446–51.
41. Bonvehi J, Coll F. Evaluation of purine alkaloids and diketopiperazines contents in processed cocoa powder. Eur Food Res Technol. 2000;210:189–95.
42. MAFF. Survey of caffeine and other methylxanthines in energy drinks and other caffeine-containing products (updated).In: Food Surveillance Information Sheet 144; 1998.
43. Smit H, Rogers P. Effects of caffeine on mood. In: Hetherington M, editor. Food cravings and addiction. Leatherhead: Leatherhead Food RA Publishing; 2001.
44. Shively C, Tarka S, Arnaud M, et al. High levels of methylxanthines in chocolate do not alter theobromine disposition. Clin Pharmacol Ther. 1985;37:415–24.
45. Mumford G, Benowitz N, Evans S, et al. Absorption rate of methylxanthines following capsules, cola and chocolate. Eur J Clin Pharmacol. 1996;51(3–4):319–25.
46. Mumford G, Evans S, Kaminski B, et al. Discriminative stimulus and subjective effects of theobromine and caffeinc in humans. Psychopharmacology. 1994;115:1 8.
47. Basheer R, Strecker R, Thakkar MM, McCarley RW. Adenosine and sleep-wake regulation. Prog Neurobiol. 2004;73(6):379–96.
48. Fredholm B, Battig K, Holmén J, et al. Actions of caffeine in the brain with special reference to factors that contribute to its widespread use. Pharmacol Rev. 1999;51(1):84–133.
49. Daly J, Butts Lamb P, Padgett W. Subclasses of adenosine receptors in the central nervous system: interaction with caffeine and related methylxanthines. Cell Mol Neurobiol. 1983;3:69–80.
50. Fredholm B, Dunwiddie T. How does adenosine inhibit transmitter release? Trends Pharmacol. 1988;9:130–4.
51. James J. Caffeine and health. London: Academic; 1991.
52. Lorist M, Snel J, Kok A. Influence of caffeine on information processing stages in well rested and fatigued subjects. Psychopharmacology. 1994;113:411–21.
53. Smit H, Gaffan E, Rogers P. Methylxanthines are the psycho-pharmacologically active constituents of chocolate. Psychopharmacology. 2004;176:412–9.
54. Porkka-Heiskanen T. Methylxanthines and sleep. In: Fredholm B, editor. Methylxanthines: handbook of experimental pharmacology, vol. 200. Heidelberg: Springer Verlag; 2011. p. 331–48.
55. Huang Z. Adenosine A2A, but not A1, receptors mediate the arousal effect of caffeine. Nature Neurosci. 2005; 8:858–9.
56. Stahl S. Essential psychopharmacology: neuroscientific basis and practical applications. 3rd ed. New York: Cambridge University Press; 2008.
57. Rozin P, Levine E, Stoess C. Chocolate craving and liking. Appetite. 1991;17:199–212.
58. Cutrufelli R, Perhrsson P. Composition of foods: snacks and foods. In: U.S. Division, Agricultural Handbook. Washington, DC: US Government Printing Office;1991; Number 8–19.
59. Bruinsma K, Douglas L. Chocolate: food or drug? J Am Diet Assoc. 1999;99(10):1249–56.
60. Ohl F, Laarakker M, Schetters D, et al. Association between plasma magnesium-ion concentration and anxiety in inbred mice. Behav Pharmacol. 2004;16:S37.
61. Mousain-Bosc M, Roche M, Rapin J, et al. Magnesium vitB6 intake reduces central nervous system hyperexcitability in children. J Am Coll Nutr. 2004;23(5):5455–85.
62. Fromm L, Pharm B, Heath D, et al. Magnesium attenuates post-traumatic depression/anxiety following diffuse traumatic brain injury in rats. J Am Coll Nutr. 2004;23(5):5295–335.
63. Rodin J, Mancuso J, Granjer J, et al. Food cravings in relation to body mass index, restraint and estradiol levels: a repeated measures study in healthy women. Appetite. 1991;17:177–85.
64. Wood G, Lass R. Cocoa. New York: Longman; 1985. p. 596.
65. Abraham G, Lubran M. Serum and red cell magnesium levels in patients with premenstrual tension. Am J Clin Nutr. 1981;34:2364–6.

66. Hill A, Heaton-Brown L. The experience of food craving: a prospective investigation in healthy women. J Psychosom Res. 1994;38:801–14.
67. Abraham G. The normal menstrual cycle. In: Givens J, editor. Endocrine causes of menstrual disorders. Chicago: Yearbook Medical Publishers; 1978. p. 15–44.
68. Rude R, Behune J. Renal tubular maximum for magnesium in normal and hypoparathyroid man. J Clin Endocrinol Metab. 1980;5:1425–31.
69. Spring B. Effects of foods and nutrients on the behavior of normal individuals. In: Wurtman R, Wurtman J, editors. Nutrition and the brain. New York: Raven; 1986. p. 1–47.
70. Liebermen H, Spring B, Garfield G. The behavioral effects of food constituents: strategies used in studies of amino acids, protein, carbohydrate and caffeine. Nutr Rev. 1986;44(Suppl):61–70.
71. Glaeser B, Maher T, Wurtman R. Changes in brain levels of acidic, basic, and neutral amino acids after consumption of single meals containing various proportions of protein. J Neurochem. 1983;41(4):1016–21.
72. Wurtman R. Ways that foods can affect the brain. Nutr Rev. 1986;44(Suppl):2–6.
73. Rose N, Koperski S, Golomb B. Mood food: chocolate and depressive symptoms in a cross-sectional analysis. Arch Intern Med. 2010;170(8):699–703.
74. Eysenck H, Eysenck M. Perspectives on individual differences. In: Eysenck H, Eysenck M, editors. Personality and individual differences: a natural science approach. New York: Plenum; 1985.
75. Macht M, Mueller J. Immediate effects of chocolate on experimentally induced mood states. Appetite. 2007;4:667–74.
76. Christensen L, Burrows R. Dietary treatment of depression. Behav Ther. 1990;21:183–93.
77. Mahler S, Smith K, Berridge K. Endocannabinoid hedonic hotspot for sensory pleasure: anandamide in nucleus accumbens shell enhances "liking" of a sweet reward. Neuropsychopharmacology. 2007;32(11):2267–78.

Chapter 32
Chocolate and Pain Tolerance

Kristina M. Eggleston and Theresa White

Key Points

- Chocolate has a reputation as a pain reliever that spans hundreds of years.
- Mechanisms of chocolate analgesia could include associative learning and direct chemical activity.
- Associative learning is likely mediated through endogenous opioid activity.
- Opioid activity in response to eating is likely an evolutionary response to promote feeding in the face of danger.
- Although several different chemical components of chocolate could directly reduce pain, the most likely compounds are flavonoids.
- Flavonoids both decrease inflammation and diminish trigeminal nerve activity.
- Increased pain tolerance from chocolate likely reflects some combination of both learning and direct chemical activity.

Keywords Pain tolerance • Chocolate • Associative learning • Sucrose analgesia • Flavonoids

Introduction

Throughout history, individuals have attributed positive health benefits to the virtues of chocolate consumption. According to one study, individuals who regularly consumed chocolate (and other candy) one to three times per month lived approximately *one entire year longer* than individuals who abstained from candy consumption. Of course, study participants who consumed chocolate and candy in larger quantities could not expect to live quite as long, but were still almost 20% more likely to survive the period of the longitudinal study than participants who completely abstained from eating

K.M. Eggleston, Integral Honors B.S. Psychology
Massachusetts Institute of Technology, Picower Institute of Learning and Memory,
46-5273 43 Vassar St. BLDG 46, Cambridge, MA 02139, USA
e-mail: eggleskm@mit.edu

T. White, M.Sc., Ph.D. (✉)
Department of Psychology, Le Moyne College,
1419 Salt Springs Rd, Syracuse, NY 13214, USA
e-mail: whitel@lemoyne.edu

R.R. Watson et al. (eds.), *Chocolate in Health and Nutrition*, Nutrition and Health 7,
DOI 10.1007/978-1-61779-803-0_32, © Springer Science+Business Media, LLC 2013

sweet treats [1]. This finding suggests that consuming chocolate may have positive health effects, an idea that is supported by the utilization and prescription of chocolate (or its various components) for an abundant variety of medicinal uses, including analgesia or pain relief, by both ancient cultures and modern medicine. Of course, chocolate also has been reputed to trigger pain, particularly migraine headaches, though this claim has been largely disputed [2]. This chapter will focus on the pain-reductive, rather than pain-inductive, qualities of chocolate. In an effort to understand its analgesic effects, this chapter will examine chocolate's compounds and their plausible effects on the human body and make a case for the most convincing mechanisms behind chocolate-related analgesia.

Evidence for Cocoa-Related Analgesia

Historical Uses of Cacao

The cultures that first identified and established cacao (the raw bean form of chocolate, originating from the *Theobroma cacao* tree) as a precious commodity were the first people to recognize its medicinal value. For example, the Mayans' use of cacao is outlined in an ancient text known as the *Badianus Manuscript* or *Badianus Codex*, dated from the middle part of the sixteenth century [3]. The text states that cacao (usually dried and crushed to a powder form and sometimes mixed with a fermented corn mash or wine to make a ritualized drink called *chocolatl)* was ordinarily used to treat painful nutrition-related issues, including angina, constipation, dental problems, dysentery, indigestion, fatigue, gout, and lactation difficulties. The codex also states that cocoa contained powerful qualities, enabling it to decrease abdominal pain and increase alertness and courage. A Mayan language codex called the *Ritual of the Bacabs* was discovered in the Yucatan peninsula around 1914. The document included specific instructions for the ritual use of cocoa to treat medical patients. Individuals with skin eruptions, fever, and seizures were chanted over, and then a mixture of cocoa, spices, honey, and tobacco juice was to be consumed by the patients [4].

The *Florentine Codex*, circa 1590, is another early text documenting the medicinal use of cocoa [5]. Though it was not written by indigenous people like the Mayan *Badianus Codex*, the *Florentine Codex* was compiled by a Spanish priest who documented 60 years of ancient Aztec agriculture, botany, cultural practices, dietary patterns, and medicinal practices. According to the *Florentine Codex*, the Aztecs used *chocolatl* to treat both the painful symptoms of gastrointestinal problems and the problems themselves. They also mixed cocoa with other plants to treat infection, fever, and cough [4].

According to the account of the Spanish botanist Francisco Hernandez [6], cacao was administered (as in indigenous cultures) to European patients for fever and liver issues. An American treatise on the plants growing in English plantations, titled *The American Physitian* [7], documents chocolate's use as an "anodyne," or pain killer, and widespread prescription in the 1600s as a "bromide,"or cure all [8]. Chocolate's reputation as a bromide led to it continuing to be prescribed throughout the eighteenth and nineteenth centuries for an abundance of ailments, primarily indigestion and diarrhea, undernutrition, fatigue, and weakness [4].

Modern Uses of Chocolate in Pain Relief

In the twentieth century and modern times, chocolate and its derivatives are still prescribed as a healthful beverage, and as a preventative measure against cardiovascular disease [9]. New data suggest chocolate may be a beneficial prescription and have pain-relieving qualities, especially for sufferers of asthma, arthritis, heart disease, migraine, abdominal pain, toothache, and tempromandibular joint disorder, or TMJ.[10–12] Although these studies were performed on animals, the analgesic properties of chocolate have benefitted people as well. Several studies have demonstrated that chocolate lowers blood pressure [13, 14], which increases acutely in pain [15]. Chocolate has also been reported to increase insulin sensitivity in people [16], which is in contrast to the insulin resistance observed during acute pain [17]. It is worth mentioning, however, that chocolate's analgesic effects have not always been observed in humans. At least one study has observed that cocoa's addition to sucrose reduced pain tolerance in a cold-pressor test [18]. Still, given the number of historical and modern reports of the pain-relieving and medicinal properties of chocolate, it is worth examining the potential mechanisms that could contribute to chocolate analgesia.

Mechanisms of Chocolate's Influence on Pain

The mechanism underlying chocolate-related analgesia may include learning, essentially experiential associations with chocolate, which many people call a comfort food [19]. Alternatively, it is possible that the chemical compounds in chocolate directly interact with human physiology and are responsible for improving pain tolerance. These alternatives are not mutually exclusive, and it is quite possible that the mechanism of chocolate analgesia actually combines both possibilities.

Experiential Associations as a Mechanism for Analgesia

Commercially available chocolate is comprised of a number of components, including fats and sugars, which are combined to produce an extremely palatable food product that is often high in calories. Over time, people are able to form associations between the ingestion of chocolate and the pleasure that comes from eating it. Pleasure in eating is associated with the opioid release that follows ingestion, a process that may have originated with the use of taste as an evolutionary signal for caloric value.

The Opioid Reward System

The opioid reward system likely plays a major role in how chocolate-based pain tolerance occurs. The most widely accepted reason for chocolate's influence on pain tolerance is its hedonic sensory appeal, which likely triggers the release of endorphins, similar to other pleasant-tasting or "liked" foods [20, 21]. These endogenous opioids not only produce a euphoric mood but also decrease sensitivity to pain, due to interaction with neurotransmitter systems in the brain [20, 22–24], and therefore could create an analgesic effect [25].

Studies with lower mammals, such as rats, have shown that consuming sweet or pleasant foods can cause the release of endorphins within the reward centers of the brain [26]. It is believed that endorphins work two ways: first, by attaching to the opiate receptors in the brain, blocking pain-related chemical messages from being received, and second, by interacting with the opiate receptors to increase the synaptic rate of dopamine release [27]. Both of these mechanisms reduce the body's perception of pain similarly to addictive drugs, such as morphine and codeine [27]. In addition, manipulating opioid levels through the administration of either agonists such as morphine or antagonists such as naltrexone has a corresponding effect on the intake of sweet and other highly palatable foods, raising or lowering them, respectively [28]. A consumer's experience of the food (whether or not it is pleasant tasting) is implicated in the release of endorphins in the brain, and thus, the effects of chocolate in an individual's body can be increased or decreased by his or her pleasant experience of the food [21].

Like other highly caloric sweet foods, chocolate consumption changes the balance of neurotransmitters in the brain, decreasing negative affect and improving mood [19]. The ability of chocolate to alleviate a negative mood, or to lift the spirits, is believed to be due at least in part to the palatability of chocolate [29] and the post-ingestive effects of endorphins. It is possible that these endogenous opioids could be directly activated by the chemicals in chocolate instead of associative learning. Yet because of the only trace amounts of chemical compounds in chocolate that could plausibly have an effect at the level of neural synapses [30] and because the elevation in mood that chocolate consumption provides is not prolonged [29], it seems more likely that opiate involvement is related to an evolutionary mechanism.

Evolutionary Signaling

The ability for palatable foods to increase endogenous opioids [26] is presumably connected to an evolutionary survival mechanism: consuming foods with high nutritive, and often high caloric, value (such as chocolate chip cookies) was necessary for survival, and the brain rewarded these actions with both a euphoric rush and a reduction in pain [31].

The sugars in commercially available chocolate are an integral part of the product, and it may be the typical association with the highly caloric sucrose that is responsible for chocolate's reported analgesic effects. Numerous studies have shown that the sweet taste of sucrose increases pain tolerance [18, 32, 33], particularly in people with high-sweet preferences [34], while the taste of bitterness decreases pain tolerance [18, 33]. The administration of opioid antagonists (in humans) reduces preference for the taste of sweetness [35], implicating the involvement of the opioid system in the consumption of sweet foods. Although it is possible that the association with sucrose is responsible for the analgesia associated with chocolate, sucrose analgesia in a cold-pressor test has been shown to be impaired by the addition of cocoa [18]. The reduction in analgesia is possibly because, in this experiment, the nutritive taste information transmitted from the sweetened cocoa beverage was complex; the bitter qualities of the cocoa were obvious and contradicted the sweet nutritive cues of sugar, decreasing cold pain tolerance as compared to a plain sucrose solution [18].

It is still possible that the learned nutritional value of chocolate, in terms of calories and other nutrients, may contribute to its success in pain reduction. Nutrition is critical for survival, yet many animals place themselves at risk in order to feed. Animals must obtain nutrients while at the same time protecting their well-being; however, feeding seems to trump escape behavior, because it is more essential for survival, and so, more immediately necessary, as opposed to merely protective [36]. Thus, the body suppresses pain in order to optimize feeding in potentially dangerous situations [36, 37]. For example, while feeding, rats' withdrawal responses from noxious heat stimuli on the paw were either completely suppressed or greatly delayed [36]. In an animal model of migraine pain, male rats subjected to painful chemical injections continued to eat chocolate chips instead of tending to their injection site [38]. Recordings from pain-modulating neurons in the ventromedial medulla (VMM) of rats during feeding and exposure to noxious stimuli demonstrated that the cells suppressed responses to external stimuli during ingestion of food, as opposed to simply biting or nibbling it [36]. In subsequent studies, ingestion has been proved to be important to pain tolerance, but not sufficient; rather, the *expectation* of positive nutritional value was necessary to affect paw withdrawal latencies [37]. Both the medullary raphe magus (RM) in the brainstem and the opioid pathway have been implicated in hedonic ingestion analgesia in rodents [37].

Chemical Compounds in Chocolate as Mechanisms for Pain Relief

The pharmacological constituents present in chocolate may also play a role in its medicinal capabilities. The compounds in pure cocoa, the form which would have been used by ancient indigenous and early European cultures, were undoubtedly in stronger form than they are in the processed and fat-integrated chocolate consumed today. The chemical compounds in cocoa of most interest for pain tolerance include the pleasure chemicals (phenylethylamine, anandamide, and tryptophan), the methylxanthines (caffeine and theobromine), and the polyphenols, including flavonoids.

The Pleasure Chemicals

Phenylethylamine (PEA), anandamide, and tryptophan are the pharmacologic constituents considered the "pleasure chemicals" in chocolate. These compounds are also endogenous biogenic amines, naturally occurring chemicals within the body, and act as sympathomimetic compounds, eliciting effects similar to the stimulation of the sympathetic nervous system through the activation of opioid systems in the brain and the resulting release of neurotransmitters [39].

The biogenic amine **PEA** is similar to speed and heroin in that it triggers the release of natural opiates in the brain, creating feelings of ecstasy [8, 40]. Phenylethylamine is assumed to stimulate the release of opiate-like endorphins in the brain, which results in a synaptic increase of dopamine [23]. However, although the pharmacological properties of PEA are similar to amphetamine, the amount of the chemical supplied in cocoa and commercially processed chocolate cannot influence mood because of the body's natural metabolism [41]. Not only is PEA broken down extremely rapidly in the bloodstream but, in order to consume the minimal active dose (1 g) of PEA, an individual would have to rapidly consume an enormous amount (15 kg or 33 lb) of chocolate [41].

Even though an individual cannot possibly obtain an active dose of PEA by consuming chocolate, there is evidence that *craving* for chocolate may be related to PEA. Phenylethylamine is a stimulant structurally similar to amphetamines and MDMA, and correlations in activation of brain areas related to addiction (primarily nucleus accumbens) between chocolate and MDMA have been observed [41] in which MDMA abusers had intense and episodic chocolate cravings and binges [42]. Furthermore, PEA's stimulation of dopamine and noradrenergic pathways may establish it as a mood modulator, as depleted levels of PEA in the bloodstream are highly correlated with depression [39]. Some hypothesize that abnormally low levels of PEA throughout the central nervous system may lead to chocolate craving (and binging or addictive behaviors) in an attempt to self-medicate against pain and regulate mood [39, 42]. Yet, this theory of chocolate craving still presents a paradox, because other foods, like cheeses and sausages, contain much higher levels of PEA but are not associated with craving or addiction [39].

Anandamide is another chemical present in chocolate that is naturally produced by the brain. It is almost identical in composition to tetrahydrocannabinol (THC), the active ingredient in marijuana [40]. Tetrahydrocannabinol and anandamide both bind to and activate the natural anandamide receptors in the brain, causing the release of pleasure chemicals and resulting in euphoria and heightened sensitivity [40].

Because of the "high" associated with consumption of anandamide, it has been suggested that an increase in brain anandamide levels could increase the sensory properties of chocolate [40]. This relationship may be related to craving, because if the positive hedonic value of chocolate is increased, the experience of chocolate consumption becomes more rewarding and endogenous opioids may also be increased. Furthermore, anandamide may interact with other biologically active constituents to produce a greater feeling of well-being. For example, the molecules N-oleoylethanolamine and N-linoleoylethanolamine, which are also found in chocolate, keep anandamide in the brain synapses longer than normal, preventing its normally rapid breakdown and prolonging the associated "high" [40, 43].

Although both marijuana and chocolate have a chemical that produces an almost identical chemical effect in the synapses of the brain's pleasure receptors, the visible physical effects of these substances differ. Chocolate seems to produce a much more mellow and longer-lasting sense of well-being than marijuana or other drugs [40]. This, however, may be attributed not only to the interactions of other chemical components but also to the small amount of anandamide in commercially available chocolate [40].

The essential amino acid **tryptophan** is another "pleasure chemical" component of chocolate. Tryptophan naturally occurs in the human body and is essential to the manufacture of neurotransmitters, especially serotonin [22]. Because the ingestion of tryptophan-containing foods leads to an increase in production of serotonin and melatonin (a neurotransmitter important to the circadian rhythm), tryptophan consumption is associated with feelings of sleepiness [22]. This relationship may explain why some physicians in ancient times viewed chocolate to have a soothing, tranquilizing effect or why the Mayan priests served it to victims just prior to their sacrifice [4]. However, the levels of tryptophan in chocolate are relatively low in comparison to other foods, like beans, turkey, or nuts, and the stimulant chemicals present in chocolate (such as caffeine) likely offset any major chemical depressant effect [40].

Methylxanthines

Methylxanthines are naturally occurring stimulant compounds that excite the central nervous system resulting in elevated alertness and energy, such as caffeine and theobromine [24]. Stimulants work by causing central nervous system nerve fibers to release stimulating neurotransmitters, such as noradrenaline, which activate the sympathetic nervous system. Methylxanthines also compete with inhibitory neuromodulators (such as adenosine), resulting in arousal [44]. Common physical effects of stimulants include diuretic effects, stimulation of cardiac muscle, and relaxation of smooth muscle (especially bronchial muscle). Although the effects of stimulants vary across individuals, most people experience increased voluntary muscle contraction, which increases an individual's capacity for work and decreases fatigue, probably due to increased release of acetylcholine [44]. In other words, stimulants commonly cause increased alertness, mood elevation, heart rate, and blood pressure, but may also cause unpleasant effects such as anxiety, jitteriness, heart palpitations, diarrhea, or increased urination [24]. Stimulants also may affect pain tolerance by increasing the speed at which pain-relieving actions in the body take place and by affecting the body's vasculature.

Caffeine, a methylxanthine stimulant, is one of the major compounds in chocolate. The behavioral effects of caffeine consumption can be positive or negative, depending upon the dosage amount and other factors, which may include genetic predispositions and expectations [44]. A dosage between 75 and 300 mg usually produces behavioral effects normally considered positive [44]: increased vigilance, alertness, arousal, and perceptual motor speed [45]. These behavioral reactions are due to the stimulatory effects of caffeine, as the compound neurologically catalyzes the release of epinephrine. Caffeine blocks the adenosine receptors in the brain that naturally slow down and relax nerve cell activity, thus causing neurons to fire more quickly and release dopamine, a neurotransmitter that produces an emotional lift [45]. The stimulation of the sympathetic nervous system that results from caffeine consumption also triggers blood pressure elevation and increased heart rate, blood sugar level, and alertness [45]. Because the sympathetic nervous system is activated, normal blood flow is redirected from the digestive system to the muscles and brain, which is a likely explanation for caffeine's association with enhanced mental and physical performance [45, 46] and chocolate's use by ancient cultures to treat stomach pain and gastrointestinal issues. Although caffeine consumption is generally correlated with elevated performance levels, excessive caffeine ingestion can produce negative behavioral effects, and doses of greater than 500 mg of caffeine commonly result in anxiety [45]. In studies of acute pain, caffeine appears to have an independent analgesic effect, with 65 mg of caffeine as effective as 648 mg of acetaminophen [47].

Another methylxanthine compound found in chocolate is **theobromine**, a stimulant compound similar to caffeine [43]. It would take about 214 mg of theobromine to equal the stimulatory effects of 21 mg of caffeine (which is the approximate amount of caffeine in a 1.5-oz bar of dark chocolate, equivalent to the stimulatory effects of a can of soda). Whereas caffeine strengthens voluntary muscle contraction, theobromine relaxes smooth muscle tissue [40]. Although theobromine and caffeine have different immediate effects on muscle, they both act as adenosine receptor antagonists in the brain, thus complementing each others' stimulatory action on the central nervous system [40].

Caffeine and theobromine have been reported to exert anti-inflammatory effects by modulating cellular function [48], so these methylxanthines might be considered as candidates for the pain reduction effects of chocolate. However, methylxanthines are found in relatively low concentrations in cocoa as compared to coffee and tea, neither of which is generally purported to have analgesic properties. So, it is therefore unlikely that methylxanthines cause an effect large enough on the vasculature to influence pain tolerance.

Flavonoids

Another set of chemicals found in chocolate, present in relatively high amounts, are antioxidants. More specifically, most of the antioxidants in cocoa, which make up 10% of chocolate composition in total, are the polyphenols [49]. Polyphenols are molecules with at least one aromatic ring and one or more hydroxyl group, divided into class by the number of phenol rings and structural binding features. Polyphenols have antioxidant, anticarcinogenic, antistress, and anti-inflammatory effects. One good-quality dark chocolate bar can have more polyphenols than two whole days' worth of fruit and vegetables [40], and more antioxidant activity than red wine [50]. Feeding cocoa polyphenols to rats has been shown to effectively reduce the effects of stress due to confinement as measured by locomotor and other activity signs [51].

The class of polyphenols most commonly found in cocoa is flavonoids. The most abundant flavonoids in cocoa are catechin, epicatechin, and procyanidins [52]. Ingestion of these compounds has been shown to have anti-inflammatory effects on the vascular system, therefore, having markedly positive effects for cardiovascular patients [52], but likely influencing pain tolerance as well. Flavonoids have also been demonstrated to reduce the activity of some sensory nerves that transmit pain information, making these chemical components of chocolate likely contributors to its analgesic effects.

Flavonoids and Inflammation

Although methylxanthines may not be present in cocoa in amounts significant enough to have an effect on vasculature, caffeine and theobromine may work in concert with antioxidant flavonoids to cause significant anti-inflammatory effects in consumers. The relationship between pain and inflammation is complex, but both are immune responses, and flavonoid effects on inflammation may be vital in determining how cocoa influences pain. Flavonoids act as immunoregulators, affecting the secretion of inflammatory mediators, specifically inhibiting the production of pro-inflammatory molecules [53]. Through competitive binding with ATP at the catalyst site, flavonoids can inhibit enzymes from forming the molecules that cause the inflammatory immune response [53].

Cocoa flavonoids have been demonstrated to be effective in decreasing inflammation. In both in vitro cell culture studies and in vivo experiments of rats fed cocoa-enriched diets (pellet food mixed with raw cocoa powder), the presence of various cocoa flavonoids show decreased activation of cytokines, chemokines, and macrophages, which are all inflammatory molecules involved in the normal immune response [54–57]. In addition, in vivo studies of various models of animal inflammatory immune response (including paw edema, granuloma, arthritis, colitis, and ear edema) demonstrate that cocoa flavonoid compounds (either ingested or injected) reduce painful general inflammation [57–60]. In a murine model of asthma, mice treated with the cocoa flavonoid, quercetin, displayed far less inflammatory immune reaction than those that were not treated [61]. Another study of the infection disseminated candidiasis in mice showed that administration of flavonoids found in cocoa actually protected the animals against the illness [62].

Flavonoids and Trigeminal Nerve Activity

Another way in which cocoa polyphenols may influence inflammation, and therefore pain, is through their effect on the trigeminal nerve. The trigeminal nerve, or cranial nerve V, connects peripheral tissues with the central nervous system and is responsible for the perception of external environmental information (temperature, touch, etc.) to the brain. More specifically, activation of trigeminal nerves releases calcitonin gene-related peptide (CGRP), which increases in level with painful experiences. Cocoa, as evidenced by its ancient uses (abdominal pain, toothache, joint pain), seems to be associated with diminished activity of the cranial nerves. In an investigation of CGRP expression in both trigeminal ganglia cultures and in an in vivo model of temporomandibular joint (TMJ) inflammation, *Theobroma cacao* extract enriched for polyphenols decreased inflammation from two stimulants that cause nerve inflammation: potassium chloride and capsaicin [11]. In vivo administration of cacao extract also decreased cellular pain response, demonstrated by decreased expression of CGRP, suggesting that pain tolerance could be enhanced by the reduction in CGRP release [11].

Another in vivo investigation of the influence of cocoa on trigeminal nerve activity incorporated raw cocoa powder (at low 1% and high 10%) into the diet of rodents and injected the animals with an inflammatory agonist (capsaicin) in the trigeminal nerve 14 days later [10]. Data showed that animals on cocoa diets had significantly suppressed neuronal expression of CGRP and upregulation of MKP (good-protein kinase enzymes that react after inflammatory stimuli to return neuropeptide levels to normal). Thus, a rodent model of chronic inflammation showed decreased trigeminal nuclear and cytosolic P-p38 (phosphorylated form of inflammatory MAPK protein) when observed in animals on a diet that included cocoa as compared to animals on normal diets. Daily consumption of cocoa seems to have reduced inflammatory and pain effects in the trigeminal nerve via a mechanism similar to glucocorticoids [10]. In other words, a cocoa-rich diet decreases trigeminal nerve activation by upregulating helper MKP proteins, downregulating inflammatory proteins (MAPK and CGRP), and therefore results in decreased pain perception. These results were replicated in an additional study using grape seed polyphenols in place of the cocoa polyphenols, thus suggesting that the observed anti-inflammatory action on the trigeminal nerve is moderated by the polyphenol content of a diet, rather than the other chemical components of cocoa [10].

Besides chronic models of pain and inflammation such as migraine and TMJ, infection, and edema, cocoa flavonoids also seem to have influence on acute models of inflammatory pain. In contrast to previous studies using less-alkaline chocolate, a recent study found that consumption of a flavonoid-rich dark chocolate beverage after exercise led to less muscle soreness and pain, likely due to a decrease in inflammation [63].

Summary

Both historic and modern reports of chocolate as an analgesic may be driven by a combination of learning and pharmacological effects on the body. Through learning, chocolate's palatability and perceived pleasantness can become associated with increased levels of endogenous opioids [22] that may reduce pain perception. Furthermore, the increase in opioids may serve as a self-preservation function, since consuming a food of high nutritional value such as chocolate may be more evolutionarily important to an animal than a brief response to pain [18, 37].

Cocoa contains a variety of pharmacological components, each of which may possibly contribute to its nourishing value and medicinal qualities in a unique way; however, more research is needed to determine the effects (beneficial or not) of cocoa consumption. It is not disputed that several different classes and families of chemicals that alter mood, physiological response, and response to pain are present in chocolate. The question is whether many of the pharmacological constituents in cocoa exist at levels that can really have an effect on behavior and physiology. Chocolate contains tryptophan (which increases serotonin levels), but the amount of protein in most chocolate inhibits the serotonin-increasing mechanism [22]. In addition, methylxanthines, PEA, and anandamide levels are all too low to have an effect [22]. A 1.65-oz bar of chocolate contains only 10% the amount of caffeine as a cup of coffee [39], and in order to get an active amount of anandamide, an individual would have to consume 25 lb of chocolate [40]. However, flavonoids are components in chocolate that could present a possible mechanism for chocolate analgesia. Cocoa's flavonoids act to reduce inflammation [12] and decrease sensory nerve activation [11, 12].

Ancient practitioners prescribed cocoa for painful medical issues, such as dental problems, dysentery, indigestion, gastrointestinal issues, fever, cough, and infection [4], all of which are related to the immune response of inflammation. Perhaps, these early healers were aware on some level of cocoa's anti-inflammatory properties, or perhaps, they were simply aware of its role as a mood enhancer or comfort food. Regardless of their awareness of the mechanisms, we can be certain that they, like us, were well aware of chocolate's effect on pain tolerance.

References

1. Lee I-M, Paffenbarger RS. Life is sweet: candy consumption and longevity. BMJ. 1998;317:19–26.
2. Marcus DA, Scharff L, Turk D, et al. A double-blind provocative study chocolate as a trigger of headache. Cephalalgia. 1997;17(8):855–62.
3. De la Cruz M. The Badianus Manuscript (Codex Barberini, Latin 241) Vatican Library; an Aztec herbal of 1552. Baltimore: The John Hopkins University Press; 1940.
4. Dillinger TL, Barriga P, Escárcega S, et al. Food of the gods: cure for humanity? A cultural history of the medicinal and ritual use of chocolate. J Nutr. 2000;130:2057S–72.
5. Sahagún B. (1590/1981) General history of the things of New Spain. The School of American Research and The University of Utah Monographs of The School of American Research and The Museum of New Mexico/Santa Fe.
6. Varey S editor. The Mexican Treasury: the writings of Dr. Francisco Hernández. Translated by Chambran R, Chamberlin CL, Varey S. Stanford: Stanford University Press; 2000.
7. Hughes W. The American Physitian, or A Treatise of the Roots, Plants, Trees, Shrubs, Fruit, Herbs etc. Growing in the English Plantations in America: Describing the Place, Time. Names, Kindes, Temperature, Vertues and Uses of Them, either for Diet, Physick, etc. whereunto Is Added a Discourse of the Cacao-Nut-Tree, and the Use of its Fruit; with All the Ways of Making of Chocolate. The Like Never Extant Before. London, UK: Printed for William Crook the Green Dragon without Temple-Bar; 1672.
8. Fuller LK. Chocolate fads, folklore, & fantasies: 1,000 chunks of chocolate information. Binghamton: The Haworth Press; 1994.
9. Keen CL. Chocolate: food as medicine/medicine as food. J Am Coll Nutr. 2001;20(5):4365–95.
10. Cady RJ, Durham PL. Cocoa-enriched diets enhance expression of phosphatases and decrease expression of inflammatory molecules in trigeminal ganglion neurons. Brain Res. 2010;1323:18–32.
11. Abbey MJ, Patil VV, Vause CV, et al. Repression of calcitonin gene-related peptide expression in trigeminal neurons by a Theobroma cacao extract. J Ethnopharmacol. 2008;115:238–48.
12. Pérez-Cano FJ, Franch A, Castellote C, et al. Cocoa and the immune system and proliferative disorders. In: Watson RR, Zibadi S, Preedy VR, editors. Dietary components and immune function, Nutrition and health, Part 5. New York: Springer; 2010. p. 469–96.
13. Fisher ND, Hughes M, Gerhard-Herman M, et al. Flavanol rich cocoa induces nitric-oxide-dependent vasodilation in healthy humans. J Hypertens. 2003;21(12):2281–6.

14. Engler MB, Engler MM, Chen CY, et al. Flavonoid-rich dark chocolate improves endothelial function and increases plasma epicatechin concentrations in healthy adults. J Am Coll Nutr. 2004;23:197–204.
15. Nordin M, Fagius J. Effect of noxious stimulation on sympathetic vasoconstrictor outflow to human muscles. J Physiol. 1995;489(Pt 3):885–94.
16. Grassi D, Lippi C, Necozione S, et al. Short-term administration of dark chocolate is followed by a significant increase in insulin sensitivity and a decrease in blood pressure in healthy persons. Am J Clin Nutr. 2005; 81:611–4.
17. Greisen J, Juhl CB, Grøfte T, et al. Acute pain induces insulin resistance in humans. Anesthesiology. 2001;95(3):578–84.
18. Eggleston K, White TL, Sheehe PR. Adding cocoa to sucrose: the effect on cold pain tolerance. Chem Senses. 2010;35:269–77.
19. Dubé L, LeBel JL, Lu J. Affect asymmetry and comfort food consumption. Physiol Behav. 2005;86:559–67.
20. Kracke G, Uthoff K, Tobias J. Sugar solution analgesia: the effects of glucose on expressed mu opioid receptors. Anesth Analg. 2005;101:64–8.
21. Small D, Zatorre R, Dagher A, et al. Changes in brain activity related to eating chocolate. Brain. 2001;124(9):1720–33.
22. Nehlig A. Coffee, tea, chocolate, and the brain. New York: CRC Press; 2004.
23. Knight I. Minerals in cocoa and chocolate. In: Knight I, editor. Chocolate and cocoa: health and nutrition. Oxford, UK: Blackwell Sciences; 1999. p. 143–52.
24. Weil AT, Rosen W. From chocolate to morphine: everything you need to know about mind-altering drugs. 3rd ed. Boston: Hougton Mifflin; 1998.
25. Segato FN, Castro-Souza C, Segato EN, Morato S, Coimbra NC. Sucrose ingestion causes opioid analgesia. Braz J Med Biol Res. 1997;30:981–4.
26. Yamamoto T, Sakoa N, Maedab S. Effects of taste stimulation on β-endorphin levels in rat cerebrospinal fluid and plasma. Physiol Behav. 2000;69(3):345–50.
27. Nevid JS, Rathus SA. Stress: what it is and how to manage it. In: Psychology and the challenges of life. 11th ed. Hoboken: Wiley; 2009: p. 90–110.
28. Lynch WC, Libby L. Naloxone suppresses intake of highly preferred saccharin solutions in food deprived and sated rats. Life Sci. 1983;33:1909–14.
29. Macht M, Mueller J. Immediate effects of chocolate on experimentally induced mood states. Appetite. 2007; 49(3):667–74.
30. Parker G, Parker I, Brotchie H. Mood state effects of chocolate. J Affect Disord. 2006;92(2–3):149–59.
31. Mercer ME, Holder MD. Antinociceptive effects of palatable sweet ingesta on human responsivity to pressure pain. Physiol Behav. 1997;61(2):311–8.
32. Haouari N, Wood C, Griffiths G, et al. The analgesic effect of sucrose in full term infants: a randomized controlled trial. Br Med J. 1995;310:1498–500.
33. Lewkowski MD, Ditto B, Roussos M, et al. Sweet taste and blood pressure-related analgesia. Pain. 2003; 106:181–6.
34. Yeomans M, Prescott J, Gould N. Acquired hedonic and sensory characteristics of odours: influence of sweet liker and propylthiouracil taster status. Q J Exp Psychol. 2009;62(8):1648–64.
35. Fantino M, Hosotte J, Apfelbaum M. An opioid antagonist, naltrexone, reduces preference for sucrose in humans. Am J Physiol. 1986;251:91–6.
36. Foo H, Mason P. Sensory suppression during feeding. Proc Natl Acad Sci. 2005;102(46):16865–9.
37. Foo H, Mason P. Analgesia accompanying food consumption requires ingestion of hedonic foods. J Neurosci. 2009;29(41):13053–62.
38. Foo H, Crabtree K, Thrasher A, et al. Eating is a protected behavior even in the face of persistent pain in male rats. Physiol Behav. 2009;97(3–4):426–9.
39. Bruinsma K, Taren DL. Chocolate: food or drug? J Am Diet Assoc. 1999;99(10):1249–56.
40. Jacobsen R. Chocolate unwrapped: the surprising health benefits of America's favorite passion. Montpelier: Invincible Cities Press; 2003.
41. Benton D. The biology and psychology of chocolate craving. In: Nehlig A, editor. Coffee, tea, chocolate, and the brain. Boca Raton: CRC Press; 2004. p. 205–18.
42. Schifano F, Magni G. MDMA ("ecstasy") abuse: psychopathological features and craving for chocolate, a case series. Biol Psychiatry. 1994;36:763–7.
43. Kelishadi R. Cacao to cocoa to chocolate: healthy food? ARYA J. 2005;1(1):29–35.
44. Apgar JL, Tarka SM. Methylxanthines. In: Knight I, editor. Chocolate and cocoa: health and nutrition. Oxford, UK: Blackwell Sciences; 1999.
45. Snel J, Lorist MM, Tieges Z. Coffee, caffeine, and cognitive performance. In: Nehlig A, editor. Coffee, tea, chocolate, and the brain. Boca Raton: CRC Press; 2004. p. 205–18.

46. Smits P, Lenders JWM, Thien T. Caffeine and theophylline attenuate adenosine-induced vasodilation in humans. Clin Pharmacol Ther. 1990;48:410–8.
47. Ward N, Whitney C, Avery D, et al. The analgesic effects of caffeine in headache. Pain. 1991;44(22):151–5.
48. Hasko G, Cronstein B. Methylxanthines and inflammatory cells. In: Fredholm BB, editor. Methylxanthines. handbook of experimental pharmacology, vol. 200. New York: Springer; 2011. p. 457–68.
49. Wollgast J, Anklam E. Review on polyphenols in Theobroma cacao: changes in composition during the manufacture of chocolate and methodology for identification and quantification. Food Res Int. 2000;33(6):423–47.
50. Waterhouse AL, Shirley JR, Donovan JL. Antioxidants in chocolate. Lancet. 1996;348(9030):834. British ed.
51. Takeda H. Utility of chocolate/cocoa: anti-stress effect of cacao mass polyphenol. Evaluation in menopausal unidentified complaint model. Shoku no Kagaku (Japanese). 1999;252:50–2.
52. Slayback DL, Watson RR. Bioflavonoids and cardiovascular health: tea, red wine, cocoa, and pycnogenol®. JANA. 2006;9(2):16–21.
53. Manthey JA. Biological properties of flavonoids pertaining to inflammation. Microcirculation. 2000;7(6):29–36.
54. Ramiro E, Franch A, Castellote C, et al. Flavonoids from *Theobroma cacao* down-regulate inflammatory mediators. J Agric Food Chem. 2005;53:8506–11.
55. Sanbongi C, Suzuki N, Sakane T. Polyphenols in chocolate, which have antioxidant activity, modulate immune functions in humans in vitro. Cell Immunol. 1997;177:129–36.
56. Hämäläinen M, Nieminen R, Vuorela P, et al. Anti-inflammatory effects of flavonoids: genistein, kaempferol, quercetin, and daidzein inhibit STAT-1 and NF-kB activations, whereas flavone, isorhamnetin, naringenin, and pelargonidin inhibit only NF-kB activation along with their inhibitory effect on iNOS expression and NO production in activated macrophages. Mediators Inflamm. 2007;45673–45683.
57. Comalada M, Camuesco D, Sierra S, et al. In vivo quercitrin anti-inflammatory effect involves release of quercetin, which inhibits inflammation through down-regulation of the NF-kB pathway. Eur J Immunol. 2005;35:584–92.
58. Matsuoka Y, Hasegawa H, Okuda S, et al. Ameliorative effects of tea catechins on active oxygen-related nerve cell injuries. J Pharmacol Exp Ther. 1995;274:602–8.
59. Rotelli AE, Guardia T, Juárez AO, et al. Comparative study of flavonoids in experimental models of inflammation. Pharmacol Res. 2003;48:601–6.
60. Kim H, Kong H, Choi B, et al. Metabolic and pharmacological properties of rutin, a dietary quercetin glycoside, for treatment of inflammatory bowel disease. Pharm Res. 2005;22:1499–509.
61. Park HJ, Lee CM, Jung ID, et al. Quercetin regulates Thl/Th2 balance in a murine model of asthma. Int Immunopharmacol. 2009;9:261–7.
62. Lee JY, Lee JH, Park JH, et al. Liquiritigenin, a licorice flavonoid, helps mice resist disseminated candidiasis due to *Candida albicans* by TH1 immune response, whereas liquiritin, its glycoside from, does not. Int Immunopharmacol. 2009;9:632–8.
63. McBrier NM, Vairo GL, Bagshaw D, et al. Cocoa-based protein and carbohydrate drink decreases perceived soreness after exhaustive aerobic exercise: a pragmatic preliminary analysis. J Strength Cond Res. 2010;24(8): 2203–10.

Chapter 33
Cocoa for Recovery

Kelly Pritchett and Phillip A. Bishop

Key Points

- Cocoa may enhance recovery from exercise owing to its antioxidant properties and partly because it usually is associated with other nutrients (e.g., carbohydrate and protein in chocolate milk).
- Chocolate milk, a cocoa-based beverage, appears comparable to more expensive recovery beverages and seems well accepted by many athletes.
- Consuming 1.0–1.5 $g/kg^{-1}/h^{-1}$ of carbohydrate, with or without cocoa, immediately after exercise and at 30-min intervals for up to 6 h postexercise, appears to be optimal for adequate glycogen resynthesis.
- A cocoa-based beverage ingested at 1.0–1.5 $g/kg^{-1}/h^{-1}$ may be optimal for recovery, and the addition of flavanol-rich cocoa may decrease perceptions of postexercise muscle soreness and decrease levels of plasma isoprostanes.
- Research should examine different types of cocoa (natural and Dutched) and differences in solid, liquid, and gel forms of cocoa-based recovery meals, the dosage of cocoa, timing, and frequency of ingestion on postexercise recovery measures, including performance, muscle soreness, and oxidative stress indices.

Keywords Dutched cocoa • Flavanols • Isoprostanes • Nutritional strategies • Postexercise carbohydrate • Postexercise muscle damage • Postexercise protein

K. Pritchett, Ph.D., RD., CSSD (✉)
Department of Nutrition, Exercise, and Health Sciences, Central Washington University,
400 E. University Way, ms 7572, Ellensburg, WA 98926, USA
e-mail: kkerr@cwu.edu

P.A. Bishop, B.S., M.S.Ed., Ed.D.
Department of Kinesiology, University of Alabama, 610 8th Street, Tuscaloosa, AL 35487-0312, USA

Department of Kinesiology, University of Alabama, 102 Moore Street, Tuscaloosa, AL 35487-0312, USA
e-mail: pbishop@bama.ua.edu

R.R. Watson et al. (eds.), *Chocolate in Health and Nutrition*, Nutrition and Health 7,
DOI 10.1007/978-1-61779-803-0_33, © Springer Science+Business Media, LLC 2013

Introduction

Athletes, coaches, and sports dietitians have sought postexercise nutritional strategies that will enhance muscle glycogen resynthesis after exercise, hasten recovery, and maintain or improve the quality of future workouts or performances [1–3]. Athletes may participate in multiple training sessions a day with as little as 6 h of recovery between workouts [3]. A principal component of preparation for an elite athlete is to maximize their training in order to increase their potential for competition. Because full and often rapid recovery is necessary for subsequent optimal performance [2], athletes should practice nutritional strategies that maximize recovery. Likewise, in some sports, athletes may compete multiple times in a single day. Enhanced recovery between competitions is important in these cases as well. Anything that enhances recovery permits more effective successive practice sessions and should improve performance [3].

Because muscle glycogen is the main fuel during intense exercise, replenishing muscle glycogen stores in the postexercise recovery period is an important factor influencing recovery and subsequent performance. In addition, the timing and composition of a postexercise meal is highly dependent upon the duration and intensity of the preceding exercise bout [4]. Research regarding postexercise nutritional strategies has focused on timing of ingestion, type (solid vs. liquid) of carbohydrate (CHO), amount of CHO, presence of other nutrients (e.g., antioxidants, protein), and frequency of postexercise feedings to determine the most effective way to enhance glycogen resynthesis [5].

Recently, research has examined the effects of carbohydrate beverages containing cocoa [6–11] on recovery measures. Research examining cocoa-based carbohydrate beverage consumption during recovery periods of 4 h or more suggests enhanced recovery [7–9, 11], which would be beneficial to athletes competing in events with short recovery periods (preliminary heats, finals) such as track and field, swimming, and multiple-day events, such as the Tour de France.

This chapter will provide an overview and discussion of the research that focuses on the effects of cocoa-based beverages and isolated cocoa on various indices of recovery including muscle damage, glycogen resynthesis, and exercise performance. Guidelines regarding the optimal timing and amount of cocoa-based carbohydrate postexercise will also be discussed in this chapter [3, 4].

Cocoa and Postexercise Recovery

Exercise has been shown to increase the production of free radicals due to the greater flux of oxygen into the cells, thereby increasing skeletal muscle damage and oxidative stress [10]. Recently, cocoa has received substantial attention in the clinical sports medicine world in its ability to reduce levels of oxidative stress [6, 10]. As a result, the efficacy of dietary supplementation with cocoa as an antioxidant has been studied.

Very few studies to date have examined the effectiveness of cocoa on exercise recovery. Wiswedel et al. (2004) suggested that cocoa-based drinks containing flavanols decreased levels of F2-isoprostane, a measure of oxidative stress, after strenuous exercise [6]. Recently, another study examined the effectiveness of cocoa-based carbohydrate-protein beverage (with a 3.5:1 carbohydrate-to-protein ratio), containing natural cocoa, on skeletal muscle damage and muscle soreness after downhill running. In a repeated-measures design, each participant ingested an experimental beverage condition immediately after exercise, 2 h postexercise, and before bed. Three beverage conditions (cocoa drink after, water, or cocoa drink before) were administered in a random order. The cocoa-based beverage had no effect on blood markers of muscle damage (CK, IL-6, IL-8, or urinary isoprostanes); however, the authors reported that the cocoa was effective in reducing perceived muscle soreness from 24 to 48 h after downhill running. The authors speculated that the cocoa-based beverage might have been effective in reducing soreness because the beverage contained natural cocoa instead of Dutched cocoa (i.e., cocoa treated with alkali). It should be noted that the actual flavanoid content of the cocoa-based beverage was not reported in this study nor was Dutched cocoa tested. Furthermore, no performance measurements were taken in this study [10].

Recently, low-fat chocolate milk has been suggested to be an effective but lower-cost recovery aid, with a carbohydrate-to-protein ratio similar to many commercial recovery and carbohydrate-replacement beverages [7–9]. Chocolate milk is composed of cocoa plus monosaccharides (glucose and fructose) and disaccharides (lactose), while most commercially available recovery beverages consist of monosaccharides (glucose and fructose) and complex carbohydrates (maltodextrin). Low-fat chocolate milk has a 4-to-1 carbohydrate: protein ratio similar to many commercial recovery beverages. In comparison to many carbohydrate-electrolyte beverages, chocolate milk packs substantially more carbohydrates per 240 mL. It also provides fluids and sodium (in rehydration, replacing sodium is crucial) [12] which needs to be replaced due to sweat loss during a workout. Chocolate milk is also high in calcium necessary for building and maintaining strong bones and a major constituent involved in muscular contraction. Based on the recommendations regarding postexercise carbohydrate intake [4], a 70-kg male would need to consume 510–810 mL (70–84 g CHO, and 19–30 g PRO) and a 60-kg female 435–690 mL (60–72 g CHO, and 16–26 g PRO) of low-fat chocolate milk per hour. These are reasonable amounts for most athletes.

Several research studies have examined the effectiveness of postexercise low-fat chocolate milk consumption on various markers of recovery [7–9, 11]. Karp et al. (2006) examined the effectiveness of consuming chocolate milk as a postexercise recovery aid between two cycling sessions in trained cyclists. After 4 h of recovery, cycling time to exhaustion at 70% of VO_{2max} was significantly ($p \leq 0.05$) longer (40.0 ± 14.7 min) for the chocolate milk trial compared to the over-the-counter recovery beverage (26.8 ± 14.7 min) [7]. The authors suggested that differences in performance demonstrated in their study may have been attributed to the different types of carbohydrate in the beverages. Because increases in muscle glycogen levels during the early hours of recovery are greater with simple versus complex carbohydrate [13], perhaps the 4-h recovery period did not allow adequate time for the complete digestion of the complex carbohydrates in the commercial recovery beverage. The authors also speculated as to whether the higher fat content of chocolate milk may have increased the levels of free fatty acids in the blood, and possibly delaying glycogen depletion during the subsequent cycling trial to exhaustion and allowing subject to cycle longer [7]. This explanation seems plausible because, when postexercise consumption of low-fat chocolate milk was compared to an isocaloric over-the-counter recovery beverage (based on 1 g $CHO.kg^{-1}$ of body weight/h postexercise for the first 2 h) after a high-intensity fatiguing trial, the authors found no differences in cycling time to exhaustion at 85% of VO_{2max} [8]. Being equivalent to a commercial recovery beverage is important because chocolate milk is well received by most athletes and is more economical than most commercial recovery beverages.

Pritchett et al. (2009) compared the postexercise consumption of low-fat chocolate milk to an over-the-counter commercial recovery beverage matched for carbohydrate and protein content and found that the increase in creatine kinase (CK), a marker of muscle damage, was significantly reduced ($p < 0.05$) from pre- to postexercise in the chocolate milk trial compared to the commercial recovery beverage (CRB) trial (increase CHOC 27.9 ± 134.8 $U.L^{-1}$, CRB 211.9 ± 192.5 $U.L^{1}$), with differences not significantly different between the two trials for CKpost (CHOC 394.8 ± 166.1 $U.L^{-1}$, CRB 489.1 ± 264.4 $U.L^{-1}$). The apparent inconsistency in the findings (i.e., a difference in change scores but not in CK postlevels) is not uncommon and was chiefly attributable to the magnitude of the mean differences. In other words, the analysis apparently was not sensitive enough to the pre- or postdifferences between the two treatments, but when expressed as a change score, relative to the error variance, a difference was detected. As always, the difference we found in the change score should be interpreted with caution. After completion of the study, participants were asked to provide feedback regarding the recovery beverages. All ($n = 10$) participants' preferred the taste of chocolate milk [8].

Similarly, Gilson et al. (2010) examined the effectiveness of low-fat chocolate milk versus a high-CHO recovery beverage consumed postexercise for a week in collegiate soccer players. The soccer players continued their normal training regimen, which was similar among subjects. This study found significantly ($p \leq 0.05$) lower creatine kinase (CHOC: 316.9±188.3 $U.L^{-1}$, CHO: 431.6±310.8 $U.L^{-1}$) levels after 1 week of supplementation with chocolate milk versus a high carbohydrate-only beverage. However, no performance differences were detected between the two beverages [11].

These findings support the efficacy of chocolate milk as a recovery aid between workouts. Low-fat chocolate milk serves as a more convenient, cheaper, premixed, and, based on current observations, more palatable recovery beverage option for many athletes. Furthermore, recent research has suggested that compared to other recovery beverages, chocolate milk may be more beneficial for decreasing muscle damage after a workout. However, it is unclear as to whether these recovery benefits seen from chocolate milk can actually be attributed to the cocoa flavanols or to other factors, such as the 3:1 carbohydrate-to-protein ratio that has been found to enhance postexercise recovery. Chocolate milk contains "Dutched cocoa," or cocoa treated with alkali, meaning that some of the flavanols would not have been preserved during this processing. The flavanol content in cocoa-based beverages and chocolate varies and is strongly dependent upon the processing [6].

Further research is needed to examine the efficacy of cocoa-based foods and beverages as a postexercise recovery aid. In addition, research needs to examine the dosage of cocoa, timing, type of coca processing, and frequency of ingestion on postexercise recovery measures, including performance, muscle soreness, and oxidative stress levels.

The Role of Carbohydrate in Postexercise Recovery

Cocoa is seldom ingested in isolation. Typically, cocoa is taken combined with carbohydrates in the form of chocolate bars or chocolate beverages. Consequently, any consideration of cocoa in recovery must also take into account the associated carbohydrates and other nutrients.

Muscle glycogen is the primary fuel source during high-intensity exercise and an important source during endurance exercise. Therefore, postexercise glycogen restoration plays a very important role in the recovery process. Glycogen resynthesis is highly dependent on the extent of glycogen depletion, as well as the type, duration, and intensity of the exercise session [14]. For up to 6 h postexercise, the rate of muscle glycogen resynthesis is accelerated compared to rest, and within 24 h postexercise, complete restoration of glycogen stores can occur if sufficient amounts of carbohydrate are consumed [1, 15].

Glycogen resynthesis, after either a glycogen-depleting exercise bout or endurance exercise, occurs in two phases (Fig. 33.1). The first phase, or the rapid phase, lasts anywhere from 30 to 60 min and is

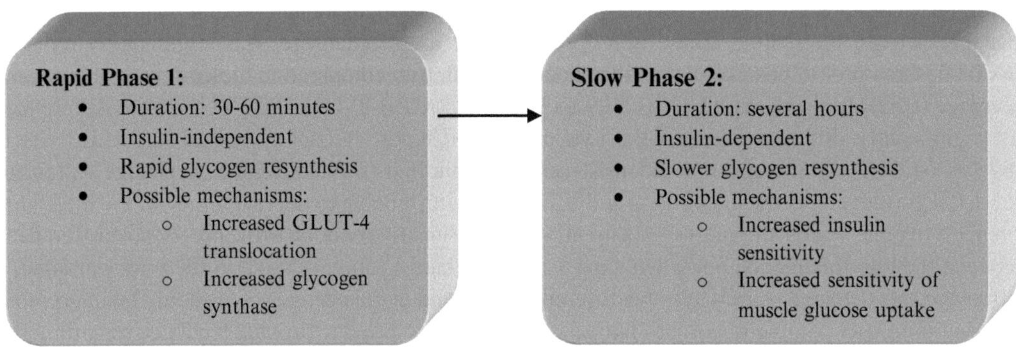

Fig. 33.1 Characteristics of postexercise glycogen resynthesis during the rapid phase (1) and the slow phase (2)

insulin independent owing to the increased permeability of the muscle cell as a result of exercise-induced translocation of GLUT-4 (glucose transporter carrier protein-4) and plausibly an upregulation of glycogen synthase. It should be noted that GLUT-4 concentrations and insulin sensitivity are higher in trained individuals, perhaps allowing for enhanced glycogen resynthesis compared to their untrained counterparts [1].

An increase in postexercise glycogen synthase and exercise-induced increases in insulin sensitivity may be the potential mechanisms responsible for the 2–4-h period of enhanced glycogen resynthesis following exercise. The literature suggests that low postexercise muscle glycogen levels may provide a stimulus for this increase in glycogen synthase. The literature indicates that a recovery meal consumed within 2 h postexercise, compared to no feeding, is more effective in improving recovery [16].

Following the rapid glycogen resynthesis phase is the slow phase where muscle glycogen resynthesis occurs at a slower rate than the initial phase. The slow phase is highly dependent on carbohydrate availability and insulin levels and can potentially last several hours [1].

Because complete muscle glycogen resynthesis can take as long as 24 h, even under optimal conditions, studies have examined methods to increase the rate of muscle glycogen resynthesis [17]. Depending on the extent of glycogen depletion, consuming 1.0–1.5 g CHO.kg^{-1} h^{-1} immediately after exercise, and at 30-min intervals for up to 6 h postexercise, appears to be optimal for adequate glycogen resynthesis [1, 4, 7, 18, 19]. On the other hand, if CHO intake is delayed by 2 h postexercise, glycogen resynthesis rates have been found to be 45% lower [3, 20].

The Addition of Protein to Carbohydrate in a Recovery Meal

As mentioned, cocoa may be consumed with milk or other protein sources. Consequently, it is important to consider the role of protein in recovery.

Protein contributes an estimated 5–15% of total energy expenditure during endurance exercise [19], which is considerably less than the contribution of carbohydrate. The literature concerning the effects of postexercise ingestion of a CHO-PRO supplement compared to a CHO supplement (i.e., sports beverage) on performance is inconsistent. Some studies show improved performance with the CHO-PRO postexercise complex versus a CHO only [7, 21–25], whereas others show no difference in performance [23, 26–28]. It should be noted that the majority of the literature has examined recovery in trained cyclists or runners when performing a time trial to exhaustion at 70–85% of VO_{2max} [7, 21–28].

The addition of protein to a postexercise recovery meal may also enhance net protein anabolism [21]. During the postexercise period, there is an increased rate of muscle protein synthesis in trained individuals [3]. The results of studies reported [14, 22, 29] seem to be in support of the addition of protein to the recovery beverage.

In addition, studies have reported decreases in muscle damage (CPK) with the addition of PRO to a recovery beverage after exercise sessions [8, 16, 19, 21, 25, 30]. Both high-intensity and prolonged endurance exercise can damage skeletal muscle, resulting in delayed-onset muscle soreness with concurrent increases in markers of muscle damage such as creatine kinase (CK), myoglobin (Mb), cortisol, and lactate dehydrogenase (LDH) [31, 32]. Elevated levels of these enzymatic markers are associated with decreased performance [32]. Due to the applied nature of recovery studies, the majority of the literature examining muscle damage has included multiple indicators of muscle damage, including blood-borne creatine kinase (CK), and subjective measures of muscle soreness (using a visual scale) [33]. However, CK has been criticized as an effective indicator of muscle damage because of poor correlations with direct measures of muscle damage.

Postexercise recovery beverages containing protein seem to be effective in improving recovery indices. However, some of the results may be due to the higher caloric content of the CHO: PRO supplements. The additional protein calories via glucogenesis may have provided additional substrate for glycogen resynthesis to occur, therefore aiding in an enhanced recovery [20]. Currently, research suggests that 20–25 g of high-quality protein during a single feeding is optimal. Future research should examine the type of protein, timing of intake, and the effects of distribution of protein throughout the day on recovery indices [3].

Limitations of Studies

Various limitations have been discussed throughout this review of studies that have examined the efficacy of cocoa-based beverages and cocoa on optimal recovery. The vast majority of the studies examined the acute effects of a postexercise cocoa-based beverage on exercise performance. Practically, it would be more beneficial to examine the effects of these nutritional strategies over a longer duration similar to a training regimen. Greatly needed are studies of consumption of chocolate in other forms, including sweetened chocolate solid foods, and comparisons of different preparations of cocoa (i.e., Dutched vs. natural).

Extraneous variables, such as dietary intake immediately before the study and sleep patterns, were often not controlled in previous cocoa studies. In order to examine the effect of a recovery beverage on performance and recovery, it is vital to control for dietary intake. It is ideal to provide a food-frequency questionnaire as well as a 3-day food record to get an accurate depiction of diet. Sleep is another variable that could influence the results of study; therefore, variations in sleep patterns should be considered when comparing trials.

Many athletes consume foods in solid or gel forms during workout or competition. For practical purposes, research should examine differences in solid, liquid, and gel forms of cocoa-based recovery meals on recovery measures and oxidative stress indices. In addition, because it is very practical to incorporate a recovery beverage into athletes' daily regimens, it would be more beneficial to the athletic population to examine the effectiveness of a postexercise recovery beverage taken on a daily basis.

Also, research should examine other measures of muscle damage. Creatine phosphokinase and subjective measures of muscle soreness have been the primary dependent variables in the literature. Due to the variable nature of CK, measuring additional blood parameters, such as LDH (lactate dehydrogenase), myoglobin, and cortisol, in conjunction with CK would enhance the quality of the studies

Summary

The available literature regarding postexercise nutritional strategies for optimal performance is evolving. The optimal timing regarding postexercise nutritional strategies for maximal glycogen resynthesis appears to be within the first 2 h postexercise [14, 22, 29, 34]. The literature suggests that 1.0–1.5 g/kg^{-1}/h^{-1} [1, 2, 7] of a cocoa-based beverage may be optimal for recovery. The additional protein (20–25 g protein) in the food or beverage may aid in muscle protein resynthesis [2, 14, 22, 29, 34]. Furthermore, the addition of flavanol-rich cocoa may decrease perceptions of postexercise muscle soreness [10] and decrease levels of plasma isoprostanes [6]. Given the limited number of studies, however, further research is warranted to examine the efficacy of cocoa flavanols on exercise recovery.

References

1. Jentjens RL, Jeukendrup AE. Determinants of post-exercise glycogen synthesis during short term recovery. Int J Sport Nutr Exerc Metab. 2003;33(2):117–44.
2. Burke L. Fasting and recovery from exercise. Br J Sports Med. 2010;44:502–8.
3. Bishop PA, Jones E, Woods K. Recovery from training: a brief review. J Strength Cond Res. 2008;229(3):1–10.
4. American College of Sports Medicine, American Dietetic Association, Dietitians of Canada. Nutrition and athletic performance. Joint position statement of the American Dietetic Association, Dietitians of Canada, and the Medicine and American College of Sports Medicine. Med Sci Sport Exerc. 2009;109:509–27.
5. Berardi JM, Price TB, Noreen EE, Lemon PWR. Postexercise muscle glycogen recovery enhanced with carbohydrate-protein supplement. Med Sci Sports Exerc. 2006;38(6):1106–13.
6. Wiswedel I, Hirsh D, Kropf S, Gruening M, Pfister E, Schewe T, Sies H. Flavanol-rich cocoa drink lowers plasma F2-ioprostane concentration in humans. Free Radic Biol Med. 2004;37(3):411–21.
7. Karp JR, Johnston JD, Tecklenburg S, Mickleborough TD, Fly AD, Stager JM. Chocolate milk as a post-exercise recovery aid. Int J Sport Nutr Exerc Metab. 2006;16:78–91.
8. Pritchett KL, Bishop PA, Pritchett RC, Green JM, Katica C. Acute effects of chocolate milk and a commercial recovery beverage on post-exercise muscle damage and cycling performance. Appl Physiol Nutr Metab. 2009;34(6):1017–22.
9. Thomas K, Morris P, Stevenson E. Improved endurance capacity following chocolate milk consumption compared with 2 commercially available sports drinks. Appl Physiol Nutr Metab. 2009;34:78–82.
10. McBrier NM, Vairo GL, Bagshaw D, Lekan JM, Bordi PL, Kris-Etherton PM. Cocoa-based protein and carbohydrate drink decreases perceived soreness after exhaustive aerobic exercise: a pragmatic preliminary analysis. J Strength Cond Res. 2010;24(8):2203–9.
11. Gilson S, Suanders MJ, Moran C, Moore R, Womack CJ, Todd K. Effects of chocolate milk consumption on markers of recovery following soccer training: a randomized cross-over study. J Int Soc Sport Nutr. 2010;7:19.
12. Clapp AJ, Bishop PA, Smith JF, Mansfield ER. Effects of carbohydrate-electrolyte content of beverages on voluntary hydration in a simulated industrial environment. Am Ind Hyg Assoc J. 2000;61:692–9.
13. Freidman JE, Neufer PD, Dohm GL. Regulation of glycogen resynthesis following exercise. Sports Med. 1991;11(4):232–43.
14. van Loon LJ, Saris WM, Ruijshoopanda MK, Wagenmakers AM. Maximizing postexercise muscle glycogen synthesis: carbohydrate supplementation and the application of amino acid or protein hydrolysate. Am J Clin Nutr. 2000;72:106–11.
15. Ryans M. Sports nutrition for endurance athletes. 2nd ed. Boulder: Velo Press; 2007.
16. Romano-Ely BC, Todd K, Saunders MJ, Laurent St T. Effect of an isocaloric carbohydrate-protein-antioxidant drink on cycling performance. Med Sci Sports Exerc. 2006;38(9):1608–16.
17. van Hall G, Shirreffs SM, Calbert JAL. Muscle glycogen resynthesis during recovery from cycle exercise: no effect of additional protein ingestion. J Appl Physiol. 2000;88:1631–6.
18. Doyle JA, Sherman WM, Strauss RL. Effects of eccentric and concentric exercise on muscle glycogen replenishment. J Appl Physiol. 1993;74:1848–55.
19. Saunders MJ, Luden ND, Herrick JE. Consumption of an oral carbohydrate-protein gel improves cycling endurance and prevents postexercise muscle damage. J Strength Cond Res. 2007;21(3):678–84.
20. Ivy JL, Katz AL, Cutler CL. Muscle glycogen synthesis after exercise: effect of time of carbohydrate ingestion. J Appl Physiol. 1988;64(4):1480–5.
21. Saunders MJ, Kane MD, Todd MK. Effects of carbohydrate-protein beverage on cycling endurance and muscle damage. Med Sci Sport Exerc. 2004;36(7):1233–8.
22. Williams MB, Raven PB, Fogt DL, Ivy JL. Effects of recovery beverages on glycogen restoration and endurance exercise performance. J Strength Cond Res. 2003;17:12–9.
23. Berardi JM, Noreen EE, Lemon PWR. Recovery from a cycling time trial is enhanced with carbohydrate-protein supplementation vs. isoenergetic carbohydrate supplementation. J Int Soc Sport Nutr. 2008;5:24.
24. Ivy JL, Res PT, Sprague RC, Widzer MO. Effect of a carbohydrate-protein supplement on endurance performance during exercise of varying intensity. Int J Sport Nutr Exerc Metab. 2003;13(3):382–95.
25. Valentine RJ, Saunders MJ, Todd MK, St Laurent TG. Influence of carbohydrate-protein beverage on cycling endurance and indices of muscle disruption. Int J Sport Nutr Exerc Metab. 2008;18:379–88.
26. Betts JA, Stevenson E, Williams C, Steppard C, Grey E, Griffin J. Recovery of endurance running capacity: effect of carbohydrate-protein mixtures. Int J Sport Nutr Exerc Metab. 2005;15:590–609.
27. Carrithers JA, Williamson DL, Gallagher PM, Godard MP, Schulze KE, Trappe SW. Effects of postexercise carbohydrate-protein feedings on muscle glycogen restoration. J Appl Physiol. 2000;88:1976–82.
28. Green MS, Corona BT, Doyle JA, Ingalls CP. Carbohydrate-protein drinks do not enhance recovery from exercise-induced muscle injury. Int J Sport Nutr Exerc Metab. 2008;18:1–18.

29. Ivy JL, Goforth HW, Damon BM, McCauley TR, Parsons EC, Price TB. Early postexercise muscle glycogen recovery is enhanced with a carbohydrate-protein supplement. J Appl Physiol. 2002;93:1337–44.
30. Ready SL, Seifert JL, Burke E. The effects of two sports drinks formulations on muscle stress and performance. Med Sci Sports Exerc. 1999;31:S119.
31. Clarkson PM, Kearns AK, Rouzier P, Rubin R, Thompson PD. Serum creatine kinase levels and renal function measures in exertional muscle damage. Med Sci Sports Exerc. 2006;38(4):623–7.
32. White JP, Wilson JM, Austin KG, Greer BK, St John N, Panton LB. Effect of a carbohydrate-protein supplement timing on acute exercise-induced muscle damage. J Int Soc Sport Nutr. 2008;5:5.
33. Luden ND, Saunders MJ, Todd K. Postexercise carbohydrate-protein-antioxidant ingestion decreases plasma creatine kinase and muscle soreness. Int J Sport Nutr Exerc Metab. 2007;17:109–23.
34. Zawadzki KM, Yaspelkis BB, Ivy JL. Carbohydrate-protein complex increases the rate of muscle glycogen storage after exercise. J Appl Physiol. 1992;72:1854–9.

Chapter 34
Chocolate and Withdrawal

Carla da Silva Benetti and Patrícia Pelufo Silveira

Key Points

- Evidence has indicated palatable foods and sugar as potential addictive substances because they induce effects on the mesolimbic reward system consistent with those promoted by drugs of abuse. Bingeing on sugar increases the release of neurotransmitters from brain structures involved in the regulation of appetite and/or addictive behaviors.
- Experimental models have demonstrated that excessive sugar consumption stimulates dopamine releasing in the nucleus accumbens as well as modifies dopamine receptors binding in the accumbens and striatum. Similar findings were detected in drug-dependent individuals.
- Abstinence of palatable food precipitates withdrawal symptoms, altering neurochemical homeostasis of the opioid pathway, and these effects, like opiate withdrawal, can be also seen when drugs of abuse are removed.
- Besides neuronal modifications produced in response to palatable food withdrawal, addiction-like behaviors have been described in some studies as well. Spontaneous withdrawal signs, like anxiety, teeth chattering, and aggressive behavior, are induced by palatable food (such as chocolate) deprivation.
- The interaction between genes and environment has a determining role in the individual tendency to eating disorders and metabolic diseases in adulthood.
- The neonatal environment seems to be a modulator of the vulnerability to palatable food addiction and withdrawal signs at the behavioral and neurochemical level in response to its deprivation.

Keywords Chocolate • Withdrawal • Food addiction • Palatable foods • Sugar

C. da Silva Benetti, Ph.D. (✉) • P.P. Silveira, M.D., Ph.D.
Hospital de Clínicas de Porto Alegre-HCPA, Laboratório de Pediatria Translacional-LPT, 2350, Ramiro Barcelos street, Largo Eduardo Zaccaro Faraco, 90035-903 Porto Alegre, Rio Grande do Sul, Brazil
e-mail: carlabenetti@yahoo.com.br; 00032386@ufrgs.br

R.R. Watson et al. (eds.), *Chocolate in Health and Nutrition*, Nutrition and Health 7,
DOI 10.1007/978-1-61779-803-0_34, © Springer Science+Business Media, LLC 2013

Introduction

The term "withdrawal" describes a group of symptoms that occurs upon the abrupt discontinuation or a decrease in dosage of the intake of an addictive substance. In order to experience the symptoms of withdrawal, it is first necessary to have developed a chemical dependency, which happens after consuming the substance for a certain period of time.

Some patterns of food consumption, such as overeating, binge eating, stress-induced eating, and emotional eating, bear a striking resemblance to substance-use disorders [1, 2]. Therefore, evidence of the existence of "food addiction" has mounted. Foods identified as having potential addictive properties include sweets, carbohydrates, fats, as well as sweet/fat combinations. One example of such a combination of these two addictives substances (sweet and fat) is chocolate, a palatable highly caloric food readily available in Western societies, where the prevalence of overweight and obesity is constantly increasing [3, 4]. It is interesting to note that since the beginning of the twentieth century, chocolate is described as the most commonly and intensely craved food in North American [5] and European countries [6]. Some investigators suggest that the popularity of this palatable food seems to be mainly due to its potential to arouse sensory pleasure and positive emotions [7].

Similarities Between Drug and Food Addiction

Drug-seeking behavior is motivated and reinforced not only by a drug's positive effects but also the negative state or "antireward" that accompanies abstinence from drug use [8]. Different neural pathways mediate the motivation, including dopaminergic, GABAergic, opioid, and serotonergic neural circuits in the striatum, amygdala, orbitofrontal cortex (OFC), and midbrain. The irritability, anhedonia, and dearth of motivation for natural rewards that characterize the state of antireward result not only from the loss of function of the brain's reward systems but also from the activation of stress systems in the amygdala [8].

Although addictive behavior is generally associated with drugs, alcohol, or sexual behavior, it is becoming apparent that certain food substances, notably chocolate, may cause similar physiological and psychological reactions in vulnerable people. Addiction, however, is a complex concept that includes several components, and some have questioned its applicability to chocolate cravings. Although there are many definitions of addiction, two characteristics are consistently included: a compulsion to use a substance that results in excessive and uncontrolled consumption, and existence of withdrawal symptoms when the substance is withheld[9].

Avena and colleagues [10] classified sugar as an addictive substance because it follows the typical addiction pathway that consists of bingeing, withdrawal, craving, and cross-sensitization. The definitions of the components involved in the addiction process are as follows:

- Bingeing: It consists of unusually large bouts of intake [11].
- Withdrawal: It is indicated by signs of anxiety and behavioral depression [12].
- Craving: It is measured during sugar abstinence as enhanced by responding to sugar [13].
- Cross-sensitization: It results from sugar to drugs of abuse [14].

More recently, this same group of researchers redefined "bingeing" as an escalation of intake with a high proportion of consumption at one time, usually after a period of voluntary abstinence or forced deprivation. And "craving" was also better defined by increased efforts to obtain a substance of abuse or its associated cues as a result of dependence and abstinence [15]. According to Liu and colleagues [16], these definitions play a major role in helping define and classify palatable food as a true additive substance in comparison to the criteria for drug dependence as observed at least in animal models [17].

Some studies reported a very high association between chocolate and sweet preference [18], but chocolate seems to be more widely craved and more highly preferred than other very sweet items [19].

Interestingly, it was recently shown that a previous experience with drugs of abuse could play a role in the desire to consume chocolate. This study shows that participants who had already experienced drug-like effects were more likely to want to consume more of the tasted chocolate than control individuals [20].

Chocolate has a set of extremely appealing sensory characteristics. In a general review of cravings, Weingarten and Elston [5] emphasize the importance of sensory characteristics in accounting for cravings. Its consumption may also serve for homeostatic regulation of certain neurotransmitters involved in the regulation of appetite, hunger, mood, and/or addictive behaviors. Several studies have described that negative mood is prominent among cravers and that they have a high prevalence of "emotional eating" [21]. Low central levels of serotonin have been associated with depression, addiction, and obsessive-compulsive disorder, while the ingestion of carbohydrates (especially chocolate) has been suggested to selectively increase tryptophan uptake and serotonin production by the brain [22]. However, although chocolate provides pleasure, for those who consider intake of this food to be excessive, any pleasure experienced is short-lived and accompanied by feelings of guilt [23], an effect that is evident even among children [24].

An interesting study shows that the personality style modulates the craving for chocolate in states of emotional dysregulation (anxiety, irritability, and depression) and suggests that craving for such food serves to promote soothing of their personality-based capacity to experience emotional dysregulation [25].

The simple availability of chocolate within sight can increase the desire for it [26], and this is associated with feelings of both craving and guilt [27]. In the presence of chocolate or related cues, self-reported "chocolate addicts" were more aroused, reported greater cravings, experienced more negative affect, and also ate more chocolate than control subjects [28]. Interestingly, visual images of chocolate from magazines influence the desire for this type of food, in the form of craving, and may induce negative affect such as guilt, anxiety, and depression, but only in women who had dieted at some point in their life [29].

Chocolate cravings, like sweet cravings, have been found to occur perimenstrually for many women [18], and it is the only high-carbohydrate food that is craved more at the menses than at other times during the menstrual cycle [30]. Some researchers have proposed that the cause for these chocolate cravings appears to be the result of a cultural rather than a physiological cause [31, 32].

Neurochemistry of Food Addiction and Withdrawal

Recent studies demonstrate that the support for the food addiction hypothesis comes from alterations in neurochemistry (dopamine) [15, 33], neuroanatomy (mesolimbic system) [34, 35], and self-medication behaviors [36]. Evidence from the literature describes that the consumption of high-sugar, high-fat food (palatable food) induces neurochemical modifications in mesolimbic-dopaminergic system, which is involved in the hedonic feeding behavior regulation [37]. Among these neurochemical changes, palatable food intake was reported to activate the mesolimbic reward system [38]. Experimental approaches demonstrate an increased mesolimbic-dopaminergic response to the exposure to sweet food in adult rats, which is measured by a higher dopamine release in the nucleus accumbens and medial prefrontal cortex, in comparison to control rats [39].

Other experimental models have also found that the consumption of sugar stimulates accumbens dopamine releasing. For instance, the intermittent access to sugar can promote sugar bingeing while dopamine is released repeatedly in the nucleus accumbens in response to the taste of sweet food [40]. It is curious to emphasize that dopamine increase can be early detected during the first hour of exposure to the sweet food [12, 41]. Moreover, another investigation found that when rats binged on sugar, the amount of dopamine released is proportional to the sucrose concentration, but not to the volume of sucrose consumption [42].

Taking into account the ability of sweet food and addictive drugs to increase accumbal dopamine release, a recent series of studies have proposed that endogenous hormones could mediate the effect of drugs of abuse on stimulating the mesolimbic dopamine system. It is well established that peripheral ghrelin controls food intake, energy balance, and appetite [43–45]. More recently, this hormone was described to activate mesolimbic dopamine system and to have a role in food-seeking behavior and in the accumbal DA release in response to alcohol [46–49]. It is known that appetite scoring (for instance, using a VAS scale) and peripheral ghrelin correlate positively. Interestingly, while this correlation disappears completely after chocolate eating, it is reversed after chocolate smelling [50]. This suggests that the different sensory proprieties of chocolate (smell, taste, texture) may diversely impact the subjective hunger sensation and ghrelin levels, which can probably lead to a peculiar mesolimbic DA response.

The excessive sugar intake modifies the binding to the dopamine receptor in some brain structures. Rats with a pattern of bingeing on palatable food exhibit an increased dopamine D1 receptor binding in the accumbens core and shell, in contrast to a decreased dopamine D2 receptor binding in the dorsal striatum [11]. Similarly, in a clinical study, obese individuals demonstrated reductions in striatal dopamine D2 receptors like those observed in drug-addicted subjects [51].

Others have demonstrated similarities in brain dopamine between obese and drug-dependent individuals, showing that both groups have lower dopamine receptor levels. Besides, this investigation also described an inverse correlation between dopamine D2 receptor density and BMI, which can suggest decreased sensitivity in the reward system in obese subjects [52]. Some years later, Stoeckel and colleagues [34] found that obese women have greater activation in a number of cerebral regions, such as the amygdala, nucleus accumbens, ventral striatum, and medial prefrontal cortex, in response to high-calorie foods. Based on that, the authors proposed that the mesolimbic system in obese women may be hyperresponsive to food rewards.

According to recent data, not only high-sugar but also high-fat food intake can alter the binding to dopamine in the brain. For example, a chronic exposure to high-fat diets modifies dopamine D2 receptor gene expression in the striatum and dopamine D4 receptor in the ventromedial nucleus of the hypothalamus, as well as increases hydroxylase tyrosine protein expression, which is an enzyme essential to dopamine production, in the ventral tegmental area [53, 54]. In addition, the early exposure to a high-fat diet in rats alters dopamine signaling markers permanently in the nucleus accumbens [55].

The striatal opioid system also seems to regulate the hedonic properties of palatable food, as infusion of an opioid agonist into the NAc stimulates the feeding behavior in rats with ad libitum access to food [56, 57]. Conversely, opioid receptor antagonists infused into the NAc decrease consumption of preferred food without affecting intake of less palatable alternatives [58]. Because m-opioid receptor activation results in the inhibition of medium spiny neuron activity in the NAc, it has been proposed that the NAc shell exerts a tonic inhibitory influence on palatable food consumption [59]. Consistent with this view, stimulation of inhibitory GABAA or GABAB receptors [60, 61], or blockade of excitatory ionotropic glutamate receptors [62] in the NAc shell increases food consumption. Similarly, excitotoxic lesion of the NAc shell also increases food consumption and enhances sensitivity to food reward [63, 64].

Considering that addiction-like behavior is related to binge consumption of palatable food, the evidence of opiate-like withdrawal is of particular interest. According to Colantuoni and colleagues [12], withdrawal symptoms can be precipitated when a palatable food is removed, inducing alterations in the opioid system. The main neurochemical modifications in food withdrawal, like opiate withdrawal, are a decrease in extracellular dopamine in the nucleus accumbens and an increase in acetylcholine release from accumbens interneurons. Such a dopamine/acetylcholine imbalance seems to be characteristic during withdrawal of several drugs of abuse [65]. It is stated that dopamine, acetylcholine, and opioid production induced by the consumption of sugar is similar to the one induced by most narcotic substances [10], concerning withdrawal symptoms.

Withdrawal from chronic drug exposure can dampen phasic and tonic dopamine levels in the absence of drug [66–68]. However, withdrawal from psychostimulants does not necessarily lead to decreased tonic levels [69]. Returning dampened dopamine levels to baseline may be one of the driving forces in relapse to drug-taking behavior [70], since evidence indicates that animals pursue drug infusion when tonic dopamine levels in the NAcc decrease past a certain threshold [71]. In humans, a positive correlation is described between dopamine plasma levels at the beginning of the withdrawal and the presence of withdrawal symptoms [72].

Although synaptic levels of dopamine show a deficiency during drug withdrawal, dopamine release shows a profound sensitization to the well-established permissive or facilitatory function of serotonin on dopamine release in the nucleus accumbens [73, 74]. Another series of studies suggest that the reward induced by ethanol, for instance, may be independent of the activation of dopamine D2 receptors and may be mediated through serotonin 5-HT1B and 5-HT2 receptors [75, 76]. In recently abstinent alcohol-dependent patients, there is a significant reduction in the availability of brainstem serotonin transporters, which is significantly correlated with lifetime alcohol consumption and with ratings of depression and anxiety during withdrawal [77].

Also in ethanol dependence, withdrawal influences gene expression in the prefrontal cortex, reflecting that this is a site of damage and neurodegeneration in addicts [78]. Besides, (gamma) γ-aminobutyric acid (GABA), the major inhibitory neurotransmitter in the CNS, and its receptor are downregulated during chronic use, and after abstention, this downregulation of the GABA receptor contributes to many of the symptoms of the withdrawal [79, 80]. A number of investigators have provided evidence for associations between elevations in (alpha) α4 subunit–containing GABAA receptors and alterations in activation of this receptor in the cerebral cortex and the hippocampus, further implicating α4-containing GABAA receptors as having a role in increased anxiety and seizure risk during withdrawal [81, 82].

Prolonged intoxication also inhibits activity in the glutamate neurotransmitter system, the major excitatory neurotransmitter in the CNS, by acting on the ion-gated N-methyl-D-aspartate (NMDA) glutamate receptor [79]. Abstinence reverses the inhibition of the NMDA receptor, producing many of the signs and symptoms of withdrawal.

During withdrawal, noradrenaline (NA) release is enhanced in animals [83] and in humans, [84] which correlates with the peripheral symptoms of hypertension tachycardia and general autonomic reactivity. Increases in the CSF correlations of NA and its major CNS metabolite (MHPG), in the central noradrenaline (NA) metabolism [84–87], as well as raised peripheral levels of NA and elevated excretion of MHPG (3-methoxy-4-hydroxyphenylglycol) have been reported [88]. The mechanism underlying this increased activity of NA neuronal systems is suggested to be alterations in α-2-adrenoceptors, which would lead to increased firing and predispose to paroxysmal episodes of neuronal discharge. Several groups have found reduced postsynaptic alpha-2-adrenoceptor function in alcoholics in acute withdrawal [89–92] and early abstinence [93].

Finally, withdrawal induces prolonged increases in glucocorticoid concentrations in specific regions of the rodent brain, while plasma concentrations may increase or remain unchanged [79, 80]. Extracellular CRF levels in the CeA, as measured by microdialysis, are substantially elevated during acute ethanol and cocaine withdrawal [94, 95] as well as during antagonist-precipitated withdrawal from chronic cannabinoid treatment [96], being implicated in the aversive and anxiogenic-like behavioral manifestations of alcohol [97] and opiate withdrawal [98]. Type II glucocorticoid receptor availability in prefrontal cortex is decreased after withdrawal, and nuclear localization of glucocorticoid receptors is increased, suggesting enhanced glucocorticoid type II receptor activation [99].

Evidence of Chocolate Withdrawal

Artificial chemical rewarding stimuli and natural rewarding activities act at the same pathways. Regarding feeding, a decrease in blood glucose level, induced by food deprivation, causes hunger and motivates the animal to seek food, since in this state eating strongly stimulates the reward system and becomes pleasurable. Then, in homeostatic conditions, when the animal has consumed enough food, satiety mechanisms suppress the reward system connected with feeding. However, these well-controlled mechanisms do not seem to be as efficient when considering highly palatable foods.

For instance, addiction-like behaviors, which are represented by signs of withdrawal, are naturally caused by the response to palatable food deprivation. Avena et al. [100] have developed a rat model of binge eating on sugar and sugar/fat solutions, showing that rats provided with 2 h of access to a sugar/fat solution will consume the majority of their daily energy intake during this window, despite having continuous access to normal laboratory chow. Moreover, sugar-bingeing rats consume more sugar solution over time (evidence of tolerance), demonstrate increased sugar consumption after a period of abstinence (deprivation effect), and exhibit signs of opiate-like withdrawal. Sugar-bingeing rats become cross-sensitized to both amphetamines – which has no effect on naive rats – and cocaine [100].

The same research group has described that spontaneous withdrawal signs, like anxiety and teeth chattering, are induced by palatable food deprivation or injection of naloxone in sugar-dependent rats. Accumbens microdialysis revealed that naloxone decreases extracellular dopamine (DA) and causes a dose-dependent increase in acetylcholine (ACh) in these animals [12]. Others described signs of aggressive behavior [101] and anxiety-like behavior [102] during periods of sugar withdrawal. This finding is in accordance with a recent experimental study [103] that has shown spontaneous increase in the frequency of headshakes induced by chocolate withdrawal. Another study shows that chocolate-olfactory priming, ineffective in inducing conditioned place preference in the control group, reactivated the extinguished place preference in the group previously conditioned to chocolate consumption [104].

In humans, an association was described between the behavior of self-medicating dysphoric mood with sweets or chocolate and hysteroid dysphoric traits, assessed by a questionnaire investigating the ingestion of sweets or chocolate in response to anxiety, depression, tension/irritability, or anger. Besides, self-medicators included a significantly higher percentage of people reporting to abuse chocolate than controls [105]. Self-defined chocolate addicts find chocolate irresistible and eat it in excess, report positive mood and pleasure during chocolate consumption, but may also report negative mood in response to chocolate cues [27, 106]. An interesting study shows that in response to chocolate images, high cravers report more pleasure and arousal but less control, having an inhibited cardiac defense but potentiated startle reflex, as compared to low cravers [107]. This suggests that under conditions of deprivation, cues that may be initially associated with reward can lead to unpleasant affect and a reversal of the pattern of reflex modulation [108].

Recent research shows that suppressing food-related thoughts can cause a subsequent increase in consumption. In agreement with this, a study performed in humans shows that participants reporting frequent use of food-thought suppression report greater chocolate cravings [109].

Developmental Origins of Chocolate Preference and Withdrawal Responses

Early environment influences the chocolate preference, metabolic response to the chronic exposure, and abdominal fat deposition in adult female rats exposed chronically to chocolate. Rats handled as neonates (brief daily separation from the mothers for 10 min/day in the first ten days of life) prefer to eat more, but are less vulnerable to the metabolic disturbances related to the chronic exposure to chocolate [110]. In addition, non-handled rats demonstrate increased frequency of headshaking after 24 h of chocolate withdrawal as well as a decreased Na^+,K^+-ATPase activity in both amygdala and hippocampus in comparison to handled ones [103]. The activity of cerebral Na^+,K^+-ATPase is reported to be modulated by drugs of abuse, such as morphine [111], and it has been proposed that the impairment of Na^+,K^+-ATPase function is related to opioid tolerance and dependence [112]. Considering the evidence that palatable food interacts with cerebral opioids and dopamine, producing effects similar to those of drugs of abuse [15], it seems that neonatal handling is associated with a decrease in the response to chocolate withdrawal in adulthood, suggesting that early life events may modulate the individual vulnerability to withdrawal after the discontinuation of a preferred food.

Other recent evidence has reported a genetic association between sweet preference and paternal alcohol dependence. It was described that individuals with a familial history of alcohol addiction are three times more probable to develop sweet preference or sugar addiction than individuals with no paternal history of alcohol dependence, independently to their alcoholism status [113]. The genetic predisposition to sugar addiction was also detected in animal studies. Rat offsprings from mothers fed with junk food (including sweet snacks) during gestation and lactation period exhibit a tendency to develop appetite alterations (e.g., sugar addiction). It is interesting to emphasize that these rat mothers displayed bingeing and overeating episodes during the period of exposure to the palatable food [114].

Finally, some beneficial effects of the early exposure to chocolate have been described. For instance, mothers who reported daily consumption of chocolate during pregnancy rated more positively the temperament of their infants at 6 months of age. Maternal prenatal stress predicted more negatively tuned ratings of the infant temperament, particularly among those mothers who reported never/seldom chocolate consumption during gestation. However, this effect was not observed among the mothers reporting weekly or daily chocolate consumption during pregnancy [115], suggesting that maternal well-being during pregnancy somehow interacts with chocolate consumption and affects the maternal perception of the children's temperament at 6 months of age.

Summary

In summary, this chapter demonstrates that excessive palatable food consumption induces behavioral and neuronal modifications, which can also be seen in drug abuse and withdrawal, indicative of pathological conditions such as addiction. Because chocolate is a highly palatable food, it is possible that this food also induces the same effects, although direct evidence investigating the induction of withdrawal by discontinuing the chocolate use is scarce. Moreover, it appears that bouts of immoderate palatable food intake are necessary for both behavioral and neurochemical alterations to be established and for bingeing to develop as well. Although the mechanisms involved in such effects remain unclear, certain investigations that have demonstrated some of the neurotransmitters, its receptors, and brain regions involved in binge-like behavior have proposed dopamine and opioid peptides as the most probable candidates. Furthermore, genetic aspects contribute notably to binge risk, in particular to sugar addiction, in a way that bingeing can induce a state that promotes perpetuation of the behavior once initiated.

Better understanding of the mechanisms implied in behavioral and neuronal outcomes consistent with either palatable food or drug addiction, as well as the development of new approaches driving to translational investigations in the future, can help to prevent and to find adequate treatment of eating disorders, including binge-type consumption of palatable food.

References

1. Corsica JA, Pelchat ML. Food addiction: true or false? Curr Opin Gastroenterol. 2010;26:165–9.
2. Ifland JR, Preuss HG, Marcus MT, et al. Refined food addiction: a classic substance use disorder. Med Hypotheses. 2009;72:518–26.
3. Flegal KM. Epidemiologic aspects of overweight and obesity in the United States. Physiol Behav. 2005;86: 599–602.
4. Wang Y, Beydoun MA. The obesity epidemic in the United States–gender, age, socioeconomic, racial/ethnic, and geographic characteristics: a systematic review and meta-regression analysis. Epidemiol Rev. 2007;29:6–28.
5. Weingarten HP, Elston D. The phenomenology of food cravings. Appetite. 1990;15:231–46.
6. Hill AJ, Heaton-Brown L. The experience of food craving: a prospective investigation in healthy women. J Psychosom Res. 1994;38:801–14.
7. Macht M, Dettmer D. Everyday mood and emotions after eating a chocolate bar or an apple. Appetite. 2006; 46:332–6.
8. Koob GF. Dynamics of neuronal circuits in addiction: reward, antireward, and emotional memory. Pharmacopsychiatry. 2009;42:S32–41.
9. Roberts AJ, Koob GF. The neurobiology of addiction: an overview. Alcohol Health Res World. 1997;21:101–6.
10. Avena NM, Rada P, Hoebel BG. Evidence for sugar addiction: behavioral and neurochemical effects of intermittent, excessive sugar intake. Neurosci Biobehav Rev. 2008;32:20–39.
11. Colantuoni C, Schwenker J, McCarthy J, et al. Excessive sugar intake alters binding to dopamine and mu-opioid receptors in the brain. Neuroreport. 2001;12:3549–52.
12. Colantuoni C, Rada P, McCarthy J, et al. Evidence that intermittent, excessive sugar intake causes endogenous opioid dependence. Obes Res. 2002;10:478–88.
13. Avena NM, Long KA, Hoebel BG. Sugar-dependent rats show enhanced responding for sugar after abstinence: evidence of a sugar deprivation effect. Physiol Behav. 2005;84:359–62.
14. Avena NM, Carrillo CA, Needham L, et al. Sugar-dependent rats show enhanced intake of unsweetened ethanol. Alcohol. 2004;34:203–9.
15. Avena NM, Bocarsly ME, Rada P, et al. After daily bingeing on a sucrose solution, food deprivation induces anxiety and accumbens dopamine/acetylcholine imbalance. Physiol Behav. 2008;94:309–15.
16. Liu Y, von Deneen KM, Kobeissy FH, et al. Food addiction and obesity: evidence from bench to bedside. J Psychoactive Drugs. 2010;42:133–45.
17. Haddock CK, Dill PL. The effects of food on mood and behavior: implications for the addictions model of obesity and eating disorders. In: Poston WSC, Haddock CK, editors. Food as a drug. New York: Haworth Press, Inc.; 2000.
18. Rozin P, Levine E, Stoess C. Chocolate craving and liking. Appetite. 1991;17:199–212.
19. Weingarten HP, Elston DA. Survey of cravings in a student sample. Appetite. 1991;17:167–75.
20. Nasser JA, Bradley LE, Leitzsch JB, et al. Psychoactive effects of tasting chocolate and desire for more chocolate. Physiol Behav. 2011;104:117–21.
21. Hill AJ, Weaver CFL. Food craving, dietary restraint and mood. Appetite. 1991;17:187–97.
22. Wurtman RJ, Wurtman JJ. Do carbohydrates affect food intake via neurotransmitter activity? Appetite. 1988;11:S42–7.
23. Macdiarmid JI, Hetherington MM. Mood modulation by food: an exploration of affect and cravings in 'chocolate addicts'. Br J Clin Psychol. 1995;34:129–38.
24. Cartwright F, Stritzke WG, Durkin K, et al. Chocolate craving among children: implications for disordered eating patterns. Appetite. 2007;48:87–95.
25. Parker G, Crawford J. Chocolate craving when depressed: a personality marker. Br J Psychiatry. 2007;191: 351–2.
26. Painter JE, Wansink B, Hieggelke JB. How visibility and convenience influence candy consumption. Appetite. 2002;38:237–8.
27. Rogers PJ, Smit HJ. Food craving and food "addiction": a critical review of the evidence from a biopsychosocial perspective. Pharmacol Biochem Behav. 2000;66:3–14.
28. Tuomisto T, Hetherington MM, Morris MF, et al. Psychological and physiological characteristics of sweet food "addiction.". Int J Eat Disord. 1999;25:169–75.
29. Fletcher BC, Pine KJ, Woodbridge Z, et al. How visual images of chocolate affect the craving and guilt of female dieters. Appetite. 2007;48:211–7.
30. Tomelleri MS, Grunewald KK. Menstrual cycle and food cravings in young college women. J Am Diet Assoc. 1987;87:311–5.
31. Zellner DA, Garriga-Trillo A, Centeno S, et al. Chocolate craving and the menstrual cycle. Appetite. 2004; 42:119–21.

32. Osman JL, Sobal J. Chocolate cravings in American and Spanish individuals: biological and cultural influences. Appetite. 2006;47:290–301.
33. Small DM, Jones-Gotman M, Dagher A. Feeding-induced dopamine release in dorsal striatum correlates with meal pleasantness ratings in healthy human volunteers. Neuroimage. 2003;19:1709–15.
34. Stoeckel LE, Weller RE, Cook 3rd EW, et al. Widespread reward-system activation in obese women in response to pictures of high-calorie foods. Neuroimage. 2008;41:636–47.
35. Pelchat ML. Food addiction in humans. J Nutr. 2009;139:620–2.
36. Corsica JA, Spring BJ. Carbohydrate craving: a double-blind, placebo-controlled test of the self-medication hypothesis. Eat Behav. 2008;9:447–54.
37. Levine AS, Kotz CM, Gosnell BA. Sugars and fats: the neurobiology of preference. J Nutr. 2003;133:831S–4.
38. Erlanson-Albertsson C. How palatable food disrupts appetite regulation. Basic Clin Pharmacol Toxicol. 2005; 97:61–73.
39. Gambarana C, Masi F, Leggio B, et al. Acquisition of a palatable-food-sustained appetitive behavior in satiated rats is dependent on the dopaminergic response to this food in limbic areas. Neuroscience. 2003;121:179–87.
40. Avena NM, Rada P, Moise N, et al. Sucrose sham feeding on a binge schedule releases accumbens dopamine repeatedly and eliminates the acetylcholine satiety response. Neuroscience. 2006;139:813–20.
41. Rada P, Avena NM, Hoebel BG. Daily bingeing on sugar repeatedly releases dopamine in the accumbens shell. Neuroscience. 2005;134:737–44.
42. Hajnal A, Smith GP, Norgren R. Oral sucrose stimulation increases accumbens dopamine in the rat. Am J Physiol Regul Integr Comp Physiol. 2004;286:31–7.
43. Cummings DE, Purnell JQ, Frayo RS, et al. A preprandial rise in plasma ghrelin levels suggests a role in meal initiation in humans. Diabetes. 2001;50:1714–9.
44. Tschöp M, Smiley DL, Heiman ML. Ghrelin induces adiposity in rodents. Nature. 2000;407:908–13.
45. Wren AM, Seal LJ, Cohen MA, et al. Ghrelin enhances appetite and increases food intake in humans. J Clin Endocrinol Metab. 2001;86:5992.
46. Abizaid A, Liu ZW, Andrews ZB, et al. Ghrelin modulates the activity and synaptic input organization of midbrain dopamine neurons while promoting appetite. J Clin Invest. 2006;116:3229–39.
47. Egecioglu E, Jerlhag E, Salomé N, et al. Ghrelin increases intake of rewarding food in rodents. Addict Biol. 2010;15:304–11.
48. Jerlhag E. Systemic administration of ghrelin induces conditioned place preference and stimulates accumbal dopamine. Addict Biol. 2008;13:358–63.
49. Jerlhag E, Landgren S, Egecioglu E, et al. The alcohol-induced locomotor stimulation and accumbal dopamine release is suppressed in ghrelin knockout mice. Alcohol. 2011;45:341–7.
50. Massolt ET, van Haard PM, Rehfeld JF, et al. Appetite suppression through smelling of dark chocolate correlates with changes in ghrelin in young women. Regul Pept. 2010;161:81–6.
51. Wang GJ, Volkow ND, Thanos PK, et al. Similarity between obesity and drug addiction as assessed by neurofunctional imaging: a concept review. J Addict Dis. 2004;23:39–53.
52. Wang GJ, Volkow ND, Logan J, et al. Brain dopamine and obesity. Lancet. 2001;357:354–7.
53. Huang X-F, Yu Y, Zavitsanou K, et al. Differential expression of dopamine D2 and D4 receptor and tyrosine hydroxylase mRNA in mice prone, or resistant, to chronic high-fat diet-induced obesity. Mol Brain Res. 2005;135:150–61.
54. Li Y, South T, Han M, et al. High-fat diet decreases tyrosine hydroxylase mRNA expression irrespective of obesity susceptibility in mice. Brain Res. 2009;1268:181–9.
55. Teegarden SL, Scott AN, Bale TL. Early life exposure to a high fat diet promotes long-term changes in dietary preferences and central reward signaling. Neuroscience. 2009;162:924–32.
56. Peciña S, Berridge KC. Hedonic hot spot in nucleus accumbens shell: where do mu-opioids cause increased hedonic impact of sweetness? J Neurosci. 2005;25:11777–86.
57. Zhang M, Gosnell BA, Kelley AE. Intake of high-fat food is selectively enhanced by mu opioid receptor stimulation within the nucleus accumbens. J Pharmacol Exp Ther. 1998;285:908–14.
58. Kelley AE, Bless EP, Swanson CJ. Investigation of the effects of opiate antagonists infused into the nucleus accumbens on feeding and sucrose drinking in rats. J Pharmacol Exp Ther. 1996;278:1499–507.
59. Kelley AE, Baldo BA, Pratt WE, et al. Corticostriatal-hypothalamic circuitry and food motivation: integration of energy, action and reward. Physiol Behav. 2005;86:773–95.
60. Basso AM, Kelley AE. Feeding induced by GABA(A) receptor stimulation within the nucleus accumbens shell: regional mapping and characterization of macronutrient and taste preference. Behav Neurosci. 1999; 113:324–36.
61. Stratford TR, Kelley AE. GABA in the nucleus accumbens shell participates in the central regulation of feeding behavior. J Neurosci. 1997;17:4434–40.
62. Maldonado-Irizarry CS, Swanson CJ, Kelley AE. Glutamate receptors in the nucleus accumbens shell control feeding behavior via the lateral hypothalamus. J Neurosci. 1995;15:6779–88.

63. Johnson PI, Parente MA, Stellar JR. NMDA-induced lesions of the nucleus accumbens or the ventral pallidum increase the rewarding efficacy of food to deprived rats. Brain Res. 1996;722:109–17.
64. Maldonado-Irizarry CS, Kelley AE. Excitotoxic lesions of the core and shell subregions of the nucleus accumbens differentially disrupt body weight regulation and motor activity in rat. Brain Res Bull. 1995;38:551–9.
65. Hoebel BG, Avena NM, Rada P. Accumbens dopamine-acetylcholine balance in approach and avoidance. Curr Opin Pharmacol. 2007;7:617–27.
66. Rossetti ZL, Melis F, Carboni S, et al. Marked decrease of extraneuronal dopamine after alcohol withdrawal in rats: reversal by MK-801. Eur J Pharmacol. 1991;200:371–2.
67. Parsons LH, Smith AD, Justice Jr JB. Basal extracellular dopamine is decreased in the rat nucleus accumbens during abstinence from chronic cocaine. Synapse. 1991;9:60–5.
68. Crippens D, Robinson TE. Withdrawal from morphine or amphetamine: different effects on dopamine in the ventral-medial striatum studied with microdialysis. Brain Res. 1994;650:56–62.
69. Paulson PE, Robinson TE. Regional differences in the effects of amphetamine withdrawal on dopamine dynamics in the striatum. Analysis of circadian patterns using automated on-line microdialysis. Neuropsychopharmacology. 1996;14:325–37.
70. Willuhn I, Wanat MJ, Clark JJ, et al. Dopamine signaling in the nucleus accumbens of animals self-administering drugs of abuse. Curr Top Behav Neurosci. 2010;3:29–71.
71. Wise RA, Newton P, Leeb K, et al. Fluctuations in nucleus accumbens dopamine concentration during intravenous cocaine self-administration in rats. Psychopharmacology. 1995;120:10–20.
72. Heinz A, Schmidt K, Baum SS, et al. Influence of dopaminergic transmission on severity of withdrawal syndrome in alcoholism. J Stud Alcohol. 1996;57:471–4.
73. Parsons LH, Koob GF, Weiss F. Serotonin dysfunction in the nucleus accumbens of rats during withdrawal after unlimited access to intravenous cocaine. J Pharmacol Exp Ther. 1995;274:1182.
74. Parsons LH, Justice JR. Perfusate serotonin increases extracellular dopamine in the nucleus accumbens as measured by in vivo microdialysis. Brain Res. 1993;606:195.
75. Risinger FO, Bormann NM, Oakes RA. Reduced sensitivity to ethanol reward, but not ethanol aversion, in mice lacking 5-HT1B receptors. Alcohol Clin Exp Res. 1996;20:1401–5.
76. Panocka I, Massi M. Long-lasting suppression of alcohol preference in rats following serotonin receptor blockade by ritanserin. Brain Res Bull. 1992;28:493–6.
77. Heinz A, Ragan P, Jones DW, et al. Reduced central serotonin transporters in alcoholism. Am J Psychiatry. 1998;155:1544–9.
78. Wiren KM, Hashimoto JG, Alele PE, et al. Impact of sex: determination of alcohol neuroadaptation and reinforcement. Alcohol Clin Exp Res. 2006;30:233–42.
79. Adinoff B. The alcohol withdrawal syndrome: neurobiology of treatment and toxicity. Am J Addict. 1994;3:277–88.
80. Glue P, Nutt DJ. Overexcitement and disinhibition: dynamic neurotransmitter interactions in alcohol withdrawal. Br J Psychiatry. 1990;157:491–9.
81. Cagetti E, Liang J, Spigelman I, et al. Withdrawal from chronic intermittent ethanol treatment changes subunit composition, reduces synaptic function and decreases behavioral responses to positive modulators of GABAA receptors. Mol Pharmacol. 2003;63:53–64.
82. Sundstrom-Poromaa I, Smith DH, Gong QH, et al. Hormonally regulated alpha(4)beta(2)delta GABA(A) receptors are a target for alcohol. Nat Neurosci. 2002;5:721–2.
83. Griffiths PJ, Littleton JM, Ortiz A. Changes in monoamine concentrations in mouse brain associated with ethanol dependence and withdrawal. Br J Pharmacol. 1974;50:489–98.
84. Hawley RJ, Major LF, Schulman EA, et al. Cerebrospinal fluid 3-methoxy-4-hydroxyphenylglycoland norepinephrine levels in alcohol withdrawal. Arch Gen Psychiatr. 1985;42:1056–62.
85. Ogata M, Mendelson JH, Mello NK, et al. Adrenal function and alcoholism. Psychosom Med. 1971;33:159.
86. Borg S, Czarnecka A, Kvande H, et al. Clinical conditions and concentrations of MOPEG in the cerebrospinal fluid and urine of male alcoholic patients during withdrawal. Alcohol Clin Exp Res. 1983;7:411–5.
87. Valverius P, Högström-Brandt AM, Borg S. Norepinephrine metabolite in CSF correlates with ethanol consumption and heredity in humans. Alcohol Clin Exp Res. 1993;10:499–503.
88. Potter JF, Bannan LT, Beevers DG. The effect of a non-selective lipophilic beta-blocker on the blood pressure and noradrenaline, vasopressin, cortisol and renin release during alcohol withdrawal. Clin Exp Hypertens. 1984;69:1147–60.
89. Nutt D, Glue P, Molyneux S, et al. Alpha-2-adrenoceptor function in alcohol withdrawal: a pilot study of the effects of i.v. clonidine in alcoholics and normals. Alcohol Clin Exp Res. 1988;12:14–8.
90. Balldin J, Berggren U, Engel J, et al. Alpha-2-adrenoceptor sensitivity in early alcohol withdrawal. Biol Psychiatry. 1992;31:712–9.
91. Fahlke C, Berggren U, Lundborg C, et al. Psychopathology in alcohol withdrawal: relationship to alpha-2-adrenoceptor function. Alcohol Alcohol. 1999;34(5):750–9.

92. Berggren U, Fahlke C, Norrby A, et al. Subsensitive alpha-2-adrenoceptor function in male alcohol-dependent individuals during 6 months of abstinence. Drug Alcohol Depend. 2000;57:255–60.
93. Glue P, Sellman JD, Nicholls MG, et al. Studies of alpha-2-adrenoceptor function in abstinent alcoholics. Br J Addict. 1989;84:97–102.
94. Merlo PE, Koob GF, Vale W, et al. Release of corticotropin releasing factor (CRF) from the amygdala of ethanol-dependent rats measured with microdialysis. Alcohol Clin Exp Res. 1994;18:522.
95. Richter RM, Weiss F. In vivo CRF release in rat amygdala is increased during cocaine withdrawal in self-administering rats. Synapse. 1999;32:254.
96. de Fonseca FR, Carrera M, Navarro M, et al. Activation of corticotropin-releasing factor in the limbic system during cannabinoid withdrawal. Science. 1997;276:2050.
97. Rassnick S, Heinrichs SC, Britton KT, et al. Microinjection of a corticotropin-releasing factor antagonist into the central nucleus of the amygdala reverses anxiogenic-like effects of ethanol withdrawal. Brain Res. 1993;05:25.
98. Heinrichs SC, Menzaghi F, Schulteis G, et al. Suppression of corticotropin-releasing factor in the amygdala attenuates aversive consequences of morphine withdrawal. Behav Pharmacol. 1995;6:74.
99. Little HJ, Croft AP, O'Callaghan MJ, et al. Selective increases in regional brain glucocorticoid: a novel effect of chronic alcohol. Neuroscience. 2008;156:1017–27.
100. Avena NM, Rada P, Hoebel BG. Sugar and fat bingeing have notable differences in addictive-like behavior. J Nutr. 2009;139:623–8.
101. Galic MA, Persinger MA. Voluminous sucrose consumption in female rats: increased "nippiness" during periods of sucrose removal and possible oestrus periodicity. Psychol Rep. 2002;90:58–60.
102. Cottone P, Sabino V, Steardo L, et al. Consummatory, anxiety-related and metabolic adaptations in female rats with alternating access to preferred food. Psychoneuroendocrinology. 2009;34:38–49.
103. da Silva Benetti CS, Silveira PP, Matté C, et al. Effects of a chronic exposure to a highly palatable diet and its withdrawal, in adulthood, on cerebral Na+, K+-ATPase and plasma S100B in neonatally handled rats. Int J Dev Neurosci. 2010;28:153–9.
104. La Mela I, Latagliata EC, Patrono E, et al. Olfactory priming reinstates extinguished chocolate-induced conditioned place preference. Appetite. 2010;54:237–40.
105. Schuman M, Gitlin MJ, Fairbanks L. Sweets, chocolate, and atypical depressive traits. J Nerv Ment Dis. 1987; 175:491–5.
106. Hetherington MM. Food cravings and addiction. Surrey: Leatherhead Publishing; 2001.
107. Rodríguez S, Fernández MC, Cepeda-Benito A, et al. Subjective and physiological reactivity to chocolate images in high and low chocolate cravers. Biol Psychol. 2005;70:9–18.
108. Drobes DJ, Miller EJ, Hillman CH, et al. Food deprivation and emotional reactions to food cues: implications for eating disorders. Biol Psychol. 2001;57:153–77.
109. Erskine JA, Georgiou GJ. Effects of thought suppression on eating behaviour in restrained and non-restrained eaters. Appetite. 2010;54:499–503.
110. da Silva Benetti CS, Silveira PP, Portella AK, et al. Could preference for palatable foods in neonatally handled rats alter metabolic patterns in adult life? Pediatr Res. 2007;62:405–11.
111. Wu ZQ, Chen J, Chi ZQ, Liu JG. Involvement of dopamine system in regulation of Na+, K+-ATPase in the striatum upon activation of opioid receptors by morphine. Mol Pharmacol. 2007;71:519–30.
112. Taylor DA, Fleming WW. Unifying perspectives of the mechanisms underlying the development of tolerance and physical dependence to opioids. J Pharmacol Exp Ther. 2001;297:11–8.
113. Fortuna JL. Sweet preference, sugar addiction and the familial history of alcohol dependence: shared neural pathways and genes. J Psychoactive Drugs. 2010;42:147–51.
114. Bayol SA, Farrington SJ, Stickland NC. A maternal 'junk food' diet in pregnancy and lactation promotes an exacerbated taste for 'junk food' and a greater propensity for obesity in rat offspring. Br J Nutr. 2007;98: 843–51.
115. Räikkönen K, Pesonen AK, Järvenpää AL, et al. Sweet babies: chocolate consumption during pregnancy and infant temperament at six months. Early Hum Dev. 2004;76:139–45.

Chapter 35
Cues Linked to the Consumption of Chocolate: Orexin and Reward-Based Feeding Behaviors

Derrick L. Choi, Jon F. Davis, and Stephen C. Benoit

Key Points

- Reward-based feeding behaviors are an important component of the overall consideration of obesity and metabolic disorders.
- Orexin is an important neurobiological regulator of brain reward circuits involved in the control of reward-based feeding behaviors.
- The paraventricular thalamic nucleus may be a critical site of integration for orexinergic signaling in reward-based feeding.
- Cues associated with the consumption of chocolate activate the orexin system, suggesting that orexin plays a role in the regulation of conditioned reward-based feeding.

Keywords Orexin • Reward • Conditioned feeding • Progressive ratio

Neurobiological regulation of energy homeostasis, the internal maintenance of a physiological equilibrium, is in constant flux and involves complex interactions among many effector molecules [1, 2]. The central nervous system is responsible for sensing nutritional status and engaging mechanisms that ultimately increase or decrease food intake to maintain energy balance. However, in certain circumstances, this central homeostatic system can be overridden even in the face of contrary signals of

D.L. Choi, Ph.D. (✉)
Department of Psychiatry and Behavioral Neuroscience, University of Cincinnati, College of Medicine, 2180 E. Galbraith Rd, Cincinnati, OH 45237-506, USA

Department of Psychiatry and Behavioral Neuroscience, University of Cincinnati, College of Medicine, UC Reading Campus, Bldg. C, Room 216-A, Cincinnati, OH 45237-506, USA
e-mail: choidl@mail.uc.edu

J.F. Davis, Ph.D.
Department of Psychiatry and Behavioral Neuroscience, University of Cincinnati, College of Medicine, 2170 Galbraith Road, Cincinnati, OH 45237, USA
e-mail: jon.davis@uc.edu

S.C. Benoit, Ph.D.
Department of Psychiatry and Behavioral Neuroscience, University of Cincinnati, 2170 E Galbraith Road, Cincinnati, OH 45237, USA
e-mail: stephen.benoit@uc.edu

R.R. Watson et al. (eds.), *Chocolate in Health and Nutrition*, Nutrition and Health 7,
DOI 10.1007/978-1-61779-803-0_35, © Springer Science+Business Media, LLC 2013

metabolic need. Non-homeostatic feeding describes food intake that is driven by the taste, texture, or palatability of the food source [3, 4]. In simple terms, eating can be driven by the rewarding properties of the food. In this chapter, the term "reward-based feeding" will refer generally to behaviors involving these mechanisms in the control of eating for pleasure or hedonic value [5]. In this chapter, we will outline the neurobiological mechanisms behind reward-based feeding and detail how chocolate and drugs of abuse share common neural pathways.

Importantly, animals and humans have evolved the ability to modulate food intake by engaging in appropriate behaviors leading to positive consequences (and suppression of behaviors associated with potentially negative outcomes). The positive attributes of a food or feeding experience, such as the taste, palatability, or simply the feeling of relieving hunger, are variables that are learned about and utilized to initiate the onset of meals based on these predicted rewarding outcomes of eating. Learned positive associations between food and environmental cues are potent inducers of feeding, even when energy demands are satisfied [6–10]. The anticipation of food rewards represents a critical point in the regulation of reward-based feeding behaviors based on environmental cues. The mesolimbic dopamine reward pathway is most commonly studied for its role in mediating the reinforcement of reward-related behaviors [11] and will be a neurobiological area of focus for this discussion.

At the center of the mesolimbic reward circuit are dopaminergic neurons of the ventral tegmental area (VTA) of the midbrain that project to limbic forebrain areas, namely, the nucleus accumbens in the ventral striatum. These structures are extensively implicated in the regulation of behaviors driven by a variety of rewarding stimuli, including palatable foods, sexual behaviors, and drugs of abuse. The lateral hypothalamus (LH) serves as a locus for receiving metabolic signals such as leptin, insulin, ghrelin, and glucose to act in the brain and can then affect reward-based components of feeding [12–16].

One feature of chocolate is its ability to activate and engage mesolimbic circuitry [5]. These observations led to the hypothesis that, similar to drugs of abuse, food sources may also take on "addictive" properties. When animals are conditioned to expect chocolate or other rewards (e.g., drugs of abuse), cues that predict the delivery of either substance lead to increased neuronal activation within the mesolimbic system [17].

Like expectation of drugs of abuse, expectation of chocolate is capable of activating brain reward regions associated with the consumption of the chocolate reward. When animals are initially exposed to chocolate, they are "neophobic" and typically sample the chocolate in small amounts. However, following successive exposures, animals increase their intake of chocolate, presumably because they develop an expectation. In fact, many antiobesity drugs have been tested for their ability to reduce chocolate intake. For example, the endocannabinoid inhibitor rimonabant dose dependently decreases chocolate consumption in rodents [18]. Rimonabant is also capable of reducing the intake of abused drugs, an effect hypothesized to occur through the reduction of pleasure associated with drug consumption. The ability of rimonabant to reduce chocolate consumption was important because it suggested that blockade of the endocannabinoid system, a system that acts partly on mesolimbic structures, may be effective at reducing the intake of palatable foods and also diet-induced obesity.

The endogenous stress system is also a potent inducer of palatable food consumption, and it is often used to study overfeeding in the context of obesity. Stress-induced feeding and the consumption of "comfort foods" are popular concepts in the lay and scientific community and have been touted as reasons for the rapid increases in obesity in adults and children [19]. There are many systems and peptides that modulate the effects of stress, but only a few that mediate the effects of stress on food reward. The gut hormone ghrelin is a potent orexigenic agent that stimulates feeding behavior in rodents [20]. Apart from mediating normal feeding behavior, ghrelin also mediates food reward [16], food anticipatory activity [21], and stress-induced feeding [22]. For example, subordinate mice that are chronically defeated by their dominant counterparts typically increase their consumption of palatable foods. In a recent study, Chuang and colleagues reported that disruption of ghrelin signaling blocked the ability of social defeat stress to induce chocolate consumption [22]. Importantly, the ability of ghrelin to mediate stress-induced consumption of chocolate was associated with increased activation of the mesolimbic dopamine system. We interpret these findings to suggest that the rewarding properties of chocolate are regulated by the gut hormone ghrelin through modification of mesolimbic dopamine signaling.

In addition to activating mesolimbic structures, chocolate is capable of entraining biological rhythms in rodents. In rodents, sleep-wake cycles are controlled by the suprachiasmatic nucleus (SCN) of the hypothalamus. Chemical lesions that destroy neurons in this nucleus disrupt the sleep-wake cycles, providing evidence that the SCN is a critical modulator of biological rhythms. The expression of the gene product period 1 (per1) in the SCN oscillates with neuronal activity in this region and has thus been referred to as a "clock gene." In this way, the expression of per1 can be used to determine which phase of the sleep-wake cycle an animal is experiencing. When rats are conditioned to expect chocolate, the expression of per1 is induced in several brain regions implicated in the control of biological rhythms and reward such as the hypothalamus, nucleus accumbens, frontal cortex, and amygdala [23]. This finding suggests that chocolate has the ability to entrain biological rhythms in rodents. One implication of this finding is that metabolic stimuli possess the ability to alter sleep-wake function. Importantly, sleep-wake function in mammals is controlled in part by the lateral hypothalamic orexin system [24].

Orexins (also called hypocretins) are neurotransmitters produced in small neuronal populations of the lateral (LH) and perifornical (PFA) areas of the hypothalamus. Orexins are known to regulate arousal, wakefulness, food intake, and reward-related behaviors [25–27]. There are two different orexin peptides, both cleaved from prepro-orexin: orexin-A (33 amino acids) and orexin-B (28 amino acids). Orexin-A can bind to both orexin-1 receptor (OX1R) and, with lower affinity, orexin-2 receptor (OX2R), while orexin-B has preferential binding affinity for OX2R. Both orexin receptors are G protein-coupled receptor subtypes [28] and are widely distributed throughout the central nervous system [29, 30]. Due to the lack of an effective and commercially available OX2R antagonist, orexin-A signaling on OX1R is more widely studied and is much better characterized. As such, several studies have made use of a common selective OX1R antagonist, SB-334867. Most studies have focused on the role for orexin-A and OX1R, as most evidence suggests that the predominant role of orexin-B is in sleep-wake regulation. Nevertheless, orexins have diverse roles in the regulation of multiple behavioral systems.

Orexin was first named and described for its ability to induce feeding, for which it was named after the Greek root word for appetite, *orexis* [28]. Numerous groups have since demonstrated the orexigenic properties of orexin-A in which intracerebroventricular (ICV) administration of orexin-A increases food intake [28, 31]. Interestingly, when given a choice, rats will selectively increase intake of a preferred high-fat diet [32]. OX1R antagonism is effective in blocking central orexin-A-induced hyperphagia and behavioral satiety [33]. Orexin neurons receive input from hypothalamic neuropeptide-Y/agouti-related peptide-expressing neurons of the arcuate nucleus and are thought to be "second-order" neurons of a hierarchy in the integration processes involved in promoting food intake [34, 35]. However, other evidence suggests that orexin neurons may also act as "first-order" neurons, sensors of metabolic status, that are directly modulated by circulating factors such as leptin, glucose, and ghrelin [15, 36–38]. Orexin neurons most likely take on both first- and second-order roles in integrating intra- and extra-hypothalamic circuits to affect consummatory behaviors.

Recent studies suggest that orexin plays a critical role in promoting reward-related behaviors [25]. The VTA is a primary component of the mesolimbic dopamine pathway and receives input from orexin neurons [39, 40]. Additionally, VTA neuron populations express both orexin receptor subtypes [29]. It is proposed that plasticity in the VTA could be regulated and enhanced by secondary input from the orexin system [41, 42]. Indeed, orexinergic projections to the VTA signal, specifically on a majority of dopamine neurons, increase dopaminergic neuron firing rates and activate the mesolimbic pathway [43, 44]. Drug-associated cues activate orexin neurons, and central application of orexin-A reinstates drug-seeking behavior [45]. Moreover, orexin is critical for learning reward-stimulus associations as VTA

OX1R antagonism attenuates acquisition of morphine-conditioned place preference [44]. These associations are likely mediated by long-term plastic changes occurring within VTA neurons. When OX1R antagonist is administered into the VTA, behavioral sensitization and neurophysiological changes that are typically caused by chronic cocaine use are blocked [46]. Similar results have been demonstrated with other substances of abuse using OX1R blockade to attenuate operant responding for alcohol and the reinstatement of alcohol and cocaine seeking [47, 48]. Furthermore, studies in male sexual behavior also suggest a role for VTA orexin in natural reward processing [49]. Other reports have also evaluated orexin and natural reward function in food reward-related behaviors. Orexin-A is implicated in operant responding for palatable food rewards [50–52] and, within the mesolimbic system, is necessary for opioid-induced high-fat intake [53]. Together, these findings suggest that orexin plays a role as a functional neuromodulator of food reward-related behaviors at the level of the VTA. Further, these studies suggest that orexin's effects on reward-related behaviors include not only those of drugs of abuse but also of natural rewards like food and sex.

In addition to the growing body of evidence supporting orexin as a modulator of reward-related behavior, substantial evidence indicates that orexin also contributes to the control of wakefulness and general arousal. Several groups have identified a neurobiological basis for narcolepsy attributed in part to genetic deficiencies in the orexin system [27, 54, 55]. Human narcoleptic patients exhibit a substantial 85–95% reduction in orexin-A measured in cerebrospinal fluid [56] and in the number of orexin-expressing neurons [57, 58]. Reported difficulties maintaining attention in narcoleptic patients further implicate the orexin system as a critical player for the regulation of not only wakefulness but also attentional arousal [59]. Furthermore, prominent and extensive orexinergic innervations of midline thalamic nuclei and the medial prefrontal cortex (mPFC) suggest a role in attentional processes and higher executive function [60]. Further functional neuroanatomical evidence revealed that excitatory thalamocortical terminals synapsing in the mPFC are receptive to orexinergic excitation and that this pathway is an important mechanism for enhancing behavioral performance during situations requiring greater attention [61–63]. In support of this notion, during novel exploratory behavior, a sustained attentional and arousal state-dependent process, orexin neurons are maximally active [64] and systemic or ICV OX1R blockade disrupts attentional performance [65]. Furthermore, dominant rats, established by social hierarchy in the visible burrow system group housing model, exhibit greater risk-taking behavior but decreased acquisition of operant responding for food rewards. These behavioral characteristics are correlated with increased orexin receptor expression in the mPFC relative to subordinate and control rats [66]. These studies provide evidence that orexin can play unique roles in goal-oriented behaviors and attentional arousal, both of which are likely to be inextricably linked.

The paraventricular nucleus of the thalamus (PVT), a midline thalamic nucleus where thalamocortical projections originate, is most closely linked to attention and arousal. The PVT receives dense inputs from LH orexinergic fibers and expresses orexin receptors [40, 67]. Orexin can synapse within the midline thalamus to directly activate cortical-projecting PVT excitatory neurons as another mechanism for which orexin can regulate arousal and attentional tone [61, 68]. Recent studies demonstrating that the PVT mediates reinstatement of extinguished alcohol-seeking and conditioned place aversion during morphine withdrawal [69, 70] have established the possibility that the PVT is also involved in reward-related behaviors. In addition, excitatory PVT projections to the nucleus accumbens, a major component of the mesolimbic reward circuit, occur on dopaminergic terminals and increase dopamine efflux independent of VTA function [71–73]. Several recent studies have also implicated PVT orexin in affecting locomotor activity, anxiety, emotionality, and addiction behaviors, all of which likely require attentional and arousal inputs to be appropriately regulated [70, 74–76]. Taken together, these studies strongly support a role for orexin as a modulator of attention and arousal, to activate the necessary alertness required to promote and enhance goal-oriented, specifically reward-related, behaviors.

In recent studies from our lab, using behavioral pharmacology in rats, experiments from two approaches were utilized to assess the hypothesis that orexin promotes reward-based feeding. The results of experiments using ICV orexin-A and systemic OX1R antagonism indicated that orexin is indeed a critical promoter of conditioned responding for chocolate rewards. These results were consistent with findings from other groups using different conditions and schedules of reinforcement [50–52]. A progressive ratio (PR) schedule of operant responding, as utilized in our study, is considered a robust test of motivation to obtain rewards as the animal must work progressively harder for each subsequent food reward. Moreover, PR responding is a dopamine-dependent behavior [77]. Thus, these data suggest that orexin promotes reward-based feeding by affecting responding for chocolate at both low- and high-response requirements. In an assessment of unlearned hedonic feeding driven by mechanisms independent of homeostatic needs, others demonstrated the necessity of orexin function in the VTA by attenuating an opioid agonist–induced high-fat feeding response following intra-VTA OX1R antagonism [53]. Similarly, we assessed the effect of systemic OX1R antagonism in a model of reward-based high-fat feeding following a large chow meal. We presume that high-fat feeding is driven by hedonics or palatability, when consumed immediately following a satiating meal of lab chow. Our data also suggest that orexin is at least partly required for this effect [5]. However, the specific sites of orexin's actions on reward-based feeding in the brain remain less clear. One primary target is the VTA, due to its well-established position as a gateway for the mesolimbic circuit and dopamine outflow. Indeed, orexin fibers do signal onto VTA neurons [29, 39, 40, 43], and this pathway is implicated in orexin's role in reward-related behaviors [44, 46, 53].

In order to elucidate the potential sites of action for orexinergic signaling, neuronal activation profiles of rats conditioned to expect chocolate were assessed. In these experiments, we hypothesized that conditioned anticipation of chocolate increases activation of orexin neurons and specific orexin target regions. Using c-Fos immunolabeling techniques, our study aimed to identify areas of the brain that expressed orexin, orexin receptors, and also c-Fos protein. This study enabled the characterization of the orexin neuron subpopulations within the overall PFA-LH region and the identification of potential target regions activated by conditioned cues. Increased activation, as measured by c-Fos immunolabeling, in the PFA-LH, PVT, mPFC, and VTA of rats expecting chocolate is consistent with several similar reports [8, 17, 45, 78].

Importantly, both chocolate-conditioned and meal-fed rats displayed increased activation in orexin-A neurons in the PFA-LH, suggesting that cues associated with not only palatable food but also daily chow availability activates orexin-A-expressing cells. One possible reason for this result in the activation profile of this region may lie in the specific activation of other PFA-LH neuropeptide systems, such as melanin-concentrating hormone, also known to be involved in the regulation of feeding- and reward-related behaviors [79–81]. The increased activation of orexin-A-expressing neurons under these conditions was further characterized to be selectively significant in the PFA but not the LH region of the orexin field. This distinction is consistent with a previous report indicating that pharmacologically induced high-fat feeding specifically activates PFA orexin neurons [53]. This is in contrast to another report that drug- and food-associated cues selectively activate LH but not PFA orexin neurons [45]. Moreover, nonspecific activation of PFA orexin neurons may be indicative of heightened general arousal. A conditioned meal-feed response to limited chow access or a conditioned reward-based feeding response both likely require a learned increase in general arousal to maximize the animal's ability to benefit from such conditions. Based on these and other findings, the ability for PFA orexin to signal arousal-based activity to selectively act on reward processing may depend more on the specific sites of action (target regions) as opposed to the origin (PFA orexin activation).

In accordance with a proposed dichotomy in orexin function [82], PFA orexin activation represents a more generalized arousal response to multiple reward-predicting cues. In this way, orexin may promote reward-based feeding indirectly, via alternate extra-mesolimbic brain regions, by serving to enhance arousal, concurrently involved in supporting reward processing. In support of this notion, increased activation in the mPFC, PVT, and VTA, all target regions of orexin fibers, in chocolate-conditioned rats suggests that orexin's role in promoting food-reward anticipation involves multiple downstream targets. The mPFC is a known regulator of cue-induced anticipation of feeding, and removal of this region attenuates cue-induced elevations in feeding [8, 9]. Importantly, the PVT, an area densely innervated by orexin fibers and PVT cells expressing OX1R exhibited increased activation in rats expecting chocolate. This is consistent with the hypothesis that the PVT is a potential primary target of orexin fibers [67] and may be a unique relay center for orexin-driven reward processing. Together, these results solidify the role for orexin as an important regulator of reward-based feeding and as a component of learning mechanisms involved in the conditioned expectation of food rewards.

From this collection of data several themes emerge. It is clear that chocolate, like drugs of abuse, is a potent inducer of brain reward circuitry. Chocolate is a rewarding substance to rodents and humans and is thus readily consumed. Given the rewarding aspects of chocolate, it is often used to measure effects of social stress on reward-based feeding in rodents. The ability of stress to induce the intake of chocolate is regulated by the gut hormone ghrelin, an effect that is modulated by mesolimbic dopamine signaling. Moreover, the rewarding properties of chocolate are capable of entraining biological rhythms in rodents, an effect that coincides with increased per1 expression in brain regions that regulate metabolic need and rewards, respectively. Furthermore, chocolate activates neurons in the LH that produce orexin, a critical regulator of biological rhythms in rodents and humans. In addition to activating orexin neurons, modulation of orexin signaling regulates a rodents "willingness" or motivation to work for chocolate rewards. Overall these data highlight the powerful properties of chocolate to regulate a variety of neurobiological circuits that are thought to underlie concepts such as addiction, reward, stress, arousal, and motivation. These data also underscore the utility of chocolate as a laboratory tool to better understand the causes and consequences of consuming diets that induce obesity.

References

1. Schwartz MW, Woods SC, Porte Jr D, Seeley RJ, Baskin DG. Central nervous system control of food intake. Nature. 2000;404:661–71.
2. Shin AC, Zheng H, Berthoud HR. An expanded view of energy homeostasis: neural integration of metabolic, cognitive, and emotional drives to eat. Physiol Behav. 2009;97:572–80.
3. Berthoud HR. Neural control of appetite: cross-talk between homeostatic and non-homeostatic systems. Appetite. 2004;43:315–7.
4. Davis JF, Choi DL, Benoit SC. Insulin, leptin and reward. Trends Endocrinol Metab. 2010;21(2):68–74.
5. Choi DL, Davis JF, Fitzgerald ME, Benoit SC. The role of orexin-A in food motivation, reward-based feeding behavior and food-induced neuronal activation in rats. Neuroscience. 2010;167:11–20.
6. Cornell CE, Rodin J, Weingarten H. Stimulus-induced eating when satiated. Physiol Behav. 1989;45:695–704.
7. Rogers PJ, Hill AJ. Breakdown of dietary restraint following mere exposure to food stimuli: interrelationships between restraint, hunger, salivation, and food intake. Addict Behav. 1989;14:387–97.
8. Petrovich GD, Holland PC, Gallagher M. Amygdalar and prefrontal pathways to the lateral hypothalamus are activated by a learned cue that stimulates eating. J Neurosci. 2005;25:8295–302.
9. Petrovich GD, Ross CA, Holland PC, Gallagher M. Medial prefrontal cortex is necessary for an appetitive contextual conditioned stimulus to promote eating in sated rats. J Neurosci. 2007;27:6436–41.
10. Weingarten HP. Conditioned cues elicit feeding in sated rats: a role for learning in meal initiation. Science. 1983;220:431–3.
11. Chen BT, Hopf FW, Bonci A. Synaptic plasticity in the mesolimbic system: therapeutic implications for substance abuse. Ann N Y Acad Sci. 2010;1187:129–39.
12. Davis JF, Choi DL, Schurdak JD, Fitzgerald MF, Clegg DJ, Lipton JW, et al. Leptin regulates energy balance and motivation through action at distinct neural circuits. Biol Psychiatry. 2011;69(7):668–74.
13. Berthoud HR, Baettig K. Effects of insulin and 2-deoxy-D-glucose on plasma glucose level and lateral hypothalamic eating threshold in the rat. Physiol Behav. 1974;12:547–56.
14. Berthoud HR, Baettig K. Effects of nutritive and nonnutritive stomach loads on plasma glucose level and lateral hypothalamic eating threshold in the rat. Physiol Behav. 1974;12:1015–9.
15. Lopez M, Seoane L, Garcia MC, Lago F, Casanueva FF, Senaris R, et al. Leptin regulation of prepro-orexin and orexin receptor mRNA levels in the hypothalamus. Biochem Biophys Res Commun. 2000;269:41–5.
16. Perello M, Sakata I, Birnbaum S, Chuang JC, Osborne-Lawrence S, Rovinsky SA, et al. Ghrelin increases the rewarding value of high-fat diet in an orexin-dependent manner. Biol Psychiatry. 2010;67:880–6.

17. Schroeder BE, Binzak JM, Kelley AE. A common profile of prefrontal cortical activation following exposure to nicotine- or chocolate-associated contextual cues. Neuroscience. 2001;105:535–45.
18. Gessa GL, Orrù A, Lai P, Maccioni P, Lecca R, Lobina C, et al. Lack of tolerance to the suppressing effect of rimonabant on chocolate intake in rats. Psychopharmacology (Berl). 2006;185(2):248–54.
19. Dallman MF, Pecoraro N, Akana SF, La Fleur SE, Gomez F, Houshyar H, et al. Chronic stress and obesity: a new view of "comfort food". Proc Natl Acad Sci U S A. 2003;100(20):11696–701.
20. Tschöp M, Smiley DL, Heiman ML. Ghrelin induces adiposity in rodents. Nature. 2000;407(6806):908–13.
21. Davis JF, Choi DL, Clegg DJ, Benoit SC. Signaling through the ghrelin receptor modulates hippocampal function and meal anticipation in mice. Physiol Behav. 2011;103(1):39–43.
22. Chuang JC, Perello M, Sakata I, Osborne-Lawrence S, Savitt JM, Lutter M, et al. Ghrelin mediates stress-induced food-reward behavior in mice. J Clin Invest. 2011;121(7):2684–92.
23. Angeles-Castellanos M, Salgado-Delgado R, Rodríguez K, Buijs RM, Escobar C. Expectancy for food or expectancy for chocolate reveals timing systems for metabolism and reward. Neuroscience. 2008;155(1):297–307.
24. Willie JT, Chemelli RM, Sinton CM, Yanagisawa M. To eat or to sleep? Orexin in the regulation of feeding and wakefulness. Annu Rev Neurosci. 2001;24:429–58.
25. Aston-Jones G, Smith RJ, Moorman DE, Richardson KA. Role of lateral hypothalamic orexin neurons in reward processing and addiction. Neuropharmacology. 2009;56(Suppl 1):112–21.
26. Cason AM, Smith RJ, Tahsili-Fahadan P, Moorman DE, Sartor GC, Aston-Jones G. Role of orexin/hypocretin in reward-seeking and addiction: implications for obesity. Physiol Behav. 2010;100:419–28.
27. Sakurai T, Mieda M, Tsujino N. The orexin system: roles in sleep/wake regulation. Ann N Y Acad Sci. 2010;1200:149–61.
28. Sakurai T, Amemiya A, Ishii M, Matsuzaki I, Chemelli RM, Tanaka H, et al. Orexins and orexin receptors: a family of hypothalamic neuropeptides and G protein-coupled receptors that regulate feeding behavior. Cell. 1998;92:1, page following 696.
29. Marcus JN, Aschkenasi CJ, Lee CE, Chemelli RM, Saper CB, Yanagisawa M, et al. Differential expression of orexin receptors 1 and 2 in the rat brain. J Comp Neurol. 2001;435:6–25.
30. Trivedi P, Yu H, MacNeil DJ, Van der Ploeg LH, Guan XM. Distribution of orexin receptor mRNA in the rat brain. FEBS Lett. 1998;438:71–5.
31. Rodgers RJ, Ishii Y, Halford JC, Blundell JE. Orexins and appetite regulation. Neuropeptides. 2002;36:303–25.
32. Clegg DJ, Air EL, Woods SC, Seeley RJ. Eating elicited by orexin-a, but not melanin-concentrating hormone, is opioid mediated. Endocrinology. 2002;143:2995–3000.
33. Rodgers RJ, Halford JC, Nunes de Souza RL, Canto de Souza AL, Piper DC, Arch JR, et al. SB-334867, a selective orexin-1 receptor antagonist, enhances behavioural satiety and blocks the hyperphagic effect of orexin-A in rats. Eur J Neurosci. 2001;13:1444–52.
34. Dube MG, Horvath TL, Kalra PS, Kalra SP. Evidence of NPY Y5 receptor involvement in food intake elicited by orexin A in sated rats. Peptides. 2000;21:1557–60.
35. Yamanaka A, Kunii K, Nambu T, Tsujino N, Sakai A, Matsuzaki I, et al. Orexin-induced food intake involves neuropeptide Y pathway. Brain Res. 2000;859:404–9.
36. Burdakov D, Gerasimenko O, Verkhratsky A. Physiological changes in glucose differentially modulate the excitability of hypothalamic melanin-concentrating hormone and orexin neurons in situ. J Neurosci. 2005; 25:2429–33.
37. Cai XJ, Widdowson PS, Harrold J, Wilson S, Buckingham RE, Arch JR, et al. Hypothalamic orexin expression: modulation by blood glucose and feeding. Diabetes. 1999;48:2132–7.
38. Toshinai K, Date Y, Murakami N, Shimada M, Mondal MS, Shimbara T, et al. Ghrelin-induced food intake is mediated via the orexin pathway. Endocrinology. 2003;144:1506–12.
39. Fadel J, Deutch AY. Anatomical substrates of orexin-dopamine interactions: lateral hypothalamic projections to the ventral tegmental area. Neuroscience. 2002;111:379–87.
40. Peyron C, Tighe DK, van den Pol AN, de Lecea L, Heller HC, Sutcliffe JG, et al. Neurons containing hypocretin (orexin) project to multiple neuronal systems. J Neurosci. 1998;18:9996–10015.
41. Carr D, Kalivas PW. Orexin: a gatekeeper of addiction. Nat Med. 2006;12:274–6.
42. Scammell TE, Saper CB. Orexin, drugs and motivated behaviors. Nat Neurosci. 2005;8:1286–8.
43. Korotkova TM, Sergeeva OA, Eriksson KS, Haas HL, Brown RE. Excitation of ventral tegmental area dopaminergic and nondopaminergic neurons by orexins/hypocretins. J Neurosci. 2003;23:7–11.
44. Narita M, Nagumo Y, Hashimoto S, Narita M, Khotib J, Miyatake M, et al. Direct involvement of orexinergic systems in the activation of the mesolimbic dopamine pathway and related behaviors induced by morphine. J Neurosci. 2006;26:398–405.
45. Harris GC, Wimmer M, Aston-Jones G. A role for lateral hypothalamic orexin neurons in reward seeking. Nature. 2005;437:556–9.
46. Borgland SL, Taha SA, Sarti F, Fields HL, Bonci A. Orexin A in the VTA is critical for the induction of synaptic plasticity and behavioral sensitization to cocaine. Neuron. 2006;49:589–601.

47. Boutrel B, Kenny PJ, Specio SE, Martin-Fardon R, Markou A, Koob GF, et al. Role for hypocretin in mediating stress-induced reinstatement of cocaine-seeking behavior. Proc Natl Acad Sci U S A. 2005;102:19168–73.
48. Lawrence AJ, Cowen MS, Yang HJ, Chen F, Oldfield B. The orexin system regulates alcohol-seeking in rats. Br J Pharmacol. 2006;148:752–9.
49. Muschamp JW, Dominguez JM, Sato SM, Shen RY, Hull EM. A role for hypocretin (orexin) in male sexual behavior. J Neurosci. 2007;27:2837–45.
50. Borgland SL, Chang SJ, Bowers MS, Thompson JL, Vittoz N, Floresco SB, et al. Orexin A/hypocretin-1 selectively promotes motivation for positive reinforcers. J Neurosci. 2009;29:11215–25.
51. Nair SG, Golden SA, Shaham Y. Differential effects of the hypocretin 1 receptor antagonist SB 334867 on high-fat food self-administration and reinstatement of food seeking in rats. Br J Pharmacol. 2008;154:406–16.
52. Thorpe AJ, Cleary JP, Levine AS, Kotz CM. Centrally administered orexin A increases motivation for sweet pellets in rats. Psychopharmacology (Berl). 2005;182:75–83.
53. Zheng H, Patterson LM, Berthoud HR. Orexin signaling in the ventral tegmental area is required for high-fat appetite induced by opioid stimulation of the nucleus accumbens. J Neurosci. 2007;27:11075–82.
54. Chemelli RM, Willie JT, Sinton CM, Elmquist JK, Scammell T, Lee C, et al. Narcolepsy in orexin knockout mice: molecular genetics of sleep regulation. Cell. 1999;98:437–51.
55. Lin L, Faraco J, Li R, Kadotani H, Rogers W, Lin X, et al. The sleep disorder canine narcolepsy is caused by a mutation in the hypocretin (orexin) receptor 2 gene. Cell. 1999;98:365–76.
56. Nishino S, Ripley B, Overeem S, Lammers GJ, Mignot E. Hypocretin (orexin) deficiency in human narcolepsy. Lancet. 2000;355:39–40.
57. Peyron C, Faraco J, Rogers W, Ripley B, Overeem S, Charnay Y, et al. A mutation in a case of early onset narcolepsy and a generalized absence of hypocretin peptides in human narcoleptic brains. Nat Med. 2000;6:991–7.
58. Thannickal TC, Moore RY, Nienhuis R, Ramanathan L, Gulyani S, Aldrich M, et al. Reduced number of hypocretin neurons in human narcolepsy. Neuron. 2000;27:469–74.
59. Rieger M, Mayer G, Gauggel S. Attention deficits in patients with narcolepsy. Sleep. 2003;26:36–43.
60. Fadel J, Bubser M, Deutch AY. Differential activation of orexin neurons by antipsychotic drugs associated with weight gain. J Neurosci. 2002;22:6742–6.
61. Huang H, Ghosh P, van den Pol AN. Prefrontal cortex-projecting glutamatergic thalamic paraventricular nucleus-excited by hypocretin: a feedforward circuit that may enhance cognitive arousal. J Neurophysiol. 2006; 95:1656–68.
62. Lambe EK, Aghajanian GK. Hypocretin (orexin) induces calcium transients in single spines postsynaptic to identified thalamocortical boutons in prefrontal slice. Neuron. 2003;40:139–50.
63. Lambe EK, Olausson P, Horst NK, Taylor JR, Aghajanian GK. Hypocretin and nicotine excite the same thalamocortical synapses in prefrontal cortex: correlation with improved attention in rat. J Neurosci. 2005;25:5225–9.
64. Mileykovskiy BY, Kiyashchenko LI, Siegel JM. Behavioral correlates of activity in identified hypocretin/orexin neurons. Neuron. 2005;46:787–98.
65. Boschen KE, Fadel JR, Burk JA. Systemic and intrabasalis administration of the orexin-1 receptor antagonist, SB-334867, disrupts attentional performance in rats. Psychopharmacology (Berl). 2009;206:205–13.
66. Davis JF, Krause EG, Melhorn SJ, Sakai RR, Benoit SC. Dominant rats are natural risk takers and display increased motivation for food reward. Neuroscience. 2009;162:23–30.
67. Kirouac GJ, Parsons MP, Li S. Orexin (hypocretin) innervation of the paraventricular nucleus of the thalamus. Brain Res. 2005;1059:179–88.
68. Ishibashi M, Takano S, Yanagida H, Takatsuna M, Nakajima K, Oomura Y, et al. Effects of orexins/hypocretins on neuronal activity in the paraventricular nucleus of the thalamus in rats in vitro. Peptides. 2005;26:471–81.
69. Hamlin AS, Clemens KJ, Choi EA, McNally GP. Paraventricular thalamus mediates context-induced reinstatement (renewal) of extinguished reward seeking. Eur J Neurosci. 2009;29:802–12.
70. Li Y, Li S, Sui N, Kirouac GJ. Orexin-A acts on the paraventricular nucleus of the midline thalamus to inhibit locomotor activity in rats. Pharmacol Biochem Behav. 2009;93:506–14.
71. Jones MW, Kilpatrick IC, Phillipson OT. Regulation of dopamine function in the nucleus accumbens of the rat by the thalamic paraventricular nucleus and adjacent midline nuclei. Exp Brain Res. 1989;76:572–80.
72. Parsons MP, Li S, Kirouac GJ. Functional and anatomical connection between the paraventricular nucleus of the thalamus and dopamine fibers of the nucleus accumbens. J Comp Neurol. 2007;500:1050–63.
73. Pinto A, Jankowski M, Sesack SR. Projections from the paraventricular nucleus of the thalamus to the rat prefrontal cortex and nucleus accumbens shell: ultrastructural characteristics and spatial relationships with dopamine afferents. J Comp Neurol. 2003;459:142–55.
74. Li Y, Li S, Wei C, Wang H, Sui N, Kirouac GJ. Changes in emotional behavior produced by orexin microinjections in the paraventricular nucleus of the thalamus. Pharmacol Biochem Behav. 2010;95:121–8.
75. Li Y, Li S, Wei C, Wang H, Sui N, Kirouac GJ. Orexins in the paraventricular nucleus of the thalamus mediate anxiety-like responses in rats. Psychopharmacology (Berl). 2010;212:251–65.

76. Li Y, Wang H, Qi K, Chen X, Li S, Sui N, et al. Orexins in the midline thalamus are involved in the expression of conditioned place aversion to morphine withdrawal. Physiol Behav. 2011;102:42–50.
77. Hubner CB, Moreton JE. Effects of selective D1 and D2 dopamine antagonists on cocaine self-administration in the rat. Psychopharmacology (Berl). 1991;105(2):151–6.
78. Schiltz CA, Bremer QZ, Landry CF, Kelley AE. Food-associated cues alter forebrain functional connectivity as assessed with immediate early gene and proenkephalin expression. BMC Biol. 2007;5:16.
79. Chung S, Hopf FW, Nagasaki H, Li CY, Belluzzi JD, Bonci A, et al. The melanin-concentrating hormone system modulates cocaine reward. Proc Natl Acad Sci U S A. 2009;106:6772–7.
80. Pissios P. Animals models of MCH function and what they can tell us about its role in energy balance. Peptides. 2009;30:2040–4.
81. Pissios P, Frank L, Kennedy AR, Porter DR, Marino FE, Liu FF, et al. Dysregulation of the mesolimbic dopamine system and reward in MCH−/− mice. Biol Psychiatry. 2008;64:184–91.
82. Harris GC, Aston-Jones G. Arousal and reward: a dichotomy in orexin function. Trends Neurosci. 2006;29: 571–7.

Chapter 36
Behavioral, Cognitive, and Affective Consequences of Trying to Avoid Chocolate

James A.K. Erskine and George J. Georgiou

Key Points

- Chocolate has become almost ubiquitous in its widespread availability.
- There has been a consistent rise in the average size of chocolate bars.
- These factors have resulted in increased average consumption and a need to attempt to restrain one's intake.
- Two common strategies to avoid consuming too much chocolate are (1) to try not to think about chocolate and (2) to try not to eat chocolate.
- Studies are now indicating that both trying not to think of chocolate and deliberately restraining oneself from eating it can result in increased rather than reduced subsequent consumption.
- Although these effects can present in most individuals, they appear to be strongest in restrained eaters or participants with a high propensity towards dieting.

Keywords Chocolate consumption • Overeating • Avoidance • Thought suppression • Behavioral avoidance • Restrained eating • Disinhibition

Introduction

> Strength is the ability to break a chocolate bar into four pieces with your bare hands - and then eat just one of those pieces.
>
> – Judith Viorst

The ability to resist the temptation to eat chocolate is a difficult endeavor. Chocolate is a food that consumed in excess would lead to a multitude of negative consequences such as increased weight, dental problems, and negative emotions, especially guilt. Furthermore, this chapter will show there is evidence that chocolate is a substance that one can become addicted to. As the widespread availability

J.A.K. Erskine, B.Sc. (Hons) (Psychology), M.Sc. (Research Methods), Ph.D. (Psychology) (✉)
School of Population Health Sciences and Education, St. George's, University of London, London, UK
e-mail: jerskine@sgul.ac.uk

G.J. Georgiou, B.Sc. (Hons), M.Sc., Ph.D.
School of Psychology, University of Hertfordshire, Hatfield, Hertfordshire, UK
e-mail: g.j.georgiou@herts.ac.uk

R.R. Watson et al. (eds.), *Chocolate in Health and Nutrition*, Nutrition and Health 7,
DOI 10.1007/978-1-61779-803-0_36, © Springer Science+Business Media, LLC 2013

of chocolate has become common, it has therefore become increasingly necessary for individuals to attempt to moderate or reduce their intake, often via avoiding thinking about chocolate or attempting to directly restrain consumption. This chapter will focus on the consequences of these processes of attempted control.

Chocolate is the most widely craved food in the western world and is disproportionately craved by women [1–6]. Food cravings are commonly defined as the intense desire to consume a certain specific food or foods [7–9]. Bruinsma and Taren [10] define craving as "an intense, periodic motivation aimed at gaining the craved substance." Chocolate cravings have been reported in 40% of females and 15% of males, and in addition, alternative substances will not satisfy their desire to eat chocolate [3, 5]. Almost 100% of women and roughly 70% of men state that they have had at least one or more food-craving episode in the past year [4].

Importantly, research on cravings relating to many foods, including chocolate, has demonstrated significant associations between food cravings and higher body mass index (BMI) scores. In short, heavier people report greater cravings for sweet foods [11, 12]. Furthermore, there is evidence that these associations also make binge eating more likely [13]. Because cravings are a common feature of addictive behaviors [14] and they also occur with reference to chocolate [15], this has prompted some researchers to suggest that chocolate could potentially be classified as a drug rather than as a food [10]. Furthermore, there is evidence that chocolate can be addictive [3, 16, 17]. Among those reporting high cravings for chocolate, research indicates there are frequent negative emotional associations, including depression, guilt, anxiety, frustration, negative body image, and ultimately eating disorders [17, 18]. Hetherington and Macdiarmid [16] also provide evidence that chocolate addicts experience intense cravings and report greater pleasure during active consumption, but their "addiction" interferes with their daily behaviors; for example, avoiding chocolate results in low mood and subsequent overconsumption. In addition, food cravings are often associated with other poor psychological outcomes such as low mood, and decreased quality of life [19–21]. Furthermore, participants reporting severe chocolate cravings often state that they try to avoid chocolate, but this results in irritability, low mood, and subsequent overeating, which is consistent with other drugs of abuse [10, 16, 17].

There is some evidence that chocolate addicts can be negatively affected by withdrawal symptoms [10, 16]. Although there are negative consequences to considering chocolate a drug, such as its addictive potential, other researchers have highlighted the positive medicinal aspects of chocolate consumption [22].

As a result of these recent findings, the status of obesity is being debated, and one suggestion is that its status changes from weight problem to psychiatric diagnosis. This is currently being debated for inclusion in the forthcoming edition of the Diagnostic and Statistical Manual of Mental Disorders [23]. In addition, research is uncovering some of the nervous system's centrally mediated hedonic mechanisms, which may govern the addictive component of certain foods [24].

Western nations, in particular Europe and the Americas, are the highest consumers of chocolate internationally. Data indicate that there has been a significant rise in the consumption of chocolate in these countries over the last two decades, with countries such as the United Kingdom rising from an average of 7.4 kg per year per capita [25] to 10.07 kg per year per capita in 2008 [26].

In tandem with increased chocolate consumption, the incidence of obesity has also risen dramatically over the past few decades. Thus, on a global scale, obesity has doubled since 1980. In 2008, over 1.5 billion adults aged over 20 years could be classified as overweight, and of these, over 200 million males and 300 million females could be classed as obese. Furthermore, almost 65% of the population of the world live in nations where overweight and obesity result in greater mortality than underweight. Reports also indicate that in 2010, almost 43 million children under 5 years of age could be classed as overweight [27–29].

In addition to growing worldwide obesity and chocolate consumption, there has also been a significant increase in the average portion size of various foods, including chocolate bars, mainly with the introduction of "king-sized" bars [30–32]. Research has consistently demonstrated that larger portion size results in increased consumption [33]. Furthermore, Vermeer et al. [32] conducted a study showing that with king-sized chocolate bars, 92% of participants intended to finish the whole bar by the end of the day, therefore indicating that if oversized chocolate bars are purchased, they are likely to be wholly consumed. Further to this, chocolate consumption has been shown to be increased when participants are asked to select chocolate (M&M's) using a large spoon as opposed to a small spoon [34].

Kemps, Tiggemann, and Hart [35] and Knight and Boland [36] suggest that people regularly attempt to control their intake of chocolate as it is commonly regarded as one of the definite foods "to be avoided" when dieting. Owing to the factors cited above, it is likely that individuals commonly attempt to regulate and control their consumption of chocolate. One common form of avoidance that people, attempting to control their food (and chocolate) intake, typically resort to is thought suppression. Therefore, if one wants to avoid overconsumption of chocolate, it is plausible that not thinking about chocolate may help. In addition, it is likely that individuals will also attempt to behaviorally avoid chocolate by choosing to evade locations where it is present and to restrain the impulse to consume it when it cannot be avoided in the environment. Critically, Geyskens, Soetens, and Roets [37] have demonstrated that cues of tempting foods (e.g., chocolate) elicited participants to direct their attention away from the stimuli. This effect was not moderated by other personal characteristics of the participants and provides evidence that when presented with tempting food, avoidance is a typical response.

Effects of Thought Suppression on Thinking About Foods

Thought suppression refers to the effortful avoidance of certain thoughts [38]. For example, an individual attempting to control his or her chocolate intake may decided to suppress thoughts about chocolate in an effort to avoid chocolate thought intrusion (including cravings) and the possibility of overeating. Research on thought suppression has consistently indicated that suppressing thoughts does not result in the desired outcome of fewer thoughts but paradoxically results in one thinking about the suppressed item more rather than less [39].

The consequences of trying to avoid certain thoughts were first studied by Wegner and colleagues [39]. They demonstrated that suppressed thoughts subsequently rebound and that individuals end up thinking about the thought more after attempted avoidance than before. This was in comparison to individuals who were instructed to actively engage in thought about the concept the suppression group was trying to avoid. This is known as the rebound effect. Admittedly, as a first demonstration, they chose to employ a neutral thought as the target of suppression. Thus, participants suppressed the neutral and largely irrelevant thought, "white bear." Since this first demonstration, several other studies have shown that suppressed thoughts subsequently return more frequently and that this process can occur with negative thoughts that are personally relevant. A meta-analysis undertaken in 2001 showed that rebound effects following thought suppression are robust [40]. Therefore, if one suppresses thoughts about chocolate, a likely outcome is that chocolate thoughts will paradoxically increase.

Importantly, a study by Harnden, McNally, and Jinserson [41] demonstrated that suppressing body weight thoughts can result in a subsequent rebound of these same thoughts – that is, an increase in thoughts regarding weight rather than the intended reduction, in comparison to control groups that were not suppressing body weight thoughts.

Another study investigating the effects of suppressing thoughts of food by Soetens, Braet, De Jonckheere, and Roets [42] asked participants to think aloud in three separate consecutive periods. In the first period, participants were free to think about anything they liked but were asked to monitor for food- or eating-related intrusions. In the second period, participants either suppressed thoughts or continued to monitor food- or eating-related thoughts. In the final period, all participants again returned to monitoring food-related intrusions. After each period, participants filled out a state scale measuring willingness and desire to eat [43]. Critically, Soetens et al. [42] also investigated participants' restrained eating status. Restrained eating has been viewed as a chronic tendency to restrict food intake for weight-related reasons and therefore has much in common with dieting [44, 45]. In addition, Soetens et al. [42] further investigated the participants' tendency towards disinhibited eating. Disinhibition is often viewed as a tendency towards overeating across a variety of situations [45]. The study also took measure of eating disorders [46, 47] and the tendency to use thought suppression in everyday life (White Bear Suppression Inventory, WBSI) [48]. Results indicated that people who were not inhibited in their eating (disinhibited restrainers) scored significantly higher than inhibited restrainers on the measure of the everyday use of thought suppression, than low restrainers. Furthermore, participants were categorized by their combination of the restrained eating scale and the disinhibited eating status. Results indicated that there was a main effect of group (inhibited restrainer versus disinhibited restrainer versus low restrainer), with disinhibited restrainers having more food-related thoughts than both other groups. Analysis indicated a significant rise in intrusive thoughts about food between the baseline period 1 and period 3 for the disinhibited restrainers that suppress. No similar effect was found for inhibited restrainers or low restrainers that suppressed. In summary, the Soetens et al. [42] study demonstrates that individuals who are naturally inclined to restrain their eating (restrainers) but are often unsuccessful (disinhibited) show significant increases in food thoughts that they had been trying to suppress. This effect was not present in groups that did not suppress food thoughts and were not disinhibited restrainers.

In a similar study, Oliver and Huon [43] had high and low disinhibitors' suppress thoughts about food and eating. Participants monitored thoughts for three 5-min periods: in baseline period 1 and period 2, participants suppressed (did not think about food and eating) versus expressed; in period 3, participants returned to monitoring. Results indicated that low disinhibitors instructed to suppress had more food-related thoughts than high disinhibitors. This pattern of results was reversed for the non-suppression condition, where high disinhibitors reported more thoughts than low disinhibitors. While the effects reported by Oliver and Huon [43] seem disparate with those reported by Soetens et al. [42], they did not examine participants' restrained eating status. The main message from this study is that food thought rebound is possible following suppression.

Further studies have indicated that suppression of naturally occurring negative thoughts can also result in rebound [49, 50], but note that some studies report that thought suppression may have no effect on personally relevant thoughts [51].

It appears that disinhibition status affects the level of rebound following suppression. One way in which studies have artificially manipulated disinhibition levels is by preloading – giving participants a high-calorie food prior to participation [52]. O'Connell, Larkin, Mizes, and Fremouw [53] had participants either take a preload or not (high-calorie chocolate or vanilla milk shake versus no milk shake). During period 1, half of the participants suppressed food and eating thoughts (try not to think about food or eating) and the other half of the participants merely monitored food and eating thoughts. During period 2, all participants monitored for food and eating intrusive thoughts. Finally, all participants took part in a taste preference task with the opportunity to try various ice creams (chocolate, vanilla, and strawberry) with instructions to at least try all three. Results indicated that restrained participants in the suppression group had more food intrusions in both thought periods but no rebound effect. Results for the taste test demonstrated a significant effect of preload – preload ate significantly less in the taste test. However, there was no effect of group or restraint on the amount of ice cream consumed.

In summary, the results regarding the suppression of food- and eating-related thoughts have provided mixed evidence. Thus, if one suppresses thoughts of food (including chocolate) or eating, there is enough evidence to suggest that these thoughts may increase following suppression but only under certain conditions or for particular individuals. This thought rebound for food appears most evident in restrained and disinhibited eaters.

Correlational Evidence of Associations Between Food-Related Thought Suppression and Issues with Eating Behavior

In view of the studies outlined above, Barnes, Fisak, and Tantleff-Dunn [54] developed the food thought suppression inventory, measuring an individual's tendency to suppress thoughts about food in everyday life. Importantly, this measure of food-related thought suppression has been found to be associated with eating pathologies, including binge eating. This is unsurprising, as the general tendency to suppress thought has been found to correlate positively with various psychopathologies, such as anxiety and depression [55]. In a further study investigating the food thought suppression inventory in males, associations were again found between the use of food thought suppression and pathological eating, including binge eating and compensatory behaviors. Furthermore, body mass index scores (BMI) were associated with more reported food thought suppression [56]. In addition to these findings, Barnes and Tanleff-Dunn [57] examined 312 overweight obese community men and women, measuring levels of food thought suppression via participants' completion of the food thought suppression inventory, weight history, and eating behaviors. The results showed that women were more likely to report using food thought suppression, as were dieters compared to non-dieters. The use of food thought suppression was shown to predict binge eating, food cravings, and other disordered eating habits.

Development of Attitudes Towards Chocolate Questionnaire

In an attempt to examine more quantitatively chocolate consumption and attitudes towards chocolate, Benton and colleagues [58] developed the attitudes towards chocolate questionnaire. Results in a large sample of 224 females and 104 males indicated that three factors accounted for the data on chocolate consumption. Factor 1 was labeled craving and was typified by intense urges to consume chocolate and the use of chocolate under stress. Factor 2 was labeled guilt and was typified by items indicating negative feelings when consuming chocolate. The third factor indicated adaptive chocolate consumption – for example, when missing a meal or to replenish energy after exercise. Importantly, the findings indicated that craving was associated with greater consumption, whereas high scores related to guilt for eating chocolate were linked to bingeing and vomiting.

In a further follow-up analysis of the attitudes towards chocolate questionnaire, Cramer and Hartleib [59] again factor analyzed the measure in a sample of 701 participants. In this study, a two-factor solution was optimal corresponding to the factors of craving and guilt. Importantly, females scored higher than males on both cravings and guilt over chocolate. Furthermore, the craving scale was associated with depression and obsession coupled with low self-esteem and social desirability. High guilt scores were associated with depression, anxiety, disordered eating, obsessions, and low self-esteem.

In summary, it would appear that the habitual suppression of thoughts related to chocolate is associated with disordered emotions and eating patterns.

Effects of Suppressing Chocolate Thoughts on Behavior

Even though there are no specific studies looking at suppression of chocolate thought on subsequent chocolate intrusions, it seems clear from other research on general food-related suppression that it is possible that suppressing thoughts about chocolate will lead to greater subsequent thought rather than less. However, the impact of suppressing chocolate-related thoughts on subsequent behavior is clearer, as will be shown in the following section.

In one of the earliest studies to explore behavioral effects of chocolate thought suppression, Johnston, Bulik, and Anstiss [60] had participants (20 chocolate cravers and 22 non-cravers) suppress or monitor their thoughts about chocolate. Participants then took part in a computerized task in which better performance was rewarded with increasing amounts of chocolate. Results indicated that suppression participants performed significantly better on the computerized task, earning more chocolate than monitoring participants. However, the outcome was not affected by whether the participants were classified as chocolate cravers or non-cravers. Therefore, although the authors did not measure actual chocolate consumption, they demonstrated that participants who had been trying to avoid thinking about chocolate subsequently worked harder in the computer task to earn greater amounts of chocolate.

Studies and research on eating behavior have often been perplexed by the relationship between heightened control of eating and overeating [44, 61–63]. In an attempt to explore this further, Mann and Ward [64] either prohibited participants from eating a food (of their own choosing with the requirement that it was something they rated as neither disliking or liking) or encouraged them to avoid it by choice or gave them no prohibition. They then assessed food intake, food thoughts, and food desire both pre- and post-forbidding the food. Results indicated that in both avoidance conditions, thoughts of the food increased. However, desire only increased among participants required to avoid the food. When examining actual consumption, no effect of the manipulations was demonstrated.

Following this study, Erskine and colleagues reasoned that if suppressed thoughts subsequently return more frequently (i.e., rebound), then perhaps suppressed thoughts regarding behaviors may also rebound [65–67]. Therefore, Erskine [65] had non-dieting participants either suppress, express (think about), or merely monitor (control group) thoughts of chocolate for a 5-min period. Participants were then introduced to an ostensibly unrelated task where they were asked to try two different brands of chocolate and answer a lengthy questionnaire about which they preferred. Participants were unaware that the amount of chocolate they consumed was being measured. Importantly, the results demonstrated that both males and females in the suppression condition consumed significantly more chocolate than the control participants, thus rebounding. Interestingly, for females, suppression participants also consumed more than the group that had been actively thinking about chocolate (this was not the case for males).

In a follow-up study, Erskine and Georgiou [66] investigated how the effects of chocolate thought suppression might be moderated by participants' restrained eating status. Restrained eating refers to a chronic tendency to restrict food intake [44]. Thus, in this study, 127 female participants were again asked to suppress, express, or merely monitor chocolate thoughts for a 5-min period. However, measures of participants' restrained eating status were taken prior to participation. After the thought manipulation period, once again all participants had the opportunity to try two brands of chocolate and answer questions regarding their preference. In this study, an interaction between the effects of the thought manipulation and participants' restrained eating status was demonstrated. Thus, restrained eaters who suppressed thoughts of chocolate ate significantly more than all other participants who ate less but comparable amounts. Erskine and Georgiou [66] also gave all participants the attitudes towards chocolate questionnaire assessing guilt, cravings, and functional use of chocolate and measures of the extent to which participants used thought suppression in everyday life [48, 58]. The correlational results indicated that the amount of chocolate consumed was related to higher body mass index scores and greater reported liking for chocolate. In addition, dietary restraint was highly positively correlated with participants having higher body mass index scores and greater guilt over chocolate consumption. However, everyday use of thought suppression was not correlated with the outcome measure of actual chocolate consumption (but this measure of thought suppression was not specific to chocolate suppression).

In several further studies examining a different behavior (smoking behavior), Erskine and colleagues have demonstrated that the post-suppression behavioral rebound previously demonstrated with chocolate consumption also occurs with smoking behavior [67]. Thus, Erskine et al. [67] show that suppression of smoking thoughts in comparison with expression or monitoring results in greater subsequent smoking. Interestingly, Erskine et al. [68] also show that suppression of smoking thoughts does not appear to elevate smoking cravings but may elevate scores on associated behaviors. Thus, they demonstrate that participants who suppressed thoughts of smoking for 5 min subsequently reported being significantly hungrier than participants who had expressed or monitored their smoking thoughts. This effect was still present 15 min after suppression. This demonstrates that the suppression of one appetitive thought can lead to increases in other related appetitive thoughts.

There is also evidence that direct behavioral control may have similar effects to thought control. Thus, Polivy et al. [69] separated 103 participants into three experimental groups. One group was deprived of chocolate for 1 week, one group was deprived of vanilla, and the final group was not deprived. After the week, participants took part in a laboratory-based taste preference task in which they had to rate their preference for certain foods. Once again, Polivy et al. [69] measured participants' restrained eating status. Results indicated that restrained eaters that were deprived of chocolate subsequently ate the most chocolate in the taste test and reported experiencing more cravings. Thus, it would appear that in addition to suppressing thoughts of chocolate, direct behavioral prohibition may have similar effects.

In another study of the effects of behavioral prohibition, Stirling and Yeomans [70] gave half of their participants a bag of 30 chocolates (Hershey's Kisses) and asked them to carry it with them for the next 24 h and were told to avoid eating any of the chocolate. Control participants were not given any chocolate or instructions to avoid chocolate for the 24-h period. Participants were also divided into restrained and non-restrained eaters. Results indicated that both restrainers and non-restrainers experienced chocolate cravings during the 24-h period. Furthermore, some of the restrained eaters that had been given the bag of chocolates consumed some of its contents, but this was not the case for the non-restrainers. Participants were then given a laboratory-based taste test where they were asked to try three bowls of chocolate and answer questions about their taste characteristics. Critically, they found two main effects. Thus, restrained eaters consistently consumed more than non-restrainers. And participants in the temptation condition consumed more than the control condition. There was no interaction between the conditions.

In a new study Georgiou, Erskine, Soetens, Craeynest, and Paphiti-Galeano [71] wanted to tease apart the effects of thought suppression and behavioral avoidance on subsequent chocolate consumption. Thus, prior research has suggested that if both behavioral prohibition and thought suppression regarding chocolate can result in increased subsequent consumption, would participants engaging in both strategies simultaneously consume even more chocolate than participants using only one individual strategy [66]. In this study [71], therefore, participants were asked to either suppress thoughts of chocolate, avoid eating chocolate without suppressing thoughts about it, or suppress chocolate thoughts and avoid eating it or merely monitor their thoughts and not attempt control. Once again, measures of restrained eating status were taken. Results indicated that only restrained eaters that were in the group that both suppressed chocolate thoughts and the behavior of eating chocolate subsequently came to consume chocolate at inflated rates. This study, therefore, suggests that rebounds may be stronger if participants use both avoidant strategies in combination. In this study, there was a trend for thought suppression alone to also result in increased subsequent consumption, but it was not significant at conventional levels.

There is also evidence that participants may show a general inhibitory deficit that can account for some of their overeating. Thus, in a study examining the effects of general inhibition of chocolate consumption by Allan, Johnston, and Campbell [72], participants were allowed to consume unlimited quantities of chocolate by being told they were part of a consumer study. All participants reported healthy dietary intentions. Importantly, the amount of chocolate consumed was predicted by a general measure of cognitive inhibition (an inhibitory stroop task). Thus, individuals who had problems suppressing irrelevant information ate more chocolate and scored higher on the body mass index.

While it is possible that individuals attempting to avoid overeating chocolate could use one or other of the above strategies, it is most probable that in everyday life, individuals will use both in tandem [71, 73]. Thus, to illustrate, imagine you are an individual who frequently experiences intense chocolate cravings and is prone to overeating at times. In addition, you currently are a few kilograms over your ideal weight for your height and therefore have decided to cut out excess chocolate in an attempt to normalize your weight. Under these circumstances, it is probable that you will seek to avoid consuming chocolate; thus, you will start to behaviorally avoid it as well as try to avoid thoughts and cravings about chocolate via suppression. Soetens, Braet, Van Vlierberghe, and Roets [74] report that a combination of suppressing chocolate thoughts and behavioral avoidance in a naturalistic context for 24 h results in enhanced subsequent laboratory chocolate consumption relative to groups not attempting control. These processes may include strategies using willpower, say when out with friends for a meal, but may also include strategies such as not keeping chocolate in the house, that is, avoiding cues that have been shown to cause cravings [75–78]. Fujuita [79] has suggested that controlling oneself does not merely involve inhibiting various impulses but is much broader. Furthermore, Fujuita suggests that it is precisely when individuals rely too heavily on the effortful inhibition of various impulses that they experience more failures of self-control [79]. Studies have also indicated that controlling oneself may consume resources that become depleted with time. Therefore, after participants engage in a period of self-control, they frequently become less successful at controlling themselves on subsequent tasks [80, 81]. Thus, for example, if one attempts to avoid eating and thinking about chocolate all day, it is probable that by evening, one will have depleted the resources that enable self-control and therefore will be highly susceptible to self-control failure, which will include eating the chocolate that one is attempting to avoid, if encountered. The depletion of self-regulatory resources has been shown to lead to poorer self-control in subsequent eating behaviors, and evidence also shows that positive mood can restore previously depleted resources. In Tice et al. [81] (study 4), participants were asked to avoid eating for 3 h prior to the study. They were then exposed to the sight and smells of freshly baked cookies, chocolate M&M's, and a bowl of radishes. Participants were informed that they were all in the radish condition and would be performing a taste test on the radishes but would have to refrain from eating the cookies or M&M's. They were then left for a period of 10 min while eating radishes and avoiding eating the cookies and chocolates. Participants were also told that they would have to refrain from eating cookies and M&M's for 24 h after the study to artificially induce desire for these items. Next, half of the participants viewed a neutral film, while the other half viewed a comedy. Participants were then instructed to perform a drawing task (that they were unaware was actually insolvable) and to notify the experimenter when they wished to end the task. Results showed that participants viewing the comedy persisted longer than participants viewing the neutral film. This study demonstrates that depleted self-control resources may be replenished by the induction of positive mood.

What is evident from the behavioral rebound literature is that chocolate avoidance (both in thought and behavior) will lead to increased subsequent consumption, although the effects seem most evident in participants who are also restrained eaters. Recent research has started to indicate methods by which one may reduce consumption without untoward consequences. Chief among these are mindfulness-based mental strategies focusing on accepting thoughts and cravings rather than suppression [82]. In addition, Morewedge and colleagues have also shown that chocolate consumption can be reduced following participants imagining consuming chocolate prior to being given the opportunity to actually consume chocolate [83]. The reduction in chocolate consumption following imagined consumption is hypothesized to result from habituation to chocolate following repeated imagined consumption.

Summary

The effects reported in this chapter may go some of the way towards explaining the often-cited yet perplexing findings that diets where avoidance is recommended do not commonly work and that the net effect of going on a diet is a subsequent increase in thoughts about food, increased consumption, and weight gain [63, 84, 85].

To conclude, although there are no studies on the effects of chocolate thought suppression on subsequent chocolate thought intrusions, the data from studies on general food-related thought suppression indicate that suppressed food thoughts can rebound, although various factors interact to make this only occur in certain individuals and certain circumstances. In contrast, the data on suppressing thoughts about chocolate on subsequent chocolate consumption are remarkably consistent. The general findings indicate that avoiding chocolate thoughts will lead to increased subsequent consumption. Furthermore, there is consistent evidence that restraining oneself from eating chocolate with or without associated thought suppression can also lead to subsequent overeating of chocolate. However, in both cases, the effects are far more pronounced in restrained eaters. It would therefore appear that as chocolate represents one of the most frequently craved and avoided substances and it is ubiquitously available, it represents a danger for common overconsumption. This chapter has indicated that avoidance of chocolate in both thought and behavior is not likely to engender abstinence. Future research needs to investigate potential strategies that might more effectively enable individuals to successfully regulate their chocolate consumption. Most promising appear mindfulness-based cognitive strategies and imagined consumption.

References

1. Michener W, Rozin P. Pharmacological versus sensory factors in the satiation of chocolate craving. Physiol Behav. 1994;56:419–22.
2. Rogers PJ, Smit HJ. Food craving and food "addiction": a critical review of the evidence from a biopsychosocial perspective. Pharmacol Biochem Behav. 2000;66:3–14.
3. Rozin P, Levine E, Stoess C. Chocolate craving and liking. Appetite. 1991;17:199–212.
4. Pelchat ML. Food cravings in young and elderly adults. Appetite. 1997;28:103–13.
5. Weingarten H, Elston D. Food cravings in a college population. Appetite. 1991;17:167–75.
6. Zellner DA, Garriga-Trillo A, Rohm E, Centeno S, Parker S. Food liking and craving: a cross-cultural approach. Appetite. 1999;33:61–70.
7. Gendall KA, Joyce PR, Sullivan PF. Impact of definition on prevalence of food cravings in a random sample of young women. Appetite. 1997;28:63–72.
8. Rozin P. The selection of food by rats, humans, and other animals. In: Rosenblatt JS, Hinde RA, Shaw E, Beer C, editors. Advances in the study of behaviour. New York: Academic; 1976. p. 21–74.
9. Weingarten H, Elston D. The phenomenology of food cravings. Appetite. 1990;15:231–46.
10. Bruinsma K, Taren DL. Chocolate: food or drug? J Am Diet Assoc. 1999;99:1249–56.
11. Martin CK, O'Neil PM, Tollefson G, et al. The association between food cravings and consumption of specific foods in a laboratory taste test. Appetite. 2008;51:324–6.
12. White MA, Whisenhunt BL, Williamson DA, et al. Development and validation of the food-craving inventory. Obes Res. 2002;10:107–14.
13. White MA, Grilo C. Ethnic differences in the prediction of eating and body image disturbances among female adolescent psychiatry inpatients. Int J Eat Disord. 2005;38:78–84.
14. Pickens RW, Johanson C. Craving: consensus of status and agenda for future research. Drug Alcohol Depend. 1992;30:127–31.
15. Pelchat ML. Of human bondage: food craving, obsession, compulsion, and addiction. Physiol Behav. 2002;76:347–52.
16. Hetherington MM, Macdiarmid JI. "Chocolate addiction": a preliminary study of its description and its relationship to problem eating. Appetite. 1993;21:233–46.
17. Macdiarmid JI, Hetherington MM. Mood modulation by food: an exploration of affect and cravings in "chocolate addicts.". Br J Clin Psychol. 1995;34:129–38.
18. Tuomisto T, Hetheringon MM, Morris MF, et al. Psychological and physiological characteristic of sweet food "addiction.". Int J Eat Disord. 1997;25:169–75.
19. Forman EM, Hoffman KL, McGrath KB, et al. A comparison of acceptance- and control-based strategies for coping with food cravings: an analog study. Behav Res Ther. 2007;45:2372–86.
20. Lafay L, Thomas F, Mennen L, et al. Gender differences in the relation between food cravings and mood in an adult community: results from the Fleurbaix Laventie Ville Santé study. Int J Eat Disord. 2001;29:195–204.

21. Vander Wal JS, Johnson KA, Dhurandhar NV. Psychometric properties of the state and trait food cravings questionnaires in overweight and obese persons. Eat Behav. 2007;8:211–23.
22. Keen CL. Chocolate: food as medicine/medicine as food. J Am Coll Nutr. 2001;20:436S–9.
23. Volkow ND, O'Brien CP. Issues for DSM-V: should obesity be included as a brain disorder? Am J Psychiatry. 2008;165:138–9.
24. Cota D, Tschöp MH, Hovath TL, et al. Cannabinoids, opioids and eating behaviour: the molecular face of hedonism? Brain Res Rev. 2006;51:85–107.
25. Seligson FH, Krummel DA, Apgar JL. Patterns of chocolate consumption. Am J Clin Nutr. 1994;60:1060S–4.
26. Ranking of consumption of chocolate. http://www.caobisco.com/doc_uploads/Charts/ranking_of_consumption_chocolate_confectionery_2008.pdf. CAOBISCO. Accessed 16 June 2011.
27. Hedley AA, Ogden CL, Johnson CL, et al. Prevalence of overweight and obesity among us children, adolescents, and adults, 1999–2002. JAMA. 2004;291:847–2850.
28. Obesity and overweight. Retrieved Nov 2005, from http://www.who.int/hpr/NPH/docs/gs_obesity.pdfS. WHO. 2003.
29. Obesity fact sheet. WHO. 2010. http://www.who.int/mediacentre/factsheets/fs311/en/. Accessed 15 Mar 2011.
30. Nielsen SJ, Popkin BM. Patterns and trends in food portion sizes, 1977–1998. JAMA. 2003;289:450–3.
31. Young LR, Nestle M. The contribution of expanding portion sizes to the US obesity epidemic. Am J Public Health. 2002;92:246–9.
32. Vermeer WH, Bruins B, Steenhuis IHM. Two pack king size chocolate bars. Can we manage our consumption? Appetite. 2010;54:414–7.
33. Rolls BJ, Roe LS, Meengs JS. The effect of large portion sizes on energy intake is sustained for 11 days. Obesity. 2007;15:1535–43.
34. Geier AB, Rozin P, Doros G. A new heuristic that helps explain the effect of portion size on food intake. Psychol Sci. 2006;17:521–5.
35. Kemps E, Tiggemann M, Hart G. Chocolate cravings are sensitive to visuo-spatial interference. Eat Behav. 2005;6:101–7.
36. Knight LJ, Boland FJ. Restrained eating: an experimental disentanglement of this disinhibiting variables of perceived calories and food type. J Abnorm Psychol. 1989;98:412–20.
37. Geyskens K, Soetens B, Roets A. Exposure makes the heart grow less fond! The effects of exposure to palatable foods on the subsequent attention processing of food cues. J Health Psychol. 2010;15:230–8.
38. Wegner DM. White bears and other unwanted thoughts: suppression, obsession, and the psychology of mental control. New York: Viking/Penguin; 1989.
39. Wegner DM, Schneider DJ, Carter S, et al. Paradoxical effects of thought suppression. J Pers Soc Psychol. 1987;53:5–13.
40. Abramowitz JS, Tolin DF, Street GP. Paradoxical effects of thought suppression: a meta-analysis of controlled studies. Clin Psychol Rev. 2001;21:683–703.
41. Harnden JL, McNally RJ, Jinserson DC. Effects of suppressing thoughts about body weight: a comparison of dieters and nondieters. Int J Eat Disord. 1997;22:285–90.
42. Soetens B, Braet C, Dejonckheere P, et al. When suppression backfires: the ironic effects of suppressing eating-related thoughts. J Health Psychol. 2006;11:655–68.
43. Oliver K, Huon G. Eating-related thought suppression in high and low disinhibitors. Int J Eat Disord. 2001;30:329–37.
44. Polivy J. Psychological consequences of food restriction. J Am Diet Assoc. 1996;96:589–92.
45. Van Strien T. Success and failure in the measurement of restraint: notes and data. Int J Eat Disord. 1999;25:441–50.
46. Fairburn CG, Beglin SJ. Assessment of eating disorders: interview or self-report questionnaire? Int J Eat Disord. 1994;16:363–70.
47. Garner DM, Olmsted MP, Polivy J. Development and validation of a multidimensional eating disorder inventory for anorexia nervosa and bulimia. Int J Eat Disord. 1983;2:15–34.
48. Wegner DM, Zanakos S. Chronic thought suppression. J Pers. 1994;62:615–40.
49. McNally RJ, Ricciardi JN. Suppression of negative & neutral thoughts. Behav Cogn Psychother. 1996;24:17–26.
50. Salkovskis PM, Campbell P. Thought suppression induces intrusion in naturally occurring negative intrusive thoughts. Behav Res Ther. 1994;32:1–8.
51. Kelly AE, Kahn JH. Effects of suppression of personal intrusive thoughts. J Pers Soc Psychol. 1994;66:998–1006.
52. Herman CP, Mack D. Restrained and unrestrained eating. J Pers. 1975;43:657–60.
53. O'Connell C, Larkin K, Mizes JS, et al. The impact of caloric preloading on attempts at food and eating-related thought suppression in restrained and unrestrained eaters. Int J Eat Disord. 2005;38:42–8.
54. Barnes RD, Fisak B, Tantleff-Dunn S. Validation of the food thought suppression inventory. J Health Psychol. 2010;15:373–81.

55. Erskine JAK, Kvavilashvili L, Kornbrot DE. The predictors of thought suppression in young and old adults: effects of anxiety, rumination and other variables. Pers Individ Differ. 2007;42:1047–57.
56. Barnes RD, White MA. Psychometric properties of the food thought suppression inventory in men. J Health Psychol. 2010;15:1113–20.
57. Barnes RD, Tantleff-Dunn S. Food for thought: examining the relationship between food thought suppression and weight-related outcomes. Eat Behav. 2010;11:175–9.
58. Benton DK, Greenfield M, Morgan M. The development of the attitudes to chocolate questionnaire. Pers Individ Differ. 1998;24:513–20.
59. Cramer KM, Hartleib M. The attitudes to chocolate questionnaire: a psychometric evaluation. Pers Individ Differ. 2001;31:931–42.
60. Johnston L, Bulik CM, Anstiss V. Suppressing thoughts about chocolate. Int J Eat Disord. 1999;26:21–7.
61. Herman CP, Polivy J. Mental control of eating: excitatory and inhibitory food thoughts. In: Wegner DM, Pennebaker JW, editors. Handbook of mental control. Englewood Cliffs: Prentice-Hall; 1993. p. 491–505.
62. Polivy J. The effects of behavioural inhibition: integrating internal cues, cognition, behaviour and affect. Psychol Inq. 1998;9:181–204.
63. Polivy J, Herman CP. Dieting and binging, a causal analysis. Am Psychol. 1985;40:193–201.
64. Mann T, Ward A. Forbidden fruit: does thinking about a prohibited food lead to its consumption? Int J Eat Disord. 2001;29:319–27.
65. Erskine JAK. Resistance can be futile: investigating behavioural rebound. Appetite. 2008;50:415–21.
66. Erskine JAK, Georgiou GJ. Effects of thought suppression on eating behaviour in restrained and non-restrained eaters. Appetite. 2010;54:499–503.
67. Erskine JAK, Georgiou GJ, Kvavilashvili L. I suppress therefore I smoke: the effects of thought suppression on smoking behaviour. Psychol Sci. 2010;21:1225–30.
68. Erskine JAK, Ussher M, Cropley M, et al. Effect of thought suppression on desire to smoke and tobacco withdrawal symptoms. Psychopharmacology. 2011;201:205–11.
69. Polivy J, Coleman J, Herman CP. The effect of deprivation on food cravings and eating behaviour in restrained and unrestrained eaters. Int J Eat Disord. 2005;38:301–9.
70. Stirling LJ, Yeomans MR. Effect of exposure to a forbidden food on eating in restrained and unrestrained women. Int J Eat Disord. 2004;35:59–68.
71. Georgiou GJ, Erskine JAK, Soetens B, et al. Effects of thought and behavioural suppression on subsequent consumption. In preparation.
72. Allan JL, Johnston M, Campbell N. Unintentional eating. What determines goal-incongruent chocolate consumption? Appetite. 2010;54:422–5.
73. Erskine JAK, Georgiou GJ. My own worst enemy: how trying not to think of an action might lead you down that very path. The Psychologist. In press.
74. Soetens B, Braet C, Van Vlierberghe L, et al. Resisting temptation: effects of exposure to a forbidden food on eating behaviour. Appetite. 2008;51:202–5.
75. Fedoroff IC, Polivy J, Herman CP. The effect of pre-exposure to food cues on the eating behaviour of restrained and unrestrained eaters. Appetite. 1997;28:33–47.
76. Fedoroff I, Polivy J, Herman CP. The specificity of restrained versus unrestrained eaters' responses to food cues: general desire to eat, or craving for the cued food? Appetite. 2003;41:7–13.
77. Fletcher BC, Pine KJ, Woodbridge Z, et al. How visual images of chocolate affect the craving and guilt of female dieters. Appetite. 2007;48:211–7.
78. Papies EK, Stroebe W, Aarts H. Pleasure in the mind: restrained eating and spontaneous hedonic thoughts about food. J Exp Soc Psychol. 2007;43:810–7.
79. Fujita K. On conceptualizing self-control as more than the effortful inhibition of impulses. Pers Soc Psychol Rev. 2011;15(4):352–66.
80. Baumeister RF, Bratslavsky E, Muraven M, et al. Ego depletion: is the active self a limited resource? J Pers Soc Psychol. 1998;74:1252–65.
81. Tice DM, Baumeister RF, Shmueli D, et al. Restoring the self: positive affect helps improve self-regulation following ego depletion. J Exp Soc Psychol. 2007;43:379–84.
82. May J, Andrade J, Batey H, et al. Less food for thought. Impact of attentional instructions on intrusive thoughts about snack foods. Appetite. 2010;55:279–87.
83. Morewedge CK, Huh YE, Vosgerau J. Thought for food: imagined consumption reduces actual consumption. Science. 2010;330:1530–3.
84. Field AE, Austin SB, Taylor CB, et al. Relation between dieting and weight change among preadolescents and adolescents. Pediatrics. 2003;112:900–6.
85. Urbszat D, Herman CP, Polivy J. Eat, drink, and be merry, for tomorrow we diet: effects of anticipated deprivation on food intake in restrained and unrestrained eaters. J Abnorm Psychol. 2002;111:396–401.

Chapter 37
Chocolate and Children's Food and Flavor Preferences

Janet E. Standen-Holmes and Djin Gie Liem

Key Points

- Chocolate is ranked among children's favorite foods in the West and increasingly in other societies.
- Children's chocolate preferences appear to be more diverse than adults'.
- Chocolate of the sort generally marketed to children contains approximately 30% fat and up to 45% sugar, giving it a very high energy density (2,200 Kj/100 g) likely to engender satiety, making it likely that children will acquire a taste for it without difficulty.
- Taste preferences are mostly learned through repeated exposure, but overexposure can lead to decreased liking, even for chocolate.
- Severe parental restriction of sweet foods may increase children's liking and desire for them, which can lead to increased consumption when parents are absent.

Keywords Chocolate • Cocoa • Snack • Children • Taste • Food choice • Preference • Sensory • Parenting

> The one thing he longed for more than anything else was … CHOCOLATE. Walking to school in the mornings, Charlie could see great slabs of chocolate piled up high in the shop windows, and he would stop and stare and press his nose against the glass, his mouth watering like mad.
>
> – *Charlie and the Chocolate Factory,* by Roald Dahl

Introduction

In his book *Charlie and the Chocolate Factory*, Roald Dahl describes a child's fantasy of chocolate waterfalls, endless chocolate bars, and the prospect of owning your own chocolate factory. Children and chocolate seem to be a natural marriage, which is celebrated across the world on various occasions throughout the year. Easter is the holiday most often associated with chocolate in the western

J.E. Standen-Holmes, B.Sc., Graduate Certificate in Human Nutrition • D.G. Liem, Ph.D., M.Sc. (✉)
School of Exercise and Nutrition Science, Deakin Sensory Science Group, Deakin University, Burwood, Australia
e-mail: janeteholmes@gmail.com; Gie.liem@deakin.edu.au

R.R. Watson et al. (eds.), *Chocolate in Health and Nutrition*, Nutrition and Health 7,
DOI 10.1007/978-1-61779-803-0_37, © Springer Science+Business Media, LLC 2013

world, 15% of annual chocolate sales occur in the 4 weeks prior, and production is ramped up over 6 months earlier in order to build up stocks to cater for this [1]. The iconic Easter rabbit is a construct that is clearly aimed at children, both as a fantasy and as a hollow, foil-covered chocolate confection, while chocolate eggs and boxes of chocolates are an appropriate gift for both children and adults.

In the Netherlands, the traditional Sinterklaas Eve party is a child-centered tradition that is always celebrated with, among other things, a gift of an initial of each child's name made out of chocolate. Originally these initials were made of bread dough, but like Easter eggs, chocolate is now the norm [2]. For the Jewish holiday of Chanukah, children were traditionally given coins, which these days are often made of foil-wrapped chocolate instead. In fact, celebrations of all sorts are so associated with chocolate that the Mars confectionery company distributes a box of chocolates called "Celebrations." No child will be surprised to receive chocolate as part of a special occasion.

This chapter will explain what is currently known about the role of chocolate in children's diets and drivers of children's chocolate consumption, such as taste preferences and parental influences.

Role of Chocolate in Children's Diets

Although the food industry presumably has a good understanding of the quantity and types of chocolate children consume, there is relatively little information publicly available. Several studies, as will be shown below, investigated the snacking behavior of children, but very few provide specific information about chocolate.

Skinner et al. (1998) found that according to their mothers, 89% of toddlers liked and ate milk chocolate [3]. A US survey found that between 1977 and 1996 there was an increase in the incidence of children's snacking behavior, although the snacks did not change much in size or energy density, the number of snacking occasions increased, and the average number of calories obtained from snack foods among 6–11-year-olds rose from 18% to 24% of daily intake. Snacks represented about 24% of the calorie intake of the average child but only 20% of intake of micronutrients [4], suggesting that snacks are a less desirable method of feeding a child whose diet is not energy depleted. The US FITS cross-sectional survey of toddlers and infants found that by 18 months of age 10% of children were being given candy as part of their afternoon snack, but specific data for chocolate was not given [3].

In a longitudinal study of 5–11-year-old girls in the United States, it was found that chocolate was ranked highest, along with Skittles, cookies, and ice cream, and preferences remained relatively stable as the children grew up. Unfortunately, it was not clear from this study what type of chocolate the children preferred [5]. The Texas-based CATCH project found that chocolate candy was a popular snack food, coming seventh on the list of snacks and representing 4% of snacks eaten (as measured per piece rather than by weight). They also found that children preferred skimmed chocolate-flavored milk over plain whole or skim milk at lunchtime; it represented 19% of dairy food choices and was chosen three times as often as chocolate candy. Thus, while American children like chocolate, they consume it even more often as milk flavoring and in cakes and cookies than as confectionery [6]. This suggests that chocolate flavoring offers a potential method of encouraging the consumption of low-fat milk.

A review of chocolate consumption published in 1994 found that snacks accounted for 66% of chocolate candy intake. The remainder occurred largely at lunch and dinner, leaving breakfast as the only meal of the day without a significant chocolate component [7]. A longitudinal study of French children reported that chocolate was nominated by 84–86% of children as one of their ten favorite foods [8]. Also, in Edinburgh [4] and Liverpool, children nominated chocolate as among their favorite foods [9]. An English survey of children's snacking behavior as reported by both their parents and themselves found chocolate to be the second most frequently eaten children's snack food [10].

In a 1994 study of the chocolate and snack food intake of 200 English and German children between 9 and 11 years of age, the children were asked about the food they brought to school and their

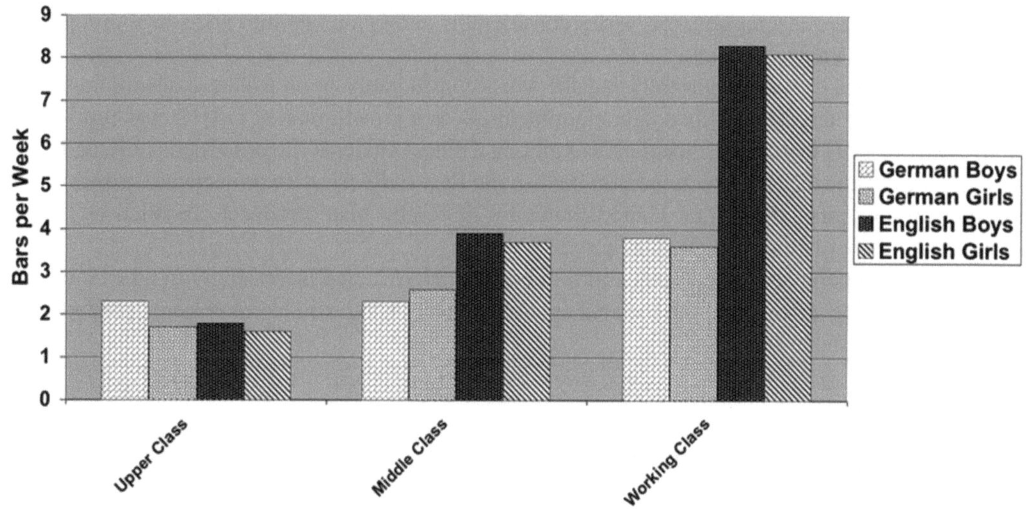

Fig. 37.1 Average chocolate consumption of English and German children in bars per week. Created with data collected from Reference [11]

weekly chocolate consumption; they were also offered a table of snack foods to choose from. There were clear national differences in chocolate consumption, which were most significant in the lower SES children. The lower SES English children ate approximately five times as much chocolate as the higher SES children from both countries, while low SES German children only reported eating twice as much chocolate as high SES children. National differences were negligible among the higher SES children and increased somewhat among middle SES children (Fig. 37.1) [11].

In another study, British children of lower SES were found to eat more of their daily energy intake in the form of snacks than higher SES children [12].

Thus, it is apparent that while the consumption of chocolate by children is widespread, it is also highly variable and probably increasing along with increasing snacking activity in general, driven at least in part because children have more disposable income and are likely to spend some of on candy, including chocolate. It may also reflect a more intensive and successful environment for marketing to children in England than in Germany [13]. A 2007 survey by the British consumer magazine *Which* found that children were most likely to spend their pocket money (allowance) on snacks, including sweets, chocolates, and crisps. These were often bought without their parents' knowledge, on the way to and from school. They found that the amount of pocket money the children received had risen approximately 600% in the previous 20 years, which is well in excess of inflation [14]. This would imply that British children's improved financial position is contributing to increased snack consumption, a significant amount of which is in the form of chocolate. Research done in Chicago on low-income families in 2008 found that the more pocket money children received, the more likely they were to have similar unhealthy eating patterns to their mothers' [15]. However, a survey of New Zealand children found that they saved most of their pocket money, and when asked what snack they would buy, only 14.3% nominated chocolate [16].

It has also been found that grandparents have a significant effect on children's pocket-money status and are thus an important contributor to children's independent snack purchases. Grandparents are often ignored in studies, which may help to explain the erratic nature of the results, since grandparents' impact on children's lives is likely to be extremely variable [17]. It would appear that while increasing discretionary income (pocket money/allowances) among children has led to increased snack and chocolate consumption, this varies considerably across nationalities and social groups, and very little is really known about it.

Although Africa and Asia currently lag the West considerably in chocolate consumption, Asia especially is being targeted by chocolate manufacturers as a growth market [18]. It was reported in *China Today* in 1993 that Swiss chocolate was among Chinese students' favorite foods [19], and retail sales volumes of chocolate confectionery in the Middle East and Africa are projected to grow by 20% over the next 5 years [20]. In the United States in 2009, chocolate products classified as "Kids" accounted for 12% of new product launches [21].

Chocolate is a popular snack among children both in the West and increasingly in other societies. Children's snacking behavior and associated chocolate intake is most likely increasing, partly due to their improved financial situation.

Drivers of Chocolate Consumption

Although children are limited in their food choices by their cultural environment, wherever the food supply is adequate, there is a wide variety of foods available, and children tend to choose the foods they like [22], regardless of their knowledge of the likely health consequences [17]. Thus, an understanding of what foods children like, why they like them, and how these preferences develop is crucial to an understanding of children's chocolate consumption.

Taste and Flavor Preferences of Children

What in common parlance is called "taste" generally refers to a combination of taste and aroma that is more properly described as "flavor." Humans can detect five distinct tastes: sweet, salty, bitter, sour, and umami, which can be perceived on all areas of the tongue [23]. Recent research suggests that humans can also detect various fatty acids, which would make fat the sixth taste [24–26]. Infants' and children's sensory perception of umami and fat taste will not be discussed in this chapter, owing to the very limited research currently available. The experience of taste is a result of the stimulation of specialized taste receptor cells grouped in small clusters known as taste buds mainly on the tongue and around the oral cavity but also on the hard and soft palate, the pharynx, the larynx, the tonsils, the esophagus, and the epiglottis. The taste buds on the tongue are located in three different types of taste papillae, the fungiform, circumvallate, and foliate papillae [27, 28].

Liking and preference tend to be conflated; however, liking in the context of food refers to a hedonic enjoyment of the flavor and process of eating (e.g., I like this chocolate cake), whereas preference is the hedonic evaluation of a stimulus relative to other stimuli (e.g., I prefer this chocolate cake over that banana cake) [29]. Although people may prefer chocolate cake over banana cake, it does not necessarily mean that they like chocolate cake; it may be that both cakes are disliked, but chocolate cake is just a little bit less disliked than banana cake.

Flavor Masking and Chocolate

In a study conducted in Utah, children were asked to rate the palatability of various flavorings in activated charcoal (an unpleasant-tasting medication that is given in large quantities to children suspected of swallowing poisons, among other things). Chocolate milk was preferred by 39% of children ahead of Coca-Cola and cherry syrup at 23% each [30]. However, a Canadian study found that 50% of children preferred cola and only 19.2% preferred chocolate milk and 15.4% preferred orange juice and plain water each, which meant that only cola was significantly preferred over plain water [31]. The children in this study were approximately 2 years younger than in the former, with a mean age of 6.5 ± 1.4 years of age, whereas the Utah group had a mean age of 8.3 years, which may explain some of the difference. The discrepancy may also perhaps be due to regional or national differences in the drinks that the children are normally given. Additionally, more than 60% of the Utah children indicated that their preferred activated charcoal flavor was not the flavor they normally preferred. More work needs to be done in order to properly understand the best use of chocolate flavor in improving the palatability of various medicines [32].

Development of Taste and Flavor Preferences

The biochemical processes leading to chocolate flavor development are extremely complex and not yet fully understood. Chocolate, like wine, starts with differing plant varieties and growing conditions and then requires elaborate processing to achieve the final product [33]. The flavor of cocoa mass is a result of both volatile and nonvolatile aromatic compounds, the most important of which are 2-methylpropanol, 2-methylbutanol, and 3-methylbutanol, but there are many hundreds of different chemicals that make a significant contribution [34]. Chocolate also contains methylxanthines, including theobromine and caffeine, which are responsible for some of the bitterness in the flavor [35]. People vary in their ability to taste different bitter compounds, and increased sensitivity is related to food aversions in children [36]. Most chocolate contains a considerable amount of sugar, with milk chocolate being approximately 45% sugar and dark chocolate going from 0% to 45% sugar, depending on brand and type. Therefore, the two basic tastants – among a wide variety of other flavors – that dominate chocolate are sweet and bitter, which will be discussed further.

Sweet Taste

At birth, human infants exhibit an innate dislike of sour and bitter tastes and a preference for sweet tastes, with a preference for salty tastes developing after a few months of age [37]. Historically, sweetness was generally associated with energy from sugars, in the case of breast milk, lactose, and galactose. Infants and children have a high need for energy because of their rapid growth and development. It has repeatedly been shown that children have a preference for higher concentrations of mono- and disaccharides than do adults [38]. Recently, it has been found that children who are in a stage of rapid growth have a higher preference for sweetness than their peers who have passed this stage of rapid growth [39]. Furthermore, it has been found that 8–9-year-old boys still have significant maturation of their taste systems to undergo and have measurably poorer ability to detect sweet, salty, sour, and bitter tastes than either girls their own age or adults, which may also help explain the preference

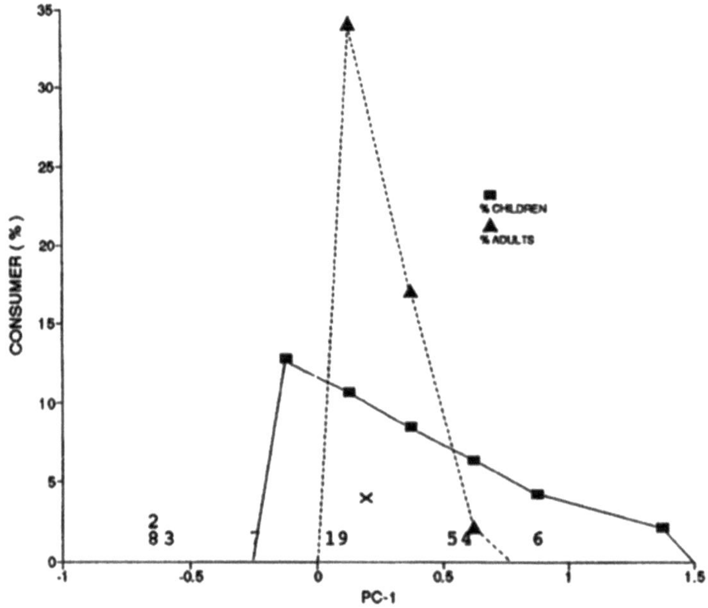

Fig. 37.2 Distribution of consumer's ideal points along the aroma/flavor first principal component using a one-dimensional ideal point model. Numbers represent chocolate milk samples. X represents the position of the average consumer (Reprinted from Food Quality and Preference, 9/4, Guillermo Hough, Ricardo Sánchez, Descriptive analysis and external preference mapping of powdered chocolate milk, 197–204, Copyright 1998, with permission from Elsevier)

among some children for more extreme flavors. Girls' gustatory systems, however, are fully mature at this age [40]. This would suggest that it is wise to distinguish children by sex when doing taste tests or food marketing, as the boys may be expected to like different, perhaps stronger, flavors than the girls, leading to the possibility that boys may prefer darker (higher cocoa mass) chocolate more than girls. This may be a fruitful area for further research.

Bitter Taste

Children's dislike for bitter tastes has been well reported on, but individual variations do exist. Anlinker measured children's sensitivity to a bitter compound, 6-n-propylthiouracil (PROP), which some people perceive as extremely bitter and others are not able to taste at all. Children who showed high sensitivity for PROP were found to be slower to eat cheese and quicker to drink milk than those who were less able to taste it, implying that they liked mild-tasting milk more than stronger-tasting cheese in contradistinction to the nontasting children who ate the cheese before they drank the milk [35]. In other studies, PROP-sensitive children were found to dislike the taste of spinach [41] and to consume smaller quantities of bitter-tasting vegetables (black olives, cucumber, and broccoli) than nontaster children [42]. Nonetheless, children's liking for bitterness seems to increase with age. Whereas children in general dislike bitter notes in foods [43], adults learn to enjoy some bitter foods such as coffee and alcohol [44]. This is likely to be related to repeated exposure and the positive post-ingestive consequences of these foods [45]. Unlike other taste receptors, which respond to a small range of stimuli, bitter receptors must respond to a vastly wider range of potentially toxic chemicals and must be able to do so at much lower concentrations in order to avoid poisoning [46]. Although most poisons are bitter tasting, in the modern food supply, most bitter-tasting substances are not toxic in the concentrations found in food. Therefore, the ability to learn to like bitter tastes in association with ingested kilojoules increases the number of different foods available, helping to forestall famine and improve micronutrient intake [47]. It has been suggested that the combination of fat and sugar makes foods, including chocolate, highly palatable [48]. The high sweetness of chocolate may decrease the perceived bitterness, which would be favorably perceived by children who prefer very sweet foods but generally show an aversion to bitter-tasting foods.

Very little is known about children's preferences regarding chocolate flavoring, but they may not necessarily be the same as adults'. In a study of preferences for powdered chocolate milk formulas, children's (11–12 years) average preferences for appearance, flavor, and texture of the various chocolate milk recipes were slightly different from adults' average preferences and considerably more variable. Figure 37.2 shows the distribution of adult and child consumers' ideal points along the aroma/flavor axis where principal component one (PC1) relates almost exclusively to chocolate flavor, with sweet/milky descriptors to the left and chocolate flavor on the right. It can be clearly seen that children's preferences for cocoa flavor are much more spread out than adults', with many children preferring a sweeter, less chocolaty recipe as well as a significant number preferring a stronger more bitter cocoa flavor. Children who preferred a strong chocolate flavor also preferred darker-colored chocolate milk, which may imply that children preferring a more intense chocolate flavor are aware that it can be assessed by color as well by taste.

A formula reflecting average preferences would satisfy most adults, while many children would prefer either considerably more or less cocoa flavor [49].

Adults' preferences have likely been conditioned by exposure over their lives to converge on flavors similar to those promoted by the large chocolate manufacturers, whereas children's preferences have not had time to do this and perhaps tend to reflect their innate variability. It is also possible that incomplete maturation of the boys' gustatory systems causes them to prefer stronger flavors in some cases. This study did not distinguish between girls and boys, so it is not possible to assess whether boys preferred stronger more chocolaty flavor than girls. It is also possible that these results would be relevant to chocolate candy preferences and that manufacturers of children's snacks are neglecting a market for darker, more bitter chocolate flavors. As dark chocolate has by definition more cocoa mass and less sugar and fat than milk chocolate and thus larger amounts of beneficial flavonoids, it may be appropriate to encourage these children's preference for it.

Repeated Exposure: Role in Forming Preferences

Children's taste preferences in general change as a function of exposure. Birch, among others, has suggested that children need at least eight exposures to an unfamiliar food before it can become liked. It is likely that significantly fewer exposures are required when the novel food is high in sugar, fat, and salt [50]. The positive effect of exposure on liking is facilitated by two basic learning mechanisms, flavor-nutrient learning and flavor-flavor learning.

Flavor-Nutrient Learning

Children's food likes and dislikes are learned and modified as a result of the post-ingestive consequences of their food intake [51]. Animals learn to avoid foods that make them feel ill [52] and to prefer foods that make them feel satiated; thus, foods and flavors associated with high energy densities become preferred [53]. The same biological principles apply to children. For example, children who were offered two differently flavored drinks with associated differences in carbohydrate content learned to prefer the drink with the higher energy density [54]. Children can also acquire preferences for foods with high energy densities based on fat content. In an experiment using yogurts with identical carbohydrate and protein content but varying fat contents, the children learned to prefer the high-fat yogurt after drinking appreciable quantities of the food. A similar group that was only supplied with a small taste of the yogurts insufficient to create satiety did not prefer either flavor, indicating that it was the first group's bodies' recognition of energy-dense food in the gut that lead to the high-fat flavor becoming preferred [55]. Chocolate of the sort generally marketed to children contains 30% fat and up to 45% sugar, giving it a very high energy density (2,200 Kj/100 g) likely to engender a feeling of satiety and making it very likely that children will acquire a taste for it without difficulty [56].

Flavor-Flavor Learning

Flavor-flavor learning is a result of the ability to learn to enjoy a new flavor when it associated with a familiar liked flavor [57]. For example, Havermans and colleagues exposed children to broccoli. The first group was asked to eat the broccoli in combination with a nonnutritive sweetener; the other group was asked to eat it plain. After only a single exposure to the sweetened broccoli, children began to develop a liking for broccoli even without the added sweetener. Chocolate is a well-liked flavor among children; by using the flavor-flavor association principle, chocolate flavor could potentially be used to help children develop liking for new foods.

Flavor Aversion and Chocolate

In the same way that flavors become liked by being combined with positive post-ingestive consequences, flavors may become disliked if they are associated with negative post-ingestive consequences, specifically nausea. Animals that have experienced nausea shortly after eating a food are likely to develop an aversion to it, especially if it is unfamiliar to them. Foods with strong tastes and odors more easily become the target of acquired food aversions [58]. Chocolate, along with coffee, tea, citrus fruit, red meat, and ice cream, is one of the most commonly reported food aversions among adults and children undergoing chemotherapy [59]. Some of the foods were associated with the treatment and thus came to be disliked; others just seemed to taste unpleasant. Retrospective surveys have suggested that between 38% and 65% of healthy individuals have at least one conditioned food aversion [60]. Caretakers of sick children who may be anticipated to suffer from nausea and anorexia may be advised to keep foods such as chocolate out of their child's diet during treatment regimens, so as to maintain its favored status until such time as the child is well enough to eat properly and gain weight again.

Persistence of Preferences

It is generally assumed that preferences learned in childhood will persist into adulthood, but there is not much evidence in support of this belief, as longitudinal studies are rare. In the Minnesota Heart Health Program study, children's food preferences were tracked over 6 years, during which the children were presented with pairs of snack foods and asked to "choose which they would usually eat when given the choice." Choices were found to be relatively stable over that period of time, with those who initially made the greatest number of healthy choices still making the greatest number of healthy food choices 6 years later, and vice versa, suggesting that dietary habits do persist. Snack foods are by design highly palatable, and such foods apparently do not fall out of favor over the course of childhood; on the contrary, they become more popular. It was rare for the snack foods to become less liked [61]. This study did not include fruits or vegetables as snacks, which, being less palatable, may have shown a different trajectory. A more general study of the persistence of children's food preferences also found that the number of foods liked did not change significantly from ages 3 to 8, though the number disliked grew as the children were introduced to new vegetables.

While exposure is necessary for liking for most foods to develop, it is also the case that overexposure can cause boredom and ennui. In a study of snack eating, those eating a chocolate bar everyday reported a decrease in pleasantness, their preference for and frequency of eating chocolate [62]. Thus, it might be expected that children's preferences for specific chocolate snack bars would decline between ages 5 and 11, if eaten very frequently. That this does not occur suggests that intake is sufficiently infrequent for the formation of monotony effects. Although the high level of new product launches in the "kids," chocolate lines would indicate that manufacturers do not wish to rely on this being the case.

Parental Influences

Parents' and children's food and snack intakes have shown significant correlation along with eating motivations and body dissatisfaction, leading to the suggestion that children learn these behaviors and attitudes from their parents, usually their mothers [10, 63].

Unfortunately, since parents generally provide both a child's genes and much of its environment in its early years, it is difficult to disentangle the one from the other. Studies of twins have found a greater similarity in food preferences in monozygotic as opposed to dizygotic twins, indicating an important though modest role for genetics in individual food preferences [64, 65].

Parenting Styles

As chocolate is a highly energy-dense food and children's liking of chocolate is well established, it can be assumed that children will eat more of it when it is available than is consonant with a healthy diet. Although in laboratory studies it has been suggested that children are reasonably able to control their energy intake [66], in real life, they are distracted by portion size and energy density. Studies found that children eat up to 30% more when larger portion sizes of energy-dense food are provided than when they are presented with smaller portions of lower energy-dense foods [67]. Thus, many parents feel called upon to control their children's intake of high-energy foods such as chocolate and other snacks by limiting their children's access to these foods and by restricting their intake in their presence [68].

It is, however, questionable whether severe restriction has the desired effects. Restricting children's access to candy has also been shown to correlate with a higher level of preferred sweetness in an orangeade beverage. While only 33% of the less restricted children preferred the highest sugar concentration, 55% of the most restricted children did [69]. In another study, a group of 5–6-year-olds who were randomly assigned to a snack prohibition or an ad libitum group were presented with M&Ms (chocolates) and crisps colored red and yellow. The prohibition group was instructed not to eat the red foods. They were then assessed for their desire to eat each of the groups of foods and watched to see which of the groups they ate more of after the prohibition was lifted. Prohibition was found to increase desire after only 5 min, so forbidding the eating of a food immediately makes it more desirable, even when an identically palatable food is offered in its place. It was also found that children, whose parents imposed either very high or very low levels of food restriction at home, ate more kilocalories during the experiment than children with a moderate level of food restriction at home [70].

These experiments suggest that restrictions on children's snack intake can be counterproductive, serving to increase their desire for and intake of the restricted foods when surveillance is lifted.

Parents whose children are heavy and who are concerned about their children's weight are more likely to attempt to restrict their children's diets, and there is some evidence that parents who are more concerned about their own weight may also be more inclined to restrict their children's food choices. Fisher and Birch reported in 1999 that mothers of overweight children imposed more restrictions on their children's snack intake, which then led to greater consumption among the girls when unsupervised, though not the boys [71]. There is evidence to show that parental restriction of children's food choices is associated with higher BMIs and overeating in general, but while it is probable that restricting access to certain highly palatable foods makes them more liked and desired by children, it is also possible that parents of overweight or overeating children feel a greater need to monitor their food intake than others [72].

It would appear that there is a cycle whereby children with inherently poor energy self-regulation resulting in heaviness are then restricted in their food intake (especially snack food) by their parents [73]. This then leads to an even greater preference for high energy density snack foods, and greater intake once parental vigilance is relaxed, as it inevitably must be when children grow up. Thus, it may be recommended that even overweight children should be moderately restrained in their snack food intake rather than being forced to endure extreme and counterproductive restrictions.

A US study of 12 focus groups of mothers of varying SES and ethnic background found all 12 groups used sweets as bribes, rewards, and pacifiers [74], which have also been shown to increase children's preference for snack foods [75].

Summary

Chocolate plays an important role in many child-oriented celebrations across the globe. The many foods of which chocolate is a part, such as confectionery, drinks, bakery products, and cereals, along with the lack of specific consumption data, make it difficult to obtain a good estimate of how much chocolate children consume and what percentage of daily energy intake can be attributed to chocolate consumption. For most western children, chocolate forms a small but well liked and probably growing part of the diet. It is also popular among children in other countries where it is a more recent addition to the diet. Children tend to eat more chocolate if they are from Northern Europe, if they are of a lower SES, if they are boys, if they are older, if they have more autonomy, and if they have more pocket money. However, it needs to be taken into account that very limited data are currently available in the public domain.

In general, children prefer high concentrations of sugar in a variety of foods and reject bitter flavors. Children can quite easily learn to appreciate flavors that are associated with high energy. Given the high sweetness and energy profile of chocolate, it is not surprising that this commodity is well liked by children. The high acceptance of chocolate and its powerful flavor make it potentially a useful tool for masking unpleasant flavors in medical applications.

Parents play an important role in children's taste development, food choice, and food consumption. Sugar-rich foods, including chocolate, are often restricted by parents in order to limit children's intake. The effectiveness of such a strategy has, however, been questioned.

Chocolate and children are, and will mostly likely always be, strongly associated with each other. The pleasure chocolate gives to children is important; however, as with all high-energy foods, moderation is the key to lifelong enjoyment.

Acknowledgments We wish to thank Professor Tony Worsley for his comments on previous drafts of the chapter and Janet's son Andrew for his generous technical assistance.

References

1. Borell J. Who dictates the Australian Easter market? Manufacturing Confectioner. 1997;3:38–40.
2. Dutch chocolate letters. http://www.stnicholascenter.org/pages/chocolate-letters/. Accessed 26 July 2011.
3. Skinner JD, Ziegler P, Pac S, Devany B. Meal and snack patterns of infants and toddlers. Am Diet Assoc. 2004;104(Suppl 1):S65–70.
4. Jahns L, Siega-Riz M, Popkin BM. The increasing prevalence of snacking among US children from 1977 to 1996. J Pediatr. 2001;138:493–8.
5. Rollins BY, Loken E, Birch LL. Stability and change in snack food likes and dislikes from 5 to 11 years. Appetite. 2010;55:371–3.
6. Ward JL, Hoelscher M, Briley ME. Food choices of third grade children in Texas. J Am Diet Assoc. 2002;102:409–12.
7. Seligson FH, Krummel DA, Apgar JL. Patterns of chocolate consumption. Am J Clin Nutr. 1994;60(Suppl): 1060s–4.
8. Bellisle F, Rolland-Cachera MF, Barthelmy L, Boucher B, Dartois AM, et al. Three consecutive (1993, 1995, 1997) surveys of food intake, nutritional attitudes and knowledge, and lifestyle in 1,000 French children, aged 9–11 years. J Hum Nutr Diet. 2000;13:101–11.
9. Johnson B, Hackett AF. Eating habits of 11–14-year-old schoolchildren living in less affluent areas of Liverpool, UK. J Hum Nutr Diet. 1997;10:135–44.
10. Brown R, Ogden J. Children's eating attitudes and behaviour: a study of the modelling and control theories of parental influence. Health Educ Res. 2004;19:261–71.
11. Kortzinger I, Neale RJ, Tilson CH. Children's snack food consumption patterns in Germany and England. Br Food J. 1994;96:10–5.
12. Ruxton CH, Kirk TR, Belton NR, Holmes MA. Relationships between social class, nutrient intake and dietary patterns in Edinburgh schoolchildren. Int J Food Sci Nutr. 1996;47:341–9.
13. European Heart Network. The marketing of unhealthy food to children in Europe. http://www.ehnheart.org. Accessed 29 July 2011.
14. Blain L. UK children's pocket money allowances rising 6 times faster than inflation. Gizmag. http://www.gizmag.com/go/7666/. Accessed 26 Jan 2011.
15. Wang Y, Li J, Caballero B. Resemblance in dietary intakes between urban low-income African-American adolescents and their mothers: the healthy eating and active lifestyles from school to home for kids study. J Am Diet Assoc. 2009;109:52–63.
16. Marshall D, O'Donhoe KS. Families, food and pester power: beyond the blame game? J Consum Behav. 2007;6:164–81.
17. Fitzgerald A, Heary C, Nixon E, Kelly C. Factors influencing the food choices of Irish children and adolescents: a qualitative investigation. Health Promot Int. 2010;25:289–98.
18. Zhang F. http://ezinearticles.com/?Chinas-Chocolate-Market-Dominated-by-Foreign-Brands&id=1013918. Accessed 27 Jan 2011.
19. Cheng C. Little emperors make big consumers. China Today. 1993;42:47–9.
20. Redruello F. Middle East and Africa confectionary: demand for chocolate picks up in 2010. http://blog.euromonitor.com/2010/10/middle-east-and-africa-confectionery-demand-for-chocolate-picks-up-in-2010.html. Accessed 27 Feb 2011.
21. News and Analysis. US Chocolate sales up, volume down. Candy Industry. 2010; Sept. http://www.candyindustry.com/Articles/General_News/BNP_GUID_9-5-2006_A_10000000000000984799. Accessed 31 July 2011.
22. Ross S. 'Do I really have to eat that?': a qualitative study of children's food choices and preferences. Health Educ J. 1995;54:312–21.
23. Cumming D, Overduin J. Gastrointestinal regulation of food intake. J Clin Invest. 2007;1:13–23.
24. Mattes RD. Fat taste and lipid metabolism in humans. Physiol Behav. 2005;86:691–7.
25. Druce M, Bloom SR. The regulation of appetite. Arch Dis Child. 2005;91:183–7.
26. Fox MK, Devaney B, Reidy C, Razafindrakoto C, Ziegler P. Relationship between portion size and energy intake among infants and toddlers: evidence of self-regulation. J Am Diet Assoc. 2006;106:S77–83.
27. Swinburn B. Obesity prevention in children and adolescents. Child Adolesc Psychiatr Clin N Am. 2009;1:209–23.
28. Birch LL, Fisher JO. Development of eating behaviors among children and adolescents. Pediatrics. 1998;3(part 2):539–49.

29. Cecil JE, Palmer CNA, Wrieden W, Murrie I, Bolton-Smith C, Watt P, et al. Energy intakes of children after preloads: adjustment, not compensation. Am J Clin Nutr. 2005;82:302–8.
30. Skokan EG, Junkins EP, Corneli HM, Schunk JE. Children rate flavoring agents used with activated charcoal. Arch Pediatr Adolesc Med. 2001;155:683–6.
31. Dagone D, Matsui D, Rieder MJ. Assessment of the palatability of vehicles for activated charcoal in pediatric volunteers. Pediatr Emerg Care. 2002;1:19–21.
32. Mennella JA, Beauchamp GK. Optimizing oral medications for children. Clin Ther. 2008;11:2120–32.
33. Afoakwa EO, Paterson A, Fowler M, Ryan A. Flavor formation and character in cocoa and chocolate: A critical review. Crit Rev Food Sci Nutr. 2008;48(9):840–57.
34. Counet C, Calleman D, Ouwerx C, Collin S. Use of gas chromatography-olfactometry to identify key odorant compounds in dark chocolate. Comparison of samples before and after conching. J Agric Food Chem. 2002;50:2385–91.
35. Schenker S. The nutritional and physiological properties of chocolate. Nutr Bull. 2000;25:303–13.
36. Anliker JA, Bartoshuk L, Ferris AM, Hooks LD. Children's food preferences and genetic sensitivity to the bitter taste of 6-n-propylthiouracil (PROP). Am J Clin Nutr. 1991;54:316–20.
37. Birch LL. Development of food preferences. Annu Rev Nutr. 1999;19:41–62.
38. De Graaf C, Zandstra EH. Sweetness intensity and pleasantness in children, adolescents, and adults. Physiol Behav. 1999;67:513–20.
39. Coldwell SE, Oswald TK, Reed DR. A marker of growth differs between adolescents with high vs. low sugar preference. Physiol Behav. 2009;96:574–80.
40. James EC, Laing DG, Oram N. A comparison of the ability of 8–9-year-old children and adults to detect taste stimuli. Physiol Behav. 1997;62:193–7.
41. Turnbull B, Matisoo-Smith E. Taste sensitivity to 6-n-propylthiouracil predicts acceptance of bitter-tasting spinach in 3-6-y-old children. Am J Clin Nutr. 2002;5:1101–5.
42. Bell KI, Tepper BJ. Short term vegetable intake by young children classified by 6-n-propylthiouracil bitter-taste phenotype. Am J Clin Nutr. 2006;1:245–51.
43. Mennella JA, Pepino MY, Reed DR. Genetic and environmental determinants of bitter perception and sweet preferences. Peds. 2005;115:e216–22.
44. Hayes JE, Wallace MR, Knopik VS, Herbstman DM, BartoshukLM DVB. Allelic variation in TAS2R bitter receptor genes associates with variation in sensations from and ingestive behaviors toward common bitter beverage in adults. Chem Senses. 2011;36:311–9.
45. Kern DL, McPhee L, Fisher J, Johnson S, Birch LL. The postingestive consequences of fat condition preference for flavors associated with high dietary fat. Physiol Behav. 1993;1:71–6.
46. Frank ME, Bouverat BP, MacKinnon BI, Hettinger TP. The distinctiveness of ionic and non-ionic bitter stimuli. Physiol Behav. 2004;80:421–31.
47. Hladik CM, Pasquet P, Simmen B. New perspectives on taste and primate evolution: the dichotomy in gustatory coding for perception of beneficent versus noxious substances as supported by correlations among human thresholds. Am J Phys Anthropol. 2002;117:342–8.
48. Benton D. Carbohydrate ingestion, blood glucose and mood. Neurosci Biobehav Rev. 2002;26:293–308.
49. Hough G, Sanchez R. Descriptive analysis and external preference mapping of powdered chocolate milk. Food Qual Pref. 1998;9:197–204.
50. Liem DG, de Graaf C. Sweet and sour preferences in young children and adults: role of repeated exposure. Physiol Behav. 2004;3:421–9.
51. Johnson SL, McPhee L, Birch LL. Conditioned preferences: young children prefer flavours associated with high dietary fat. Physiol Behav. 1991;50:1245–51.
52. Garcia J, Kimldorf DJ, Koelling RA. Conditioned aversion to saccharin resulting from exposure to gamma radiation. Science. 1955;122(3160):157–8.
53. Capaldi ED, Campbell DH, Sheffer JD, Bradford JP. Conditioned flavor preferences based on delayed caloric consequences. J Exp Psychol Anim Behav Process. 1987;2:150–5.
54. Booth DA, Mather P, Fuller J. Starch content of ordinary foods associatively conditions human appetite and satiation indexed by intake and eating pleasantness of starch paired flavours. Appetite. 1982;3:163–84.
55. Kern DL, McPhee L, Fisher J, Johnson S, Birch LL. The postingestive consequences of fat condition preferences for flavors associated with high dietary fat. Physiol Behav. 1993;54:71–6.
56. Jones S, Phillipson L, McVie D, Reid A, Lynch M, et al. Food marketing to children in Australia. Centre for Health Initiatives. http://www.cancer.org.au/File/PolicyPublications/FoodMarketingtoChildreninAustralia.pdf. Accessed 29 Jan 2011.
57. Logan FA. Hybrid theory of classical conditioning. In: Bower GH, Spence KW, editors. The psychology of learning and motivation: advances in research and theory. London: Academic; 1977.

58. Yamamoto T, Shimura T, Sako N, Yasoshima Y, Sakai N. Neural substrates for conditioned taste aversion in the rat. Behav Brain Res. 1994;2:123–37.
59. Skolin I, Wahiln YB, Hursti UK, Larsson MV, Hernell O. Altered food intake and taste perception in children with cancer after start of chemotherapy: perspective of children, parents and nurses. Support Care Cancer. 2006;14:369–78.
60. Bernstein IL. Taste aversion learning: a contemporary perspective. Nutrition. 1999;15:229–34.
61. Skinner JD, Carruth BR, Bounds W, Ziegler PJ. Children's food preferences: a longitudinal analysis. J Am Diet Assoc. 2002;102:1638–47.
62. Hetherington MM, Bell A. Effects of repeat consumption on pleasantness, preference and intake. Br Food J. 2000;102:507–21.
63. Vereecken CA, Keulelier E, Maes L. Influence of mother's educational level on food parenting practices and food habits of young children. Appetite. 2004;43:93–103.
64. Breen FM, Plomin R, Wardle J. Heritability of food preferences in young children. Physiol Behav. 2006;88:443–7.
65. Falciglia GA, Norton PA. Evidence for a genetic influence on preference for some foods. J Am Diet Assoc. 1994;94:154–8.
66. Birch LL, Johnson SL, Andresen G, Peters JC, Schulte MC. The variability of young children's energy intake. N Engl J Med. 1991;324:232–5.
67. Mrdjenovic G, Levitsky DA. Children eat what they are served: the imprecise regulation of energy intake. Appetite. 2005;44:273–82.
68. Johnson SL, Birch LL. Parents' and children's adiposity and eating style. Pediatrics. 1994;94:653–61.
69. Liem DG, Mars M, De Graaf C. Sweet preferences and sugar consumption of 4- and 5-year-old children: role of parents. Appetite. 2004;43:235–45.
70. Jansen E, Mulkens S, Jansen A. Do not eat the red food!: prohibition of snacks leads to their relatively higher consumption in children. Appetite. 2007;49:572–7.
71. Fisher JO, Birch LL. Restricting access to foods and children's eating. Appetite. 1999;32:405–19.
72. Browne M. A comparative study of parental behaviors and children's eating habits. Infant Child Adolesc Nutr. 2011;1:11–4.
73. Powers SW, Chamberlin LA, van Shaick KB. Maternal feeding strategies, child eating behaviors, and child BMI in low-income African-American preschoolers. Obesity. 2006;14:2026–33.
74. Sherry B, McDivitt J, Birch LL, Cook FH, Sanders S, Prish JL, et al. Attitudes, practices, and concerns about child feeding and child weight management status among socioeconomically diverse White, Hispanic and African-American mothers. J Am Diet Assoc. 2004;104:215–21.
75. Birch LL, Zimmerman SI, Hind H. The influence of social-affective context on the formation of children's food preferences. Child Dev. 1980;51:856–61.

Chapter 38
Chocolate Use and Knowledge in Oral and Dental Health in Children

Emine Efe and Selma Öncel

Key Points

- Dental health and dental caries are important worldwide health issues that are common in childhood.
- Sugars are the most important dietary etiological cause of dental caries.
- World Health Organization and Food and Agriculture Organization of the United Nations emphasize the importance of reducing the consumption, frequency, and amount of carbohydrates that could be easily fermented for the prevention of dental caries, which is considered among the chronic degenerative diseases that could be prevented by improving individual lifestyle.
- Nutrition education for the purposes of reducing caries incidence in children is aimed at informing parents about the importance of reducing dietary exposures to sweet foods and hidden sugars.
- Implementation of community-based preventive oral health programs focusing on healthy diet and adequate oral hygiene practices should be promoted in schools through integration into the school curriculum and services to combat the growing problem of dental caries among schoolchildren.

Keywords Sugar • Chocolate • Oral health • Childhood caries • Education

Dental Health

Dental health and dental caries are important worldwide health problems that are common in childhood [1]. Dental caries and related oral diseases like gingivitis and periodontitis are the most common oral diseases throughout the world. The prevalence of these diseases is continuously increasing in parallel with people's changes in dietary habits and high consumption of sugar [2].

E. Efe, Ph.D. (✉)
Child Health Nursing Department, Akdeniz University, School of Health, Antalya, Turkey
e-mail: eefe@akdeniz.edu.tr

S. Öncel, Ph.D.
Public Health Nursing Department, Akdeniz University, School of Health, Antalya, Turkey
e-mail: oselma@akdeniz.edu.tr

R.R. Watson et al. (eds.), *Chocolate in Health and Nutrition*, Nutrition and Health 7,
DOI 10.1007/978-1-61779-803-0_38, © Springer Science+Business Media, LLC 2013

In contrast to improvements seen in many parts of the world, recent epidemiological surveys suggest that caries experienced by young children in Middle Eastern countries has remained rather high considering the FDI/WHO goals for the year 2000 (50% of the children would be free of caries for the primary dentition at the age of 5–6 years) [3].

In children under 6 years of age, early childhood caries (ECC) is a condition characterized by the presence of one or more decayed, missing, or filled primary teeth [4].

Prevalence

Dental caries is known as a chronic and multifactorial disease. The prevalence and severity of this disease historically have been very high, particularly among preschool- and school-age children, and have been considered as one of the most important global oral health burdens [5].

Dental caries is accepted as a disease related to diet regime, and considering its prevalence and severity, there is a reduction in the developed countries, while an increase is noted in the developing countries due to the increasing sugar consumption and inadequate fluorine intake [6].

Although there have been significant developments in oral and dental health in recent years, 19% of children between the ages of 2 and 5 years in the United States have dental caries, and this rate increases up to 52% in children between 5 and 9 years of age [7]. A prevalence study in Brazil reported the incidence of caries to be 43.8% for the same age group [8]. However, 83% of 6-year-old children in Turkey have caries in their milk teeth and/or permanent teeth. In the 6–8-year-old group, there is an average of 4.5–5 cavities in milk teeth and the prevalence of caries reaches up to 80% [9]. About 90% of school children throughout the world and most adults have experienced caries, and it is most prevalent in Asian and Latin American countries [5]. In Malaysia, the annual impact indicators, percentage of DMFT (decay, missing, filled, extraction) among 12-year-old children, had improved from 47.4% with three or fewer decayed, missing, or filled teeth in 1990 to 91.2% in 2000. In Sarawak, the mean DMFT score for 12-year-old decreased over the period of 10 years from 2.5 in 1997 [10] to 1.4 in 2007 [11]. Although this decreasing trend indicates an improvement in oral health status, environmental risk factors such as diet, nutrition, oral hygiene, tobacco, and alcohol [5] can play an important role in the prevention of dental caries.

A linear correlation between total sugar consumption and caries prevalence has been well documented, and increased sugar consumption has long been considered as the most important risk factor for the disease [5].

Reasons for Dental Caries

Dental caries is a multifactorial disease influenced by many factors, including age, sex, dietary habits, mother's education, oral hygiene level, oral health behavior, microorganisms, trace elements, saliva, genetic predisposition, and tooth morphology [12–14].

Early childhood caries has a complex etiology. It is affected by the amount of cariogenic bacteria, the presence of cariogenic foods, and the susceptibility of host tooth [15, 16]. Other environmental and behavioral factors such as oral hygiene habits, socioeconomic status, enamel hypoplasia, fluoride exposure, previous caries experience, access to dental care, and other dietary factors also affect caries development [15–17]. Early childhood caries may lead to pain, improper speech development, reduced ability to chew, low weight for age, low self-esteem, and sleep disturbances; however, there is only weak or moderate evidence for many of these associations [18, 19]. It is likely that the arctic nutrition transition has been exacerbated by other oral health risk factors common in these regions, including the practice of giving bottles to children at bedtime, lack of fluoridated water, inadequate access to dental health care, and lack of education about preventative measures [20–22].

In the recent studies, the rate of dental caries was reported higher in children of families with low socioeconomic level [23–26].

Extensive dental decay in preschool children or early childhood caries (ECC) is a troubling child health problem. Parents' beliefs and attitudes toward dental health influence the way they practice oral health for their children [27]. Therefore, parental belief systems and practices are important factors that mediate the influences of culture on children's oral health. Despite this gap in the knowledge, substantial literature has emerged related to the prevention and management of ECC and caries relapse [28].

Relationship Between Dental Caries and Diet

Sugars are the most important dietary etiological cause of dental caries [29–32]. The authors concluded that the relationship between sugar consumption and caries is much weaker in the modern age of fluoride exposure than it used to be. However, controlling the consumption of sugar remains a justifiable part of caries prevention, if not always the most important aspect. The role of sugar consumption in the development of caries has been established for decades. Based on a large body of worldwide epidemiological studies, sugar – particularly sucrose – is considered as one of the most important etiological factors for dental caries. It has almost become a universal rule to include the control of sugar consumption in all caries management and prevention programs and guidelines across nations, age, as well as cultural and socioeconomic groups. However, an exact relationship still remains unclear between sugar consumption and dental caries [33].

The relationships between nutrition and health were approved by the Food and Drug Administration (FDA), which gives approval to drugs to be used by humans in America. One of them was the relation between sugar alcohols in foods and dental caries [34, 35].

Sugars given in sticky form between meals have shown to significantly increase the activity of caries [36]. Most dietary sugars are easily metabolized into acid by organisms, while some forms of dietary sugar can be slowly broken down in the mouth by salivary amylase. These sugars can then be metabolized by plaque organisms, such as *Streptococcus mutans,* which facilitate the establishment of dental caries [37]. This might be due to the increasing availability and open sale in school canteens and streets of high-sugar-containing snacks such as chips, cheese curls and other air-filled "sitsiryas," candies, carbonated/synthetic juice drinks/beverages, and artificially colored water-/sugar-based ice candies. These are further explained by the fact that children stay for about 10 h in the school, and with their small budgets, they could only afford to buy hard candies or soft drinks for their snacks. These factors might have contributed to the high consumption of these foods and the resulting high prevalence of dental carries among the students. The level of sugar consumption by children in the Yabao et al. [38] study (21 kg/year) was higher than that reported for Sri Lanka (18.10), Thailand (15.34), Pakistan (16.02), Myanmar (1.44), Nigeria (10.15), and Senegal (14.26), but lower than that reported for Japan (23.32), Malaysia (35.10), Singapore (49.56), Saudi Arabia (38.66), Syria (40.18), the United States (32.14), the United Kingdom (43.95), Switzerland (44.98), the Netherlands (53.54), and Sweden (45.6) [39].

In a study investigating the relationship between nutrition and dental caries, the importance of frequency of sugar intake was emphasized and it was concluded that consumption of sugary foods more than four times a day in snacks increased the risk of dental caries [29].

Numerous studies have identified the popularity of certain snack foods among children. Confectionery, chocolate, and sweets are clearly the favorite choices of most children, and crisps, biscuits, and fruit are also commonly preferred [40, 41]. Since most researchers [42] agree that limiting sugar intake is still the most important factor in the prevention of caries, studies identifying the role of snacks on overall sugar consumption are important. The role of soft drinks on dental erosion is already well established [19]. It has been reported that people of lower socioeconomic status have worse oral health. In particular, children from lower social groups, families receiving benefits, or from ethnic minority groups are more likely to be at risk of severe caries. Studies have shown that socioeconomic status affects dietary intake [43, 44]; hence, it is likely that differences in diet, particularly sugar intakes, may be a significant determinant of the variation in caries observed between social classes [45]. Maliderou et al. [46] found that there was no significant difference between boys and girls considering the amount of sugar consumption, although girls did consume slightly more on average (145 ± 32 g/day compared to the boys 134 ± 26 g/day).

According to the results of the study implemented by Bozkurt et al. [47], periodontal health of adolescents in rural areas is better than those in urban areas owing to high consumption rates of chocolate, sugar, and sugary foods in urban areas.

In socioeconomically developing countries, transition from a traditional lifestyle to a Western lifestyle has, among other things, led to an increase in sugar consumption from food and beverages and in the form of chocolate and candies [48].

Female students consumed more sugar than male students through sweet food, sweet snacks, and soft drinks. Studies have shown that girls usually have greater preference for sweet foods, whereas boys opt for higher fat and salt content fast foods and snacks [49]. Lian et al. [50] found that female students have a higher daily consumption of sweet food compared to male students. A similar pattern was found for the frequency of taking candy/chocolate/sweets and the consumption of soft drinks.

Dental caries is a major dental health problem among schoolchildren. As in other developing countries, this may be due to an increased availability of refined sugar products without a concurrent rise in oral health awareness. Implementation of community-based preventive oral health programs on healthy diet and practices of adequate oral hygiene should be promoted in schools through integration into the school curriculum and services to combat the growing problem of dental caries among schoolchildren. The high level of untreated caries in all age groups is a cause for concern.

Abdullah et al. [51] found that 45.6% of the subjects disliked chocolates, whereas 34% of respondents desired to taste chocolates only sometimes.

The World Health Organization's recommendation for sugar consumption designates 27.40–30 g of sugar/day or 10 kg of sugar [38]. Yabao et al. [38] found that students' sugar intake was twice the WHO recommendation, with a mean daily total intake of 59 g per person. Most common sources of dietary sugar were hard candies (89%) and chocolate (78.4%).

In the study of Şanlıer and Özgen [52], it was reported that nearly half of the students (46.7%) consumed a considerable amount of chocolate, coke, sugary gum, and so forth. In the study of Özgen [53], consumption of sugary foods and carbonated beverages was found to be quite high in students attending college and university.

The National Diet and Nutrition Survey [54] reported that for young people aged 4–18 years, the intake of non-milk intrinsic sugar was 85 ± 38.7 and 69 ± 29.0 g/day in boys and girls, respectively. Other studies [45] failed to demonstrate clear social class differences in terms of the consumption of sugar-containing foods. However, Inchley et al. [55] investigated dietary trends among Scottish schoolchildren and concluded that the consumption of high-sugar foods was higher in children from lower socioeconomic groups.

In the case of dental caries, there is a consensus about the relationship between the frequency of intake of non-milk extrinsic sugars and incidence of tooth decay [56, 57]. Evidence suggests that commercialized sugar products have become easily available and are highly preferred and frequently consumed, particularly among the higher socioeconomic status groups and urban residents [58]. Eating habits during childhood tend to persist into adulthood [59, 60]. The policy of promoting restricted sugar consumption requires information about the patterns of sugared snack intake at an early age and its effective factors. Dietary studies on children have additional problems because they must address children's cognitive ability to record and remember intake as well as their limited knowledge of food and food preparation [61].

Children are a group urgently requiring diet-related dental health promotion and intervention since food habits derived in childhood often remain into adulthood [62].

Television advertising is, however, only one way in which manufacturers are able to influence children's behavior in terms of dietary habits and preferences. Children are now being targeted using "viral marketing" and "underground communication."

Pacey et al. [63] found that 93.2% consumed chocolate or candy, and mean intake frequency was 1.1 times per day. When asked what their child's preferred brand of breakfast cereal was, 72.6% reported a high-sugar brand.

Freeman et al. [64] found that 27% of the entire population of adolescents stated that they consume chocolate and confectionery several times a day. On the other hand, 25% stated that they drink carbonated drinks with sugar, and a slightly lower proportion of adolescents admitted to consuming cakes, hot drinks with sugar, and fruit squashes several times a day.

Formation of Dental Caries

In the formation of dental caries, there is a dynamic balance between pathological and protecting factors. It was determined that organic acids resulting from the anaerobic metabolism of carbohydrates that could be fermented, taken in diet, and considered among important pathological factors within the context of this balance could pose a local risk factor for dental caries by demineralizing enamel and dentine [65, 66].

Dental caries is initiated via the demineralization of tooth hard tissue by organic acids resulting from the fermentation of dietary sugar by dental plaque odontopathogenic bacteria [67]. In the recent Australian oral health survey, 61% of 12-year-olds showed signs of caries with a mean decayed, missing, and filled tooth index (DMFT) of 1.8 [68].

Kara et al. [69] carried out a study on students of dentistry faculty and reported that sugary and chocolate foods are the most demanded nutrition (76.6%) in snacks. This situation poses a risk for the formation of dental caries.

The previous studies reported that people prefer sugar-rich foods when they skip meals, and this situation was determined to increase the formation of dental caries [70–72].

Properties of foods consumed in the daily diet and individual characteristics like daily consumption frequency, consumption type, and sugar-clearing speed of mouth are considered factors that affect the incidence rate of caries [73].

In the studies performed on humans and animals, physical form of sugar, consumption frequency, consumption amount, and stickiness of sugar were found to be related to dental caries, while cariogenicity of sugary food was reported to cause excessive acid formation due to the prolonged duration in mouth and decreased plate pH [19].

Öncel et al. [26] stated that sugar, chocolate, sweet, hot/cold foods, acid foods, fruits/vegetables, and meat could be harmful to dental health of children [26]. Efe et al. [74] found that more than 60% of the children spent their allowance on chocolate, chips, cola, candy, and other acidic drinks.

Prevention of Dental Caries

Oral health is fundamental and an essential part of general health and well-being [75]. World Health Organization and Food and Agriculture Organization of the United Nations emphasize the importance of reducing the consumption frequency and amount of carbohydrates that could be easily fermented for the prevention of dental caries, which are considered among the chronic degenerative diseases that could be prevented by improving an individual's lifestyle [76].

The overall health, education, and development of children, families, and communities could be enhanced through oral health [77]. Treating the tooth with fluoride, decreasing the carbohydrate source of bacteria, reducing the levels of cariogenic bacteria, and lessening the ability of bacteria to produce acid are among the preventive approaches that can remineralize the initial carious lesion. It is necessary to reduce sugar intake in snacks and maintain good oral health for the prevention of dental caries [78, 79]. The most important habit of mouth hygiene is to brush teeth with fluoride-containing toothpaste twice a day [78, 79].

A number of studies have demonstrated that individuals with high levels of education and income are more likely to engage in preventive oral hygiene behaviors and have better eating habits [80, 81]. Compared to males, females are generally less likely to pay more attention to their diet and engage in more preventive oral health behavior [82].

Evidence showed that strong knowledge of oral health demonstrates better oral care practice [83]. Similarly, people with a more positive attitude toward oral health are influenced by greater knowledge in taking care of their teeth. Studies showed that appropriate oral health education can help to cultivate healthy oral health practices [84]. Transition to healthy attitude and practice can be triggered by adequate information, motivation, and practice of the measures to the subjects [83].

A high-carbohydrate sweet diet has clearly been implicated in the etiology and epidemiology of dental caries [85]. When efforts are invested in oral health education, it is a rule of thumb that emphasis should include dietary and oral hygiene improvement [85, 86]. Livny et al. [87] found that before the program, 37.7% of all the children reported bringing sandwiches with sweetened spreads (chocolate spread, jelly, etc.). This level decreased to 33.2% after the program (not statistically significant). A significant decrease in sweet sandwiches was found separately for boys, from 46.8% to 29.6% ($p=0.003$). Most children in this study (88.7%) reported that their parents were the major source of oral hygiene guidance. Programs should, therefore, include the involvement of parents who are integrally related to children's health behavior. Only few children reported having been taught by dentists, and none by dental hygienists or other health personnel [87].

Maintaining Oral and Dental Health

In addition to controlling sugar consumption, a number of other actions have been taken to control the disease. The proper use of systemic and topical fluoride, perhaps, is one of the most cost-effective approaches and has resulted in major reductions in dental caries. The Centers for Disease Control and Prevention has named fluoridation of water as one of the ten most important public health measures of the twentieth century [88].

The schools may serve as the best platform for promoting oral health care among teenagers. The oral health education programs should be intensified to promote oral health care as a lifelong practice. The incorporation of oral health education activities into the school's curriculum has already taken place, but more efforts are needed in the form of educational materials and health promotion activities. Undoubtedly, the support from the parents and teachers in various ways – both organization and participation – is essential. The school health policies are already in place with the introduction of healthy eating behavior that bans sugar foods and drinks in the school premises. However, the implementation still needs to be strengthened. Healthy foods must be made available in the school canteens, while the canteens should be prohibited from selling food and drinks that contain high sugar levels. Although parents play a role of influence on the eating behavior of their children, a more effective idea would be to allow children to make healthy food choices. At this age, these students, as compared to primary school children, are more likely to purchase food on their own [50].

There is no standard form in studies related to oral health. Problems occur in the comparison of the obtained results [89].

Importance of Education

In order to organize such health education programs, the assessment of knowledge, attitude, and practice is essential [90].

Ehizele et al. [91] showed that a great percentage of teachers know that dental caries and gingival bleeding are not normal, but only a few of them know about the etiological factors. The national school curriculum sets out a clear and statutory entitlement of learning for all pupils from all levels of schooling. Oral health education is part of the primary school curriculum in many countries of the world [92, 93]. The high level of involvement of teachers in oral health education reflected by this study is commendable and should be encouraged as contextualized oral health educational activities in the schools, which is said to have positive effects on oral hygiene, gingival bleeding score, oral health behavior, and information level about oral health [94].

Güngör et al. [95] reported in their study that teeth-brushing habit increases in parallel with the education level; therefore, these individuals have better oral health.

Roles of Parents in Maintaining Oral and Dental Health

Family and friends seems to play a minor role in advising a dental visit. Barker and Horton [96] carried out a study on preschool children in California and showed that parents played a major role in influencing their children's oral health and access to care.

The previous studies determined a significant relation between the education level of parents and the dental health of children. Children of parents with a high education level have higher knowledge about dental health [26, 97]. Families with good economic status can take their children to the dentist at any time; in addition, they can easily afford toothbrushes and toothpaste.

Blay et al. [98] found that females reported daily higher consumption of soda, sugared coffee/tea, and chocolate/sweets than males, both in urban and rural areas. There was no statistically significant difference in the consumption of soda, sugared coffee/tea, cakes/biscuits, chocolate/sweets, and tooth brushing among urban respondents with less highly educated and more highly educated mothers. Among the rural residents, people with a highly educated father and mother were significantly more likely to take soda, cakes/biscuits, and chocolate/sweets daily as compared to their counterparts with less educated parents. Blay et al. [98] suggested that having a highly educated father and mother generally is the most important influence on adolescents' sugar and oral hygiene habits in Ghana.

Nutrition Education and Counseling

Nutrition education for the purpose of reducing caries incidence in children is aimed at informing parents about the importance of reducing dietary exposures to sweet foods and hidden sugars. Education is necessary, but not sufficient, to change eating behaviors. Diet counseling aims to help parents change their and their children's dietary behaviors so that they could choose diets with low or noncariogenic snacks, limit sweet foods to mealtimes, and perform tooth brushing after sugar exposures. Dietary recommendations must be realistic and always based on current dietary behaviors of the family. It is pointless to prescribe changes that a patient cannot or will not implement [99].

Peker and Alkurt [100] found that oral and dental health behaviors and attitudes and also their knowledge about oral and dental health care of dental students improved with increasing level of education, while oral and dental health care of female students were better than males.

Recommendations

The World Health Organization cited the following items in its twenty-first century dental health targets: (1) 80% of children aged 6 years will have no dental caries until 2010, and DMFT will be maximum 1.5 in children aged 12 years, and (2) health-protecting and health-improving programs will be established in at least 50% of nursery schools and 95% of all schools [101].

The WHO Oral Health Report 2003 [102] noted that the prevalence and incidence of dental caries can be controlled by the joint action of communities, professionals, and individuals. In many developing countries, however, access to oral health services is very limited, while in developed countries, significant portions of the population are underserved. For these reasons, professionally applied fluorides were not considered to be relevant to this review. Rather, this review focused on public health approaches, including the following:

- Water fluoridation
- Salt fluoridation
- Milk fluoridation
- Development of affordable fluoride toothpastes

The advantages of topical fluoride in a variety of forms have been firmly established [103]. In 2001, the Centers for Disease Control and Prevention advised that it was beneficial for patients of all ages to drink water with optimal fluoride concentration and brush with fluoridated toothpaste twice a day [104].

References

1. Simon C, Tesfaye F, Berhane Y. Assessment of the oral health status of school children in Addis Ababa. Ethiop Med J. 2003;41:245–56.
2. Saini S, Aparna, Gupta N, Mahajan A, Arora DR. Microbial flora in orodental infections. Indian J Med Microbiol. 2003;21:111–4.
3. Fédération Dentaire Internationale. Goals for the oral health in the year 2000. Fédération Dentaire Internationale. Int Dent J. 1982;32:74–7.
4. American Academy of Pediatric Dentistry (AAPD). Early childhood caries (ECC): classifications, consequences, and preventative strategies. Retrieved 3 Jan 2010, from http://www.aapd.org/media/policies_Guidelines/P_ECCClassifications.pdf
5. Petersen PE, Bourgeois D, Ogawa H, Estupinan-Day S, Ndiaye C. The global burden of oral diseases and risks to oral health. Bull World Health Org. 2005;83:661–9.
6. Global Oral Health Data Bank. Geneva: World Health Organization; 2001.
7. Edelstein BL. Evidence-based dental care for children and the age 1 dental visit. Pediatr Ann. 1998;27:569–74.
8. Pattussi MP, Marcenes W, Croucher R, Sheiman A. Social deprivation, income inequality, social cohesion and dental caries in Brazilian school children. Soc Sci Med. 2001;53:915–25.
9. Çubukçu ÇE. Reason for primary care dentistry. Hacettepe Toplum Hekimliği Bülteni. 1. 2003. (http://www.thb.hacettepe.edu.tr2003/20034.shtml)
10. Oral Health Division. National oral health survey of schoolchildren 1997 (NOHSS'97). Kuala Lumpur: Ministry of Health; 1998.
11. Sarawak Health Department. Annual Report 2008: Oral Health Division. Kuching: Ministry of Health, Malaysia; 2008.
12. Lulic-Dukic O, Juric H, Dukie W, Glavina D. Factors predisposing to early childhood caries (ECC) in children of pre-school age in the city of Zagreb, Croatia. Coll Antropol. 2001;25:297–302.
13. Dash JK, Sahoo PK, Bhuyan SK, Sahoo SK. Prevalence of dental caries and treatment needs among children of Cuttack (Orissa). J Indian Soc Pedod Prev Dent. 2002;20:139–43.
14. Kumar M, Chandu GN, Shafiulla MD. Oral health status and treatment needs in institutionalized psychiatric patients: one year descriptive cross sectional study. Indian J Dent Res. 2006;17:171–7.

15. Seow W. Biological mechanisms of early childhood caries. Community Dent Oral Epidemiol. 1998;26(Suppl 1): 8–27.
16. Reisine S, Douglass J. Psychosocial and behavioral issues in early childhood caries. Community Dent Oral Epidemiol. 1998;26(Suppl 1):32–44.
17. Schroth R, Harrison R, Moffatt M. Oral health of Indigenous children and the influence of early childhood caries on childhood health and well-being. Pediatr Clin North Am. 2009;56(6):1481–99.
18. Weinstein P, Smith W, Fraser-Lee N, Shimono T, Tsubouchi J. Epidemiologic study of 19-month-old Edmonton, Alberta children: caries rates and risk factors. ASDC J Dent Child. 1996;63(6):426–33.
19. Moynihan PJ, Petersen PE. Diet, nutrition and prevention of dental diseases. Public Health Nutr. 2004;7:201–26.
20. Leake J, Jozzy S, Uswak G. Severe dental caries, impacts and determinants among children 2–6 years of age in Inuvik Region, Northwest Territories, Canada. J Can Dent Assoc. 2008;74(6):519.
21. Tiberia M, Milnes A, Feigal R, Morley K, Richardson D, Croft W, et al. Risk factors for early childhood caries in Canadian preschool children seeking care. Pediatr Dent. 2007;29(3):201–8.
22. Lawrence H, Binguis D, Douglas J, McKeown L, Switzer B, Figueiredo R, et al. A 2-year community randomized controlled trial of fluoride varnish to prevent early childhood caries in Aboriginal children. Community Dent Oral Epidemiol. 2008;36:503–16.
23. Usmen E. Ankara ve köylerdeki ilkokul çocuklarında diş çürüğü sıklığı ve bunu etkileyen faktörler. Diş Hek Fak Derg. 1971;5:84–6.
24. Yazıcıoğlu N, Yazıcıoğlu B, Ulusoy M, Bumin Ç, et al. Ankara'da ilkokul çocuklarında diş sağlığı konusunda bir prevelans araştırması. Bölüm I. TED Ankara Koleji ilk kısım öğrencilerinde diş sağlığı. AÜ Diş Hek Fak Derg. 1985;12:693–706.
25. Saraçoğlu A, Kümbüloğlu Ö, Hatipoğlu HA, User A. Relationship between social status, caries and periodontal disorder prevalence in dental students (An epidemiological study). Cumhuriyet Üniversitesi Diş Helimliği Fakültesi Dergisi. 2007;10:10–5.
26. Oncel S, Akcan A, Keser I. Study of the dental health of primary school students. Sağlık ve Toplum. 2007;17:75–84.
27. Wong D, Perez-Spiess S, Julliard K. Attitudes of Chinese parents toward the oral health of their children with caries: a qualitative study. Pediatr Dent. 2005;27(6):505–12.
28. Amin MS, Harrison RL. Understanding parent's oral health behaviors for their young children. Qual Health Res. 2011;19:116–27.
29. Sheiham A. Dietary effects on dental diseases. Public Health Nutr. 2001;4:569–91.
30. Moynihan PJ. Dietary advice in dental practice. Br Dent J. 2002;193:563–8.
31. Sayegh A, Dini EL, Holt RD, Bedi R. Oral health, sociodemographic factors, dietary and oral hygiene practices in Jordanian children. J Dent. 2002;33:379–88.
32. Sanders I. Diet and general health dietary counseling. Caries Res. 2004;38(Suppl l):3–8.
33. Li Y. Controlling sugar consumption still has a role to play in the prevention of dental caries. J Evid Based Dent Pract. 2011;11:24–6.
34. Korver O. "Healthy" developments in the food industry. Cancer Lett. 1997;114:19–23.
35. Questions and answers about functional foods. http://ific.org/publications/qa/funcfdsqa.cfm?renderforprint=1
36. Krasse B. The Vipeholm Dental Caries Study: recollections and reflections 50 years later. J Dent Res. 2001;80:1785–8.
37. Wells JE, Reed WM, Country VM. Review of basic science and clinical dentistry, vol. 2. Cambridge: Harper and Row; 1980.
38. Yabao RN, Duante CA, Velandria FV, Lucas M, Kassu A, Nakamori M, et al. Prevalence of dental caries and sugar consumption among 6-12-y-old schoolchildren in La Trinidad, Benguet, Philippines. Eur J Clin Nutr. 2005;59:1429–38.
39. Woodward M, Walker ARP. Sugar consumption and dental caries: evidence from 90 countries. Br Dent J. 1994;176:297–302.
40. Howard S, Reeves S. The snacking habits of adolescents: is snack food necessary to meet dietary recommendations. Health Educ J. 2005;64:51–7.
41. Hackett A, Kirby F, Howie MA. National Survey of the diet of children aged 13–14 years living in urban areas of the United Kingdom. J Hum Nutr Diet. 1997;10:37–51.
42. Rayner J, Holt R, Blinkhorn F, Duncan K. British Society of Paediatric Dentistry: a policy on oral health care in pre-school children. Int J Paediatr Dent. 2003;13:279–81.
43. Shelton N. What not to eat: inequalities in healthy eating behaviour, evidence from 1998 Scottish Heart Survey. J Public Health. 1998;27:36–44.
44. Shohoumi S, Bingham S, Welch AA, Luben R, Day N, Wareham N, et al. Occupational social class, educational level, area deprivation, independently predict plasma ascorbic acid concentration. Eur J Clin Nutr. 2004;58:1432–5.

45. Hobdell MH, Oliveira ER, Bautista R, Myburgh NG, Lalloo R, Narendran S, et al. Oral disease and socio-economic status. Br Dent J. 2003;194:91–6.
46. Maliderou M, Reeves S, Noble C. The effect of social demographic factors, snack consumption and vending machine use on oral health of children living in London. Br Dent J. 2006;201:441–4.
47. Bozkurt FY, Fentoğlu Ö, Kıran M. Isparta il merkezi ve kırsalında yaşayan adölesanlarda ağız bulgularının değerlendirilmesi. T Klin Diş Hek Bil. 2002;8:25–30.
48. Murray CJL, Lopez AD. Global mortality, disability and the contribution of risk factors: Global Burden of Disease Study. Lancet. 1997;349:1498–504.
49. Oogarah-Pratap B. Dietary habits of Mauritian school adolescents. Nutr Food Sci. 2007;37:442–51.
50. Lian CW, Phing TS, Chat CS, Shin BC, Baharuddin LH, Che'Jalil ZBJ. Oral health knowledge, attitude and practice among secondary school students in Kuching, Sarawak. Arch Orofac Sci. 2010;5:9–16.
51. Abdullah M, Ali H, Rahiman S. Knowledge, practice and awareness of oral hygiene among three different age populations of same ethnic group-a community based study. Braz J Oral Sci. 2010;9:481–7.
52. Sanlıer N, Ozgen L. Ogrencilere farklı yöntemlerle verilen eğitim ağız-diş sağlığı ve beslenme bilgisi üzerine etkisi. Türk Eğitim Bilimleri Dergisi. 2005;3:1–14.
53. Özgen L. Farklı eğitim düzeyindeki öğrencilerin hızlı hazır yiyecek (fast food) tüketim durumlarının ve alışkanlıklarının saptanması üzerine bir araştırma. G. Ü. Sosyal Bilimler Enstitüsü. Ankara. Bilim Uzmanlık Tezi; 1998.
54. Gregory J, Lowe S, Bates CJ, Prentice A, Jackson LV, Smithers G, et al. National diet and nutrition survey: young people aged 4 to 18 years, Report of the diet and nutrition survey, vol. 1. London: The Stationery Office; 2000.
55. Inchley IJ, Todd C, Bryce C, Currie C. Dietary trends among Scottish school children in the 1990's. J Hum Nutr Diet. 2001;14:207.
56. Sreebny LM. Sugar availability, sugar consumption and dental caries. Community Dent Oral Epidemiol. 1982;10:1–7.
57. van Loveren C, Duggal MS. Experts' opinion on the role of diet in caries prevention. Caries Res. 2004;38:16–23.
58. Jamel HA, Sheiham A, Cowell CR, Watt RG. Taste preference for sweetness in urban and rural populations in Iraq. J Dent Res. 1996;75:1879–84.
59. Kelder SH, Perry CL, Klepp KL, Lytle LL. Longitudinal tracking of adolescent smoking, physical activity and food choice behaviour. Am J Public Health. 1994;84:1121–6.
60. Astrom AN. Comparative risk judgements for oral health hazards among Norwegian adults: a cross sectional study. BMC Oral Health. 2004;2:1–7.
61. Rockett HRH, Breitenbach M, Frazier AL, Witschi J, Wolf AM, Field AE, et al. Validation of a youth/adolescent food frequency questionnaire. Prev Med. 1997;26:808–16.
62. Walker A, Gregory J, Bradnock G, et al. National Diet and Nutrition Survey: young people aged 4 to 18 years, Report of the oral health survey, vol. 2. London: HMSO; 2002.
63. Pacey A, Nancarrow T, Egeland GM. Prevalence and risk factors for parental-reported oral health of Inuit pre-schoolers: Nunavut Inuit child health survey, 2007–2008. Rural Remote Health. 2010;10:1368.
64. Freeman R, Heimonen H, Speedy P, Tuutti H. Determinants of cariogenic snacking in adolescents in Belfast and Helsinki. Eur J Oral Sci. 2000;108:504–10.
65. Lingstrom P, Moynihan P. Nutrition, saliva, and oral health. Nutrition. 2003;19:576–9.
66. Featherstone JDB. The continuum of dental caries-evidence for a dynamic disease process. J Dent Res. 2004;83(Spec Issue C):39–42.
67. Loesche WJ. Role of *Streptococcus mutans* in human dental decay. Microbiol Rev. 1986;59:353.
68. Barnard P. National oral health survey Australia 1987–88. Department of Health, Housing, Local Government and Community Services. Canberra: AGPS; 1993.
69. Kara M, Gürbüz E, Mete A, Sahin T, Celik C, Yamanel K. The relationship between dietary habits and oral health status of dental students. Atatürk Üniv Diş Hek Fak Derg. 2009;19:161–7.
70. Bruno-Ambrosius K, Swanholm G, Twetman S. Eating habits, smoking and tooth brushing in relation to dental caries: a 3-year study in Swedish female teenagers. Int J Paediatr Dent. 2005;15:190–6.
71. Dweyr JT, Evans M, Stone EJ, Feldman HA, Lytle L, Hoelscher D, et al. Adolescents' eating patterns influence their nutrient intakes. J Am Diet Assoc. 2001;101:798–802.
72. Summerbell CD, Moody RC, Shanks J, Stock MJ, Geissler C. Sources of energy from meals versus snacks in 220 people in four age groups. Eur J Clin Nutr. 1995;49:33–41.
73. Hefferren JJ, editor. Scientific consensus conference on methods for the assessment of the cariogenic potential of foods. University of Texas Health Science Center, San Antonio, Texas, November 17–21;1985. Sponsored by Food, Nutrition and Dental Health Program, American Dental Association. J Dent Res. 1986;65:1473–543.
74. Efe E, Sarvan S, Kukulu K. Self-reported knowledge and behaviors related to oral and dental health in Turkish children. Issues Compr Pediatr Nurs. 2007;30:133–46.
75. Sheiham A. Oral health, general health and quality of life. Bull World Health Organ. 2005;83:641–720.

76. Peker K, Bermek G. The importance of fermentable carbohydrates in the etiology and prevention of dental caries. Istanbul Univ Dishekim Fak Derg. 2008;42:1–9.
77. Kwan S, Peterson PE. Evaluation of community-based oral health promotion and oral disease prevention. WHO recommendations for improved evidence in public health practice. Commun Dent Health. 2004;21:319–29.
78. Nguyen L, Hakkinen U, Knuttila M, Jarvelin MR. Should we brush twice a day? Determinants of dental health among young adults in Finland. Health Econ. 2008;17:267–86.
79. Puy CL, Navarro LF. Evidence concerning the medical management of caries. Med Oral Patol Oral Cir Bucal. 2008;13(5):E325–30.
80. Payne BJ, Locker D. Relationship between dental and general health behaviors in a Canadian population. J Public Health Dent. 1996;56:198–204.
81. Rosin P, Peters J, Robling MR. Social class and preventive behavior: a British example. J Epidemiol Community Health. 1995;49:28–32.
82. Sakki TK, Knuutila MLE, Anttila SS. Lifestyle, gender and occupational status as determinants of dental health behavior. J Clin Periodontol. 1998;25:566–70.
83. Smyth E, Caamano F, Fernandez-Riveiro P. Oral health knowledge, attitudes and practice in 12-year-old schoolchildren. Med Oral Patol Oral Cir Bucal. 2007;12:E614–20.
84. Ab-Murat N, Watt RG. Chief dentists' perceived strengths and weaknesses of oral health promotion activities in Malaysia. Annal Dent Univ Malaya. 2006;13:1–5.
85. Selwitz RH, Ismail AI, Pitts NB. Dental caries. Lancet. 2007;369(9555):51–9.
86. Sgan-Cohen HD. Oral hygiene: past history and future recommendations. Int J Dent Hyg. 2005;3(2):54–8.
87. Livny A, Vered Y, Slouk L, Sgan-Cohen HD. Oral health promotion for schoolchildren-evaluation of a pragmatic approach with emphasis on improving brushing skills. BMC Oral Health. 2008;8:1–6.
88. Morbidity Mortality World Report (MMWR). Achievements in public health, 1900–1999: fluoridation of drinking water to prevent dental caries. 1999; 48: 933–940. http://www.cdc.gov/mmwr/
89. Owens J. Barriers to oral health promotion in the Republic of Ireland. Scand J Public Health. 2011;39:93–7.
90. Al-Omiri MK, Al-Wahadni AM, Saeed KN. Oral health attitudes, knowledge, and behavior among school children in North Jordan. J Dent Educ. 2006;70:179–87.
91. Ehizele A, Chiwuzie J, Ofili A. Oral health knowledge, attitude and practices among Nigerian primary school teachers. Int J Dent Hyg. 2011;9:254–60.
92. Abdellatif H. Oral health knowledge and sources of information of fluoride among Saudi parents in Riyadh. Saudi Dent J. 2004;16:3–8.
93. Chapman A, Copestake SJ, Duncan K. An oral health education programme based on the national curriculum. Int J Paediatr Dent. 2006;16:40–4.
94. Petersen PE, Peng B, Tai B, Bian Z, Fan M. Effect of a school-based oral health education programme in Wuhan city, peoples republic of China. Int Dent J. 2004;54:33–41.
95. Güngör K, Tüter G, Bal B. Eğitim düzeyi ile ağız sağlığı arasındaki ilişkinin değerlendirilmesi. GÜ Diş Hek Fak Derg. 1996;16(1):21–5.
96. Barker JC, Horton SB. An ethnographic study of Latino preschool children's oral health in rural California: Intersections among family, community, provider and regulatory sectors. BMC Oral Health. 2008;8:8.
97. Altıntaş H, Altıntaş SC, Telatar TG, Hürmüz P, et al. The study of determining some of the knowledge and behaviors of fifth grade students, attending to elementary schools in Ankara batıkent II health center region, about dental health. GÜ Diş Hek Fak Derg. 2004;21:97–105.
98. Blay D, Astrom AN, Haugejorden O. Oral hygiene and sugar consumption among urban and rural adolescents in Ghana. Community Dent Oral Epidemiol. 2000;28:443–50.
99. Tinanoff N, Palmer CA. Dietary determinants of dental caries and dietary recommendations for preschool children. J Public Health Dent. 2000;60:197–206.
100. Peker I, Alkurt MT. Oral health attitudes and behavior among a group of Turkish dental students. Eur J Dent. 2009;3:24–31.
101. Aktı A. Okul sağlığı hizmet uygulamalarında ağız diş sağlığı, deneyimler ve öneriler. Klinik Çocuk Forumu. 2005: 17–25.
102. The World Oral Health Report 2003. Geneva: WHO; 2003. Available from: http://www.who.int/oral_health
103. American Dental Association Council on Scientific Affairs. Professionally applied topical fluoride: evidence-based clinical recommendations. J Am Dent Assoc. 2006;137:1151–9.
104. Centers for Disease Control and Prevention. Recommendations for using fluoride to prevent and control dental caries in the United States. MMWR Recomm Rep. 2001;50(RR-14):1–42.

Chapter 39
Acne and Chocolate: Is There Any Evidence of Their Association?

Adilson Costa and Maria Carolina Fidelis

Key Points

- Acne is a common dermatosis and affects millions of people around the world.
- For several years, acne has been associated with some types of foods. Popular wisdom claims that chocolate is the most important food that causes acne.
- Recent observational studies have shown a possible relation between a Westernized dietary pattern (based on a high glycemic index that can promote a hyperinsulinemia scenario) and acne establishment.
- Modern nutritional studies have supported the idea that chocolate has biologically active compounds that can affect acne.

Keywords Acne • Chocolate • High glycemic index • Hyperinsulinemia • Growth-factor insulin-simile 1 (IGF-1)

Acne Vulgaris

Acne vulgaris is a common dermatological condition in the United States [1] that affects over 17 million Americans of all ages [1, 2]. Its overall prevalence varies between 35% and 90% in adolescents. This percentage is higher in the West [3, 4]. Its development and prevalence are higher among males, owing to androgenic influence [3, 5, 6]. Severe acne is associated with low self-esteem, distorted self-image, social withdrawal, and depression [2, 7].

This chronic dermatosis attacks the pilosebaceous follicle and presents a multifactorial etiopathogeny, including sebaceous hyperproduction, follicular hyperkeratinization, increase of *Propionibacterium acnes* colonization, and periglandular dermal inflammation [3, 4, 8–10]. Other factors have also been indicated as contributors to acne prevalence and intensity, such as physiological factors (menstrual cycle, pregnancy, anxiety, and depression) and external factors (humidity and high temperatures, lack or excess of skin cleaning, use of certain cosmetics, smoking, and diet) [8].

A. Costa, M.D., M.Sc., Ph.D. (✉) • M.C. Fidelis, M.D.
Dermatology Department, Catholic University of Campinas, Campinas, SP, Brazil
e-mail: adilson_costa@hotmail.com; carolfidelis@hotmail.com

R.R. Watson et al. (eds.), *Chocolate in Health and Nutrition*, Nutrition and Health 7,
DOI 10.1007/978-1-61779-803-0_39, © Springer Science+Business Media, LLC 2013

Hormones may increase sebum production and cause the hyperproliferation of follicular cells, with the consequent obstruction of the follicular opening and formation of a comedo [1, 2]. The full obstruction of the follicular opening causes closed comedones ("whiteheads"), whereas the partial obstruction causes open comedones ("blackheads") [1, 2].

Acne may appear as noninflammatory comedones, superficial inflammatory lesions (papules and pustules), and deep inflammatory lesions (nodules and cysts) [1, 2, 8]. Inflammatory lesions are mostly associated with the colonization of the pilosebaceous follicle by *Propionibacterium acnes*, which causes the follicular rupture and a neutrophilic cascade [1]. Acne rarely presents as a result of nonbacterial causes [1].

Chocolate

Chocolate derives from seeds of a tropical plant, the *Theobroma cacao* [11]. A thick paste known as chocolate liquor is produced through the processing of these seeds and serves as base for all chocolate products [11]. Chocolate liquor has over 50% of lipids in its content. Lipids, known as cocoa butter, may be separated from the liquor through pressure and used commercially [11]. Cocoa is rich in polyphenols, which act as antioxidants [12, 13]; however, a part of its high antioxidant capacity is lost in the manufacturing of chocolate [12].

Chocolate Liquor Composition

The chemical compounds of chocolate liquor are basically lipids, theobromine, caffeine, and tannins.

Lipids

Contrary to other natural fats, cocoa butter is almost entirely composed of triglycerides of only three fatty acids: palmitic, stearic, and oleic acids [11].

Theobromine and Caffeine

Xanthine alkaloids are known for their pharmacologic properties and used as stimulants and diuretics. Their precise action mechanism is obscure. It is, however, acknowledged that small dosages may increase sugar in the blood as well as basal metabolic rates. Chocolate is the only food product in which xanthine alkaloids are found together with a large quantity of lipids [11].

Tannins

Chocolate liquor contains two tannins in particular, the purple and brown cacao, which give it the common color of chocolate [11].

Composition of Processed Chocolate

The industrial processing of chocolate basically depends on the mixture of sugar (sucrose) with chocolate liquor and extra-virgin cocoa butter. Milk solids, small quantities of lecithin, and secondary flavoring agents may be added. In dark chocolate, 33% of the lipid content derives from cocoa butter, while lipids of milk chocolate, also of approximately 33%, are produced through 27% of cocoa butter and 6% of milk fat. Chocolate contains about 50% of sugar, which makes it a distinct product that combines a large amount of sugar with a high level of lipids [11].

Diet and Acne

Food Perception and Acne

Many recent studies have evaluated the knowledge, beliefs, and perceptions involving the causes of acne in patients suffering from it. No great differences have been noticed in the beliefs and perceptions of the acne pathogenesis in patients of different societies and cultures, but the role of the diet has been uniformly reported in these populations [10].

In a study conducted in Australia, evaluators questioned last-year medical students about their perceptions on acne. Almost one-half of them believed that the diet was an etiological factor; as expected, chocolate was at the top of the list of foods that would cause it [14].

This epidemiological example leads us to reflect on the reason why certain works, for many decades, have tried to confirm an ancient theory, surrounded by many myths and folktales: the association of acne and diet [3, 15–17].

Westernized Dietary Patterns and Acne Development

A great difference is observed in the prevalence of acne between non-Western and totally modernized societies. In this scenario, dietary patterns were suspected to be the reason [10, 18].

Studies conducted with the Inuit Eskimo population have demonstrated that acne was absent in such population until the introduction of Westernized food habits [10]. In the same manner, studies involving the natives of Kitava Island, the Ache Indians of Paraguay, and communities of Papua New Guinea attribute the absence of acne in these populations to their diet [3, 10].

Other studies carried out with populations from rural villages, such as those in Kenya, Zambia, and Bantu, in South Africa, have concluded that these people have significantly less acne than their descendants who migrated to the United States and United Kingdom [3, 19–21] – descendants who, in other words, are now exposed to Westernized diets.

Based on these epidemiological findings, we may postulate that, contrary to the traditional food habits of these populations, Westernized diets, which present a high glycemic index, exposed these populations to hyperinsulinemia and therefore, influenced follicular epithelium growth, keratinization, and sebaceous secretion, leading predisposed individuals to develop acne.

The Role of Hyperinsulinemia

One of the characteristics of Westernized diets is the chronic adoption of high glycemic index foods. This situation leads to hyperglycemia and hyperinsulinemia [22] and may cause a series of hormonal alterations, in addition to increasing the ratio between growth-factor insulin-simile 1 (IGF-1) and insulin-like growth factor-binding protein-3 (IGFBP-3). Consequently, we have an increase of the synthesis of ovarian and testicular androgens, together with a decrease in the production of sex hormone-binding globulins (SHBG), which may result in polycystic ovary syndrome (PCOS), acne, and cancer of the epithelial cells, among other diseases [22].

Authors of a randomized clinical study evaluated the effect of low glycemic index diets on the risk of acne and insulin sensitivity [2]. Individuals who followed a low glycemic index diet presented an improvement in the average number of acne lesions, when compared to those of the control group [2]. In addition, individuals of the group of the diet under study presented a decrease in their average weight as well as an increase of insulin sensitivity and SHBG levels. With the increase of SHBG levels, a decrease of free circulating androgens in the same proportion is expected. Therefore, an increase of SHBG levels enables a direct decrease in lesion counting [2].

In this situation, a possible hyperinsulinemia state, associated with the secondary presence of IGF-1, would stimulate the synthesis of androgens, which stimulates sebum production [3, 17, 23], and also influence follicular epithelial growth and keratinization, thus contributing to the acne pathogenesis [3, 24].

Milk and Acne

Can milk cause acne? A study in 1949 monitored 1,925 patients through daily diet journals and found out that milk was the food most commonly involved in the development of acne lesions [10].

In more recent reports that also support the association between the ingestion of milk and acne, we find the retrospective evaluation of data from an emblematic study that presents a positive association between severe acne in adolescents and the ingestion of whole and skim milk [1–4, 10].

In the same manner, another prospective study for cutoff points in adolescents confirmed the relationship between milk ingestion and acne, being the association stronger between skim milk and the development of lesions in boys [2–4], a fact that points to the protein fraction of milk in the development of the lesions [1–4]. This demonstrates that the association does not result from the fat content of milk and, therefore, reinforces the IGF-1 level theory [3].

Milk, yogurt, and some cheeses containing lactose, despite having a low glycemic load, obtain a very high insulin response [22, 25], which induces an increase of IGF-1 levels and facilitates the development and/or aggravation of acne, which is particularly higher in individuals who ingest skim milk [3, 26, 27]. This is because IGF-1 stimulates the synthesis of ovarian, testicular, and adrenal androgens, which act on the pilosebaceous follicles and stimulate the production of sebum [10]. In addition, milk contains progesterone and dihydrotestosterone precursors, which are also implicated in the comedogenesis [3, 28, 29].

Another case that explains the association between acne and milk is the iodine content of milk [3, 10]. It results from the diet supplementation offered to animals and from the use of iodine-based solutions in dairy and milking equipment [3]. What helps to reinforce this relation is the result of a study conducted with 1,066 adolescents who were evaluated through questionnaires aimed at determining whether the levels of iodine contained in water or salt may affect the prevalence or seriousness of acne [3]. Patients that lived in the region with habits of higher consumption of salt presented more prevalence to severe acne [3, 29].

Therefore, does milk cause acne? We may conclude, based on the presented evidence, that there is a strong relation between acne and milk ingestion [10].

Chocolate and Acne

Chocolate has especially been blamed as an aggravating factor of acne. Patients often report the development of pustules a few days after the ingestion of this food [3]. After all, what is the role of chocolate in the development of acne?

In the double-blind study conducted by Fulton et al., 65 individuals ingested a chocolate bar (112 g) that was rich in chocolate liquor and cocoa butter, daily for 4 weeks. After this period, the regimen was changed and patients ingested a control bar, with the same weight, without chocolate liquor and cocoa butter, for 4 more weeks. Lesions were classified in three categories: aggravated, improved, and unaltered. Since there was no significant difference between the ingestion of chocolate and control bars in the three categories, the authors concluded that the ingestion of a great amount of chocolate does not interfere in the course of acne vulgaris or in sebum composition [3, 15].

The conclusion of the Fulton et al. study buried for decades the ultimate association between chocolate ingestion and acne. However, reports from patients in dermatology clinics always pointed to an opposite evidence, and this has intrigued many scholars. Science remained bothered.

Years later, however, the control bar employed in the Fulton et al. study was evaluated. It was a bar designed to imitate the classic chocolate, which did not contain cacao. Nevertheless, such bar contained 28% of lipids, such as hydrogenated vegetable fat, and 53% of sucrose; the chocolate bar that was adopted for the study contained 34% of lipids, such as cocoa butter, and 44.3% of sucrose [3, 11, 17]. Therefore, from a quantitative standpoint, there was no actual difference between the two bars with respect to lipid and sugar contents. Thus, the two bars predisposed to hyperglycemia and hyperinsulinemia, factors involved in the development of acne. Perhaps, due to this fact, Fulton et al. did not find differences in the course of acne vulgaris among the patients who ingested a chocolate bar and those who ingested the control bar [3, 11, 17].

Many of those who supported the Fulton et al. theory, even in light of contrary evidence presented years later, did so because of a study conducted 4 years before by Grant and Anderson, a study in which methodological flaws were evident. In this study, eight patients (only eight patients!) were selected, among whom three affirmed that the ingestion of chocolate aggravated their acne condition. Such patients ingested a great amount of chocolate for 7 days and at the end of the study, no alteration in the number or severity of lesions was observed [30]. However, pre- and post-experimental lesions were not considered in this study, and there was no control group or statistical analysis, in addition to having reduced sampling.

Decades passed until new research in health science started to be ruled by robust methodological criteria. The new criteria justified new studies on this possible association.

Therefore, an Australian study was carried out to test the postprandial insulin of food with chocolate. This study compared the plasmatic profile of healthy patients after the ingestion of food with and without chocolate. Despite the absence of significant variation in the glycemic index of patients in each group, the insulin index was on average 28% higher in products with chocolate. The greater difference occurred in the category of aromatized milk; the chocolate-flavored version caused an insulinemia 45% higher than the strawberry-flavored one [31].

An explanation for the findings of the Australian group may be that chocolate is rich in biologically active compounds, such as caffeine, theobromine, serotonin, phenylethylamine, triglycerides, and cannabinoid-like fatty acids, which increase secretion of and peripheral resistance to insulin [32]. Moreover, the amino acids present in chocolate (such as arginine, leucine, and phenylalanine) are extremely insulinotropic when ingested with carbohydrates; other amino acids (valine, lysine, and isoleucine), found in other types of food, especially those rich in lactose, can also cause this plasmatic behavior [3].

Based on recent findings, new light has been shed on the research of the relation between acne and the ingestion of chocolate. Nowadays, supported by reports published in this literature, we may indeed venture to say that the lines of this research have a reason to exist, once the ingestion of chocolate-based food seem to somewhat influence the development of clinical cases of acne vulgaris.

New findings may be added to this historical binominal. A group of Norwegian specialists recently carried out an epidemiological study with 3,775 adolescents from Oslo, ages 18 and 19, and discovered a strong relation between acne and the ingestion of chocolate among male adolescents [33].

As previously seen, industrial chocolate bars contain about 50% of sugars and most of them contain milk derivatives in their composition. Both profiles, alone or associated, lead to an increase of glycemic index with frequent increase of insulinemia, therefore triggering an increase of IGF-1 and of the ratio between IGF-1 and IGFBP-3, a decrease of SHBG production, and an increase of androgens – all factors implicated in this comedogenesis [1–4, 10, 17, 22, 23, 25–27, 29].

Summary

For ages, the old folktale says, "The ingestion of chocolate causes acne." Doctors always ask themselves, "Does it really?"

For a long time, the modern community of specialists has avoided the evolution of scientific evidence in this practical field of medicine. Based on premature conclusions from studies with weak methodology and on the voice of skeptical opinion formers who were not tendentious to promote the development of researches that would provide clarifying answers, whether positive or negative, to support this proposition, "Is there any relationship between acne and chocolate?"

Modern findings with respect to the etiopathogeny of acne vulgaris show us that hyperinsulinemia is a crucial element for the development and/or aggravation of this dermatosis. The clinical evolution of this disease was stimulated by food. In this context, chocolate took its role.

Obviously, it is still premature to precisely affirm and confirm the relation between acne and chocolate. The most recent findings pull us closer to a favorable conclusion to this relation, but more studies are mandatory. We must carry out well-designed, placebo-controlled studies, with closed populations, with or without acne vulgaris, to reach a definitive explanation.

Specialists in health science, however, must be prepared for the final conclusions, whatever they may be, with respect to the binomial acne-chocolate. The worst of all sciences is the science that disregards the facts that are revealed. To this date, we have been induced not to render any response to this relation.

References

1. Spencer EH, Ferdowsian HR, Barnard ND. Diet and acne: a review of the evidence. Int J Dermatol. 2009;48:339–47.
2. Ferdowsian HR, Levin S. Does diet really affect acne? Skin Therapy Lett. 2010;15(3):1–5.
3. Costa A, Lage D, Moisés TA. Acne e dieta: verdade ou mito? An Bras Dermatol. 2010;85(3):346–53.
4. Melnik BC. Evidence for acne-promoting effects of milk and other insulinotropic dairy products. Nestle Nutr Workshop Ser Pediatr Program. 2011;67:131–45.
5. Dreno B, Poli F. Epidemiology of acne. Dermatology. 2003;206:7–10.
6. Sobral Filho JF, Nunes Maia HGS, Fonseca ESVB, et al. Aspectos epidemiológicos da acne vulgar em universitários de João Pessoa – PB. An Bras Dermatol. 1993;68:225–8.

7. Smithard A, Glazebrook C, Williams HC. Acne prevalence, knowledge about acne and psychological morbidity in mid-adolescence: a community-based study. Br J Dermatol. 2001;145:274–9.
8. El-Akawi Z, Abdel-Latif Nemr N, Abdul-Razzac K, et al. Factors believed by Jordanian acne patients to affect their acne condition. East Medit Health J. 2006;12(6):840–6.
9. Magin P, Adams J, Heading G, et al. The causes of acne: a qualitative study of patient perceptions of acne causation and their implications for acne care. Dermatol Nurs. 2006;18(4):344–9.
10. Davidovici BB, Wolf R. The role of diet in acne: facts and controversies. Clin Dermatol. 2010;28:12–6.
11. Mackie BS, Mackie LE. Chocolate and acne. Aust J Dermatol. 1974;15:103–9.
12. Williams S, Tamburic S, Lally C. Eating chocolate can significantly protect the skin from UV light. J Cosmet Dermatol. 2009;8:169–73.
13. Mursu J, Voutilainen S, Nurmi T, et al. Dark chocolate consumption can increases HDL cholesterol concentration and chocolate fatty acids may inhibit lipid peroxidation in healthy humans. Free Radic Biol Med. 2004; 37(9):1351–9.
14. Green J, Sinclair RD. Perceptions of acne vulgaris in final year medical student written examination answers. Aust J Dermatol. 2001;42:98–101.
15. Fulton JE, Plewig G, Kligman AM. Effect of chocolate on acne vulgaris. JAMA. 1969;2010:2071–4.
16. Anderson PC. Foods as the cause of acne. Am Fam Physician. 1971;3:102–3.
17. Cordain L. Implications for the role of diet in acne. Sem Cutan Med Surg.. 2005;24:84–91.
18. Halvorsen JA, Dalgard F, Thorsen M, et al. BMC Public Health. 2009;9:340–7.
19. Verhagen A, Koten J, Chaddah V, et al. Skin diseases in Kenya. A clinical and histopathological study of 3,168 patients. Arch Dermatol. 1998;98:577–86.
20. Ratnan A, Jayaraju K. Skin disease in Zambia. Br J Dermatol. 1979;101:449–53.
21. Park R. The age distribution of common skin disorder in the Bantu of Pretoria, Transvaal. Br J Dermatol. 1968;80:758–61.
22. Carrera-Bastos P, Fontes-Villalba M, O'Keefe JH, et al. The western diet and lifestyle and diseases of civilization. Res Rep Clin Cardiol.. 2011;2:15–35.
23. Edmondson SR, Thumiger SP, Werther GA, et al. Epidermal homeostasis: the hole of the growth hormone and insulin-like growth factor systems. Endocr Rev. 2003;24:737–64.
24. Cordain L, Lindeberg S, Hurtado M, et al. Acne vulgaris: a disease of Western civilization. Arch Dermatol. 2002;138:1584–90.
25. Melnik BC. Milk signalling in the pathogenesis of type 2 diabetes. Med Hypotheses. 2011;76(4):553–9.
26. Adebamowo CA, Spiegelman D, Danby FW, et al. High school dietary intake and teenage acne. J Am Acad Dermatol. 2005;52:207–14.
27. Adebamowo CA, Spiegelman D, Berekey CS, et al. Milk consumption and acne in adolescent girls. Dermatol Online J. 2006;3012:1.
28. Danby FW. Acne and milk, the diet myth, and beyond. J Am Acad Dermatol. 2005;52:360–2.
29. Hitch JM, Greenburg BG. Adolescent acne and dietary iodine. Arch Dermatol. 1961;84:898–911.
30. Grant JD, Anderson PC. Chocolate as a cause of acne: a dissenting view. Mo Med. 1965;62:459–60.
31. Brand-Miller J, Holt SHA, De Jong V, et al. Cocoa powder increases postprandial insulinemia in lean young adults. J Nutr. 2003;133:3149–52.
32. Herraiz T. Tetrahydro-β-carbolines, potential neuroactive alkaloids, in chocolate and cocoa. J Agric Food Chem. 2000;48(10):4900–4.
33. Halvorsen JA, Dalgard F, Thoresen M, et al. Is the association between acne and mental distress influenced by diet? Results from a cross-sectional population study among 3775 late adolescents in Oslo, Norway. BMC Public Health. 2009;16(9):340.

Chapter 40
Cocoa and Chocolate in Humans: The Rationale for Clinical Studies on Healthy Volunteers

Ario Conti, Maria Teresa Pinorini-Godly, and Maria Laura Colombo

Key Points

- To help to maintain a good psychoneuroimmunological balance between large and small components, it might be possible to contribute with natural products.
- It is necessary to be aware of the big difference in action between a synthetic product represented by one molecule and naturally derived products represented by a hundred molecules.
- Cacao and chocolate products have generated significant interest owing to their association with various protective and therapeutic activities, and at least 45 human studies have been performed.
- Studies showing opposite effects among different cacao flavonoids suggested that other compounds might contribute to cacao's immune effects.
- This study on immunological properties of cacao and chocolate represents the first 28-day long clinical study on healthy young volunteers.
- It is a fact that dark chocolate helps lower systolic blood pressure.
- White chocolate, in normal human healthy volunteers, increases interleukin 10 production by activated peripheral blood mononuclear cells, plasmatic serotonin, and dopamine.
- Cacao and chocolate are good candidates for modulating neuroimmune interactions and help in preventing immune-based degenerative diseases.
- To better understand the effects of natural products, cocoa, and chocolate, both dark and white, we should invest much more in clinical research.

Keywords *Theobroma cacao* • Chocolate • Flavonoids • Immune system • Blood pressure • Interleukin 10 • Tumor necrosis factor alpha • Serotonin • 5-HIAA • 5-HTOL • Dopamine

---

A. Conti, Ph.D. (✉) • M.T. Pinorini-Godly, Dr.For.Sci.
Alpine Institute of Chemistry and Toxicology, Alpine Foundation for Life Sciences,
Olivone, Ticino 6718, Switzerland
e-mail: ario.conti@ti.ch; pinorini.alpisti@bluewin.ch

M.L. Colombo, Ph.D.
Department of Drug and Science Technology, School of Pharmacy, University of Torino,
Via Pietro Giuria 9, Torino 10125, Italy
e-mail: marialaura.colombo@unito.it

R.R. Watson et al. (eds.), *Chocolate in Health and Nutrition*, Nutrition and Health 7,
DOI 10.1007/978-1-61779-803-0_40, © Springer Science+Business Media, LLC 2013

Introduction

The research work on immunomodulating drugs is a relatively new field in pharmacology, which was developed only in the last 20 years. It will provide both theoretical significance and therapeutic values in many diseases, including cancer, viral infections, autoimmune diseases, organ transplantation, and aging, caused by deficiency or imbalance of the immune function of patients. On the other hand, the old concept of psychoneuroimmunomodulation has been rediscovered and modernized by scientists and integrated in many fields of medical sciences [1]. In fact, we have all experienced a sense of impending illness prior to its actual onset. Indeed, a lot of us have felt that either before or during sickness, our emotions may have influenced the course of the disease. Such anecdotes have existed since ancient times and, in the last decade, have been the initial foundation for a new and rapidly developing area of medical research. Basic to these sensations, as well as the new field of neuroendocrine and immune interactions, is the realization that our brain and endocrine system can influence our immunity and that the immune system serves as a sensory organ that ultimately signals the brain. During the last decades, conceptual shifts in biological sciences have provided new evidence to support intuitive beliefs regarding the connection between the mind–body unit, external and/or internal stimuli such as viruses and bacteria, and primordial environmental stimuli such as light–dark cycle, moon cycle, tides, magnetic forces, and humidity. Moreover, many new factors such as climate changes; air pollution; the rise in world population, particularly in developing countries; the rise of poverty in developed countries; and their social and environmental effects are becoming increasingly sophisticated. Consequently, the role of human management of the ecosystem has been reconsidered by each and every one of us, scientists, politicians, and the lay public. On the other hand, methodological communication was not a major problem during the early days of medicine. Treatment modalities were based on a gift from mother nature: plants and extracts thereof. Herbal medicine, which must be integrated into this vision, has long been an accepted treatment for various diseases in European countries, South/North America, Africa, Australia, and Asia, which includes China, the new emergent country. One key to eventually understand and use this complex communication network, which implies external and/or internal human environments and psycho–immune–nervous–endocrine system for the treatment of human diseases, would seem to be in a complete knowledge of the involved cellular and molecular processes. However, to help this network to maintain a good balance between all small and big components, it might be possible to contribute with natural products. Before going on with this approach and therapeutic philosophy, it is necessary to be aware of the big difference in action between the synthetic product represented by one molecule and the naturally derived products represented by tens of hundred molecules [1, 2]. The new concept of phytotherapy is based on knowledge and experience collected by many world and ethnic populations, including those of China, America (mainly South), Africa, and also Europe. In these traditions, herbs used for thousands of years are considered as tonics for the improvement of general health. In more recent time, animal experiments and clinical trials have shown that quite a few herbs are immunologically active, and most of the tonics are excellent immunopotentiating and immunomodulating agents. In this chapter, we reviewed some of our results on the clinical effect of a very special plant and fruit, that is, cocoa and chocolate, on human healthy volunteers.

Cocoa, Chocolate, and Derivatives: Between Ethnobotanic and History

In the last decades, the international scientific community has become aware of the therapeutic potentiality of cocoa and/or cocoa derivatives and by-products. A scientific paper review has recently been published [3] reporting a short history of cocoa and chocolate, their implications for public health, and their importance for the international market. Currently, the characteristics of cocoa seeds, beans, and different cultivar groups of cacao beans used to make cocoa and chocolate, as well as chemical content of cacao and chocolate and/or derivatives, are well known [4, 5].

Going back in history, chromatographic analyses of residues extracted from pottery vessels show that cacao beverages were made before 1,000 B.C., extending the confirmed use of cacao in Mesoamerica back to at least 500 years earlier. Cocoa came to Europe in the sixteenth century, and in 1737, Linnaeus named the cocoa tree Theobroma (food of God) [3]. In 1590, the Florentine Codex suggested a remedy made out of cocoa beans, maize, and the herb *tlacoxochitl* (*Calliandra anomala*) to alleviate fever, shortness of breath, and heart conditions. Manuscripts produced in Europe and New Spain from the sixteenth to early twentieth century revealed more than 100 medicinal uses for cocoa and/or chocolate [3]. Basically, at the beginning, cocoa beverages were manufactured as a complex mixture of cocoa and spices diluted with water, milk, beer, or wine and sweetened with honey. In Europe, chocolate was sold by a traveling salesman, and the art of making chocolate was learned in Italy. There are three main cultivar groups of cacao beans used to make cocoa and chocolate: Criollo, the cocoa tree used by the Mayans – highly prized and rare – less bitter and more aromatic than other beans, from which only 5–10% of chocolate is made. Latin America and Asia are the two continents with highest production. Forastero trees, which include several subvarieties, are significantly hardier than Criollo trees and produce cheaper cocoa beans; in fact, they are used for 80% of world chocolate production. The Arriba variety is considered the best one. Countries that produce the quality Forastero are Africa, Brazil, and Ecuador. The last cultivar is represented by Trinitario, a hybrid of Criollo and Forastero, used in about 10–15% of chocolate production [3] and produced where the growing conditions are satisfactory. The cocoa beverage was a product that everybody enjoyed over the centuries, and it became a popular drink.

A common worldwide saying is that "Switzerland is the country of clocks and chocolate." Is that true? When it comes to chocolate, the saying is not completely wrong, but it should be seen in the right perspective. The pioneers of Swiss chocolate helped to develop and ameliorate the quality of chocolate. François Luis Cailler (1796–1852) in Corsier, State of Vaud, built the first factory with specific machinery useful for manufacturing chocolate. Later, Daniel Peter (1836–1919) invented milk chocolate, while Rudolph Lindt (1855–1909) invented plain chocolate. Moreover, Henri Nestlè (1814–1890) invented a special milk-based product defined as "child's flour." Nestlè became the first big world entrepreneur. Beginning in the last century (1904), his company started reselling chocolate produced by a third party and began its world expansion.

The clinical study reported in this chapter was conceived in Olivone, Blenio Valley, Switzerland, because at the beginning of the last century (1906), at the same time as Nestlè, a chocolate factory was born in Blenio Valley. A special chocolate was produced there until 1968: the fierce competition with other chocolate firms in Switzerland and abroad ended the life of this Blenio factory, where, since 2003, the Alpine Institute of Chemistry and Toxicology is still active in research and analytical services.

Cocoa, Chocolate, and Derivatives: Clinical Studies

In 1996, the first human clinical study with chocolate was run by Kondo et al., who found that 35 g of defatted cacao decreased LDL oxidation between 2 and 4 h after ingestion. Since 1996, cacao and chocolate products have generated significant interest for the pharmaceutical and nutraceutical industry, owing to their association with various health-protective and therapeutic activities, and at least 45 human studies involving the use of cacao in different forms have been performed [3–9].

Effect on Biological Systems

The first available scientific data on the relationship between cacao, reactive oxygen species (ROS), and cytokine production show that cacao liquor polyphenols inhibit ROS and reduce the expression of interleukin 4 (IL-4) mRNA in human lymphocytes [10]. Excluding some very interesting studies [9–15], there is a paucity of information regarding the potential immunoregulatory effects of cacao and chocolate on human peripheral blood mononuclear cells (PBMCs) and polymorphonuclear cells (PMNs). On the other hand, recent findings suggest immunomodulatory functions like regulation of cytokine production in vivo [16–18]. Chronic and acute inflammation underlies the molecular basis of many pathologies: cacao products with flavonoids (flavonols, as demonstrated by in vitro and ex vivo studies, might represent an interesting dietary support to condition and modulate the inflammatory pathway). Indeed, proinflammatory cytokines [11–14], activation of platelets [11], nitric oxide-mediated mechanisms, and plasma leukotriene/prostacyclin ratio, a measure of the proinflammatory/anti-inflammatory eicosanoid balance, speak in favor of an anti-inflammatory role of cacao [19]. In particular, activity of cocoa molecules seems to act on Th1/Th2 balance by increasing IL-4 secretion and therefore Th2 response [20]. Degree of flavonoid polymerization might be involved on Th2 cytokines: short-chain cacao procyanidins increase IL-4 and IL-5, while long-chain reduce both of them [13, 21]. The mechanism by which cacao exerts its opposing effects on Th1/Th2 cytokines remains to be established [22, 23]. Other effects, such as cacao macrophage and lymphocyte downregulation in vitro, have been reported. In vivo effects of cocoa-rich compounds exert beneficial effects on blood pressure, insulin resistance, vascular damage, and oxidative stress [24]. Given their powerful antioxidant activity, flavonoids seem to be the perfect candidates for immune regulation; nonetheless, studies showing opposite effects among different cacao flavonoid fractions suggest that other compounds may contribute to cacao's immune effects. In any case, as the profile of flavonoids absorbed in vivo differs from the one present in crude cacao extract, the physiological relevance of these data is limited [9].

28-Day Action and Effects of Black and White Chocolate on Neuroimmune Systems of Healthy Human Volunteers

This study comprises three phases: the first phase characterized each chemical characteristic of black and white chocolate used in the study, the second phase included the clinical study, and the third was focused on results analysis and perspectives on the real potentiality of chocolate treatment of some human disturbances and/or diseases (Table 40.1).

A 28-day treatment with and/or without chocolate has been planned to lay the basis for further specific clinical studies. The aim of the project was to understand whether a relatively long-term chocolate treatment might have some effects on psychoneuroimmunological parameters in healthy adult human volunteers. Since it was not possible to characterize 380 molecules, a chocolate with higher polyphenol, theobromine, and epicatechin (dark) or without (white) content was used. All obtained results are in accordance with data reported in literature by Tomas-Barberan et al. [25].

Both dark and white chocolate presented the following characteristics: energetic values (Kcal/100 g: 550 dark, 561 white) and proteins (g/100: 8 dark, 6 white) are similar between dark and white, while a notable difference was noted for carbohydrates (g/100 g: 13 dark, 57 white) and fats (g/100 g: 63 dark, 34 white). The most relevant difference between dark and white chocolate has been reported for total polyphenols (25.03 mgGAE/g), theobromine (9.25 mg/g), and epicatechin (0.930 mg/g) in dark chocolate 78% versus not detected values in white chocolate. The 28-day long study design involved 21 healthy adult men aged between 25 and 30, investigator blinded and a parallel-group trial in accordance with ethical guidelines. Chocolate doses –25 g per patient – were wrapped in aluminum foil and

Table 40.1 Cocoa and chocolate in humans: the rationale for clinical studies on healthy volunteers

Third phase (4 months)		Analysis of results, conclusions, and perspectives of clinical studies with cocoa and chocolate in humans
Second phase (1 year)		Blinded clinical study
	T_{28}	Parameter monitoring and physical examination no. 2 by a doctor
	T_{21}	Telephone interview (at home)
	T_{14}	Telephone interview (at home)
	T_7	Parameter monitoring and physical examination no.1 by a doctor
	T_0	Parameter monitoring
	T_{-7}	Screening of participants by doctors, nutritionist, and scientist. Definition of study groups :
		Dark chocolate group
		White chocolate group
First phase (1.5 year)		Definition and selection of the participants to the study
		Application and authorization of the ethical committee
		Definition of the plan of clinical and experimental studies
		Chemical analysis of black and white chocolate
		Definition of the project of search, the sponsor's search

provided in nontransparent bags that transferred no information about the content. After a cacao-free run-in period of at least 4 days and an overnight fast of 10 or more hours, participants were allocated to receive over 4 weeks 75 g/day of commercially available polyphenol-rich dark chocolate or a matching dose per day of white chocolate, containing similar energy. Participants were instructed to ingest the total daily amount of chocolate in three doses of 25 g. Each dose was eaten at three different moments: at 4.00, 6.00, and 8.00 P.M. According to the second phase of the study, all plasma parameters were assessed, while each participant was in the 12-h fasting state between 6 and 10 A.M. Peripheral venous blood was taken after the run-in period (T_0), after 1 (T_7) and 4 (T_{28}) weeks of treatment. Adverse events were monitored every week via interview, on T_7 and T_{28} throughout physical examinations and laboratory testing, while on T_{14} and T_{21}, via telephonic interview. All personal data, for each participant in the study, were sent to their respective physicians.

The aim of this study was to evaluate on venous peripheral blood whether the immune system cell subpopulation and derived cytokines are affected upon chocolate intake. More than 40 basic clinical parameters, T and B lymphocyte cell markers (different T cell subsets activated or not), neurotransmitters such as serotonin (5-HT), 5-hydroxytryptophan (5-HTOL), 5-hydroxyindoleacetic acid (5-HIAA), dopamine (DA), homovanillic acid (HVA), adrenaline (A), and noradrenaline (NA) were measured. The functionality in vitro of some peripheral blood cells has been verified by assessing three specific cytokines, that is, interleukin 10 (IL-10), interleukin 8 (IL-8), and tumor necrosis factor alpha (TNF-α). This 28-day clinical study, aimed at assessing the effect of dark and white chocolate on healthy human volunteers, has allowed us to corroborate some previously published evidences by other authoritative scientists. On the other hand, we have been surprised by some new effects on neuroimmune parameters. No adverse events and/or side effects occurred in both groups of treatment, dark and white chocolate. Fifty percent of the participants allocated to the chocolate treatment group found it hard to eat 75 g of chocolate every day for 4 weeks. Supplementation with dark or white chocolate was well tolerated. Baseline data, at time T_0, of both treated groups did not differ, either in terms of anthropometric and hemodynamic measures or in terms of systolic and diastolic blood pressure. Finally, 21 participants were randomized. Nineteen completed the study and attended all monitoring sessions. None of the data were lost. According to the participants' self-reports, the weekly monitoring of the returned empty bags and the personal interview during each week, all chocolate portions were eaten and no other cacao products were consumed. Moreover, all chocolate portions were consumed at the requested time with maximum 1 h of difference. Participants showed no

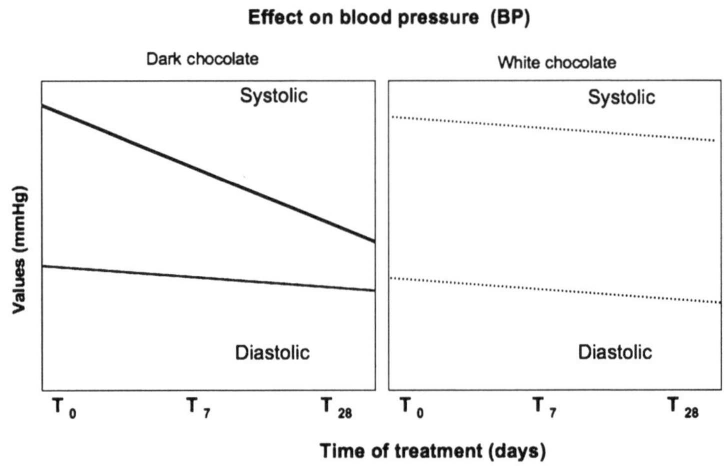

Fig. 40.1 Effect of dark and white chocolate on diastolic and systolic blood pressure. Systolic pressure decreased in both groups during treatment, but only in the dark chocolate treatment group was the decrease statistically significant

significant difference to the reported habitual frequency of food intake and physical activity during treatment, and from the participants' food diaries, no important changes in nutritional composition of diets from the run-in phase to the end of the intervention were observed. Due to the energy, nutrient, and electrolyte contribution of the daily dark or white chocolate doses to the total diet, body weight and clinical parameters did not change with clinical relevance during the study. Also, the body mass index (calculated as weight in kilograms divided by height in square meters) did not show any modification and remain between 18.5 and 25. Basic clinical parameters monitored along the 28 days remained in the normal range. The cytofluorimetric analysis to understand any possible changes in lymphocyte subset populations as, for example, B lymphocytes, T helper, and T suppressor lymphocytes, activated or not, shows no statistical differences between the group treated with dark chocolate and the group treated with white chocolate. But the main question at this point was to know whether black and/or white chocolate in healthy human volunteers showed any activity on other parameters. First of all, dark and white chocolates influence blood pressure (Fig. 40.1) where only systolic blood pressure decreased in both groups of treatment. Only in the dark chocolate group this result is statistically significant and all data confirmed which is also reported in other studies [26–28].

Other interesting results have been obtained on parameters connected with the immunologic system, that is, C-reactive protein (CRP), the first marker of inflammation; IL-10; and TNF-α. C-reactive protein concentration, although not statistically significant, showed a decreasing trend in the dark chocolate group, whereas no difference seems to be assessed in the white chocolate one. Interleukin 10 is a regulator of lymphoid and myeloid cell function produced by B and T cells, activated mast cells, macrophages, and keratinocytes. Due to the ability of IL-10 to block cytokine synthesis and several accessory cell functions of macrophages, this cytokine is a potent suppressor of the effector functions of macrophages, T cells, and natural killer cells. In addition, IL-10 participates in regulating proliferation and differentiation of B cells, mast cells, and thymocytes. Moreover, IL-10 was recently identified as the cytokine produced by the T helper subpopulation cells (Th2) that inhibit the synthesis of immunostimulatory cytokines by the Th1 cells. IL-10 exhibits inhibitory effects on monocytes including downregulation of MHC class II antigen expression and suppression of IL-1α (alpha) and β (beta), IL-6, IL-8, GM-CSF, G-CSF, and TNF-α production; synergizes with IL-2 and IL-4 to promote the proliferation of thymocytes; and synergizes with IL-3 and IL-4 to enhance the survival of mast cells. IL-10 production can be induced by mitogenic lectins and LPS, while IL-4 and IFN-γ inhibit this production. The suppressive effects of IL-10 on monocytes and Th1 cytokine synthesis

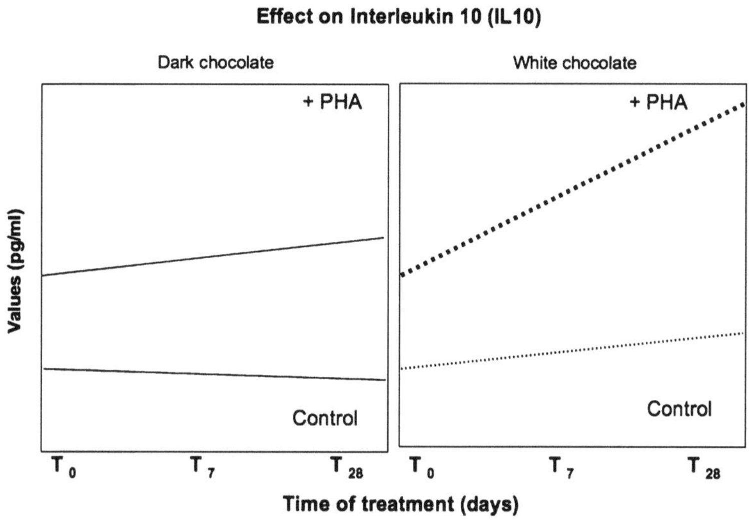

Fig. 40.2 Effect of dark and white chocolate treatment on in vitro production of IL-10 by cultured resting and phytohematoagglutin (PHA) cultured resting and PHA-activated peripheral blood mononuclear cells (PBMCs) collected from human venous blood. No significant differences in IL-10 concentration in cultured resting and PHA-activated PBMCs were observed in the dark chocolate treatment group, whereas IL-10 production *increased* in PHA-activated PBMCs collected from the white chocolate treatment group after 1 and 4 weeks of treatment ($p<0.01$)

suggest that IL-10 may have utility as a general suppressor of immune function. The targets for such immunosuppressive drugs include infectious disease, transplantation, induction of tolerance, and possibly cancer. IL-10 is currently in preclinical studies to evaluate its potential in various disease states [29]. No statistically significant differences in IL-10 concentration in cultured resting and PHA-activated PBMCs were observed in the dark chocolate group, whereas a statistically significant ($p<0.01$) increase in IL-10 production by PHA-activated PBMCs after 1 and 4 weeks of treatment was assessed in the white chocolate group (Fig. 40.2).

TNF-α (alpha), produced by monocytes and macrophages throughout mediation by lymphokines and endotoxines and secreted upon stimulation by IFN-γ (gamma), is a multipotent modulator of immune response [8, 28]. TNF-α responds to stimuli such as infectious agents or tissue injury by activating neutrophils, altering properties of vascular endothelial cells, regulating metabolic activities of other tissues, as well as exhibiting tumoricidal activity by inducing localized blood clotting. Activation of B cells by the Epstein–Barr virus can be inhibited by TNF-α [30]. Moreover, TNF-α may play a role in the pathogenesis of many disease states such as inflammatory disease of joints and other tissues, septicemia and meningococcal disease, parasitic infections, myocarditis, and AIDS. No statistically significant modulation by 28 days after chocolate administration is observed: to be critical, a decrease in both groups, dark and white chocolate, on the TNF-α production of this proinflammatory cytokine by cultured resting and activated PBMCs has been observed (Fig. 40.3).

The last group of molecule assessed by this study is connected with indolamines and catecholamines, where 5-HT (serotonin) and 5-HIAA increased during treatment in both dark chocolate and white chocolate groups. Serotonin increase is statistically significant on T_{28} in the WCG, whereas its metabolite 5-HIAA increased significantly in the DCG (Fig. 40.4). The amount of catecholamines in the extracellular space is a function of the balance between their vesicular release and their reuptake by the monoamine transporter system.

A significant increase of dopamine plasmatic concentration has been detected in the white chocolate group, whereas there has been a very small increase in the dark chocolate group (Fig. 40.5).

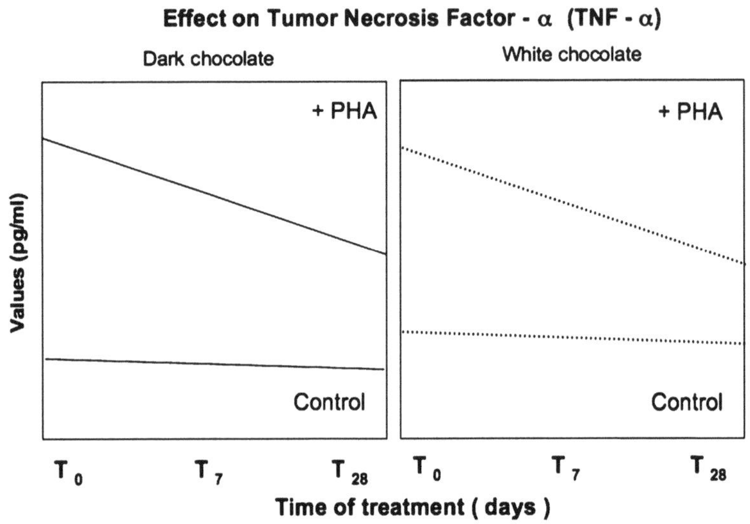

Fig. 40.3 Production of TNF-α by cultured resting and PHA-activated PBMCs. No apparently significant modulation has been reported, although it has been observed a trend to TNF-α decrease in both dark and white chocolate treatment groups

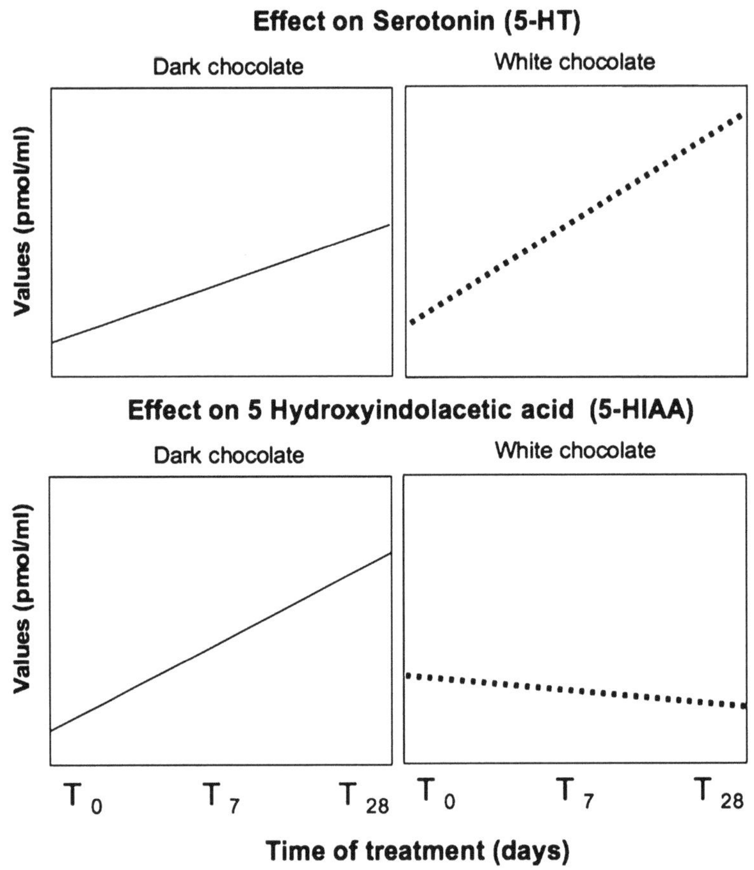

Fig. 40.4 Plasmatic concentration of serotonin (5-HT) and 5-hydroxyindoleacetic acid (5-HIAA); 5-HT increased in both treatment groups after 28 days of treatment; the increase is statistically significant only in the white chocolate group ($p<0.01$). The metabolite 5-HIAA increased only in the group treated with dark chocolate

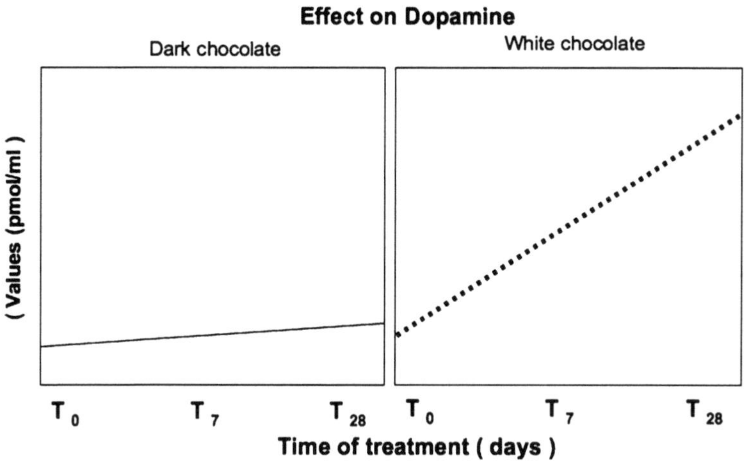

Fig. 40.5 Effect of dark and white chocolate treatment on dopamine plasmatic concentration. A significant increase is detected in the white chocolate–treated group ($p<0.01$), whereas no changes in dopamine concentration have been reported in the dark chocolate–treated group

The decrease of adrenaline (A) is statistically significant on T_{28} versus T_0 in the dark chocolate group. Noradrenaline (NA) increased on T_7 (+51.3% vs. T_0), returning to baseline values at T_{28}. On the other hand, in the white chocolate increased +2.7% on T_7 and decreased −32% on T_{28} versus baseline, whereas differences are not statistically significant. Noradrenaline increase in the white group is almost linear, showing +34% on T_{28}. Also, in this case, differences are not significant (Fig. 40.6).

Summary

This clinical study is one of the first reports demonstrating a neuroimmunomodulatory role of chocolate in healthy human volunteers and, surprisingly, not only does dark chocolate seem to have special effects but also the white one. In fact, the latter increases (+29.6% vs. baseline) IL-10 production from PHA-activated PBMCs and causes a decrease in monocyte population. This may indicate suppressive effects of IL-10 on monocytes. At T_{28}, both IL-10 and monocytes are increased. It is not easy to explain this data. In fact, other active compounds, like biogenic amines present in cacao and chocolate, may also affect the stimulation capacity of PBMCs. Moreover, this study confirms the systolic blood pressure-lowering effect of dark chocolate with high concentration of polyphenols in subjects with recently diagnosed and untreated stage one mild isolated systolic hypertension and adults with untreated upper-range prehypertension or stage one hypertension without concomitant risk factors [8, 28]. On the other hand, these data are controversial and need to be confirmed in other models [31, 32], although, taken together, all this indicates that cacao and dark chocolate may play a role in normalizing elevated blood pressure. In people who have normal blood pressure, a dietary supplement of chocolate may help to maintain pressure in the normal range. How to explain such interesting action of chocolate? The decrease of blood pressure might be related to a decrease of adrenaline production, which plays a role in limiting the production of angiotensin II precursor, renin, which in turn is responsible for the increase in blood pressure. Moreover, chocolate contains moderate concentrations of potassium (dark chocolate, ca. 263 mg; white chocolate, ca. 287 mg) [33]. It is extensively recognized that reduction of blood pressure correlates with increased intake of potassium, probably

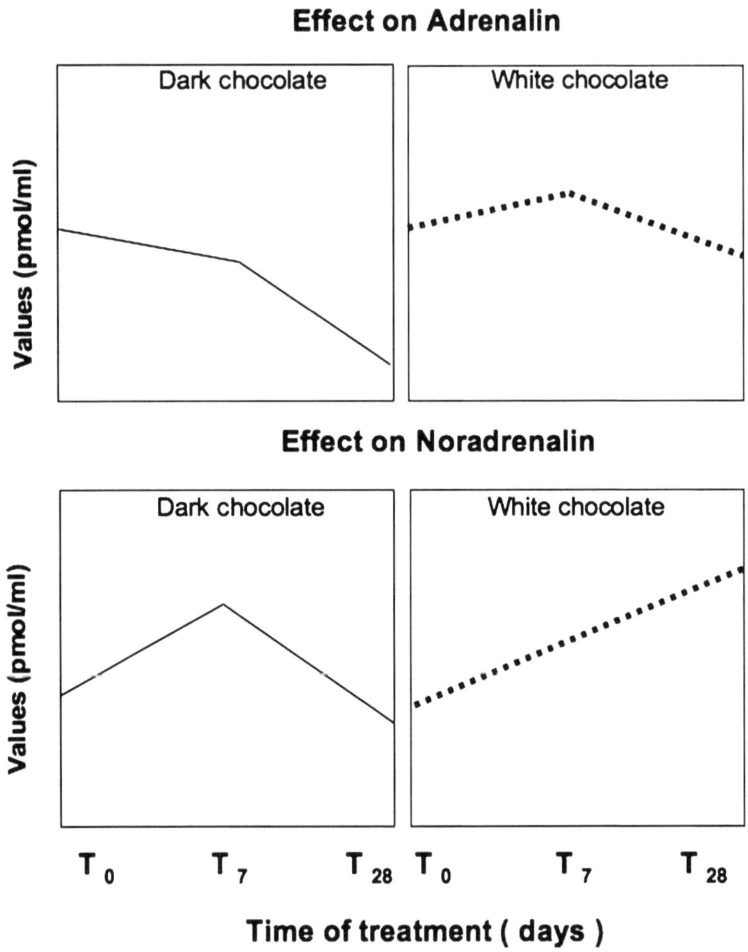

Fig. 40.6 Change in plasmatic concentration of adrenaline and noradrenaline. The decrease of adrenaline in the dark chocolate group is statistically significant after 28 days of treatment, whereas noradrenaline, in the same group, increased significantly after 7 days of treatment and returned to basal level after 28 days. In the white chocolate group, no differences have been observed on adrenaline production, while a clear trend to increase has been noted on noradrenaline

because potassium increases the excretion of sodium and for its activity on blood vessels. Nevertheless, certain cacao flavanol fractions are able to induce synthesis of prostacyclin and relaxation of isolated aortic rings with increased nitric oxide synthesis. Finally, the biological activity attributed mainly to flavonoids seems to be most effective on endothelium and, in turn, on BP decrease. In fact, the decrease of BP is observed in the dark chocolate group only. Moreover, the white chocolate group did not take polyphenols. Epicatechin, catechin, and dimeric flavanols might exhibit a dose-dependent accumulation within the nuclei, and this might also be possible for metabolites. To get any presumably antioxidant effect, the first dose of flavanol-rich cacao is not sufficient to reach the maximum effect. The response becomes interesting after many boluses have been administered and a new higher plateau will be reached: data indicate that 5–7 days are required to reach a new steady state [31, 34]. In fact, Engler et al. [31] and Fisher et al. [32] were unable to document any blood pressure change in young, healthy, normotensive persons after consumption of flavonoid-rich dark chocolate. The reason is more likely to be linked with the period of treatment (5 days). Results obtained with the present long-period study seem to confirm this hypothesis. Moreover, the most abundant flavonoid compounds may not necessarily lead to the highest concentrations of biologically active metabolites in target tissues nor may they be the most biologically active in relation to specific health outcomes [35]. Moreover, and possibly related, a residual vascular response is evident 15 h after the last dose, at a time when pharmacokinetic studies indicate that the flavanols and their known metabolites have largely disappeared from the circulation. Thus, one possible mechanistic sequence would involve activation at nitric oxide synthase gene level, as a first step. As a second possibility, the responsible agent may be a metabolite of the flavanols that gradually accumulates [34]. However, metabolites of dietary phenolics, which appear in the circulatory system in very low concentrations (nmol/L to low mmol/L), might exert modulatory effects in cells through actions on components of the intracellular signaling cascades. Such events are important for cellular growth, proliferation, and apoptosis. In addition, the intracellular concentrations required to affect cell signaling pathways are considerably lower than those required to impact on antioxidant capacity.

Our clinical study seems to confirm the preliminary in vitro data showing that plasma concentration of 5-HT (serotonin) and 5-HIAA increased during dark and white chocolate treatments [15]. Interestingly and surprisingly enough, our data seem not to be correlated with the nonfat cacao percentage, that is, polyphenols, epicatechin, and theobromine, which are not present in white chocolate. Chocolate craving was reported to be a form of self-medication in atypical depression, in seasonal affective disorder, as well as an interesting impact on brain neurotransmitters with antidepressant benefits [36, 37]. Accordingly, several psychoactive constituents, including anandamide, caffeine, or phenylethylamine, with also neural and immunologic activity, have been identified in cacao. Moreover, at least in the gastrointestinal tract, one can assume the existence of all compounds present in cacao at effectual concentrations where they could also increase the availability of tryptophan and production of serotonin [15]. A great proportion of serotonin is synthesized and stored in the gastrointestinal tract where it plays a paracrine secretion and motility role. On this basis, ingestion of cacao products like polyphenols, which are present in cacao, might play an important role in tryptophan metabolism and serotonin availability. Furthermore, antioxidant capacity of cacao products may locally shift the redox equilibrium in the gastrointestinal tract, which could be of benefit for the intestine and the whole organism and might improve quality of life [15]. An increase in plasmatic serotonin concentration has been observed in the white chocolate control group. This unexpected increase could be linked with an enhanced consumption of carbohydrates and fats, which, in turn, may increase serotonin synthesis and modulate neurochemical imbalance. On the other hand, some component present in chocolate diminished tryptophan degradation, most probably by the inhibition of indoleamine 2,3-dioxygenase (IDO) activity. Consequently, this fact might increase the production of serotonin (5-HT). Serotonin (5-HT) and 5-hydroxyindoleacetic acid (5-HIAA) increased in both groups but reached smaller values of plasmatic concentration in the dark chocolate one. It might be believed that some active compound, and/or combination of two or more different molecules present in both dark and white chocolate, could be responsible for the observed effects.

This clinical study on the effect of chocolate shows that C-reactive protein, a specific marker of inflammation, remains within the normal range in both groups, thus confirming previous data in other clinical studies [28]. On the other hand, a decrease of CRP concentration after 30 days' intervention with flavanol-rich cacao containing 321 mg flavanols per dose/day on 41 medicated diabetic patients was reported [38]. Popular medicine perceives chocolate and cacao as food cravings that may create dependence. Interestingly, within this clinical study, it has been observed that many volunteers, at the end of the 28-day protocol, manifested a moderate aversion to chocolate, in particular for the dark one. Discussion about craving is a little bit complex because chocolate is a complex of cocoa molecules including sugars and psychoactive molecules [39–41]. Because chocolate craving has some features of addiction, attempts have been made to identify any psychoactive ingredient. Several candidates have been identified (the biogenic stimulant amines caffeine, theobromine, tyramine, and phenylethylamine), but up to now, their concentrations are considered too low to have a significant psychoactive effect, and they are also present in higher concentrations in non-craved foods [42].

Comparisons of subjects who ingested milk chocolate, dark chocolate, white chocolate, and cacao powder (powdered cacao mass with some cacao butter extract) have demonstrated that milk chocolate is the favorite one. If psychoactive substances were involved, then cacao powder should equally satisfy the craving and dark chocolate should be the most preferred. Chocolate contains two analogues of anandamide that are similar to the cannabinoid responsible for euphoria caused by cannabis [43]. Our project was mainly aimed at evaluating whether intake of chocolate by human volunteers might influence some neuroimmunological parameters and represents one of the few studies focused on acquiring data on the effects of long-term administration of chocolate on the neuroimmune system of healthy volunteers. The clinical study has been conducted on healthy human volunteers living a normal lifestyle. Our basic results, as well as in the paucity of studies published in the last decade, might constitute a sufficient rationale for a more complete and multidisciplinary study on selected human volunteers suffering from specific pathologies. With this preliminary clinical study, many other relevant conclusions might be formulated and will need to be confirmed in a larger multicenter study. No changes in body mass index or other adverse effects were detected. In fact, all clinical parameters tested during the study protocol remained within the normal range. We could nevertheless observe an increase of the total leukocyte and eosinophil subpopulation in the dark chocolate group (DCG). Data remained within the normal clinical range, a slight allergic answer to the dark chocolate can consequently be assumed. Possibly, an increased eosinophil population can be hypothesized to be related to the Th2 immune response. An increase in monocytes and basophil plasmatic concentration was detected in the white control group (WCG). Also, in this case, all values remained within the normal range, and this clinical evidence could be associated with an increased consumption of triglycerides and fats [42]. Finally, our study has made us think of two different ways of action of cacao and chocolate in vivo. The antioxidant effect seems to be linked to the blood pressure-lowering effect and positively related to the content of particular polyphenols such as the flavanols (catechin, epicatechin, and polymeric procyanidins). Concentrations of flavanol in cacao and cacao-derived products like chocolate depend on the geographical origin of the beans and their different manufacturing steps, such as roasting and fermentation, which can, in turn, decrease the polyphenol content by more than 80%.

The presence of these molecules is linked to the nonfat cacao mass. Cacao powder and some dark chocolates contain very high concentration of polyphenols, and their effect on the endothelium is now recognized. In contrast, white chocolate does not contain this type of antioxidant molecules and, consequently, no decrease in blood pressure was found after white chocolate consumption. White chocolate contains much more cacao butter, fat, fatty acids, sterols, and sugars than dark chocolate. In addition, some lipophilic substances such as tryptophan, serotonin, dopamine, and anandamide possibly act synergistically to elicit a human immunomodulating effect in vivo and ex vivo. These substances could be present in the fat–cacao fraction, also in dark chocolate but at lower concentration than in the white one. This might be the reason why a similar, but less potent, effect on cytokine production in both resting and activated PBMCs was observed in the dark chocolate group in the present study. Moreover, (beta)β-sitosterol and stigmasterol have been shown to be safe and effective in lowering circulating cholesterol levels [44]. Recent studies have linked the consumption of plant sterols and stanol esters with a reduction in risk of cardiovascular disease in those who do not adopt a Mediterranean diet [45]. As largely reported in related literature, the cacao extract used in vitro, in vivo, and ex vivo is depleted from the fat component (defatting cacao and chocolate is a crucial step in obtaining a flavanol extract). As a consequence, lipophilic substances, which are also not easily water soluble, are not present in the extract. For this reason, the immunomodulating effect of this extract could be related to the presence of very high concentrations of an antioxidant substance that is not the same as those present in chocolate. Moreover, the concentration used for cell treatment in vitro is difficult, if not impossible, to reach in vivo.

A more general conclusion lets us postulate that traditional folk medicine, such as European natural medicine, traditional Chinese medicine, and American folk medicine, is a great treasure of the

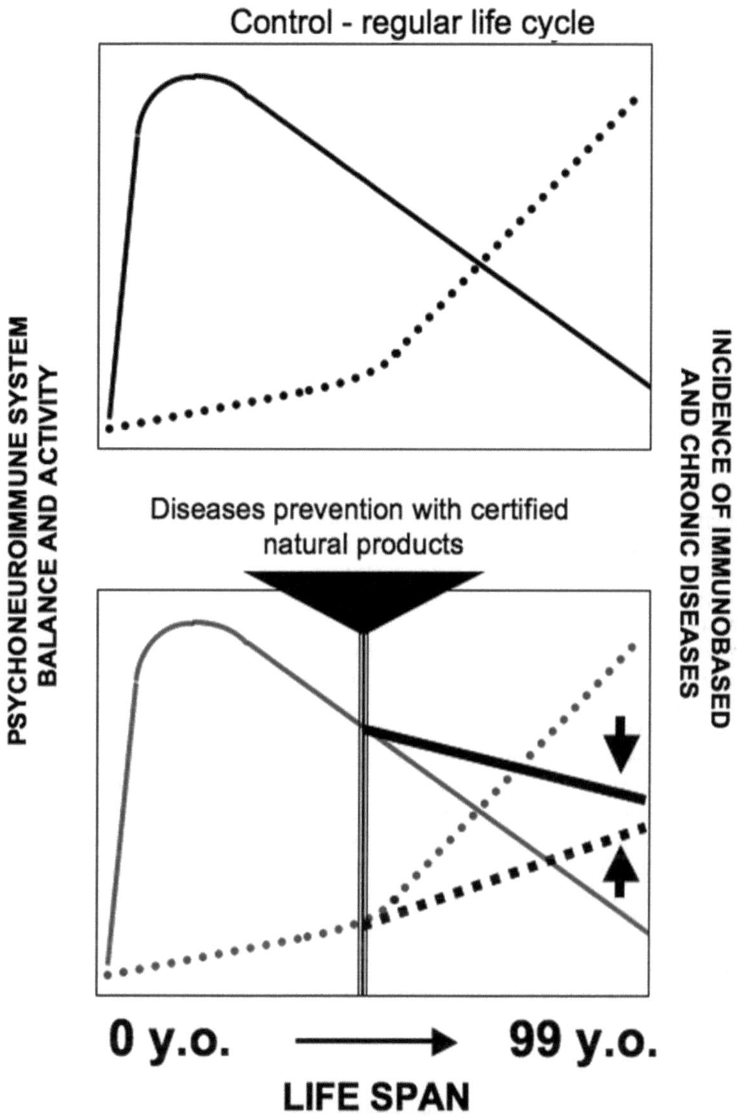

Fig. 40.7 During the normal life cycle, the psychoneuroimmune balance and activity undergo normal involution. This physiologic mechanism could be disturbed as a result of altered hormonal and immunological molecule production (cytokines, lymphokines, neurotransmitters, etc.) in physiopathologic situations such as depression, neuropathies, psychiatric diseases, aging, metabolic and/or endocrinologic diseases, and exposure to unnatural photoperiod and electromagnetic fields. To summarize, on the one hand, there is an involution; during the life span, the neuroimmune system is prone to developing diseases. With the use of certified natural products (i.e., chocolate), it might be possible to halt the decline of the neuroimmune system and prevent certain immune-based and chronic diseases

people. It offers significant potential for new drug discovery, and cocoa/chocolate contain a large number of molecules.

Our Alpine Institute of Chemistry and Toxicology is willing to cooperate with scientists from other countries to do joint research. We believe that new drugs that could help in the care of immune-based degenerative diseases such as cancer and AIDS (Fig. 40.7) and make human longevity related to immunomodulators will be developed in the next century and will have great contribution to the health of mankind.

Acknowledgments This work was supported by a private sponsor – Confiserie Teuscher Zürich, Switzerland – and by the doctoral (Ph.D.) research work done by Manuel Rusconi, collaborator of the Alpine Institute of Chemistry and Toxicology. The authors gratefully acknowledge the sponsor and Dr. M. Rusconi.

References

1. McCann SM, Lipton JM, et al., editors. Neuroimmunomodulation: molecular aspects, integrative systems and clinical advances. In: Proceedings of the 3rd congress of the International Society of NeuroImmunoModulation (ISNIM). Bethesda; 13–15 Nov 1996, Ann NY Acad Sci 1998;840:1–866.
2. Savino W, Silva PO, Besedowsky H, editors. Neuromodulation: from fundamental biology to therapy. In: Proceedings of the VII congress of the International Society of Neuroimmunomodulation (ISNIM). 14–27 April 2008. Rio de Janeiro. Ann NY Acad Sci 2009;1153:xi, 1–271.
3. Rusconi M, Conti A. *Theobroma cacao* L., the food of the gods: a scientific approach beyond myths and claims. Pharmacol Res. 2010;61:5–13.
4. Kondo K, Hirano R, Matsumoto A, Igarashi O, Itakura H. Inhibition of LDL oxidation by cacao. Lancet. 1996;348:1514.
5. Visioli F, Bernaert H, Corti R, Ferri C, Heptinstall S, Molinari E, et al. Chocolate lifestyle and health. Crit Rev Food Sci Nutr. 2009;49:299–312.
6. Di Giuseppe R, Di Castelnuovo A, Centritto F, Zito F, De Curtis A, Costanzo S, et al. Regular consumption of dark chocolate is associated with low serum concentration of C-reactive protein in a healthy Italian population. J Nutr. 2008;138:1939–45.
7. Faridi Z, Njike VY, Dutta S, Ali A, Katz DL. Acute dark chocolate and cocoa ingestion and endothelial function: a randomized controlled crossover trial. Am J Clin Nutr. 2008;88:58–63.
8. Taubert D, Roesen R, Lehmann C, Jung N, Schömig E. Effects of low habitual cocoa intake on blood pressure and bioactive nitric oxide. JAMA. 2007;298:49–60.
9. Ramiro-Puig E, Castell M. Cacao: antioxidant and immunomodulator. Br J Nutr. 2009;101(7):931–40.
10. Sanbongi C, Suzuki N, Sakane T. Polyphenols in chocolate, which have antioxidant activity, modulate immune functions in humans in vitro. Cell Immunol. 1997;177:129–36.
11. Mao TK, van de Water J, Keen CL, Schmitz HH, Gershwin ME. Modulation of TNF-alpha secretion in peripheral blood mononuclear cells by cacao flavanols and procyanidins. Dev Immunol. 2002;9(3):135–41.
12. Mao TK, Van De Water J, Keen CL, Schmitz HH, Gershwin ME. Cacao flavonols and procyanidins promote transforming growth factor-beta1 homeostasis in peripheral blood mononuclear cells. Exp Biol Med. 2003;228(1):93–9.
13. Mao TK, Van De Water J, Keen CL, Schmitz HH, Gershwin ME. Effect of cacao flavanols and their related oligomers on the secretion of interleukin-5 in peripheral blood mononuclear cells. J Med Food. 2002;5(1):17–22.
14. Mao TK, Powell J, Van de Water J, Keen CL, Schmitz HH, Hammerstone JF, et al. The effect of cacao procyanidins on the transcription and secretion of interleukin 1 beta in peripheral blood mononuclear cells. Life Sci. 2000;66(15):1377–86.
15. Jenny M, Santer E, Klein A, Ledochowski M, Schennach H, Ueberall F, et al. Cacao extracts suppress tryptophan degradation of mitogen-stimulated peripheral blood mononuclear cells. J Ethnopharmacol. 2009;122:261–7.
16. Ramiro-Puig E, Urpí-Sardà M, Pérez-Cano FJ, Franch A, Castellote C, Andrés-Lacueva C, et al. Cocoa-enriched diet enhances antioxidant enzyme activity and modulates lymphocyte composition in thymus from young rats. J Agric Food Chem. 2007;55:6431–8.
17. Ramiro-Puig E, Pérez-Cano FJ, Ramírez-Santana C, Castellote C, Izquierdo-Pulido M, Permanyer J, et al. Spleen lymphocyte function modulated by a cocoa enriched diet. Clin Exp Immunol. 2007;149:535–42.
18. Ramiro-Puig E, Pérez-Cano FJ, Ramos-Romero S, Pérez-Berezo T, Castellote C, Permanyer J, et al. Intestinal immune system of young rats influenced by cocoa enriched diet. J Nutr Biochem. 2008;19:555–65.
19. Schramm DD, Wang JF, Holt RR, Ensunsa JL, Gonsalves JL, Lazarus SA, et al. Chocolate procyanidins decrease the leukotriene-prostacyclin ratio in humans and human aortic endothelial cells. Am J Clin Nutr. 2001; 73:36–40.
20. Ramiro E, Franch A, Castellote C, Andrés-Lacueva C, Izquierdo-Pulido M, Castell M. Effect of Theobroma cacao flavonoids on immune activation of a lymphoid cell line. Br J Nutr. 2005;93:859–66.
21. Mao T, Van De Water J, Keen CL, Schmitz HH, Gershwin ME. Cacao procyanidins and human cytokine transcription and secretion. J Nutr. 2000;130:2093S–9.
22. Mowen KA, Glimcher LH. Signaling pathways in Th2 development. Immunol Rev. 2004;202:203–22.
23. Ramiro-Puig E, Casadeus G, Lee HG, Zhu X, McShea A, Perry G, et al. Neuroprotective effect of cacao flavonoids on in vitro oxidative stress. Eur J Nutr. 2009;48:54–61.

24. Selmi C, Cocchi CA, Lanfredini M, Keen CL, Gershwin ME. Chocolate at heart: the anti-inflammatory impact of cacao flavanols. Mol Nutr Food Res. 2008;52:1340–8.
25. Tomas-Barberan F, Cienfuegos-Jovellanos E, Marín A, Muguerza B, Gil-Izquierdo A, Cerda B, et al. A new process to develop a cacao powder with higher flavonoid monomer content and enhanced bioavailability in healthy humans. J Agric Food Chem. 2007;55:3926–35.
26. Hollenberg NK, et al. Cacao, flavanols and cardiovascular risk. Br J Cardiol. 2004;11(5):379–86.
27. Taubert D, Berkels R, Roesen R, Klaus W. Chocolate and blood pressure in elderly individuals with isolated systolic hypertension. JAMA. 2003;290(8):1029–30.
28. Grassi D, Lippi C, Necozione S, Desideri G, Ferri C. Short-term administration of dark chocolate is followed by a significant increase in insulin sensitivity and a decrease in blood pressure in healthy persons. Am J Clin Nutr. 2005;81(3):611–4.
29. Moore KW, O'Garra A, de Waal Malefyt R, Vieira P, Mosmann TR. Interleukin 10. Ann Rev Immunol. 1993;11:165–90.
30. Janssen O, Kaebelitz D. Tumor necrosis factor selectively inhibits activation of human B cells by Epstein Barr virus. J Immunol. 1998;140:125.
31. Engler MB, Engler MM, Chen CY, Malloy MJ, Browne A, Chiu EY, et al. Flavonoid-rich dark chocolate improves endothelial function and increases plasma epicatechin concentrations in healthy adults. J Am Coll Nutr. 2004; 23:197–204.
32. Fisher ND, Hughes M, Gerhard-Herman M, Hollenberg NK. Flavanol-rich cacao induces nitric-oxide-dependent vasodilation in healthy humans. J Hypertens. 2003;21:2281–6.
33. Beutler B, Cerami A. Tumor necrosis factor: a macrophage hormone governing cellular metabolism and inflammatory response. Endocr Rev. 1998;9:57.
34. Hollenberg NK. Vascular action of cacao flavanols in humans: the roots of the story. J Cardiovasc Pharmacol. 2006;47:S99–102.
35. Hooper L, Kroon PA, Rimm EB, Cohn JS, Harvey I, Le Cornu KA, et al. Flavonoids, flavonoid-rich foods, and cardiovascular risk: a meta-analysis of randomized controlled trials. Am J Clin Nutr. 2008;88:38–50.
36. Wurtman RJ, Wurtman JJ. Carbohydrates and depression. Sci Am. 1989;260:68–75.
37. Parker G, Parker I, Brotchie H. Mood state effects of chocolate. J Affect Disord. 2006;92:149–59.
38. Balzer J, Rassaf T, Heiss C, Kleinbongard P, Lauer T, Merx M, et al. Sustained benefits in vascular function through flavanol-containing cacao in medicated diabetic patients a double-masked, randomized, controlled trial. J Am Coll Cardiol. 2008;51:2141.
39. Weingarten HP, Elston D. Food cravings in a college population. Appetite. 1991;17:167–75.
40. Rozin P, Levine E, Stoess C. Chocolate craving and liking. Appetite. 1991;17:199–212.
41. Drewnowski A. Food preferences and the opioid peptide system. Trends Food Sci Tech. 1992;3:97–9.
42. Facchini F, Hollenbeck CB, Chen YN, Chen YD, Reaven GM. Demonstration of a relationship between white blood cell count, insulin resistance, and several risk factors for coronary heart disease in women. J Intern Med. 1992;232:267–72.
43. DiTomaso E, Beltramo M, Piomelli D. Brain cannabinoids in chocolate. Nature. 1996;382:677–8.
44. Thompson GR, Grundy SM. History and development of plant sterol and stanol esters for cholesterol-lowering purposes. Am J Cardiol. 2005;96:3–9.
45. Athyros VG, Kakafika AI, Papageorgiou AA, Tziomalos K, Peletidou A, Vosikis C, et al. Effect of a plant stanol ester-containing spread, placebo spread, or Mediterranean diet on estimated cardiovascular risk and lipid, inflammatory and haemostatic factors. Nutr Metab Cardiovasc Dis. 2011;21(3):213–21.

About the Editors

Ronald R. Watson, Ph.D., attended the University of Idaho but graduated from Brigham Young University in Provo, Utah, with a degree in chemistry in 1966. He received his Ph.D. in biochemistry from Michigan State University in 1971. His postdoctoral schooling in nutrition and microbiology was completed at the Harvard School of Public Health, where he gained 2 years of postdoctoral research experience in immunology and nutrition.

From 1973 to 1974, Dr. Watson was assistant professor of immunology and performed research at the University of Mississippi Medical Center in Jackson. He was assistant professor of microbiology and immunology at the Indiana University Medical School from 1974 to 1978 and associate professor at Purdue University in the Department of Food and Nutrition from 1978 to 1982. In 1982, Dr. Watson joined the faculty at the University of Arizona Health Sciences Center in the Department of Family and Community Medicine of the School of Medicine. He currently serves as professor of health promotion sciences in the Mel and Enid Zuckerman Arizona College of Public Health.

Dr. Watson is a member of several national and international nutrition, immunology, cancer, and alcoholism research societies. Among his patents, he has one on a dietary supplement, passion fruit peel extract, with more pending. He had done DHEA research on its effects on mouse AIDS and immune function for 20 years. He has previously edited a book on melatonin (Watson RR. *Health Promotion and Aging: The Role of Dehydroepiandrosterone (DHEA)*. Harwood Academic Publishers, 1999, 164 pages). For 30 years, Dr. Watson has been funded by Wallace Research Foundation to study dietary supplements in health promotion. He has edited more than 100 books on nutrition, dietary supplements and over-the-counter agents, and drugs of abuse, as well as scientific reference books. He has also published more than 500 research and review articles.

Professor Victor R. Preedy currently serves as professor of nutritional biochemistry in the Department of Nutrition and Dietetics, King's College London, and professor of clinical biochemistry in the Department of Clinical Biochemistry, King's College London. He is also director of the Genomics Centre, King's College London. Professor Preedy graduated in 1974 with a degree in biology and physiology with pharmacology. He received his Ph.D. in 1981 in the field of nutrition and metabolism, specializing in protein turnover. In 1992, he received his membership of the Royal College of Pathologists, based on his published works, and in 1993 he received his D.Sc. for outstanding contribution to protein metabolism. At the time, he was one of the university's youngest recipients of this distinguished award. Professor Preedy was elected as a fellow to the Royal College of Pathologists in 2000. Since then, he has been elected as a fellow to the Royal Society for the Promotion of Health (2004) and the Royal Institute of Public Health (2004). In 2009, he was elected as a fellow of the

R.R. Watson et al. (eds.), *Chocolate in Health and Nutrition*, Nutrition and Health 7,
DOI 10.1007/978-1-61779-803-0, © Springer Science+Business Media, LLC 2013

Royal Society for Public Health. Professor Preedy has written or edited over 550 articles, which include over 160 peer-reviewed manuscripts based on original research, 85 reviews, and 30 books. He has a wide interest in health-related matters, particularly nutrition and diet.

Dr. Sherma Zibadi received her Ph.D. in nutrition from the University of Arizona and is a graduate of the Mashhad University of Medical Sciences, where she also received her M.D. She recently completed her postdoctoral research fellowship, awarded by the American Heart Association. Dr. Zibadi engages in the research field of cardiology and complementary medicine. Her main research interests include maladaptive cardiac remodeling and heart failure, study of the underlying mechanisms, and potential mediators of the remodeling process, which help in identifying new targets for treatment of heart failure. Dr. Zibadi's research interest also extends into alternative medicine, exploring the preventive and therapeutic effects of natural dietary supplements on heart failure and its major risk factors in both basic animal and clinical studies, translating lab research finding into clinical practice. Dr. Zibadi is an author of multiple research papers published in peer-reviewed journals and books, as well as coeditor of several books.

About the Series Editor

Dr. Adrianne Bendich has recently retired as Director of Medical Affairs at GlaxoSmithKline (GSK) Consumer Healthcare where she was responsible for leading the innovation and medical programs in support of many well-known brands, including TUMS and Os-Cal. Dr. Bendich had primary responsibility for GSK's support for the Women's Health Initiative (WHI) intervention study. Prior to joining GSK, she was at Roche Vitamins Inc. and was involved with the groundbreaking clinical studies showing that folic acid–containing multivitamins significantly reduced major classes of birth defects. Dr. Bendich has coauthored over 100 major clinical research studies in the area of preventive nutrition. She is recognized as a leading authority on antioxidants, nutrition and immunity, and pregnancy outcomes, vitamin safety, and the cost-effectiveness of vitamin/mineral supplementation.

Dr. Bendich, who is now President at Consultants in Consumer Healthcare LLC, is the Editor of ten books, including *Preventive Nutrition: The Comprehensive Guide for Health Professionals, Fourth Edition*, coedited with Dr. Richard Deckelbaum, and is Series Editor of Nutrition and Health for Springer/Humana Press (www.springer.com/series/7659). The series contains 40 published volumes – major new editions in 2010–2011 include *Vitamin D, Second Edition,* edited by Dr. Michael Holick; *Dietary Components and Immune Function,* edited by Dr. Ronald Ross Watson, Dr. Sherma Zibadi, and Dr. Victor R. Preedy; *Bioactive Compounds and Cancer,* edited by Dr. John A. Milner and Dr. Donato F. Romagnolo; *Modern Dietary Fat Intakes in Disease Promotion,* edited by Dr. Fabien DeMeester, Dr. Sherma Zibadi, and Dr. Ronald Ross Watson; *Iron Deficiency and Overload,* edited by Dr. Shlomo Yehuda and Dr. David Mostofsky; *Nutrition Guide for Physicians*, edited by Dr. Edward Wilson, Dr. George A. Bray, Dr. Norman Temple, and Dr. Mary Struble; *Nutrition and Metabolism,* edited by Dr. Christos Mantzoros; and *Fluid and Electrolytes in Pediatrics,* edited by Leonard Feld and Dr. Frederick Kaskel. Recent volumes include *Handbook of Drug-Nutrient Interactions,* edited by Dr. Joseph Boullata and Dr. Vincent Armenti; *Probiotics in Pediatric Medicine,* edited by Dr. Sonia Michail and Dr. Philip Sherman; *Handbook of Nutrition and Pregnancy,* edited by Dr. Carol Lammi-Keefe, Dr. Sarah Couch, and Dr. Elliot Philipson; *Nutrition and Rheumatic Disease,* edited by Dr. Laura Coleman; *Nutrition and Kidney Disease,* edited by Dr. Laura Byham-Grey, Dr. Jerrilynn Burrowes, and Dr. Glenn Chertow; *Nutrition and Health in Developing Countries,* edited by Dr. Richard Semba and Dr. Martin Bloem; *Calcium in Human Health,* edited by Dr. Robert Heaney and Dr. Connie Weaver; and *Nutrition and Bone Health,* edited by Dr. Michael Holick and Dr. Bess Dawson-Hughes.

R.R. Watson et al. (eds.), *Chocolate in Health and Nutrition*, Nutrition and Health 7,
DOI 10.1007/978-1-61779-803-0, © Springer Science+Business Media, LLC 2013

Dr. Bendich served as Associate Editor for "Nutrition," the International Journal; served on the Editorial Board of the *Journal of Women's Health and Gender-based Medicine*; and was a member of the Board of Directors of the American College of Nutrition.

Dr. Bendich was the recipient of the Roche Research Award, is a *Tribute to Women and Industry* Awardee, and was a recipient of the Burroughs Wellcome Visiting Professorship in Basic Medical Sciences, 2000–2001. In 2008, she was given the Council for Responsible Nutrition's (CRN) Apple Award in recognition of her many contributions to the scientific understanding of dietary supplements. Dr. Bendich holds academic appointments as Adjunct Professor in the Department of Preventive Medicine and Community Health at UMDNJ, has an adjunct appointment at the Institute of Nutrition, Columbia University P&S, and is an Adjunct Research Professor, Rutgers University, Newark Campus. She is also listed in *Who's Who of American Women*.

Index

A
Acetic acid bacteria (AAB), 50–52
Acne vulgaris
    chocolate
        amino acids, 521
        ingestion, 518
        liquor composition, 518
        milk derivatives, 522
        postprandial insulin, 521
    closed and open comedones, 518
    food perception, 519
    hyperinsulinemia, 520
    milk, 520–521
    prevalence and intensity, 517
    processed chocolate composition, 519
    westernized dietary patterns, 519
Acute myocardial infarction (AMI), 307
Afflatoxins, 68–69
Aging, 392
Alata samina, 32–33
Albumin, 192
Alzheimer's disease, 382–383
Analgesia
    evidence for, 438–439
    evolutionary signaling, 440
    opioid reward system, 439–440
Animal Research Institute (ARI), 24
Anxiety
    cocoa-containing food effects, 398–400
    definition, 395
Arachidic acid, 150, 152
Area under the plasma pharmacokinetic curve (AUC), 122
Aromatic amino acids, 194
Arterial disease prevention
    chocolate flavonoids
        active components, bioavailability, 293–294
        antioxidant function, 294–295
        atherosclerosis reduction, 296–297
        cardioprotectiverole, evidence, 297, 298
        C6-C3-C6 backbone structure, 293
        CVD progression, 295
        cyclooxygenase inactivation, 295
        history, 291
        hypertension control, 296
        insulin resistance, 297
        lipids, 292
        minerals, 292
        NF-kappa B pathway modulation, 296
        nitric oxide pathway induction, 296
        oxygen radical absorbance capacity values, 293
        platelet modulation, 297
        polyphenols classification, 293
        processing, 291–292
        procyanidin, 293
    diet, 290
    environmental factors, 290
    fruit and vegetable consumption, 290
    plant-based foods, 290
Aspartic endoprotease, 94, 97
Atherosclerosis, 392
Atomic spectrometric techniques
    chocolate-and cocoa-derived products
        elements in, 131–132
        metal determination, 127–129
    CPE, sample pretreatment, 130
    emulsions and microemulsions, 130, 132
    FAAS, 127, 130
    GFAAS, 127, 130
    ICP-AES, 127
    inorganic analysis, solid samples, 125, 126
    OES, 127

B
Bacterial fermentation, 196
Bingeing, 458
Blood–brain barrier (BBB), 392–393

C
Cacao beans
    acetic acid, 114
    amino acid profiles, 108, 109
    antinutritional compounds, 114
    biogenic amines, 115–116
    carbohydrates, 111–112
    chemical compositions, geographical origin, 106, 108
    chlorogenic acid, 115
    cotyledon proteins, 108

R.R. Watson et al. (eds.), *Chocolate in Health and Nutrition*, Nutrition and Health 7,
DOI 10.1007/978-1-61779-803-0, © Springer Science+Business Media, LLC 2013

Cacao beans (*cont.*)
glutelins and prolamins, 108
hydrophobic amino acids, 108
lipid content
composition, 109
phospholipids, 110, 111
triglycerides, 110
oxalic acid, 114
phytic acid, 114
pods, 106, 107
polyphenols, 112, 113
purinic alkaloids, 115
quinic and caffeic acid, 115
seed proteins, 108, 109
Theobroma cacao L, 106
Cacao husk
LCC
antibacterial activity, 254
anti-HIV activity, 251–253
anti-influenza virus activity, 253
biological activities, 248
cigarette smoke-induced cytotoxicity, 255, 256
cytokine production, 258–259
formation, 248
LPS contamination effect, 259–260
macrophage activation, 256, 257
NO radical scavenging activity, 258
preparation method, 248, 249
synergistic action, 254–255
yield, 249–251
ligin fractions, 248, 249
Caco-2 cellular models, 184–185
Caffeine, 390
California's Safe Drinking Water and Toxic Enforcement Act, 74
Calliandra anomala, 8
Cancer prevention
biomarkers, 362
dietary assessment, 361–362
epidemiologic studies
case-control studies, 359–360
catechins, 359
cocoa products intake estimation, 359
cohort studies, 360–361
flavanols intake estimation, 359
genistein, 359
Kuna Indians and mainland Panama, 359
procyanidins intake estimation, 359
nutritional databases, 362
potential biologic mechanisms
anti-inflammatory effects, 357–358
antioxidative activity, 356–357
antiproliferative action, 358
gut microbiota, 358–359
Candida krusei. *See Pichia kudriavzevii*
Carboxypeptidase, 94, 96
Cardiovascular disease. *See* Coronary heart disease
(+)-Catechin, 195, 390, 391
metabolism
biliary and urinary excretion, 213–214
liver and biliary circulation, 215–216
methylation, 214–215
monomeric polyphenol biotransformation, 213
MS and NMR analysis, 216
in small intestine and liver, 214
in vitro digestion models, cocoa matrices, 181–182
Catecholamines, 193
CB. *See* Cocoa butter (CB)
Check-all-that-apply (CATA), 169
Chemistry and biochemistry of cocoa, 390–392
Children
chocolate consumption
bitter taste, 496–497
English and German children, 493
exposure, 497
flavor-flavor learning, 498
flavor masking, 495
flavor-nutrient learning, 497–498
national differences, 493
pocket money/allowance, 493–494
quantity and types, 492
sweet taste, 495–496
taste and flavor preferences, 494, 495
US FITS cross-sectional survey, 492
dental caries
chronic and multifactorial disease, 506
and diet, 507–509
formation, 509
prevalence and severity, 506
prevention, 506, 509–510
reasons, 506–507
dental health
early childhood caries, 506
maintenance, 510
parents role, 511
flavor aversion, 498
health education programs, 511
nutrition education and counseling, 511
oral health, 510, 511
parental influences, 499–500
persistence of preferences, 498–499
public health approaches, 512
Chlorogenic acid, 115
Chocolate
addiction, 423–424
medical use
beneficial effects, 16
chocolate rehabilitation, 20
deficiencies and failures, 19
Florence, 17
fluid and body damage, 18
food, 13
heat substance, 17–18
Hippocratic-Galenic approach, 14–15
hot/cold properties, 14
hypochondriac melancholy, 15
life regime, 12
liver and kidney disease, 14
metaphor, 12
milky spirit, 18
nourishment and therapeutic substance, 19–20
Regimina Sanitatis, 12–13

restore energy, 15
therapeutic purpose, 13–14
volatile substances property, 16
types, 421, 422
withdrawal
chocolate preference, 462–463
definition, 458
drug and food addiction, 458–459
evidence, 462
neurochemistry, 459–461
Chocolate bars
aquatic biotechnology, 146
edible nanotubes, 144
energy density, 143, 144
high-protein diet, 146
human food production, 144
nanocellulose, 144
nutrient density, 143, 145
vitamin A, 146
Chocolate craving, 394–395, 423–424, 535
Chocolate thought suppression
attitude development, 483
food thought suppression inventory, 483
low mood and subsequent overconsumption, 480
negative consequences, 480
negative emotional associations, 480
obesity, 480
psychiatric diagnosis, 480
rebound effect, 481–482
restrained eating status, 484, 485
self-control, 486
Chromosome length polymorphism (CLP), 49–50
Chronic fatigue syndrome (CFS), 269, 400–403
Chronic inflammation, 357
Chronic stress, 194
Cloud point extraction (CPE), 130
CLP. *See* Chromosome length polymorphism
Cochrane Program Review Manager version 5.1, 315
Cocoa
aroma
cocoa beans (*see* Cocoa beans)components, 90–91
clinical benefits
antioxidant properties, 266–267
cancer, 271
cardiovascular health, 267–268
inflammatory response and immune functions, 270–271
nervous system, 268–270
oxidative stress prevention, 266
fermentation
AAB, 50–52
aeration, 46
Bacillus spp. role, 47–48, 52–53
box fermentation, 41–42
curing, 39
ethanol production, 46–47
heap fermentation system, 40
inocula origin, 45
LAB, 47, 50, 51
laborious process, 41
mechanical drying, 43
microbial growth, 55
microorganism inoculation, 40
mold, 55
mycotoxin production, 48
pectinolytic filamentous fungi and AAB, 47
pod storing, 40
pulp, microbiological activity, 39–40
rewetting protection, 43, 44
starter culture, 55–56
storage and transportation, 43
sun drying, 43, 44
terminating time, 42–43
tray fermentation, 41, 42
turning heap, 41
yeast (*see* Yeast, fermentation)pre-hispanic use
archeological evidence, 3
biomarker, 3–4
cascading waterfall-like effect, 6
CB, 9
Chaco Canyon, cylinder jar, 6–7
chocolate drink consumption, 5
coprolites, 3
cultivation, 3, 4
economic partnerships and conflict, 8
luxury foods, 4–5
nutritional and healthful properties, 8
pochteca, 8
serving, chocolate drink, 7–8
Theobroma cacao, 5
Cocoa beans, 201–202, 247
acid-induced proteolytic process, 89, 92–93
aspartic endoprotease, 94, 97
carboxypeptidase, 94
fermentation, 91–92
globular storage protein (*see* Globular storage protein)
shell, 33
TcAP2 gene product, 94
vicilin (7S)-class globulin, 96–97
Cocoa butter (CB), 33
CBEs, 152
CBRs, 150, 152
fatty acid composition, 150, 152
TAGs
composition, 150, 151
oil fractionation technology (*see* Oil fractionation technology)
Cocoa butter equivalents (CBEs), 152
Cocoa butter replacers (CBRs), 150, 152
Cocoa by-products
cocoa bean shell, 33
cocoa butter, 33
CPH (*see* Cocoa pod husk)
economic viability assessment, 24
farmer income and alleviating farmer poverty, 34
pulp juice
alcohol production, 28–29
beans, mucilaginous pulp, 25

Cocoa by-products (*cont.*)
fermentary transportation, 25, 26
fermentation, 25
pectin, 26, 28
perforated PVC assembly, 25–27
physicochemical analyses, 26, 30
polytank, 25–27
soft drink development, 28
sweating, 25–26
raw/waste material collection, 34
relative distribution, 24
Cocoa pod husk (CPH)
animal feed ingredient, 31–32
bean processing, 29
chemical composition, 31
monogastric diet, 35
nutrient analysis and antinutritional factors, 31
nylon bag technique, 32
potash fertilizer, 33
preparation and storage, 30
production and utilization, 24
soap making, 32–33
wet feeding, 32
Cocoa polyphenols
animal studies, pharmacokinetic aspects, 224–226
bioavailability in vivo, 185–186
(+)-catechin metabolism
biliary and urinary excretion, 213–214
liver and biliary circulation, 215–216
methylation, 214–215
monomeric polyphenol biotransformation, 213
MS and NMR analysis, 216
in small intestine and liver, 214
chocolate matrix characteristics and type
(-)-epicatechin bioavailability, 235–236
plasma (+)-catechin temporal change, 236
plasma (+)-epicatechin temporal change, 236–237
plasma pharmacokinetic data, 238
procyanidins bioavailability, 238–239
protein, fat, and petroleum ether content, 238
in vitro digestion models, 235
diet components role
AUC value plot, 234
coingestion of dark chocolate, 230–231
flow-mediated dilation (FMD), 231
milk protein, 231–232
normolipidemic fasted subjects, 233–234
plasma (+)-catechin temporal change, 232
plasma (+)-epicatechin temporal change, 233
presence and absence of milk, 229–230
(–)-epicatechin metabolism
absorption, 219
intragastrical dose, 216
metabolic conjugation, 219
metabolites of, 218
MS and NMR analysis, 217
oral absorption, 217
urinary excretion, 218
fermentation and roasting, 120
flavan-3-ols/catechins, 179
flavonoids (*see* Flavonoids)FMD, 119
free-radical scavengers/chelators, 180
glutathione metabolism, 180
human studies, pharmacokinetic aspects, 226–227
intestinal flora interaction
anaerobic incubation, 209
colonic microbiota, 207
gut microbes, 208, 211
metabolic breakdown, 208
microbial metabolic products, 209–210
microbial stereoselectivity, 209
phenolic acid metabolites, 208–209
procyanidin B2, 210–211
monomeric catechins
metabolism, 219–221
structure, 202
monomeric flavanol
metabolism, 211–212
profile, 204–205
ORAC, 205–206
passage and stability, gastrointestinal tract, 206–207
phenolic and flavonoid content, 205
proanthocyanidins and anthocyanins, 179
procyanidins
metabolism, 221–224
structure, 203–204
stereochemical features role, 227–229
theobromine, 179
in vitro digestion models
bioaccessibility, 181
caco-2 cellular models, 184–185
cocoa matrices, 181–183
colonic fermentation, 183–184
digested and nondigestion fraction, 180
food matrix effect, 180–181
static and dynamic models, 181
Cocoa Research Institute of Ghana (CRIG), 24
Cognitive functions
ACTICOA powder, 401
CFS, 400
flavonoid-rich chocolate
ACTICOA powder, 383
Alzheimer's disease, 382–383
anti-inflammatory polyphenolic compounds, 383
antioxidant activity, 383
beneficial effects, 386
blood oxygenation level-dependent response, 383–384
caffeine and theobromine, 385
cerebral blood flow, 384
cerebral infarction, 382–383
clinical studies, 382
clinical trial, 384
dementia risk factors, 382
episodic memory, 384
global cognition, 384
health benefits, 382
H_2O_2 induced apoptosis, 385
mental fatigue, 384
neuroinflammation suppression, 386

perceptual speed and executive function, 384
semantic memory, 384
sensorimotor speed, 384
vasodilatation, 383
visuospatial skills, 384
neuropsychological functioning test, 402
Colon microorganisms, 193
Common Fund for Commodities (CFC), 24
Coronary heart disease (CHD)
biological mechanisms, 309–310
chocolate consumption
catechin consumption, 304
cross-check dietary history method, 304
inverse dose-dependent association, 307
mortality, 304, 307
prospective cohort studies, 304–306
methodological limitations, 308–309
randomized controlled trial, 304
4-Cresol, 193
Cross-sensitization, 458

D
Dark chocolate
diabetes mellitus (*see* Diabetes mellitus)
hypertension (*see* Hypertension)
multiple health benefits, 390
Decreased LDL oxidation, 295
Depression, 395
cerebrovascular blood flow, 396–397
cocoa-containing food effects, 398–400
Diabetes
caffeine, 332
cocoa and dark chocolate, 332
epidemiological studies, 328–329
hyperglycemia, 327
magnesium, 332
obesity, 331
polyphenol-rich cocoa
beneficial effects, 329
early human studies, 332, 334
(-)-epicatechin, 329
glucose metabolism, 329, 332, 333
glucose tolerance, 330
hypoglycemic effects, 329
insulin-resistant and diabetic patients, 334–335
macrovascular and microvascular complications, 330, 331
normoglycemic and hyperglycemic rats, 330
overweight and obese subjects, 335–336
pro-inflammatory factors, 330
and reduced-calorie milk chocolates
acceptance means, 173
cocoa drinks, 169
conjoint analysis, 171, 173
consumer perception, 171
vs. conventional milk chocolates, 169, 170
descriptive analysis and consumer affective testing, 167
formulations, 167, 168
lipid, and carbohydrate-based fat replacers, 169
polydextrose and lactitol, 168
preference mapping, 169
sensory properties, 167
storage time, 169, 171, 172
Sucra and Sucra/WPC chocolates, 174
WPC, 168
theobromine, 332
type I and type II diabetes, 327–328, 331
Diabetes mellitus
chocolate
atherosclerosis and lipid profle, 346–347
blood pressure, 344
carbohydrate metabolism, 343
definition, 342–343
endothelium function, 345–346
glucocorticoid hormones, 347
inflammatory markers, 348
vascular health, 345
insulin resistance, 342
obesity, 341
Diaminoxidase (DAO) inhibitors, 115
Dopamine, 459–461
Drosophila melanogaster, 45
Dutched cocoa, 451, 452
Dynamic digestion models, 181

E
Emotional eating strategy, 431–432
(-)-Epicatechin, 181–183, 390, 391
metabolism
absorption, 219
intragastrical dose, 216
metabolic conjugation, 219
metabolites of, 218
MS and NMR analysis, 217
oral absorption, 217
urinary excretion, 218
European Food Safety Authority (EFSA), 66

F
FAAS. *See* Flame atomic absorption spectrometry (FAAS)
Fatigue, 402–403
Federation for Cocoa Commerce (FCC), 66
Flame atomic absorption spectrometry (FAAS), 127, 130
Flavanols, 180, 190, 195, 196
Flavonoids
antioxidants, 309–310
bioavailability
aglycones, 120–121
Enterococcus casseliflavus, 122
Eubacterium ramulus, 122
milk protein, 121–122
procyanidin absorption, 121
non-fermented cocoa nibs, 120
Flow-mediated dilation (FMD), 119, 231, 267, 345

Fungi
  chocolate quality, 61–62
  cocoa pod, 62, 63
  fermentation process, 62
  microbial growth and spoilage, 65
  physiologic condition, 64
  roasting and alkalization, 65
  sensory property, alteration, 62
  sun drying, 64
  water content and microbial growth, 64

G
Gastrointestinal enzymes, 196
Gingivitis, 506
Globular storage protein
  aspartic endoprotease, 95
  carboxypeptidase, 96
  hydrophilic peptides and free amino acids, 95–96
  21-kDa albumin, 94
  proteolytic formation, 94, 95
  vicilin (7S)-class globulin, 94, 96
Glucocorticoids, 193
Graphite furnace atomic absorption spectrometry (GFAAS), 127, 130
Gut microbiota
  chocolate and cocoa bioactives, 194–195
  energy and stress-related metabolism, 193–194

H
Hanseniaspora guilliermondii, 49
Hedonic effects, 423
High-oleic high-stearic (HOHS) sunflower oil. *See* Oil fractionation technology
Hippocratic-Galenic approach, 14–15
^{1}H NMR spectroscopy, 193
Human and gut microbial metabolic interactions
  chocolate and cocoa bioactives, 194–195
  cocoa, 190, 192
  dietary preferences and habits, 189–190
  energy and stress-related metabolism, 193–194
  food matrix, 190
  HDL cholesterol, 190, 192
  individuals specific food habits, 192–193
  modulation, host metabolism, 195–196
  nutrimetabonomics, 190, 191
  polyphenols, 190, 192
  systolic blood pressure, 192
Human healthy volunteers
  biological systems effect, 528
  clinical studies, 527–259
  dark chocolate group, 536
  ethnobotanic and history, 526–527
  immune-based degenerative diseases, 537
  neuroimmune systems, black and white chocolate
    adrenaline and noradrenaline, plasmatic concentration, 533, 534
    chemical characteristics, 528
    C-reactive protein, 530
    cytofluorimetric analysis, 530
    diastolic and systolic blood pressure, 530
    dopamine plasmatic concentration, 531, 533
    indolamines and catecholamines, 531
    interleukin 10, 530, 531
    plasma parameters assessment, 529
    psychoneuroimmunological parameters, 528
    serotonin (5-HT) and 5-hydroxyindoleacetic acid (5-HIAA), plasma concentration, 531, 532
    TNF-α production, 531, 532
  white chocolate group, 536
Hydrophobic amino acids, 92–93
Hydroxybenzoic and hydroxycinnamic acids, 112
4-Hydroxyphenylacetate, 193
3-Hydroxyphenylpropionic acid, 196
Hyperinsulinemia, 520
Hypertension
  baseline blood pressure, 320
  blinding, 319
  blood pressure-lowering mechanisms, 315
  chocolate and cocoa products, 316, 318
  cocoa consumption, 314
  epicatechin and catechin, 320
  flavanol dosage, 319
  Kuna Indians, 313–314
  meta-analysis, 316, 317
    age, 321–322
    blood pressure-reducing effect, 316, 318
    Cochrane Program Review Manager version 5.1, 315
    duration, 321
    flow-mediated dilation, 322
    intervention trials, 315
    systolic and diastolic blood pressure, 316
  sugar, 320–321
  theobromine, 320
Hypochondriacal melancholy, 382
Hypothalamic–pituitary–adrenal metabolic axis, 193
Hysteroid dysphoria, 426

I
Increased antioxidant capacity, 295
Increased chocolate consumption worldwide, 297
Inductively coupled plasma-atomic emission spectrometry (ICP-AES), 127
International Cocoa Organisation (ICCO), 24
Intestinal flora interaction
  anaerobic incubation, 209
  colonic microbiota, 207
  gut microbes, 208, 211
  metabolic breakdown, 208
  microbial metabolic products, 209–210
  microbial stereoselectivity, 209
  phenolic acid metabolites, 208–209
  procyanidin B2, 210–211
In vitro digestion models
  bioaccessibility, 181
  caco-2 cellular models, 184–185

cocoa matrices
epicatechin and catechin, 181–182
food matrix composition, 182
macronutrient typology, 182
micellarization, 182
polysaccharide moiety, 183
procyanidin oligomers, 181
water-insoluble cocoa fraction, 183
colonic fermentation, 183–184
digested and nondigestion fraction, 180
food matrix effect, 180–181
static and dynamic models, 181
In vitro insulin, 192
Isoelectric points (IEPs), 97

J

Joint FAO/WHO Expert Committee on Food Additives (JECFA), 66

L

Lactic acid bacteria (LAB), 47, 50, 51
Lactobacillus fermentum, 50
LDL, 190, 192
Lignin-carbohydrate complex (LCC)
antibacterial activity, 254
anti-HIV activity, 251–253
anti-influenza virus activity, 253
biological activities, 248
cigarette smoke-induced cytotoxicity, 255, 256
cytokine production, 258–259
formation, 248
LPS contamination effect, 259–260
macrophage activation, 256, 257
NO radical scavenging activity, 258
preparation method, 248, 249
synergistic action, 254–255
yield, 249–251
Low-calorie chocolates
flavonoids, 165
food cravings, 163–164
milk chocolates (*see* Milk chocolates)
obesity, 164
sensory characteristics, 164

M

Maillard reactions, 92, 106
MAO inhibitors. *See* Monoamine oxidase (MAO) inhibitors
Medline and Cochrane databases, 315
Metabolic syndrome, 341
Methylxanthine, 179
Microencapsulation, 196
Milk chocolates. *See also* Diabetes
formulation acceptance, 165
formulation B, 166
magnitude estimation and category scaling, 165
rebaudioside, 167
sensory attributes, 165
sucralose and stevioside, 166–167
time-intensity method and curves, 165–166
Monoamine oxidase (MAO) inhibitors, 115, 116
Monomeric catechins
metabolism, 219–221
structure, 202
Monomeric flavanol
metabolism, 211–212
profile, 204–205
Monosaccharides and oligosaccharides, 112
Mood disorder, 394, 395
Mood effects of chocolate, 431–432
adverse effects, 416
benefits, 416
depression, 410
elucidation, 410
emotion-inducing film clip, 410
in humans, 410, 414–415
oxidative stressors, 416
palatability, 410
potential mechanisms, animal studies, 410–413
Moreishness, 424
Mycotoxin
afflatoxins, 68–69
chocolate occurence, 67
fungal toxin, 65
human exposure, 65
inhibitory effect, 66
investigative study, 66
OTA, 65–67
regulation of, 69
shell fraction and toxic contaminants, 67

N

Neurocognitive functioning, cocoa-and chocolate-related product effects
animal studies, 373–374
blood oxygenation level–dependent response, 371
clinical trial, 371
dose-dependent associations, 372
mechanisms, 374–375
methylxanthines in chocolate, 370, 371
multivariate models, 372
neuropsychological tests, 371
psychopharmacological activity of cocoa powder, 370
psychostimulant effects, 371
State-Trait Anxiety Inventory state scale, 372
visual contrast sensitivity, 373
Neurotransmitters, nutrient-dependent chocolate effects, 430–431
Nonnutritive constituents
agricultural practice
and cocoa bean processing, 78
disease, 77
distribution, 77
harvesting, 78
nitrogen and phosphorous application, 77

Nonnutritive constituents (*cont.*)
  cadmium, 81–82
  chemical fumigation, 83
  cocoa and chocolate, 73–74
  contamination pathway
    airborne fallout, 79
    chocolate manufacture contamination, 80
    grinding equipment contamination, 80
    ion adsorption, 79
    mycotoxin, 80
    pesticides, 80
    physiological uptake, 79
    shell fragments, 80
    soil contamination, 79
  insecticides and fungicides, 74
  lead, 81
  market basket survey, 74
  mycotoxins
    Aflatoxins, 76
    ochratoxin A, 76–77
  nickel, 82
  ochratoxin A, 83
  Pb-contaminated candy, 74
  Pb isotopic characteristics, 83
  pesticides, 82
  toxic effects, and regulatory value
    cadmium, 75
    lead, 75
    nickel, 75–76
    pesticides, 76

O
Ochratoxin A (OTA)
  chocolate occurrence, 67
  inhibitory effect, 66
  investigative study, 66
  shell fraction and toxic contaminants, 67
Oil fractionation technology, 152–153
  dry fractionation
    characteristics, 154–156
    crystal growth, 153
    crystallization kinetics, 153, 154
    nucleation process, 153
  solvent fractionation
    chemical composition, 157
    N17 oil, 156, 157
    SHS, 156, 158–159
    vacuum-filtered fat crystals, 154
Oligosaccharides, 196
Opioid reward system, 439–440
Optical emission spectrometry (OES), 127
Orexins, 471–474
Orosensory properties, 422–423
Oxalic acid, 114
Oxygen radical absorbance capacity (ORAC),
    205–206

P
Pain tolerance
  analgesia
    evidence for, 438–439
    evolutionary signaling, 440
    opioid reward system, 439–440
  chemical compounds
    anandamide, 441–442
    flavonoids, 443–444
    methylxanthines, 442–443
    phenylethylamine, 441
    tryptophan, 442
Palatable food, 458
Parkinson's disease (PD), 270, 416
Periodontitis, 506
Personality traits, 394
Phenylacetylglutamine, 194
2-Phenylethylamine (PEA)
  anandamide, 427
  antipsychotic effects, 426
  depression and dopamine, 426
  endogenous amphetamine and mania, 426
  methylxanthines, 428–430
  noradrenaline and MAO-B activity, 426
  salsolinol, 427–428
  serotonin, 426
  tetrahydro-β-carbolines, 427–428
  tyramine, 426
Phytic acid, 114
Pichia kudriavzevii, 45
Pichia membranifaciens, 45
Polydextrose, 164
Polyphenol-free acetone-dry powder, 93
Postexercise recovery
  carbohydrate, 452–453
  cocoa, 450–452
  limitations of studies, 454
  muscle glycogen, 450
  postexercise nutritional strategies, 450
  protein addition to recovery meal, 453–454
Proanthocyanidins, 195
Procyanidins, 180, 382
  metabolism, 221–224
  structure, 203–204
Propionibacterium acnes, 517, 518
Provisional tolerable weekly intake (PTWI), 75
Psychoactive components, 422, 424–426

R
Regimina Sanitatis, 12–13
Reward-based feeding behaviors
  c-Fos immunolabeling techniques, 473
  endocannabinoid inhibitor rimonabant dose, 470
  endogenous stress system, 470
  ghrelin, 470–471
  mesolimbic circuitry, 470

mesolimbic dopamine reward pathway, 470, 471
narcoleptic patients, 472
orexins, 471–474
paraventricular nucleus of thalamus, 472
rimonabant, 470
sleep-wake function, 471
social stress, 474
suprachiasmatic nucleus, 471
Rimonabant, 470
Rustic cacao systems, 136

S
Saccharomyces theobromae, 48
Salivary habituation, 164
SHS. *See* Sunflower hard stearins (SHS)
SSS. *See* Sunflower soft stearins (SSS)
Static digestion models, 181
Stearic acid, 150, 152, 310
Stevioside, 164
Striatal opioid system, 460
Sunflower hard stearins (SHS)
characteristics, 154–156
physical properties, 158–159
Sunflower soft stearins (SSS), 154–156
Sympathoadrenal system, 193

T
TAG. *See* Triacylglycerol (TAG)
Theobroma cacao, 20, 39, 136
Theobromine, 179, 390, 391
Triacylglycerol (TAG)
composition, 150, 151
oil fractionation technology, 152–153
dry fractionation, 153–156
solvent fractionation, 154–159
Two-dimensional gel electrophoresis, 96–97
Type 2 diabetes. *See* Diabetes mellitus

U
United Nations Codex Alimentarius Commission, 74
US Food Safety Modernization Act, 82

V
Vascular tone, endothelial control
blood fluidity and control, 277, 278
cardiovascular diseases, 279
endothelial dysfunction, 279
endothelial NO synthase, 278
flavonoid classes, 279
health and disease, humans, 282–283
platelet adhesion and activation prevention, 278
polyphenol-rich sources
animal models, endothelial function, 281–282
endothelium-dependent relaxations, 279–281
Vasodilation, 392, 396
Vicilin (7S)-class globulin, 98–99
aspartic endoprotease and carboxypeptidase, 96
hydrophilic peptides and free amino acids, 95–96
proteolytic formation, 94, 95
structure, 96–97

W
Whey protein concentrate (WPC), 168
Wild and domestic animals
animal feed
cocoa by-products, 136–137
theobromine content, 137–138
cacao plantations and biodiversity, 136
chocolate toxicity, 138–139
Witches'broom, 77

Y
Yeast, fermentation
box and heap fermentation, 48
CLP, 49–50
ethanol production, 46–47
growth, 45–46
Hanseniaspora guilliermondii/opuntiae complex, 49
molecular biology–based method, 48–49

Z
Zymomonas mobilis, 54

MIX
Papier aus verantwortungsvollen Quellen
Paper from responsible sources
FSC® C105338

If you have any concerns about our products,
you can contact us on
ProductSafety@springernature.com

In case Publisher is established outside the EU,
the EU authorized representative is:
Springer Nature Customer Service Center GmbH
Europaplatz 3, 69115 Heidelberg, Germany

Printed by Libri Plureos GmbH
in Hamburg, Germany